T0190326

Structural Analysis with the Finite Element Method
Linear Statics

Volume 2. Beams, Plates and Shells

Lecture Notes on Numerical Methods in Engineering and Sciences

Aims and Scope of the Series

This series publishes text books on topics of general interest in the field of computational engineering sciences.

The books will focus on subjects in which numerical methods play a fundamental role for solving problems in engineering and applied sciences. Advances in finite element, finite volume, finite differences, discrete and particle methods and their applications are examples of the topics covered by the series.

The main intended audience is the first year graduate student. Some books define the current state of a field to a highly specialised readership; others are accessible to final year undergraduates, but essentially the emphasis is on accessibility and clarity.

The books will be also useful for practising engineers and scientists interested in state of the art information on the theory and application of numerical methods.

Titles:

Structural Analysis with the Finite Element Method
Linear Statics

Volume 2. Beams, Plates and Shells

Eugenio Oñate

International Center for Numerical Methods in Engineering (CIMNE)
School of Civil Engineering
Universitat Politècnica de Catalunya (UPC)
Barcelona, Spain

ISBN: 978-94-017-7703-2
ISBN: 978-1-4020-8743-1 (e-book)
DOI: 10.1007/978-1-4020-8743-1
Depósito legal: B-18335-2012

Lecture Notes Series Manager: **Mª Jesús Samper,** CIMNE, Barcelona, Spain

Cover page: **Pallí Disseny i Comunicació,** www.pallidisseny.com

Printed by: **Artes Gráficas Torres S.A.,**
Morales 17, 08029 Barcelona, España
www.agraficastorres.es

Printed on elemental chlorine-free paper

Structural Analysis with the Finite Element Method. Linear Statics.
Volume 2. Beams, Plates and Shells
Eugenio Oñate

First edition, 2013

© International Center for Numerical Methods in Engineering (CIMNE), 2013
Gran Capitán s/n, 08034 Barcelona, Spain
www.cimne.com
Softcover re-print of the Hardcover 1st edition 2013

To my family

Foreword

The present volume continues the objective of Volume 1: to present the Finite Element Method (FEM) for solid and structural mechanics from a balanced perspective that interweaves theory, formulation, physical modeling and computer implementation. The challenging balance is made possible by the three-decade practical experience of the author in teaching FEM courses while building and leading a large research center (CIMNE) that conducts advanced research in computational mechanics for a wide variety of engineering applications. That experience is put to good use to support of the educational goals addressed by this book series. A major difference between the first and second volume is in the level and detail of coverage.

The goal of the first volume was to offer an introductory coverage of FEM. The material concentrated on solid and structural mechanics to maintain a clear application focus. The problem class is restricted to linear static analysis. This volume addressed the fundamentals steps of the Direct Stiffness Method (DSM) version of FEM, while introducing related mathematical considerations such as consistency, accuracy and convergence. These are in turn translated into practical modeling rules. Emphasis is on the physical interpretation of the method as a "divide and conquer" technique. Accordingly, the exposition level is that appropriate to a first course in FEM at the master level. The assumed preparation level of a student taking such a course is expected to include multivariate calculus, linear algebra, and a basic knowledge of structural analysis at the Mechanics of Materials level, as well as some familiarity with computer programming concepts. Because of the inevitable space and time limitations of a first-course treatment, the contents of Volume 1 were restricted to simple structural elements, notably axially loaded members and two-dimensional solids. Those models are sufficient, however, to illustrate the primary steps of DSM, as well as to provide specific examples that teach the most important modeling rules.

A key virtue of DSM is that its steps are applicable to any finite element model formulated within the framework of the stiffness equations. Consequently, those steps need not be repeated in a more advanced treatment such as that presented in this volume. The author is free to directly proceed to more difficult problems by focusing on more advanced element formulation techniques and associated modeling rules. The problems addressed in the volume involve the classical structural components that carry bending actions: beams, plates and shell, as well as combinations such as stiffened plates and shells. The coverage still targets on linear static problems.

Consideration of bending effects brings about a new set of formulation and modeling difficulties. Foremost is locking: a pathological overstiffness endemic to certain "thin" configurations, and which must be overcome to avoid unsafe designs. Remedies to those difficulties, however, can in turn produce undesirable side effects. For example reduced integration to cure locking may give rise to numerical instabilities; e.g., mesh "hourglassing". The delicate interplay between diagnosing and curing requires more advanced mathematical tools, which go beyond those deemed sufficient for Volume 1. Accordingly, the present treatment is intended for follow-up courses that cover more advanced models in structural mechanics, as well as a more detail coverage of the computer implementation. The following Chapter summary give a more specific idea of the contents of the present volume.

Chapters 1-4 cover structural beam members. The Euler-Bernoulli and Timoshenko models of plane beams are presented in Chapters 1 and 2, respectively. Chapter 3 deals with composite laminate farbrications of plane beans. Chapter 4 addresses 3D beams, focusing on composite fabrication and the problem of cross section warping under torsion and shear. Rotation-free beam element models are introduced as advanced topic.

Chapters 5-7 deal with flat plate structural components. The Kirchhoff and Reissner-Mindlin Model are covered in Chapters 5 and 6 respectively, while Chapter 7 addresses composite laminates fabrication of plate walls. Rotation free plate elements are covered again; at the plate level this represents a still ongoing research topic.

Chapters 8 through 10 deal with shell structures. The coverage includes facet element models, axisymmetric configurations, doubly curved shell models, culminating with the treatment of stiffened shells.

Chapter 11 covers prismatic structures. This is a specialized form of shells (e.g., folded thin roofs) which deserves special treatment on account of its industrial importance as well as ubiquity in various engineering branches (Aerospace, Civi, Marine and Mechanical).

Chapter 12 present a computer implementation dubbed MATfem, supported by the MATLAB high level programming language. This may be viewed as a "unification chapter" that brings together models formulated in the previous chapters, as well as in Volume 1, into a MATLAB framework. This material was organized in collaboration with Professor Zárate.

Each of the formulation oriented chapters (1 through 11) covers its title material in three stages. First, classical theories for the pertinent structural configuration is summarized in a form suitable for FEM formulation. Second, the construction of finite element models based on those theories. Third, advanced material pertaining to the topic is included; for example, how to overcome difficulties such as locking. Some of the advanced material is still the matter of ongoing research by the author's team at CIMNE. An important example are rotation-free elements for beam, plates and shells, as well as the treatment of prismatic structures.

The volume concludes with seven Appendices that summarize relevant reference material for the benefit of the reader.

Several features that clearly distinguish this volume from other texts at a similar level should be noted. The rich treatment of composite fabrication in Chapters 3, 4 and 7 does not have a counterpart in other FEM textbooks. The treatment of rotation-free beam, plate and shell elements in Chapters 1, 2, 5 and 9 reflects the long-term involvement of the author in that research thrust, which offers significant promise in its extension to extremely large deformations as occur in important fabrication processes. Finally, the MATfem implementation in Chapter 12 stands out for the careful attention to modularity and completeness.

While as noted Volume 1 was specifically oriented to an introductory course, the present volume can be used in two contexts:

1. A textbook that support advanced FEM courses. Since the overall content is too extensive to be covered in a one-semester course, the instructor will likely need to select specific presentation topics, and perhaps designate others as launching pads for course team projects. The instructor will have to provide exercitation problems that enhance the student comprehension of the material, and well as exam problems that test that understanding.

2. As a reference monograph for advanced topics that are not adequately treated aside from the specialized literature. An instance would be rotation free elements for complex plates and shell assemblies.

In summary, the present volume ably complements and supplements the first one with a wealth of material that provide both instructors and researchers with a wealth of possibilities.

Carlos Felippa
Professor in the University of Colorado at Boulder
November 2012

Preface

This book complements the content of the first volume: *Structural Analysis wit the FEM. Basis and solids* (Springer/CIMNE, 2009). The scope of the second volume covers the finite element analysis of "structural elements" such as beams, plates and shells. Similarly, as in Volume 1, the study is restricted to linear static analysis only.

The book is addressed to undergraduate students and readers that are exposed to the FEM analysis of beams, plates and shells for the first time. No previous experience on the FEM is strictly necessary. However, some knowledge of basic FEM concepts, such as mesh discretization, displacement interpolation, shape functions, numerical integration, element matrices and vectors, assembly of the stiffness equations, etc., as described in Volume 1, will facilitate the understanding of the underlying ideas behind the FEM.

Throughout the text emphasis has been put in the study of beam, plate and shell structures with composite laminated material, this being a possibility of increasing interest for practical applications. For didactic reasons, however, the homogeneous and isotropic material case is considered first, and in most cases in a separate chapter, in order to facilitate the study to an inexperienced reader.

As in Volume 1, the study of each structural model is introduced with the detailed description of the underlying theory. This includes the general structural mechanics assumptions, the kinematic description, the stress and stress fields, the constitutive relationship and the expression of the virtual work principle in terms of stress resultants and generalized strains defined over the "reference geometry" of the element, i.e. a line for a beam, a plane for a plate or a flat shell and a curved surface for a general shell structure.

Chapter 1 introduces the FEM analysis of two-dimensional (2D) plane slender beams following the classical Euler-Bernoulli beam theory. The popular two-noded Euler-Bernoulli beam element is derived in some detail. The chapter concludes with the formulation of two *rotation-free* beam elements. These elements have the vertical deflection as the only nodal variable and are an interesting alternative to standard slender beam elements.

Chapter 2 describes the formulation of beam elements based on the more advanced Timoshenko beam theory. This theory accounts for the effect of shear deformation and is therefore applicable to slender and thick beams. Timoshenko beam elements suffer from the so-called shear locking defect that leads to overstiff situations for slender beams. A number of procedures to avoid shear locking are described.

Chapter 3 is devoted to the FEM analysis of beams with composite laminated material. Emphasis is put in studying the particularities introduced by the composite material in the mechanical behaviour of the beam, the more important one being the introduction of the axial deformation mode in addition to the standard bending mode typical of homogeneous plane beams. Higher order beam theories that are able to reproduce the complexity of the in-plane displacement field across the thickness are described. An effective 2-noded composite laminated Timoshenko beam element based on the so-called refined zigzag theory is detailed.

Chapter 4 studies three-dimensional (3D) Timoshenko and Euler-Bernoulli beams under arbitrary loading. Both the classical Saint-Venant theory allowing for a uniform torsion field and the more complex theory for non-uniform torsion of beams with thin-walled open section are described. Both theories are generalized for homogeneous and composite material. A number of two-noded beam elements adequate for each theory considered are presented.

Chapter 5 focusses on the FEM analysis of homogeneous thin plates following the classical Kirchhoff plate theory. The difficulties for satisfying the continuity condition for the deflection slopes across the element sides are detailed. A number of conforming and non-conforming quadrilateral and triangular thin plate elements are presented. Part of the chapter is devoted to the study of two families of rotation-free thin plate triangles.

Chapter 6 extends the FEM analysis to thick and thin homogeneous plates that follow the more advanced Reissner-Mindlin plate theory. This can be viewed as an extension of Timoshenko beam theory and, thus, it accounts for transverse shear deformation effects. Different techniques to avoid the shear locking defect are detailed. A collection of Reissner-Mindlin quadrilateral and triangular plate elements is presented.

Chapter 7 is devoted to composite laminated plates. The effect of the axial displacement induced by the changes in the material properties across the thickness in the governing equations of the plate problem are detailed. An accurate 4-noded composite laminated plate quadrilateral based on the refined zigzag theory is presented.

Chapter 8 introduces the FEM study of shell structures via flat shell elements. The basic equations of Reissner-Mindlin flat shell theory expressed in the local axes of the element are identical to those of a composite laminated plate as studied in the previous chapter. The equations for a flat shell element are formulated for homogeneous and composite laminated material. The particularity of the assembly of the element equations in global axes are detailed. The formulation of thin shell elements following Kirchhoff theory is also explained. The chapter concludes with the description of some higher order theories for composite laminated flat shell elements, including the refined zigzag theory.

Chapter 9 deals with axisymmetric shell structures. Reissner-Mindlin assumptions for the shell kinematics are studied first. Both homogeneous and composite laminated materials are considered. A simple locking-free 2-noded troncoconical axisymmetric shell element is presented in some detail. The formulation of curved axisymmetric shell elements is also studied and the combined effect of shear and membrane locking is discussed. The derivation of thin axisymmetric shell elements following Kirchhoff assumptions is explained and two interesting rotation-free thin axisymmetric shell elements are presented. The axisymmetric formulation is particularized for circular plates, arches and shallow shells. The derivation of axisymmetric shell elements via the degeneration of 3D axisymmetric solid elements is also explained. The chapter concludes with the study of higher order theories for composite laminated axisymmetric shells.

Chapter 10 studies the analysis of 3D shell structures of arbitrary shape using degenerated solid elements. The degeneration process is explained and the element matrices and vectors are obtained. Different techniques for deriving degenerated shell elements free of shear and membrane locking are presented. Several procedures for the explicit integration of the element stiffness matrix across the thickness are explained. The basic concepts of the isogeometry approach and the derivation of isogeometric shell elements are studied. The chapter concludes with the formulation of stiffened shell elements by coupling beam and shell elements.

Chapter 11 presents the derivation of finite strip and finite prism methods for analysis of prismatic plate/shell and 3D solid structures, respectively. These elements combine a finite element approximation across the transverse section of the structure with Fourier series expansions for representing the longitudinal response. Both procedures can therefore be considered as a particular class of reduced order models. The finite strip formulation is detailed for plates, folded plates with straight and circular plant and axisymmetric shells with non arbitrary loading. The simple 2-noded strip allows one to analyze complex prismatic shell structures, such as box-girder bridges, in an extremely simple form. The finite prism formulation is derived for prismatic solids with straight and circular plant and axysymmetric solids with arbitrary loading.

Chapter 12 explains the programming of some of the elements studied in the book for beam, plate and shell analysis using MATLAB. The elements are implemented in the MAT-Fem code that can be freely downloaded from the web. The general structure of the code is explained as well as the basic concepts for programming the stiffness matrix and the equivalent nodal force for the element. The essential parts of the MAT-Fem code are listed for each element considered.

The book concludes with a number of appendices with details of topics of general interest such as the properties of selected materials, the equilibrium equations for a solid, some numerical integration quadratures for triangular, quadrilateral and hexahedral elements, the derivation of the shear correction parameters, the shear center and the warping functions in beams, the stability conditions for beam and plate elements based on the assumed strain technique, the analytical solution for circular plates and the expression of the shape functions for some C^0 triangular and quadrilateral elements.

I want to express my gratitude to Dr. Francisco Zárate who was responsible for writing the computer program Mat-fem presented in Chapter 12 and also undertook the task of writing this chapter.

The content of the book is an expanded version of the course on Finite Element Analysis of Structures which I have taught at the School of Civil Engineering in the Technical University of Catalonia (UPC) since 1979. I want to express my thanks to my colleagues in the Department of Continuum Mechanics and Structural Analysis at UPC for their support and cooperation over many years. Special thanks to Profs. Benjamín Suárez, Miguel Cervera and Juan Miquel, Drs. Francisco Zárate and Daniel di Capua and Mr. Miguel Angel Celigueta with whom I have shared the teaching of the mentioned course during many years.

Many examples included in the book are the result of problems solved by academics and research students at UPC and CIMNE in cooperation with companies which are acknowledged in the text. I thank all of them for their contributions. Special thanks to the GiD team at CIMNE for providing many pictures shown in the book.

Many thanks also to my colleagues and staff at CIMNE for their cooperation and support during so many years that has made possible the publication of this book.

Finally, my special thanks to Mrs. María Jesús Samper from CIMNE for her excellent work in the typing and editing of the manuscript.

Eugenio Oñate
Barcelona, November 2012

Contents

1

SLENDER PLANE BEAMS. EULER-BERNOULLI THEORY

1.1 INTRODUCTION

This chapter studies the bending of slender plane beams using the classical Euler-Bernoulli beam theory and the FEM. Many readers will ask themselves why we are applying the FEM to a simple structural problem that can be solved by standard Strength of Materials techniques [Ti2]. The answer is that the study of beam elements is of great interest as it allows us to introduce concepts which will be applied for the formulation of thin plate and shell elements in the subsequent chapters.

The matrices and vectors for the 2-noded Euler-Bernoulli beam element are in many cases identical to those obtained via standard matrix analysis methods [Li,Pr]. This coincidence illustrates the analogy between the FEM and classical matrix techniques for structural analysis.

The organization of the chapter is as follows. First, the classical Euler-Bernoulli theory for plane beams which neglects the effect of shear deformation is presented. This is followed by the formulation of the 2-noded Euler-Bernoulli beam element. This element introduces the C^1 continuous Hermite shape functions. The formulation of rotation-free beam elements using the deflection as the only nodal variable is also described. Thick beam elements based on Timoshenko beam theory which takes shear deformation into account will be derived in Chapter 2.

The formulation of Timoshenko and Euler-Bernoulli beam elements for plane composite laminated beams and 3D beams will be studied in Chapters 3 and 4, respectively.

E. Oñate, *Structural Analysis with the Finite Element Method. Linear Statics: Volume 2: Beams, Plates and Shells*, Lecture Notes on Numerical Methods in Engineering and Sciences, DOI 10.1007/978-1-4020-8743-1_1,
© International Center for Numerical Methods in Engineering (CIMNE), 2013

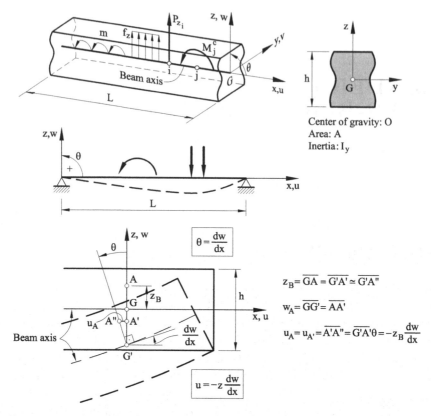

Fig. 1.1 Euler-Bernoulli plane beam. Definition of loads and displacements

1.2 CLASSICAL BEAM THEORY

1.2.1 Basic assumptions

Let us consider a beam of length L and cross-sectional area A under vertical point loads and moments acting on the plane xz which is assumed to be a principal plane of inertia (Figure 1.1). We will also accept that the beam axis x coincides with the line joining the gravity centers G of the cross-sections. The material properties are assumed to be *isotropic and homogeneous*, so that the beam axis coincides with the neutral axis (Section 3.6).

The classical Euler-Bernoulli plane beam theory is based on the following hypotheses [Ti2]:

1. The vertical displacement (deflection) w of the points contained on a cross-section are small and equal to the deflection of the beam axis.

2. The lateral displacement v (along the y axis in Figure 1.1) is zero.
3. Cross-sections normal to the beam axis remain plane and orthogonal to the beam axis after deformation (normal orthogonality condition).

1.2.2 Displacement field

Following the above hypotheses the displacement field is written as

$$\begin{aligned} u(x, y, z) &= -z\theta(x) \\ v(x, y, z) &= 0 \\ w(x, y, z) &= w(x) \end{aligned} \tag{1.1}$$

Hypothesis 3 implies that the rotation is equal to the slope of the beam axis (Figure 1.1); i.e.

$$\boxed{\theta = \frac{dw}{dx}} \quad \text{and} \quad \boxed{u = -z\frac{dw}{dx}} \tag{1.2}$$

1.2.3 Strain and stress fields

Starting from the strain field for a 3D solid [On4] we find

$$\varepsilon_x = \frac{du}{dx} = -z\frac{d^2w}{dx^2} \quad , \quad \varepsilon_y = \varepsilon_z = \gamma_{xy} = \gamma_{xz} = \gamma_{yz} = 0 \tag{1.3}$$

i.e. the beam is under a pure axial strain (ε_x) state. The axial stress σ_x (Figure 1.2) is related to ε_x by Hook law [Ti2] as

$$\sigma_x = E\varepsilon_x = -zE\frac{d^2w}{dx^2} \tag{1.4}$$

where E is the Young modulus. The values of E for some typical materials are given in Appendix A.

1.2.4 Bending moment-curvature relationship

The bending moment for a cross section is defined as (Figure 1.2)

$$M = -\iint_A z\sigma_x\, dA = \left(\iint_A z^2 dA\right) E\frac{d^2w}{dx^2} = EI_y\frac{d^2w}{dx^2} = EI_y\kappa \tag{1.5}$$

where $I_y = \iint_A z^2\, dA$ is the moment of inertia (or inertia modulus) of the section with respect to the y axis and $\kappa = \dfrac{d^2w}{dx^2}$ is the curvature of the beam axis.

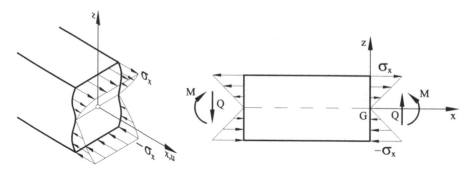

Fig. 1.2 Sign criteria for axial stress σ_x, bending moment M and shear force Q

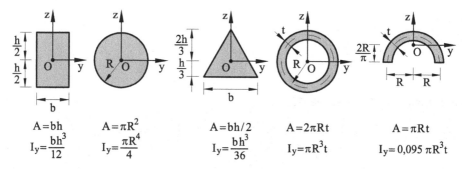

$A=bh$
$I_y=\dfrac{bh^3}{12}$

$A=\pi R^2$
$I_y=\dfrac{\pi R^4}{4}$

$A=bh/2$
$I_y=\dfrac{bh^3}{36}$

$A=2\pi Rt$
$I_y=\pi R^3t$

$A=\pi Rt$
$I_y=0{,}095\,\pi R^3t$

Fig. 1.3 Center of gravity G, area and inertia modulus around the y axis (I_y) for some beam cross sections

Figure 1.3 shows the value of I_y for some cross sections. The position of the center of gravity G and the inertia moduli for thin wall sections and solid sections of arbitrary shape can be found in specialized texts [PCh,Yo].

Eq.(1.5) assumes homogeneous material. Composite laminated plane beams are studied in Chapter 3. The general case of 3D beams with heterogeneous material properties is treated in Chapter 4.

1.2.5 Principle of Virtual Work

The principle of virtual work (PVW) is written as

$$\iiint_V \delta\varepsilon_x\sigma_x\,dV = \int_0^L \left[\delta w f_z + \delta\left(\frac{dw}{dx}\right)m\right]dx + \sum_i \delta w_i P_{z_i} + \sum_j \delta\left(\frac{dw}{dx}\right)_j M_j^c \tag{1.6a}$$

where the sums in the r.h.s. of Eq.(1.6a) extend over the number of points with point loads P_{z_i} and concentrated moments M_j^c. The upper index in M_j^c distinguishes the concentrated moment acting at a point j from the internal couple M_i at the ith element node.

External vertical point loads P_{z_i} and distributed loads $f_z(x)$ are taken as positive if they act in the direction of the global z axis. The external concentrated moments M_j^c acting at beam points and the distributed moment $m(x)$ are taken as positive if they act anticlockwise, in consistency with the definition of the rotation θ (Figure 1.1).

The integral in the l.h.s. of Eq.(1.6a) represents the virtual strain work (also called internal virtual work) and it can be simplified as follows. Using Eqs.(1.3)–(1.5)

$$
\iiint_V \delta\varepsilon_x \sigma_x \, dV = \int_0^L \delta\left(\frac{d^2w}{dx^2}\right)\left[\iint_A z^2 \, dA\right] E\left(\frac{d^2w}{dx^2}\right) dx
$$

$$
= \int_0^L \delta\left(\frac{d^2w}{dx^2}\right) EI \frac{d^2w}{dx^2} \, dx = \int_0^L \delta\kappa M \, dx \qquad (1.6b)
$$

The virtual strain work is therefore expressed as the integral along the beam axis of the product between the bending moment and the virtual curvature.

Substituting Eq.(1.6a) into (1.6b) gives

$$
\int_0^L \delta\kappa M \, dx = \int_0^L \left[\delta w f_z + \delta\left(\frac{dw}{dx}\right) m\right] dx + \sum_i \delta w_i P_{z_i} + \sum_j \delta\left(\frac{dw}{dx}\right)_j M_j^c
$$

$$
(1.7)
$$

The reduction of the virtual work expression to integrals over the "reference" geometry of the body (i.e. the beam axis, the mid-plane in plates and the mid-surface in shells) is a distinct characteristic of beam, plate and shell theories.

The only unknown in Euler-Bernoulli beam theory is the vertical deflection w. However, the PVW involves second derivatives of w. Therefore, C^1 continuity is required (w and dw/dx must be continuous) as explained in Section 3.8.3 of [On4]. This requirement has a physical explanation. As the rotation dw/dx coincides with the slope of the beam axis, the slope has to be continuous to ensure a smooth deflection field.

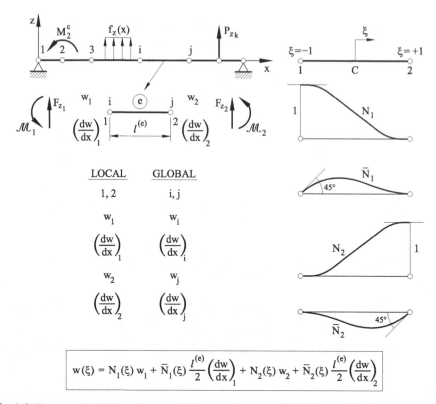

Fig. 1.4 Two-noded Euler-Bernoulli beam element. Nodal variables, equilibrating nodal forces and Hermite shape functions

1.3 THE 2-NODED EULER-BERNOULLI BEAM ELEMENT

1.3.1 Approximation of the deflection, curvature and bending moment fields

The simplest C^1 continuous beam element is the 2-noded beam element shown in Figure 1.4. The continuity of the beam slope across adjacent elements requires that dw/dx be a nodal variable. Therefore, the element has four degrees of freedom (DOFs): w_i and $(dw/dx)_i$ at each node. This allows us to define a cubic expansion for the deflection as:

$$w = \alpha_0 + \alpha_1 x + \alpha_2 x^2 + \alpha_3 x^3 \qquad (1.8)$$

The parameters α_i are obtained by substituting the deflection value and its derivatives (i.e. the rotation $\frac{dw}{dx}$) at the nodes in Eq.(1.8). This yields the following system of equations

$$w_1 = \alpha_0 + \alpha_1 x_1 + \alpha_2 x_1^2 + \alpha_3 x_1^3$$

$$\left(\frac{dw}{dx}\right)_1 = \alpha_1 + 2\alpha_2 x_1 + 3\alpha_3 \ x_1^2$$

$$w_2 = \alpha_0 + \alpha_1 x_2 + \alpha_2 x_2^2 + \alpha_3 x_2^3 \qquad (1.9)$$

$$\left(\frac{dw}{dx}\right)_2 = \alpha_1 + 2\alpha_2 x_2 + 3\alpha_3 x_2^2$$

Eq. (1.8) can be rewritten, after substituting the α_i parameters obtained from the solution of Eqs.(1.9), as

$$w = N_1 w_1 + \overline{N}_1 \frac{l^{(e)}}{2} \left(\frac{dw}{dx}\right)_1 + N_2 w_2 + \overline{N}_2 \frac{l^{(e)}}{2} \left(\frac{dw}{dx}\right)_2 \qquad (1.10)$$

where the element shape functions are

$$N_1 = \frac{1}{4} (2 - 3\xi + \xi^3) \qquad ; \qquad N_2 = \frac{1}{4} (2 + 3\xi - \xi^3)$$

$$\overline{N}_1 = \frac{1}{4} (1 - \xi - \xi^2 + \xi^3) \qquad ; \qquad \overline{N}_2 = \frac{1}{4} (-1 - \xi + \xi^2 + \xi^3) \qquad (1.11a)$$

with the local coordinate ξ defined as (Figure 1.4)

$$\xi = \frac{2}{l^{(e)}} (x - x_c) \qquad \text{and} \qquad x_c = \frac{x_1 + x_2}{2} \qquad (1.11b)$$

Eq.(1.10) can be written in compact form as

$$w = \mathbf{N} \ \mathbf{a}^{(e)} \qquad (1.12a)$$

where

$$\mathbf{N} = \left[N_1, \overline{N}_1, N_2, \overline{N}_2\right] \qquad \text{and} \qquad \mathbf{a}^{(e)} = \left[w_1, \left(\frac{dw}{dx}\right)_1, w_2, \left(\frac{dw}{dx}\right)_2\right]^T \qquad (1.12b)$$

are respectively the shape functions matrix and the nodal displacement vector for the element, including the nodal deflections w_1 and w_2 and the nodal rotations $\left(\frac{dw}{dx}\right)_1$ and $\left(\frac{dw}{dx}\right)_2$.

The shape functions (1.11) coincide with Hermite polynomials [AS]. Figure 1.4 shows that N_1 and N_2 take a unit value at a node, zero at the other node and their first derivatives are zero at both nodes, while the opposite occurs with $\bar{N}_1$ and $\bar{N}_2$. We will see in Chapter 5 that choosing the deflection and the rotations as the nodal variables for thin plate bending elements leads, in general, to non-conforming situations where the rotations are discontinuous along the common element boundaries. This is not so for Euler-Bernoulli beam elements for which the rotation takes a single value at the nodes shared by two adjacent elements.

It is deduced from Eq.(1.11b) that $\dfrac{dx}{d\xi} = \dfrac{l^{(e)}}{2}$, and thus

$$dx = \frac{l^{(e)}}{2}d\xi \quad ; \quad \frac{dw}{dx} = \frac{2}{l^{(e)}}\frac{dw}{d\xi} \quad \text{and} \quad \frac{d^2w}{dx^2} = \frac{4}{(l^{(e)})^2}\frac{d^2w}{d\xi^2} \qquad (1.13)$$

The rotation can be expressed in terms of the nodal displacements from Eqs.(1.10), (1.11) and (1.13) as

$$\frac{dw}{dx} = \left[\frac{dN_1}{dx}, \frac{d\bar{N}_1}{dx}, \frac{dN_2}{dx}, \frac{d\bar{N}_2}{dx}\right]\mathbf{a}^{(e)} = \frac{2}{l^{(e)}}\left[(-3+3\xi^2), (-1-2\xi+3\xi^2),\right.$$

$$\left.(3-\xi^2), (-1+2\xi+3\xi^2), (-1+2\xi+3\xi^2)\right]\mathbf{a}^{(e)} = \hat{\mathbf{N}}\mathbf{a}^{(e)}$$

$$(1.14)$$

The virtual deflection and virtual rotation fields are interpolated in term of the respective nodal variables as

$$\delta w = \mathbf{N}\delta\mathbf{a}^{(e)} \quad , \quad \delta\left(\frac{dw}{dx}\right) = \hat{\mathbf{N}}\delta\mathbf{a}^{(e)} \qquad (1.15a)$$

where

$$\delta\mathbf{a}^{(e)} = \left[\delta w_1, \delta\left(\frac{dw}{dx}\right)_1, \delta w_2, \delta\left(\frac{dw}{dx}\right)_2\right]^T \qquad (1.15b)$$

and $\mathbf{N}$ and $\hat{\mathbf{N}}$ are defined in Eqs.(1.12b) and (1.14), respectively.

The curvature at a point within the element is obtained in terms of the nodal DOFs using Eqs.(1.10) and (1.13) by

$$\kappa = \frac{d^2w}{dx^2} = \frac{4}{(l^{(e)})^2}\left(\frac{d^2N_1}{d\xi^2}w_1 + \frac{l^{(e)}}{2}\frac{d^2\bar{N}_1}{d\xi^2}\left(\frac{dw}{dx}\right)_1 + \frac{d^2N_2}{d\xi^2}w_2 + \frac{l^{(e)}}{2}\frac{d^2\bar{N}_2}{d\xi^2}\left(\frac{dw}{dx}\right)_2\right)$$

$$= \left[\frac{6\xi}{(l^{(e)})^2}, \frac{(-1+3\xi)}{l^{(e)}}, \frac{-6\xi}{(l^{(e)})^2}, \frac{(1+3\xi)}{l^{(e)}}\right]\begin{Bmatrix} w_1 \\ \left(\dfrac{dw}{dx}\right)_1 \\ w_2 \\ \left(\dfrac{dw}{dx}\right)_2 \end{Bmatrix} = \mathbf{B}_b\mathbf{a}^{(e)} \qquad (1.16a)$$

where $\mathbf{B}_b$ is the bending strain (or curvature) matrix for the element.

From Eq. (1.16a) we deduce the relationship between the virtual curvature and the virtual nodal displacement as

$$\delta\kappa = \mathbf{B}_b\delta\mathbf{a}^{(e)} \tag{1.16b}$$

The bending moment is expressed in terms of the nodal displacements from Eqs.(1.5) and (1.16a) as

$$M = (EI_y)\mathbf{B}_b\mathbf{a}^{(e)} \tag{1.16c}$$

Note that the curvature and the bending moment vary linearly within the element.

1.3.2 Discretized equilibrium equations for the element. Element stiffness matrix

The PVW for an individual element of length $l^{(e)}$ is written as (Eq.(1.7))

$$\int_{l^{(e)}} \delta\kappa M dx = \int_{l^{(e)}} \left(\delta w f_z + \delta\left(\frac{dw}{dx}\right)m\right) dx + \sum_{i=1}^{2}\left[\delta w_i F_{z_i} + \delta\left(\frac{dw}{dx}\right)_i \mathcal{M}_i\right] \tag{1.17}$$

where F_{z_i} and $\mathcal{M}_i$ are the equilibrating nodal forces transmitted to node i by the adjacent elements (Figure 1.4).

Substituting Eqs.(1.12a), (1.15a) and (1.16b) into (1.17) gives after simplification of the virtual nodal displacements

$$\int_{l^{(e)}} \mathbf{B}_b^T M dx = \int_{l^{(e)}} (\mathbf{N}^T f_x + \hat{\mathbf{N}}^T m)dx + \mathbf{q}^{(e)} \tag{1.18a}$$

where
$$\mathbf{q}^{(e)} = [F_{z_1}, \mathcal{M}_1, F_{z_2}, \mathcal{M}_2]^T \tag{1.18b}$$

is the *equilibrating nodal force vector* for the element.

Substituting the constitutive expression for M (Eq.(1.16c)) into (1.18a) and using Eqs.(1.13) gives

$$\left(\int_{-1}^{+1} \mathbf{B}_b^T\mathbf{B}_b\frac{EI_y l^{(e)}}{2} d\xi\right)\mathbf{a}^{(e)} - \int_{-1}^{+1}(\mathbf{N}^T f_x + \hat{\mathbf{N}}^T m)\frac{l^{(e)}}{2}d\xi = \mathbf{q}^{(e)} \tag{1.19a}$$

Eq.(1.19a) can be written in compact matrix form as

$$\mathbf{K}^{(e)}\mathbf{a}^{(e)} - \mathbf{f}^{(e)} = \mathbf{q}^{(e)} \tag{1.19b}$$

where $\mathbf{K}^{(e)}$ and $\mathbf{f}^{(e)}$ are, respectively, the *stiffness matrix* and the *equivalent nodal force vector* for the element.

The element stiffness matrix has the general form

$$\mathbf{K}^{(e)} = \int_{-1}^{+1} \mathbf{B}_b^T \mathbf{B}_b \, \frac{EI_y l^{(e)}}{2} \, d\xi \tag{1.19c}$$

Integrals in beam elements can be computed either analytically or by numerical integration [Ral,WR]. In practice the Gauss integration rule (also called Gauss quadrature) is chosen. For 1D integrals, this is generically written for a function $f(\xi)$ as

$$\int_{-1}^{+1} f(\xi)d\xi = \sum_{p=1}^{n_p} f(\xi_p)W_p \tag{1.19d}$$

where n_p is the number of quadrature points, $f(\xi_p)$ is the value of function $f(\xi)$ at the ξ_p quadrature point and W_p is the corresponding weight. For details of the 1D Gauss quadrature see Appendix C and [On4,ZTZ].

The stiffness matrix of the 2-noded Euler-Bernoulli beam element can be computed analytically or using a 2-point Gauss quadrature as

$$\mathbf{K}^{(e)} = \left(\frac{EI_y}{l^3}\right)^{(e)} \begin{bmatrix} 12 & 6l^{(e)} & -12 & 6l^{(e)} \\ \ddots & 4(l^{(e)})^2 & -6l^{(e)} & 2(l^{(e)})^2 \\ & \ddots & 12 & -6l^{(e)} \\ \text{sym.} & & \ddots & 4(l^{(e)})^2 \end{bmatrix} \tag{1.20}$$

Those readers familiar with matrix analysis of structures will recognize the coincidence of Eq.(1.20) with the expression derived from the standard slope-deflection relationships of Strength of Materials (see Eq.(1.37b) of [On4]) [Li,Pr,ZTZ]. The reason is that the cubic deflection field assumed in Eq.(1.10) coincides with that obtained by integrating the differential equations of equilibrium for a beam segment (Example 1.1).

1.3.3 Equivalent nodal force vector for the element

The equivalent nodal force vector for a distributed vertical loading f_z and a distributed bending moment m is deduced from Eq.(1.19a) as

$$\mathbf{f}^{(e)} = \begin{Bmatrix} f_{z_1} \\ m_1 \\ f_{z_2} \\ m_2 \end{Bmatrix} = \int_{-1}^{+1} \left(\mathbf{N}^T f_z + \hat{\mathbf{N}}^T m\right) \frac{l^{(e)}}{2} \, d\xi \tag{1.21a}$$

a) Sign for equilibrating nodal forces ($F_{z_i}, \mathcal{M}_i$) and equivalent nodal forces (f_{z_i}, m_i)

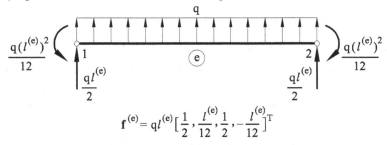

b) Equivalent nodal forces for uniform load q

$$\mathbf{f}^{(e)} = ql^{(e)}\left[\frac{1}{2}, \frac{l^{(e)}}{12}, \frac{1}{2}, -\frac{l^{(e)}}{12}\right]^{T}$$

Fig. 1.5 Two-noded Euler-Bernoulli beam element a) Sign criterion for equilibrating forces and equivalent nodal forces. b) Equivalent nodal forces for uniformly distributed loading

The signs for the components of $\mathbf{f}^{(e)}$ coincide with those of $\mathbf{q}^{(e)}$ (Figure 1.5).

For a uniformly distributed loading of intensity $f_z = q$ (Figure 1.5) and $m = 0$ then

$$\mathbf{f}^{(e)} = ql^{(e)}\left[\frac{1}{2}, \frac{l^{(e)}}{2}, \frac{1}{2}, -\frac{l^{(e)}}{2}\right]^{T} \tag{1.21b}$$

For a uniformly distributed moment m (with $f_z = 0$), then

$$\mathbf{f}^{(e)} = m[-1, 0, 1, 0]^{T} \tag{1.21c}$$

Hence, a distributed vertical load generates nodal bending moments m_i in $\mathbf{f}^{(e)}$. Conversely, a uniformly distributed bending moment m yields vertical equivalent nodal forces with opposite sign and zero nodal bending moments. These results are a consequence of the link between the deflection field and the nodal rotations via Eq.(1.10). It is interesting that the terms of $\mathbf{f}^{(e)}$ in Eq.(1.21b) coincide (with opposite sign) with the vertical reactions and the couples at the ends of a clamped beam under uniformly distributed loading (Figure 1.5b). This coincidence is fortuitous and can not be generalized to other loading types (Example 1.2).

The equivalent nodal force vector for *external concentrated loads* acting directly at a node i is

$$\mathbf{p}_i = [P_{z_i}, M_i^c]^T \tag{1.21d}$$

where P_{z_i} and M_i^c are the vertical force and the bending moment acting at node i (Figure 1.1). The components of $\mathbf{p}_i$ are directly assembled in the global nodal force vector $\mathbf{f}$.

1.3.4 Global equilibrium equations

The global stiffness equations $\mathbf{Ka} = \mathbf{f}$ are obtained by setting the equilibrium of nodal forces at the beam nodes in the standard manner [On4]. The stiffness matrix $\mathbf{K}$ and the equivalent nodal force vector $\mathbf{f}$ for the beam are obtained by assembling the element contributions, as it is usual in the FEM [On4,ZT2,ZTZ].

Nodal reactions at prescribed nodes can be treated as external concentrated loads and assembled into the global equivalent nodal force vector (Examples 1.3 and 1.4). Solution of the global system of equations for the nodal deflections and rotations requires prescribing the deflection and/or the rotation at the constrained nodes [On4].

The reactions at the constrained nodes can be obtained a posteriori from the computed displacement field as

$$\mathbf{r} = \mathbf{Ka} - \mathbf{f}^{\text{ext}} \tag{1.22}$$

where $\mathbf{r}$ contains the vertical force and the moment reactions at the constrained nodes and $\mathbf{f}^{\text{ext}}$ is obtained by assembling the equivalent nodal force vector $\mathbf{f}^{(e)}$ due to external loads only. Other alternatives for computing the reactions are explained in [On4].

We note that the signs for the reactions coincide with those for the equivalent nodal forces; i.e. couples are positive if they act anticlockwise and vertical forces are positive if they point towards the vertical z axis.

Once the nodal deflections and the rotations have been obtained, the bending moment at any point of an element is computed as

$$M = EI\ \kappa = EI_y\ \mathbf{B}_b\ \mathbf{a}^{(e)} \tag{1.23}$$

The shear force distribution over the element can be obtained from the equilibrium equations (Example 1.1), the moment-curvature relationship (Eq.(1.5)) and Eqs.(1.10), (1.11a) and (1.13) as

$$Q = -\frac{dM}{dx} = -EI_y\frac{d^3w}{dx^3} = \frac{EI_y}{(l^{(e)})^3}[12, 6l^{(e)}, -12, 6l^{(e)}]\mathbf{a}^{(e)} \tag{1.24}$$

Figure 1.2 shows the sign criterium for the shear force Q. This force is constant over the element. This is consequence of the cubic interpolation chosen for the deflection field.

The shear force and the couple at the element nodes can be computed by noting that (Figures 1.2 and 1.5)

$$Q_1 = -F_{z_1} , \; M_1 = \mathcal{M}_1 , \; Q_2 = -F_{z_2} , \; M_2 = \mathcal{M}_2 \qquad (1.25a)$$

and, hence,

$$[-Q_1, -M_1, Q_2, M_2]^T = \mathbf{q}^{(e)} = \mathbf{K}^{(e)} \mathbf{a}^{(e)} - \mathbf{f}^{(e)} \qquad (1.25b)$$

where Q_1, M_1 and Q_2, M_2 are the shear forces and the couples at nodes 1 and 2 of the element, respectively. The equivalence between the components of $\mathbf{q}^{(e)}$ and the shear forces and the couples at the element nodes is useful in practice.

Example 1.1: Derive the displacement field for an unloaded beam segment by integrating the differential equation of equilibrium.

- Solution

Let us consider the beam segment of length l of Figure 1.6 with a shear force and a couple acting at each end.

The equilibrium of an unloaded infinitesimal beam segment of length dx leads to the well known differential equations (Figure 1.7) [Ti2]:

Equilibrium of couples: $\dfrac{dM}{dx} = -Q$

Equilibrium of vertical forces: $\dfrac{dQ}{dx} = 0$

Differentiating the first equation and making use of the second one and of the bending moment-curvature relationship of Eq.(1.5) gives

$$\frac{d^4 w}{dx^4} = 0$$

The solution of this differential equation is the cubic polynomical

$$w(x) = a_1 + a_2 + a_3 x^2 + a_4 x^3$$

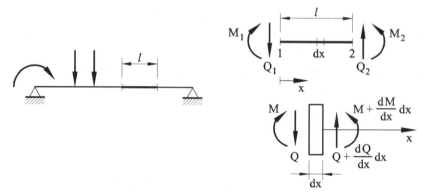

Fig. 1.6 Forces and couples acting at the ends of beam segments of finite and infinitesimal lengths

The conditions at the unloaded beam segment ends are (Figure 1.6)

$$w = w_1 \quad \text{and} \quad \frac{dw}{dx} = \left(\frac{dw}{dx}\right)_1 \quad \text{at} \quad x = 0$$

$$w = w_2 \quad \text{and} \quad \frac{dw}{dx} = \left(\frac{dw}{dx}\right)_2 \quad \text{at} \quad x = l$$

which lead to the following equations system

$$\left\{ \begin{array}{c} w_1 \\ \left(\dfrac{dw}{dx}\right)_1 \\ w_2 \\ \left(\dfrac{dw}{dx}\right)_2 \end{array} \right\} = \begin{bmatrix} 1 & 0 & 0 & 0 \\ 0 & 1 & 0 & 0 \\ 1 & l & l^2 & l^3 \\ 0 & 1 & 2l & 3l^2 \end{bmatrix} \left\{ \begin{array}{c} a_1 \\ a_2 \\ a_3 \\ a_4 \end{array} \right\}$$

from which the parameters a_i can be obtained. Substituting these into the cubic deflection field gives

$$w(x) = f_1(x)\, w_1 + f_2(x)\, \frac{l}{2}\left(\frac{dw}{dx}\right)_1 + f_3(x)\, w_2 + f_4(x)\, \frac{l}{2}\left(\frac{dw}{dx}\right)_2$$

where

$$f_1(x) = 1 - 3\left(\frac{x}{l}\right)^2 + 2\left(\frac{x}{l}\right)^3 \quad ; \quad f_2(x) = \frac{2x}{l} - 4\left(\frac{x}{l}\right)^2 + 2\left(\frac{x}{l}\right)^3$$

$$f_3(x) = 3\left(\frac{x}{l}\right)^2 - 2\left(\frac{x}{l}\right)^3 \quad ; \quad f_4(x) = -2\left(\frac{x}{l}\right)^2 + 2\left(\frac{x}{l}\right)^3$$

The coincidence of functions $f_i(\xi)$ with the Hermite shape functions (1.11a) can be recognized after making a simple transformation to the natural coordinate system, i.e. changing x by $\frac{l}{2}(1 + \xi)$ (Eq.(1.11b)).

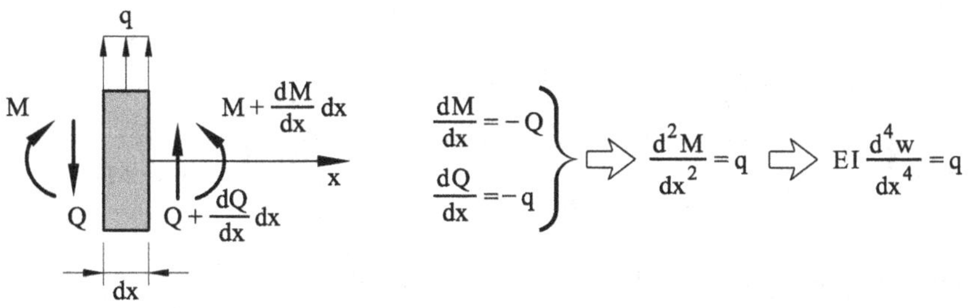

Fig. 1.7 Differential equations for an infinitesimal beam segment under distributed loading

Using the PVW it is easy to deduce that the stiffness matrix for the unloaded beam segment of length l of Figure 1.6 coincides precisely with that for the 2-noded Hermite beam element (Eq.(1.20)).

Example 1.2: Obtain the reactions in a clamped beam of length L under uniformly distributed loading $f_z = q$ by integrating the differential equation of equilibrium.

- Solution

The deflection in this case satisfies the differential equation (Figure 1.7)

$$EI_y \frac{d^4 w}{dx^4} = q$$

where q is the intensity of the distributed load acting in the direction of the vertical axis.

The solution of the above equation is the 4th order polynomial

$$w = a_0 + a_1 x + a_2 x^2 + a_3 x^3 + a_4 x^4$$

with $a_4 = \dfrac{q}{24EI_y}$. The parameters a_0, a_1, a_2 and a_3 are obtained from the boundary conditions at the clamped ends

$$w = \frac{dw}{dx} = 0 \quad \text{at} \quad x = 0 \quad \text{and} \quad x = L$$

which gives $a_0 = a_1 = 0$ and the system

$$a_2 + a_3 L = \frac{qL}{24EI_y}$$

$$2a_2 + 3a_3 L = \frac{qL^2}{6EI_y}$$

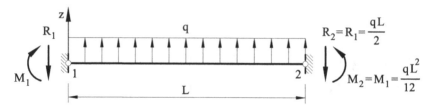

Fig. 1.8 End reactions for a clamped beam under uniformly distributed loading

giving $a_2 = \dfrac{qL^2}{24EI_y}$ and $a_3 = -\dfrac{qL}{12EI_y}$.

The expression for the deflection field is

$$w(x) = \frac{1}{EI_y}\left(\frac{qL^2}{24}x^2 - \frac{qL}{12}x^3 + \frac{qx^4}{24}\right)$$

The bending moment reactions at the clamped ends are

$$M_1 = -(M)_{x=0} = -EI_y\left(\frac{dw}{dx^2}\right)_{x=0} = -\frac{qL^2}{12}$$

$$M_2 = (M)_{x=L} = EI_y\left(\frac{d^2w}{dx^2}\right)_{x=L} = \frac{qL^2}{12}$$

and the vertical force reactions

$$R_1 = -(Q)_{x=0} = EI_y\left(\frac{d^3w}{dx^3}\right)_{x=0} = -\frac{qL}{2}$$

$$R_2 = (Q)_{x=L} = -EI_y\left(\frac{d^3w}{dx^3}\right)_{x=L} = -\frac{qL}{2}$$

Note that the signs for the reactions coincide with those for the equilibrating nodal forces (Figure 1.5). The reactions are displayed in Figure 1.8.
The forces acting on the supports are equal to the above reactions, *but with the opposite sign*. Grouping these forces in a vector $\mathbf{f}$ gives

$$\mathbf{f} = \left[\frac{qL}{2}\ ,\ \frac{qL^2}{12}\ ,\ \frac{qL}{2}\ ,\ -\frac{qL^2}{12}\right]^T$$

Vector $\mathbf{f}$ coincides with the equivalent nodal force vector $\mathbf{f}^{(e)}$ for the 2-noded Euler-Bernoulli beam element of length $l^{(e)} = L$ (Figure 1.5 and Eq.(1.21b)). This coincidence is fortuitous and can not be generalized to other loading cases. However, it allows us to interpret $\mathbf{f}^{(e)}$ as a nodal force system which *equilibrates* the external loads.

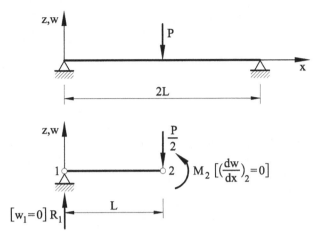

Fig. 1.9 Simple supported beam under central point loading. Analysis with a single 2-noded Euler-Bernoulli element

Example 1.3: Obtain for the simply supported beam of Figure 1.9 the deflection and the rotation at the center and the vertical reaction at the supports using a single 2-noded Euler-Bernoulli beam element.

- Solution

The global equilibrium equations for the beam are written in matrix form as (taking into account symmetry conditions)

$$
\left(\frac{EI_y}{L^3}\right)
\begin{bmatrix}
12 & 6L & -12 & 6L \\
\ddots & 4L^2 & -6L & 2L^2 \\
 & & 12 & -6L \\
\text{Symm.} & & \ddots & 4L^2
\end{bmatrix}
\begin{Bmatrix}
w_1 \\
\left(\dfrac{dw}{dx}\right)_1 \\
w_2 \\
\left(\dfrac{dw}{dx}\right)_2
\end{Bmatrix}
=
\begin{Bmatrix}
R_1 \\
0 \\
-\dfrac{P}{2} \\
M_2
\end{Bmatrix}
$$

with $w_1 = 0$ and $\left(\dfrac{dw}{dx}\right)_2 = 0$.

Eliminating the prescribed DOFs gives

$$
\left(\frac{EI_y}{L^3}\right)
\begin{bmatrix}
4L^2 & -6L \\
-6L & 12
\end{bmatrix}
\begin{Bmatrix}
\left(\dfrac{dw}{dx}\right)_1 \\
w_2
\end{Bmatrix}
=
\begin{Bmatrix}
0 \\
-\dfrac{P}{2}
\end{Bmatrix}
$$

which yields

$$w_2 = -\frac{PL^3}{6EI_y} \quad ; \quad \left(\frac{dw}{dx}\right)_1 = -\frac{PL^2}{4EI_y}$$

$$R_1 = \frac{P}{2} \quad \text{and} \quad M_2 = \frac{PL}{2}$$

The nodal solution coincides with the exact values [Ti2]. The exact deflection field for this problem is a cubic polynomial (Example 1.1) and, therefore, the finite element solution coincides with the exact one throughout the beam.

Example 1.4: Obtain the end displacements and the reactions for the cantilever beam of Figure 1.10 using a single 2-noded Euler-Bernoulli element for
a) A uniformly distributed loading $f_z = q$,
b) A uniformly distributed bending moment m.

- Solution

a) Uniformly distributed loading q

Using Eqs.(1.19b), (1.20) and (1.21b) the following global system is obtained

$$\left(\frac{EI_y}{L^3}\right) \begin{bmatrix} 12 & 6L & -12 & 6L \\ \ddots & 4L^2 & -6L & 2L^2 \\ & & 12 & -6L \\ \text{Sym.} & & \ddots & 4L^2 \end{bmatrix} \begin{Bmatrix} w_1 \\ \left(\dfrac{dw}{dx}\right)_1 \\ w_2 \\ \left(\dfrac{dw}{dx}\right)_2 \end{Bmatrix} = \begin{Bmatrix} R_1 + \dfrac{qL}{2} \\ M_1 + \dfrac{qL^2}{12} \\ \dfrac{qL}{2} \\ -\dfrac{qL^2}{12} \end{Bmatrix}$$

with $w_1 = 0$ and $\left(\dfrac{dw}{dx}\right)_1 = 0$.

Eliminating the first two rows and columns corresponding to the prescribed DOFs at the clamped end and solving the system gives

$$w_2 = \frac{qL^4}{8EI_y} \quad ; \quad \left(\frac{dw}{dx}\right)_2 = \frac{qL^3}{6EI_y}$$

$$R_1 = -qL \quad \text{and} \quad M_1 = -\frac{qL^2}{2}$$

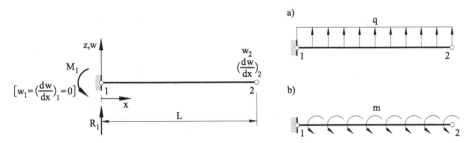

Fig. 1.10 Cantilever beam under uniformly distributed vertical loading and moment. Analysis with one 2-noded Euler-Bernoulli element

which coincide with the exact values. This coincidence occurs only at the nodes since the exact solution is a quartic polynomial (Example 1.3) whereas the finite element solution is a cubic. The exact nodal values are a consequence of the coincidence of the element stiffness matrix and the equivalent nodal force vector with the expressions of standard matrix analysis (Example 1.2).

b) Uniformly distributed bending moment m

The global stiffness equations coincide with those of case a) with the r.h.s. changed to (Eq.(1.21c))

$$\mathbf{f} = [R_1 - m, M_1, m, 0]^T$$

Solution of the global equation system gives

$$w_2 = \frac{mL^3}{3EI_y} \quad , \quad \left(\frac{dw}{dx}\right)_2 = \frac{mL^2}{2EI_y} \quad , \quad R_1 = 0 \quad , \quad M_1 = -mL$$

The solution coincides with the exact one throughout the beam. The reason is that the exact solution is a cubic polynomial which can be exactly matched by the cubic approximation chosen. The reader can verify this coincidence.

The coincidence of the nodal values with the exact ones in the two previous examples is not common in finite element analysis and, in fact, only happens for some 1D problems for which the shape functions satisfy the homogeneous form of the equilibrium differential equations [ZTZ] and Section 2.3.5 of [On4].

It is easy to verify that this condition is fulfilled for the two problems considered. The homogeneous differential equation for the beam is

$$EI_y \frac{d^4w}{dx^4} = 0 \tag{1.26}$$

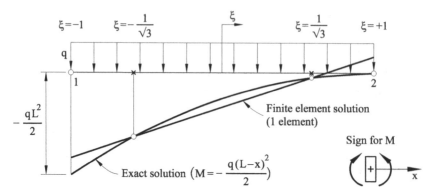

Fig. 1.11 Cantilever beam of length L under uniformly distributed loading. Bending moment distribution

which is satisfied by the cubic approximation used. Unfortunately, this property does not hold for plate problems.

Figure 1.11 shows the bending moment distribution obtained in Example 1.4a and the exact one. Both solutions coincide at the two Gauss points for $\xi = \pm \frac{1}{\sqrt{3}}$ that integrate exactly the element stiffness matrix (Section 1.3.2 and Appendix C). What may look as a coincidence is in fact due to the property of the Gauss points which *are the optimal sampling points for evaluating the strains and stresses*. In fact, the numerical results at the Gauss points have a higher order approximation than if sampled at any other element point (Section 6.7 of [On4] and [Hu,ZTZ]). For the example considered here, the exact bending moment field is a parabola which intersects the linear finite element distribution at the two Gauss points. Consequently, the bending moment values are exact at these points.

1.4 ROTATION-FREE EULER-BERNOULLI BEAM ELEMENTS

An alternative family of Euler-Bernoulli beam elements can be derived by approximating the curvature over selected control domains in terms of the differences in the slopes at appropriate nodal points. The slopes are in turn expressed as the differences in the nodal deflections. *This leads to an equilibrium equation where the nodal deflections are the only variables.*

In the following sections we will study two families of the so-called *rotation-free beam elements*: the cell-centred beam element where the control domain coincides with a 2-noded element, and the cell-vertex beam element for which the control domain is centred around a node.

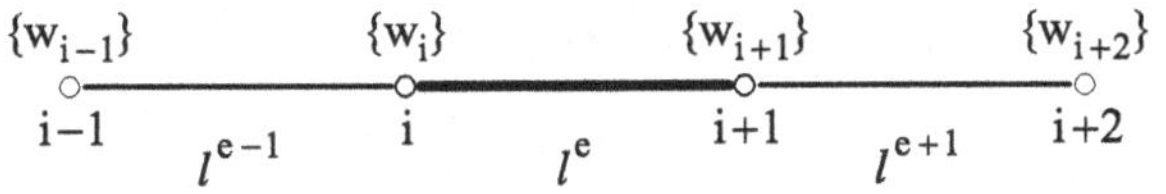

{w$_{i-1}$}   {w$_i$}   {w$_{i+1}$}   {w$_{i+2}$}

i-1   i   i+1   i+2

l^{e-1} l^e l^{e+1}

Fig. 1.12 Patch of 3 elements for computing the curvature over the central element e with nodes $i, i+1$

1.4.1 Cell-centred beam (CCB) element

Let us consider the patch of three beam elements shown in Figure 1.12. From the definition of the curvature $\left(\kappa = d^2w/dx^2\right)$ we write for the central element e with nodes $i, i+1$

$$\int_{le} \left(\kappa - \frac{d^2w}{dx^2}\right) dx = 0 \qquad (1.27)$$

where l^e is the length of element e.

Eq.(1.27) simply states that the curvature is equal to the second derivative of the deflection field in an average sense over the element.

Let us assume that the curvature is *constant* over element e and equal to κ^e. From Eq.(1.27) we obtain

$$\kappa^e = \frac{1}{l^e} \int_{le} \frac{d^2w}{dx^2} dx = \frac{1}{l^e} \left[\left(\frac{dw}{dx}\right)_{i+1} - \left(\frac{dw}{dx}\right)_i\right] \qquad (1.28)$$

Eq.(1.28) is the standard way for computing the curvature at the element center as the difference between the slopes at the two end nodes. This procedure is common in centred finite difference methods [Ral,PFTV].

Eq.(1.25) defines a constant bending moment over the element of value

$$M^e = (EI_y)^e \kappa^e = \left(\frac{EI_y}{l}\right)^e \left[\left(\frac{dw}{dx}\right)_{i+1} - \left(\frac{dw}{dx}\right)_i\right] \qquad (1.29)$$

Let us also assume a linear interpolation for the deflection within each 2-noded element, i.e.

$$w = \sum_{i=1}^{2} N_i w_i \qquad (1.30)$$

where N_i are the standard C^o continuous linear shape functions for the 2-noded Lagrange element $(N_i = \frac{1}{2}(1 + \xi\xi_i)$, Figure 2.4 and [On4]).

The deflection gradients are *discontinuous* between elements and therefore the nodal slopes are not uniquely defined at the element edges. This ambiguity is overcome by a simple averaging of the nodal slopes, giving

$$\left(\frac{dw}{dx}\right)_i = \frac{1}{2}\left[\left(\frac{dw}{dx}\right)_i^{(e)} + \left(\frac{dw}{dx}\right)_i^{(e-1)}\right] = \frac{1}{2}\left[\frac{w_{i+1} - w_i}{l^e} + \frac{w_i - w_{i-1}}{l^{e-1}}\right]$$
(1.31a)

$$\left(\frac{dw}{dx}\right)_{i+1} = \frac{1}{2}\left[\left(\frac{dw}{dx}\right)_{i+1}^{(e)} + \left(\frac{dw}{dx}\right)_{i+1}^{(e+1)}\right] = \frac{1}{2}\left[\frac{w_{i+1} - w_i}{l^e} + \frac{w_{i+2} - w_{i+1}}{l^{e+1}}\right]$$
(1.31b)

Substituting the above equations into Eq.(1.28) yields

$$\kappa^e = \mathbf{B}_b \bar{\mathbf{w}}^{(e)}$$
(1.31c)

with

$$\mathbf{B}_b = \frac{1}{2l^e l^{e+1} l^{e-1}}\left[l^{e+1}, -l^{e+1}, -l^{e-1}, l^{e-1}\right]$$
(1.31d)

$$\bar{\mathbf{w}}^{(e)} = [w_{i-1}, w_i, w_{i+1}, w_{i+2}]^T$$
(1.31e)

Substituting Eq.(1.31c) into the PVW and following the standard process of Section 1.3.2 yields the element stiffness matrix as

$$\mathbf{K}^{(e)} = l^e \mathbf{B}_b^T (EI_y)^e \mathbf{B}_b$$
(1.32)

In Eq.(1.32) we have assumed that EI_y is constant in the element.

The equilibrium equations for the element (termed CCB for Cell Centred Beam element) involve the deflections at the four nodes belonging to the three-element patch associated to element e. This increases the bandwidth of the global stiffness matrix. What is relevant here is *that the rotations have been eliminated as nodal variables* and the nodal deflection are the only DOFs. This explains why these elements are termed "rotation free" beam elements [JO,OZ,OZ2].

The equivalent nodal force vector is obtained in the standard way via the linear displacement field. For a uniformly distributed load $f_z = q$

$$\mathbf{f}^{(e)} = \frac{ql^e}{2}[0, 1, 1, 0]^T$$
(1.33)

Point loads acting at a node are directly assigned to the node, as usual. An external bending moment can not be directly accounted for in this formulation. For a mesh of equal length elements the solution is to replace the moment by two equivalent vertical loads of value $\frac{M_i}{l^e}$ with opposite sign acting at the two adjacent nodes to node i.

The stiffness equations are assembled as usual leading to the global system $\mathbf{Kw} = \mathbf{f}$ with $\mathbf{w} = [w_1, w_2, \cdots, w_N]$ where N is the total number of nodes in the mesh. Solution of the system yields the deflections at all nodes. This requires prescribing the constrained DOFs as explained below.

Boundary conditions

The prescribed nodal deflection values are imposed when solving the global system of equations in the standard manner. Prescribed rotations must be introduced when building the element curvature matrix as detailed next.

Free or simply supported node

We consider that the left-end node of the beam is free or simply supported (Figure 1.13a). The rotation at the prescribed node i is computed as

$$\left(\frac{dw}{dx}\right)_i^{(e)} = \frac{w_{i+1} - w_i}{l^e} \tag{1.34}$$

The rotation at node $i + 1$ is expressed by Eq.(1.31b). Substituting Eqs.(1.34) and (1.31b) into (1.28) gives

$$\kappa^e = \frac{1}{2(l^e)^2 l^{e+1}} [0, l^{e+1}, -(l^e + l^{e+1}), l^e] \begin{Bmatrix} w_{i-1} \\ w_i \\ w_{i+1} \\ w_{i+2} \end{Bmatrix} = \mathbf{B}_b \bar{\mathbf{w}}^{(e)} \tag{1.35}$$

The element stiffness matrix is computed by Eq.(1.32). Note that a fictitious node $i - 1$ has been introduced into $\bar{\mathbf{w}}^{(e)}$ in Eq.(1.35). This is convenient in order to keep a 4×4 dimension for all the element stiffness matrices. The stiffness associated to the nodal deflection w_{i-1} is obviously zero. This does not affect the equation solution process as the value of w_{i-1} is prescribed as zero.

We follow a similar approach for a free or simply supported node at the right-end of the beam. The element curvature is computed as

$$\kappa^e = \frac{1}{2(l^e)^2 l^{e-1}} [l^e, -(l^e + l^{e-1}), l^{e-1}, 0] \begin{Bmatrix} w_{i-1} \\ w_i \\ w_{i+1} \\ w_{i+2} \end{Bmatrix} = \mathbf{B}_b \bar{\mathbf{w}}^{(e)} \tag{1.36}$$

The fictitious deflection w_{i+2} is now prescribed as zero.

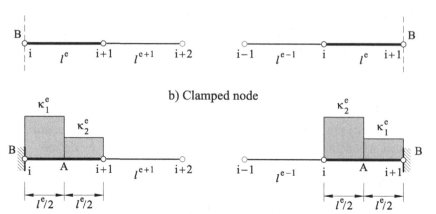

Fig. 1.13 CCB element. (a) Free or simply supported node B at left and right ends of a beam. (b) Clamped or symmetry node B at left and right beam ends. Definition of domains 1 and 2 splitting the element containing the clamped node

Clamped or symmetry edge

Let us consider that the left-end node i is clamped or lays on a symmetry axis. An obvious possibility is to introduce the condition of zero rotation when computing the curvature in the element adjacent to the clamped edge. Thus, from Eqs.(1.28) and (1.31b) we obtain

$$\kappa^e = \frac{1}{l^e}\left(\frac{dw}{dx}\right)_{i+1} = \frac{1}{2(l^e)^2 l^{e+1}}\,[0, -l^{e+1}, (l^e + l^{e+1}), -l^e]\begin{Bmatrix} w_{i-1} \\ w_i \\ w_{i+1} \\ w_{i+2} \end{Bmatrix} = \mathbf{B}_b \bar{\mathbf{w}}^e$$

$$(1.37)$$

Matrix $\bar{\mathbf{B}}_b$ substitutes $\mathbf{B}_b$ for computing the stiffness matrix in Eq.(1.33). A similar method is used for a right-end clamped edge [JO,OZ2].

A more accurate solution is obtained by splitting the element adjacent to the clamped node into two domains of equal size $\frac{l^e}{2}$ (Figure 1.13b). For a left-end clamped node, the curvature in the first domain is expressed as

$$\kappa_1^e = \frac{2}{l^e}\left[\frac{w_{i+1} - w_i}{l^e}\right] = \frac{2}{(l^e)^2}[0, -1, 1, 0]\begin{Bmatrix} w_{i-1} \\ w_i \\ w_{i+1} \\ w_{i+2} \end{Bmatrix} = \mathbf{B}_1 \bar{\mathbf{w}}^{(e)} \qquad (1.38a)$$

The stiffness matrix corresponding to the first domain is

$$\mathbf{K}_1^e = \frac{l^e}{2}\mathbf{B}_1^T (EI_y)^e \mathbf{B}_1 \tag{1.38b}$$

The curvature in the second domain is

$$\kappa_2^e = \frac{2}{l^e}\left[\frac{1}{2}\left(\frac{w_{i+1}-w_i}{l^e} + \frac{w_{i+2}-w_{i+1}}{l^{e+1}}\right) - \frac{w_{i+1}-w_i}{l^e}\right] =$$

$$= \frac{1}{(l^e)^2 l^{e+1}}[0, l^{e+1}, -(l^e + l^{e+1}), l^e]\left\{\begin{array}{c} w_{i-1} \\ w_i \\ w_{i+1} \\ w_{i+2} \end{array}\right\} = \mathbf{B}_2\bar{\mathbf{w}}^{(e)} \tag{1.39a}$$

and the corresponding stiffness matrix is

$$\mathbf{K}_2^{(e)} = \frac{l^e}{2}\mathbf{B}_2^T (EI_y)^e \mathbf{B}_2 \tag{1.39b}$$

The element stiffness matrix $\mathbf{K}^{(e)}$ is obtained by the sum of $\mathbf{K}_1^{(e)}$ and $\mathbf{K}_2^{(e)}$. Again the fictitious nodal rotation w_{i-1} is prescribed to a zero value during the solution process.

The same procedure applies for a right-end clamped or symmetry node (Figure 1.3b). The curvatures at the two splitting domains that split the element are

$$\kappa_1^e = \frac{2}{(l^e)^2}[0, -1, 1, 0]\left\{\begin{array}{c} w_{i-1} \\ w_i \\ w_{i+1} \\ w_{i+2} \end{array}\right\} = \mathbf{B}_1\bar{\mathbf{w}}^{(e)} \tag{1.40}$$

$$\kappa_2^e = \frac{1}{(l^e)^2 l^{e-1}}[-l^e, (l^e + l^{e-1}), -l^{e-1}, 0]\left\{\begin{array}{c} w_{i-1} \\ w_i \\ w_{i+1} \\ w_{i+2} \end{array}\right\} = \mathbf{B}_2\bar{\mathbf{w}}^{(e)} \tag{1.41}$$

Matrices $\mathbf{K}_1^{(e)}$ and $\mathbf{K}_2^{(e)}$ are computed as described above and their sum gives $\mathbf{K}^{(e)}$. The fictitious nodal deflection w_{i+2} is prescribed as zero.

Further details on the derivation of the CCB element can be found in [JO,OZ2]. The example given next shows the assembly of the global stiffness equations for a typical node and the comparison of the CCB element formulation with a centred finite difference scheme.

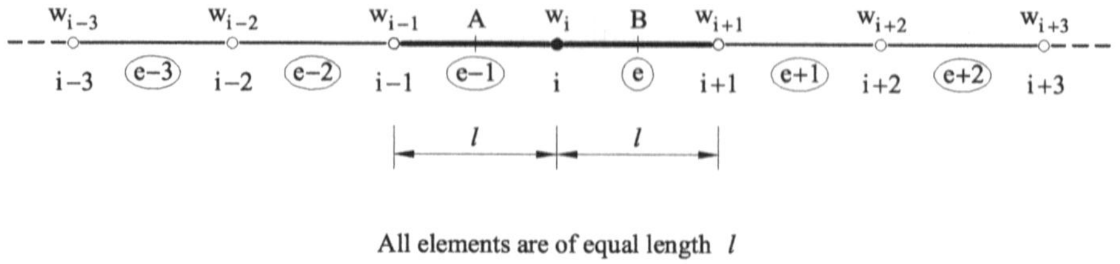

All elements are of equal length l

Fig. 1.14 Regular mesh of rotation-free beam elements

Example 1.5: Obtain the assembled stiffness equation for a node i in a mesh of CCB elements of equal length l. Verify the analogy with a centred finite difference scheme.

- Solution

Figure 1.14 shows a regular mesh of rotation-free beam element with the deflection as the only nodal variable. A uniformly distributed load of intensity q is assumed.

The stiffness matrix for the CCB element e in Figure 1.14 is deduced from Eq.(1.32) as

$$
\mathbf{K}^{(e)} = \frac{EI_y}{4l^3}
\begin{array}{c}
\begin{array}{cccc} w_{i-1} & w_i & w_{i+1} & w_{i+2} \end{array} \\
\left[
\begin{array}{cccc}
1 & -1 & -1 & 1 \\
-1 & 1 & 1 & -1 \\
-1 & 1 & 1 & -1 \\
1 & -1 & -1 & 1
\end{array}
\right]
\begin{array}{c} w_{i-1} \\ w_i \\ w_{i+1} \\ w_{i+2} \end{array}
\end{array}
$$

The global stiffness for node i is the assembly of the stiffness contributions from elements $e-2$, $e-1$, e and $e+1$. The assembly of $\mathbf{K}^{(e-2)}$, $\mathbf{K}^{(e-1)}$, $\mathbf{K}^{(e)}$ and $\mathbf{K}^{(e+1)}$ gives

$$
\mathbf{K} = \frac{EI_y}{4l^3}
\begin{array}{c}
\begin{array}{ccccccc} w_{i-3} & w_{i-2} & w_{i-1} & w_i & w_{i+1} & w_{i+2} & w_{i+3} \end{array} \\
\left[
\begin{array}{ccccccc}
1 & -1 & -1 & 1 & \cdot & \cdot & \cdot \\
-1 & 2 & 0 & -2 & 1 & \cdot & \cdot \\
-1 & 0 & 3 & -1 & 1 & 1 & \cdot \\
1 & -2 & -1 & 4 & -1 & -2 & 1 \\
\cdot & 1 & -2 & -1 & 3 & 0 & -1 \\
\cdot & \cdot & 1 & -2 & 0 & 2 & -1 \\
\cdot & \cdot & \cdot & 1 & -1 & -1 & 1
\end{array}
\right]
\begin{array}{c} w_{i-3} \\ w_{i-2} \\ w_{i-1} \\ w_i \\ w_{i+1} \\ w_{i+2} \\ w_{i+3} \end{array}
\end{array}
$$

All other terms outside the central band in $\mathbf{K}$ are zero.

The stiffness equation for node i (i.e. $K_{ij}w_j = f_i$) for a uniformly distributed load q is deduced from the row corresponding to w_i in $\mathbf{K}$ and Eq.(1.33) as

$$\frac{EI_y}{4l^3}(w_{i-3} - 2w_{i-2} - w_{i-1} + 4w_i - w_{i+1} - 2w_{i+2} + w_{i+3}) - ql = 0 \quad (1.42)$$

We will verify next that Eq.(1.42) coincides with the expression derived via a finite difference scheme centred at the element midpoint. The starting point is the differential equation of equilibrium (Figure 1.7)

$$EI_y\frac{d^4w}{dx^4} - q = 0 \quad (1.43)$$

The fourth derivative term is sampled at node i and computed as follows

$$\left(\frac{d^4w}{dx^4}\right)_i = \frac{1}{2l}\left[\left(\frac{d^3w}{dx^3}\right)_{i+1} - \left(\frac{d^3w}{dx^3}\right)_{i-1}\right]$$

The third derivatives are computed as

$$\left(\frac{d^3w}{dx^3}\right)_{i+1} = \frac{1}{2l^2}\left[\left(\frac{d^2w}{dx^2}\right)_{i+2} - \left(\frac{d^2w}{dx^2}\right)_i\right] ; \left(\frac{d^2w}{dx^3}\right)_{i-1} = \frac{1}{2l^2}\left[\left(\frac{d^2w}{dx^2}\right)_i - \left(\frac{d^2w}{dx^2}\right)_{i-2}\right]$$

Substituting these expressions into the previous one gives

$$\left(\frac{d^4w}{dx^4}\right)_i = \frac{1}{4l^2}\left[\left(\frac{d^2w}{dx^2}\right)_{i+2} - 2\left(\frac{d^2w}{dx^2}\right)_i + \left(\frac{d^2w}{dx^2}\right)_{i-2}\right] \quad (1.44)$$

with

$$\left(\frac{d^2w}{dx^2}\right)_{i+2} = \frac{1}{l^2}(w_{i+3} - 2w_{i+2} + w_{i+1}) ; \left(\frac{d^2w}{dx^2}\right)_{i-2} = \frac{1}{l^2}(w_{i-3} - 2w_{i-2} + w_{i-1})$$

$$\left(\frac{d^2w}{dx^2}\right)_i = \frac{1}{l^2}(w_{i+1} - 2w_i + w_{i-1})$$

Substituting the expressions of the second derivatives into Eq.(1.44) and this one into Eq.(1.43) gives

$$\frac{EI_y}{4l^4}(w_{i-3} - 2w_{i-2} - w_{i-1} + 4w_i - w_{i+1} - 2w_{i+2} + w_{i+3}) - q = 0$$

which coincides with Eq.(1.42). The CCB formulation is therefore equivalent to a finite difference scheme centred at the node, using the third derivative of w at nodes $i+1$ and $i-1$ for computing the fourth derivative at node i. Clearly, the global stiffness stencil for node i involves the deflection values at node i and the six adjacent nodes.

Above coincidence has a tutorial value. The analogy with the finite difference scheme does not necessarily hold for non regular meshes or for the treatment of the boundary conditions.

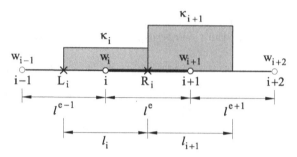

Fig. 1.15 CVB element. Control domains around two nodes i and $i+1$

1.4.2 Cell-vertex beam (CVB) element

An alternative rotation-free beam element can be derived by computing the curvature *at each node* using a finite difference scheme. The resulting element is termed CVB (for Cell-Vertex Beam element).

Let us consider the control domain formed by half of the lengths of the elements adjacent to a node. The curvature at node i is computed as

$$\kappa_i = \frac{1}{l_i}\left[\left(\frac{\partial w}{\partial x}\right)_{R_i} - \left(\frac{\partial w}{\partial x}\right)_{L_i}\right] \tag{1.45}$$

where $l_i = \frac{1}{2}(l^e + l^{e-1})$ and subscripts R_i and L_i denote the midpoints of the elements located at the right and left of node i (Figure 1.15). The curvature κ_i is assumed to be constant in the control domain l_i assigned to node i (Figure 1.15).

The rotations $\left(\dfrac{\partial w}{\partial x}\right)_{R_i}$ and $\left(\dfrac{\partial w}{\partial x}\right)_{L_i}$ are expressed in terms of the nodal deflections as

$$\left(\frac{\partial w}{\partial x}\right)_R = \frac{w_{i+1} - w_i}{l^e} \quad , \quad \left(\frac{\partial w}{\partial x}\right)_L = \frac{w_i - w_{i-1}}{l^{e-1}} \tag{1.46}$$

Substituting Eqs.(1.46) into (1.45) gives

$$\kappa_i = \frac{2}{l^e l^{e-1}(l^e + l^{e-1})}[l^e, -(l^e + l^{e-1}), l^{e-1}, 0]\begin{Bmatrix} w_{i-1} \\ w_i \\ w_{i+1} \\ w_{i+2} \end{Bmatrix} = \mathbf{B}_i \bar{\mathbf{w}}^{(e)} \tag{1.47}$$

where

$$\mathbf{B}_i = \frac{2}{l^e l^{e-1}(l^e + l^{e-1})}[l^e, -(l^e + l^{e-1}), l^{e-1}, 0] \quad , \quad \bar{\mathbf{w}}^{(e)} = \begin{Bmatrix} w_{i-1} \\ w_i \\ w_{i+1} \\ w_{i+2} \end{Bmatrix} \quad (1.48)$$

Similarly, the curvature at node $i+1$ is found as

$$\kappa_{i+1} = \frac{2}{(l^e + l^{e+1})}\left[\frac{w_{i+2} - w_{i+1}}{l^{e+1}} - \frac{w_{i+1} - w_i}{l^e}\right] = \mathbf{B}_{i+1}\bar{\mathbf{w}}^{(e)} \qquad (1.49a)$$

with

$$\mathbf{B}_{i+1} = \frac{1}{l^e l^{e+1}(l^e + l^{e+1})}[0, l^{e+1}, -(l^e + l^{e+1}), l^e] \qquad (1.49b)$$

The internal virtual work over the element e is obtained by adding up the contributions from the two control domains l_i and l_{i+1} as

$$\delta U^{(e)} = \int_{\frac{l_i}{2}} \delta\kappa_i(EI_y)\delta\kappa_i dx + \int_{\frac{l_{i+1}}{2}} \delta\kappa_{i+1}(EI_y)\delta\kappa_{i+1} dx \qquad (1.50)$$

Substituting Eqs.(1.47) and (1.49a) into (1.50) and following the standard process yields the element stiffness matrix as

$$\mathbf{K}^{(e)} = \mathbf{K}_i^{(e)} + \mathbf{K}_{i+1}^{(e)} \qquad (1.51a)$$

where

$$\mathbf{K}_i^{(e)} = \int_{\frac{l_i}{2}} \mathbf{B}_i^T(EI_y)\mathbf{B}_i dx = \frac{l_i}{2}\mathbf{B}_i^T(EI_y)^{(e)}\mathbf{B}_i$$

$$\mathbf{K}_{i+1}^{(e)} = \int_{\frac{l_{i+1}}{2}} \mathbf{B}_{i+1}^T(EI_y)\mathbf{B}_{i+1} dx = \frac{l_{i+1}}{2}\mathbf{B}_{i+1}^T(EI_y)^{(e)}\mathbf{B}_{i+1} \qquad (1.51b)$$

In the computation of the above integrals we have assumed EI_y to be constant over the control domains.

Similarly as for the CCB element, the stiffness matrix for a CVB element involves the nodal deflections of the adjacent elements.

The global stiffness matrix for a mesh of CVB elements can be directly assembled from the so-called *nodal stiffness matrices*, given for an arbitrary node i by

$$\mathbf{K}_i = l_i\bar{\mathbf{B}}_i^T(EI_y)\bar{\mathbf{B}}_i \qquad (1.52a)$$

with

$$\bar{\mathbf{B}}_i = \frac{2}{l^e l^{e-1}(l^e + l^{e-1})}[l^e, -(l^e + l^{e-1}), l^{e-1}] \qquad (1.52b)$$

Matrix $\mathbf{K}_i$ links the deflections of nodes $i-1$, i and $i+1$ and plays the role of an element matrix. As such it can be assembled into the global stiffness matrix in the usual manner (Example 1.6).

The global equivalent nodal force f_i is computed for the simplest case as

$$f_i = P_{z_i} + f_{z_i} \frac{(l^e + l^{e+1})}{2} \tag{1.53a}$$

with

$$f_{z_i} = \frac{1}{2}\left[f_z^{(e)} + f_z^{(e+1)} \right] \tag{1.53b}$$

where P_{z_i} is the external nodal force acting at node i and and $f_z^{(e)}$ and $f_z^{(e+1)}$ are uniformly distributed loads acting over elements e and $e+1$, respectively. Indeed, more sophisticated loading types can be considered.

Boundary conditions

The prescribed values for the deflection are imposed when solving the global system of equations, as usual. We detail next how to compute the curvature at a free end and at simply supported and clamped nodes.

The curvature at a *free end node and at a simply supported node* is *zero*. This condition is implemented by neglecting the contribution of the boundary node to the element stiffness matrix. For a boundary element with a free or simply supported node the stiffness matrix is (Figure 1.16a)

$$\mathbf{K}^{(e)} = \mathbf{K}_{i+1}^{(e)} \quad \text{if} \quad \kappa_i = 0 \tag{1.54a}$$

$$\mathbf{K}^{(e)} = \mathbf{K}_i^{(e)} \quad \text{if} \quad \kappa_{i+1} = 0 \tag{1.54b}$$

The curvature at a *clamped or symmetry left-end node i* is computed as (Figure 1.16b)

$$\kappa_i = \frac{2}{l^e}\frac{(w_{i+1} - w_i)}{l^e} = \frac{2}{(l^e)^2}[0, -1, 1, 0] \begin{Bmatrix} w_{i-1} \\ w_i \\ w_{i+1} \\ w_{i+2} \end{Bmatrix} = \mathbf{B}_i \bar{\mathbf{w}}^{(e)} \tag{1.55}$$

where w_{i-1} is an auxiliary fictitious deflection.

Similarly, for a right-end clamped or symmetry node (Figure 1.16b)

$$\kappa_{i+1} = \frac{2}{l^e}\frac{(w_{i+1} - w_i)}{l^e} = \frac{2}{(l^e)^2}[0, -1, 1, 0] \begin{Bmatrix} w_{i-1} \\ w_i \\ w_{i+1} \\ w_{i+2} \end{Bmatrix} = \mathbf{B}_{i+1} \bar{\mathbf{w}}_i \tag{1.56}$$

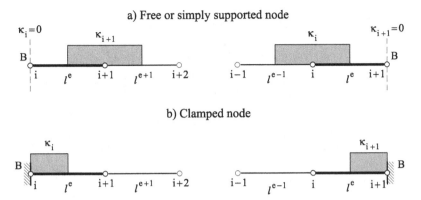

Fig. 1.16 CVB element. (a) Boundary condition of zero curvature at a free or simply supported node B. (b) Control domain for a clamped or symmetry node

where w_{i+2} is an auxiliary fictitious deflection.

The element stiffness matrix is computed in both cases by Eq.(1.51b) with $\mathbf{B}_i$ or $\mathbf{B}_{i+1}$ equal to zero, as appropriate. The fictitious nodal deflections w_{i-1} or w_{i+2} (and indeed w_i) are prescribed to a zero value, when solving the global system of equations.

Further details of the CVB formulation are given in [JO,OZ3].

The example presented next shows the process for assembling the stiffness equations for a typical node i in a mesh of CVB elements using the element stiffness matrix and the nodal stiffness matrix. The equivalence of both procedures and the coincidence of the assembled stiffness equations for the node with a centred finite difference scheme are shown.

Example 1.6: Obtain the assembled stiffness equation for a node i in a mesh of equal length CVB elements via the element and nodal stiffness matrices. Verify that the assembled nodal equation coincides with a centred finite difference scheme

- Solution

We will consider the mesh of equal length CVB elements surrounding a node i as shown in Figure 1.14.

Assembly of the element stiffness matrices

The stiffness matrix for a CVB element with nodes i and $i + 1$ in a regular mesh is deduced from Eqs.(1.51) as

$$
\mathbf{K}^{(e)} = \frac{EI_y}{2l^3}
\begin{array}{c}
\begin{array}{cccc} w_{i-1} & w_i & w_{i+1} & w_{i+2} \end{array} \\
\begin{bmatrix}
1 & -2 & 1 & 0 \\
-2 & 5 & -4 & 1 \\
1 & -4 & 5 & -2 \\
0 & 1 & -2 & 1
\end{bmatrix}
\begin{array}{c} w_{i-1} \\ w_i \\ w_{i+1} \\ w_{i+2} \end{array}
\end{array}
$$

The assembly of $\mathbf{K}^{(e-2)}$, $\mathbf{K}^{(e-1)}$, $\mathbf{K}^{(e)}$, $\mathbf{K}^{(e+1)}$ and $\mathbf{K}^{(e+2)}$ gives

$$
\mathbf{K} = \frac{EI_y}{2l^3}
\begin{array}{c}
\begin{array}{ccccccc} w_{i-3} & w_{i-2} & w_{i-1} & w_i & w_{i+1} & w_{i+2} & w_{i+3} \end{array} \\
\begin{bmatrix}
1 & -2 & 1 & 0 & \cdot & \cdot & \cdot \\
-2 & 6 & -6 & 2 & 0 & \cdot & \cdot \\
1 & -6 & 11 & -8 & 2 & 0 & \cdot \\
0 & 2 & -8 & 12 & -8 & 2 & 0 \\
\cdot & 0 & 2 & -8 & 11 & -6 & 1 \\
\cdot & \cdot & 0 & 2 & -6 & 6 & -2 \\
\cdot & \cdot & \cdot & 0 & 1 & -2 & 1
\end{bmatrix}
\begin{array}{c} w_{i-3} \\ w_{i-2} \\ w_{i-1} \\ w_i \\ w_{i+1} \\ w_{i+2} \\ w_{i+3} \end{array}
\end{array}
$$

The stiffness stencil for node i in the global equation system for a uniformly distributed load q is deduced from the ith row in the above matrix as

$$
\frac{EI_y}{l^3}[w_{i-2} - 4w_{i-1} + 6w_i - 4w_{i+1} + w_{i+2}] - ql = 0 \tag{1.57}
$$

Assembly of the nodal stiffness matrices

The general expression for the nodal stiffness matrix $\mathbf{K}_i$ for equal length CVB elements is deduced from Eqs.(1.52) as

$$
\mathbf{K}_i = \frac{EI_y}{l^3}
\begin{array}{c}
\begin{array}{ccc} w_{i-1} & w_i & w_{i+1} \end{array} \\
\begin{bmatrix}
1 & -2 & 1 \\
-2 & 4 & -2 \\
1 & -2 & 1
\end{bmatrix}
\begin{array}{c} w_{i-1} \\ w_i \\ w_{i+1} \end{array}
\end{array}
$$

The global stiffness equation for node i is obtained by assembling the contributions from nodes $i - 1$, i and $i + 1$. The assembly of $\mathbf{K}_{i-1}$, $\mathbf{K}_i$ and $\mathbf{K}_{i+1}$ gives

$$
\mathbf{K}^{(e)} = \frac{EI_y}{l^3}
\begin{array}{c}
\begin{array}{ccccc} w_{i-2} & w_{i-1} & w_i & w_{i+1} & w_{i+2} \end{array} \\
\begin{bmatrix}
1 & -2 & 1 & 0 & 0 \\
-2 & 5 & -4 & 1 & 0 \\
1 & -4 & 6 & -4 & 1 \\
0 & 1 & -4 & 5 & -2 \\
0 & 0 & 1 & -2 & 1
\end{bmatrix}
\begin{array}{c} w_{i-2} \\ w_{i-1} \\ w_i \\ w_{i+1} \\ w_{i+2} \end{array}
\end{array}
$$

It can be clearly seen that the global stiffness stencil for node i coincides with Eq.(1.57). Both assembly procedures therefore lead to the same system of equations.

Let us verify the coincidence of Eq.(1.57) with the expression obtained from a centred finite difference scheme using the nodes as sampling points.
The starting point is the differential equation of equilibrium for the beam under uniformly distributed loading given by (Figure 1.7)

$$EI_y \frac{d^4 w}{dx^4} - q = 0 \tag{1.58}$$

The fourth derivative is sampled at node i and computed as follows (Figure 1.14)

$$\left(\frac{d^4 w}{dx^4}\right)_i = \frac{1}{l}\left[\left(\frac{d^3 w}{dx}\right)_B - \left(\frac{d^3 w}{dx^3}\right)_A\right]$$

$$\left(\frac{d^3 w}{dx}\right)_B = \frac{1}{l}\left[\left(\frac{d^2 w}{dx^2}\right)_{i+1} - \left(\frac{d^2 w}{dx^2}\right)_i\right] ; \left(\frac{d^3 w}{dx^3}\right)_A = \frac{1}{l}\left[\left(\frac{d^2 w}{dx^2}\right)_i - \left(\frac{d^2 w}{dx^2}\right)_{i-1}\right]$$

Substituting the last two expressions into the previous one gives

$$\left(\frac{d^4 w}{dx^4}\right)_i = \frac{1}{l^2}\left[\left(\frac{d^2 w}{dx^2}\right)_{i+1} - 2\left(\frac{d^2 w}{dx^2}\right)_i + \left(\frac{d^2 w}{dx^2}\right)_{i-1}\right] \tag{1.59}$$

The second derivatives are now computed as

$$\left(\frac{d^2 w}{dx^2}\right)_{i+1} = \frac{w_i - 2w_{i+1} + w_{i+1}}{l^2}, \left(\frac{d^2 w}{dx^2}\right)_i = \frac{w_{i-1} - 2w_i + w_{i+1}}{l^2}$$

$$\left(\frac{d^2 w}{dx^2}\right)_{i-1} = \frac{w_{i-2} - 2w_{i+1} + w_i}{l^2} \tag{1.60}$$

Substituting Eqs.(1.60) into (1.59) and this one into Eq.(1.58) gives

$$\frac{EI_y}{l^4}(w_{i-2} - 4w_{i-1} + 6w_i - 4w_{i+1} + w_{i+2}) - q_i = 0$$

which coincides with Eq.(1.57). The CVB formulation is therefore analogous to a centred finite difference scheme if the fourth derivative of w at a node is computed from the third derivatives at the midpoint of the adjacent elements. The global stiffness equation for node i involves five nodal deflection values (w_i and the deflection at the four adjacent nodes w_{i-2}, w_{i-1}, w_{i+1} and w_{i+2}). The bandwith of the global stiffness matrix is therefore smaller than for the CCB element, as this one involves seven nodal deflections in the stiffness equation for node i (Eq.(1.42)).
Similarly as for the CCB element, the analogy with the finite difference formulation does not necessarily hold for the treatment of the boundary conditions or for non-regular meshes.

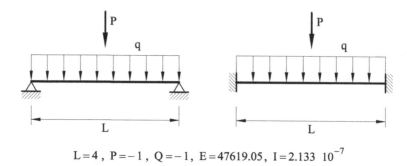

$$L=4\,,\ P=-1\,,\ Q=-1\,,\ E=47619.05\,,\ I=2.133\ \ 10^{-7}$$

Fig. 1.17 Simply supported and clamped beams under a central point load P and a uniformly distributed load q analyzed with the CCB and CVB rotation-free beam elements. Units are in the International System

1.4.3 Examples of application of CCB and CVB rotation-free beam elements

We present a study of the performance of the CCB and CVB rotation-free beam elements for a simple supported beam and a fully clamped beam under a central point load and a uniformly distributed load (Figure 1.17).

Table 1.1 shows the convergence of the central deflection error versus the analytical value [Ti2] using uniform and non-uniform meshes. Only half of the beam has been discretized in all cases, taken advantage of the symmetry of the problem.

Figures 1.18 shows the distribution of the deflection and the bending moment along the beam for the uniformly distributed load, using different regular meshes of CCB and CVB elements.

In general, both elements yield excellent results for relatively coarse meshes. Typically, the CVB element performs better than the CCB one. The accuracy of both elements deteriorates slightly for non-uniform meshes.

More examples of the performance of the CCB and CVB rotation-free Euler-Bernoulli beam elements can be found in [JO,OZ3].

The CCB element can be extended to account for shear deformation effects. This requires introducing the shear angle as an additional nodal variable. Details can be found in Section 2.10 and [OZ2].

The extension of the CCB and CVB elements to derive rotation-free thin plate and shell triangles will be studied in Chapters 5, 8 and 10.

CCB element							
Simply supported				Both ends clamped			
Central Point Load		Uniformly Distributed Load		Central Point Load		Uniformly Distributed Load	
U	NU	U	NU	U	NU	U	NU
No. of elements*							
4	12.50	25.47	5.00	12.48	50.00	58.2	31.25	58.06
8	3.12	7.23	1.25	5.35	12.50	28.9	7.81	32.42
16	0.78	1.81	0.32	1.34	3.12	7.22	1.95	8.10
32	0.19	0.45	0.08	0.33	0.78	1.79	0.49	2.02

CVB element							
Simply supported				Both ends clamped			
Central Point Load		Uniformly Distributed Load		Central Point Load		Uniformly Distributed Load	
U	NU	U	NU	U	NU	U	NU
No. of elements*							
4	3.12	7.23	1.25	6.23	12.50	28.90	7.80	45.60
8	0.78	1.81	0.31	1.56	3.12	7.23	1.90	11.40
16	0.19	0.45	0.08	0.39	0.78	1.81	0.49	2.85
32	0.05	0.11	0.02	0.09	0.19	0.45	0.12	0.71

*Only half beam is discretized due to symmetry

Table 1.1 Simply supported and clamped beams analyzed with CCB and CVB elements. Convergence of central deflection for uniform (U) and non-uniform (NU) meshes. Numbers show percentage error versus the analytical solution

1.5 CONCLUDING REMARKS

Euler-Bernoulli beam elements require C^1 continuous Hermite shape functions. The stiffness matrix for the simple 2-node beam element is identical to that obtained via standard matrix analysis techniques. This coincidence also occurs for the equivalent nodal force vector for some loads.

Understanding Euler-Bernoulli beam elements is useful as an introduction to the study of plate elements based on Kirchhoff thin plate theory in Chapter 5.

Rotation-free beam elements combine a finite difference approximation for the curvature over a control domain with standard finite element methods. Rotation-free beam elements just have deflection DOFs. These concepts will be extended for deriving rotation-free plate and shell triangles in other chapters.

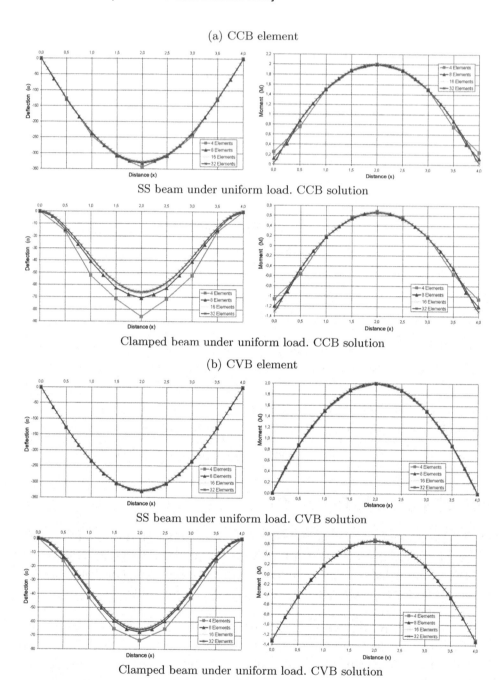

Fig. 1.18 Symply supported (SS) and clamped beams under uniform load analyzed with the CCB element (a) and CVB element (b). Deflection and bending moment distributions for different meshes. Number of elements denotes the discretization in half of the beam due to symmetry. Results for the whole beam shown have been projected from the symmetric solution

2

THICK/SLENDER PLANE BEAMS. TIMOSHENKO THEORY

2.1 INTRODUCTION

This chapter studies Timoshenko plane beam elements. Timoshenko beam theory accounts for the effect of transverse shear deformation. Timoshenko beam elements are therefore applicable for "thick" beams $\left(\lambda = \frac{L}{h} < 10\right)$ where transverse shear deformation has an influence in the solution, as well as for slender beams ($\lambda > 100$) where this influence is irrelevant [Ti].

Timoshenko beam elements have also advantages for the analysis of *composite laminated beams*, as the effect of transverse shear deformation is relevant in these cases (Chapters 3 y 4).

Timoshenko beam elements require C^0 continuity for the deflection and rotation fields and, therefore, are simpler than Euler-Bernoulli beam elements. Unfortunately, they suffer generally from the so-called *shear locking* defect which yields unrealistically stiffer solutions for slender beams. Felippa [Fel] has written an entertaining review of Timoshenko beam elements.

In the following sections, Timoshenko plane beam theory is described. Next, the formulation of 2 and 3-noded Timoshenko beam elements is presented. Shear locking is explained and some techniques to overcome it via reduced integration, linked interpolations and assumed shear strain fields are described. After this we derive an exact 2-noded Timoshenko beam element by integrating the equations of equilibrium. An extension of the rotation-free slender beam element studied in Chapter 1 to account for transverse shear deformation effects is presented. The treatment of beams on an elastic foundation is also described.

E. Oñate, *Structural Analysis with the Finite Element Method. Linear Statics: Volume 2: Beams, Plates and Shells*, Lecture Notes on Numerical Methods in Engineering and Sciences, DOI 10.1007/978-1-4020-8743-1_2,
© International Center for Numerical Methods in Engineering (CIMNE), 2013

The study of this chapter is important as an introduction to the formulation of thick plate and shell elements later in the book. 3D Timoshenko beam elements will be studied in Chapter 4.

2.2 TIMOSHENKO PLANE BEAM THEORY

2.2.1 Basic assumptions

Timoshenko plane beam theory shares hypotheses 1 and 2 of conventional Euler-Bernoulli theory for the vertical and lateral motion of a beam (Section 1.2.1). Hovever, hypothesis 3 for the normal kinematics now reads as follows: "cross sections normal to the beam axis before deformation remain plane *but not necessarily orthogonal* to the beam axis after deformation". This assumption represents a better approximation of the true deformation of the cross section in deep beams. As the beam slenderness (length/thickness ratio) diminishes, the beam cross sections do not remain plane. Timoshenko hypothesis is equivalent to assuming an average rotation for the deformed cross section which is kept plane (Figure 2.1).

The rotation of the cross section is deduced from Figure 2.1 as

$$\theta = \frac{dw}{dx} + \phi \tag{2.1}$$

where $\frac{dw}{dx}$ is the slope of the beam axis and ϕ is an additional rotation due to the distortion of the cross-section. Note that the rotation θ does not coincide with the slope $\frac{dw}{dx}$, as it happened in Euler-Bernoulli theory.

2.2.2 Strain and stress fields

The strain field is obtained by combining Eqs.(1.1), (1.3) and (2.1) to give the following non-zero strains

$$\varepsilon_x = \frac{du}{dx} = -z\frac{d\theta}{dx} \quad ; \quad \gamma_{xz} = \frac{dw}{dx} + \frac{du}{dz} = \frac{dw}{dx} - \theta = -\phi \tag{2.2}$$

Hence, Timoshenko theory introduces a transverse shear deformation γ_{xz}, which absolute value coincides with the rotation ϕ.

The axial and shear stresses σ_x and τ_{xz} at a point of the beam cross section are related to the corresponding strains by

$$\sigma_x = E\varepsilon_x = -zE\frac{d\theta}{dx} \tag{2.3a}$$

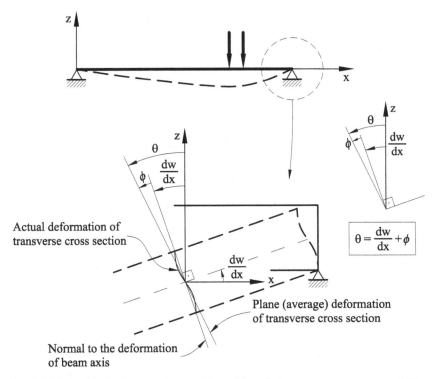

Fig. 2.1 Timoshenko beam theory. Rotation of the transverse cross section

$$\tau_{xz} = G\gamma_{xz} = G\left(\frac{dw}{dx} - \theta\right) \tag{2.3b}$$

where G is the shear modulus $G = \frac{E}{2(1+\nu)}$ and ν is the Poisson ratio [On4].

2.2.3 Resultant stresses and generalized strains

The bending moment M and the shear force Q are defined with the sign criterion of Figure 2.2, as

$$M = -\iint_A z\sigma_x\, dA \quad , \quad Q = \iint_A \tau_{xz}\, dA \tag{2.4a}$$

Substituting σ_x and τ_{xz} from Eqs.(2.3) into (2.4a) gives

$$M = \hat{D}_b \frac{d\theta}{dx} = \hat{D}_b\kappa \quad , \quad Q = \hat{G}\gamma_{xz} \tag{2.4b}$$

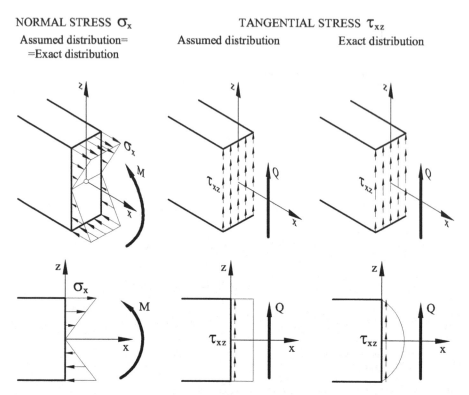

NORMAL STRESS σ_x

Assumed distribution=
=Exact distribution

TANGENTIAL STRESS τ_{xz}

Assumed distribution   Exact distribution

Fig. 2.2 Timoshenko beam theory. Distribution of the normal and tangential stresses

with

$$\hat{D}_b = \iint_A Ez^2 dA \quad \text{and} \quad \hat{G} = \iint_A G dA \qquad (2.5a)$$

In the following, a "hat" on the constitutive parameters denotes integrated (also called *resultant* or *generalized*) values over the section (in beams) or the thickness (in plates and shells).

For homogeneous material

$$\hat{D}_b = EI_y \quad \text{and} \quad \hat{G} = GA \qquad (2.5b)$$

In Eq.(2.4b) $\kappa = \frac{d\theta}{dx}$ is the bending strain (sometimes called incorrectly the *curvature*). κ and γ_{xz} are termed "generalized strains" as they are sectional quantities which depend on the axial coordinate x only.

Eq.(2.3a) tell us that the normal stress σ_x varies linearly through the thickness, and this can be considered "exact" according to classical beam theory [Ti2]. On the other hand, Eq.(2.3b) shows that the shear stress τ_{xz} is constant across the thickness. This is in contradiction with the exact quadratic distribution for a rectangular beam (Figure 2.2) [Ti2]. This problem can be overcome by modifying the internal energy dissipated by the constant shear stresses in the PVW to match the exact shear stress energy deduced from beam theory [Co6,Ti2]. Thus, we take

$$\tau_{xz} = k_z \, G \, \gamma_{xz} \tag{2.6a}$$

and from Eq.(2.4b)

$$Q = k_z \, \hat{G} \, \gamma_{xz} = \hat{D}_s \, \gamma_{xz} \quad \text{with} \quad \hat{D}_s = k_z \hat{G} \tag{2.6b}$$

The shear correction parameter $k_z (k_z \leq 1)$ takes into account the distortion of the cross section [Co6]. This distortion is shown in Figure 2.1.

For homogeneous material

$$Q = k_z GA \, \gamma_{xz} = GA^* \, \gamma_{xz} \quad \text{and, hence,} \quad \hat{D}_s = GA^* \tag{2.6c}$$

where $A^* = k_z A$ is the *reduced cross sectional area* [Ti2].

2.2.3.1 Computation of the shear correction parameter

The value of k_z can be computed by assuming cylindrical bending in the xz plane (i.e. $\tau_{xy} = 0$) and matching the exact transverse shear strain energy (U_s) with that given by Timoshenko beam theory (U_s^T) corrected by the coefficient k_z. The values of U_s and U_s^T are [Ti2]

$$U_s = \frac{1}{2} \iint_A \frac{\tau_{xz}^2}{G} dA \quad , \quad U_s^T = \frac{1}{2} \frac{Q^2}{k_z \hat{G}} \tag{2.7a}$$

where τ_{xz} is the exact transverse shear stress. Equaling U_s and U_s^T yields

$$k_z = \frac{Q^2}{2\hat{G}U_1} = \frac{Q^2}{\hat{G}} \left[\iint_A \frac{\tau_{xz}^2}{G} dA \right]^{-1} \tag{2.7b}$$

A general approach for computing τ_{xz}, and hence k_z is presented in Appendix D. Figure 2.3 shows the value of k_z for different sections. The computation of the shear correction parameter for composite laminated plane beams is detailed in Section 3.8.

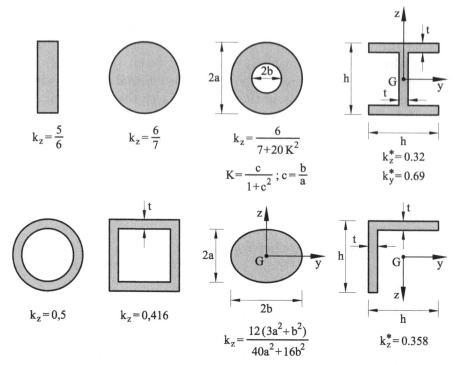

$$k_z = \frac{5}{6} \qquad k_z = \frac{6}{7} \qquad k_z = \frac{6}{7+20\,K^2}$$

$$K = \frac{c}{1+c^2} \; ; \; c = \frac{b}{a}$$

$$k_z^* = 0.32$$
$$k_y^* = 0.69$$

$$k_z = 0.5 \qquad k_z = 0.416 \qquad k_z = \frac{12\,(3a^2 + b^2)}{40a^2 + 16b^2} \qquad k_z^* = 0.358$$

Fig. 2.3 Shear correction parameter k_z for some cross sections. Asterisk denotes values computed with the FEM [BD5,Bo,Co6]

2.2.4 Principle of virtual work

We consider a beam under the loads shown in Figure 1.1. The internal virtual work involves the axial and shear stresses and the PVW is written as

$$\iiint_V (\delta\varepsilon_x\sigma_x + \delta\gamma_{xz}\tau_{xz})\,dV = \int_0^L (\delta w f_z + \delta\theta m)\,dx + \sum_i \delta w_i P_{z_i} + \sum_j \delta\theta_j M_j^c \tag{2.8a}$$

The virtual internal work in the l.h.s. of Eq.(2.8a) can be modified using Eqs.(2.2)-(2.6) as

$$\iiint_V \left[-z\sigma_x\delta\left(\frac{d\theta}{dx}\right) + \tau_{xz}\delta\left(\frac{dw}{dx} - \theta\right)\right] dV =$$
$$= \int_0^l \left[\delta\left(\frac{d\theta}{dx}\right)\left(\iint_A -z\sigma_x\,dA\right) + \delta\gamma_{xz}\left(\iint_A \tau_{xz}dA\right)\right] dx = \tag{2.8b}$$
$$= \int_0^l \left[\delta\left(\frac{d\theta}{dx}\right)M + \delta\gamma_{xz}Q\right] dx$$

Substituting Eq.(2.8b) into (2.8a) yields the PVW in terms of integrals along the beam axis as

$$\int_0^L \left[\delta\left(\frac{d\theta}{dx}\right)M + \delta\gamma_{xz}Q\right]dx = \int_0^L (\delta w f_z + \delta\theta m)\,dx + \sum_i \delta w_i P_{z_i} + \sum_j \delta\theta_j M_j^c$$

(2.9)

The first integral is the internal virtual work induced by the bending moment and the transverse shear force while the r.h.s. is the virtual work of the applied loads. Eqs.(2.8b) and (2.9) show that *the PVW involves just the first derivatives of the deflection and the rotation.* As a consequence just C^o continuity for w and θ is required to satisfy the integrability condition (Section 3.8.3 of [On4] and [Hu,ZT2,ZTZ]). Eq.(2.9) is the basis for the finite element discretization presented in the next section.

2.3 TWO-NODED TIMOSHENKO BEAM ELEMENT

2.3.1 Approximation of the displacement field

Let us consider first the simple 2-noded Timoshenko beam element (Figure 2.4). The deflection w and the rotation θ are now independent variables and each one is linearly interpolated using C^o shape functions as

$$\begin{aligned}w(\xi) &= N_1(\xi)w_1 + N_2(\xi)w_2 \\ \theta(\xi) &= N_1(\xi)\theta_1 + N_2(\xi)\theta_2\end{aligned}$$

(2.10)

or

$$\mathbf{u} = \left\{\begin{array}{c} w \\ \theta \end{array}\right\} = \sum_{i=1}^2 \mathbf{N}_i \mathbf{a}_i = \mathbf{N}^{(e)}\mathbf{a}^{(e)}$$

(2.11a)

with

$$\mathbf{a}^{(e)} = \left\{\begin{array}{c} \mathbf{a}_1^{(e)} \\ \mathbf{a}_2^{(e)} \end{array}\right\} \quad ; \quad \mathbf{a}_i^{(e)} = \left\{\begin{array}{c} w_i \\ \theta_i \end{array}\right\}$$

(2.11b)

$$\mathbf{N}^{(e)} = [\mathbf{N}_1, \mathbf{N}_2] \quad ; \quad \mathbf{N}_i = \begin{bmatrix} N_1 & 0 \\ 0 & N_2 \end{bmatrix}$$

(2.11c)

In the above, $\mathbf{a}^{(e)} = [w_1, \theta_1, w_2, \theta_2]^T$ is the nodal displacement vector for the element, w_1, θ_1 and w_2, θ_2 are the deflection and the rotation of nodes 1 and 2, respectively, and $N_1(\xi)$ and $N_2(\xi)$ are the standard C^o linear shape functions (Figure 2.4).

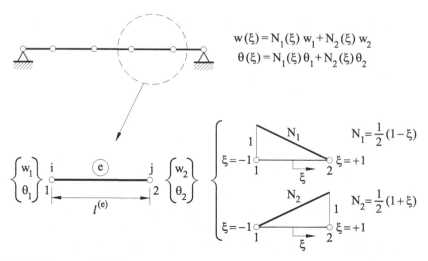

$$w(\xi) = N_1(\xi)\, w_1 + N_2(\xi)\, w_2$$
$$\theta(\xi) = N_1(\xi)\, \theta_1 + N_2(\xi)\, \theta_2$$

Fig. 2.4 Two-noded Timoshenko beam element. Displacement interpolation

Note the difference between the approximation (2.10) and Eq.(1.10) for the 2-noded Euler-Bernoulli beam element for which the deflection and the rotation were depending variables due to the C^1 continuity requirement.

2.3.2 Approximation of the generalized strains and the resultant stresses

The bending strain κ and the transverse shear strain γ_{xz} are expressed in terms of the nodal DOFs using Eq.(2.10) as

$$\kappa = \frac{d\theta}{dx} = \frac{d\xi}{dx}\frac{d\theta}{d\xi} = \frac{d\xi}{dx}\left[\frac{dN_1}{d\xi}\theta_1 + \frac{dN_2}{d\xi}\theta_2\right] \tag{2.12}$$

$$\gamma_{xz} = \frac{dw}{dx} - \theta = \frac{d\xi}{dx}\left[\frac{dN_1}{d\xi}w_1 + \frac{dN_2}{d\xi}w_2\right] - (N_1\theta_1 + N_2\theta_2) \tag{2.13}$$

The element geometry is interpolated in terms of the coordinates of the two nodes in the standard isoparametric manner as $x = \sum\limits_{i=1}^{2} N_i(\xi)x_i$ [On4]. From this we deduce $\frac{dx}{d\xi} = \frac{l^{(e)}}{2}$. Substituting the inverse of this expression in Eqs.(2.12) and (2.13) and using a matrix notation we can write

$$\kappa = \frac{d\theta}{dx} = \mathbf{B}_b\, \mathbf{a}^{(e)} \qquad , \qquad \gamma_{xz} = \frac{dw}{dx} - \theta = \mathbf{B}_s\, \mathbf{a}^{(e)} \tag{2.14}$$

where

$$\mathbf{B}_b = \left[0, \frac{2}{l^{(e)}}\frac{dN_1}{d\xi}, 0, \frac{2}{l^{(e)}}\frac{dN_2}{d\xi}\right] = \left[0, -\frac{1}{l^{(e)}}, 0\ \frac{1}{l^{(e)}}\right]$$

$$\mathbf{B}_s = \left[\frac{2}{l^{(e)}}\frac{dN_1}{d\xi}, -N_1, \frac{2}{l^{(e)}}\frac{dN_2}{d\xi}, -N_2\right] = \left[-\frac{1}{l^{(e)}}, \frac{-(1-\xi)}{2}, \frac{1}{l^{(e)}}, \frac{-(1+\xi)}{2}\right]$$

(2.15)

are the bending and transverse shear strain matrices for the element.

The virtual displacement and the virtual strain fields are expressed in terms of the virtual nodal displacements via Eqs.(2.10) and (2.14) as

$$\delta\mathbf{u} = \mathbf{N}\delta\mathbf{a}^{(e)} \quad , \quad \delta\kappa = \mathbf{B}_b\delta\mathbf{a}^{(e)} \quad , \quad \delta\gamma_{xz} = \mathbf{B}_s\delta\mathbf{a}^{(e)} \qquad (2.16)$$

with $\delta\mathbf{a}^{(e)} = [\delta w_1, \delta\theta_1, \delta w_2, \delta\theta_2]^T$.

The bending moment and the shear force (Figure 1.2) are obtained from the nodal displacements using Eqs.(2.4b), (2.6b) and (2.14) as

$$M = \hat{\mathbf{D}}_b\mathbf{B}_b\mathbf{a}^{(e)} \qquad , \qquad Q = \hat{\mathbf{D}}_s\mathbf{B}_s\mathbf{a}^{(e)} \qquad (2.17)$$

Clearly M is constant while Q has a linear distribution within the element. We will see later that, in practice, the value of Q at the element mid-point should be taken.

2.3.3 Discretized equations for the element

The PVW for an individual element can be written as (see Eq.(2.9) and Figure 2.4)

$$\int_{l^{(e)}} [\delta\kappa M + \delta\gamma_{xz}Q]\,dx = \int_{l^{(e)}} \delta\mathbf{u}^T \left\{\begin{matrix} f_z \\ m \end{matrix}\right\}\,dx + \left[\delta\mathbf{a}^{(e)}\right]^T\mathbf{q}^{(e)} \qquad (2.18a)$$

where

$$\mathbf{q}^{(e)} = [F_{z_1}, \mathcal{M}_1, F_{z_2}, \mathcal{M}_2]^T \qquad (2.18b)$$

is the equilibrating nodal force vector for the element. The signs for the components of $\mathbf{q}^{(e)}$ are shown in Figure 1.4.

Substituting Eqs.(2.16) into (2.18a) gives, after simplifying the virtual displacements

$$\int_{l^{(e)}} \left[\mathbf{B}_b^T M + \mathbf{B}_s^T Q\right]dx - \int_{l^{(e)}} \mathbf{N}^T \left\{\begin{matrix} f_z \\ m \end{matrix}\right\}\,dx = \mathbf{q}^{(e)} \qquad (2.19)$$

Substituting the constitutive equations for M and Q (Eqs.(2.17) and using Eqs.(2.14) gives

$$\left(\int_{l^{(e)}} \left[\mathbf{B}_b^T (\hat{D}_b) \mathbf{B}_b + \mathbf{B}_s^T (\hat{D}_s) \mathbf{B}_s \right] dx \right) \mathbf{a}^{(e)} - \int_{l^{(e)}} \mathbf{N}^T \left\{ \begin{matrix} f_z \\ m \end{matrix} \right\} dx = \mathbf{q}^{(e)}$$

(2.20a)

In compact matrix form

$$\underbrace{\left[\mathbf{K}_b^{(e)} + \mathbf{K}_s^{(e)} \right]}_{\mathbf{K}^{(e)}} \mathbf{a}^{(e)} - \mathbf{f}^{(e)} = \mathbf{q}^{(e)}$$

(2.20b)

where the element stiffness matrix is

$$\mathbf{K}^{(e)} = \mathbf{K}_b^{(e)} + \mathbf{K}_s^{(e)}$$

(2.21a)

and

$$\mathbf{K}_b^{(e)} = \int_{l^{(e)}} \mathbf{B}_b^T (\hat{D}_b) \mathbf{B}_b \, dx \quad ; \quad \mathbf{K}_s^{(e)} = \int_{l^{(e)}} \mathbf{B}_s^T (\hat{D}_s) \mathbf{B}_s \, dx$$

(2.21b)

are respectively the bending and shear stiffness matrices for the element,

$$\mathbf{f}^{(e)} = \left\{ \begin{matrix} \mathbf{f}_1^{(e)} \\ \mathbf{f}_2^{(e)} \end{matrix} \right\} \quad \text{with} \quad \mathbf{f}_i^{(e)} = \left\{ \begin{matrix} f_{z_i} \\ m_i \end{matrix} \right\} = \int_{l^{(e)}} N_i \left\{ \begin{matrix} f_z \\ m \end{matrix} \right\} dx$$

(2.22)

is the equivalent nodal force vector due to the distributed loading f_z and the distributed moment m.

The above integrals can be expressed in the natural coordinate system. Recalling that $dx = \frac{l^{(e)}}{2} d\xi$, the matrices and vectors of Eqs.(2.20)–(2.22) are rewritten as

$$\mathbf{K}_b^{(e)} = \int_{-1}^{+1} \mathbf{B}_b^T (\hat{D}_b) \, \mathbf{B}_b \, \frac{l^{(e)}}{2} d\xi \quad ; \quad \mathbf{K}_s^{(e)} = \int_{-1}^{+1} \mathbf{B}_s^T (\hat{D}_s) \, \mathbf{B}_s \frac{l^{(e)}}{2} \, d\xi \quad (2.23)$$

and

$$\mathbf{f}_i^{(e)} = \int_{-1}^{+1} N_i \left\{ \begin{matrix} f_z \\ m \end{matrix} \right\} \frac{l^{(e)}}{2} \, d\xi$$

(2.24a)

For a uniformly distributed values of f_z and m then

$$\mathbf{f}_i^{(e)} = \frac{l^{(e)}}{2} \left\{ \begin{matrix} f_z \\ m \end{matrix} \right\}$$

(2.24b)

i.e. the total distributed vertical force and the bending moment are equally split between the two nodes of the element. The external vertical forces and bending moments give uncoupled contributions to vector $\mathbf{f}^{(e)}$. This is due to the independent C^o interpolation for w and θ (Eq.(2.10)). Recall that in Euler-Bernoulli beam elements a vertical load induces nodal couples due to the C^1 interpolation for the deflection (Section 1.3.3).

The integrals can be evaluated numerically using a 1D Gauss quadrature as

$$\mathbf{K}_a^{(e)} = \sum_{p=1}^{n_p} (\mathbf{B}_a^T \hat{D}_a \mathbf{B}_a)_p W_p \frac{l^{(e)}}{2} \quad , \quad \text{with } a = b, s \qquad (2.25)$$

where n_p is the number of integration points in the beam element and W_p are the quadrature weights (Appendix C and [On4]).

The element stiffness matrix can also be computed as

$$\mathbf{K}^{(e)} = \int_{l^{(e)}} \mathbf{B}^T \hat{\mathbf{D}} \mathbf{B} \, dx \qquad (2.26a)$$

where $\mathbf{B}$ and $\hat{\mathbf{D}}$ are generalized strain and constitutive matrices, respectively with

$$\mathbf{B} = \begin{Bmatrix} \mathbf{B_b} \\ \mathbf{B_s} \end{Bmatrix} \quad \text{and} \quad \mathbf{D} = \begin{bmatrix} \hat{D}_b & 0 \\ 0 & \hat{D}_s \end{bmatrix} \qquad (2.26b)$$

The split of the element stiffness matrix via Eq.(2.21a) is more convenient as it allows us to identify the bending and shear contributions. This is also of interest for using different quadrature rules for $\mathbf{K}_b$ and $\mathbf{K}_s$ ir order to avoid shear locking as shown in the next section.

The global stiffness matrix and the global equivalent nodal force vector $\mathbf{f}$ are assembled from the element contributions as usual. Point loads $\mathbf{p}_i = [P_{z_i}, M_i^c]^T$ acting at nodes are directly assembled into vector $\mathbf{f}$.

The reactions at prescribed nodes can be obtained "a posteriori" once the nodal displacements have been found, as described in Section 1.3.4.

2.4 LOCKING OF THE NUMERICAL SOLUTION

From Eqs.(2.15) and (2.23) we deduce that the exact evaluation of the bending stiffness matrix $\mathbf{K}_b^{(e)}$ requires a single Gauss integration point, as all the terms in the integrand are constant (Appendix C). Exact integration gives (for homogeneous material)

$$\mathbf{K}_b^{(e)} = \left(\frac{\hat{D}_b}{l} \right)^{(e)} \begin{bmatrix} 0 & 0 & 0 & 0 \\ 0 & 1 & 0 & -1 \\ 0 & 0 & 0 & 0 \\ 0 & -1 & 0 & 1 \end{bmatrix} \qquad (2.27a)$$

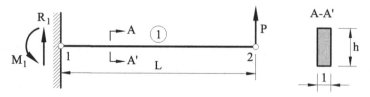

Fig. 2.5 Cantilever beam under end point load. Analysis with one 2-noded Timoshenko beam element

The exact integration of the shear stiffnes matrix $\mathbf{K}_s^{(e)}$ requires two Gauss integration points, as quadratic terms in ξ are now involved, due to the products $N_i N_j$ (Appendix C). For homogeneous material

$$
\mathbf{K}_s^{(e)} = \left(\frac{\hat{D}_s}{l}\right)^{(e)}
\begin{bmatrix}
1 & \dfrac{l^{(e)}}{2} & -1 & \dfrac{l^{(e)}}{2} \\[2mm]
 \ddots & \dfrac{\left(l^{(e)}\right)^2}{3} & -\dfrac{l^{(e)}}{2} & \dfrac{\left(l^{(e)}\right)^2}{6} \\[2mm]
 & & 1 & -\dfrac{l^{(e)}}{2} \\[2mm]
\text{Symm.} & & \ddots & \dfrac{\left(l^{(e)}\right)^2}{3}
\end{bmatrix}
\tag{2.27b}
$$

The performance of the 2-noded Timoshenko beam element with exact integration can be assessed in the analysis of an homogeneous cantilever beam under an end point load. A single element is used first (Figure 2.5). The global equilibrium equation is

$$
\left[\mathbf{K}_b^{(1)} + \mathbf{K}_s^{(1)}\right]\mathbf{a}^{(1)} = \mathbf{f}
\tag{2.28}
$$

Substituting Eqs.(2.27a) for $\hat{D}_b = EI_y$, $\hat{D}_s = GA^*$ and $l^{(e)} = L$ gives

$$
\begin{bmatrix}
\dfrac{GA^*}{L} & \dfrac{GA^*}{2} & -\dfrac{GA^*}{L} & \dfrac{GA^*}{2} \\[2mm]
 & \left(\dfrac{GA^*}{3}L + \dfrac{EI_y}{L}\right) & -\dfrac{GA^*}{2} & \left(\dfrac{GA^*}{6}L - \dfrac{EI_y}{L}\right) \\[2mm]
 & & \dfrac{GA^*}{L} & -\dfrac{GA^*}{2} \\[2mm]
\text{Symm.} & & \ddots & \left(\dfrac{GA^*}{3}L + \dfrac{EI_y}{L}\right)
\end{bmatrix}
\begin{Bmatrix} w_1 \\ \theta_1 \\ w_2 \\ \theta_2 \end{Bmatrix}
=
\begin{Bmatrix} R_1 \\ M_1 \\ P \\ 0 \end{Bmatrix}
\quad
\begin{matrix} w_1 = 0 \\ \theta_1 = 0 \end{matrix}
\tag{2.29}
$$

Once the clamped DOFs have been eliminated, the following simplified system is obtained

$$
\begin{bmatrix}
\dfrac{GA^*}{L} & -\dfrac{GA^*}{2} \\[3mm]
-\dfrac{GA^*}{2} & \left(\dfrac{GA^*}{3}L + \dfrac{EI_y}{L}\right)
\end{bmatrix}
\left\{ \begin{matrix} w_2 \\ \theta_2 \end{matrix} \right\}
= \left\{ \begin{matrix} P \\ 0 \end{matrix} \right\}
\tag{2.30}
$$

The solution is

$$
\left\{ \begin{matrix} w_2 \\ \theta_2 \end{matrix} \right\}
= \mathbf{F}\,\mathbf{f}
= \dfrac{\beta}{\beta+1}
\begin{bmatrix}
\left(\dfrac{L}{GA^*} + \dfrac{L^3}{3EI}\right)\dfrac{L^2}{EI_y} & \dfrac{L^2}{EI_y} \\[3mm]
\dfrac{L^2}{EI_y} & \dfrac{L}{EI_y}
\end{bmatrix}
\left\{ \begin{matrix} P \\ 0 \end{matrix} \right\}
\tag{2.31}
$$

where $\mathbf{F} = \mathbf{K}^{-1}$ is the flexibility matrix and

$$
\beta = \frac{12\,EI_y}{GA^*L^2}
\tag{2.32}
$$

The parameter β characterizes the influence of the transverse shear strain in the numerical solution. A small value of β indicates that shear shear strain effects are negligible. β dependes on the geometry and the material properties of the transverse cross section. For a rectangular beam of unit width, height h, homogeneous material and $I_y = \frac{h^3}{12}$,

$$
\beta = \frac{12EI_y}{L^2GA^*} = \frac{E}{k_zG}\left(\frac{h}{L}\right)^2 = \frac{E}{k_zG\lambda^2}
\tag{2.33}
$$

where $\lambda = L^2/h$ is the *beam slenderness ratio*. Therefore, β tends to zero for very slender beams $(\lambda \to \infty)$ as expected. For an homogeneous isotropic rectangular section with $\nu = 0.25$ and $\alpha = 5/6$, then $\beta = \frac{3}{\lambda^2}$. For the same section with $\frac{E}{k_zG} = 50$, then $\beta = \frac{50}{\lambda^2}$. The value of β for some composite laminated sections is given in Table 3.3.

The deflection and the rotation at the free end are found from Eq.(2.31) as

$$
w_2 = \frac{\beta}{\beta+1}\left(\frac{L}{GA^*} + \frac{L^3}{3EI_y}\right)P \quad , \quad
\theta_2 = \frac{\beta L^2}{(\beta+1)EI_y}P
\tag{2.34a}
$$

The reactions are obtained from the first two rows of Eq.(2.29) as

$$
R_1 = -P \quad , \quad M_1 = -Pl
\tag{2.34b}
$$

Let us study the influence of λ on the numerical solution.

The flexibility matrix giving *exact nodal results* for this problem (after eliminating the prescribed DOFs) using conventional Euler-Bernoulli beam theory (via Eq.(1.20)) and Timoshenko beam theory (via Eq.(2.101c); see also Example 2.10) is

a) Euler-Bernoulli theory   **b) Timoshenko theory**

$$
\mathbf{F} = \begin{bmatrix} \dfrac{L^3}{3EI_y} & \dfrac{L^2}{2EI_y} \\ \dfrac{L^2}{2EI_y} & \dfrac{L}{EI_y} \end{bmatrix} \quad ; \quad \mathbf{F} = \begin{bmatrix} \left(\dfrac{L}{GA^*} + \dfrac{L^3}{3EI_y}\right) & \dfrac{L^2}{2EI_y} \\ \dfrac{L^2}{2EI_y} & \dfrac{L}{EI_y} \end{bmatrix} \tag{2.35}
$$

The "exact" end displacements for each theory are

$$
w_2^{EB} = \frac{L^3}{3EI_y} P \quad ; \quad w_2^T = \left(\frac{L}{GA^*} + \frac{L^3}{3EI_y}\right) P
$$

$$
\theta_2^{EB} = \frac{L^2}{2EI_y} P \quad , \quad \theta_2^T = \frac{L^2}{2EI_y} P \tag{2.36}
$$

where upper indices EB and T refer to Euler-Bernoulli and Timoshenko beam theories respectively. Note that the end rotations are the same for both theories.

The effect of transverse shear deformation is negligible for a slender beam (i.e. for a large value of λ). Hence, Timoshenko solution should coincide for this case with that of conventional Euler-Bernoulli theory. The ratio between the end deflection value using the 2-noded Timoshenko beam element and the "exact" Euler-Bernoulli solution is deduced from Eqs.(2.34a) and (2.36) as

$$
r_w = \frac{w_2}{w_2^{EB}} = \frac{\beta}{\beta+1} \frac{\left(\dfrac{L}{GA^*} + \dfrac{L}{3EI_y}\right) P}{\left(\dfrac{L^3}{3EI_y}\right) P} = \frac{3(4\lambda^2 + 3)}{4\lambda^2(\lambda^2 + 3)} \tag{2.37}
$$

Clearly, the ratio r_w should tend to one as λ increases.

Figure 2.6 shows the change in r_w with λ. For very slender beams ($\lambda \to \infty$) r_w tends to zero. Thus, as the beam slenderness increases the numerical solution is progressively stiffer than the exact one. This means that the 2-noded Timoshenko beam element is unable to reproduce the conventional solution for slender beams. This phenomenon, known as *shear*

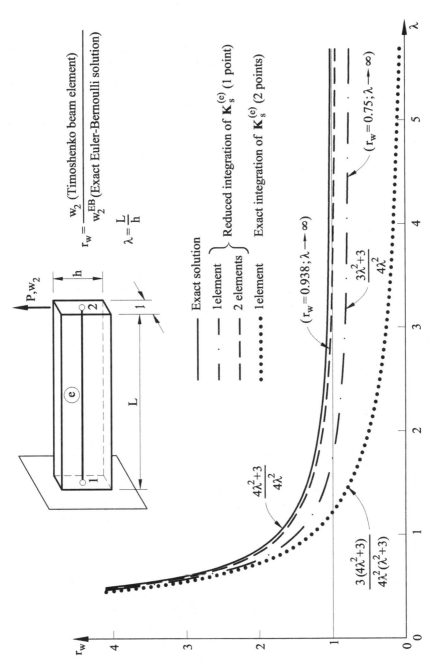

Fig. 2.6 Cantilever beam analyzed with one 2-noded Timoshenko beam element. Change in the ratio r_w between the end deflection for the 2-noded Timoshenko beam element and the exact Euler-Bernoulli solution with the beam slenderness ratio λ. Influence of the integration order for $\mathbf{K}_s^{(e)}$

locking, in principle disqualifies Timoshenko beam elements for analysis of slender beams.

Many procedures to eliminate shear locking in Timoshenko beam elements have been proposed. A popular method is to reduce the influence of the transverse shear stiffness by under-integrating the terms in $\mathbf{K}_s^{(e)}$ using a quadrature of one order less than is needed for exact integration (the so-called *reduced integration*). The terms of $\mathbf{K}_b^{(e)}$ are still integrated exactly.

For homogeneous material, the computation of $\mathbf{K}_s^{(e)}$ with a *single integration point* gives

$$
\mathbf{K}_s^{(e)} = \left(\frac{\hat{D}_s}{l}\right)^{(e)}
\begin{bmatrix}
1 & \dfrac{l^{(e)}}{2} & -1 & \dfrac{l^{(e)}}{2} \\[4pt]
& \dfrac{\left(l^{(e)}\right)^2}{4} & -\dfrac{l^{(e)}}{2} & \dfrac{\left(l^{(e)}\right)^2}{4} \\[4pt]
& & 1 & -\dfrac{l^{(e)}}{2} \\[4pt]
& & & \dfrac{\left(l^{(e)}\right)^2}{4} \\
\text{Symm.} & & &
\end{bmatrix}
\tag{2.38}
$$

The element stiffness matrix with *a uniform one-point integration* for $\mathbf{K}_b^{(e)}$ and $\mathbf{K}_s^{(e)}$ is therefore

$$
\mathbf{K}^{(e)} =
\begin{bmatrix}
\left(\dfrac{\hat{D}_s}{l}\right)^{(e)} & \dfrac{\hat{D}_s^{(e)}}{2} & -\left(\dfrac{\hat{D}_s}{l}\right)^{(e)} & \dfrac{\hat{D}_s^{(e)}}{2} \\[10pt]
& \left(\dfrac{\hat{D}_s l}{4} + \dfrac{\hat{D}_b}{l}\right)^{(e)} & \dfrac{\hat{D}_s^{(e)}}{2} & \left(\dfrac{\hat{D}_s l}{4} - \dfrac{\hat{D}_b}{l}\right)^{(e)} \\[10pt]
& & \left(\dfrac{\hat{D}_s}{l}\right)^{(e)} & -\dfrac{\hat{D}_s^{(e)}}{2} \\[10pt]
\text{Symm.} & & \left(\dfrac{\hat{D}_s}{l}\right)^{(e)} & \left(\dfrac{\hat{D}_s l}{4} + \dfrac{\hat{D}_b}{l}\right)^{(e)}
\end{bmatrix}
\tag{2.39}
$$

Using Eq.(2.39) instead of Eq.(2.27b), the stiffness and flexibility matrices for the single Timoshenko beam element of Figure 2.5 with uniform

	Number of elements				
Limit end deflection ratio r_w	1	2	4	8	16
$r_w = \dfrac{w}{w^{EB}}$ for $\lambda \to \infty$	0.750	0.938	0.984	0.996	0.999

Table 2.1 Cantilever beam under end point load. Convergence of the end deflection ratio r_w with the number of 2-noded Timoshenko beam elements for very slender beams $(\lambda \to \infty)$, using uniform one-point integration

one-point integration, after eliminating the prescribed DOFs, are

$$\mathbf{K} = \begin{bmatrix} \dfrac{GA^*}{L} & -\dfrac{GA^*}{2} \\ -\dfrac{GA^*}{2} & \left(\dfrac{GA^*}{4}L + \dfrac{EI}{L}\right) \end{bmatrix} \quad ; \quad \mathbf{F} = \begin{bmatrix} \left(\dfrac{L}{GA^*} + \dfrac{L^3}{4EI}\right) & \dfrac{L^2}{2EI} \\ \dfrac{L^2}{2EI} & \dfrac{L}{EI} \end{bmatrix} \quad (2.40)$$

Note that $\mathbf{F}$ now coincides with the "exact" expression (2.35), except for the term F_{11}. Solving for the end displacements gives

$$w_2 = F_{11} \, P = \left(\dfrac{L}{GA^*} + \dfrac{L^3}{4EI_y}\right) P \qquad (2.41)$$

$$\theta_2 = F_{12}P = \dfrac{L^2}{2EI_y}P$$

It is interesting that the exact Timoshenko solution for the end rotation has been obtained (see Eq.(2.36)).

The end deflection ratio r_w is now

$$r_w = \dfrac{w_2}{w_2^{EB}} = \dfrac{3\lambda^2 + 3}{4\lambda^2} \qquad (2.42)$$

The new distribution of r_w with λ is plotted in Figure 2.6. Now $\lambda \to$ 0.75 for $r_w \to \infty$ and, therefore, shear locking has been avoided. Obviously, the limit solution is not exact due to the coarse mesh used. We can check that the limit value of r_w (for $\lambda \to \infty$) converges rapidly to the unity as the mesh is refined (Table 2.1). For a two element mesh, r_w tends to 0.938 and the solution practically coincides with the exact one for all values of the slenderness ratio λ (Figure 2.6).

Analyzing the single beam element under uniformly distributed loading $(f_z = q)$ leads to similar conclusions as for the point load case. The exact quadrature leads to shear locking, whereas the reduced one point quadrature for $\mathbf{K}_s^{(e)}$ gives the following end displacements

$$w_2 = q\left(\frac{L}{GA^*} + \frac{L^3}{8EI}\right) \quad, \quad \theta_2 = \frac{qL^2}{4EI} \qquad (2.43)$$

Now the end rotation is not exact, while the end deflection for the limit slender case ($\lambda \to \infty$) coincides with the exact value of $\frac{qL^3}{8EI}$ (Example 1.4). The end rotation value obviously improves as the mesh is refined.

A similar example is presented next for the same beam under a uniformly distributed moment.

Example 2.1: Solve the cantilever beam of Figure 2.5 under a uniformly distributed moment m using one 2-noded Timoshenko beam element.

- Solution

The solution with exact integration is obtained from Eq.(2.29) by substituting the nodal force vector in the r.h.s. by

$$\mathbf{f} = \left[R_1, M_1 + \frac{mL}{2}, 0, \frac{mL}{2}\right]^T$$

The values for w_2 and θ_2 are deduced from Eq.(2.31) with the r.h.s. given by $\mathbf{f} = [0, mL/2]^T$ as

$$w_2 = \frac{\beta}{\beta+1}\frac{mL^3}{2EI} \quad , \quad \theta_2 = \frac{\beta}{\beta+1}\frac{mL^2}{2EI}$$

For very slender beams, $\beta \to 0$ and the solution locks giving $w_2 = \theta_2 = 0$. Also from Eq.(2.29) we find $R_1 = 0$ and $M_1 = mL$.

The solution with one-point reduced integration of the shear stiffness terms is obtained via matrix $\mathbf{F}$ of Eq.(2.40) giving

$$w_2 = \frac{mL^3}{4EI} \quad , \quad \theta_2 = \frac{mL^2}{2EI} \quad , \quad R_1 = 0 \quad , \quad M_1 = -mL$$

Note that the values of w_2 and θ_2 are independent of the shear modulus. The value for θ_2 coincides with the exact solution of Euler-Bernoulli theory (Example 1.4b). The value for w_2 converges fast to the "exact" slender solution of $\frac{mL^3}{3EI}$ as the mesh is refined, similarly as for the end point load case. It is interesting that the "exact" end displacements for a slender beam coincide with those for the end point load case for $P = m$.

We conclude that *the one point reduced quadrature for $\mathbf{K}_s^{(e)}$ yields a 2-noded Timoshenko beam element valid for both thick and slender beams.* Once the nodal displacements have been obtained, the bending moment and the shear force are computed at the element mid-point which is "optimal" for the evaluation of stresses (Figure 6.12 of [On4]).

2.4.1 Substitute transverse shear strain matrix

Matrix $\mathbf{K}_s^{(e)}$ in Eq.(2.38) can be obtained by the following expression

$$\mathbf{K}_s^{(e)} = (\overline{\widehat{D}}_s)\overline{\mathbf{B}}_s^T\overline{\mathbf{B}}_s l^{(e)} \tag{2.44}$$

where $\overline{(\cdot)}$ denotes values computed at the single quadrature point located at the element center.

Matrix $\overline{\mathbf{B}}_s$ of Eq.(2.44) is

$$\overline{\mathbf{B}}_s = \left[-\frac{1}{l^{(e)}}, -\frac{1}{2}, \frac{1}{l^{(e)}}, -\frac{1}{2}\right] \tag{2.45}$$

Matrix $\overline{\mathbf{B}}_s$ is the *substitute transverse shear strain matrix* and it leads to a locking-free 2-noded Timoshenko beam element. Matrix $\overline{\mathbf{B}}_s$ can also be obtained by the procedures to avoid shear locking described in Section 2.8.

2.5 MORE ON SHEAR LOCKING

The effect of shear locking can be also explained by studing the behaviour of the global system $\mathbf{Ka} = \mathbf{f}$ as the beam slenderness increases. For a single element mesh this system can be written making use of Eqs.(2.21a) (assuming homogeneous geometrical and material properties) as

$$\left(\frac{EI_y}{L^3}\overline{\mathbf{K}}_b + \frac{GA^*}{L}\overline{\mathbf{K}}_s\right) \mathbf{a} = \mathbf{f} \tag{2.46}$$

The "exact" solution for slender beams is proportional to $\frac{L^3}{3EI_y}$ (see Eq.(2.36)). Multiplying Eq.(2.46) by this value we obtain

$$\left(\overline{\mathbf{K}}_b + \frac{4}{\beta}\overline{\mathbf{K}}_s\right) \mathbf{a} = \frac{L^3}{3EI_y}\,\mathbf{f} = \overline{\mathbf{f}} \tag{2.47}$$

where β is given by Eq.(2.33) and $\overline{\mathbf{f}}$ is a vector *of the same order of magnitude* as the exact slender beam solution. For slender beams β decreases and, therefore, the factor multiplying $\overline{\mathbf{K}}_s$ in Eq.(2.47) is much larger than the terms of $\overline{\mathbf{K}}_b$. Consequently, Eq.(2.47) tends for very slender beams to

$$\frac{4}{\beta}\overline{\mathbf{K}}_s\,\mathbf{a} = \overline{\mathbf{f}} \qquad (2.48)$$

In the slender limit for $h \to 0$, then $\beta \to 0$ and

$$\overline{\mathbf{K}}_s\,\mathbf{a} = \frac{\beta}{4}\overline{\mathbf{f}} \to \mathbf{0} \qquad (2.49)$$

Consequently, as the beam slenderness increases the finite element solution progressively stiffens (*locks*) and the limit slender solution is *infinitely stiffer than the correct Euler-Bernoulli solution*. Furthermore, from Eq.(2.49) we deduce that the trivial solution $\mathbf{a} = \mathbf{0}$ can be avoided if $\overline{\mathbf{K}}_s$ (or $\mathbf{K}_s$) *is a singular matrix*. The singularity of $\mathbf{K}_s$ appears as a necessary (though not always sufficient) condition for the existence of the correct solution in the analysis of slender beams using Timoshenko elements [ZT2].

Singularity of the shear stiffness matrix $\mathbf{K}_s$ can be induced by reduced integration. It can be proved that the numerical integration of the stiffness matrix introduces s independent relationships at each integration point, s being the number of strains involved in the computation of the stiffness matrix [ZT2]. Thus, if p is the total number of integration points and j the number of free DOFs (after eliminating the prescribed values), the stiffness matrix will be singular if the total number of independent relationships introduced can not balance the total number of unknowns, i.e. if

$$\boxed{j - s \times p > 0} \qquad (2.50)$$

The proof of this inequality is given in Appendix E. Eq.(2.50) allows us to study the singularity of the shear stiffness matrix $\mathbf{K}_s$ and also that of the global stiffness matrix $\mathbf{K}$, both for an individual element and for a patch of elements. In all cases we find that $\mathbf{K}_s$ becomes singular by using a reduced quadrature. The subintegration must however preserve the proper rank of the global matrix $\mathbf{K}$ to avoid instabilities in the solution (Section 3.10.3 of [On4]). As an example let us consider the beam in Figure 2.23. The number of free DOFs is two (w_2 and θ_2) and only the transverse shear strain is involved in the computation of $\mathbf{K}_s$ (i.e. $s = 1$). Using exact integration for $\mathbf{K}_s$ ($p = 2$) gives

$$2 - 1 \times 2 = 0$$

and, consequently, condition (2.50) is not satisfied.

We can verify that $\mathbf{K}_s$ is not singular in this case. Eliminating the prescribed values in Eq.(2.27b) gives

$$\left| \mathbf{K}_s \right| = \begin{vmatrix} \dfrac{GA^*}{l} & -\dfrac{GA^*}{2} \\ -\dfrac{GA^*}{2} & \dfrac{GA^*}{3}l \end{vmatrix} = \dfrac{l}{12} GA^* \tag{2.51}$$

Computing $\mathbf{K}_s$ with a single quadrature point ($p = 1$), the rule (2.50) gives

$$2 - 1 \times 1 = 1 > 0$$

and, therefore, $\mathbf{K}_s$ should be singular. This can be verified by using the expression of $\mathbf{K}_s$ from Eq.(2.38), i.e.

$$\left| \mathbf{K}_s \right| = \begin{vmatrix} \dfrac{GA^*}{l} & -\dfrac{GA^*}{2} \\ -\dfrac{GA^*}{2} & \dfrac{GA^*}{4}l \end{vmatrix} = 0 \tag{2.52}$$

It is very important to check always that the global stiffness matrix $\mathbf{K}$ is not singular. The number of strains involved is two (κ and γ_{xz}). Using a single integration point for $\mathbf{K}_b$ and $\mathbf{K}_s$ the rule (2.50) gives

$$2 - 2 \times 1 = 0$$

which guarantees the non-singularity of $\mathbf{K}$ and the existence of a correct solution, as shown in the example of the previous section.

The need for the singularity of the transverse shear strain matrix to avoid shear locking can be argued on different grounds. For instance, consider the total internal energy of the beam written as

$$U = \frac{1}{2}\mathbf{a}^T \mathbf{K}_b \mathbf{a} + \frac{1}{2}\mathbf{a}^T \mathbf{K}_s \mathbf{a} = U_b + U_s \tag{2.53}$$

where U_b and U_s represent the bending and shear contributions to the internal energy, respectively. Timoshenko beam elements are able to reproduce the Euler-Benouilli solution if the shear strain energy U_s tends to zero as the beam slenderness ratio increases. In the limit thin case, U_s should vanish and this justifies the need for the singularity of $\mathbf{K}_s$.

There are other procedures to avoid shear locking which are related to the singularity of $\mathbf{K}_s$. Some of these methods are discussed in Section 2.8.

2.6 SUBSTITUTE SHEAR MODULUS FOR THE TWO-NODED TIMOSHENKO BEAM ELEMENT

The behaviour of the 2-noded Timoshenko beam element with reduced integration for $\mathbf{K}^{(e)}$ can be enhanced by using a "substitute shear modulus" $\overline{GA}^*$. This is defined such that the "exact" flexibility matrix coincides with that obtained using a single Timoshenko beam element. Equaling F_{11} in Eqs.(2.35) and (2.40) gives

$$\frac{l^{(e)}}{\overline{GA}^*} + \frac{(l^{(e)})^3}{4EI_y} = \frac{l^{(e)}}{GA^*} + \frac{(l^{(e)})^3}{3EI_y} \tag{2.54a}$$

and

$$\frac{1}{\overline{GA}^*} = \frac{1}{GA^*} + \frac{(l^{(e)})^2}{12EI} \tag{2.54b}$$

Introducing $\overline{GA}^*$ into the expression (2.40) for $\mathbf{K}_s^{(e)}$ and using $\mathbf{K}_b^{(e)}$ from Eq.(2.27a) we obtain an enhanced stiffness matrix for the 2-noded Timoshenko beam element as

$$K_{11}^{(e)} = K_{33}^{(e)} = -K_{13}^{(e)} = 12\left(\frac{K_1}{K_2}\right)^{(e)} \quad ; \quad K_{22}^{(e)} = K_{44}^{(e)} = K_1^{(e)}\left(1 + \frac{3}{K_2^{(e)}}\right)$$

$$K_{24}^{(e)} = K_1^{(e)}\left(\frac{3}{K_2^{(e)}} - 1\right) \quad ; \quad K_{12}^{(e)} = K_{14}^{(e)} = -K_{34}^{(e)} = -K_{23}^{(e)} = \frac{l^{(e)}}{2}K_{11}^{(e)}$$

$$K_1^{(e)} = \left(\frac{EI_y}{l^3}\right)^{(e)} \qquad \text{and} \qquad K_2^{(e)} = \left[1 + \beta^{(e)}\right]$$

$$\tag{2.55}$$

where $\beta^{(e)}$ is deduced from Eq.(2.33) changing L by $l^{(e)}$. For very slender beams $\frac{r}{\beta^{(e)}} \to 0$ and the terms in Eq.(2.55) coincide with those of the stiffness matrix for the 2-noded Euler-Bernoulli beam element (Eq.(1.20)).

Matrix (2.55) yields *nodally exact results* for thick and slender beams under uniformly distributed loads and nodal point loads. The exact solution for different loads requires modifying the equivalent nodal force vector. This is detailed in Section 2.9 where the stiffness matrix of (2.55) and the equivalent nodal force vector for an "exact" 2-noded Timoshenko beam element are obtained by integrating the equilibrium equations.

2.7 QUADRATIC TIMOSHENKO BEAM ELEMENT

Let us will consider the 3-noded Timoshenko beam element with quadratic Lagrange shape functions shown in Figure 2.7. The deflection and the

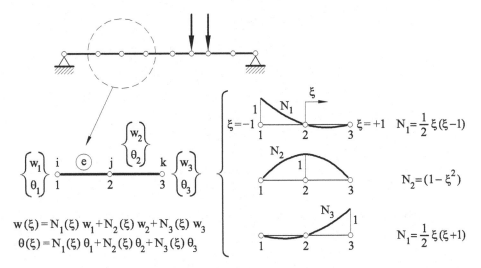

Fig. 2.7 3-noded quadratic Timoshenko beam element. Nodal displacements and quadratic shape functions

rotation are independently interpolated as

$$w(\xi) = N_1(\xi)w_1 + N_2(\xi)w_2 + N_3(\xi)w_3$$
$$\theta(\xi) = N_1(\xi)\theta_1 + N_2(\xi)\theta_2 + N_3(\xi)\theta_3 \tag{2.56}$$

The geometry is interpolated in an isoparametric form, similarly as for the 3-noded rod element of Section 3.3.4 of [On4], i.e.

$$x = N_1 x_1 + N_2 x_2 + N_3 x_3 \tag{2.57}$$

For simplicity we assume that node 2 is at the center of the element. This gives $\frac{dx}{d\xi} = \frac{l^{(e)}}{2}$ (Section 3.3.4 of [On4]).

The bending strain is obtained by

$$\kappa = \frac{d\theta}{dx} = \mathbf{B}_b \mathbf{a}^{(e)} \tag{2.58}$$

where

$$\mathbf{B}_b = \left[0, \frac{dN_1}{d\xi}\frac{d\xi}{dx}, 0, \frac{dN_2}{d\xi}\frac{d\xi}{dx}, 0, \frac{dN_3}{d\xi}\frac{d\xi}{dx}\right] = \frac{2}{l^{(e)}}\left[0, \xi - \frac{1}{2}, 0, -2\xi, 0, \xi + \frac{1}{2}\right] \tag{2.59}$$

and

$$\mathbf{a}^{(e)} = \left\{\begin{matrix} \mathbf{a}_1 \\ \mathbf{a}_2 \\ \mathbf{a}_3 \end{matrix}\right\}, \quad \text{with } \mathbf{a}_i = \left\{\begin{matrix} w_i \\ \theta_i \end{matrix}\right\} \tag{2.60}$$

The transverse shear strain is expressed as

$$\gamma_{xz} = \frac{dw}{dx} - \theta = \mathbf{B}_s \mathbf{a}^{(e)} \tag{2.61}$$

with

$$\mathbf{B}_s = \left[\frac{dN_1}{d\xi}\frac{d\xi}{dx}, -N_1, \frac{dN_2}{d\xi}\frac{d\xi}{dx}, -N_2, \frac{dN_3}{d\xi}\frac{d\xi}{dx}, -N_3 \right] =$$

$$= \frac{2}{l^{(e)}} \left[\xi - \frac{1}{2}, -\frac{l^{(e)}}{4}(\xi^2 - \xi), -2\xi, -\frac{l^{(e)}}{2}(1 - \xi^2), \xi + \frac{1}{2}, -\frac{l^{(e)}}{4}(\xi^2 + \xi) \right] \tag{2.62}$$

The element stiffness matrix is obtained as explained for the 2-noded beam element and it can also be split as

$$\mathbf{K}^{(e)} = \mathbf{K}_b^{(e)} + \mathbf{K}_s^{(e)} \tag{2.63}$$

where $\mathbf{K}_b^{(e)}$ and $\mathbf{K}_s^{(e)}$ are the bending and shear stiffness matrices, respectively, given by Eqs.(2.21b). The equivalent nodal force vector is

$$\mathbf{f}^{(e)} = \left\{ \begin{array}{c} \mathbf{f}_1^{(e)} \\ \mathbf{f}_2^{(e)} \\ \mathbf{f}_3^{(e)} \end{array} \right\} \quad \text{with} \quad \mathbf{f}_i^{(e)} = \int_{-1}^{+1} N_i \left\{ \begin{array}{c} f_z \\ m \end{array} \right\} \frac{l^{(e)}}{2} \, d\xi \tag{2.64}$$

The terms $\frac{dN_i}{d\xi}\frac{dN_j}{d\xi}$ in $\mathbf{K}_b^{(e)}$ are quadratic in ξ and the exact integration requires a two-point Gauss quadrature (Appendix C).

On the other hand, the terms N_iN_j in $\mathbf{K}_s^{(e)}$ are quartic in ξ and a three-point Gauss quadrature is needed to integrate them exactly (Appendix C). Unfortunately the exact integration of $\mathbf{K}_s^{(e)}$ leads to shear locking in many situations. This problem disappears if a reduced two-point quadrature is used for $\mathbf{K}_s^{(e)}$. As an example let us consider a beam clamped at one end and simply supported at the other (Figure 2.8). A single 3-noded beam element is used. The number of available DOFs is just three, and the rule (2.50) for a 3 point quadrature for $\mathbf{K}_s^{(e)}$ gives

$$j - s \times p = 3 - 1 \quad \text{shear strain} \times 3 \text{ point} = 0$$

i.e., $\mathbf{K}_s^{(e)}$ is not singular and the solution will lock for slender beams. Singularity is guaranteed by using a reduced two-point quadrature for $\mathbf{K}_s^{(e)}$. In this case $3 - 1 \times 2 = 1 > 0$ and $\mathbf{K}_s^{(e)}$ is singular. Figure 2.9 shows $\mathbf{K}_b^{(e)}$ and $\mathbf{K}_s^{(e)}$ using a two-point Gauss quadrature for both matrices.

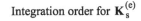

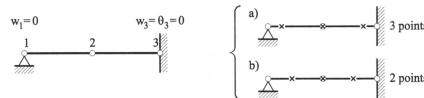

Available degrees of freedom, $j = 6 - 3 = 3$
Shear strain components, $s = 1$

$$\text{Singularity condition} \begin{cases} \text{a) } 3 - 1 \times 3 = 0 \\ \text{b) } 3 - 1 \times 2 = 1 > 0 \ \underline{\text{Singular}} \end{cases}$$

Fig. 2.8 Singularity rule for $\mathbf{K}_s^{(e)}$ in a simply supported/clamped beam analyzed with a single 3-noded Timoshenko beam element.

$$\mathbf{K}_b^{(e)} = \left(\frac{\hat{D}_b}{3l}\right)^{(e)} \begin{bmatrix} 0 & 0 & 0 & 0 & 0 & 0 \\ 0 & 7 & 0 & -8 & 0 & 1 \\ 0 & 0 & 0 & 0 & 0 & 0 \\ 0 & -8 & 0 & 16 & 0 & -8 \\ 0 & 0 & 0 & 0 & 0 & 0 \\ 0 & 1 & 0 & -8 & 0 & 7 \end{bmatrix}$$

$$\mathbf{K}_s^{(e)} = \left(\frac{\hat{D}_s}{9l}\right)^{(e)} \begin{bmatrix} 21 & -\frac{9}{2}l^{(e)} & -24 & -6l^{(e)} & 3 & \frac{3}{2}l^{(e)} \\ -\frac{9}{2}l^{(e)} & (l^{(e)})^2 & 6l^{(e)} & (l^{(e)})^2 & -\frac{3}{2}l^{(e)} & -\frac{(l^{(e)})^2}{2} \\ -24 & 6l^{(e)} & 48 & 0 & -24 & -6l^{(e)} \\ -6l^{(e)} & (l^{(e)})^2 & 0 & 4(l^{(e)})^2 & 6l^{(e)} & (l^{(e)})^2 \\ 3 & -\frac{3}{2}l^{(e)} & -24 & 6l^{(e)} & 21 & \frac{9}{2}l^{(e)} \\ \frac{3}{2}l^{(e)} & -\frac{(l^{(e)})^2}{2} & -6l^{(e)} & (l^{(e)})^2 & \frac{9}{2}l^{(e)} & (l^{(e)})^2 \end{bmatrix}$$

Fig. 2.9 $\mathbf{K}_b^{(e)}$ and $\mathbf{K}_s^{(e)}$ matrices for the 3-noded Timoshenko beam element obtained with a uniform two-point Gauss quadrature

Reduced integration is not strictly necessary for analysis of the cantilever beam in Figure 2.5 using a single 3-noded Timoshenko element. Here the exact three-point integration of $\mathbf{K}_s^{(e)}$ satisfies the singularity rule (2.50). This is an exception and, in practice, the reduced quadrature for $\mathbf{K}_s^{(e)}$ is recommended. Also, the bending moment and the shear force should be computed at the two integration points which are the optimal sampling points (Figure 6.12 of [On4]).

The reduced integration of the shear stiffness matrix appears to be a "panacea" which yields an improved solution at a lower computing cost. However, as for 2D and 3D solid elements [On4], despite its potential benefits, reduced integration should be used with extreme care in order not to perturb the proper rank of the global stiffness matrix. This is not the case for the linear and quadratic Timoshenko beam elements for which reduced integration of $\mathbf{K}_s^{(e)}$ is recommended for practical purposes.

2.8 ALTERNATIVES FOR DERIVING LOCKING-FREE TIMOSHENKO BEAM ELEMENTS

2.8.1 Reinterpretation of shear locking

A detailed inspection of Timoshenko beam elements shows that the equal order interpolation for the deflection and the rotation leads to the limit condition of zero shear strain not being satisfied, which in turn leads to shear locking. Let us consider, as an example, the simple 2-noded Timoshenko beam element. The linear displacement approximation yields the following transverse shear strain field

$$\gamma_{xz} = \frac{\partial w}{\partial x} - \theta = \alpha_1 + \alpha_2 \xi \qquad (2.65)$$

with

$$\alpha_1 = \frac{w_2 - w_1}{l^{(e)}} - \frac{1}{2}(\theta_1 + \theta_2) \quad ; \quad \alpha_2 = \frac{1}{2}(\theta_1 - \theta_2) \qquad (2.66)$$

The limit Euler-Bernoulli condition of vanishing transverse shear strain for slender beams ($\gamma_{xz} = 0$) requires

$$\begin{array}{ll} \alpha_1 \to 0 & \text{i.e.} \quad \dfrac{w_2 - w_1}{l^{(e)}} = \dfrac{\theta_1 + \theta_2}{2} \\[2mm] \alpha_2 \to 0 & \text{i.e.} \quad \theta_1 = \theta_2 \end{array} \qquad (2.67)$$

The condition for α_1 expresses the coincidence of the average element rotation and the element slope (which obviously should be identical for slender beams). However, the condition $\theta_1 = \theta_2$ for α_2 has not a physical meaning and leads to a zero curvature field (as $\frac{d\theta}{dx} = \frac{1}{l^{(e)}}(\theta_2 - \theta_1) = 0$) and, hence, to zero flexural stiffness. This originates locking of the numerical solution.

Therefore, shear locking can be seen as *a consequence of imposing a non-physical relationship on the nodal displacements* in order to satisfy the condition of zero transverse shear strain. It is then obvious that the linear term in Eq.(2.65) must be eliminated so that the condition $\gamma_{xz} = 0$ can be satisfied naturally without introducing spureous constraints. A simple way to cancel this term is to evaluate γ_{xz} at the element midpoint ($\xi = 0$). This gives $\gamma_{xz} = \alpha_1$ and the element then behaves correctly in the limit slender case. *This is equivalent to using a single point quadrature for* $\mathbf{K}_s^{(e)}$. Reduced integration appears here as an effective procedure for eliminating the spureous contribution in the discretized transverse shear strain field which is the source of locking.

There are other procedures to avoid shear locking. Following the above arguments it is reasonable to assume that the physical conditions of the problem will not be violated if the coefficients of the polynomial representing γ_{xz} are linear functions of both the nodal displacements and the rotations. This can be achieved if the polynomial terms originating from the slope $\dfrac{dw}{dx}$ are *of the same degree* as those contributed by θ. This is satisfied if the polynomial interpolation for w is one degree higher than that used for θ. This technique is studied in the next two sections.

Another alternative for eliminating shear locking is to assume a priori a "good" transverse shear strain field over the element (i.e. $\gamma_{xz} = \alpha_1$ in Eq.(2.65)). Thus, the spurious terms are omitted from the onset and the source of locking disappears. This is the basis of the *assumed shear strain* technique studied in a next section. This procedure has been widely used for deriving locking-free beam, plate and shell elements, and some of them will be studied in the following chapters. There are interesting analogies between the different procedures for eliminating shear locking which in some cases are completely equivalent.

2.8.2 Use of different interpolations for deflection and rotation

The key to this approach is to use an interpolation for the deflection that is one degree higher than the one used for the rotation. Hence, the condition $\gamma_{xz} = \dfrac{dw}{dx} - \theta = 0$ can be naturally fulfilled in the limit.

This technique has been used by different authors [Cr, DL, Ma, TH] for deriving thick beam and plate elements. It can be verified that the simplest option of using a linear approximation for the deflection and a

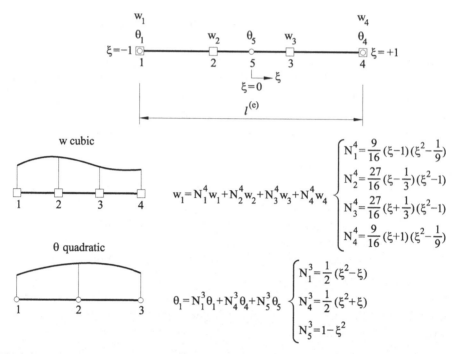

Fig. 2.10 Timoshenko beam element with cubic deflection and quadratic rotation

constant one for the rotation is equivalent to using one point integration for the transverse shear stiffness matrix. As an alternative we can choose a quadratic interpolation for the deflection and a linear one for the rotation. A more interesting option is to use a cubic approximation for the deflection and a quadratic one for the rotation leading to a parabolic distribution of the transverse shear strain over the element (Figure 2.10).

Some caution should be taken as the resulting cubic/quadratic element is sometimes unable to reproduce a constant shear distribution. Consider, for instance, a simple supported beam under a central point load analyzed with one cubic/quadratic element (Figure 2.11). The computed shear force distribution is quadratic and continuous and this is quite different from the "exact" constant discontinuous solution.

Example 2.3 shows how the cubic/quadratic Timoshenko beam element can be constrained to give a beam element with linear bending and constant transverse shear strain fields. Also by constraining the transverse shear strains to be zero the cubic/quadratic beam elements "degenerates" into the 2-noded Euler-Bernoulli beam element of previous chapter (Example 2.5).

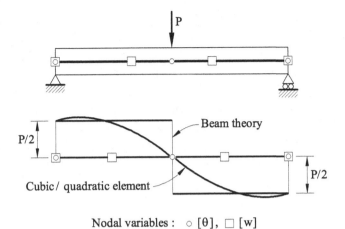

Nodal variables : ○ [θ], □ [w]

Fig. 2.11 Simply supported beam analyzed with one cubic/quadratic Timoshenko beam element. Computed and exact distributions of the shear force

2.8.3 Linked interpolation

Shear locking can be avoided by enhancing the interpolation for the deflection field with additional higher-order polynomial terms involving the nodal rotations. The aim is to obtain a transverse shear strain field that can satisfy the limit Euler-Bernoulli condition of vanishing shear strain.

The following interpolation is locking-free for the 2-noded Timoshenko beam element

$$w = \frac{1}{2}(1 - \xi)w_1 + \frac{1}{2}(1 + \xi)w_2 + (1 - \xi^2)\frac{l^{(e)}}{8}(\theta_1 - \theta_2) \qquad (2.68)$$

The above interpolation *links* the nodal rotations and the deflections. The resulting transverse shear strain field is

$$\gamma_{xz} = \frac{dw}{dx} - \theta = \frac{w_2 - w_1}{l^{(e)}} - \frac{\theta_1 + \theta_2}{2} = \left[-\frac{1}{l^{(e)}}, -\frac{1}{2}, \frac{1}{l^{(e)}}, -\frac{1}{2} \right] \begin{Bmatrix} w_1 \\ \theta_1 \\ w_2 \\ \theta_2 \end{Bmatrix} = \overline{\mathbf{B}}_s \mathbf{a}^e$$

$$(2.69)$$

The transverse shear strain vanishes for

$$\frac{w_2 - w_1}{l^{(e)}} = \frac{\theta_1 + \theta_2}{2} \qquad (2.70)$$

Eq.(2.70) states that the average rotation of the element equals the slope of the deflection field. This condition is satisfied for slender beams.

Matrix $\overline{\mathbf{B}}_s$ in Eq.(2.69) coincides with the substitute shear strain matrix of Eq.(2.45) obtained by sampling $\mathbf{B}_s$ of Eq.(2.15) at the element center. The shear stiffness matrix is given by Eq.(2.44). The resulting 2-noded beam element is therefore free from shear locking.

The equivalent nodal force vector for nodal point loads coincides with that of the standard displacement formulation as the "linking" shape function $(1 - \xi^2)$ vanishes at the end nodes. However, a distributed loading introduces nodal bending moment components to the linked interpolation. The equivalent nodal force vector for a uniform loading $f_z = q$, is

$$\mathbf{f}^{(e)} = ql^{(e)} \left[\frac{1}{2}, \frac{(l^{(e)})^2}{12}, \frac{1}{2}, -\frac{(l^{(e)})^2}{12} \right]^T \tag{2.71}$$

Let us consider, for example, a cantilever beam of length L under uniformly distributed loading analyzed with just one linked beam element. The end displacement values are readily obtained using the flexibility matrix of Eq.(2.40) and the force vector of Eq.(2.71) as

$$w_2 = \frac{qL}{2GA^*} + \frac{qL^3}{12EI_y} \quad \text{and} \quad \theta_2 = \frac{qL^2}{6EI_y} \tag{2.72}$$

The end rotation is now exact. The end deflection for a slender beam has approximately 33% error versus the exact value of $qL^3/8EI_y$. This error diminishes rapidly as the mesh is refined. The original beam element with reduced integration and $\mathbf{f}^{(e)}$ given by Eq.(2.24b) yields an exact end deflection and an approximate value for the end rotation (Eq.(2.43)).

Fraejis de Veubeke [FdV2] was the first to use linked interpolations for beam analysis. Tessler et al. [TD,Te] have used similar interpolations for beams, shallow arches and plates that they call "anisoparametric". Other applications of linked interpolations for beams can be found in [Cr,SCB].

The derivation of the displacement field of Eq.(2.68) is shown in the next example.

Example 2.2: Derive the linked interpolation of Eq.(2.68) for the 2-noded Timoshenko beam element.

- Solution

The starting point is the 3-noded Timoshenko beam element with node numbers 1,3,2 where node number 3 corresponds to the mid-node. The original quadratic interpolation for the deflection and the rotation is

$$w = \frac{1}{2}(\xi - 1)\xi w_1 + (1 - \xi^2)w_3 + \frac{1}{2}(\xi + 1)\xi w_2$$

$$\theta = \frac{1}{2}(\xi - 1)\theta_1 + (1 - \xi^2)\theta_3 + \frac{1}{2}(\xi + 1)\xi\theta_2$$

The transverse shear strain is obtained by

$$\gamma_{xz} = \frac{dw}{dx} - \theta = \frac{2}{l^{(e)}}\left(\xi - \frac{1}{2}\right)w_1 - \frac{\xi}{2}w_3 + \frac{2}{l^{(e)}}\left(\xi + \frac{1}{2}\right)w_2 - \frac{1}{2}(\xi - 1)\xi\theta_1 -$$

$$-(1 - \xi^2)\theta_3 - \frac{1}{2}(\xi + 1)\xi\theta_2 = \frac{w_2 - w_1}{l^{(e)}} + \frac{2}{l^{(e)}}\xi(w_1 - 2w_3 + w_2) - \theta_3 -$$

$$-\frac{\xi}{2}(\theta_1 - \theta_2) - \frac{\xi^2}{2}(\theta_1 - 2\theta_3 + \theta_2) = 0$$

Clearly, γ_{xz} should vanish for slender (Euler-Bernoulli) beams. This is achievable for the above interpolation if the linear and quadratic terms in ξ are zero and γ_{xz} is simply expressed as

$$\gamma_{xz} = \frac{w_2 - w_1}{l^{(e)}} - \theta_3$$

The condition $\gamma_{xz} = 0$ implies $\frac{w_2 - w_1}{l^{(e)}} = \theta_3$, i.e. the average slope equals the rotation at the mid-node, which is a physical condition for slender beams.
The vanishing of the linear and quadratic terms in the original quadratic expression for γ_{xz} yields the following two conditions

$$\frac{2}{l^{(e)}}(w_1 - 2w_3 + w_2) - \frac{\theta_1 - \theta_2}{2} = 0$$

$$\theta_1 - 2\theta_3 + \theta_2 = 0$$

From the above we obtain

$$w_3 = \frac{w_1 + w_3}{2} - \frac{\theta_1 - \theta_2}{8}l^{(e)}$$

$$\theta_3 = \frac{\theta_1 + \theta_2}{2}$$

Substituting these values into the original quadratic interpolation gives, after some algebra

$$w = \frac{1}{2}(1 - \xi)w_1 + \frac{1}{2}(1 + \xi)w_2 + (1 - \xi^2)\frac{l^{(e)}}{8}(\theta_1 - \theta_2)$$

$$\theta = \frac{1}{2}(1 - \xi)\theta_1 + \frac{1}{2}(1 + \xi)\theta_2$$

which is the linked interpolation (2.68) we are looking for. It can be verified that the above interpolation yields the constant transverse shear strain field of Eq.(2.68).

2.8.4 Assumed transverse shear strain approach

As previously explained, Timoshenko beam elements should be able to satisfy the condition of vanishing transverse shear strain for slender beams. Hence the following condition must be satisfied for $(\lambda \to \infty)$

$$\gamma_{xz} = \mathbf{B}_s \, \mathbf{a} = \alpha_1(w_i, \theta_i) + \alpha_2(w_i, \theta_i)\xi + \alpha_3(w_i, \theta_i)\xi^2 + \cdots + \alpha_n(w_i, \theta_i)\xi^n = 0$$

$$(2.73)$$

which necessarily leads to

$$\alpha_j(w_i, \theta_i) = 0 \quad ; \quad j = 1, n \qquad (2.74)$$

Eq.(2.74) imposes a linear relationship between the nodal displacements and the rotations which can usually be interpreted on physical grounds. Elements satisfying Eq.(2.74) are therefore able to reproduce naturally the limit slender beam condition without locking. However, Timoshenko beam elements typically have some α_j coefficients in Eq.(2.73) which are a function of the nodal rotations only. The condition $\alpha_j(\theta_i) = 0$ is too strong (and even non-physical) and this leads to locking.

A consequence of the above argument is that shear locking can be avoided by assuming "a priori" a polynomial transverse shear strain field of the form (2.73). The assumed transverse shear strain can be written as

$$\gamma_{xz} = \sum_{k=1}^{m} N_{\gamma_k} \, \gamma_k = \mathbf{N}_\gamma \, \boldsymbol{\gamma}^{(e)} \qquad (2.75)$$

where γ_k are the transverse shear strains sampled at m discrete points, and N_{γ_k} are the transverse shear interpolation functions. Eq. (2.75) is rewritten after expressing the transverse shear strains γ_k in terms of the nodal displacements as

$$\gamma_{xz} = \sum_{k=1}^{m} N_{\gamma_k} \bar{\mathbf{B}}_{s_k} \, \mathbf{a}_k^{(e)} = \bar{\mathbf{B}}_s \mathbf{a} \qquad (2.76)$$

Matrix $\bar{\mathbf{B}}_s$ is the *substitute shear strain matrix* mentioned in Section 2.4.1 ($\bar{\mathbf{B}}_s$ is also called *B-bar shear strain matrix*) [Cr,Hu]. The expression of $\bar{\mathbf{B}}_s$ for the 2-noded Timoshenko beam element coincides with Eq.(2.45). The coincidence is explained in the next section.

Eq.(2.76) can be written in the form (2.73) which guarantees the absence of locking.

The PVW (Eq.(2.9))can be written using (2.75), (2.4b) and (2.6b)

$$\int_0^l \left[\delta(\frac{\partial \theta}{\partial x}) \hat{D}_b \frac{\partial \theta}{\partial x} + \delta(\mathbf{N}_\gamma \, \boldsymbol{\gamma}^{(e)})^T \, \hat{D}_s (\mathbf{N}_\gamma \, \boldsymbol{\gamma}^{(e)}) \right] dx = \text{EVW} \qquad (2.77)$$

where EVW denotes the virtual work performed by the external loads (this is equal to the r.h.s. of Eq.(2.18a).

Eq.(2.77) shows that only C^o continuity for the rotation is required, whereas the deflection and the transverse shear strain can be discontinuous. This allows one the choice of independent interpolations for the rotation, the deflection and the transverse shear strain as

$$w = \mathbf{N}_w \, \mathbf{w}^{(e)} \quad ; \quad \theta = \mathbf{N}_\theta \, \boldsymbol{\theta}^{(e)} \quad ; \quad \gamma_{xz} = \mathbf{N}_\gamma \, \boldsymbol{\gamma}^{(e)} \qquad (2.78)$$

The nodal variables $\mathbf{w}^{(e)}, \boldsymbol{\theta}^{(e)}$ and $\boldsymbol{\gamma}^{(e)}$ must satisfy the following conditions to guarantee the convergence of the element (Appendix G)

$$n_\theta + n_w \geq n_\gamma \quad ; \quad n_\gamma \geq n_w \qquad (2.79)$$

where n_w, n_θ and n_γ are the number of variables involved in the interpolation of the deflection, the rotation and the transverse shear strain, respectively, disregarding the prescribed DOFs.

Eqs.(2.79) must be satisfied for each individual element and also for any patch of elements as a necessary (though not always sufficient) condition for convergence [ZQTN,ZTZ]. Eqs.(2.79), therefore, provide a fast and simple procedure to asses "a priori" the viability of a new element. The final assessment of the element performance must be verified via the patch test in all cases.

Selection of the assumed transverse shear strain field

The assumed transverse shear strain field can be obtained directly by observing the original field, bearing in mind that Eq.(2.79) must be satisfied. Hence, for the 2-noded Timoshenko beam element it is reasonable to assume a priori the following constant shear strain field (Section 2.8.1)

$$\gamma_{xz} = \alpha_1 (w_i, \theta_i) \qquad (2.80)$$

The parameter α_1 can be obtained by "sampling" γ_{xz} at the element midpoint. This leads to

$$\alpha_1 = (\gamma_{xz})_{\xi=0} = \frac{w_2 - w_1}{l^{(e)}} - \frac{\theta_1 + \theta_2}{2} \qquad (2.81)$$

and the *substitute shear strain matrix* $\overline{\mathbf{B}}_s$ is deduced from

$$\gamma_{xz} = \underbrace{\left[-\frac{1}{l^{(e)}}, -\frac{1}{2}, \frac{1}{l^{(e)}}, -\frac{1}{2} \right]}_{\overline{\mathbf{B}}_s} \mathbf{a}^{(e)} \tag{2.82}$$

Matrix $\overline{\mathbf{B}}_s$ coincides with the the original shear strain matrix $\mathbf{B}_s$ of Eq.(2.15) sampled at the element center, i.e. using a one-point quadrature, as well as with the expression of Eq.(2.69) obtained via a linked interpolation. The analogy between assumed transverse shear strain, reduced integration and linked interpolation procedures has been verified in this case.

The same argument evidences that the assumed transverse shear strain should vary linearly for the 3-noded quadratic Timoshenko beam element.

Figure 2.12 shows that the 2- and 3-noded Timoshenko beam elements with constant and linear assumed transverse shear strain fields satisfy Eqs.(2.79).

The condition $n_w + n_\theta > n_\gamma$ of Eq.(2.79) is equivalent to the singularity rule (2.50) (Appendix G). This is another explanation for the good performance of Timoshenko beam elements based on assumed transverse shear strain fields for analysis of slender beams. These concepts are of relevance for deriving locking-free thick plate and shell elements. A methodology for the systematic derivation of the substitute transverse shear strain matrix for thick plate elements is presented in Chapter 6.

The assumed transverse shear strain technique can be used to derive Euler-Bernoulli beam elements starting from Timoshenko elements. The assumed transverse shear strain field is chosen so that γ_{xz} vanishes at a number of points within the element. In this way, its behaviour approximates that of Euler-Bernoulli beam theory.

Some examples of the above concepts are presented next. Examples 2.2 and 2.3 describe two alternatives for deriving a Timoshenko beam element with constant transverse shear strain by constraining the original displacement field. In Examples 2.4 and 2.5 the 2-noded Euler-Bernoulli beam element is derived by imposing a zero transverse shear strain at selected points in two different Timoshenko beam elements. Imposing the condition of vanishing transverse shear strain at a number of discrete points within the element is also the basis of the so-called Discrete-Kirchhoff plate elements studied in Section 6.11. Examples 2.6–2.8 finally show the equivalence between reduced integration and assumed transverse shear strain techniques for linear and quadratic Timoshenko beam elements.

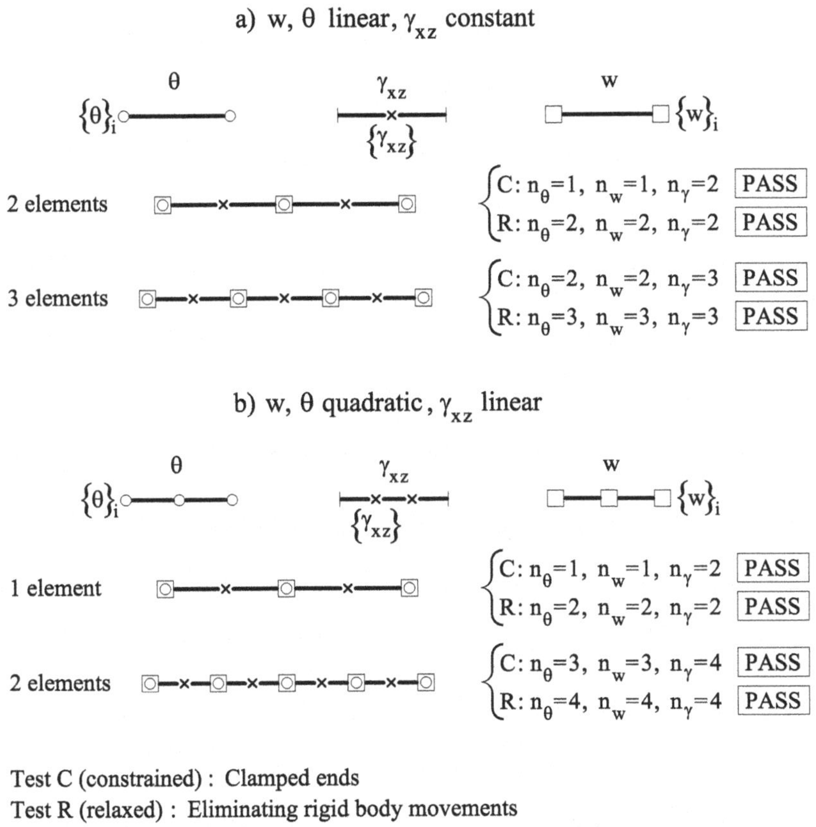

a) w, θ linear, γ_{xz} constant

$$\begin{cases} C: n_\theta=1, n_w=1, n_\gamma=2 & \boxed{PASS} \\ R: n_\theta=2, n_w=2, n_\gamma=2 & \boxed{PASS} \end{cases}$$

$$\begin{cases} C: n_\theta=2, n_w=2, n_\gamma=3 & \boxed{PASS} \\ R: n_\theta=3, n_w=3, n_\gamma=3 & \boxed{PASS} \end{cases}$$

b) w, θ quadratic, γ_{xz} linear

$$\begin{cases} C: n_\theta=1, n_w=1, n_\gamma=2 & \boxed{PASS} \\ R: n_\theta=2, n_w=2, n_\gamma=2 & \boxed{PASS} \end{cases}$$

$$\begin{cases} C: n_\theta=3, n_w=3, n_\gamma=4 & \boxed{PASS} \\ R: n_\theta=4, n_w=4, n_\gamma=4 & \boxed{PASS} \end{cases}$$

Test C (constrained) : Clamped ends
Test R (relaxed) : Eliminating rigid body movements

Fig. 2.12 Verification of Eqs.(2.79) for 2- and 3-noded Timoshenko beam elements with constant and linear assumed transverse shear strain fields respectively

---

Example 2.3: Derive a beam element with linear bending and constant transverse shear fields starting from the cubic/quadratic Timoshenko beam element of Figure 2.10.

- Solution

Figure 2.10 shows the original element and the cubic N_i^4 and quadratic N_i^3 shape functions for the deflection and the rotation, respectively. The quadratic rotation field automatically guarantees a linear bending field. The constant transverse shear field is obtained as follows. From the original displacement approximation the transverse shear strain is found as

$$\gamma_{xz} = \frac{\partial w}{\partial x} - \theta = \sum_{i=1}^{4} \frac{\partial N_i^4}{\partial x} w_i - N_1^3 \theta_1 - N_4^3 \theta_4 - N_5^3 \theta_5 = A + B\xi + c\xi^2$$

with

$$A = \frac{1}{8l^{(e)}}\,(w_1 - 27w_2 + 27w_3 - w_4 - 8l^{(e)}\theta_5)$$

$$B = \frac{9}{4l^{(e)}}\left[\,w_1 - w_2 - w_3 + w_4 + \frac{2l^{(e)}}{9}(\theta_1 - \theta_4)\,\right]$$

$$C = \frac{27}{8l^{(e)}}\left[\,-w_1 + 3w_2 - 3w_3 + w_4 - \frac{4l^{(e)}}{27}(\theta_1 + \theta_4) + \frac{8l^{(e)}}{27}\theta_5\,\right]$$

For γ_{xz} to be constant it is required that

$$B = C = 0$$

These conditions lead to two equations from which two nodal DOFs can be eliminated. Selecting the intermediate deflections w_2 and w_3 gives

$$w_2 = \frac{2w_1 + w_4}{3} + \frac{l^{(e)}}{81}(11\theta_1 - 7\theta_4 - 4\theta_5)$$

$$w_3 = \frac{w_1 + 2w_4}{3} + \frac{l^{(e)}}{81}(7\theta_1 - 1\theta_4 + 4\theta_5)$$

Substituting w_2 and w_3 into the original cubic field for w yields

$$w = \frac{1}{2}(1 - \xi)w_1 + \frac{1}{2}(1 + \xi)w_4 + \frac{l^{(e)}}{24}(2\xi^3 - 2\xi - 3\xi^2 + 3)\theta_1 +$$

$$+\frac{l^{(e)}}{24}(2\xi^3 - 2\xi + 3\xi^2 - 3)\theta_4 + \frac{l^{(e)}}{6}\xi(1 - \xi^2)\theta_5 = \sum_{i=1}^{5} N_i a_i = \mathbf{N}^{(e)}\mathbf{a}^{(e)}$$

with

$$\mathbf{a}^{(e)} = [w_1, \theta_1, w_4, \theta_4, \theta_5]^{(e)}$$

whereas the original quadratic interpolation is kept for the rotation. Note that the new interpolation for the deflection involves also the nodal rotations. This is another example of "linked" interpolation similar to those described in Section 2.8.3. It can be verified that the rigid body condition $(\sum_{i=1}^{5} N_i = 1)$ still holds in this case.

We can easily check that the required conditions are satisfied, i.e.

$$\kappa = \frac{d\theta}{dx} = \frac{2}{l^{(e)}}(2\xi - 1)\theta_1 - \frac{4\xi}{l^{(e)}}\theta_4 + \frac{1}{l^{(e)}}(2\xi + 1)\theta_5 = \frac{1}{l^{(e)}}(\theta_5 - \theta_1) +$$

$$+\frac{2}{l^{(e)}}(\theta_1 - 2\theta_4 + \theta_5)\xi = \frac{1}{l^{(e)}}\,[0, (2\xi - 1), 0, -4, (2\xi + 1)]\,\mathbf{a}^{(e)} = \mathbf{B}_b\mathbf{a}^{(e)}$$

$$\gamma_{xz} = \frac{dw}{dx} - \theta = \frac{w_4 - w_1}{l^{(e)}} - \frac{\theta_1 + 4\theta_5 + \theta_4}{6} = \left[-\frac{1}{l^{(e)}}, \frac{1}{6}, \frac{1}{l^{(e)}}, -\frac{1}{6}, \frac{2}{3}\right]\mathbf{a}^{(e)} = \mathbf{B}_s\mathbf{a}^{(e)}$$

Example 2.4: Derive the beam element of the previous example starting from the standard quadratic Timoshenko beam element in Section 2.7.

- Solution

The curvature and transverse shear strain fields for the 3-noded quadratic Timoshenko beam element of Figure 2.7 are

$$\kappa = \frac{d\theta}{dx} = \frac{\theta_3 - \theta_1}{l^{(e)}} + \frac{2}{l^{(e)}}(\theta_1 - 2\theta_2 + \theta_3)\xi$$

$$\gamma_{xz} = \frac{\partial w}{\partial x} - \theta = \frac{1}{l^{(e)}}(w_3 - w_1 - l^{(e)}\theta_2) + \frac{1}{l^{(e)}}(2w_1 - 4w_2 + 2w_3 + \frac{l^{(e)}}{2}\theta_1 -$$

$$-\frac{l^{(e)}}{2}\theta_3)\xi + \left(\theta_2 - \frac{\theta_1}{2} - \frac{\theta_3}{2}\right)\xi^2 = A + B\xi + C\xi^2$$

Note that the bending strain field coincides with that obtained in the previous example.

The condition $\gamma_{xz} = constant$ requires $B = 0$ and $C = 0$. From the later we deduce $\theta_2 = \frac{\theta_1 + \theta_3}{2}$. Substituting this value into the interpolation for the bending strain gives $\frac{d\theta}{dx} = \frac{\theta_3 - \theta_1}{l^{(e)}}$, and therefore the required linear bending distribution can not be obtained.

The alternative is to impose $\gamma_{xz} = A + C\xi^2$. The condition that must be satisfied now is

$$B = 0 \Rightarrow w_2 = \frac{w_1 + w_3}{2} + \frac{l^{(e)}}{8}(\theta_1 - \theta_3)$$

The deflection field is

$$w = \frac{1}{2}(1 - \xi)w_1 + \frac{1}{2}(1 + \xi)w_3 + \frac{l^{(e)}}{8}(\theta_1 - \theta_3) =$$

$$= \left[\frac{1}{2}(1 - \xi), \frac{l^{(e)}}{8}, \frac{1}{2}(1 + \xi), -\frac{l^{(e)}}{8}, 0\right] \mathbf{a}^{(e)}$$

with $\mathbf{a}^{(e)} = [w_1, \theta_1, w_3, \theta_3, \theta_2]^T$, and the shear strain distribution is

$$\gamma_{xz} = \frac{w_3 - w_1}{l^{(e)}} - \left(\frac{\theta_1 + \theta_3}{2}\right)\xi^2 - (1 - \xi^2)\theta_2$$

It is interesting that at $\xi = \pm\frac{1}{\sqrt{3}}$ the transverse shear strain is

$$(\gamma_{xz})_{\xi=\pm\frac{1}{\sqrt{3}}} = \frac{w_3 - w_1}{l^{(e)}} - \frac{\theta_1 + 4\theta_2 + \theta_3}{6}$$

which coincides with the constant transverse shear strain obtained in the previous example. Therefore, an "effective" constant transverse shear field can be obtained by using a two-point Gauss quadrature for the shear stiffness terms. This coincidence is a consequence of the properties of the Gauss points; i.e. the constant transverse shear distribution of Example 2.3 is the least square approximation of the quadratic field chosen here and both fields take the same value at the two Gauss points (Section 6.7 of [On4]).

Example 2.5: Derive the 2-noded Euler-Bernoulli beam element by imposing the condition of zero transverse shear strain over the cubic/quadratic Timoshenko beam element of Figure 2.10.

- Solution

The initial steps are similar to those of Example 2.3. The transverse shear strain is obtained as

$$\gamma_{xz} = \frac{dw}{dx} - \theta = A + B\xi + C\xi^2$$

where A, B and C coincide with the expressions given in Example 2.3. The condition $\gamma_{xz} = 0$ over the element is satisfied if

$$A = B = C = 0$$

This leads to a system of three equations which allows us to eliminate the internal nodal DOFs w_2, w_3 and θ_5 as

$$A = 0 \quad \Rightarrow \quad \theta_5 = \frac{1}{8l^{(e)}}(w_1 - 27w_2 + 27w_3 - w_4)$$

$$B = 0 \quad \Rightarrow \quad w_2 = w_1 - w_3 + w_4 + \frac{2l^{(e)}}{9}(\theta_1 - \theta_4)$$

Substituting these values into the equation for $C = 0$ gives

$$w_3 = \frac{1}{27}\left[7w_1 + 20w_4 + 2l^{(e)}(\theta_1 - 2\theta_4)\right]$$

and consequently

$$w_2 = \frac{1}{27}\left[20w_1 - 7w_4 + 2l^{(e)}(\theta_1 - 2\theta_4)\right]$$

Substituting w_2 and w_3 into the original cubic deflection field gives

$$w = \frac{1}{4}(\xi^3 - 3\xi + 2)w_1 + \frac{1}{4}(2 + 3\xi - \xi^3)w_4 + \frac{l^{(e)}}{8}(\xi^2 - 1)(\xi - 1)\theta_1 +$$

$$+ \frac{l^{(e)}}{8}(\xi^2 - 1)(\xi + 1)\theta_4 = N_1\, w_1 + \overline{N}_1\theta_1 + N_4 w_4 + \overline{N}_4\,\theta_4$$

where $N_1, \overline{N}_1, N_4, \overline{N}_4$ coincide with the Hermite shape functions for the 2-noded Euler-Bernoulli beam element (Eqs.(1.11a)). On the other hand, the condition $\gamma_{xz} = 0$ implies $\theta = \frac{dw}{dx}$ over the element and the deflection field can be written as

$$w = N_1\, w_1 + \overline{N}_1\left(\frac{dw}{dx}\right)_1 + N_4 w_4 + \overline{N}_4\left(\frac{dw}{dx}\right)_4$$

It can clearly be seen that the element has C^1 continuity and it coincides with the 2-noded Euler-Bernoulli beam element of Section 1.3.

Example 2.6: Derive a 2-noded beam element by imposing the condition of zero transverse shear strain at the two Gauss points $\xi = \pm\frac{1}{\sqrt{3}}$ in the quadratic Timoshenko beam element. Verify that the stiffness matrix of the new element coincides with that of the 2-noded Euler-Bernoulli beam element.

- Solution

The transverse shear strain distribution for the quadratic Timoshenko beam element can be seen in Example 2.4 and in Eq.(2.62). The condition of zero transverse shear strain at the two Gauss points $\xi = \pm\frac{1}{\sqrt{3}}$ is written as

$$(\gamma_{xz})_{\xi=\frac{1}{\sqrt{3}}} = 0 \qquad \text{and} \qquad (\gamma_{xz})_{\xi=-\frac{1}{\sqrt{3}}} = 0$$

Substituting the expression for γ_{xz} from Example 2.3, the following two equations are obtained

$$-\frac{(1 + 2a)}{l^{(e)}}w_1 + \frac{4a}{l^{(e)}}w_2 + \frac{(1 - 2a)}{l^{(e)}}w_3 - \frac{a(a + 1)}{2}\theta_1 -$$

$$-(1 - a^2)\theta_2 - \frac{a(a - 1)}{2}\theta_3 = 0$$

$$-\frac{(1 - 2a)}{l^{(e)}}w_1 - \frac{4a}{l^{(e)}} + w_2 + \frac{(1 + 2a)}{l^{(e)}}w_3 + \frac{a(1 - a)}{2}\theta_1 -$$

$$-(1 - a^2)\theta_2 - \frac{a(a + 1)}{2}\theta_1 - \frac{a(a + 1)}{2}\theta_3 = 0$$

with $a = \frac{1}{\sqrt{3}}$. Eliminating w_2 and θ_2 from the above equations gives

$$\theta_2 = -\frac{1}{4}(\theta_1 + \theta_3) + \frac{3}{2l^{(e)}}(w_3 - w_1)$$

$$w_2 = \frac{1}{2}(w_1 + w_3) + \frac{l^{(e)}}{8}(\theta_1 - \theta_3)$$

Substituting these values into the original displacement field gives

$$\theta = \frac{1}{4}(3\xi^2 - 2\xi - 1)\theta_1 + \frac{1}{4}(3\xi^2 + 2\xi - 1)\theta_3 - $$
$$- \frac{3}{2l^{(e)}}(1 - \xi^2)w_1 + \frac{3}{2l^{(e)}}(1 - \xi^2)w_3$$

$$w = \frac{1}{2}(1 - \xi)w_1 + \frac{1}{2}(1 + \xi)w_3 + \frac{l^{(e)}}{8}(1 - \xi^2)\theta_3$$

The bending strain and the transverse shear strain fields are obtained from the new displacement field as

$$\kappa = \frac{d\theta}{dx} = \frac{6\xi}{l^{(e)2}}w_1 + \frac{(3\xi - 1)}{l^{(e)}}\theta_1 - \frac{6\xi}{(l^{(e)})^2}w_3 + \frac{(3\xi + 1)}{l^{(e)}}\theta_3$$

$$\gamma_{xz} = \frac{\partial w}{\partial x} - \theta = \frac{(1 - 3\xi^2)}{2l^{(e)}}w_1 + \frac{(3\xi^2 - 1)}{4}\theta_1 + \frac{(3\xi^2 - 1)}{2l^{(e)}}w_3 + \frac{(3\xi^2 - 1)}{4}\theta_3$$

The bending strain field is identical to the curvature field for the 2-noded Euler-Bernoulli beam element (Eq.(1.16a)). It can also be verified that $\gamma_{xz} = 0$ at $\xi = \pm\frac{1}{\sqrt{3}}$. The resultant generalized strain matrix is

$$\mathbf{B} = \left\{ \begin{matrix} \mathbf{B}_b \\ \mathbf{B}_s \end{matrix} \right\} = \left[\begin{matrix} \dfrac{6\xi}{(l^{(e)})^2} & \dfrac{(3\xi - 1)}{l^{(e)}} & -\dfrac{6\xi}{(l^{(e)})^2} & \dfrac{(3\xi + 1)}{l^{(e)}} \\ \dfrac{(1 - 3\xi^2)}{2l^{(e)}} & \dfrac{(3\xi^2 - 1)}{4} & \dfrac{(3\xi^2 - 1)}{2l^{(e)}} & \dfrac{(3\xi^2 - 1)}{4} \end{matrix} \right]$$

and the element stiffness matrix is

$$\mathbf{K}^{(e)} = \int_{-1}^{+1} \left[\mathbf{B}_b^T(\hat{D}_b)\mathbf{B}_b + \mathbf{B}_s^T(\hat{D}_s)\mathbf{B}_s \right] \frac{l^{(e)}}{2} d\xi = \mathbf{K}_b^{(e)} + \mathbf{K}_s^{(e)}$$

The expression for $\mathbf{K}_b^{(e)}$ can be integrated exactly using a two-point Gauss quadrature as

$$\mathbf{K}_b^{(e)} = \left(\frac{\hat{D}_b}{l^3} \right)^{(e)} \left[\begin{matrix} 12 & 6l & -12 & 6l \\ & 4l^2 & -6l & 2l^2 \\ & & 12 & -6l \\ \text{Symm.} & & & 4l^2 \end{matrix} \right]^{(e)}$$

which coincides with the stiffness matrix for the 2-noded Euler-Bernoulli beam element (Eq.(1.20)).

The exact integration of $\mathbf{K}_s^{(e)}$ requires a three-point Gauss quadrature (Appendix C). For the two-point reduced quadrature, $\mathbf{B}_s$, and therefore $\mathbf{K}_s^{(e)}$, vanish and the stiffness matrix is the same as for the 2-noded Euler-Bernoulli beam element.

This coincidence is logical as the quadratic rotation field obtained yields a linear bending field, as for the 2-noded Euler-Bernoulli element. Also $\gamma_{xz} = 0$ at the two Gauss points that integrate the bending stiffness exactly. Therefore, the shear stiffness contribution is zero and the stiffness matrix coincides with that for the 2-noded Euler-Bernoulli beam element.

Example 2.7: Figure 2.13 shows a quadratic/linear Timoshenko beam element. Derive the stiffness matrix for the 2-noded Timoshenko beam element by eliminating the central deflection imposing a constant transverse shear strain field over the element.

- Solution

The transverse shear strain field in the original quadratic/linear element is

$$
\gamma_{xz} = \frac{\partial w}{\partial x} - \theta = (2\xi - 1)\frac{w_1}{l^{(e)}} - 4\xi\frac{w_2}{l^{(e)}} + (2\xi + 1)\frac{w_3}{l^{(e)}} - \frac{1}{2}(1 - \xi)\theta_1 - \frac{1}{2}(1 + \xi)\theta_3
$$

$$
= \frac{w_3 - w_1}{l^{(e)}} - \frac{\theta_1 + \theta_3}{2} + \xi\left(2\frac{w_1}{l^{(e)}} - \frac{4w_2}{l^{(e)}} + \frac{2w_3}{l^{(e)}} - \frac{\theta_3 - \theta_1}{2}\right)
$$

For γ_{xz} to be constant, it is necessary to cancel out the term in brackets, i.e.

$$
2\frac{w_1}{l^{(e)}} - \frac{4w_2}{l^{(e)}} + \frac{2w_3}{l^{(e)}} - \frac{\theta_3 - \theta_1}{2} = 0
$$

which gives $w_2 = \frac{w_1 + w_3}{2} - \frac{l^{(e)}}{8}(\theta_3 - \theta_1)$.

Substituting this value into the original deflection field yields

$$
w = \frac{w_2 + w_1}{2} + \left(\frac{w_3 - w_1}{2}\xi + \frac{l^{(e)}}{8}(1 - \xi^2)(\theta_1 - \theta_3)\right)
$$

Hence, the resulting displacement field is expressed in terms of four nodal variables: w_1, θ_1, w_3 and θ_3.

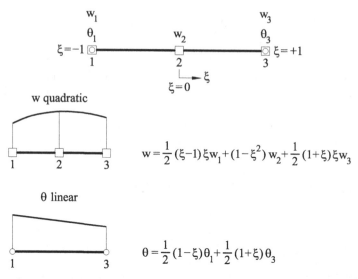

Fig. 2.13 Quadratic/linear Timoshenko beam element of length $l^{(e)}$

The bending strain and the transverse shear strain in the new 2-noded element are

$$\kappa = \frac{\partial \theta}{\partial x} = -\frac{\theta_1}{l^{(e)}} + \frac{\theta_3}{l^{(e)}}$$

$$\gamma_{xz} = \frac{\partial w}{\partial x} - \theta = \frac{w_3 - w_1}{l^{(e)}} - \frac{\theta_1 + \theta_3}{2}$$

The generalized strain matrix is

$$\mathbf{B} = \begin{bmatrix} 0 & -\dfrac{1}{l^{(e)}} & 0 & \dfrac{1}{l^{(e)}} \\[2mm] -\dfrac{1}{l^{(e)}} & -\dfrac{1}{2} & \dfrac{1}{l^{(e)}} & -\dfrac{1}{2} \end{bmatrix} = \left\{ \begin{matrix} \mathbf{B}_b \\ \mathbf{B}_s \end{matrix} \right\}$$

The exact expression of the stiffness matrix for the new element is

$$\mathbf{K}^{(e)} = \begin{bmatrix} \left(\dfrac{\hat{D}_s}{l}\right)^{(e)} & \dfrac{\hat{D}_s^{(e)}}{2} & -\left(\dfrac{\hat{D}_s}{l}\right)^{(e)} & \dfrac{\hat{D}_s^{(e)}}{2} \\[4mm] & \left(\dfrac{\hat{D}_s l}{4} + \dfrac{\hat{D}_b}{l}\right)^{(e)} & \dfrac{\hat{D}_s^{(e)}}{2} & \left(\dfrac{\hat{D}_s l}{4} - \dfrac{\hat{D}_b}{l}\right)^{(e)} \\[4mm] & & \left(\dfrac{\hat{D}_s}{l}\right)^{(e)} & -\dfrac{\hat{D}_s^{(e)}}{2} \\[4mm] \text{Symm.} & & & \left(\dfrac{\hat{D}_s l}{4} + \dfrac{\hat{D}_b}{l}\right)^{(e)} \end{bmatrix}$$

The above matrix coincides with that obtained in Eq.(2.39) for the 2-noded Timoshenko beam element using a one point reduced quadrature for $\mathbf{K}_s^{(e)}$. The reasons for the coincidence are: a) the constant curvature field is the same in both cases; and b) the one point quadrature implies the evaluation of the terms in $\mathbf{B}^{(e)}$ in Eq.(2.15) at $\xi = 0$. This is equivalent to using the constant transverse shear strain field chosen in this example.

Example 2.8: Derive the stiffness matrix for the 2-noded Timoshenko beam element by imposing a constant transverse shear strain field equal to the value of the original linear field at the element mid-point.

- Solution

The linear interpolation for the deflection and the rotation is written as

$$w = \frac{1}{2}(1-\xi)w_1 + \frac{1}{2}(1+\xi) \qquad ; \qquad \theta = \frac{1}{2}(1-\xi)\theta_1 + \frac{1}{2}(1+\xi)\theta_2$$

The condition of constant transverse shear strain is

$$\gamma_{xz} = (\gamma_{xz})_{\xi=0}$$

The original linear transverse shear strain field is

$$\gamma_{xz} = \frac{\partial w}{\partial x} - \theta = \frac{w_2 - w_1}{l^{(e)}} - \frac{\theta_1 + \theta_2}{2} - \frac{\theta_2 - \theta_1}{2}\xi$$

Therefore

$$(\gamma_{xz})_{\xi=0} = \frac{w_2 - w_1}{l^{(e)}} - \frac{\theta_1 + \theta_2}{2}$$

The substitute transverse shear strain matrix is

$$\gamma_{xz} = (\gamma_{xz})_{\xi=0} = \left[-\frac{1}{l^{(e)}}, -\frac{1}{2}, \frac{1}{l^{(e)}}, -\frac{1}{2} \right] \begin{Bmatrix} w_1 \\ \theta_1 \\ w_2 \\ \theta_1 \end{Bmatrix} = \bar{\mathbf{B}}_s \mathbf{a}^{(e)}$$

Above expression coincides with that deduced in the previous example. Clearly, the resulting stiffness matrix coincides with that obtained in Section 2.4 using a single integration point for $\mathbf{K}_s^{(e)}$ and $\mathbf{K}_b^{(e)}$.

Example 2.9: Derive the stiffness matrix for the 3-noded quadratic Timoshenko beam element with an assumed transverse shear strain field varying linearly between the two Gauss points at $\xi = \pm \frac{1}{\sqrt{3}}$.

- Solution

The linear interpolation for γ_{xz} can be written as

$$\gamma_{xz} = \frac{1}{2a}(a - \xi)(\gamma_{xz})_{\xi=-a} + \frac{1}{2a}(a + \xi)(\gamma_{xz})_{\xi=a}$$

with $a = \frac{1}{\sqrt{3}}$. From the original expression of γ_{xz} for the quadratic Timoshenko beam element (Section 2.7 and Example 2.4) we have

$$(\gamma_{xz})_{\xi=-a} = \left[-\frac{(1 + 2a)}{l^{(e)}}, \frac{a(1 + a)}{2}, \frac{4a}{l^{(e)}}, (1 - a^2), \frac{(1 - 2a)}{l^{(e)}}, \frac{a(1 - a)}{2} \right] \mathbf{a}^{(e)}$$

$$(\gamma_{xz})_{\xi=+a} = \left[\frac{(2a - 1)}{l^{(e)}}, \frac{a(a - 1)}{2}, \frac{4a}{l^{(e)}}, -\frac{a(1 + a)}{2} \right] \mathbf{a}^{(e)}$$

with $\mathbf{a}^{(e)} = [w_1, \theta_1, w_2, \theta_2, w_3, \theta_3]^T$.
Substituting these expressions into the above linear transverse shear strain field gives after some algebra

$$\gamma_{xz} = \left[\frac{(2\xi - 1)}{l^{(e)}}, \frac{(a^2 - \xi)}{2}, -\frac{4a}{l^{(e)}}\xi, (1 - a^2), \frac{(1 + 2\xi))}{l^{(e)}}, \frac{(a^2 - \xi)}{2} \right] \mathbf{a}^{(e)} = \bar{\mathbf{B}}_s \mathbf{a}^{(e)}$$

Matrix $\mathbf{B}_b$ coincides with that given in Eq.(2.59) for the quadratic beam element. The element stiffness matrix is

$$\mathbf{K}^{(e)} = \int_{-1}^{+1} \left(\mathbf{B}_b^T(\hat{D}_b)\mathbf{B}_b + \bar{\mathbf{B}}_s^T(\hat{D}_s)\bar{\mathbf{B}}_s \right) \frac{l^{(e)}}{2} d\xi = \mathbf{K}_b^{(e)} + \mathbf{K}_s^{(e)}$$

The exact integration of $\mathbf{K}_b^{(e)}$ requires a two-point quadrature and this yields the expression of Figure 2.9. Similarly, since $\bar{\mathbf{B}}_s$ is linear in ξ the exact integration of $\mathbf{K}_s^{(e)}$ also requires a two-point quadrature giving

$$\mathbf{K}_s^{(e)} = \left(\frac{\hat{D}_s}{9l} \right)^{(e)} \begin{bmatrix} 21 & -\frac{9}{2}l^{(e)} & -24 & -6l^{(e)} & 3 & \frac{3}{2}l^{(e)} \\ & (l^{(e)})^2 & 6l^{(e)} & (l^{(e)})^2 & -\frac{3}{2}l^{(e)} & -\frac{(l^{(e)})^2}{2} \\ & & 48 & 0 & -24 & -6l^{(e)} \\ & & & 4(l^{(e)})^2 & 6l^{(e)} & (l^{(e)})^2 \\ \text{Symm.} & & & & 21 & \frac{9}{2}l^{(e)} \\ & & & & & (l^{(e)})^2 \end{bmatrix}$$

Above expression coincides with that shown in Figure 2.9 obtained with the original quadratic $\mathbf{B}_s$ matrix (Eq.(2.62)) and a two-point reduced quadrature. The assumed linear shear strain field therefore coincides precisely with that introduced by the two-point reduced quadrature. This shows again the analogy between the assumed shear strain technique and reduced integration.

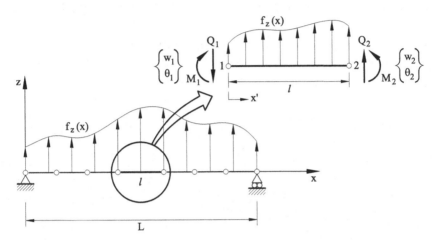

Fig. 2.14 Two-noded beam element under distributed loading

2.9 EXACT TWO-NODED TIMOSHENKO BEAM ELEMENT

We derive in this section a 2-noded Timoshenko beam element that yields nodally exact results. The method is based on the integration of the beam differential equations over a beam element.

Let us consider a 2-noded beam element of length l in a beam of constant cross-section, length L and uniform material properties along the beam, subjected to a distributed loading $f_z(x)$ with x being the neutral axis with the origin at the left-hand node of the element (Figure 2.14). The equilibrium equations are (noting that $\frac{d(\cdot)}{x} \equiv \frac{d(\cdot)}{x'}$ and $dx = dx'$)

$$\frac{dM}{dx'} + Q = 0 \quad , \quad \frac{dQ}{dx'} + f_z(x') = 0 \qquad (2.83)$$

The boundary conditions are

$$\begin{aligned} w = w_1 \; ; & \quad \theta = \theta_1 \quad \text{at } x' = 0 \\ w = w_2 \; ; & \quad \theta = \theta_2 \quad \text{at } x' = l \end{aligned} \qquad (2.84)$$

We recall the constitutive equations of Timoshenko beam theory for the bending moment M and the shear force Q (Eqs.(2.4b) and (2.6b))

$$M = \hat{D}_b \frac{d\theta}{dx'} \quad , \quad Q = \hat{D}_s \left(\frac{dw}{dx'} - \theta \right) \tag{2.85}$$

where $\hat{D}_b$ and $\hat{D}_s$ are given by Eqs.(2.5a) and (2.6b), respectively.

Substituting the constitutive equations (2.85) into Eqs.(2.83) gives

$$
\begin{aligned}
\hat{D}_b \frac{d^2\theta}{dx'^2} + \hat{D}_s \left(\frac{dw}{dx'} - \theta \right) &= 0 \\
\hat{D}_s \left(\frac{d^2w}{dx'^2} - \frac{d\theta}{dx'} \right) + f_z(x') &= 0
\end{aligned}
\tag{2.86}
$$

Eqs.(2.86) show that w and θ vary as a cubic and a quadratic polynomial, respectively for the exact solution of Timoshenko beam theory.

We will express next the resultant stresses M_1, Q_1 and M_2, Q_2 at the two ends of the beam element in terms of the end displacements w_1, θ_1 and w_2, θ_2. These equations will automatically yield the force-displacement relationship for the 2-noded beam element sought.

Let us obtain first the shear force-displacement relationship. Integration of Eq.(2.83) for Q gives

$$Q(x) = Q_1 - F(x') \quad ; \quad F(x) = \int_0^{x'} f_z(x')dx' \tag{2.87}$$

Taking the mean value of Q along the element in Eq.(2.87) yields

$$Q_1 = \frac{1}{l} \int_0^l Q(x')dx' + \frac{1}{l} \int_0^l F(x')dx' \tag{2.88}$$

Substituting the constitutive equation for the shear force (Eq.(2.85)) into (2.88) gives

$$Q_1 = -\frac{\hat{D}_s}{l} \int_0^l \theta dx' + \frac{\hat{D}_s}{l}(w_2 - w_1) + \frac{1}{l} \int_0^l F(x')dx' \tag{2.89}$$

Note that dw/dx' has been approximated by $(w_2 - w_1/l)$ in the derivation of Eq.(2.89).

From Eqs.(2.83), (2.85) and (2.87) we obtain

$$\frac{d^2\theta}{dx'^2} = -\frac{Q}{\hat{D}_b} = -\frac{1}{\hat{D}_b}(Q_1 - F(x')) \tag{2.90}$$

We will choose the following quadratic expansion for θ

$$\theta = N_1\theta_1 + N_2\theta_2 + N_m\theta_m \tag{2.91}$$

where $N_1 = 1 - \frac{x'}{l}$, $N_2 = \frac{x'}{l}$, $N_m = \left(1 - \frac{x'}{l}\right)\frac{x'}{l}$, where θ_m is the rotation at the element midpoint.

Eq.(2.90) can be written in integral form after weighting with function N_m as

$$\int_0^l N_m \frac{d^2\theta}{dx'^2} dx' = -\frac{Q_1}{\hat{D}_b}\int_0^l N_m dx' + \frac{1}{\hat{D}_b}\int_0^l N_m F(x') dx' \tag{2.92}$$

Integrating twice by parts the integral in the l.h.s. of Eq.(2.92) gives

$$\int_0^l \theta dx' = \frac{l}{2}(\theta_1 + \theta_2) + \frac{Q_1 l^3}{12\hat{D}_b} - \frac{l^2}{2\hat{D}_b}\int_0^l N_m F(x') dx' \tag{2.93}$$

Substituting above expression into (2.89) yields

$$Q_1 = \frac{\hat{D}_b}{(1+\beta)}\left[\frac{12}{l^3}(w_2 - w_1) - \frac{6}{l^2}(\theta_1 + \theta_2)\right] + f_1 \tag{2.94a}$$

with

$$\beta = \frac{12\hat{D}_b}{l^2 \hat{D}_s} \quad \text{and} \quad f_1 = \frac{1}{l(1+\beta)}\int_0^l (6N_m + \beta)F(x') dx' \tag{2.94b}$$

The above expression for the parameter β holds for composite beams and is a generalization of the form of Eq.(2.32) for homogeneous material. From the equilibrium of the element (Figure 2.14) we have

$$Q_2 = Q_1 - \int_0^l f_z(x') dx' \tag{2.95}$$

Substituting Eq.(2.94a) into (2.95) gives

$$Q_2 = \frac{\hat{D}_b}{(1+\beta)}\left[\frac{12}{l^3}(w_2 - w_1) - \frac{6}{l^2}(\theta_1 + \theta_2)\right] + f_2 \tag{2.96a}$$

with

$$f_2 = f_1 - \int_0^l f_z(x') dx' \tag{2.96b}$$

Let us derive now the moment-displacement relationship. Using N_1 as the weighting function we write the weighted integral form of the moment-shear force relationship (Eq.(2.83)) as

$$\int_0^l N_1 \frac{dM}{dx'} dx' + \int_0^l N_1 Q(x') dx' = 0 \tag{2.97}$$

Integrating by parts the first integral gives

$$M_1 = \frac{1}{l} \int_0^l M dx' + \int_0^l N_1 Q(x') dx' = 0 \tag{2.98}$$

Substituting M and Q from Eqs.(2.85) and (2.87) leads to

$$M_1 = \frac{\hat{D}_b}{l}(\theta_1 - \theta_2) + \frac{l}{2} Q_1 - \int_0^l N_1 F(x') dx' \tag{2.99}$$

Using N_2 instead of N_1 in Eq.(2.97) gives

$$M_2 = \frac{\hat{D}_b}{l}(\theta_2 - \theta_1) - \frac{l}{2} Q_1 + \int_0^l N_2 F(x') dx' \tag{2.100}$$

The final moment-displacement relationship is obtained substituting Q_1 from Eq.(2.94a) into Eqs.(2.99) and (2.100). Eqs.(2.94a), (2.96a), (2.98) and (2.100) can be grouped as

$$\mathbf{K} \mathbf{a}^{(e)} - \mathbf{f}^{(e)} = \mathbf{q}^{(e)} \tag{2.101a}$$

where

$$\mathbf{q}^{(e)} = [F_{z_1}, \mathcal{M}_1, F_{z_2}, \mathcal{M}_2]^T = [-Q_1, -M_1, Q_2, M_2]^T \tag{2.101b}$$

$$\mathbf{K}^{(e)} = \frac{\hat{D}_b}{(1+\beta)l^3} \begin{bmatrix} 12 & 6l & -12 & 6l \\ & (4+\beta)l^2 & -6l & (2-\beta)l^2 \\ & & 12 & -6l \\ \text{Symm.} & & & (4+\beta)l^2 \end{bmatrix} \tag{2.101c}$$

$$\mathbf{f}^{(e)} = \left[f_1, \frac{l}{2} f_1 - \int_0^l N_1 F dx', -f_1 + \int_0^l f_z(x') dx', \frac{l}{2} f_1 - \int_0^l N_2 F dx' \right]^T \tag{2.101d}$$

where f_1 and f_2 are defined in Eqs.(2.94b) and (2.96b), respectively.

Matrix $\mathbf{K}^{(e)}$ and vector $\mathbf{f}^{(e)}$ in Eqs.(2.101) are respectively the stiffness matrix and the equivalent nodal force vector for a 2-noded beam element of length l *which is nodally exact*, i.e. the nodal results coincide with those obtained by integrating the equilibrium equations (2.83).

Note the sign criteria for the equilibrating nodal forces $\mathbf{q}^{(e)}$ and the end shear forces and bending moments (Figures 1.5 and 2.14).

The expression of $\mathbf{f}^{(e)}$ for a uniformly distributed loading is independent of β and it coincides with the equivalent nodal force vector for the 2-noded Euler-Bernoulli beam element (Eq.(1.21b)). Table 2.2 shows the expression of $\mathbf{f}^{(e)}$ for different loading types [BD5].

f_z	f_{z_1}	m_1	f_{z_2}	m_2
	$\dfrac{ql}{2}$	$\dfrac{ql^2}{12}$	$\dfrac{ql}{2}$	$-\dfrac{ql^2}{12}$
	$\dfrac{ql\,(9+10\beta)}{60\,(1+\beta)}$	$f_{z_1}\dfrac{l}{2}-\dfrac{ql^2}{24}$	$\dfrac{ql}{2}-f_{z_1}$	$f_{z_1}\dfrac{l}{2}-\dfrac{ql^2}{8}$
$a=\dfrac{x_0}{l}$	$\dfrac{Pb}{(1+\beta)}$ $b=1-3a^2+2a^3+\beta(1-a)$	$f_{z_1}\dfrac{l}{2}-\dfrac{Pl}{2}(1-a^2)$	$P-f_{z_1}$	$\dfrac{f_{z_1}l}{2}-\dfrac{Pl}{2}(1-a^2)$

$$\mathbf{f}^{(e)}=[f_{z_1},\,m_1,\,f_{z_2},\,m_2]^T$$

Table 2.2 Equivalent nodal forces for an "exact" 2-noded Timoshenko beam element

For $\beta=0$ the stiffness matrix coincides with the expression for the 2-noded Euler Bernoulli beam element (Eq.(1.20)).

The exact 2-noded Timoshenko beam element can also be derived using a mixed formulation with a linear interpolation for w, a quadratic one for θ and a constant field for Q. Eliminating the shear force at element level leads to the expressions for $\mathbf{K}^{(e)}$ and $\mathbf{f}^{(e)}$ of Eqs.(2.101) [BD5].

The reader can verify the coincidence of Eq.(2.101c) $\mathbf{K}^{(e)}$ with the form of Eq.(2.55) for homogeneous material derived via a substitute shear modulus. Clearly, nodally exact results are only obtained if the expression for $\mathbf{f}^{(e)}$ of Eq.(2.101d) is used.

We present next an application of this element to the analysis of a cantilever beam. Exact nodal results are obtained using just one element.

Example 2.10: Obtain the exact nodal values for the end deflection and the end rotation for the cantilever beam of Figure 2.5 using the exact 2-noded Timoshenko beam element of Section 2.9.

- Solution

For a single beam element of rectangular cross-section under an end point load, the "exact" stiffness equation is (Eqs.(2.101))

$$\frac{EI_y}{(1+\beta)l^3} \begin{bmatrix} 12 & 6l & -12 & 6l \\ & (4+\beta)l^2 & -6l & (2-\beta)l^2 \\ & & 12 & -6l \\ & & & (4+\beta)l^2 \end{bmatrix} \begin{Bmatrix} w_1 \\ \theta_1 \\ w_2 \\ \theta_2 \end{Bmatrix} = \begin{Bmatrix} R \\ M_1 \\ P \\ 0 \end{Bmatrix} \tag{2.102}$$

Eliminating the DOFs at the clamped node gives

$$\frac{EI_y}{(1+\beta)l^3} \begin{bmatrix} 12 & -6l \\ -6l & (4+\beta)l^2 \end{bmatrix} \begin{Bmatrix} w_2 \\ \theta_2 \end{Bmatrix} = \begin{Bmatrix} P \\ 0 \end{Bmatrix} \tag{2.103}$$

Inverting above system yields

$$\begin{Bmatrix} w_2 \\ \theta_2 \end{Bmatrix} = \mathbf{F} \begin{Bmatrix} P \\ 0 \end{Bmatrix} \tag{2.104}$$

where the "exact" flexibility matrix is

$$\mathbf{F} = \begin{bmatrix} \left(\frac{1}{GA^*} + \frac{l^3}{3EI_y}\right) & \frac{l^2}{2EI_y} \\ \frac{l^2}{2EI_y} & \frac{l}{EI_y} \end{bmatrix} \quad \text{with} \quad A^* = \alpha A, \alpha = \frac{5}{6} \tag{2.105}$$

From Eq.(2.104)

$$w_2 = P\left(\frac{1}{GA^*} + \frac{l^3}{3EI_y}\right) \quad , \quad \theta_2 = \frac{l^2}{2EI_y}P \tag{2.106}$$

For a very slender beam $G/E \to \infty$ and the "exact" nodal solution is

$$w_2 = \frac{Pl^3}{3EI_y} \quad , \quad \theta_2 = \frac{l^2}{2EI_y} \tag{2.107}$$

Above results coincide with those given in Section 2.4 (Eq.(2.36)).

2.10 ROTATION-FREE BEAM ELEMENT ACCOUNTING FOR TRANSVERSE SHEAR DEFORMATION EFFECTS

The CCB rotation-free beam element (Section 1.4.1) can be extended to account for transverse shear deformation effects. The nodal deflection of the original rotation-free beam element is enhanced with the shear deformation angle. This allows us to compute the bending and shear deformation contributions to the PVW. The method is summarized next for the so-called CCB+1 element. Further details can be found in [ZO3].

The rotation is expressed as the sum of the slope and the shear deformation angle ϕ (hereafter termed the shear angle) in the standard manner for Timoshenko theory, i.e.

$$\theta = \frac{dw}{dx} + \phi \tag{2.108}$$

The bending strain (curvature) is expressed in terms of w and ϕ as

$$\kappa = \frac{d\theta}{dx} = \frac{d^2 w}{dx^2} + \frac{d\phi}{dx} = \kappa_w + \kappa_\phi \tag{2.109}$$

where $\kappa_w = \frac{d^2 w}{dx^2}$ and $\kappa_\phi = \frac{d\phi}{dx}$. κ_w is termed the "geometrical" curvature.

The transverse shear strain is given by (minus) the shear angle, i.e.

$$\gamma_{xz} = \frac{dw}{dx} - \theta = -\phi \tag{2.110}$$

The PVW for a single element (Eq.2.18a) is expressed in terms of w and ϕ (disregarding distributed and concentrated bending moments) as

$$\int_{l^{(e)}} [(\delta \kappa_w + \delta \kappa_\phi)M - \delta \phi Q] \, dx - \int_0^l \delta w f_z dx = \sum_{i=1}^2 \delta w_i F_{z_i} \tag{2.111}$$

where $\delta \kappa_w = \delta \left(\frac{d^2 w}{dx^2} \right)$ and $\delta \kappa_\phi = \delta \left(\frac{d\phi}{dx} \right)$.

The PVW is split into the following two independent equations

$$\int_{l^{(e)}} \delta \kappa_w M \, dx - \int_{l^{(e)}} \delta w f_z dx = \sum_{i=1}^2 \delta w_i F_{z_i}$$
$$\int_{l^{(e)}} [\delta \kappa_\phi M - \delta \phi Q] \, dx = 0 \tag{2.112}$$

The constitutive relationship between the bending moment and the curvature and between the shear force and the shear strain are written as

$$M = \hat{D}_b(\kappa_w + \kappa_\phi) \quad ; \quad Q = -\hat{D}_s \phi \tag{2.113}$$

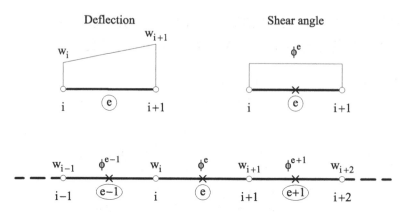

Fig. 2.15 CCB+1 rotation-free beam element accounting for transverse shear deformation effects. The deflection is linearly interpolated in terms of the nodal values and the shear angle is constant within the element

We express the deflection and shear angle within the element in terms of a linear interpolation for w and a constant field for ϕ as (Figure 2.15)

$$w = \sum_{i=1}^{2} N_i w_i \quad ; \quad \phi = \phi^e \tag{2.114}$$

where N_i are the standard linear shape functions (Figure 2.4).

An alternative procedure is to interpolate linearly both the deflection and the shear angle. This choice was used by Oñate and Zárate [OZ2] who derived an accurate rotation-free beam element (termed CCB+) accounting for shear deformation effects. The constant approximation for the shear angle of Eq. (2.114), however, simplifies the treatment of the clamped boundary conditions and also yields excellent results. This option is used for deriving the element matrices as explained below.

A *constant geometrical curvature field* κ_w^e is defined over an element with nodes $i, i+1$, as is typical in rotation-free beam elements, giving

$$\kappa_w^e = \frac{1}{l^e} \int_i^{i+1} \frac{d^2 w}{dx^2} = \frac{1}{l^e} \left[\frac{dw}{dx} \right]_i^{i+1} = \frac{1}{l^e} \left[\left(\frac{dw}{dx} \right)_{i+1} - \left(\frac{dw}{dx} \right)_i \right] = \mathbf{B}_b \bar{\mathbf{w}}^{(e)} \tag{2.115}$$

and, hence

$$\delta \kappa_w^e = \mathbf{B}_b \delta \bar{\mathbf{w}}^{(e)} \tag{2.116}$$

Matrix $\mathbf{B}_b$ coincides with Eq.(1.31d) and $\bar{\mathbf{w}}^{(e)} = [w_{i-1}, w_i, w_{i+1}, w_{i+2}]^T$ collects the deflection at the four nodes that belong to element e and the two adjacent elements (see Eq.(1.31e) and Figure 2.15).

An average curvature κ_ϕ is defined over the element as

$$\kappa_\phi^e = \frac{1}{l^e}\int_{l^e}\frac{d\phi}{dx}dx = \frac{1}{l^e}[\phi_{i+1} - \phi_i] \qquad (2.117)$$

The shear angles at the element nodes are computed as the average of the constant values for the elements adjacent to each node, i.e.

$$\phi_i = \frac{\phi^e + \phi^{e-1}}{2} \ , \ \phi_{i+1} = \frac{\phi^e + \phi^{e+1}}{2} \qquad (2.118)$$

Substituting Eqs.(2.118) into (2.117) gives

$$\kappa_\phi^e = \frac{1}{2l^e}[-1,0,1]\left\{\begin{array}{c}\phi^{e-1}\\\phi^e\\\phi^{e+1}\end{array}\right\} = \mathbf{B}_\phi\bar{\boldsymbol{\phi}}^{(e)} \qquad (2.119a)$$

with

$$\mathbf{B}_\phi = \frac{1}{2l^e}[-1,0,1] \ , \ \bar{\boldsymbol{\phi}}^{(e)} = [\phi^{e-1},\phi^e,\phi^{e+1}]^T \qquad (2.119b)$$

This technique introduces the fictitious shear angle variables ϕ^0 and ϕ^{N+1} (N being the number of elements in the mesh) which are prescribed to a zero value.

The virtual fields δw and $\delta\phi$ are expressed as

$$\delta w = [0, N_1, N_2, 0]\left\{\begin{array}{c}\delta w_{i-1}\\\delta w_i\\\delta w_{i+1}\\\delta w_{i+2}\end{array}\right\} = \mathbf{N}_w\delta\bar{\mathbf{w}}^e \qquad (2.120a)$$

$$\delta\phi = [0, 1, 0]\left\{\begin{array}{c}\delta\phi^{e-1}\\\delta\phi^e\\\delta\phi^{e+1}\end{array}\right\} = \mathbf{N}_\phi\delta\bar{\boldsymbol{\phi}}^{(e)} \qquad (2.120b)$$

Eqs.(2.113), (2.115) and (2.119a) define constant bending moment and shear force fields over the element in terms of $\bar{\mathbf{w}}^{(e)}$ and $\bar{\boldsymbol{\phi}}^{(e)}$ as

$$M = \hat{D}_b[\kappa_w^e + \kappa_\phi^e] = \hat{D}_b[\mathbf{B}_b\bar{\mathbf{w}}^{(e)} + \mathbf{B}_\phi\bar{\boldsymbol{\phi}}^{(e)}] \qquad (2.121)$$

$$Q = -\hat{\mathbf{D}}_s\bar{\boldsymbol{\phi}}^{(e)} = -\hat{\mathbf{D}}_s\mathbf{N}_\phi\bar{\boldsymbol{\phi}}^{(e)}$$

Substituting Eqs.(2.115), (2.117), (2.120) and (2.121) into (2.112) yields (assuming homogeneous material properties)

$$[\delta\bar{\mathbf{w}}^{(e)}]^T[\mathbf{B}_b^T\hat{D}_b\mathbf{B}_b\bar{\mathbf{w}}^{(e)}+\mathbf{B}_b^T\hat{D}_b\mathbf{B}_\phi\bar{\boldsymbol{\phi}}^{(e)}]l^{(e)}-[\delta\bar{\mathbf{w}}^{(e)}]^T\int_{l^{(e)}}\mathbf{N}_w^Tf_zdx=[\delta\bar{\mathbf{w}}^{(e)}]^T\mathbf{q}_w^{(e)}$$

$$\left[\delta\bar{\boldsymbol{\phi}}^{(e)}\right]^T\left[\mathbf{B}_\phi^T\hat{D}_b\mathbf{B}_b\bar{\mathbf{w}}^{(e)}+\mathbf{B}_\phi^T\hat{D}_b\mathbf{B}_\phi\bar{\boldsymbol{\phi}}^{(e)}+\mathbf{N}_\phi^T\mathbf{N}_\phi\hat{D}_sdxl^{(e)}\bar{\boldsymbol{\phi}}^{(e)}\right]=0 \quad (2.122)$$

where

$$\mathbf{q}_w^{(e)}=[0,F_{z_1},F_{z_2},0]^T \tag{2.123}$$

are the equilibrating nodal forces for the element.

After simplification of the virtual variables and following the usual global assembly process we obtain the system of equilibrium equations

$$\mathbf{K}_w\mathbf{w}+\mathbf{K}_w\boldsymbol{\phi}=\mathbf{f}_w \quad ; \quad \mathbf{K}_w^T\mathbf{w}+\mathbf{K}_\phi\boldsymbol{\phi}=\mathbf{0} \tag{2.124}$$

where $\mathbf{w}$ lists the deflection of all nodes and $\boldsymbol{\phi}$ contains the shear angles for all the elements (including the auxiliary variables at the first and last element which are prescribed to zero).

The matrices and vectors in Eq.(2.124) are formed by assembling the element contributions, as usual. The element stiffness matrices are

$$\mathbf{K}_w^{(e)}=\mathbf{B}_b^T\mathbf{B}_b\hat{D}_bl^{(e)} \quad , \quad \mathbf{K}_{w\phi}^{(e)}=\mathbf{B}_w^T\mathbf{B}_\phi\hat{D}_bl^{(e)}$$

$$\mathbf{K}_\phi^{(e)}=\mathbf{B}_\phi^T\mathbf{B}_b\hat{D}_bl^{(e)}+\mathbf{N}_\phi^T\mathbf{N}_\phi\hat{D}_sl^{(e)} \tag{2.125}$$

The equivalent nodal force vector for the element for a uniformly distributed load $f_z=q$ is

$$\mathbf{f}_w^{(e)}=\frac{ql^{(e)}}{2}[0,1,1,0]^T \tag{2.126}$$

Boundary conditions

Prescribed deflection: A zero deflection at a node is directly imposed when solving the system of equations.

Simply supported (SS) node: Matrix $\mathbf{B}_b$ is modified as explained in Section 1.4.1 (Eqs.(1.35)–(1.37)) whereas for $\mathbf{B}_\phi$ the following procedure is used.

The curvature κ_ϕ in the element $i,i+1$ adjacent to a simple support at a left-end node i (is expressed as (see Eq.(2.117) and Figure 2.16a)

$$\kappa_\phi=\frac{1}{l^e}(\phi_{i+1}-\phi_i)=\frac{1}{l^e}\left(\frac{\phi^{e+1}+\phi^e}{2}-\phi^e\right)=\frac{1}{2l^e}[0,-1,1]\left\{\begin{array}{c}\phi^{e-1}\\\phi^e\\\phi^{e+1}\end{array}\right\}=\mathbf{B}_\phi\bar{\boldsymbol{\phi}}^e$$

$$(2.127a)$$

a) Free or simply supported node

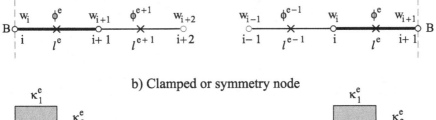

b) Clamped or symmetry node

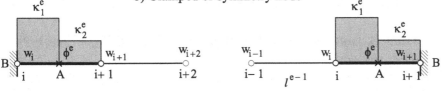

Fig. 2.16 CCB+1 element. (a) Free or SS node B at left and right ends of the beam. (b) Clamped or symmetry node B at left and right ends. Shaded areas show the domains 1 and 2 that split the element containing the prescribed node

with

$$\mathbf{B}_\phi = \frac{1}{2l^e}[0, -1, 1] \tag{2.127b}$$

Note that in Eq.(2.127a) we have assumed that $\phi_i = \phi^e$.

A similar procedure for a right-end node $i+1$ (Figure 2.16a) gives

$$\mathbf{B}_\phi = \frac{1}{2l^e}[-1, 1, 0] \tag{2.127c}$$

Clamped or symmetry node: The element adjacent to a clamped node i, or to a node on a symmetry axis, is split in two domains (Figure 2.16b).

The curvature in the first domain is expressed as

$$\kappa_1^e = \frac{2}{l^e}\int_i^A \frac{d\phi}{dx}dx = \frac{2}{l^e}[\theta_A - \underbrace{\vec{\theta_i}}_{=0}] = \frac{2}{l^e}\theta_A = \frac{2}{l^e}\left[\left(\frac{dw}{dx}\right)_A + \phi_A\right] =$$

$$= \frac{2}{l^e}\left[\frac{w_{i+1} - w_i}{l^e} + \phi^e\right] = \frac{2}{(l^{(e)})^2}[0, -1, 1, 0]\begin{Bmatrix} w_{i-1} \\ w_i \\ w_{i+1} \\ w_{i+2} \end{Bmatrix} +$$

$$+ \frac{2}{l^e}[0, 1, 0]\begin{Bmatrix} \phi^{e-1} \\ \phi^e \\ \bar{\phi}^{e+1} \end{Bmatrix} = \mathbf{B}_{b_1}\bar{\mathbf{w}}^{(e)} + \mathbf{B}_{\phi_1}\bar{\phi}^{(e)} \tag{2.128}$$

For the second domain we obtain

$$
\kappa_2^e = \frac{2}{l^e}[\theta_{i+1} - \theta_A] = \frac{2}{l^e}\left[\frac{1}{2}\left(\frac{w_{i+1} - w_i}{l^e} + \frac{w_{i+2} - w_{i+1}}{l^{e+1}}\right) - \frac{w_{i+1} - w_i}{l^e}\right.
$$

$$
\left. + \frac{\phi^{e+1} + \phi^e}{2} - \phi^e\right] = \frac{1}{(l^e)^2 l^{e+1}}[0, l^{e+1}, -(l^e + l^{e+1}), l^e]\bar{\mathbf{w}}^e +
$$

$$
+ \frac{1}{l^e}[0, -1, 1]]\bar{\boldsymbol{\phi}}^e = \mathbf{B}_{b_2}\bar{\mathbf{w}}^e + \mathbf{B}_{\phi_2}\bar{\boldsymbol{\phi}}^e \tag{2.129}
$$

Note that $\mathbf{B}_{b_1}$ and $\mathbf{B}_{b_2}$ coincide with matrices $\mathbf{B}_1$ and $\mathbf{B}_2$ of Eq.s(1.39a) and (1.40a), respectively.

The two domains in element e are treated as *two different elements* with nodes (i, A) and $(A, i+1)$, respectively. Matrices $\mathbf{K}_w^{(e)}, \mathbf{K}_{w\phi}^{(e)}$ and $\mathbf{K}_\phi^{(e)}$ of Eq.(2.125) are computed for each of the two domains using $\mathbf{B}_{b_i}$ and $\mathbf{B}_{\phi_i}$, $i = 1, 2$ instead of $\mathbf{B}_b$ and $\mathbf{B}_\phi$, respectively. The stiffness matrices for the two domains are assembled in the global stiffness matrix as usual.

The same procedure is followed for a right-end clamped node (Figure 2.18b). Matrices $\mathbf{B}_{b_1}$ and $\mathbf{B}_{\phi_1}$ for "element" $(A, i+i)$ coincide precisely with those of Eq.(2.128). For "element" (i, A) $\mathbf{B}_{b_2}$ coincides with matrix $\mathbf{B}_2$ of Eq.(1.41) and $\mathbf{B}_{\phi_2}$ is

$$
\mathbf{B}_{\phi_2} = \frac{1}{l^e}[-1, 1, 0] \tag{2.130}
$$

We recall that the fictitious shear angles ϕ^0 and ϕ^{N+1} are prescribed to a zero value.

2.10.1 Iterative computation of the nodal deflections and the element shear angles

The following iterative algorithm can be implemented for computing $\mathbf{w}$ and $\boldsymbol{\phi}$ via Eq.(2.124)

$$
\begin{aligned}
\mathbf{K}_w\mathbf{w}^i &= \mathbf{f}_w - \mathbf{K}_{w\phi}\boldsymbol{\phi}^{i-1} \rightarrow \mathbf{w}^i \\
\mathbf{K}_\phi\boldsymbol{\phi}^i &= -\mathbf{K}_{w\phi}^T\mathbf{w}^i \rightarrow \boldsymbol{\phi}^i
\end{aligned} \tag{2.131}
$$

where superindex i denotes the number of iterations. The iterations continue until convergence of the nodal deflections and the shear deformation angles is achieved. Convergence is typically measured by the L_2 norm of vectors $\mathbf{w}$ and $\boldsymbol{\phi}$ [OZ2]. The advantage of the iterative scheme versus the monolithic solution of Eqs.(2.124) is that for $i = 1$ and $\boldsymbol{\phi}^0 = \mathbf{0}$, $\mathbf{w}^1$ *coincides with the solution for Euler-Bernoulli beam theory*. The effect of

Simple supported beam. Uniform load q

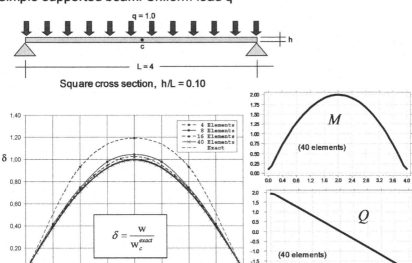

Fig. 2.17 SS thick beam under uniform load. Normalized deflection field for different meshes of CCB+1 elements. Bending moment and shear force diagrams for a 40 element mesh

transverse shear deformation is introduced progressively via the shear angle variables ϕ with the number of iterations. This effect is negligible for slender beams. Convergence of the iterative scheme (2.131) is quite fast (2–3 iterations), as the shear angle values are relatively small compared to the deflections, even for thick beams [ZO3].

For visualization purposes the nodal values of the shear angles can be computed a posteriori by a simple nodal averaging of the constant values of the two elements adjacent to a node.

2.10.2 Performance of the CCB+1 element

The accuracy of the CCB+1 beam element was tested in the analysis of simple supported (SS) and cantilever thick beams respectively loaded under distributed and end point forces. Figures 2.17 and 2.18 show the normalized distribution of the deflection (for the SS and cantilever beams) and the convergence of end deflection (for the cantilever beam) with the number of elements. The bending moment and shear force diagrams for the 40 element mesh are also shown. Only half of the beam was discretized taking advantage of the symmetry.

Cantilever beam. End point load P

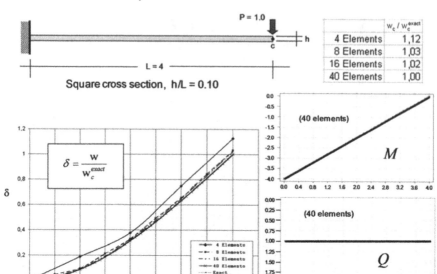

Fig. 2.18 Cantilever thick beam under end point load. Convergence of end deflection. Normalized deflection field for different meshes of CCB+1 elements. Bending moment and shear force diagrams for a 40 element mesh

Table 2.3 shows the central deflection values and the percentage error versus the thick solution for thick and slender SS beams under a uniformly distributed load. It is remarkable that $\simeq 5\%$ error is obtained with the simple four element mesh in all cases.

Other examples of the good behaviour of the CCB+1 element for slender and thick beams can be found in [ZO3].

The bending moment and shear force distributions along the beam are also plotted for the 40 element mesh. Results are practically coincident with the analytical values. Similar good behaviour was obtained for other thick and slender beam problems studied with the CCB+ element [OZ2].

2.11 BEAMS ON ELASTIC FOUNDATION

The treatment of 2D solid structures on an elastic foundation was studied in Section 9.7 of [On4]. The method can be easily extended to beams.

The virtual work includes now the work of the reactions at the nodes

No. of elements*	Central deflection w_c							
	h/L=0.01		h/L=0.05		h/L=0.10		h/L=0.20	
	w_c	% error	w_c	% error	w_c	% error	w_c	% error
2	-3,94E+02	20.02%	-6,33E-01	19.75%	-4,00E-02	19.57%	-2,63E-03	18.32%
4	-3,45E+02	5.02%	-5,54E-01	4.85%	-3,51E-02	4.87%	-2,32E-03	4.47%
8	-3,32E+02	1.27%	-5,34E-01	1.12%	-3,39E-02	1.20%	-2,24E-03	1.00%
16	-3,29E+02	0.33%	-5,29E-01	0.19%	-3,36E-02	0.28%	-2,22E-03	0.14%
32	-3,28E+02	0.10%	-5,28E-01	-0.05%	-3,35E-02	-0.05%	-2,22E-03	-0.08%
Slender solution	-3,28E+02		-5,25E-01		-3,28E-02		-2,05E-03	
Thick solution	-3,28E+02		-5,28E-01		-3,35E-02		-2,22E-03	

* Number of elements in half beam due to symmetry

Table 2.3 Simple supported beam under uniformly distributed load. Central deflection values and percentage error versus the thick beam solution for slender and thick beams analyzed with the CCB+1 rotation-free beam element

on the foundation (Figure 2.19) as

$$IVW = EVW + \int_{l_f} \delta w t_z dx \qquad (2.132)$$

where l_f is the length of the beam in contact with the foundation and IVW and EVW denote the internal and external virtual work terms given by the l.h.s. and r.h.s. of Eq.(2.9), respectively.

Noting that $t_z = -kw(x)$ where k is the elastic modulus of the foundation Eq.(2.132) can be written as

$$IVW + \int_{l_f} \delta w k w dx = EVW \qquad (2.133)$$

Introducing now the standard discretization for 2-noded Timoshenko beam elements leads to the equilibrium equation for the element

$$[\mathbf{K}^{(e)} + \mathbf{H}^{(e)}]\mathbf{a}^{(e)} - \mathbf{f}^{(e)} = \mathbf{q}^{(e)} \qquad (2.134)$$

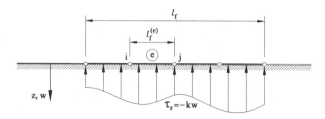

Fig. 2.19 Beam laying on an elastic foundation

where all the matrices and vectors have the usual form (Section 2.33) and

$$
\mathbf{H}^{(e)} = \begin{bmatrix} \mathbf{H}_{11} & \mathbf{H}_{12} \\ \mathbf{H}_{21} & \mathbf{H}_{22} \end{bmatrix}^{(e)} \quad \text{with} \quad \mathbf{H}_{ij}^{(e)} = \int_{l_f^{(e)}} k \begin{bmatrix} (N_i N_j) & 0 \\ 0 & 0 \end{bmatrix} dx \qquad (2.135)
$$

where $l_f^{(e)}$ is the length of the element in contact with the foundation. A simple integration gives

$$
\mathbf{H}_{ij}^{(e)} = \frac{(kl_f)^{(e)}}{3} \begin{bmatrix} \gamma & 0 \\ 0 & 0 \end{bmatrix} \qquad (2.136)
$$

with $\gamma = 1$ for $i + j = 2, 4$ and $\gamma = 1/2$ for $i + j = 3$.

$\mathbf{H}^{(e)}$ is a full matrix and this introduces a coupling between the deflection of nodes 1 and 2. A simplification is to diagonalize $\mathbf{H}$ by lumping the coefficients of each row. This is equivalent to assuming a spring of elastic constant $\frac{(kl_f)^{(e)}}{2}$ attached to each node. For equal length elements ($l_f^{(e)} = l$) matrix $\mathbf{H}$ is simply

$$
\mathbf{H}^{(e)} = kl \begin{bmatrix} \mathbf{H}_1 & \mathbf{0} \\ \mathbf{0} & \mathbf{H}_1 \end{bmatrix} \quad \text{with} \quad \mathbf{H}_1 = \begin{bmatrix} 1 & 0 \\ 0 & 0 \end{bmatrix} \qquad (2.137)
$$

The process is the same for Euler-Bernoulli beam elements. Matrix $\mathbf{H}$ must account now for the coupling between the nodal deflections and the nodal rotation (Eq.(1.10)). The expression for $\mathbf{H}_{ij}^{(e)}$ is

$$
\mathbf{H}_{ij}^{(e)} = \int_{l_f^{(e)}} k \begin{bmatrix} N_i N_j \,, & N_i \bar{N}_j \\ \bar{N}_i N_j \,, & \bar{N}_j \bar{N}_i \end{bmatrix} dx \quad , \quad i, j = 1, 2 \qquad (2.138)
$$

where $N_i, \bar{N}_i, \ i = 1, 2$ are given in Eqs.(1.11a). The lumping of $\mathbf{H}^{(e)}$ is also possible, although it is not recommended in this case.

2.12 CONCLUDING REMARKS

This chapter has focused on plane beam elements based on Timoshenko beam theory. Unlike Euler-Bernoulli beam theory, beam cross-sections in Timoshenko theory do not remain necessarily orthogonal to the beam axis. This introduces the effect of transverse shear deformation and it allows us to analize thick and slender beams using simple C^o continuous elements. Timoshenko beam elements are also more suitable for analysis of composite laminated beams (Chapters 3 and 4). Practically, the only drawback of Timoshenko beam elements is their tendency to lock for slender beams. This deficiency can be overcome in a number of ways: by reduced integration of the shear stiffness terms, by compatible interpolations for the deflection and the rotation, and by imposing "a priori" a correct transverse shear strain field over the element. An "exact" 2-noded beam element has been presented that is useful for analysis of thick and slender beams.

The extended rotation-free beam element with transverse shear deformation can be applied to the analysis of slender and thick beams in an adaptive manner. The shear angle variables are introduced only in the elements where transverse shear deformation effects are important [OZ2,ZO3].

Many of the concepts studied in this chapter will be revisited again when dealing with plates and shells.

3

COMPOSITE LAMINATED PLANE BEAMS

3.1 INTRODUCTION

Strength performance and weight advantages of composite materials versus traditional concrete and steel have led to their sustained and increased applications in aircraft and aerospace vehicles, automotive, naval and civil structures. The design of efficient and reliable composite structures however requires improved computational methods that accurately incorporate the key mechanical effects [AB,Bar2,BC].

Composite beams are typically formed by a piling of layers of composite material. The finite element analysis of the so called *composite laminated beams* has to account for the non-uniform distribution of the material properties along the beam thickness direction. Timoshenko beam theory is particularly suited to these problems as the heterogeneity of the material increases the importance of transverse shear deformation. The formulation is also applicable to mixed steel-concrete beams and standard reinforced concrete beams, where the concrete and steel bars are modelled by layers with different material properties.

In the following sections we present a finite element formulation for the analysis of *composite laminated plane beams* using 2-noded Timoshenko beam elements. This includes the formulation of an "exact" 2-noded Timoshenko composite laminated beam element. The derivation of Euler-Bernoulli composite laminated beam elements is also briefly described. The chapter concludes with an overview of composite laminated beam elements based on higher order approximations for the in-plane displacement across the thickness.

This chapter introduces the main ideas that we will extend in the subsequent chapters when dealing with 3D beams, plates and shells with composite laminated material.

E. Oñate, *Structural Analysis with the Finite Element Method. Linear Statics:*
Volume 2: Beams, Plates and Shells, Lecture Notes on Numerical Methods
in Engineering and Sciences, DOI 10.1007/978-1-4020-8743-1_3,
© International Center for Numerical Methods in Engineering (CIMNE), 2013

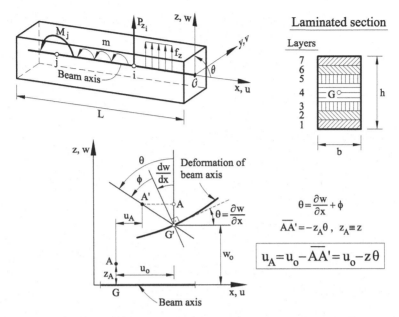

Fig. 3.1 Composite laminated Timoshenko beam. Forces and displacement field

3.2 KINEMATICS OF A PLANE LAMINATED BEAM

Let us consider a straight beam of length L and axis x linking the gravity centers G of all cross-sections with xz being a principal plane of inertia. The cross-section is formed by a piling (stacking) of layers of composite material. Hence, in general the beam axis does not coincide with the neutral axis. The loads are vertical forces and bending moments contained in the xz plane as usual for plane beams (Figure 3.1). Bending on the plane yz will not be considered here. The general bending problem will be studied in Chapter 4 when dealing with 3D beams.

We will assume Timoshenko hypothesis for the rotation of the normal to hold (Section 2.2). Under these assumptions the axial and vertical displacements of an arbitrary point A of the beam section are expressed as

$$u(x, z) = u_0(x) - z\theta(x) \qquad ; \qquad w(x, z) = w_0(x) \qquad (3.1)$$

where $(\cdot)_0$ denotes the displacements of the beams axis (Figure 3.1). The axial displacement u_0 is now accounted for in the beam kinematics due to the effect of non-uniform material properties over the beam section. This is a key difference with the homogeneous beam theories studied in the previous two chapters in which $w_0(x)$ was denoted as $w(x)$, for simplicity.

The axial and transverse shear strains are deduced from Eqs.(3.1) as

$$\varepsilon_x = \frac{\partial u}{\partial x} = \frac{\partial u_0}{\partial x} - z \frac{\partial \theta}{\partial x} \tag{3.2a}$$

$$\gamma_{xz} = \frac{\partial w}{\partial x} + \frac{\partial u}{\partial z} = \frac{\partial w_0}{\partial x} - \theta \tag{3.2b}$$

Eqs.(3.2) can be written in matrix form as

$$\varepsilon = \left\{ \begin{array}{c} \varepsilon_x \\ \gamma_{xz} \end{array} \right\} = \begin{bmatrix} 1 & -z & 0 \\ 0 & 0 & 1 \end{bmatrix} \left[\frac{\partial u_0}{\partial x}, \frac{\partial \theta}{\partial x}, \frac{\partial w_0}{\partial x} - \theta \right]^T = \mathbf{S}\hat{\varepsilon} \tag{3.3a}$$

with

$$\mathbf{S} = \begin{bmatrix} 1 & -z & 0 \\ 0 & 0 & 1 \end{bmatrix} \quad , \quad \hat{\varepsilon} = \left[\frac{\partial u_0}{\partial x}, \frac{\partial \theta}{\partial x}, \frac{\partial w_0}{\partial x} - \theta \right]^T \tag{3.3b}$$

where ε is the strain vector, $\hat{\varepsilon}$ is the generalized strain vector containing the elongation of the beam axis $\left(\frac{\partial u_0}{\partial x} \right)$, the curvature $\left(\frac{\partial \theta}{\partial x} \right)$ and the transverse shear strain $\left(\frac{\partial w_0}{\partial x} - \theta \right)$ and $\mathbf{S}$ is a strain-displacement transformation matrix depending on the thickness coordinate z.

The assumption $\theta = \frac{\partial w_0}{\partial x}$ (Euler-Bernoulli theory) leads to the transverse shear strain vanishing. As mentioned above, Timoshenko theory is preferable for composite laminated beams due to the relevance of transverse shear deformation in these structures.

The assumption of a linear displacement field across the thickness can be enhanced by using a higher order approximation based on linear or quadratic distributions for the in-plane displacement within each layer. A description of some higher order laminated beam theories is presented in the last part of the chapter. Despite the increased accuracy of these theories, the simpler Timoshenko beam formulation described in the first part of this chapter yields sufficiently good results for many composite and sandwich beams found in practice.

3.3 BASIC CHARACTERISTICS OF COMPOSITE MATERIALS

Composite materials for structural applications are typically manufactured by embedding in a matrix material fibers that are all aligned in a single direction. Appendix A lists the basic properties of the more usual fiber and matrix (resins and polymers) materials. We present next a comparative discussion of the properties of the fibers, the matrix and the composite.

	E_f (GPa)	ρ_f (Kg/m^3)	σ_{fu}^t (MPa)
E-Glass	72	2550	3400
S-Glass	86	2500	4800
Carbon	190	1410	1700
Boron	400	2600	3400
Graphite	250	1410	1700

Table 3.1 Young modulus (E), density (ρ) and ultimate tensile stress (σ_{fu}^t) of some fibers [BC]

	E_f (GPa)	ρ (Kg/m^3)	σ_u^t (MPa)
Aluminium	73	2700	620
Titanium	115	4700	1900
Steel	210	7700	4100

Table 3.2 Young modulus (E), density (ρ) and ultimate tensile stress (σ_u^t) of several metals

3.3.1 About the properties of fibers

Table 3.1 lists the Young modulus (E_f), the density ρ_f and the ultimate allowable tensile stress (σ_{fu}^t) (also called "failure " stress) of several typical fibers. Indices f and m in the material properties in the following denote fiber and matrix properties, respectively.

Table 3.2 shows the same physical properties for three commonly used metals: aluminium, titanium and steel. Clearly, the ultimate tensile stress and the Young modulus of steel are far superior to those of titanium and aluminium. However, Table 3.2 shows that while steel is far stronger and stiffer, it is also much heavier that the other two metals.

Comparing the properties of these three metals with those of the fibers in Table 3.1 we clearly see that fibers have a remarkable lighter ultimate strength than the three chosen metals. In addition, the fiber densities are comparable (and even smaller) to those of the lighter metal (aluminium), while the Young modulus for the fibers are within the same range than for the three metals.

Clearly, the various metals chosen can be directly used as structural materials, whereas the fibers can not be used, as such, as structural materials. However, the remarkable high strength/density ratio of the fibers justifies their popularity for use in structural applications.

3.3.2 About the properties of the matrix

A number of polymeric materials can be used as matrix materials. Thermosct matcrials, such as epoxy, arc commonly used as matrices for composite materials. The mechanical properties of epoxy are

$$E_m = 3.5 \text{GPa} \ , \quad \rho_m = 1300 \text{Kg/m}^3 \ , \quad \sigma^t_{mu} = 50 \text{MPa} \ \text{ and } \ \sigma^c_{mu} = 140 \text{MPa}$$
$$(3.4)$$

where E_m and ρ_m are the Young modulus and the density, respectively and $(\sigma^c_{mu}, \sigma^t_{mu})$ are the ultimate allowable stresses in tension and compression, respectively for the epoxy matrix.

3.3.3 Approximation of the properties of the composite

A crude way of approximating the material properties of a composite material consisting of fibers all aligned in a single direction embedded in a matrix is to homogenize the material properties of the composite by using a rule of mixture defined as [BC]

$$E_c = V_f E_f + (1 - V_f) E_m \ , \quad \rho_c = V_f \rho_f + (1 - V_f) \rho_m$$
$$\sigma^t_{cu} = V_f \sigma^t_{fu} + (1 - V_f) \sigma^t_{mu}$$
$$(3.5)$$

where E_c, ρ_c and σ^t_{cu} are the Young modulus, the density and the ultimate tensile stress of the composite and V_f is the volume fraction of fibers in the composite.

As for the ultimate compression stress for the composite, the value for the matrix (i.e. $\sigma^c_{cu} = \sigma^c_{mu}$) can be used as a good approximation.

Consider, for instance, a composite material consisting of graphite fibers $(V_f = 0.6)$ embedded in an epoxy matrix. The physical properties of the composite are (see Table 3.1 and Eqs.(3.4) and (3.5))

$$E_c = 250 \times 0.6 + 3.5 \times 0.4 = 150 + 1.4 = 151.4 \text{GPa}$$
$$\rho_c = 1410 \times 0.6 + 1300 \times 0.4 = 846 + 520 = 1366 \text{Kg/m}^3$$
$$\sigma^t_{cu} = 1700 \times 0.6 + 50 \times 0.4 = 1020 + 20 = 1040 \text{MPa}$$
$$\sigma^c_{cu} = \sigma^c_{mu} = 140 \text{MPa}$$
$$(3.6)$$

Note that the intrinsic properties of the matrix contribute little to the Young modulus and the ultimate tensile strength of the composite. However, the role of the matrix is essential to keep all the fibers together and to provide an adequate surface finish. A less obvious, but yet important role of the matrix is to diffuse the stresses among the individual fibers [Bar2,BC].

In Section 7.2.3 we will study a more accurate definition of the material properties and the constitutive equations for thin sheets of composite material made of unidirectional fibers embedded in a matrix.

3.4 STRESSES AND RESULTANT STRESSES

The axial and shear stresses are expressed from Eqs.(3.2) as

$$\sigma_x = E\varepsilon_x = E\left(\frac{\partial u_0}{\partial x} - z\frac{\partial \theta}{\partial x}\right) \tag{3.7a}$$

$$\tau_{xz} = G\gamma_{xz} = G\left(\frac{\partial w_0}{\partial x} - \theta\right) \tag{3.7b}$$

where $E = E(x, z)$ and $G = G(x, z)$ are the longitudinal Young modulus and the shear modulus of the beam composite material.

Eqs.(3.4) assume that one of the principal material axes is coincident with the beam axis, as it is usual in plane beam theory.

Eqs.(3.4) can be written in matrix form using Eq.(3.3) as

$$\boldsymbol{\sigma} = \left\{\begin{matrix} \sigma_x \\ \tau_{xz} \end{matrix}\right\} = \begin{bmatrix} E & 0 \\ 0 & G \end{bmatrix} \left\{\begin{matrix} \varepsilon_x \\ \gamma_{xz} \end{matrix}\right\} = \mathbf{D}\boldsymbol{\varepsilon} = \mathbf{DS}\hat{\boldsymbol{\varepsilon}} \tag{3.8}$$

where $\mathbf{D}$ is the standard constitutive matrix relating stresses and strains at a point in the transverse cross section.

The axial force N, the bending moment M and the shear force Q in a beam section (Figure 3.2) are obtained as

$$\hat{\boldsymbol{\sigma}} = \left\{\begin{matrix} N \\ M \\ Q \end{matrix}\right\} = \iint_A \left\{\begin{matrix} \sigma_x \\ -z\sigma_x \\ \tau_{xz} \end{matrix}\right\} dA = \iint_A \mathbf{S}^T\boldsymbol{\sigma}\, dA \tag{3.9}$$

where $\hat{\boldsymbol{\sigma}}$ is the resultant stress vector, $\mathbf{S}$ is the transformation matrix of Eq.(3.3b) and A is the area of the cross-section.

3.5 GENERALIZED CONSTITUTIVE MATRIX

Substituting Eq.(3.8) into (3.9) gives

$$\hat{\boldsymbol{\sigma}} = \left(\iint_A \mathbf{S}^T\mathbf{DS}\, dA\right)\hat{\boldsymbol{\varepsilon}} = \hat{\mathbf{D}}\hat{\boldsymbol{\varepsilon}} \tag{3.10}$$

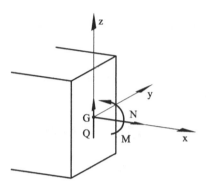

Fig. 3.2 Sign convenion for resultant stresses N, M and Q

where $\hat{\varepsilon}$ is the generalized strain vector defined in Eq.(3.3b) and $\hat{\mathbf{D}}$ is the generalized constitutive matrix. The terms of $\hat{\mathbf{D}}$ are computed as

$$\hat{\mathbf{D}} = \iint_A \mathbf{S}^T \mathbf{D} \mathbf{S} \, dA = \begin{bmatrix} \hat{D}_a & \hat{D}_{ab} & 0 \\ \hat{D}_{ab} & \hat{D}_b & 0 \\ 0 & 0 & \hat{D}_s \end{bmatrix} \qquad (3.11a)$$

with

$$\hat{D}_a = \iint_A E(x,z) \, dA \quad ; \quad \hat{D}_{ab} = -\iint_A E(x,z)z \, dA$$

$$\hat{D}_b = \iint_A E(x,z)z^2 \, dA \quad ; \quad \hat{D}_s = k_z \hat{G} \ \text{ with } \hat{G} = \iint_A G(x,z) \, dz$$

$$(3.11b)$$

where $\hat{D}_a$ is the axial stiffness, $\hat{D}_b$ is the bending stiffness, $\hat{D}_{ab}$ is the coupling axial-bending stiffness, $\hat{D}_s$ is the shear stiffness and k_z is the shear correction parameter for bending around the y axis. The computation of k_z is explained in the next section.

For a general composite material with an arbitrary distribution of the material properties over the beam section, the integrals of Eq.(3.9) can be computed by dividing the section into a grid of 2D triangular o quadrilateral elements and using numerical integration within each 2D element.

For a laminated beam with n_l layers of isotropic material with modulae E^k, G^k, thickness h_k and width b_k we have

$$\hat{D}_a = \sum_{k=1}^{n_l} (z_{k+1} - z_k) b_k E^k = \sum_{k=1}^{n_l} b_k h_k E^k$$

$$\hat{D}_{ab} = -\sum_{k=1}^{n_l} \frac{1}{2} (z_{k+1}^2 - z_k^2) b_k E^k = -\sum_{k=1}^{n_l} h_k b_k \bar{z}_k E^k$$

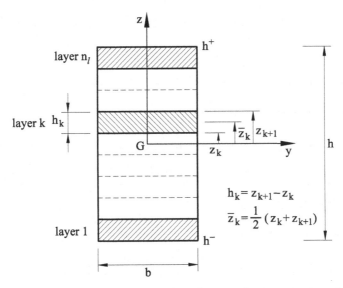

Fig. 3.3 Rectangular laminated beam. Coordinates for integrating the material properties across the layers

$$\hat{D}_b = \frac{1}{3}\sum_{k=1}^{n_l}(z_{k+1}^3 - z_k^3)b_k E^k$$

$$\hat{D}_s = k_z\sum_{k=1}^{n_l}(z_{k+1} - z_k)b_k G^k = k_z\sum_{k=1}^{n_c}h_k b_k G^k$$

(3.12)

where $\bar{z}_k$ is the vertical coordinate of the midpoint of the kth layer. If the section width is constant its value can be taken out of the sums in Eq.(3.12). Figure 3.3 shows an example of a rectangular laminated beam.

The material parameters E^k and G^k for each layer correspond to the properties of the composite along the principal material axes, i.e. the longitudinal direction (the fiber direction) or the transverse direction, as adequate. It is assumed that the fiber direction is aligned in (or orthogonal to) the beam axis direction.

3.6 AXIAL-BENDING COUPLING AND NEUTRAL AXIS

The off-diagonal term $\hat{D}_{ab}$ in matrix $\hat{\mathbf{D}}$ originates a coupling between the axial and bending effects. Thus, an axial force yields a curvature and a bending moment induces an elongation of the beam axis. This coupling term vanishes in certain circumstances for which the x axis is the so-called *neutral axis* that are studied next.

For homogeneous material

$$\hat{D}_a = EA \ , \quad \hat{D}_{ab} = -E\bar{S} \ , \quad \hat{D}_b = EI_y \ , \quad \hat{D}_s = k_z GA \qquad (3.13)$$

where A is the area of the transverse section, I_y is the moment of inertia with respect to the y axis and $\bar{S} = \iint_A z\,dA = z_g A$ where z_g is the vertical coordinate of the center of gravity of the section G. If the x axis is placed at point G then $\bar{S} = 0$ and, hence, $\hat{D}_{ab} = 0$ which means that, for homogeneous material, the x axis is the *neutral axis* and the axial and bending effects are uncoupled at a section level.

If the material properties (and the section geometry) are symmetrical with respect to the reference axis x, then x is also the neutral axis.

The position of the *neutral axis* for an arbitrary composite laminated section can be found as follows. Let us define the relative vertical coordinate $z' = z - d$ where d is the vertical distance between the beam axis x and the neutral axis. If the x axis is placed at point O defining the neutral axis (Figure 3.4), then

$$\hat{D}_{ab} = -\iint_A Ez'dA = -\iint_A E(z - d)dA = 0 \qquad (3.14a)$$

From Eqs.(3.14a) and (3.11b) we can obtain

$$d = -\frac{\hat{D}_{ab}}{\hat{D}_a} \qquad (3.14b)$$

In conclusion, the axial and bending moment effects can be decoupled at section level by simply placing the origin of the x axis at point O (Figure 3.4) and changing z by z' in all the equations. This does not affect the expressions for $\hat{D}_a$ and $\hat{D}_s$ (as they do not depend on z) and of $\hat{D}_b$ (as $\iint_A Ez'^2dA = \iint_A Ez^2dA$). The change of z by z' does affects the axial displacement u, as $u = u_0 - z'\theta$ (Eq.3.1) and the computation of the normal stress σ_x via Eq.(3.7a). However, the results for the vertical deflection w_0, the rotation θ and the transverse shear stresses are independent of the origin of the beam axis.

3.7 THERMAL STRAINS AND INITIAL STRESSES

An initial axial strain due to thermal effects (ε_x^o) and initial stresses (σ_x^o, τ_{xz}^o) can easily be accounted for in the present formulation. The strain-

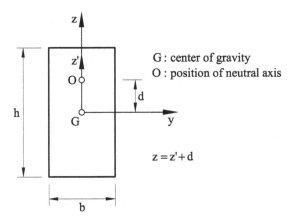

Fig. 3.4 Position of the neutral axis O in a rectangular beam section

stress relationship of Eq.(3.8) is modified as

$$\sigma_x = E(\varepsilon_x - \varepsilon_x^o) + \sigma_x^o \quad ; \quad \tau_{xz} = G\gamma_{xz} + \tau_{xz}^o \qquad (3.15)$$

where $\varepsilon_x^o = \alpha\Delta T$, α being the thermal expansion coefficient and ΔT the temperature increment. Recall that the initial tangential stresses due to a thermal expansion are zero (Section 4.2.4 of [On4]).

The relationship between resultant forces and generalized strains (Eqs. (3.7)) is modified as

$$\hat{\sigma} = \hat{\mathbf{D}}\hat{\varepsilon} + \hat{\sigma}^o \qquad (3.16a)$$

where $\hat{\sigma}^o$ is the initial resultant stress vector given by

$$\hat{\sigma}^o = [N^o, M^o, Q^o]^T \qquad (3.16b)$$

with

$$N^o = \iint_A [-E\varepsilon_x^o + \sigma_x^o]dA \ , \ M^o = \iint_A [E\varepsilon_x^o - \sigma_x^o]z \, dA \ , \ Q^o = \iint_A \tau_{xz}^o dA$$
$$(3.16c)$$

3.8 COMPUTATION OF THE SHEAR CORRECTION PARAMETER

We recall the expression for the shear correction parameter given in Eq.(2.7b)

$$k_z = \frac{Q^2}{\hat{G}} \left[\iint_A \frac{\tau_{xz}^2}{G(z)} \right]^{-1} dA \qquad (3.17)$$

This equation was found assuming cylindrical bending around the y axis and, consequently, $\tau_{xy} = 0$. What it remains now is to find an expression for the distribution of τ_{xz} over the cross section.

We will consider again for simplicity the case of cylindrical bending around the y axis. The more general case is explained in Appendix D.

The laminate beam theory chosen yields a discontinuous distribution of the normal stress σ_x across the beam layers, despite the fact that the strain ε_x varies linearly in the depth direction. This is a consequence of the different material properties for each layer (Figure 3.5). Inverting Eq.(3.10) gives

$$\frac{\partial u_0}{\partial x} = \frac{1}{\hat{D}}[\hat{D}_b N - \hat{D}_{ab} M] \tag{3.18a}$$

$$\frac{\partial \theta}{\partial x} = \frac{1}{\hat{D}}[-\hat{D}_{ab} N + \hat{D}_a M] \tag{3.18b}$$

with

$$\hat{D} = \hat{D}_a \hat{D}_b - \hat{D}_{ab}^2 \tag{3.18c}$$

Substituting Eqs.(3.18a,b) into (3.7a) gives σ_x at each layer in terms of N and M by

$$\sigma_x = \frac{E}{\hat{D}}\left[\hat{D}_b N - \hat{D}_{ab} M - z(-\hat{D}_{ab} N + \hat{D}_a M)\right] \tag{3.19}$$

Eq.(3.7b) shows that the shear stress τ_{xz} is *constant* across the beam depth (as it is usual in Timoshenko beam theory). The "correct" distribution of τ_{xz} which satisfies the equilibrium equations of elasticity can be computed a posteriori once the displacements have been obtained. From the equilibrium equation along the x direction (Appendix B and [ZTZ]) we have (recalling that $\tau_{xy} = 0$)

$$\frac{\partial \tau_{xz}}{\partial z} + \frac{\partial \sigma_x}{\partial x} = 0 \quad \rightarrow \quad \tau_{xz}(z) = -\int_{h^-}^{z} \frac{\partial \sigma_x}{\partial x} dz \tag{3.20}$$

Substituting Eq.(3.19) into Eq.(3.20) and accepting that $\dfrac{\partial N}{\partial x} = 0$ and using $\dfrac{\partial M}{\partial x} = -Q$ (Figure 1.7) gives

$$\tau_{xz}(z) = \frac{-Q}{\hat{D}} F(z) \tag{3.21}$$

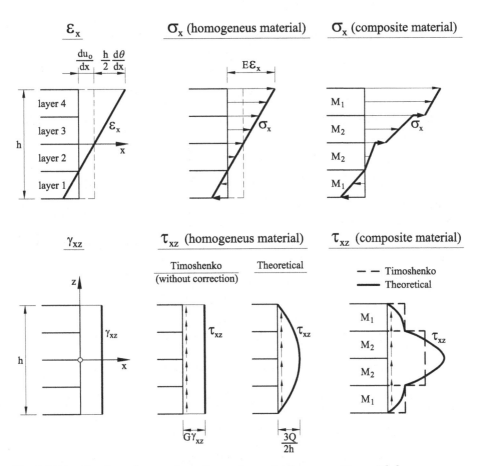

Fig. 3.5 Distribution of σ_x and τ_{xz} for rectangular beam section with homogeneous material and 4-layered composite material (symmetric). M_i denotes material type

with

$$F(z) = \hat{D}_a S(z) + \hat{D}_{ab} \int_{h^-}^{z} E(z)dz \quad , \quad S(z) = \int_{h^-}^{z} zE(z)\,dz \qquad (3.22)$$

where $S(z)$ is the static moment of the Young modulus with respect to the fiber of coordinate z. For a composite laminated section point G defining the x axis is generally not equidistant from the lower and upper fibers with coordinates h^- and h^+, respectively ($|h^-| + h^+ = h$, Figure 3.3). Both $S(z)$ and $F(z)$ are continuous functions. For composite laminated beams, τ_{xz} has a continuous distribution across the thickness with a steep gradient at the interfaces between layers with different Young modulus (Figure 3.5).

If x is the neutral axis, then $\hat{D}_{ab} = 0$, and, hence, $\hat{D} = \hat{D}_a \hat{D}_b$ and

$$\tau_{xz}(z) = \frac{-Q}{\hat{D}_b} S(z) \tag{3.23}$$

For a rectangular section with homogeneous material and $h^- = -h/2$ Eq.(3.23) simplifies to the classical parabolic distribution of elasticity theory (with $\nu = 0$) [BD4,Ti3], i.e.

$$\tau_{xz}(z) = \frac{3}{2} \frac{Q}{bh} \left(1 - \frac{4z^2}{h^2} \right) \tag{3.24}$$

Figure 3.5 shows the distribution of τ_{xz} for two beam sections of homogeneous and composite laminated material, respectively.

Substituting Eq.(3.21) into (3.17) gives (noting that $\hat{D}$ is a property of the transverse cross section)

$$k_z = \frac{\hat{D}^2}{\hat{G}} \left[\iint_A \frac{F^2(z)}{G(z)} \, dA \right]^{-1} \tag{3.25}$$

If x is the neutral axis, then $F = \hat{D}_a S(z)$, $\hat{D} = \hat{D}_a \hat{D}_b$ and

$$k_z = \frac{\hat{D}_b^2}{\hat{G}} \left[\iint_A \frac{S^2(z)}{G(z)} \, dA \right]^{-1} \tag{3.26}$$

For *homogeneous material*

$$k_z = \frac{I_y^2}{A} \left[\iint_A M_e^2 \, dA \right]^{-1} \tag{3.27}$$

where $M_e = \int_{h^-}^{z} z dz$ is the *static moment* with respect to the y axis of the area between the coordinates h^- and z.

For a *homogeneous rectangular section* of dimensions $b \times h$

$$M_e = \int_{-h/2}^{z} z dz = \frac{z^2}{2} - \frac{h^2}{8} \quad ; \quad \iint_A M_e^2 \, dA = \frac{bh^5}{120} \tag{3.28}$$

$$\text{and} \quad k_z = \frac{\left[\frac{1}{12} bh^3 \right]^2}{bh \frac{bh^5}{120}} = \frac{5}{6} \tag{3.29}$$

which is the value of k_z for homogeneous rectangular beams in Timoshenko theory (Figure 2.3).

An identical process can be followed for computing k_y for bending around the z axis.

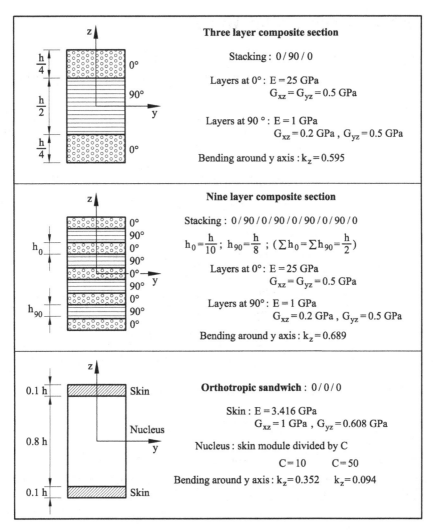

Fig. 3.6 Shear correction parameter k_z for different laminated beam sections. The stacking sequence indicates the orientation of the main orthotropic direction of each laminate with respect to the beam axis (Section 5.2.3)

Figure 3.6 shows the values of k_z for different laminated composite beams. The values of G_{yz} shown are useful for studying the bending around the z axis.

Clearly, the general expression for k_z allows for arbitrary distributions of the Young modulus and the shear modulus within the cross-section. For more information on the computation of the shear correction parameters in composite beams see [Ban,BD5,Be2,Be3,BG,DM,Ga3,Ste,TH3].

3.9 PRINCIPLE OF VIRTUAL WORK

We study the PVW for distributed loads $\mathbf{t}$ only. Other load types (i.e. point loads) can be easily taken into account as explained in Chapter 2. The expression of the PVW is

$$\iiint_V \delta\boldsymbol{\varepsilon}^T \boldsymbol{\sigma} \, dV = \int_L \delta\mathbf{u}^T \mathbf{t} \, dx \tag{3.30}$$

where $\delta\mathbf{u} = [\delta u_0, \delta w_0, \delta\theta]^T$ is the virtual displacement vector, $\delta\boldsymbol{\varepsilon}$ and $\boldsymbol{\sigma}$ are the virtual strain vector and the stress vector, respectively, and $\mathbf{t} = [f_x, f_z, m]^T$ is the vector of external forces acting over the beam axis due to distributed axial and vertical loads f_x and f_z, respectively and a distributed moment m.

As usual, the integral in the l.h.s. of Eq.(3.30) represents the internal virtual work. This can be expressed in terms of the work of the resultant stresses on the generalized strains along the beam axis as follows.

Making use of Eqs.(3.3) and (3.8), Eq.(3.30) can be written as

$$\iiint_V \delta\boldsymbol{\varepsilon}^T \hat{\boldsymbol{\sigma}} \, dV = \int_l \delta\hat{\boldsymbol{\varepsilon}}^T \left[\iint_A \mathbf{S}^T \mathbf{D} \mathbf{S} \, dA \right] \hat{\boldsymbol{\varepsilon}} \, dx = \int_L \delta\hat{\boldsymbol{\varepsilon}}^T \hat{\mathbf{D}} \hat{\boldsymbol{\varepsilon}} \, dx = \int_L \delta\hat{\boldsymbol{\varepsilon}}^T \hat{\boldsymbol{\sigma}} \, dx \tag{3.31}$$

The PVW can therefore be expressed in terms of integrals along the beam axis as

$$\int_L \delta\hat{\boldsymbol{\varepsilon}}^T \hat{\boldsymbol{\sigma}} \, dx = \int_L \delta\mathbf{u}^T \mathbf{t} \, dx \tag{3.32}$$

Note that all the derivatives appearing in the PVW are of first order. This allows us using C° continuous interpolations for the axial displacement u_0, the vertical deflection w_0 and the rotation θ.

3.10 TWO-NODED COMPOSITE LAMINATED TIMOSHENKO BEAM ELEMENT

The beam is discretized into 2-noded elements of length $l^{(e)}$. A standard linear approximation is chosen for u_0, w_0 and θ as (Figure 3.7)

$$\mathbf{u} = \begin{Bmatrix} u_0 \\ w_0 \\ \theta \end{Bmatrix} = \sum_{i=1}^{2} N_i(\xi) \mathbf{a}_i^{(e)} \quad \text{with} \quad \mathbf{a}_i^{(e)} = \begin{Bmatrix} u_0 \\ w_0 \\ \theta \end{Bmatrix}_i \tag{3.33}$$

where $(\cdot)_i$ denotes nodal values, as usual. The expression for the linear shape functions $N_i(\xi)$ is shown in Figure 2.4.

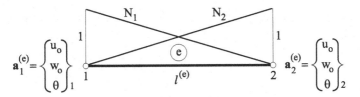

Fig. 3.7 Two-noded composite laminated Timoshenko beam element. Nodal variables and shape functions

Substituting the approximation (3.33) into the generalized strain vector of Eq.(3.3) gives

$$\hat{\boldsymbol{\varepsilon}} = \left\{ \begin{array}{c} \dfrac{\partial u_0}{\partial x} \\[2mm] \dfrac{\partial \theta}{\partial x} \\[2mm] \dfrac{\partial w_0}{\partial x} - \theta \end{array} \right\} = \sum_{i=1}^{2} \mathbf{B}_i \mathbf{a}_i^{(e)} = \mathbf{B} \mathbf{a}^{(e)} \tag{3.34}$$

with

$$\mathbf{a}^{(e)} = \left\{ \begin{array}{c} \mathbf{a}_1^{(e)} \\[1mm] \mathbf{a}_2^{(e)} \end{array} \right\} \quad \text{and} \quad \mathbf{B}_i = \left\{ \begin{array}{c} \mathbf{B}_{a_i} \\ \cdots \\ \mathbf{B}_{b_i} \\ \cdots \\ \mathbf{B}_{s_i} \end{array} \right\} = \left[\begin{array}{ccc} \dfrac{\partial N_i}{\partial x} & 0 & 0 \\ \cdots & \cdots & \cdots \\ 0 & 0 & \dfrac{\partial N_i}{\partial x} \\ \cdots & \cdots & \cdots \\ 0 & \dfrac{\partial N_i}{\partial x} & -N_i \end{array} \right] \tag{3.35}$$

where $\mathbf{B}_{m_i}, \mathbf{B}_{b_i}$ and $\mathbf{B}_{s_i}$ are the generalized strain matrices corresponding to axial, bending and transverse shear deformation effects.

Substituting the constitutive relationship (3.16a) into the PVW (Eq.(3.32)) and using Eqs.(3.33) and (3.34) leads, after standard algebra, to the system of equations $\mathbf{Ka} = \mathbf{f}$ where the stiffness matrix and the equivalent nodal force vector are assembled from the element contributions given by

$$\mathbf{K}_{ij}^{(e)} = \int_{l^{(e)}} \mathbf{B}_i^T \hat{\mathbf{D}} \mathbf{B}_j \, dx$$

$$\mathbf{f}_i^{(e)} = \left\{ \begin{array}{c} f_{x_i} \\ f_{z_i} \\ m_i \end{array} \right\} = \int_{l^{(e)}} N_i^{(e)} \mathbf{t} \, dx - \int_{l^{(e)}} \mathbf{B}_i^T \hat{\boldsymbol{\sigma}}^o dx \qquad i, j = 1, 2 \tag{3.36}$$

The second integral in the expression of $\mathbf{f}_i^{(e)}$ accounts for the effect of the initial (thermal) strain and the initial stresses.

The element stiffness matrix can be written using the components of $\mathbf{B}_i$ and $\hat{\mathbf{D}}$ as

$$\mathbf{K}_{ij}^{(e)} = \mathbf{K}_{a_{ij}}^{(e)} + \mathbf{K}_{b_{ij}}^{(e)} + \mathbf{K}_{s_{ij}}^{(e)} + \mathbf{K}_{ab_{ij}}^{(e)} + [\mathbf{K}_{ab_{ij}}^{(e)}]^T \tag{3.37a}$$

where

$$\mathbf{K}_{r_{ij}}^{(e)} = \int_{l^{(e)}} \mathbf{B}_{r_i}^T \hat{D}_r \mathbf{B}_{r_j} \, dx \qquad r = a, b, s \tag{3.37b}$$

and

$$\mathbf{K}_{ab_{ij}}^{(e)} = \int_{l^{(e)}} \mathbf{B}_{a_i}^T \hat{D}_{ab} \mathbf{B}_{b_i} \, dx \tag{3.37c}$$

In the above expressions indexes a, b, s and ab denote respectively the contribution of the axial, bending, shear and coupling axial-bending terms to the element stiffness matrix.

If x is the neutral axis, then $\hat{D}_{ab}$ is zero (Section 3.6) and so it is matrix $\mathbf{K}_{ab}^{(e)}$. This leads to a decoupling of the axial, bending and transverse shear effects at the element level. The element stiffness matrix in this case can be written as

$$\mathbf{K}^{(e)} = \begin{bmatrix} \mathbf{K}_a^{(e)} & \mathbf{0} \\ \mathbf{0} & \mathbf{K}_f^{(e)} \end{bmatrix} \tag{3.38}$$

with the nodal displacement vector for the element ordered as

$$\mathbf{a}^{(e)} = \begin{bmatrix} u_{o_1}, u_{o_2}, w_{o_1}, \theta_1, w_{o_2}, \theta_2 \end{bmatrix}^T \tag{3.39}$$

In Eq.(3.38)

$$\mathbf{K}_a^{(e)} = \frac{\hat{D}_a}{l^{(e)}} \begin{bmatrix} 1 & -1 \\ -1 & 1 \end{bmatrix} \tag{3.40}$$

is the stiffness matrix for the 2-noded axially loaded rod element studied in Chapter 2 of [On4] and $\mathbf{K}_f^{(e)}$ (f for "flexural" effects) is the 4×4 stiffness matrix of the 2-noded Timoshenko beam element incorporating the bending and shear contributions (Eq.(2.21a)).

3.11 SHEAR LOCKING IN COMPOSITE LAMINATED BEAMS

We saw in Section 2.4 that the relative value of the shear stiffness terms versus the bending terms affects the finite element solution for the Timoshenko beam problem. For thick beams, the shear terms dominate the bending ones in the stiffness matrix and this leads to unrealistically stiff results (locking). The relative influence of the shear terms over

Isotropic rectangular section $\nu = 0.25$	$k_z = \dfrac{5}{6}$; $\beta = \dfrac{3}{\lambda^2}$
Homogeneous rectangular section with $\dfrac{E}{k_z G} = 50$	$k_z = \dfrac{5}{6}$; $\beta = \dfrac{50}{\lambda^2}$
Three layer composite section (Figure 3.6)	$k_z = 0.595$; $\beta = \dfrac{105}{\lambda^2}$
Nine layer composite section (Figure 3.6)	$k_z = 0.689$; $\beta = \dfrac{65}{\lambda^2}$
Sandwich section with $C = 50$ (Figure 3.6)	$k_z = 0.094$; $\beta = \dfrac{84}{\lambda^2}$

Table 3.3 Values of the β parameter and the shear correction factor k_z for beams of rectangular section with different materials

the bending terms can be quantified by the parameter β of Eq.(2.94b):

$$\beta = \frac{12\hat{D}_b}{L^2 \hat{D}_s} \tag{3.41}$$

A small value of β indicates that the influence of transverse shear deformation is negligible in the solution. Parameter β depends on the geometrical and mechanical properties of the section. For a rectangular beam of length L, depth h and homogeneous isotropic material, $\beta = E(k_z G \lambda^2)^{-1}$ with $\lambda = L/h$ being the beam slenderness ratio.

Table 3.3 shows the value of β for different beam sections of isotropic and composite materials.

For a relatively "thick" isotropic beam ($\lambda = 4$), the ratio $\frac{E}{k_z G} \simeq 2$ and $\beta = 0.125$. It is interesting that for a slender composite beam with $\lambda = 20$ and $\frac{E}{k_z G} = 50$ the value of β is also 0.125. *Therefore, the influence of transverse shear deformation is the same for a thick isotropic beam and a slender composite beam*, both leading to a small value of β. This justifies using Timoshenko theory for composite laminated beams.

Shear locking appearing for small values of β can be eliminated by any of the methods explained in the previous chapter. For the 2-noded composite Timoshenko beam element, the simplest procedure is to evaluate all integrals in the stiffness matrix *using a single Gauss integration point*. The expression for $\mathbf{K}_{ij}^{(e)}$ in this case is

$$\mathbf{K}_{ij}^{(e)} = l^{(e)} [\mathbf{B}_i^T \hat{\mathbf{D}} \mathbf{B}_j]_c \tag{3.42}$$

where $[\cdot]_c$ denotes values at the element center. The expression of $[\mathbf{B}_i]_c$ is readily obtained from Eq.(3.35) as

$$
[\mathbf{B}_i]_c = \begin{bmatrix} \dfrac{(-1)^i}{l^{(e)}} & 0 & 0 \\[2ex] 0 & 0 & \dfrac{(-1)^i}{l^{(e)}} \\[2ex] 0 & \dfrac{(-1)^i}{l^{(e)}} & -1/2 \end{bmatrix} \tag{3.43}
$$

Similarly as for homogeneous beam element, the single point reduced quadrature is equivalent to using a constant assumed transverse shear strain field (Section 2.8.4).

3.12 EXACT TWO-NODED TIMOSHENKO BEAM ELEMENT WITH COMPOSITE LAMINATED SECTION

The derivation of an exact 2-noded composite laminated Timoshenko beam element is simple *if x is the neutral axis*. Matrix $\mathbf{K}_{ab}$ is zero in this case and the bending and shear stiffness matrices coincide precisely with those derived in Section 2.9. The stiffness matrix has the form shown in Eq.(3.38) with matrices $\mathbf{K}_a^{(e)}$ and $\mathbf{K}_f^{(e)}$ given by Eqs.(3.40) and Eq.(2.101c), respectively.

The equivalent nodal force vector is split into axial $(\mathbf{f}_a^{(e)})$ flexural $(\mathbf{f}_f^{(e)})$ component as

$$
\mathbf{f}^{(e)} = \left\{ \begin{array}{c} \mathbf{f}_a^{(e)} \\ \mathbf{f}_f^{(e)} \end{array} \right\} \tag{3.44}
$$

with

$$
\mathbf{f}_{a_i}^{(e)} = \int_{l^{(e)}} N_i f_x dx \quad , \quad i = 1, 2 \tag{3.45}
$$

and $\mathbf{f}_f^{(e)}$ is given by Eq.(2.101d).

Note that the DOFs in the displacement vector are now ordered as shown in Eq.(3.39).

The axial and flexural forces at the nodes can be computed separately in terms of the nodal displacements as

$$
\mathbf{q}_m^{(e)} = \left\{ \begin{array}{c} -\mathcal{N}_1^{(e)} \\ \mathcal{N}_2^{(e)} \end{array} \right\}^{(e)} = \mathbf{K}_m^{(e)} \left\{ \begin{array}{c} u_{o_1} \\ u_{o_2} \end{array} \right\} - \mathbf{f}_m^{(e)} \tag{3.46}
$$

$$\mathbf{q}_f^{(e)} = \begin{Bmatrix} -Q_1 \\ -M_1 \\ Q_2 \\ M_2 \end{Bmatrix}^{(e)} = \mathbf{K}_f^{(e)} \begin{Bmatrix} w_{o_1} \\ \theta_1 \\ w_{o_2} \\ \theta_2 \end{Bmatrix}^{(e)} - \mathbf{f}_f^{(e)} \tag{3.47}$$

The derivation of an exact composite laminated Timoshenko beam element when x is not the neutral axis is more complex but is still possible using mixed or hybrid formulations [BD5]. In practice it is desirable to find the vertical position of the neutral axis via Eq.(3.14b) and then change the coordinate z by z', so that x is the neutral axis and the expressions for $\mathbf{K}^{(e)}$ and $\mathbf{f}^{(e)}$ given above are applicable.

3.13 COMPOSITE LAMINATED EULER-BERNOULLI BEAM ELEMENTS

The derivation of composite laminated beam elements satisfying Euler-Bernoulli beam theory follows very similar arguments as for Timoshenko elements.

The displacement interpolation for the 2-noded composite laminated Euler-Bernoulli beam element combines a standard linear C° interpolation for the axial displacement with a cubic Hermite interpolation for the vertical deflection (Eq.(1.10)).

The generalized strain matrix contains only the axial and bending contributions $\mathbf{B}_a$ and $\mathbf{B}_b$. The expression for $\mathbf{B}_{a_i}$ coincides with that of Eq.(3.35), while $\mathbf{B}_b$ is deduced from Eq.(1.16a).

The stiffness matrix is obtained by neglecting the shear stiffness terms in Eq.(3.37a) (i.e. $\mathbf{K}_{s_{ij}}^{(e)} = \mathbf{0}$) while the rest of the stiffness matrices (i.e. $\mathbf{K}_a^{(e)}$, $\mathbf{K}_b^{(e)}$ and $\mathbf{K}_{ab}^{(e)}$) are computed by Eqs.(3.37b,c). The coupling of the axial and bending stiffness follows the rules explained in Section 3.6. The equivalent nodal force vector for a distributed loading is deduced from Eqs.(1.21) for the Euler-Bernoulli beam element.

The reader is invited to write down the formulation for the 2-noded Euler-Bernoulli composite laminated beam element, as an exercise.

3.14 HIGHER ORDER COMPOSITE LAMINATED BEAM THEORIES

Timoshenko beam theory (TBT) produces inadequate predictions when applied to relatively thick composite laminated beams with material layers that have highly dissimilar stiffness characteristics. Even with a judiciously chosen shear correction factor, Timoshenko theory tends to underestimate the axial stress at the top and bottom outer fibers of a beam. Also, along the layer interfaces of a laminated beam the transverse shear stresses predicted often exhibit erroneous discontinuities. These difficulties are due to the higher complexity of the "true" variation of the in-plane displacement field across a highly heterogeneous beam cross-section.

Indeed to achieve accurate computational results, 3D finite element analyses are often preferred over beam, plate and shell models that are based on first order shear deformation theories, such as the Timoshenko and Euler-Bernoulli theories. For composite laminates with hundred of layers, however, 3D modelling becomes prohibitely expensive, especially for non linear and progressive failure analyses.

The need for composite laminated beam, plate and shell theories with better predictive capabilities has led to the development of the so-called *higher order* theories. A review can be found in [LL,Red2]. In those beam theories higher-order kinematic terms with respect to the beam depth are added to the expression for the in-plane displacement and, in some cases, to the expressions for the deflection.

In the following sections we describe two popular higher order beam theories for composite laminated beams; namely the layer-wise theory and the zigzag theory. A detailed description of the refined zigzag theory (RZT) proposed by Tessler *et al.* [TDG] is presented and an interesting 2-noded composite laminated beam element based on the RZT is described.

3.15 LAYER-WISE THEORY

An enhancement in the prediction of the correct shear and axial stresses for thick and highly heterogenous composite laminated and sandwich beam, plate and shell structures can be achieved by using the so-called *layer-wise* theory. In this theory the thickness coordinate is split into a number of *analysis layers* that may or not coincide with the number of laminate plies. The kinematics are independently described within each layer and certain physical continuity requirements are enforced [BOM,LL2,OL,OL2,Red2]. In the more general setting the 3D displacement field in layer-wise theory is written as a linear combination of some function of the thickness coordinate and independent functions of the position within each analysis layer as

$$u_i(x, y, z) = u_i^0(x, y) + \sum_{k=1}^{N_i} u_i^k(x, y)\phi_k(z) \qquad (3.48)$$

where N_i is the number of analysis layers taken across the laminate thickness to model the ith displacement component, $u_i^k(x, y)$ are the displacements at each layer interface k and ϕ_j are known functions of the thickness coordinate z. Generally N_i coincides with the number of actual material layers (plies) n_l. The ϕ_k functions are typically piecewise and continuous within each layer. They are only defined over two adjacent layers and can be interpreted as a global C° Lagrange interpolation associated to the common interface j. Due to the local definition of $\phi_j(z)$, the displacements are continuous across the thickness but their derivatives with respect to z are not. Hence, the transverse shear strains are discontinuous at the interfaces and, consequently, the transverse shear stress can be enforced to be continuous for the case of layers with different mechanical properties.

For laminated beams Eq.(3.48) is particularized as

$$u(x, y, z) = \sum_{j=0}^{N} u^j(x) N_j(z) \quad , \quad w(x, y, z) = w_0(x) \qquad (3.49)$$

where N is the number of analysis layers, u and w are the horizontal (axial) and vertical displacements respectively and N_j is the linear shape function for each layer. Figure 3.8 shows a representation of the axial displacement field for a three-layered beam where N is equal to three.

A drawback of layer-wise theory is that the number of kinematic variables depends on the number of analysis layers. However, the layer displacements u^j can be condensed at each section in terms of the axial displacement for the top layer during the equation solution process. The method is described in Section 7.7.3 for laminated plates.

3.16 ZIGZAG THEORIES

The so-called zigzag theories are a sub-class of the general layer-wise theory. They assume a zigzag pattern for the axial displacements and enforce continuity of the transverse shear stresses across the entire laminate depth. Importantly, the number of kinematic variables in zigzag theories is *independent of the number of layers*.

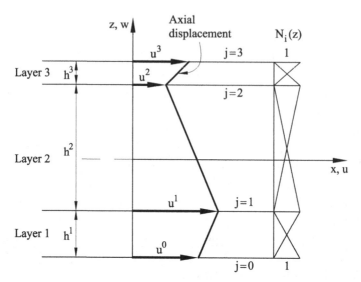

Fig. 3.8 Distribution of the axial displacement field for a three-layered beam ($n_l = 3$) in layer-wise theory

The kinematic field in zigzag beam theories is written as

$$u^k(x, z) = u_0(x) - z\theta(x) + \bar{u}^k(x, z) \quad ; \quad w(x, z) = w_0(x) \tag{3.50a}$$

where

$$\bar{u}^k = \phi^k(z)\Psi(x) \tag{3.50b}$$

is the *zigzag displacement function*.

In Eqs.(3.50), superscript k indicates quantities within the kth layer with $z_k \leq z \leq z_{k+1}$ and z_k is the vertical coordinate of the kth interface. The uniform axial displacement $u_0(x)$, the rotation $\theta(x)$ and the transverse deflection $w_0(x)$ are the primary kinematic variables of the underlying single-layer Timoshenko beam theory studied in the previous sections. Function $\phi^k(z)$ denotes *a piecewise linear zigzag function*, yet to be established, and $\Psi(x)$ is a primary kinematic variable that defines the amplitude of the zigzag function along the beam.

Zigzag theories differ in the way they define the zigzag function $\phi^k(z)$. In the early zigzag theories for plates, Di Sciuva [DiS2,DiS3] and Murakami [Mu,TM] enforced piecewise linear zigzag displacement fields that satisfy *a priori* the transverse shear stress and displacement continuity conditions at the layer interface. This model was enhanced by adding a cubic in-plane displacement to the zigzag function [Dis3]. Many zigzag theories, such as Di Sciuva theory, require C^1 continuity for the deflection field, which is a drawback versus simpler C° continuous approximations. Also many zigzag theories run into theoretical difficulties to satisfy equilibrium of forces at clamped supports.

Averill *et al.* [AA2,Av,AY] developed the piecewise linear, quadratic and cubic zigzag theories for beams and overcame the need for C^1 continuity by enforcing the continuity of the transverse shear stress across the laminate depth via a penalty method. However, Averill theory is unable to model correctly clamped conditions.

Alam and Upadhyay [AU] proposed a 2-noded beam element based on an extension of Averill's zigzag theory including a cubic in-plane displacement field within each layer. Good results were reported for cantilever and clamped composite and sandwich beams. An assessment of different zigzag theories can be found in [AA,Ca,Sa,KDJ].

Tessler *et al.* [TDG] developed a refined zigzag theory (RFT) for composite laminated beams starting from the standard Timoshenko kinematic assumptions. The zigzag functions chosen have the property of vanishing on the top and bottom surfaces of a laminate. A particular feature of this theory is that the transverse shear stresses are not required to be continuous at the layer interfaces. This results in simple piewice-constant functions that approximate the true shear stress distribution. This theory also provides good results for clamped supports.

Gherlone *et al.* [GTD] developed three and two-noded C° beam elements based on the RFT for analysis of multilayered composite and sandwich beams. Locking-free elements are obtained by using special *anisoparametric* interpolations [TD,Te] that are adapted to approximate the four independent kinematic variables modeling the beam deformation. A family of two-noded beam elements is achieved by imposing different constraints on the original displacement approximation. The constraint conditions requiring a constant variation of the transverse shear force provide an accurate 2-noded beam element [GTD].

Oñate *et al.* [OEO,OEO2] have taken a different route for deriving a simple 2-noded composite laminated beam element based on the RZT. A standard linear displacement field is used to model the four variables of the so called LRZ element. Shear locking is avoided by using reduced integration on selected terms of the shear stiffness matrix.

In the next section we describe the RZT. Then we detail the formulation of the 2-noded LRZ beam element. Its good behaviour is demonstrated in the analysis of SS and clamped composite laminated beams. An example showing its capabilities to model delamination is also presented.

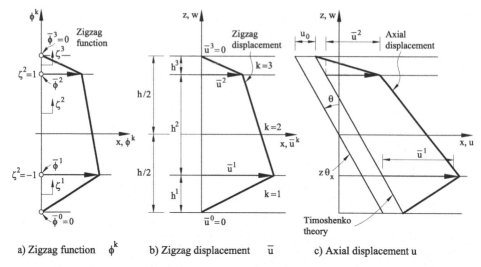

a) Zigzag function ϕ^k **b) Zigzag displacement** $\bar{u}$ **c) Axial displacement u**

Fig. 3.9 Refined zigzag theory. Distribution of zigzag function ϕ^k (a), zigzag displacement u^k (b) and axial displacement u (c) for a 3 layered section ($n_l = 3$)

3.17 REFINED ZIGZAG THEORY (RZT)

3.17.1 Zigzag displacement field

The key attributes of the RZT are, first, *the zigzag function vanishes at the top and bottom surfaces of the beam* section and does not require full shear-stress continuity across the laminated-beam depth. Second, all boundary conditions can be modelled adequately. And third, C° continuity is only required for the FEM approximation of the kinematic variables.

Within each layer the zigzag function is expressed as

$$\phi^k = \frac{1}{2}(1 - \zeta)\bar{\phi}^{k-1} + \frac{1}{2}(1 + \zeta)\bar{\phi}_k^k = \frac{\bar{\phi}^k + \bar{\phi}^{k-1}}{2} + \frac{\bar{\phi}_k^k - \bar{\phi}^{k-1}}{2}\zeta^k \quad (3.51)$$

where $\bar{\phi}^k$ and $\bar{\phi}^{k-1}$ are the zigzag functions of the k and $k - 1$ interface, respectively with $\bar{\phi}^0 = \bar{\phi}^{n_l} = 0$ and $\zeta^k = \frac{2(z - z^{k-1})}{h^k} - 1$ (Figure 3.9a).

Note that the zigzag displacement $\bar{u}^k$ (Eq.(3.50b) also vanishes at the top and bottom layers (Figure 3.9b). The axial displacement field is plotted in Figure 3.9c.

The form of ϕ^k of Eq.(3.51) yields a constant distribution of its gradient within each layer β^k, i.e.

$$\beta^k = \frac{\partial \phi^k}{\partial z} = \frac{\bar{\phi}^k - \bar{\phi}^{k-1}}{h^k} \quad (3.52a)$$

From Eq.(3.52a) and the conditions $\bar{\phi}^0 = \bar{\phi}^N = 0$ we deduce

$$\iint_A \beta^k dA = 0 \qquad (3.52b)$$

The β^k parameter is useful for computing the zigzag function as shown in the next section.

3.17.2 Strain and stress fields

The strain-displacement relations are derived from Eqs.(3.2) and (3.50) as

$$\varepsilon_x^k = \frac{\partial u_0}{\partial x} - z\frac{\partial \theta}{\partial x} + \phi^k\frac{\partial \Psi}{\partial x} = [1, -z, \phi^k] \left\{ \begin{array}{c} \dfrac{\partial u_0}{\partial x} \\ \dfrac{\partial \theta}{\partial x} \\ \dfrac{\partial \Psi}{\partial x} \end{array} \right\} = \mathbf{S}_p \hat{\varepsilon}_p \qquad (3.53a)$$

$$\gamma_{xz}^k = \gamma + \beta^k \Psi = [1, \beta^k] \left\{ \begin{array}{c} \gamma \\ \Psi \end{array} \right\} = \mathbf{S}_t^k \hat{\varepsilon}_t \qquad (3.53b)$$

with

$$\begin{array}{ll} \mathbf{S}_p = [1, -z, \phi^k] & , \hat{\varepsilon}_p = \left[\dfrac{\partial u_0}{\partial x}, \dfrac{\partial \theta}{\partial x}, \dfrac{\partial \Psi}{\partial x} \right]^T \\ \mathbf{S}_t^k = [1, \beta^k] & , \quad \hat{\varepsilon}_t = [\gamma, \Psi]^T \end{array} \qquad (3.53c)$$

where $\hat{\varepsilon}_p$ and $\hat{\varepsilon}_t$ are the generalized in-plane (axial-bending) and transverse shear strain vectors, respectively.

In Eq.(3.53b), $\gamma = \frac{\partial w_0}{\partial x} - \theta$. Integrating Eq.(3.53b) over the cross section and using Eq.(3.52b) and the fact that Ψ is independent of z yields

$$\gamma = \frac{1}{A} \iint_A \gamma_{xz}^k dA \qquad (3.54)$$

i.e. γ represents the average transverse shear strain of the cross section. Hooke stress-strain relations for the kth orthotropic layer have the standard form (Eqs.(3.7))

$$\sigma_x^k = E^k \varepsilon_x^k = E^k \mathbf{S}_p^k \hat{\varepsilon}_p \qquad (3.55a)$$

$$\tau_{xz}^k = G^k \gamma_{xz}^k = G^k \mathbf{S}_t^k \hat{\varepsilon}_t \qquad (3.55b)$$

where E^k and G^k are the axial and shear moduli for the kth layer, respectively.

Note that in the above equations we have distinguished all variables within a layer with superscript k.

3.17.3 Computation of the zigzag function

The shear strain-shear stress relationship of Eq.(3.53b) is written as

$$\tau_{xz}^k = G^k \eta + G^k(1 + \beta^k)\Psi \tag{3.56}$$

where $\eta = \gamma - \Psi$ is a difference function.

Clearly the distribution of τ_{xz}^k within each layer is constant, as η is independent of the zigzag function and β^k is constant (Eq.(3.52a)).

The distribution of τ_{xz}^k is enforced to be independent of the zigzag function. This can be achieved by constraining the term multiplying Ψ in Eq.(3.56) to be constant, i.e.

$$G^k(1 + \beta^k) = G^{k+1}(1 + \beta^{k+1}) = G, \quad \text{constant} \tag{3.57}$$

This is equivalent to enforcing the interfacial continuity of the second term in the r.h.s. of Eq.(3.56).

From Eq.(3.57) we deduce

$$\beta^k = \frac{G}{G^k} - 1 \tag{3.58}$$

Substituting β^k in the integral of Eq.(3.52b) gives

$$G = \left[\frac{1}{A}\iint_A \frac{dA}{G^k}\right]^{-1} = \left[h\sum_{k=1}^{n_l}\frac{h^k}{G^k}\right]^{-1} \tag{3.59}$$

which is the *equivalent shear modulus* for the laminate.

Substituting Eq.(3.52a) into Eq.(3.52b) gives the following recursion relation for the zigzag function values at the layer interfaces

$$\bar{\phi}_k = \sum_{i=1}^{k} h^i \beta^i \quad \text{with} \quad u^0 = u^{n_l} = 0 \tag{3.60}$$

with β^i given by Eq.(3.58).

Introducing Eq.(3.60) into (3.51) gives the expression for the zigzag function as

$$\phi^k = \frac{h^k \beta^k}{2}(\zeta^k - 1) + \sum_{i=1}^{k} h^i \beta^i \tag{3.61}$$

We recall that this zigzag theory does not enforce the continuity of the transverse shear stresses across the section. This is consistent with the kinematic freedom inherent in the lower order kinematic approximation of the underlying beam theory.

For homogeneous material $G^k = G$ and $\beta^k = 0$. Hence, the zigzag function ϕ^k vanishes and we recover the kinematic and constitutive expressions of the standard Timoshenko composite laminated beam theory.

We finally note that function Ψ can be interpreted as a weighted-average shear strain angle [TDG]. The value of Ψ should be prescribed to zero at a clamped edge and left unprescribed at a free edge.

3.17.4 Generalized constitutive relationship

The resultant stresses are defined as

$$\hat{\sigma}_p = \begin{Bmatrix} N \\ M \\ M_\phi \end{Bmatrix} = \iint_A [\mathbf{S}_p^k]^T \sigma_x^k dA = \left(\iint_A [\mathbf{S}_p^k]^T \mathbf{S}_p^k E^k dA \right) \hat{\varepsilon}_p = \hat{\mathbf{D}}_p \hat{\varepsilon}_p \quad (3.62)$$

$$\hat{\sigma}_t = \begin{Bmatrix} Q \\ Q_\phi \end{Bmatrix} = \iint_A [\mathbf{S}_t^k]^T \tau_{xz}^k dA = \left(\iint_A [\mathbf{S}_t^k]^T \mathbf{S}_t^k G^k dA \right) \hat{\varepsilon}_t = \hat{\mathbf{D}}_t \hat{\varepsilon}_t \quad (3.63)$$

In vectors $\hat{\sigma}_p$ and $\hat{\sigma}_t$, N, M and Q are respectively the axial force, the bending moment and the shear force of standard beam theory, whereas M_ϕ and Q_ϕ are an additional bending moment and an additional shear force which are conjugate to the new generalized strains $\frac{\partial \Psi}{\partial x}$ and Ψ, respectively.

The generalized constitutive matrices $\hat{\mathbf{D}}_b$ and $\hat{\mathbf{D}}_t$ are

$$\hat{\mathbf{D}}_p = \iint_A E^k \begin{bmatrix} 1 & -z & \phi^k \\ -z & z^2 & -z\phi^k \\ \phi^k & -z\phi^k & (\phi^k)^2 \end{bmatrix} dA \quad , \quad \hat{\mathbf{D}}_t = \begin{bmatrix} D_s & -\delta \\ -\delta & \delta \end{bmatrix} \quad (3.64a)$$

with

$$D_s = \iint_A G^k dA \quad , \quad \delta = D_s - GA \quad (3.64b)$$

In the derivation of the expression for $\hat{\mathbf{D}}_t$ we have used the definition of β^k of Eq.(3.58).

The generalized constitutive equation can be written as

$$\hat{\sigma} = \begin{Bmatrix} \hat{\sigma}_p \\ \hat{\sigma}_t \end{Bmatrix} = \hat{\mathbf{D}}\hat{\varepsilon} = \hat{\mathbf{D}} \begin{Bmatrix} \hat{\varepsilon}_p \\ \hat{\varepsilon}_t \end{Bmatrix} \quad \text{with} \quad \hat{\mathbf{D}} = \begin{bmatrix} \hat{\mathbf{D}}_p & \mathbf{0} \\ \mathbf{0} & \hat{\mathbf{D}}_t \end{bmatrix} \quad (3.65)$$

It is interesting that this formulation does not require a shear correction parameter k_z.

3.17.5 Virtual work expression

The virtual work expression for a distributed load $f_z = q$ is

$$\iiint_V (\delta\varepsilon_x^k \sigma_x^k + \delta\gamma_{xz}^k \tau_{xz}^k) dV - \int_L \delta w q ds = 0 \qquad (3.66)$$

The l.h.s. of Eq.(3.66) contains the internal virtual work performed by the axial and tangential stresses over the beam volume V and the r.h.s. is the external virtual work carried out by the distributed load.

Substituting Eqs.(3.53a,b) into the expression for the virtual internal work and using Eqs.(3.62) and (3.63) gives

$$\iiint_V \left(\delta\varepsilon_x^k \sigma_x^k + \delta\gamma_{xz}^k \tau_{xz}^k \right) dV = \iiint_V \left(\delta\hat{\boldsymbol{\varepsilon}}_p^T [\mathbf{S}_p^k]^T \sigma_x^k + \delta\hat{\boldsymbol{\varepsilon}}_t^T [\mathbf{S}_t^k]^T \tau_{xz}^k \right) dV =$$
$$= \int_L \left(\delta\hat{\boldsymbol{\varepsilon}}_p^T \hat{\boldsymbol{\sigma}}_p + \delta\hat{\boldsymbol{\varepsilon}}_t^T \hat{\boldsymbol{\sigma}}_t \right) dx \qquad (3.67)$$

The virtual work is therefore written as

$$\int_L \left(\delta\hat{\boldsymbol{\varepsilon}}_p^T \hat{\boldsymbol{\sigma}}_p + \delta\hat{\boldsymbol{\varepsilon}}_t^T \hat{\boldsymbol{\sigma}}_t \right) dx - \int_L \delta w q dx = 0 \qquad (3.68)$$

This expression is the basis for deriving a 2-noded zigzag beam element as explained in the next section.

3.18 TWO-NODED LRZ COMPOSITE LAMINATED BEAM ELEMENT

The kinematic variables are u_0, w_0, θ and Ψ. They are discretized using 2-noded linear C° beam elements of length $l^{(e)}$ (Figure 3.10) as

$$\mathbf{u} = \begin{Bmatrix} u_0 \\ w_0 \\ \theta \\ \Psi \end{Bmatrix} = \sum_{i=1}^{2} N_i \mathbf{a}_i^{(e)} = \mathbf{N}\mathbf{a}^{(e)} \qquad (3.69)$$

with

$$\mathbf{N} = [N_1 \mathbf{I}_4, N_2 \mathbf{I}_4] \quad , \quad \mathbf{a}^{(e)} = \begin{Bmatrix} \mathbf{a}_1^{(e)} \\ \mathbf{a}_2^{(e)} \end{Bmatrix} \quad , \quad \mathbf{a}_i^{(e)} = \begin{Bmatrix} u_{0_i} \\ w_{0_i} \\ \theta_i \\ \Psi_i \end{Bmatrix} \qquad (3.70)$$

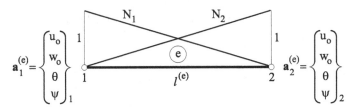

Fig. 3.10 Two-noded LRZ composite laminated beam element. Nodal variables and shape function

where N_i are the standard 1D linear shape functions (Figure 2.4), $\mathbf{a}_i^{(e)}$ is the vector of nodal DOFs and $\mathbf{I}_4$ is the 4×4 unit matrix.

Substituting Eq.(3.69) into the generalized strain vectors of Eq.(3.53c) gives

$$\hat{\varepsilon}_p = \mathbf{B}_p \mathbf{a}^{(e)} \quad , \quad \hat{\varepsilon}_t = \mathbf{B}_t \mathbf{a}^{(e)} \tag{3.71}$$

The generalized strain matrices $\mathbf{B}_p$ and $\mathbf{B}_t$ are

$$\mathbf{B}_p = [\mathbf{B}_{p_1}, \mathbf{B}_{p_2}] \quad , \quad \mathbf{B}_t = [\mathbf{B}_{t_1}, \mathbf{B}_{t_2}] \tag{3.72a}$$

with

$$\mathbf{B}_{p_i} = \begin{bmatrix} \dfrac{\partial N_i}{\partial x} & 0 & 0 & 0 \\[2mm] 0 & 0 & \dfrac{\partial N_i}{\partial x} & 0 \\[2mm] 0 & 0 & 0 & \dfrac{\partial N_i}{\partial x} \end{bmatrix} \quad , \quad \mathbf{B}_{t_i} = \left[\begin{array}{ccc|c} 0 & \dfrac{\partial N_i}{\partial x} & -N_i & 0 \\[2mm] \hline 0 & 0 & 0 & N_i \end{array} \right] = \begin{bmatrix} \mathbf{B}_{s_i} \\ \hline \mathbf{B}_{\psi_i} \end{bmatrix}$$

$$\tag{3.72b}$$

where $\mathbf{B}_{p_i}$ and $\mathbf{B}_{t_i}$ are the in-plane and transverse shear strain matrices for node i.

The virtual displacement and the generalized strain fields are expressed in terms of the virtual nodal DOFs as

$$\delta \mathbf{u} = \mathbf{N} \delta \mathbf{a}^{(e)} \quad , \quad \delta \hat{\varepsilon}_p = \mathbf{B}_p \delta \mathbf{a}^{(e)} \quad , \quad \delta \hat{\varepsilon}_t = \mathbf{B}_t \delta \mathbf{a}^{(e)} \tag{3.73}$$

The discretized equilibrium equations are obtained by substituting Eqs.(3.62), (3.63), (3.69), (3.71) and (3.73) into the virtual work expression (3.68). After simplification of the virtual nodal DOFs, the following standard matrix equation is obtained

$$\mathbf{K} \mathbf{a} = \mathbf{f} \tag{3.74}$$

where $\mathbf{a}$ is the vector of nodal DOFs for the whole mesh.

The stiffness matrix $\mathbf{K}$ and the equivalent nodal force vector $\mathbf{f}$ are obtained by assembling the element contributions $\mathbf{K}^{(e)}$ and $\mathbf{f}^{(e)}$ given by

$$\mathbf{K}^{(e)} = \mathbf{K}_p^{(e)} + \mathbf{K}_t^{(e)} \tag{3.75}$$

with

$$\mathbf{K}_{p_{ij}}^{(e)} = \int_{l^{(e)}} \mathbf{B}_{p_i}^T \hat{\mathbf{D}}_p \mathbf{B}_{p_j} dx \quad , \quad \mathbf{K}_{t_{ij}}^{(e)} = \int_{l^{(e)}} \mathbf{B}_{t_i}^T \hat{\mathbf{D}}_t \mathbf{B}_{t_j} dx \tag{3.76}$$

and

$$\mathbf{f}^{(e)} = \int_{l^{(e)}} N_i q [1, 0, 0, 0]^T dx \tag{3.77}$$

Matrix $\mathbf{K}_p^{(e)}$ is integrated with a one-point numerical quadrature which is exact in this case. Full integration of matrix $\mathbf{K}_t^{(e)}$ requires a two-point Gauss quadrature (Appendix B). This however leads to shear locking for slender composite laminated beams.

Shear locking can be eliminated by reduced integration of all (or some) of the terms of $\mathbf{K}_t^{(e)}$. For this purpose we split this matrix as

$$\mathbf{K}_t^{(e)} = \mathbf{K}_s^{(e)} + \mathbf{K}_\psi^{(e)} + \mathbf{K}_{s\psi}^{(e)} + [\mathbf{K}_{s\psi}^{(e)}]^T \tag{3.78a}$$

with

$$\mathbf{K}_{s_{ij}}^{(e)} = \int_{l^{(e)}} D_s \mathbf{B}_{s_i}^T \mathbf{B}_{s_j} dx \quad , \quad \mathbf{K}_{\psi_{ij}}^{(e)} = \int_{l^{(e)}} \delta \mathbf{B}_{\psi_i}^T \mathbf{B}_{\psi_j} dx \tag{3.78b}$$

$$\mathbf{K}_{s\psi_{ij}}^{(e)} = \int_{l^{(e)}} (-\delta) \mathbf{B}_{s_i}^T \mathbf{B}_{\psi_j} dx$$

where $\mathbf{B}_{s_i}$ and $\mathbf{B}_{\psi_i}$ are defined in Eq.(3.72b) and D_s and δ are given in Eq.(3.64b).

This beam element is termed LRZ (for **L**inear Timoshenko **Z**igzag element).

A study of the accuracy of the LRZ beam element for analysis of laminated beams of different slenderness using one and two-point quadratures for integrating $\mathbf{K}_s^{(e)}$, $\mathbf{K}_\psi^{(e)}$ and $\mathbf{K}_{s\psi}^{(e)}$ is presented in the next section.

3.19 STUDY OF SHEAR LOCKING AND CONVERGENCE FOR THE LRZ COMPOSITE LAMINATED BEAM ELEMENT

3.19.1 Shear locking in the LRZ beam element

We study the performance of the LRZ beam element for the analysis of a cantilever beam of length L under an end point load of value $F = 1$ (Figu-

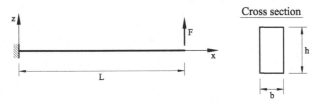

Fig. 3.11 Cantilever beam with rectangular section under point load

Composite material properties			
	Layer 1 (bottom)	Layer 2 (core)	Layer 3 (top)
h [mm]	6.6667	6.6667	6.6667
E [MPa]	2.19E5	2.19E3	2.19E5
G [MPa]	0.876E5	8.80E2	0.876E5

Table 3.4 Symmetric 3-layered cantilever beam. Material properties

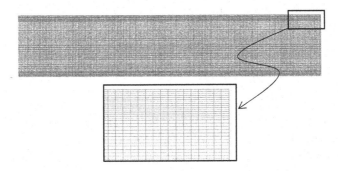

Fig. 3.12 Mesh of 27000 4-noded plane stress rectangular elements for analysis of cantilever and simple supported beams

re 3.11). The beam has a rectangular section ($b \times h$) formed by a symmetric three-layered material whose properties are listed in Table 3.4. The analysis is performed for four span-to-thickness ratios: $\lambda = 5, 10, 50, 100$ ($\lambda = L/h$) using a mesh of 100 LRZ beam elements.

The same beam was analized using a mesh of 27000 four-noded plane stress rectangles [On4] for comparison purposes (Figure 3.12).

Figure 3.13 shows the ratio r between the end node deflection obtained with the LRZ element (w_{zz}) and with the plane stress quadrilateral (w_{ps}) (i.e. $r = \frac{w_{zz}}{w_{ps}}$) versus the beam span-to-thickness ratio. Results for the LRZ element have been obtained using *exact* two-point integration for all terms of matrix $\mathbf{K}_t^{(e)}$ (Eq.(3.76)) and a one-point *reduced* integration for

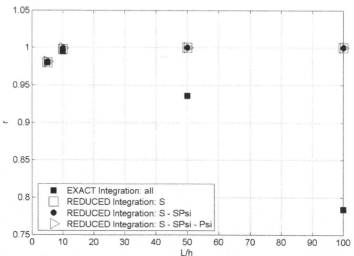

Fig. 3.13 r ratio $\left(r = \frac{w_{zz}}{w_{ps}}\right)$ versus L/h for cantilever beam under point load analyzed with the LRZ element. Labels *"all"*, S, $SPsi$ and Psi refer to matrices $\mathbf{K}_t^{(e)}$, $\mathbf{K}_s^{(e)}$, $\mathbf{K}_{s\psi}^{(e)}$ and $\mathbf{K}_\psi^{(e)}$, respectively

the following three groups of matrices: $\mathbf{K}_s^{(e)}$; $\mathbf{K}_s^{(e)}$ and $\mathbf{K}_{s\psi}^{(e)}$; and $\mathbf{K}_s^{(e)}$, $\mathbf{K}_{s\psi}^{(e)}$ and $\mathbf{K}_\psi^{(e)}$ (Eqs.(3.78b)).

Results in Figure 3.13 show that the exact integration of $\mathbf{K}_t^{(e)}$ leads to shear locking as expected. Good (locking-free) results are obtained by one-point reduced integration of the three groups of matrices considered.

The influence of reduced integration in the distribution of the transverse shear stress was studied in [OEO,OEO2] for the three groups of matrices. The conclusion is that for small values of λ the reduced or exact reduced integration of matrix $\mathbf{K}_t^{(e)}$ leads to similar results.

It is also recommended using a reduced one-point integration for matrices $\mathbf{K}_s^{(e)}$ and $\mathbf{K}_{s\psi}^{(e)}$, while matrix $\mathbf{K}_\psi^{(e)}$ should be integrated with a 2-point quadrature [OEO2].

3.19.2 Convergence of the LRZ beam element

The same three-layered cantilever beam of Figure 3.11 was studied in [OEO] for three different set of thickness and material properties for the three layers as listed in Table 3.5. Material A is the more homogeneous one, while material C is clearly the more heterogeneous.

The problem was studied with six meshes of LRZ elements ranging from 5 to 300 elements. Table 3.6 shows the convergence with the number of

	Material properties		
	Layer 1(bottom)	**Layer 2** (core)	**Layer 3** (top)
Composite A h [mm]	6.66	6.66	6.66
E [MPa]	4.40E5	2.19E4	2.19E5
G [MPa]	2.00E5	8.80E3	8.76E4
Composite B h [mm]	6.66	6.66	6.66
E [MPa]	2.19E5	2.19E3	2.19E5
G [MPa]	8.76E4	8.80E2	8.76E4
Composite C h [mm]	2	16	2
E [MPa]	7.30E5	7.30E2	2.19E5
G [MPa]	2.92E5	2.20E2	8.76E4

Table 3.5 Non symmetric 3-layered cantilever beams. Material properties

elements for the deflection and function Ψ at the beam end, the maximum axial stress σ_x at the end section and the maximum shear stress τ_{xz} at the mid section.

Convergence is measured by the relative error defined as

$$e_r = \left| \frac{v_6 - v_i}{v_6} \right| \tag{3.79}$$

where v_6 and v_i are the magnitudes of interest obtained using the finest grid (300 elements) and the ith mesh ($i = 1, 2, \cdots 5$), respectively.

Results clearly show that convergence is always slower for the heterogeneous material case, as expected.

For a mesh of 25 elements the errors for all the magnitudes considered are less than 1% for materials A and B. For material C the maximum error does not exceed 5% (Table 3.6). For the 50 element mesh errors of the order of 1% or less were obtained in all cases.

Results for a 10 element mesh are good for material A (errors $< 0.4\%$), relatively good for material B (errors $< 5\%$) and unacceptable for material C (errors ranging from around 8% to 20%) (Table 3.6).

3.20 EXAMPLES OF APPLICATION OF THE LRZ COMPOSITE LAMINATED BEAM ELEMENT

3.20.1 Three-layered laminated thick cantilever beam under end point load

We present results for a laminated thick cantilever rectangular beam under an end point load. The material properties are those of Composite C in

	(a)

	(b)

$e_r\% - w$ at $x = L$				$e_r\% - \Psi$ at $x = L$		
Number of	Composite			Number of	Composite	
elements	A	B	C	elements	A	B	C
5	1.800	9.588	42.289	5	0.040	8.563	36.113
10	0.506	2.901	19.277	10	0.003	1.814	8.042
25	0.0860	0.499	4.913	25	0.000	0.259	0.328
50	0.0191	0.123	1.406	50	0.000	0.063	0.033
100	0.0048	0.031	0.339	100	0.000	0.016	0.007
300	0.0000	0.000	0.000	300	0.000	0.000	0.000

	(c)

	(d)

$e_r\% - (\sigma_x)_{\max}$ at $x = L$				$e_r\% - (\tau_{xz})_{\max}$ at $\frac{L}{2}$		
Number of	Composite			Number of	Composite	
elements	A	B	C	elements	A	B	C
5	0.568	6.923	18.239	5	7.020	19.283	50.938
10	0.076	2.704	12.437	10	0.352	5.176	20.602
25	0.013	0.568	4.266	25	0.052	0.888	3.408
50	0.003	0.131	1.095	50	0.010	0.210	0.707
100	0.001	0.029	0.250	100	0.003	0.049	0.147
300	0.000	0.000	0.000	300	0.000	0.000	0.000

Table 3.6 Non symmetric 3-layered cantilever thick beams under end point load ($\lambda = 5$). Convergence study. (a) Relative error e_r for the maximum value of σ_x at $x = L$ and (b) idem for τ_{xz} at $x = L/2$

Table 3.5. The span-to-thickness ratio is $\lambda = 5$.

For the laminated sandwich considered the core is eight times thicker than the face sheets. In addition, the core is three orders of magnitude more compliant than the bottom face sheet. Moreover, the top face sheet has the same thickness as the bottom face sheet, but is about three times stiffer. This laminate does not possess material symmetry with respect to the mid-depth reference axis. The high heterogeneity of this stacking sequence is very challenging for the beam theories considered herein to model adequately.

The legend caption PS denotes the *reference* solution obtained with the structured mesh of 27000 four-noded plane stress quadrilaterals shown in Figure 3.12. TBT denotes the solution obtained with a mesh of 300 2-noded beam elements based on standard laminated Timoshenko beam theory. LRZ-300, LRZ-50, LRZ-25, LRZ-10 refer to the solution obtained with the LRZ beam element using meshes of 300, 50, 25 and 10 elements, respectively.

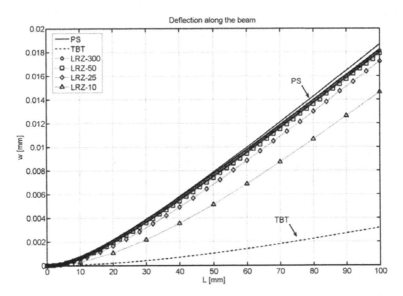

Fig. 3.14 Non symmetric 3-layered cantilever thick beam under end point load ($\lambda = 5$). Distribution of the vertical deflection w for different theories and meshes

Figure 3.14 shows the deflection values along the beam length. Very good agreement with the plane stress solution is obtained already for the LRZ-50 mesh as expected from the conclusions of the previous section.

TBT results are considerable stiffer. The difference with the reference solution is about *six times stiffer* for the end deflection value.

Figure 3.15 shows the thickness distribution of the axial displacements at two beams sections. Excellent results are again obtained with the 50 element mesh. The TBT results are far from the correct ones.

Figure 3.16 shows the distribution along the beam length of the axial stress σ_x at the bottom surface of the beam cross section. Very good agreement between the reference PS solution and the LRZ-50 and LRZ-300 results is obtained. Results for the LRZ-25 mesh compare reasonably well with the PS solution except in the vicinity of the clamped edge. The TBT results yield a linear distribution of the axial stress along the beam, as expected. This introduces large errors in the axial stress values in the vicinity of the clamped edge (Figure 3.16).

Figure 3.17 shows the thickness distribution of the axial stress (σ_x) at the clamped end. The accuracy of the LRZ results is again remarkable.

LRZ and TBT results for the distribution of the (constant) tangential shear stress τ_{xz} for the bottom layer (layer 1) along the beam length are

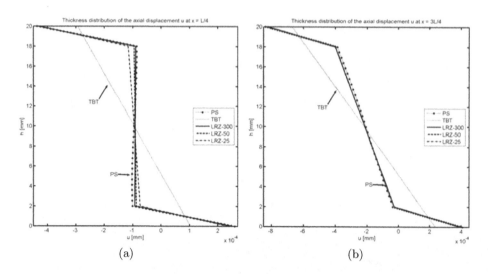

Fig. 3.15 Non symmetric 3-layered cantilever thick beam under end point load ($\lambda = 5$). Thickness distribution of the axial displacement u at $x = L/4$ (a) and $x = 3L/4$ (b)

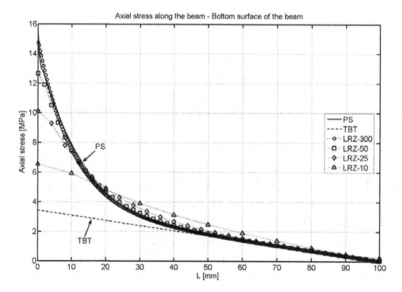

Fig. 3.16 Non symmetric 3-layered cantilever thick beam under end point load ($\lambda = 5$). Axial stress σ_x at the bottom surface of the cross section along the beam length

shown in Figure 3.18. TBT results are clearly inaccurate (except for the value at the clamped edge).

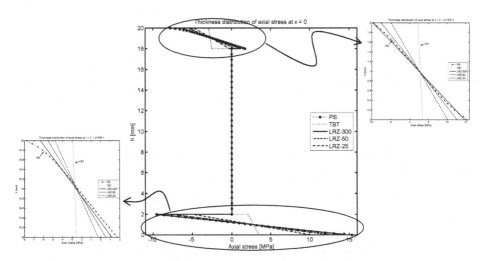

Fig. 3.17 Non symmetric 3-layered cantilever thick beam under end point load ($\lambda = 5$). Thickness distribution of the axial stress σ_x at the clamped end ($x = 0$)

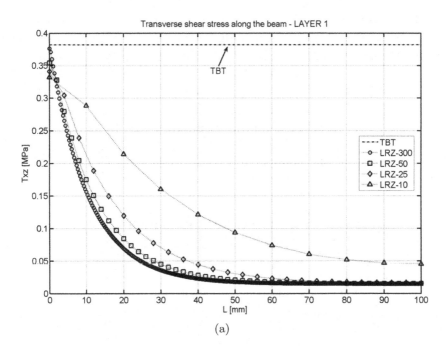

(a)

Fig. 3.18 Non symmetric 3-layered cantilever thick beam under end point load ($\lambda = 5$). Transverse shear stress τ_{xz} along the beam for the bottom layer (layer 1). Similar results are obtained for layers 2 and 3 are similar

Figure 3.19 shows the thickness distribution for the transverse shear stress τ_{xz} at two sections ($\frac{L}{20}$ and $\frac{L}{2}$). LRZ results provide an accurate estimate of the average transverse shear stress value for each layer. The distribution of τ_{xz} across the thickness can be substantially improved by using the equilibrium equations for computing τ_{xz} "a posteriori" as explained in the next section.

3.20.2 Non-symmetric three-layered simple supported thick beam under uniform load

The next example is the analysis of a three-layered simple supported thick rectangular beam under a uniformly distributed load of unit value ($q = 1$). The material properties and the thickness for the three layers are shown in Table 3.7. *The material has a non symmetric distribution* with respect to the beam axis. An unusually low value for the shear modulus of the core layer has been taken, thus reproducing the effect of a damaged material in this zone. The span-to-thickness ratio is $\lambda = 5$. Results obtained with the LRZ element are again compared to those obtained with a mesh of 300 2-noded TBT elements and with the mesh of 4-noded plane stress (PS) rectangles of Figure 3.12. The PS solution has been obtained by fixing the vertical displacement of all nodes at the end sections and the horizontal displacement of the mid-line node at $x = 0$ and $x = L$ to a zero value. This way of approximating a simple support condition leads to some discrepancies between the PS results and those obtained with beam theory.

No advantage of the symmetry of the problem for the discretization has been taken.

	Thickness and material properties		
	Layer 1 (bottom)	Layer 2 (core)	Layer 3 (top)
h [mm]	6.6666	6.6666	6.6666
E [MPa]	$2.19E^5$	$5.30E^5$	$7.30E^5$
G [MPa]	$8.76E^4$	$2.90E^2$	$2.92E^5$

Table 3.7 Thickness and material properties for 3-layered non-symmetric SS rectangular thick beam

Figure 3.20 shows the distribution of the vertical deflection for the different methods. The error in the "best" maximum central deflection value versus the "exact" PS solution is $\simeq 12\%$. The discrepancy is due

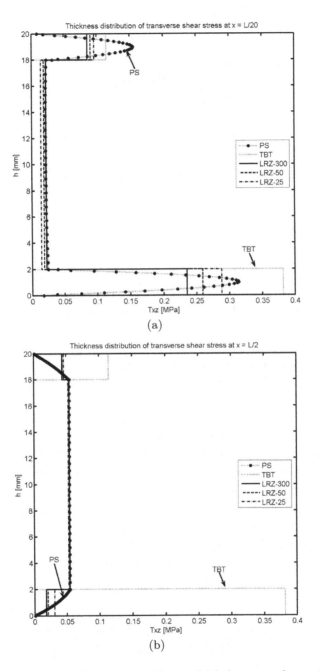

Fig. 3.19 Non symmetric 3-layered cantilever thick beam under end point load ($\lambda = 5$). Thickness distribution of τ_{xz} at $L/20$ (a) and $L/2$ (b)

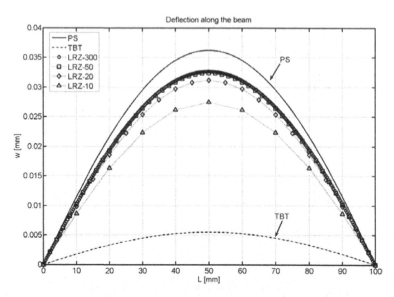

Fig. 3.20 Non symmetric 3-layered SS thick beam under uniformly distributed load ($\lambda = 5$). Distribution of vertical deflection w along the beam length

to the difference in the way the simple support condition is modelled in beam and PS theories, as well as to the limitations of beam theory to model accurately very thick beams. TBT results are quite inaccurate, as expected [OEO,OEO2].

Figure 3.21 shows the thickness distribution for the axial stress σ_x at two beam sections ($x = 0$ and $L/2$). The accuracy of the LRZ results is remarkable with a maximum error of 10% despite of the modeling limitations mentioned above. TBT results are incorrect.

Figure 3.22 shows the thickness distribution of the shear stress at the section close to the support ($x = L/20$).

LRZ results can be much improved by computing τ_{xz} "a posteriori" from the axial stress field using the equilibrium equation (Appendix B)

$$\frac{\partial \sigma_x}{\partial x} + \frac{\partial \tau_{xz}}{\partial z} = 0 \qquad (3.80)$$

The transverse shear stress at a point across the thickness with coordinate z is computed by integrating Eq.(3.80) as

$$\tau_{xz}(z) = -\int_{h^-}^{z} \frac{\partial \sigma_x}{\partial x} dz = -\frac{\partial N_z}{\partial x} \quad \text{where} \quad N_z = \int_{h^-}^{z} \sigma_x dz \qquad (3.81)$$

In Eq.(3.81) N_z is the axial force (per unit width) resulting from the thickness integration of σ_x between the coordinates h^- and z.

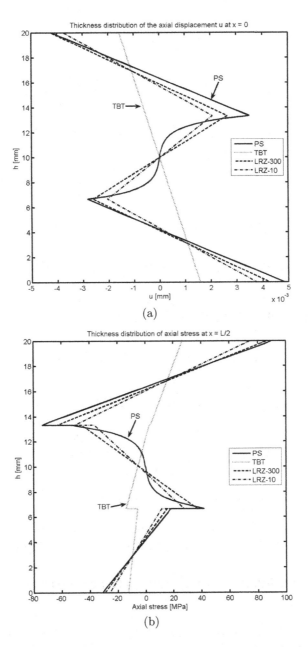

Fig. 3.21 Non symmetric 3-layered SS thick beam under uniformly distributed load ($\lambda = 5$). (a) Thickness distribution of axial displacement at $x = 0$. (b) Thickness distribution of axial stress σ_x at $x = L/2$

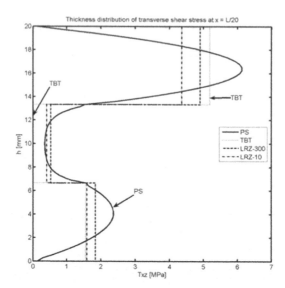

Fig. 3.22 Non symmetric 3-layered SS thick beam under uniformly distributed load ($\lambda = 5$). Thickness distribution of τ_{xz} at $x = L/20$

The space derivative of N_z in Eq.(3.80) is computed *at a node i* as

$$\frac{\partial N_z}{\partial x} = \frac{2}{l^e + l^{e-1}}(N_z^e - N_z^{e-1}) \tag{3.82}$$

where (l^e, N_z^e) and (l^{e-1}, N_z^{e-1}) are the element length and the value of N_z at elements e and $e - 1$ adjacent to node i, respectively. A value of $\tau_{xz}(h^-) = 0$ is taken. It is remarkable that the method yields automatically $\tau_{xz}(h^+) \simeq 0$.

Results for τ_{xz} obtained with this procedure are termed LRZ-10-N_z and LRZ-300-N_z in Figure 3.25. We note the accuracy of the "recovered" thickness distribution for τ_{xz}, even for the coarse mesh of 10 LRZ elements.

3.20.3 Non-symmetric ten-layered clamped slender beam under uniformly distributed loading

We present results for a ten-layered clamped slender rectangular beam ($L = 100$ mm, $h = 5$ mm, $b = 1$ mm, $\lambda = 20$) under uniformly distributed loading ($q = 1$ KN/mm). The composite material has the non-symmetric distribution across the thickness shown in Table 3.8.

Figure 3.23 shows results for the deflection along the beam for LRZ meshes with 10 and 300 elements (LRZ-10 and LRZ-300). Results obtained

(a)		
Layer	h_i	Material
1	0.5	IV
2	0.6	I
3	0.5	V
4	0.4	III
5	0.7	IV
6	0.1	III
7	0.4	II
8	0.5	V
9	0.3	I
10	1	II

(b)		
Material	E [MPa]	G [MPa]
I	2.19e5	0.876e5
II	7.3e5	2.92e5
III	0.0073e5	0.0029e5
IV	5.3e5	2.12e5
V	0.82e5	0.328e5

Table 3.8 10-layered clamped slender rectangular beam under uniformly distributed loading. (a) Thickness and material number for each of the 10 layers. (b) Properties of each material

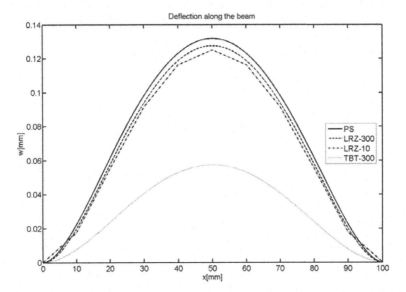

Fig. 3.23 10-layered clamped slender beam under uniform loading. Distribution of the deflection along the beam

with a mesh of 27.000 4-noded plane stress quadrilaterals and with a mesh of 300 TBT elements are also shown for comparison. Note the accuracy of the coarse LRZ-10 mesh and the erroneous results of the TBT solution.

Figure 3.24 shows the thickness distribution of the axial displacement and the axial stress (σ_x) for the section at $x = \frac{L}{4}$. The accuracy of the LRZ results is once more noticeable.

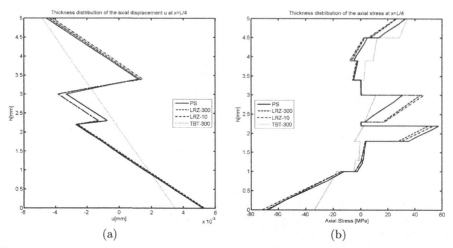

(a)   (b)

Fig. 3.24 10-layered clamped slender beam under uniform loading. Thickness distribution of axial displacement (a) and axial stress σ_x (b) for $x = \frac{L}{4}$

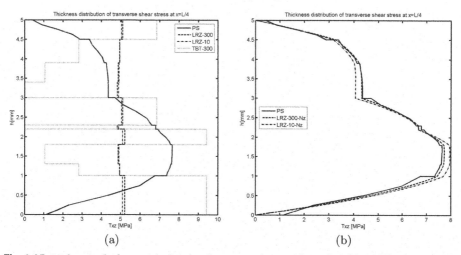

(a)   (b)

Fig. 3.25 10-layered clamped slender beam under uniform loading. Thickness distribution of τ_{xz} at $x = \frac{L}{4}$. (a) Comparison of LRZ-10 and LRZ-300 results with plane stress (PS) and TBT solutions. (b) PS solution and LRZ-10-N_z and LRZ-300-N_z results for τ_{xz} obtained by thickness integration of the equilibrium equation using the LRZ-10 and LRZ-300 results (Eq.(3.81))

Figure 3.25 shows the thickness distribution of the transverse shear stress at $x = \frac{L}{4}$. Results in Figure 3.25a show the values directly obtained with the LRZ-10 and LRZ-300 meshes. These results are clearly better than those obtained with the TBT element but only coincide in an average sense with the plane stress FEM solution.

3.20.4 Modeling of delamination with the LRZ element

Prediction of delamination in composite laminated beams is a challenge for all beam models. A method for predicting delamination in beams using a Hermitian zigzag theory was presented in [DG2,3]. A sub-laminate approach is used for which the number of kinematic unknowns depends of the number of physical layers. This increases the number of variables but it yields the correct an accurate transverse shear stress distribution without integrating the equilibrium equations.

Delamination effects in composite laminated beams can be effectively reproduced with the LRZ element *without introducing additional kinematic variables*. The delamination model simply implies *introducing a very thin "interface layer" between adjacent material layers* in the actual composite laminated section. Delamination is produced when the material properties of the interface layer are drastically reduced to almost a zero value in comparison with those of the adjacent layers due to interlamina failure. This simple delamination model allows the LRZ element to take into account the reduction of the overall beam stiffness due to the failure of the interface layer leading to an increase in the deflection and rotation field. Moreover, the LRZ element can also accurately represent the jump in the axial displacement field across the interface layer and the change in the axial and tangential stress distributions over the beam sections as delamination progresses.

Figures 3.26–3.30 show an example of the capabilities of the LRZ beam element to model delamination. The problem represents the analysis of a cantilever thick rectangular beam ($\lambda = 5$) under an end point load. The beam section has three layers of composite material with properties shown in Table 3.9. Delamination between the upper and core layers has been modelled by introducing a very thin interface layer ($h = 0.01$ mm) between these two layers (Figure 3.26). The initial properties of the interface layer coincide with those of the upper layer. Next, the shear modulus value for the interface layer has been progressively reduced up to 11 orders of magnitude from $G2 = 8.76 \times 10^4$ MPa (Model 1) to $G2 = 8.76 \times 10^{-7}$ MPa (Model 12) (Table 3.10).

We note that the reduction of the shear modulus has been applied over the whole beam length in this case. However it can applied in selected beam regions as appropriate.

Figure 3.27 shows results for the end deflection value in terms of the shear modulus of the interface layer for the LRZ-100 mesh. Note that the deflection increases one order of magnitude versus the non-delaminated

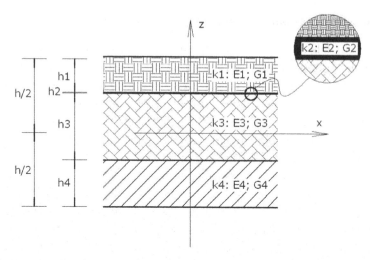

Fig. 3.26 Modeling of interface layer for delamination study in 3-layered thick cantilever beam ($\lambda = 5$) under end point load

	Composite material			
	Layer 1	Layer 2	Layer 3	Layer 4
h [mm]	2	0.01	16	2
E [MPa]	2.19E5	2.19E5	0.0073E5	7.30E5
G [MPa]	0.876E5	G2	0.0029E5	2.92E5

Table 3.9 Thickness and layer properties for delamination study in a 3-layered cantilever beam under end point load. Layer 2 is the interface layer. G2 values are given in Table 3.10

Model	G2	Model	G2	Model	G2
1	8.76E+004	5	8.76E+000	9	8.76E-004
2	8.76E+003	6	8.76E-001	10	8.76E-005
3	8.76E+002	7	8.76E-002	11	8.76E-006
4	8.76E+001	8	8.76E-003	12	8.76E-007

Table 3.10 Shear modulus values for the interface layer for delamination study in a 3-layered cantilever beam. Values of G2 in MPa

case. It is also interesting that the end deflection does not change much after the shear modulus of the interface layer is reduced beyond eight orders of magnitude (results for Model 9 in Figure 3.27). Results agree reasonably well (error $\simeq 10\%$) with those obtained with the plane stress model of Figure 3.12 introducing a similar reduction in the shear modulus of an *ad hoc* interface layer.

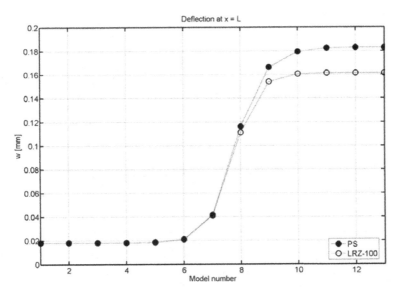

Fig. 3.27 Delamination study in 3-layered cantilever beam under end point load. Evolution of end deflection with the shear modulus value for the interface layer LRZ-100 results and PS solution

Figure 3.28 shows the thickness distribution for the axial displacement at the mid section for four decreasing values of the shear modulus at the interface layer: $G2 = 8.76, 8.76 \times 10^{-1}, 8.76 \times 10^{-3}$ and 8.76×10^{-6} MPa. The jump of the axial displacement across the thickness at the interface layer during delamination is well captured. We again note that the displacement jump at the interface layer remains stationary after a reduction of the material properties in that layer of six orders of magnitude. Results agree well with the plane stress solution also shown in the figure.

Figure 3.29 shows the thickness distribution of the axial stress (σ_x) for the same four decreasing values of G_2 in the interface layer. The effect of delamination in the stress distribution is clearly visible. Once again the LRZ-100 results agree well with the plane stress solution.

Figure 3.30 finally shows the thickness distribution for the transverse shear stress at $x = \frac{L}{2}$ for the same four values of G_2 at the interface layer. The three graphs show the PS results, the LRZ-100 results and the solution obtained by integrating the equilibrium equation (via Eqs.(3.80)–(3.83)) using the LRZ-100 results. Note the accuracy of the later solution versus the standard LRZ-100 results as delamination develops and τ_{xz} progressively vanishes at the interface layer.

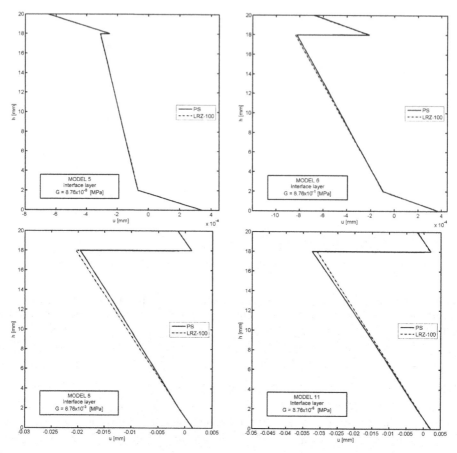

Fig. 3.28 Delamination study in 3-layered cantilever beam under end point load. Thickness distribution of axial displacement at $x = \frac{L}{2}$ for four decreasing values of the shear modulus at the interface layer (Models 5, 6, 8 and 11, Table 3.10)

Similar good results for predicting the delamination and the thickness distribution of the axial and transverse shear stresses are obtained over the entire beam length.

The example shows the capability of the LRZ element to model a complex phenomenon such as delamination in composite laminated beams *without introducing additional kinematic variables*. More evidences of the good behaviour of the LRZ beam element for predicting delamination in beams are reported in [OEO3].

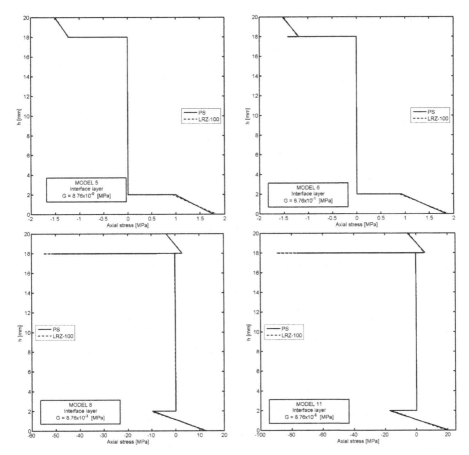

Fig. 3.29 Delamination study in 3-layered cantilever beam under end point load. Thickness distribution of σ_x at $x = \frac{L}{2}$ for four decreasing values of the shear modulus at the interface layer (Models 5, 6, 8 and 11, Table 3.10)

3.21 CONCLUDING REMARKS

The formulation of composite laminated beam elements introduces the axial elongation of the beam axis, the axial force and the corresponding axial stiffness matrix into the classical bending theory of beams. Axial and bending effects are generally coupled to each other except for a symmetric distribution of the material properties in the cross-section or, more generally, when the beam axis coincides with the neutral axis. A composite laminated material increases the importance of transverse shear deformation effects, even for slender beams. This makes Timoshenko theory more appropriate for composite laminated beams.

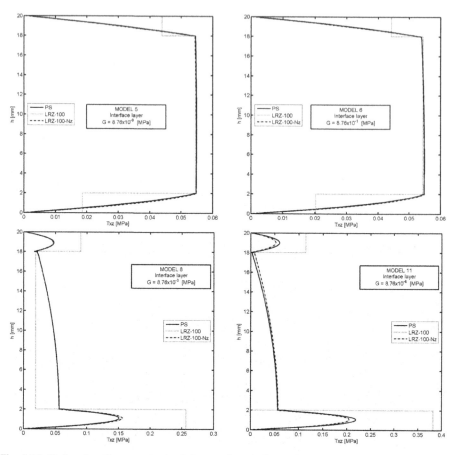

Fig. 3.30 Delamination study in 3-layered cantilever beam under end point load. Thickness distribution of τ_{xz} at $x = \frac{L}{2}$ for four values of G at the interface layer (Models 5, 6, 8 and 11, Table 3.10). LRZ-100 results, plane stress (PS) solution and LRZ-100-Nz results obtained by integrating the equilibrium equation (Eq.(3.81)) using the LRZ-100 results

Timoshenko composite laminated beam elements suffer from shear locking, although this defect can be eliminated via the techniques studied in Chapter 2. For the 2-noded beam element, the simplest procedure is the one point reduced integration of all the stiffness matrix terms. An exact 2-noded Timoshenko composite laminated beam element has been presented for the case of uncoupled axial-bending effects.

Euler-Bernoulli composite laminated beam elements share the key features of Timoshenko theory with regard to the axial-bending coupling, while they neglect transverse shear deformation effects.

Higher order beam theories, such as the refined zigzag theory described in the last part of the chapter, provide a more accurate description of the axial displacement and the transverse shear stress field across the cross-section depth. The 2-noded LRZ beam element is an excellent candidate for practical analysis of composite laminated beams. The ability of this element for reproducing delamination effect in a simple manner is remarkable.

The concepts explained in this chapter are the basis for the study of finite elements for 3D beams, plates and shells with composite material in the subsequent chapters.

4

3D COMPOSITE BEAMS

4.1 INTRODUCTION

A three-dimensional (3D) beam, also called a *rod*, is a member that carries axial, flexural (shear and bending) and torsion force resultants. Structures containing 3D beams are found in frames of buildings and industrial
constructions, arches, stiffened shells, structural parts in land transport
vehicles, fusselages of airplanes and spacecrafts, ships hulls, mechanical
parts, etc. Figure 4.1 shows schematic examples of structures formed by
an assembly of straight rods.

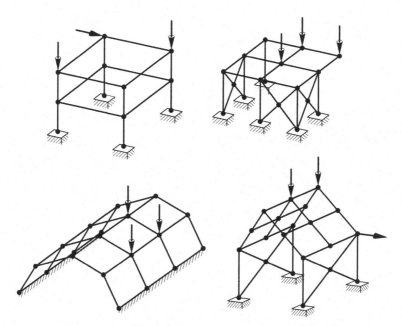

Fig. 4.1 Schematic representation of 3D beam structures with members carrying
axial, flexural and torsion effects. Points denote beam joints

E. Oñate, *Structural Analysis with the Finite Element Method. Linear Statics:*
Volume 2: Beams, Plates and Shells, Lecture Notes on Numerical Methods
in Engineering and Sciences, DOI 10.1007/978-1-4020-8743-1_4,
© International Center for Numerical Methods in Engineering (CIMNE), 2013

In this chapter we will study finite element methods for 3D beams with *arbitrary cross section and general composite material*. Many concepts are an extension of those studied for plane beams in Chapters 1–3 and several more advanced topics are introduced here. For the sake of completeness the key theoretical concepts of 3D beam theory are explained, although in a concise manner. Readers not familiar with 3D beam analysis are recommended the prior study in classical books of Strength of Materials and Structural Analysis [Li,OR,SJ,Ti2,3].

In the first part of the chapter we derive 3D composite beam elements via an extension of the Timoshenko and Euler-Bernoulli plane beam theories studied in Chapters 2 and 1, respectively. The formulation of 2-noded straight and curved 3D beam elements is described. The constitutive relationship between the three significant stresses and strains is assumed to have a diagonal form. This restricts the applicability of the formulation to a specific (but wide) range of composite materials. The free torsion (Saint Venant theory) is studied first. This theory assumes that the strains and stresses induced by the warping of the section are zero. This assumption is exact when the internal torque is constant along the beam length and warping is not restricted at any point. Saint Venant theory is also a good approximation when the torque is not uniform in beams with solid sections (rectangular, square, circular, etc.), in sections formed by thin rectangular members meeting at a point (angular section, T-type section, etc.) and in hollow cellular sections (tubes, box-type sections with width/length ≤ 4, etc.). A particularization of the 2-noded 3D beam element based on Saint-Venant theory to plane grillages is explained.

For other types of sections subjected to a non-uniform torque or to a constrained uniform torsion, warping strains and stresses must be taken into account [MB,OR,Vl]. A torsion theory adequate for thin-walled open beams exhibiting strong warping effects is presented in some detail. A refined version of this theory accounting for the transverse shear deformation induced by torsion is also briefly described.

In the last part of the chapter, we present a procedure for deriving curved 3D beam elements based on a degeneration of 3D solid elements. This formulation is applicable to composite beams and is an alternative to the traditional methods for deriving beam elements using "classical" beam theories. The extension of 3D beam elements to be used as stiffners in shells is studied in Chapter 10.

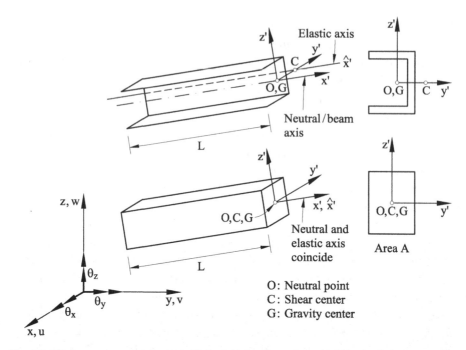

Fig. 4.2 3D homogeneous beam. Global (x, y, z) and local (x', y', z') reference systems. Neutral/beam axis and elastic axis. Global displacements and rotations

4.2 BASIC DEFINITIONS FOR A 3D COMPOSITE BEAM

4.2.1 Local and global axes

A 3D beam is a prismatic solid of length L and transverse area A, oriented in the longitudinal direction x' which dimensions in the plane $y'z'$ orthogonal to x' are relatively small with respect to the longitudinal direction. Point O defining the origin of the beam axis x' within the cross-section *will be assumed to be located at the neutral point*. Therefore, the x' axis will be called hereafter indistinguishably *beam axis* or *neutral axis*.

The beam geometry is defined in a global orthogonal coordinate system x, y, z (Figure 4.2).

In the following we will consider the beam axis to be straight and assume *constant geometrical and material properties* along x'.

The definition of the beam axis x' as the neutral axis decouples bending and axial effects (Section 3.6). Bending and torsional effects are also decoupled when the external forces along the y' and z' directions act at the *shear center* C. Point C is also called *bending center*. The straight line

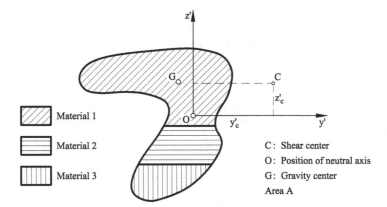

Fig. 4.3 Gravity and shear centers (G and C) and position of neutral axis (O) in a composite section

connecting points C of all sections is the *elastic axis* denoted by $\hat{x}'$. In our derivations we will assume that the neutral and elastic axes (x' and $\hat{x}'$) are *parallel*, although they may or not be coincident (Figure 4.2).

The neutral point O coincides with the gravity center of the section G for homogeneous sections (Figure 4.2 and Section 3.6).

For homogeneous solid sections, closed thin-walled sections and open thin-walled sections with double symmetry, points O, G and C coincide (Figures 4.2 and 4.6).

For composite beams, the gravity center (G), the shear center (C) and the neutral axis position (O) typically do not coincide (Figure 4.3).

4.2.2 Constitutive behaviour

The three local stresses in a section ($\sigma_{x'}, \tau_{x'y'}, \tau_{x'z'}$) (Figure 4.4) are related to the conjugate strains ($\varepsilon_{x'}, \gamma_{x'y'}, \gamma_{x'z'}$) by the constitutive relationship

$$\boldsymbol{\sigma}_{x'} = \left\{ \begin{array}{c} \sigma_{x'} \\ \tau_{x'y'} \\ \tau_{x'z'} \end{array} \right\} = \mathbf{D}' \left\{ \begin{array}{c} \varepsilon_{x'} \\ \gamma_{x'y'} \\ \gamma_{x'z'} \end{array} \right\} = \mathbf{D}'\boldsymbol{\varepsilon}' \tag{4.1}$$

The 3×3 constitutive matrix $\mathbf{D}'$ is deduced from the general 3D constitutive equation [On4] by making zero the strains that are neglected in 3D beam theory (i.e. $\varepsilon_{y'} = \varepsilon_{z'} = \gamma_{y'z'} = 0$). For a general heterogeneous material $\mathbf{D}'$ is a full matrix. For the beam theories studied in this chapter we will assume that the material is *orthotropic* with the orientation of one

Beam cross-section

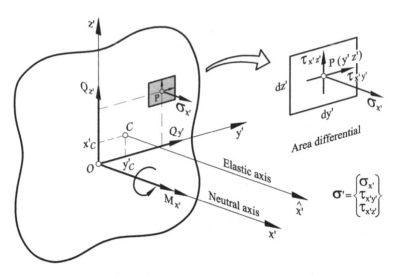

Fig. 4.4 Stresses in a 3D beam

of the principal material axes coincident with the beam axis. Under this assumption matrix $\mathbf{D}'$ has a simple (and standard) diagonal form

$$\mathbf{D}' = \begin{bmatrix} E & 0 & 0 \\ 0 & G_{y'} & 0 \\ 0 & 0 & G_{z'} \end{bmatrix} \tag{4.2}$$

where E is the longitudinal Young modulus ($E = E_{x'}$) and $G_{y'}$ and $G_{z'}$ are the transverse shear moduli ($G_{y'} = G_{x'y'}$; $G_{z'} = G_{x'z'}$). For isotropic material

$$G_{y'} = G_{z'} = G = \frac{E}{2(1+\nu)} \tag{4.3}$$

The material parameters can vary at each point of the section, providing the previous assumption holds.

The simple form of $\mathbf{D}'$ of Eq.(4.2) will allow us using many concepts of classical beam theory that are probably familiar to many readers.

Composite laminated section

For a composite laminated section (Figure 4.5a) we will assume that the constitutive parameters are constant within each layer. The values of E, $G_{y'}$ and $G_{z'}$ for the layer can be computed by rotating the constitutive equations for the layer material (assumed to be in plane stress state, i.e.

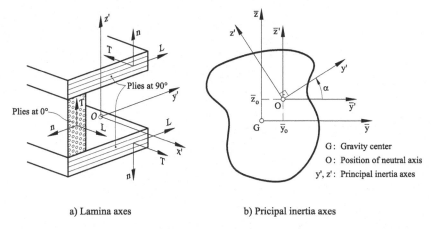

a) Lamina axes   b) Pricipal inertia axes

Fig. 4.5 (a) Lamina axes (L, T, n) in a composite laminated section. (b) Principal inertia axes (y', z') of a section

$\sigma_n = 0$) from the lamina axes (L, T, n) to the beam local axes (x', y', z') following a procedure similar as explained in Section 8.3.3. The resulting $\mathbf{D}'$ matrix is a full matrix that can be diagonalized first by eliminating $\varepsilon_{y'}$ and $\gamma_{y'z'}$, that are assumed to be zero, and then imposing that the constitutive relationships for $\sigma_{x'}$, $\tau_{x'y'}$ and $\tau_{y'z'}$ are decoupled.

An alternative procedure is to preserve the full expression of the $\mathbf{D}'$ matrix and then simplify its generalized form $(\hat{\mathbf{D}}')$ relating the resultant stresses and the generalized strains [Va,VOO].

In our derivations we will accept that $\mathbf{D}'$ is diagonal which invariably occurs when the angle between the longitudinal lamina axis L and the x' axis is 0° or 90° (Figure 4.5a). Despite this simplification, this model is applicable to a wide number of composite laminated beams.

Under the previous assumptions it is useful to define x' as the neutral axis and y' and z' as the principal axes of inertia of the section.

4.2.3 Resultant constitutive parameters and neutral axis

Let us choose an arbitrary orthogonal coordinate system $\bar{x}, \bar{y}, \bar{z}$ attached to the gravity center G with $\bar{x}$ parallel to x'. A simple translation of this system to the neutral axis O (whose position is still unknown) gives the system $x', \bar{y}', \bar{z}'$ attached to point O (Figure 4.5b). The resultant (generalized) axial and bending constitutive parameters are defined as

$$\hat{D}_a = \iint_A E \, dA \quad , \quad \hat{D}_{b_{\bar{y}'\bar{z}'}} = \iint_A E\bar{y}'\bar{z}' dA$$

$$\hat{D}_{b_{\bar{y}'}} = \iint_A E\bar{z}'^2 \, dA \quad , \quad \hat{D}_{b_{\bar{z}'}} = \iint_A E\bar{y}'^2 \, dA$$

$$\hat{D}_{ab_{\bar{y}'}} = \iint_A E\bar{z}' \, dA \quad , \quad \hat{D}_{ab_{\bar{z}'}} = \iint_A E\bar{y}' \, dA \qquad (4.4)$$

For 3D isotropic and composite beams the local axis x' of system $x', \bar{y}', \bar{z}'$ is defined as the *neutral* axis if

$$\hat{D}_{ab_{\bar{y}'}} = \iint_A E(\bar{y}', \bar{z}')\bar{z}' \, dA = 0 \qquad (4.5a)$$

$$\hat{D}_{ab_{\bar{z}'}} = \iint_A E(\bar{y}', \bar{z}')\bar{y}' \, dA = 0 \qquad (4.5b)$$

The neutral axis in the coordinate system $x', \bar{y}', \bar{z}'$ satisfies

$$\iint_A E\bar{z}' \, dA = \iint_A E(\bar{z} - \bar{z}_0) \, dA = 0$$

$$\iint_A E\bar{y}' \, dA = \iint_A E(\bar{y} - \bar{y}_0) \, dA = 0 \qquad (4.6)$$

Eqs.(4.6) give the position of the neutral point $O(\bar{y}_0, \bar{z}_0)$ in the $x', \bar{y}, \bar{z}$ system as (Figure 4.5b)

$$\bar{z}_0 = \frac{\iint_A E\bar{z} dA}{\iint_A E \, dA} = \frac{\hat{D}_{ab_{\bar{y}}}}{\hat{D}_a} \quad , \quad \bar{y}_0 = \frac{\iint_A E\bar{y} dA}{\iint_A E \, dA} = \frac{\hat{D}_{ab_{\bar{z}}}}{\hat{D}_a} \qquad (4.7)$$

The position of the neutral axis does not change if the local system $x', \bar{y}', \bar{z}'$ is rotated around x' to give the final local reference system x', y', z' where y', z' are the principal inertia axes.

4.2.4 Principal inertia axes

The principal inertia axes y', z' defining the local coordinate system x', y', z' (at the neutral point O) are obtained in terms of the resultant constitutive parameters in the $x', \bar{y}', \bar{z}'$ system (also at O) as follows.

The coordinates y', z' of an arbitrary point of the beam section are expressed in terms of the coordinates $\bar{y}', \bar{z}'$, i.e.

$$\begin{aligned} y' &= C\bar{y}' + S\bar{z}' \\ z' &= -S\bar{y}' + C\bar{z}' \end{aligned} \quad \text{with } C = \cos\alpha, S = \sin\alpha \qquad (4.8)$$

where α is the angle between the principal axis y' and the $\bar{y}'$ axis (Figure 4.5b). Using Eqs.(4.4) we obtain

$$\hat{D}_{b_{y'z'}} = \iint_A E y' z' dA = CS(\hat{D}_{b_{\bar{y}'}} - \hat{D}_{b_{\bar{z}'}}) + (C^2 - S^2)\hat{D}_{b_{\bar{y}'\bar{z}'}} \qquad (4.9)$$

Axes y' and z' are principal inertia axes if $\hat{D}_{b_{y'z'}} = 0$, i.e. if

$$\tan 2\alpha = \frac{-2\hat{D}_{b_{\bar{y}'\bar{z}'}}}{\hat{D}_{b_{\bar{y}'}} - \hat{D}_{b_{\bar{z}'}}} \qquad (4.10)$$

The principal bending constitutive parameters $\hat{D}_{b_{z'}}$ and $\hat{D}_{b_{y'}}$ are obtained in terms of the constitutive parameters in the $x', \bar{y}', \bar{z}'$ coordinate system (at point O) as

$$\hat{D}_{b_{z'}} = \iint_A E y'^2 \, dA = C^2 \hat{D}_{b_{\bar{z}'}} + 2CS\hat{D}_{b_{\bar{y}'\bar{z}'}} + S^2 \hat{D}_{b_{\bar{y}'}}$$

$$\hat{D}_{b_{y'}} = \iint_A E z'^2 \, dA = S^2 \hat{D}_{b_{\bar{z}'}} - 2CS\hat{D}_{b_{\bar{y}'\bar{z}'}} + C^2 \hat{D}_{b_{\bar{y}'}} \qquad (4.11a)$$

Using Eq.(4.10) gives

$$(\hat{D}_{b_{y'}}, \hat{D}_{b_{z'}}) = \frac{1}{2}(\hat{D}_{b_{\bar{y}'}} + \hat{D}_{b_{\bar{z}'}}) \pm \frac{1}{2}\left[\left(\hat{D}_{b_{\bar{y}'}} - \hat{D}_{b_{\bar{z}'}}\right)^2 + 4\hat{D}_{b_{\bar{y}'\bar{z}'}}^2\right]^{1/2} \qquad (4.11b)$$

4.2.5 Summary of steps for defining the local coordinate system

The steps for defining the local coordinate system x', y', z', attached to the neutral point O (Figure 4.5b), for a 3D beam are the following.

1. Find the center of gravity G of the section and define the orthogonal system $\bar{x}, \bar{y}, \bar{z}$ attached to point G, where $\bar{x}$ is parallel to x'.
2. Find the position $\bar{y}_0, \bar{z}_0$ of the neutral point O defining the system $x', \bar{y}', \bar{z}'$ attached to point O (Eqs.(4.7)). Compute the constitutive parameters $\hat{D}_a, \hat{D}_{b_{\bar{y}'}}, \hat{D}_{b_{\bar{z}'}}$ and $\hat{D}_{b_{\bar{y}'\bar{z}'}}$ by Eqs.(4.4).
3. Compute angle α defining the position of the local coordinate system x', y', z' attached to point O (Eq.(4.10)).
4. Compute the bending parameters in the principal inertia axes $\hat{D}_{b_{y'}}$, $\hat{D}_{b_{z'}}$ by Eqs.(4.11b).

For *homogeneous material* the neutral axis coincides with the center of gravity and

$$\bar{z}_0 = \frac{1}{A} \iint_A \bar{z} \, dA \quad , \quad \bar{y}_0 = \frac{1}{A} \iint_A \bar{y} \, dA \tag{4.12}$$

The principal directions of inertia y', z' are defined in this case by

$$\tan 2\alpha = -\frac{2I_{\bar{y}'\bar{z}'}}{I_{\bar{y}'} - I_{\bar{z}'}} \tag{4.13}$$

and the principal moments of inertia are

$$(I_{y'}, I_{z'}) = \frac{1}{2}(I_{\bar{y}'} + I_{\bar{z}'}) \pm \frac{1}{2} \left[(I_{\bar{y}'} - I_{\bar{z}'})^2 + 4I_{\bar{y}'\bar{z}'} \right]^{1/2} \tag{4.14a}$$

with

$$(I_{\bar{y}'}, I_{\bar{z}'}, I_{\bar{y}'\bar{z}'}) = \iint_A (\bar{z}'^2, \bar{y}'^2, \bar{y}'\bar{z}') dA \tag{4.14b}$$

If either $\bar{y}'$ or $\bar{z}'$ are symmetry axis, then $I_{\bar{y}'\bar{z}'} = 0$ and $I_{y'} = I_{\bar{y}'}$, $I_{z'} = I_{\bar{z}'}$.

4.2.6 Computation of the shear center

Let x' be the neutral axis and y', z' the principal inertia axes of a cross section. The stress field is defined by the stresses $\sigma_{x'}$, $\tau_{x'y'}$ and $\tau_{x'z'}$ (Figure 4.4). The shear stresses $\tau_{x'y'}$ and $\tau_{x'z'}$ define the shear forces $Q_{y'}$ and $Q_{z'}$ and the torque $M_{x'}$

$$Q_{y'} = \iint_A \tau_{x'y'} \, dA \quad , \quad Q_{z'} = \iint_A \tau_{x'z'} \, dA \quad , \quad M_{x'} = \iint_A (\tau_{x'z'} y' - \tau_{x'y'} z') \, dA \tag{4.15}$$

The torque with respect to the elastic axis $\hat{x}'$ passing by point C with coordinates y_c', z_c' is

$$M_{\hat{x}'} = \iint_A \left[\tau_{x'z'}(y' - y_c') - \tau_{x'y'}(z' - z_c') \right] dA = M_{x'} - y_c' Q_{z'} + z_c' Q_{y'} \tag{4.16}$$

Point C is the shear center *if the shear stresses due to bending effects* satisfy

$$M_{\hat{x}'} = M_{x'} - y_c' Q_{z'} + z_c' Q_{y'} = 0 \tag{4.17}$$

which gives

$$y_c' = \frac{M_{x'}}{Q_{z'}} \qquad \text{for } Q_{y'} = 0 \quad \text{and } Q_{z'} \neq 0 \tag{4.18a}$$

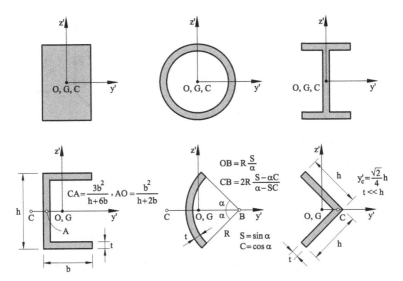

Fig. 4.6 Position of the gravity center (G), the shear center (C) and the neutral axis (O) for some homogeneous sections

$$z'_c = -\frac{M_{x'}}{Q_{y'}} \qquad \text{for } Q_{z'} = 0 \quad \text{and} \quad Q_{y'} \neq 0 \tag{4.18b}$$

The procedure for computing y'_c for a composite beam is the following:

- Compute the distribution of $\tau_{x'y'}$ and $\tau_{x'z'}$ over the section for a shear force $Q_{z'}$ following the procedure explained in Appendix D for the general case and in Section 3.7 for cylindrical bending.
- Compute $M_{x'}$ by Eq.(4.15) and then y'_c by Eq.(4.18a).
  Similar steps are repeated for computing z'_c using the shear force $Q_{y'}$. Clearly, the method is also applicable to homogeneous beams.

Figure 4.6 shows the position of the center of gravity O and the shear center C for some sections. For other sections see [PCh,Yo].

For convenience, a new local reference system $\hat{x}', \hat{y}', \hat{z}'$ is defined at the shear center C so that $\hat{x}', \hat{y}'$ and $\hat{z}'$ are parallel to x', y' and z', respectively (Figure 4.7).

4.2.7 Properties of the shear center

The properties of the shear center are the following:

- External forces $F_{\hat{y}'}$ and $F_{\hat{z}'}$ applied at the shear center do not produce a twist of the section. This means that the shear stresses $\tau_{x'y'}$ and $\tau_{x'z'}$ due to $F_{y'}$ and $F_{z'}$ are associated to a bending-shear state only.

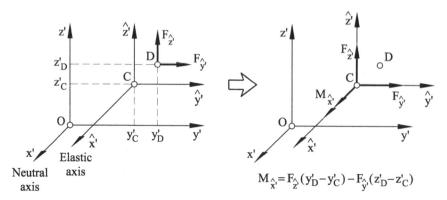

Fig. 4.7 External forces $F_{x'}$, $F_{z'}$ and torque $M_{\hat{x}'}$ acting at the shear center C

- External forces $F_{y'}$ and $F_{z'}$ applied at an arbitrary point D are balanced by shear stresses due to the shear forces $Q_{z'} = F_{\hat{z}'}$ and $Q_{y'} = F_{\hat{y}'}$ and the torque $M_{\hat{x}'} = F_{z'}(y_D' - y_c') - F_{y'}(z_D' - z_c')$.
- Consequently, any external load contained on the $y'z'$ plane can be reduced to two loads $(F_{\hat{y}'}, F_{\hat{z}'})$ acting at the shear center and a torque $M_{\hat{x}'}$ around the elastic axis $\hat{x}'$. The two forces originate displacements along the y' and z' axis and the corresponding bending states (in planes $x'y'$ and $x'z'$, respectively), while the torque induces a twist of the section ($\theta_{\hat{x}'}$) around the elastic axis (Figure 4.7).
- The shear stresses due to bending in the plane $x'y'$ satisfy

$$Q_{y'} = \iint_A \tau_{x'y'} \, dA \neq 0 \quad , \quad Q_{z'} = 0 \quad , \quad M_{\hat{x}'} = 0 \qquad (4.19a)$$

those due to bending in the plane $x'z'$ ($\tau_{x'z'}$) satisfy

$$Q_{z'} = \iint_A \tau_{x'z'} \, dA \neq 0 \quad , \quad Q_{y'} = 0 \quad , \quad M_{\hat{x}'} = 0 \qquad (4.19b)$$

and those due to a torque $M_{\hat{x}'}$ ($\tau_{x'y'}$, $\tau_{x'z'}$) satisfy

$$Q_{y'} = Q_{z'} = 0 \quad , \quad M_{\hat{x}'} = \iint_A [\tau_{x'z'}(y' - y_c') - \tau_{x'y'}(z' - z_c')] \, dA \neq 0$$
$$(4.20)$$

- For unconstrained uniform torsion a torque acting at the shear center induces just shear stresses in the section. For a non-uniform torque or a constrained uniform torsion, a torque also induces axial (warping) stresses. For all cases the axial force and the bending moments induced by a torque acting at the shear center are zero (Appendix I).

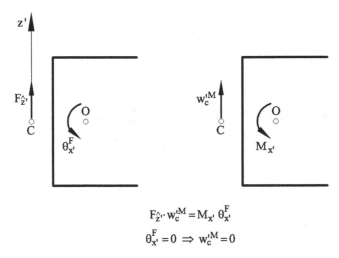

$$F_{\hat{z}'} \cdot w_c'^M = M_{x'} \cdot \theta_{x'}^F$$

$$\theta_{x'}^F = 0 \Rightarrow w_c'^M = 0$$

Fig. 4.8 The shear center remains fixed under a torque $M_{x'}$ acting at the neutral point O

- The shear center remains fixed when the section is subjected to a torque acting at the neutral point O. This property follows from the Maxwell-Betti reciprocity theorem (Figure 4.8), i.e.

$$F_{\hat{z}'} w_c'^M = M_{x'} \theta_{x'}^F \qquad (4.21)$$

If $F_{\hat{z}'}$ passes by C then the rotation $\theta_{z'}^F$ due to the force $F_{\hat{z}'}$ is zero. Hence, the vertical displacement of the shear center $w_c'^M$ due to $M_{x'}$ is also zero. The same applies for a force $\theta_{y'}^F$ act C. This explains why the shear center is also called *twist center*.
- If the section is symmetric with respect to y', point C is placed over this axis and $M_{\hat{x}'} = 0$. Also if O is a point of double symmetry, points O and C coincide. Finally, if the section is defined by an assembly of thin walls intersecting at a single point, the shear center practically coincides with that point (Figure 4.6).
- For homogeneous material the position of C is typically a geometric property of the section. The position of the shear center for different homogeneous sections can be found in many publications [PCh,Yo].
- An exception of the above statement is the case of very deformable thin walled beams with open cross-section. Here the shear center C is not longer a property of the section and it depends on the boundary conditions and the external forces (as the transverse shear stresses due to torsion are not zero on the middle line) [Hy].

- For a composite beam the position of the shear center C depends on the geometry and the material properties for each layer. Appendix I describes a procedure for computing the position of C in thin-walled open composite sections.

4.3 3D SAINT-VENANT COMPOSITE BEAMS

We study first the finite element formulation for 3D composite beams satisfying *the free torsion assumption* of Saint-Venant, i.e. the beam cross section can deform freely due to a torque. As a consequence, *the axial strains and stresses due to torsion (warping effects) are zero*, or of little importance. As mentioned earlier this assumption is satisfied by *solid sections, T-type sections and closed thin-walled sections* (including multicellular sections). The formulation is also applicable to thin-walled open sections *if the torque is uniform and warping is not constrained* along the beam. A refined formulation for thin-walled open sections accounting for warping effects is presented in Section 4.10. In all cases Timoshenko assumptions for the bending of the section are assumed, i.e. plane sections remain plane but not necessarily orthogonal to the beam axis. This introduces shear deformation effects in the bending modes, as described in Chapter 2 for plane beams.

The formulation of 3D Saint-Venant beams following Euler-Bernoulli assumptions is briefly explained in Section 4.7.

4.3.1 Displacement and strain fields. Timoshenko theory

We will assume that *the shear center C and the neutral axis O do not coincide*. Hence, a coupling between axial, bending/shear and torsion effects will exist if the kinematic variables are chosen either at O or at C.

A decoupling of these effects can be however obtained by choosing the following kinematic variables [BD5]:

- The axial displacement u' sampled at the neutral point $O(u'_0)$,
- the displacements v' and w' in the directions of axes $\hat{y}'$ and $\hat{z}'$, respectively sampled at the shear center C (v'_c, w'_c),
- the twist rotation $\theta_{\hat{x}'}$, and
- the rotations $\theta_{y'}$ and $\theta_{z'}$ around the y' and z' axes at O (Figure 4.9).

Note that, differently from previous chapters, we have chosen a vectorial representation for the rotations (Figure 4.9).

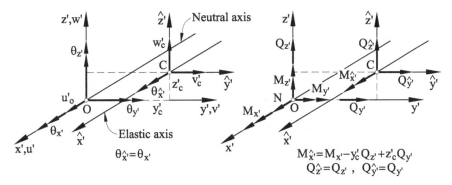

Fig. 4.9 Sampling points for the kinematic variables and the resultant stresses

Note also that as the elastic axis is parallel to the neutral axis the twisting rotations $\theta_{x'}$ and $\theta_{\hat{x}'}$ have the same value (Figure 4.9). In the following we will keep the rotation $\theta_{\hat{x}'}$ as the kinematic variable for convenience.

The kinematic variables at point C are expressed in terms of their values at point O prior to their transformation to global axes.

The displacement field over the beam cross section due to *bending effects*, following Timoshenko beam theory (Chapter 2) is written as

$$u' = u'_0 + z'\theta_{y'} - y'\theta_{z'} \quad ; \quad v' = v'_c \quad ; \quad w' = w'_c \quad (4.22)$$

where indexes 0 and c denote the sampling point for each displacement.

The displacement field is completed with the motion due to torsion. Following Saint-Venant theory we assume that the beam sections twist along the x' axis following a twist angle $\theta_{\hat{x}'}$ *which varies linearly* along $\hat{x}'$ (and also along x'). The displacements along the y' and z' axes are $-(z' - z'_c)\theta_{\hat{x}'}$ and $(y' - y'_c)\theta_{\hat{x}'}$, respectively where y'_c and z'_c are the coordinates of the shear center C. The axial displacement introduced by the twist is $\omega\phi_\omega$ where ω is the *warping function* and ϕ_ω is the twist angle. It is also assumed that the twist angle coincides with the change of the twist rotation along the beam length, i.e

$$\phi_\omega = \frac{\partial \theta_{\hat{x}'}}{\partial x'} \quad (4.23)$$

Figure 4.10 shows the free torsion displacements of an arbitrary point P within the beam section.

Assumption (4.23) is equivalent to neglecting the effect of the shear deformations induced by torsion in the analysis. The way for accounting for the shear deformation terms due to torsion in thin-walled open beams is studied in Section 4.11.

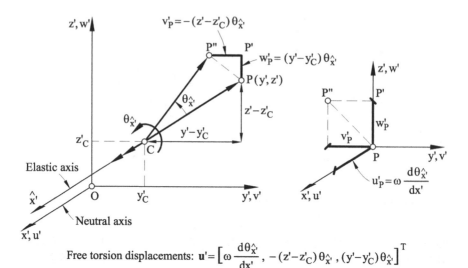

Free torsion displacements: $\mathbf{u}' = \left[\omega \dfrac{d\theta_{\hat{x}'}}{dx'}, -(z'-z'_C)\theta_{\hat{x}'}, (y'-y'_C)\theta_{\hat{x}'} \right]^T$

Fig. 4.10 Graphical representation of the free torsion displacements of a point P in local axes x', y, z'

The twisting displacement is superposed to that induced by the axial and bending motion and the resulting displacement field can be written as

$$\mathbf{u}' = \begin{Bmatrix} u' \\ v' \\ w' \end{Bmatrix} = \underbrace{\begin{Bmatrix} u'_0 \\ 0 \\ 0 \end{Bmatrix}}_{\text{axial}} + \underbrace{\begin{Bmatrix} z'\theta_{y'} - y'\theta_z \\ v'_c \\ w'_c \end{Bmatrix}}_{\text{bending}} + \underbrace{\begin{Bmatrix} \omega \frac{d\theta_{\hat{x}'}}{dx'} \\ -(z' - z'_c)\theta_{\hat{x}'} \\ (y' - y'_c)\theta_{y'} \end{Bmatrix}}_{\text{free torsion}} \quad (4.24)$$

The local displacement vector is

$$\mathbf{u}' = [u'_0, v'_c, w'_c, \theta_{\hat{x}'}, \theta_{y'}, \theta_{z'}]^T \quad (4.25)$$

The (local) strain field can be deduced from Eqs.(4.1) and (4.24) as

$$\boldsymbol{\varepsilon}' = \begin{Bmatrix} \varepsilon_{x'} \\ \gamma_{x'y'} \\ \gamma_{x'z'} \end{Bmatrix} = \underbrace{\begin{Bmatrix} \frac{\partial u'_0}{\partial x'} \\ 0 \\ 0 \end{Bmatrix}}_{\text{axial}} + \underbrace{\begin{Bmatrix} z'\frac{\partial \theta_{y'}}{\partial x'} \\ 0 \\ \frac{\partial w'_c}{\partial x'} + \theta_{y'} \end{Bmatrix}}_{\substack{\text{bending in} \\ \text{plane } x'z'}} + \underbrace{\begin{Bmatrix} -y'\frac{\partial \theta_{z'}}{\partial x'} \\ \frac{\partial v'_c}{\partial x'} - \theta_{z'} \\ 0 \end{Bmatrix}}_{\substack{\text{bending in} \\ \text{plane } x'y'}} +$$

$$+\left\{\begin{array}{c} 0 \\ \left(\frac{\partial \omega}{\partial y'} - (z' - z'_c)\right)\frac{\partial \theta_{\hat{x}'}}{\partial x'} \\ \left(\frac{\partial \omega}{\partial z'} + (y' - y'_c)\right)\frac{\partial \theta_{\hat{x}'}}{\partial x'} \end{array}\right\} \qquad (4.26)$$

free torsion

Eqs.(4.24) and (4.26) clearly show the superposition of the displacement and strains due to axial, bending and free torsion effects.

As $\theta_{\hat{x}'}$ is assumed to vary linearly along the elastic axis $\hat{x}'$, then $\frac{\partial^2 \theta_{\hat{x}'}}{\partial x'^2} = 0$, and the twisting angle does not contribute to the axial strain $\varepsilon_{x'}$. *This is not the case for thin-walled beams with open section* (Section 4.10).

Eq.(4.26) can be written as

$$\varepsilon' = \mathbf{S}_1 \hat{\varepsilon}' \qquad (4.27a)$$

where $\hat{\varepsilon}'$ is the generalized local strain vector given by

$$\boxed{\hat{\varepsilon}' = \left[\frac{\partial u'_0}{\partial x'}, \left(\frac{\partial v'_c}{\partial x'} - \theta_{z'}\right), \left(\frac{\partial w'_c}{\partial x'} + \theta_{y'}\right), \frac{\partial \theta_{y'}}{\partial x'}, \frac{\partial \theta_{z'}}{\partial x'}, \frac{\partial \theta_{\hat{x}'}}{\partial x'}\right]^T} \qquad (4.27b)$$

and $\mathbf{S}_1$ is the strain transformation matrix

$$\mathbf{S}_1 = \begin{bmatrix} 1 & 0 & 0 & z' & -y' & 0 \\ 0 & 1 & 0 & 0 & 0 & \left[\frac{\partial \omega}{\partial y'} - (z' - z'_c)\right] \\ 0 & 0 & 1 & 0 & 0 & \left[\frac{\partial \omega}{\partial z'} + (y' - y'_c)\right] \end{bmatrix} \qquad (4.28)$$

We should remember that if the shear center C and the neutral axis O coincide then $y'_c = z'_c = 0$.

4.3.2 Stresses, resultant stresses and generalized constitutive matrix

The resultant stress vector is defined as

$$\hat{\sigma}' = \left\{\begin{array}{c} N \\ Q_{y'} \\ Q_{z'} \\ M_{y'} \\ M_{z'} \\ M_{\hat{x}'} \end{array}\right\} = \iint_A \left\{\begin{array}{c} \sigma_{x'} \\ \tau_{x'y'} \\ \tau_{x'z'} \\ z'\sigma_{x'} \\ -y'\sigma_{x'} \\ \left[(y' - y'_c)\tau_{x'z'} - (z' - z'_c)\tau_{x'y'}\right] \end{array}\right\} dA \qquad (4.29)$$

where N is the axial force, $Q_{y'}$ and $Q_{z'}$ are the shear forces along the y' and z' axes, respectively, $M_{y'}$ and $M_{z'}$ are the bending moments around the y' and x' axes, respectively and $M_{\hat{x}'}$ is the torque around the $\hat{x}'$ axis (Figure 4.9).

Eq.(4.29) can be rewritten as

$$\hat{\sigma}' = \iint_A \mathbf{S}_2 \sigma' \, dA \quad \text{where} \quad \mathbf{S}_2 = \begin{bmatrix} 1 & 0 & 0 \\ 0 & 1 & 0 \\ 0 & 0 & 1 \\ z' & 0 & 0 \\ -y' & 0 & 0 \\ 0 & -(z'-z_c') & (y'-y_c') \end{bmatrix} \tag{4.30}$$

Note that

$$\mathbf{S}_1 = \mathbf{S}_2^T + \begin{bmatrix} 0 & 0 & 0 & 0 & 0 & 0 \\ 0 & 0 & 0 & 0 & 0 & \dfrac{\partial \omega}{\partial y'} \\ 0 & 0 & 0 & 0 & 0 & \dfrac{\partial \omega}{\partial x'} \end{bmatrix} \tag{4.31}$$

Recall that under a bending state the shear forces $Q_{y'}$ and $Q_{z'}$ are non zero while $M_{\hat{x}'} = 0$. Conversely, under a pure torsional state $Q_{y'} = Q_{z'} = 0$ while $M_{\hat{x}'}$ is non zero (Eqs.(4.19) and (4.20)).

4.3.3 Generalized constitutive matrix

The decoupling between the flexural and torsional effects implies that the shear forces $Q_{y'}$ and $Q_{z'}$ are computed from the tangential stresses induced by bending effects only, while the torque $M_{\hat{x}'}$ is computed from the tangential stresses due to torsion. Taking this into account, the relationship between the resultant stresses and generalized strains can be derived from Eq.(4.29), using Eqs.(4.1) and (4.26), as

$$\hat{\sigma}' = \iint_A \left\{ \begin{array}{c} E\varepsilon_{x'} \\ G_{y'}\left(\dfrac{\partial v_c'}{\partial x'} - \theta_{x'}\right) \\ G_{z'}\left(\dfrac{\partial w_c'}{\partial z'} + \theta_{y'}\right) \\ z'E\varepsilon_{x'} \\ -y'E\varepsilon_{x'} \\ D_t \end{array} \right\} dA \tag{4.32}$$

where

$$D_t = \left[G_{z'} \left(\frac{\partial w}{\partial z'} + y' - y'_c \right) (y' - y'_c) - G_{y'} \left(\frac{\partial w}{\partial y'} - z' + z'_c \right) (z' - z'_c) \right] \frac{\partial \theta_{\hat{x}'}}{\partial x'} \tag{4.33}$$

Substituting into (4.32) the expression for $\varepsilon_{x'}$ of Eq.(4.26) and recalling that y', z' are principal inertia axes we obtain

$$\hat{\sigma}' = \hat{\mathbf{D}}' \hat{\varepsilon}' \tag{4.34}$$

The generalized constitutive matrix $\hat{\mathbf{D}}'$ has the following diagonal form

$$\hat{\mathbf{D}}' = \begin{bmatrix} \hat{D}_a & \vdots & 0 & 0 & 0 & 0 & \vdots & 0 \\ \cdots & \cdots & \cdots & \cdots & \cdots & \cdots & \cdots & \cdots \\ 0 & \vdots & \hat{D}_{s_{y'}} & 0 & 0 & 0 & \vdots & 0 \\ 0 & \vdots & 0 & \hat{D}_{s_{z'}} & 0 & 0 & \vdots & 0 \\ 0 & \vdots & 0 & 0 & \hat{D}_{b_{y'}} & 0 & \vdots & 0 \\ 0 & \vdots & 0 & 0 & 0 & \hat{D}_{b_{z'}} & \vdots & 0 \\ \cdots & \cdots & \cdots & \cdots & \cdots & \cdots & \cdots & \cdots \\ 0 & \vdots & 0 & 0 & 0 & 0 & \vdots & \hat{D}_t \end{bmatrix} = \begin{bmatrix} \hat{D}_a & \mathbf{0} & \mathbf{0} \\ \mathbf{0} & \hat{\mathbf{D}}_f & \mathbf{0} \\ \mathbf{0} & \mathbf{0} & \hat{D}_t \end{bmatrix} \tag{4.35a}$$

where $\hat{D}_a, \hat{\mathbf{D}}'_f$ and $\hat{D}_t$ denote the axial, flexural and torsion contributions to matrix $\hat{\mathbf{D}}'$ with

$$\hat{D}_a = \iint_A E \, dA \; ; \quad \hat{D}_{s_{y'}} = k_{y'} \iint_A G_{y'} dA \; ; \quad \hat{D}_{s_{z'}} = k_{z'} \iint_A G_{z'} dA$$

$$\hat{D}_{b_{y'}} = \iint_A E z'^2 \, dA \; ; \quad \hat{D}_{b_{z'}} = \iint_A E y'^2 \, dA$$

$$\hat{D}_t = \iint_A \left[G_{z'} \left(\frac{\partial w}{\partial z'} + y' - y'_c \right) (y' - y'_c) - G_{y'} \left(\frac{\partial w}{\partial y'} - z' + z'_c \right) (z' - z'_c) \right] dA \tag{4.35b}$$

where $k_{y'}$ and $k_{z'}$ are the shear correction parameters accounting for a non uniform distribution of the shear stresses. These parameters can be computed as described in Section 3.8 and in Appendix D.

The diagonal form of $\hat{\mathbf{D}}'$ of Eq.(4.35a) is obtained only if x' is the neutral axis and y' and z' are principal inertia axes. Otherwise $\hat{\mathbf{D}}'$ is a full matrix (Example 4.1).

The integrals of Eq.(4.35b) for composite beams are performed taking into account the distribution of the material properties over the section. For a beam with heterogeneous material properties (satisfying Eq.(4.1)) it is convenient to divide the section in a collection of cells with specified material properties. The integration is performed by adding up the contributions of each cell to the integrals of Eq.(4.35b).

For *composite laminated beams* with a material orientation as in Figure 3.3, the axial, bending and shear constitutive parameters can be computed using an extension of Eqs.(3.12) as

$$\hat{D}_a = \sum_{k=1}^{n_l} b_k h_k E^k \ , \ \hat{D}_{s_{y'}} = k_{y'} \sum_{k=1}^{n_l} b_k h_k G_{y'}^k \ , \ \hat{D}_{s_{z'}} = k_{z'} \sum_{k=1}^{n_l} b_k h_k G_{z'}^k$$

$$\hat{D}_{b_{y'}} = \sum_{k=1}^{n_l} \frac{b_k}{3}(z_{k+1}'^3 - z_k'^3)E^k \ , \ \hat{D}_{b_{z'}} = \sum_{k=1}^{n_l} \frac{h_k b_k^3}{12} E^k$$

$$(4.36)$$

where $h_k = z_{k+1}' - z_k'$ and b_k are the depth and width of the kth layer. n_l is the number of layers and $(\cdot)^k$ denotes values for the kth layer.

The computation of the torsional stiffness $\hat{D}_t$ depends on the warping function ω. This function can be obtained as explained in Section 4.3.4.

Table 4.1 shows the average value of $\hat{D}_t$ for two beam sections with composite laminated material obtained by computing w (via Eqs.(4.44)) and $\hat{D}_t$ with the FEM using different meshes of 3-noded triangles [DB5,Bo].

Expressions (4.35b) simplify for homogeneous material to

$$\hat{D}_a = EA \quad , \quad \hat{D}_{s_{y'}} = k_{y'} G_{y'} A \quad , \quad \hat{D}_{s_{z'}} = k_{z'} G_{z'} A$$

$$\hat{D}_{b_{y'}} = EI_{y'} \quad , \quad D_{b_{z'}} = EI_{z'} \ , \ \hat{D}_t = GJ$$

$$(4.37a)$$

where $I_{y'}$ and $I_{z'}$ are the principal moments of inertia, $k_{y'}$ and $k_{z'}$ depend on the geometry of the section ($k_{y'} = k_{z'} = 5/6$ for a rectangular section) and J is the *torsional inertia* given by

$$J = \iint_A \left(\hat{y}\frac{\partial \omega}{\partial z'} - \hat{z}\frac{\partial \omega}{\partial y'} + \hat{y}^2 + \hat{z}^2 \right) dA \qquad (4.37b)$$

where $\hat{y} = y' - y_c'$, $\hat{z} = z' - z_c'$.

Figure 4.11 shows the values of J for some homogeneous sections [BD5,PCh,Yo].

The strains and stresses at a point of a section can be computed from the resultant stresses. Taking into account the decoupling between flexural and torsional stresses we deduce from Eqs.(4.27a), (4.34) and (4.1)

$$\varepsilon' = S_1 \hat{\varepsilon}' = S_1 \hat{D}'^{-1} \hat{\sigma}'$$

$$\sigma' = D'\varepsilon' = D'\bar{S}_1 \hat{D}'^{-1} \hat{\sigma}'$$

$$(4.38a)$$

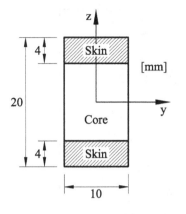

	Material 1	Material 2
Skin	Aluminium	Aluminium
	G=23664 MPa	G=23664 MPa
Core	Polystyrene Foam	Araldite
	G=7.7 MPa	G=1362 MPa

	N° of 3-noded triangles per layer	8	32	72	128
	N° of nodes	15	45	91	153
Material 1	$\hat{D}_t \times 10^{-6}$ N/m^2	7.23	7.55	7.61	7.62
Material 2	$\hat{D}_t \times 10^{-6}$ N/m^2	28.05	26.30	26.00	25.86

Table 4.1 Torsional stiffness $\hat{D}_t$ for two composite laminated beams. Results show the average value of $\hat{D}_t$ in the section obtained by solving Eqs.(4.44) and (4.35b) with the FEM using different meshes of 3-noded triangles [DB5,Bo]

where

$$\bar{\mathbf{S}}_1 = \begin{bmatrix} 1 & 0 & 0 & -z' & y' & 0 \\ 0 & 1 & 0 & 0 & 0 & 0 \\ 0 & 0 & 1 & 0 & 0 & 0 \end{bmatrix} \tag{4.38b}$$

for the strains and stresses induced by axial and flexural effects, and

$$\bar{\mathbf{S}}_1 = \begin{bmatrix} 0 & 0 & 0 & 0 & 0 & 0 \\ 0 & 0 & 0 & 0 & 0 & \dfrac{\partial w}{\partial y'} - (z' - z'_c) \\ 0 & 0 & 0 & 0 & 0 & \dfrac{\partial w}{\partial z'} + (y' - y'_c) \end{bmatrix} \tag{4.38c}$$

for the shear strains and shear stresses induced by the free torsion.

The computation of the shear stresses due to torsion presents some particular features, as it requires the knowledge of the warping function. This topic is discussed in the next sections for the general case and for thin-walled closed sections. Thin-walled open sections are studied in Section 4.10.

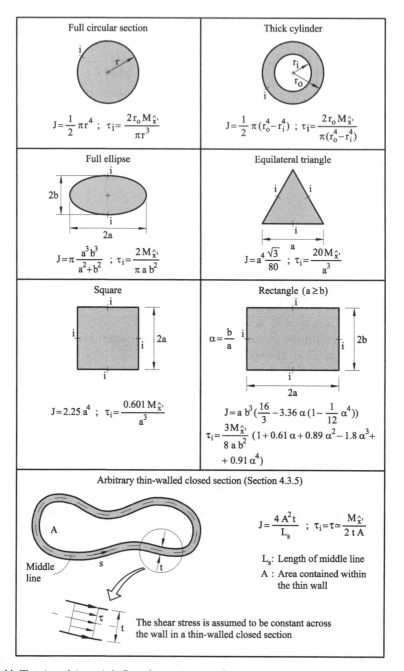

Fig. 4.11 Torsional inertial J and maximum shear stress τ_i for some homogeneous sections. Point i shows the position of τ_i

---

Example 4.1: Derive the generalized constitutive matrix $\hat{\mathbf{D}}'$ for an arbitrary position of the beam axis.

Let us assume that the x' axis does not coincide with the neutral axis and y', z' are not principal inertia axes. Substituting the expression for the axial stress $\varepsilon_{x'}$ (Eq.(4.26)) into the integral for the resultant stress vector (Eq.(4.32)) gives

$$\hat{\sigma}' = \iint_A \left\{ \begin{array}{c} E\left(\dfrac{\partial u'_0}{\partial x'} + z'\dfrac{\partial \theta_{x'}}{\partial x'} - y'\dfrac{\partial \theta_{x'}}{\partial x'}\right) \\[2mm] G_{y'}\left(\dfrac{\partial v'_c}{\partial x'} - \theta_{x'}\right) \\[2mm] G_{z'}\left(\dfrac{\partial w'_c}{\partial x'} + \theta_{y'}\right) \\[2mm] z'E\left(\dfrac{\partial u'_0}{\partial x'} + z'\dfrac{\partial \theta_{y'}}{\partial x'} - y'\dfrac{\partial \theta_{x'}}{\partial x'}\right) \\[2mm] -y'E \\[2mm] D_t \end{array} \right\} dA$$

where D_t is defined in Eq.(4.33). Simple algebra gives

$$\hat{\sigma}' = \hat{\mathbf{D}}'\hat{\varepsilon}' \quad \text{with} \quad \hat{\mathbf{D}}' = \iint_A \begin{bmatrix} E & 0 & 0 & z'E & -y'E & 0 \\ 0 & G_{y'} & 0 & 0 & 0 & 0 \\ 0 & 0 & G_{z'} & 0 & 0 & 0 \\ z'E & 0 & 0 & z'^2E & -y'z'E & 0 \\ -y'E & 0 & 0 & -y'z'E & y'^2E & 0 \\ 0 & 0 & 0 & 0 & 0 & D_t \end{bmatrix} dA$$

with $\hat{\varepsilon}'$ given by Eq.(4.27b).

We see that $\hat{\mathbf{D}}'$ is now a full (still symmetric) matrix. This implies that axial and flexural effects are coupled, i.e. an axial force induces flexural (bending and shear) effects and viceversa. The off-diagonal terms in $\hat{\mathbf{D}}'$ vanish if x' is the neutral axis and y', z' are the principal inertial axes (Section 4.2.3).

In practice, either the full form of $\hat{\mathbf{D}}'$ given above or the diagonal form of Eq.(4.35a) can be used and *both yield identical results*. The simpler diagonal form requires the "a priori" computation of the position of the neutral axis and the principal inertia axes.

Above considerations do not affect the value of the torsional stiffness $\hat{D}_t$.

4.3.4 Computation of shear stresses due to torsion and the warping function

The shear stresses due to torsion in Saint-Venant beams are expressed in terms of the displacements as (see Eqs.(4.1) and (4.26))

$$
\begin{aligned}
\tau_{x'y'} &= G_{y'}\gamma_{x'y'} = G_{y'}\frac{\partial\theta_{\hat{x}'}}{\partial x'}\left[\frac{\partial\omega}{\partial y'} - (z' - z_c')\right] \\
\tau_{x'z'} &= G_{z'}\gamma_{x'z'} = G_{z'}\frac{\partial\theta_{\hat{x}'}}{\partial x'}\left[\frac{\partial\omega}{\partial z'} + (y' - y_c')\right]
\end{aligned}
\tag{4.39}
$$

The equilibrium equation along x' (noting that $\sigma_{x'}$ is zero under pure torsion) is (Appendix B)

$$
\frac{\partial\tau_{x'y'}}{\partial y'} + \frac{\partial\tau_{x'z'}}{\partial z'} = 0 \qquad \text{in } A
\tag{4.40}
$$

where A is the beam section. Substituting Eqs.(4.39) into (4.40) gives

$$
\frac{\partial}{\partial y'}\left(G_{y'}\frac{\partial\omega}{\partial y'}\right) + \frac{\partial}{\partial z'}\left(G_{z'}\frac{\partial\omega}{\partial z'}\right) = 0 \quad \text{in } A
\tag{4.41}
$$

Eq.(4.41) is a Laplace equation which must satisfy the following condition at the cross-section boundary Γ [ZTZ]

$$
\tau_n = \tau_{x'y'}n_{y'} + \tau_{x'z'}n_{z'} = 0 \quad \text{on } \Gamma
\tag{4.42}
$$

noting that $n_{y'} = -\frac{\partial z'}{\partial s}$ and $n_{z'} = \frac{\partial y'}{\partial s}$ and using Eqs.(4.39) we have

$$
G_{y'}\frac{\partial\omega}{\partial y'}n_{y'} + G_{z'}\frac{\partial\omega}{\partial z'}n_{z'} + G_{y'}(z' - z_c')\frac{\partial z'}{\partial s} + G_{z'}(y' - y_c')\frac{\partial y'}{\partial s} = 0 \qquad \text{on } \Gamma
\tag{4.43}
$$

Eqs.(4.41) and (4.43) simplify for homogeneous material to

$$
\frac{\partial^2\omega}{\partial y'^2} + \frac{\partial^2\omega}{\partial z'^2} = 0 \quad \text{in} \quad A
\tag{4.44a}
$$

$$
\frac{\partial\omega}{\partial n} + \frac{1}{2}\frac{\partial}{\partial s}\left[(z' - z_c')^2 + (y' - y_c')^2\right] = 0 \quad \text{on } \Gamma
\tag{4.44b}
$$

Solution of the above differential equations yields the distribution of ω over the beam section. The solution can be obtained analytically for simple sections with homogeneous material. For the general case, Eqs.(4.44) are typically solved using finite differences (FD) or finite element (FE) methods [BD4,Bo,OR,PCh,Yo,ZTZ].

Eqs.(4.44) yield a non unique distribution of ω over the section, as their solution is not affected by adding a constant to ω. This is not a problem as

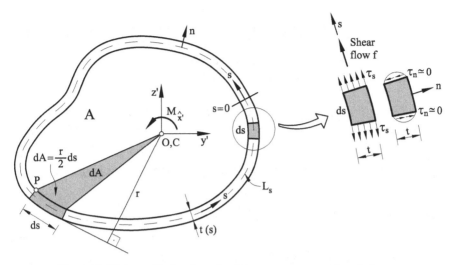

Fig. 4.12 Thin-walled tube of arbitrary cross-sectional shape

the shear stresses depend on the derivatives of ω (Eq.(4.39)) and, hence, the results are not influenced by the value of the constant.

Once the distribution of ω in the beam section has been found, the torsional stiffness $\hat{D}_t$ can be computed via Eq.(4.35b). This is typically done using the same FD or FE mesh used for solving Eqs.(4.44) [BD4,Bo]. Table 4.1 shows an example of this procedure for computing $\hat{D}_t$ in two composite laminated beams using the FEM.

Figure 4.11 shows the position of the maximum shear stress in some homogeneous sections [BD5,PCh,Yo].

4.3.5 Thin-walled closed sections

Given the geometry of thin-walled closed sections it is convenient to resolve the shear stress into its components tangential and normal to the central wall line $\tau_{x's}$ and $\tau_{x'n}$, denoted hereafter as τ_s and τ_n for simplicity (Figure 4.12). Furthermore, it is typically assumed that the normal shear stress τ_n *vanishes through the wall thickness*, as the outer surfaces of the beam are stress-free and the wall thickness is small. The flow of the tangential stress τ_s (hereafter denoted as shear flow) is defined as

$$f(s) = t\tau_s(s) \qquad (4.45)$$

From the local equilibrium equation for a differential element of the thin-walled beam (Eq.(4.40)) we deduce (assuming $\tau_n = 0$) [BC]

$$\frac{\partial \tau_s}{\partial s} + \frac{\partial \tau_n}{\partial n} = \frac{\partial \tau_s}{\partial s} = \frac{1}{t}\frac{\partial f}{\partial s} = 0 \quad \rightarrow \quad f(s) = f = \text{constant} \qquad (4.46)$$

This constant shear flow distribution generates a torque $M_{\hat{x}'}$ about the neutral point O (coinciding with the shear center C) given by

$$M_{\hat{x}'} = \int_{L_s} frds = f \int_{L_s} rds = 2Af \qquad (4.47)$$

where A is the area enclosed by the central wall line with perimeter L_s. From Eqs.(4.45) and (4.47) the tangential stress resulting from the torque is found as

$$\tau_s = \frac{M_{\hat{x}'}}{2At} \qquad (4.48)$$

This equation is completed with the relationship between the twist rate and the applied torque (Eq.(4.23))

$$\phi_w := \frac{\partial \theta_{\hat{x}'}}{\partial x'} = \frac{M_{\hat{x}'}}{\hat{D}_t} \qquad (4.49)$$

The torsional stiffness $\hat{D}_t$ can be obtained by equaling the *external complementary virtual work* [ZTZ,BC] done by the torque and the *internal complementary virtual work* performed by the tangential stress and using Eq.(4.48), i.e.

$$\phi_w \delta M_{\hat{x}'} = \int_{L_s} \gamma_s \delta\tau_s ds = \int_{L_s} \tau_s \frac{\delta\tau_s}{G} tds = \left[\int_{L_s} \frac{M_{\hat{x}'}}{4A^2Gt} ds\right] \delta M_{\hat{x}'} \qquad (4.50)$$

From Eq.(4.50) we find after simplification and noting that $M_{\hat{x}'}$ is constant within L_s

$$\phi_w = \frac{1}{\hat{D}_t} M_{\hat{x}'} \quad \text{with} \quad \hat{D}_t = 4A^2 \left[\int_{L_s} \frac{ds}{Gt}\right]^{-1} \qquad (4.51)$$

For an arbitrary closed section of constant wall thickness

$$\hat{D}_t = \frac{4GtA^2}{L_s} = GJ \quad \text{with} \quad J = \frac{4tA^2}{L_s} \qquad (4.52)$$

Eq.(4.52) shows that the cross-section of maximal torsional stiffness is the thin-walled circular tube. The tangential stress is deduced from Eq.(4.48) as

$$\tau_s = \frac{M_{\hat{x}'}}{2\pi R_m^2 t} \qquad (4.53)$$

where R_m is the mean radius of the tube.

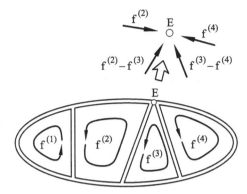

Fig. 4.13 Shear flows in each cell of a thin-walled multi-cellular section [BC]

4.3.5.1 Torsion of multi-cellular sections

For multi-cellular sections equilibrium arguments require the shear flow to remain constant along each wall. Also at each connection point the sum of the flows going into the joint must vanish [AMR,BC,OR].

These continuity requirements are automatically satisfied if constant shear flows are assumed to act in each cell. For the four-cell section shown in Figure 4.13 shear flows circulating around each cell are denoted $f^{(1)}$, $f^{(2)}$, $f^{(3)}$ and $f^{(4)}$, and their assumed positive direction is indicated [BC]. Figure 4.13 also illustrates the shear flows converging to joint E: the continuity condition is satisfied because $(f^{(4)}) + (f^{(3)} - f^{(4)}) + (f^{(2)} - f^{(3)}) + (-f^{(2)}) = 0$.

The solution of the problem requires the computation of the constant shear flows, one around each cell. The total torque, $M_{\hat{x}'}$, carried by the section equals the sum of the torques carried by each individual cell, $M_{\hat{x}'}^{(i)}$, where i indicates the cell number, i.e.

$$M_{\hat{x}'} = \sum_{i=1}^{N_{cells}} M_{\hat{x}'}^{(i)} = 2 \sum_{i=1}^{N_{cells}} A^{(i)} f^{(i)} \qquad (4.54)$$

where N_{cells} is the number of cells and $A^{(i)}$ the area enclosed by the ith cell with perimeter $C^{(i)}$. This single equation does not allow the determination of the shear flows in the N_{cells} cells.

Additional equations can be obtained by expressing the *compatibility conditions* requiring the twist rates of the various cells to be identical. In response to the shear flow, $f^{(i)}$, acting within the cell, a twist rate, $\phi_\omega^{(i)}$, develops in the cell. Compatibility of the deformations of all cells provides

N_{cells} - 1 additional equations

$$\phi_\omega^{(1)} = \phi_\omega^{(2)} = \cdots = \phi_\omega^{(i)} = \cdots = \phi_\omega^{(N_{cells})} \tag{4.55}$$

The relationship between the twist rate and the torque carried by the cell is deduced from Eq.(4.50) as

$$\phi_\omega^{(i)} = \int_{L_s^{(i)}} \frac{M_{\hat{x}'}^{(i)}}{4(A^{(i)})^2} \frac{ds}{Gt} = \int_{L_s^{(i)}} \frac{2A^{(i)}f}{4(A^{(i)})^2} \frac{ds}{Gt} = \frac{1}{2A^{(i)}} \int_{L_s^{(i)}} \frac{f}{Gt} ds \tag{4.56}$$

Eqs.(4.55) and (4.56) provide the N_{cells} equations needed to solve for the N_{cells} shear flows in the cells of a multi-cellular section under torsion.

Example 4.2: Two-cell cross-section.

The thin-wall cross-section shown in Figure 4.14 taken from [BC] represents a highly idealized air-foil structure for which the curved portion is the leading edge, the thicker vertical web is the spar, and the trailing straight segments form the aft portion of the airfoil. Eq.(4.54) gives

$$M_{\hat{x}'} = 2 \sum_{i=1}^{N_{cell}} A^{(i)} f^{(i)} = \pi R^2 f^{(1)} + 6R^2 f^{(2)} \tag{4.57}$$

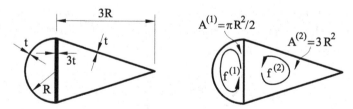

Fig. 4.14 A two-cell thin-walled section under torsion [BC]

The compatibility condition requires twist rates for the two cells to be identical. Eq.(4.56) yields the twist rate for the front cell as

$$\begin{aligned}
\phi_\omega^{(1)} &= \frac{1}{2A^{(1)}} \int_{C^{(1)}} \frac{f}{Gt(s)} ds = \frac{1}{G\pi R^2/2} \left[\frac{f^{(1)}}{t} \pi R + \frac{f^{(1)} - f^{(2)}}{3t} 2R \right] \\
&= \frac{1}{G\pi Rt} \left[\pi f^{(1)} + \frac{2}{3}(f^{(1)} - f^{(2)}) \right]
\end{aligned} \tag{4.58}$$

and the twist rate for the aft cell is

$$
\begin{aligned}
\phi_\omega^{(2)} &= \frac{1}{2A^{(2)}} \int_{C^{(2)}} \frac{f}{Gt(s)} ds = \frac{1}{6G3R^2} \left[\frac{f^{(2)} - f^{(1)}}{3t} 2R + f^{(2)} 2\sqrt{10} \frac{R}{t} \right] \\
&= \frac{1}{6GRt} \left[\frac{2}{3}(f^{(2)} - f^{(1)}) + 2\sqrt{10} f^{(2)} \right]
\end{aligned}
$$

$$(4.59)$$

Equating the two twist rates yields the second equation for the shear flows

$$
\frac{1}{\pi} \left[\pi f^{(1)} + \frac{2}{3}(f^{(1)} - f^{(2)}) \right] = \frac{1}{6} \left[\frac{2}{3}(f^{(2)} - f^{(1)}) + 2\sqrt{10} f^{(2)} \right] \qquad (4.60)
$$

which simplifies to $f^{(1)} = 1.04 f^{(2)}$.

This result, along with Eq.(4.57), can be used to solve for $f^{(1)}$ and $f^{(2)}$ giving

$$
R^2 f^{(1)} = 1.04 M_{\hat{x}'}/(6 + 1.04\pi) \quad \text{and} \quad R^2 f^{(2)} = M_{\hat{x}'}/(6 + 1.04\pi)
$$

Note that the shear slow in the front cell, $f^{(1)}$, is only about 4% greater than that in the aft cell, $f^{(2)}$, and hence, the shear flow in the spar, $R^2(f^{(1)} - f^{(2)}) = 0.04 M_1/(6 + 1.04\pi)$, nearly vanishes.

Because the torsional stiffness of a closed section is proportional to the square of the enclosed area, the largest contribution to the torsional stiffness comes from the outermost closed section, which is the union of the front and aft cells. Consequently, the largest shear flow circulates in this outermost section, leaving the spar nearly unloaded.

The torsional stiffness is computed as the ratio of the torque to the cell twist rate (Eq.(4.49)). Since the twist rates of the two cells are equal, either $\phi_\omega^{(1)}$ or $\phi_\omega^{(2)}$ can be used. For instance, using $\phi_\omega^{(1)}$ yields

$$
\hat{D}_t = \frac{M_{\hat{x}'}}{\phi_\omega^{(1)}} = \frac{(\pi 1.04 + 6)R^2 f^{(2)}}{1/\pi GRt[1.04\pi + 2/3(1.04 - 1)]f^{(2)}} = 2.81\pi GR^3 t \qquad (4.61)
$$

4.3.6 Virtual work expression

The PVW is written as

$$
\iiint_V [\delta\varepsilon_{x'}\sigma_{x'} + \delta\gamma_{x'y'}\tau_{x'y'} + \delta\gamma_{x'z'}\tau_{x'z'}] dV = \int_L \delta\mathbf{u}'^T \mathbf{t}' dx' + \sum_i \delta\mathbf{u}_i'^T \mathbf{p}_i'
$$

$$(4.62)$$

where $\mathbf{t}'$ and $\mathbf{p}_i'$ are distributed forces along the beam axis and point loads, respectively, V is the beam volume and L is the beam length.

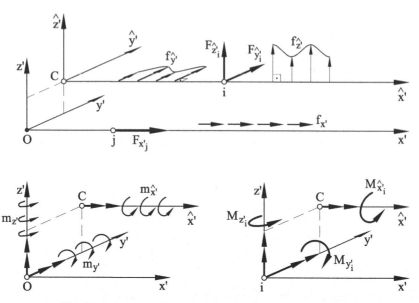

Fig. 4.15 Forces and moments acting on a 3D beam

The components of the virtual displacement vector $\delta\mathbf{u}'$ and the external force vectors $\mathbf{t}'$ and $\mathbf{p}'_i$ are expressed in the local coordinate system as (Figure 4.15)

$$
\begin{aligned}
\delta\mathbf{u}' &= [\delta u'_0,\ \delta v'_c,\ \delta w'_c,\ \delta\theta_{\hat{x}'},\ \delta\theta_{y'},\ \delta\theta_{z'}]^T \\
\mathbf{t}' &= [f_{x'},\ f_{\hat{y}'},\ f_{\hat{z}'},\ m_{\hat{x}'},\ m_{y'},\ m_{z'}]^T \\
\mathbf{p}'_i &= [F_{x'_i},\ F_{y'_i},\ F_{\hat{z}'_i},\ M_{\hat{x}'_i},\ M_{y'_i},\ M_{z'_i}]^T
\end{aligned}
\tag{4.63}
$$

Note that the distributed and point forces along the $\hat{y}'$ and $\hat{z}'$ axes $(f_{y'}, f_{z'}, F_{y'_i},\ F_{z'_i})$ as well as the torques $(m_{\hat{x}'}, M_{\hat{x}'_i})$ *act on the elastic axis* $\hat{x}'$, while the axial forces $(f_{x'}, F_{x'_i})$ and the bending moments $(m_{y'}, m_{z'}, M_{y'_i},\ M_{z'_i})$ *act along the neutral axis* x' (Figure 4.15).

The internal virtual work can be written in terms of the resultant stresses and the virtual generalized strains. Using a matrix notation we can write the l.h.s. of Eq.(4.62) using Eqs.(4.27a) and (4.31) as

$$
\iiint_V \delta\boldsymbol{\varepsilon}'^T \boldsymbol{\sigma}' dV = \iiint_V \delta\hat{\boldsymbol{\varepsilon}}'^T \mathbf{S}_1^T \boldsymbol{\sigma}'\, dV = \iiint_V \delta\hat{\boldsymbol{\varepsilon}}'^T \mathbf{S}_2 \boldsymbol{\sigma}' dV +
$$
$$
+ \iiint_V \frac{\partial\theta_{\hat{x}'}}{\partial\hat{x}'}\left(\frac{\partial\omega}{\partial y'}\tau_{x'y'} + \frac{\partial\omega}{\partial z'}\tau_{x'z'}\right) dV \tag{4.64}
$$

The first integral in the r.h.s. can be expressed using Eq.(4.30) as

$$
\iiint_V \delta\hat{\boldsymbol{\varepsilon}}'^T \mathbf{S}_2\boldsymbol{\sigma}'\, dV = \int_L \delta\hat{\boldsymbol{\varepsilon}}'^T\left(\iint_A \mathbf{S}_2\boldsymbol{\sigma}' dA\right) dx' = \int_L \delta\hat{\boldsymbol{\varepsilon}}'^T \hat{\boldsymbol{\sigma}}' dx' \tag{4.65}
$$

The second integral in the r.h.s. of Eq.(4.64) vanishes as shown next. Recalling that $\dfrac{\partial \theta_{\hat{x}'}}{\partial x'}$ is constant and using integration by parts we have

$$I = \frac{\partial \theta_{\hat{x}'}}{\partial x'} \iiint_V \left(\frac{\partial \omega}{\partial y'} \tau_{x'y'} + \frac{\partial \omega}{\partial z'} \tau_{x'z'} \right) dV =$$

$$\frac{\partial \theta_{\hat{x}'}}{\partial x'} \left[-\iiint_V \omega \left(\frac{\partial \tau_{x'y'}}{\partial y'} + \frac{\partial \tau_{x'z'}}{\partial z'} \right) dV + \iint_S \omega(\tau_{x'y'} n_{y'} + \tau_{x'z'} n_{z'}) \, ds dx' \right]$$

$$(4.66)$$

The surface integral in the r.h.s. of Eq.(4.66) is zero as $\tau_{x'y'} n_{y'} + \tau_{x'z'} n_{z'} = 0$ on the beam boundary in absence of other traction forces (Appendix B). Also, for shear stresses originating from torsion effects $\sigma_{x'} = 0$, and from the equilibrium equations of 3D elasticity (Appendix B)

$$\frac{\partial \sigma_{x'}}{\partial x'} + \frac{\partial \tau_{x'y'}}{\partial y'} + \frac{\partial \tau_{x'z'}}{\partial z'} = \frac{\partial \tau_{x'y'}}{\partial y'} + \frac{\partial \tau_{x'z'}}{\partial z'} = 0 \qquad (4.67)$$

and, therefore, $I = 0$.

For shear stresses originating from bending effects, then from the previous equilibrium equation

$$\frac{\partial \tau_{x'y'}}{\partial y'} + \frac{\partial \tau_{x'z'}}{\partial z'} = -\frac{\partial \sigma_{x'}}{\partial x'} \qquad (4.68a)$$

and, hence,

$$I = -\frac{\partial \theta_{\hat{x}'}}{\partial x'} \iiint_V \omega \frac{\partial \sigma_{x'}}{\partial x'} dV \qquad (4.68b)$$

From Eqs.(4.68) and (4.26) we deduce

$$\frac{\partial \sigma_{x'}}{\partial x'} = E \frac{\partial \varepsilon_{x'}}{\partial x'} = E \left[\frac{\partial u_0'}{\partial x'} + z' \frac{\partial \theta_{y'}}{\partial x'} - y' \frac{\partial \theta_{z'}}{\partial x'} \right] \qquad (4.69)$$

Substituting (4.69) into (4.68b) gives

$$I = -\frac{\partial \theta_{\hat{x}'}}{\partial x'} \iiint_V \omega E \left(\frac{\partial u_0'}{\partial x'} + z' \frac{\partial \theta_{y'}}{\partial x'} - y' \frac{\partial \theta_{z'}}{\partial x'} \right) dV =$$

$$= -\frac{\partial \theta_{\hat{x}'}}{\partial x'} \int_L \left[\frac{\partial u_0'}{\partial x'} \iint_A E\omega dA + \frac{\partial \theta_{y'}}{\partial x'} \iint_A z' E\omega dA - \frac{\partial \theta_{z'}}{\partial x'} \iint_A y' E\omega dA \right] dx'$$

$$(4.70)$$

This integral will vanish if

$$\iint_A E\omega \, dA = \iint_A z' E\omega \, dA = \iint_A y' E\omega \, dA = 0 \qquad (4.71)$$

These conditions can be proved by noting that the axial force and the bending moments induced by a torque acting at the shear center are zero (Section 4.2.7). In conclusion, the PVW can be written as

$$\boxed{\int_L \delta\hat{\boldsymbol{\varepsilon}}'^T \hat{\boldsymbol{\sigma}}' dx' = \int_L \delta\mathbf{u}'^T \mathbf{t}' \, dx' + \sum_i \delta\mathbf{u}_i'^T \mathbf{p}_i'}$$ (4.72)

4.4 FINITE ELEMENT DISCRETIZATION. 2-NODED TIMOSHENKO 3D BEAM ELEMENT

4.4.1 Definition of neutral axis and element matrices

The reference (neutral) line is discretized in C° continuous 1D straight finite elements of length $l^{(e)}$, such as the 2 and 3-noded beam elements shown in Figure 4.16. The coordinates of a point on the reference line are obtained by the isoparametric interpolation [On4]

$$\mathbf{x} = [x_0, y_0, z_0]^T = \sum_{i=1}^{n} \mathbf{N}_i \mathbf{x}_i$$ (4.73)

with $\mathbf{N}_i = N_i(\xi)\mathbf{I}_3$, where $N_i(\xi)$ is the 1D shape function of node i (Figure 2.4) and n is the number of nodes per element. The unit tangent vector to the beam axis along x' is obtained as

$$\mathbf{e}_1 = \frac{\mathbf{x}_1 - \mathbf{x}_2}{|\mathbf{x}_1 - \mathbf{x}_2|}$$ (4.74)

where $\mathbf{x}_1$ and $\mathbf{x}_2$ are the coordinate vectors of the element end nodes. For a straight beam $\mathbf{e}_{1_i} = \mathbf{e}_1$. For a curved beam modeled with straight segments $\mathbf{e}_{1_i}$ is taken as the average of the tangent vectors of the two elements meeting at node i.

Unit vectors $\mathbf{e}_{2_i}$ and $\mathbf{e}_{3_i}$ are defined along the principal directions y' and z' at each node, respectively. For beams with non uniform section the principal directions may vary at each point. The position of vectors $\mathbf{e}_1$, $\mathbf{e}_2$ are interpolated within the element from the nodal values as

$$\mathbf{e}_a = \sum_{i=1}^{n} N_i \mathbf{e}_{a_i} \quad , \quad a = 1, 2, 3$$ (4.75)

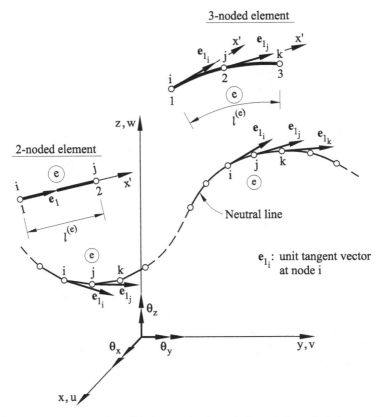

Fig. 4.16 Discretization of a 3D beam in 2-noded and 3-noded (curved) beam elements. Definition of unit tangent vectors at nodes

Above expressions are particularly useful for curved elements (Section 4.6).

The local displacements are interpolated as

$$\mathbf{u}' - \sum_{i=1}^{n} N_i(\xi)\mathbf{I}_6\mathbf{a}_i'^{(e)}; \quad \mathbf{a}_i'^{(e)} = [u_{0_i}', v_{c_i}', w_{c_i}', \theta_{\hat{x}_i'}, \theta_{y_i'}, \theta_{z_i'}]^T \qquad (4.76)$$

where $\mathbf{I}_6$ is the 6×6 unit matrix. Substituting Eq.(4.76) into the generalized strain vector $\hat{\boldsymbol{\varepsilon}}'$ of (4.27b) gives

$$\hat{\boldsymbol{\varepsilon}}' = \left[\frac{\partial u_0'}{\partial x'}, \left(\frac{\partial v_c'}{\partial x'} - \theta_{z'} \right), \left(\frac{\partial w_c'}{\partial x'} + \theta_{y'} \right), \frac{\partial \theta_{y'}}{\partial x'}, \frac{\partial \theta_{z'}}{\partial x'}, \frac{\partial \theta_{\hat{x}'}}{\partial x'} \right]^T = \sum_{i=1}^{n} \mathbf{B}_i'\mathbf{a}_i'^{(e)}$$

$$(4.77)$$

with

$$\mathbf{B}'_i = \begin{bmatrix} \dfrac{\partial N_i}{\partial x'} & 0 & 0 & 0 & 0 & 0 \\ \cdots & \cdots & \cdots & \cdots & \cdots & \cdots \\ 0 & \dfrac{\partial N_i}{\partial x'} & 0 & 0 & 0 & -N_i \\ 0 & 0 & \dfrac{\partial N_i}{\partial x'} & 0 & N_i & 0 \\ 0 & 0 & 0 & 0 & \dfrac{\partial N_i}{\partial x'} & 0 \\ 0 & 0 & 0 & 0 & 0 & \dfrac{\partial N_i}{\partial x'} \\ \cdots & \cdots & \cdots & \cdots & \cdots & \cdots \\ 0 & 0 & 0 & \dfrac{\partial N_i}{\partial x'} & 0 & 0 \end{bmatrix} = \begin{bmatrix} \mathbf{B}_{a_i} \\ \cdots \\ \mathbf{B}_{f_i} \\ \cdots \\ \mathbf{B}_{t_i} \end{bmatrix} \qquad (4.78)$$

where $\mathbf{B}_{a_i}$, $\mathbf{B}_{f_i}$ and $\mathbf{B}_{t_i}$ are the axial, flexural and torsion contributions to the generalized strain matrix of node i.

Substituting the expression of $\hat{\boldsymbol{\varepsilon}}'$ of Eq.(4.77) into the PVW (Eq.(4.72)) and using Eqs.(4.34) and (4.76) yields the element stiffness matrix and the equivalent nodal force vector for distributed forces in local axes as

$$\mathbf{K}_{ij}^{\prime(e)} = \int_{l^{(e)}} \mathbf{B}_i^{\prime T}\hat{\mathbf{D}}'\mathbf{B}'_j \, dx' \quad , \quad \mathbf{f}_i^{\prime(e)} = \int_{l^{(e)}} N_i \mathbf{t}' \, dx' \quad , \quad i,j = 1,2 \quad (4.79)$$

Introducing Eqs.(4.35a) and (4.78) into $\mathbf{K}_{ij}^{\prime(e)}$ gives

$$\mathbf{K}'_{ij} = \int_{l^{(e)}} \left[\mathbf{B}_{a_i}^T \hat{D}_a \mathbf{B}_{a_j} + \mathbf{B}_{f_i}^T \hat{\mathbf{D}}_f \mathbf{B}_{f_j} + \mathbf{B}_{t_i}^T \hat{D}_t \mathbf{B}_{t_j} \right] dx' \qquad (4.80)$$

The element matrices are typically evaluated using a numerical quadrature. Shear locking can be avoided by under integrating the shear contributions in $\mathbf{K}_{ij}^{\prime(e)}$. The simplest 3D beam element is *the linear 2-noded element with one point uniform quadrature*. Its local stiffness matrix can be obtained explicitly as

$$\mathbf{K}_{ij}^{\prime(e)} = [\mathbf{B}_i^{\prime T}\hat{\mathbf{D}}\mathbf{B}'_j]_c l^{(e)} = \left[\mathbf{B}_{a_i}^T \hat{D}_a \mathbf{B}_{a_j} + \mathbf{B}_{f_i}^T \hat{\mathbf{D}}_f \mathbf{B}_{f_j} + \mathbf{B}_{t_i}^T \hat{D}_t \mathbf{B}_{t_j} \right]_c l^{(e)}$$
$$(4.81a)$$

where $(\cdot)_c$ denotes values at the element center. The expression of $[\mathbf{B}'_i]_c$ is

$$[\mathbf{B}'_i]_c = \begin{bmatrix} \mathbf{B}_{a_i} \\ \cdots \\ \mathbf{B}_{f_i} \\ \cdots \\ \mathbf{B}_{t_i} \end{bmatrix} = \begin{bmatrix} a_i & 0 & 0 & 0 & 0 & 0 \\ \cdots & \cdots & \cdots & \cdots & \cdots & \cdots \\ 0 & a_i & 0 & 0 & 0 & -1/2 \\ 0 & 0 & a_i & 0 & 1/2 & 0 \\ 0 & 0 & 0 & 0 & a_i & 0 \\ 0 & 0 & 0 & 0 & 0 & a_i \\ \cdots & \cdots & \cdots & \cdots & \cdots & \cdots \\ 0 & 0 & 0 & a_i & 0 & 0 \end{bmatrix} \quad \text{with} \quad a_i = (-1/l^{(e)})^i$$

(4.81b)

Matrix $\mathbf{K}'^{(e)}_{ij}$ for the 3D 2-noded beam element is shown in Box 4.1. Note that it is an extension of the stiffness matrix for the 2-noded Timoshenko plane beam element with one-point quadrature (Eq.(2.39)).

4.4.2 Stiffness and force transformations

Before assembly it is necessary to refer all nodal variables to the neutral point O. From Eqs.(4.24) we obtain (for $y' = z' = 0$)

$$v'_c = v'_0 - z'_c \theta_{\hat{x}'}$$
$$w'_c = w'_0 + y'_c \theta_{\hat{x}'}$$

(4.82)

The relationship between the nodal displacement vector $\mathbf{a}'_i$ and the vector containing the DOFs at point O ($\bar{\mathbf{a}}'_i$) is

$$\mathbf{a}'_i = \mathbf{L}_i \bar{\mathbf{a}}'_i$$

(4.83a)

where

$$\bar{\mathbf{a}}'_i = [u'_{0_i}, v'_{0_i}, w'_{0_i}, \theta_{x'_i}, \theta_{y'_i}, \theta_{z'_i}]^T \quad \text{and} \quad \mathbf{L}_i = \begin{bmatrix} 1 & 0 & 0 & 0 & 0 & 0 \\ 0 & 1 & 0 & -z'_{c_i} & 0 & 0 \\ 0 & 0 & 1 & y'_{c_i} & 0 & 0 \\ 0 & 0 & 0 & 1 & 0 & 0 \\ 0 & 0 & 0 & 0 & 1 & 0 \\ 0 & 0 & 0 & 0 & 0 & 1 \end{bmatrix}$$

(4.83b)

In Eq.(4.83b) we have taken $\theta_{\hat{x}'} = \theta_{x'}$, as mentioned in Section 4.3.1. The nodal point forces are transformed as

$$\bar{\mathbf{p}}'_i = \mathbf{L}_i^T \mathbf{p}'_i$$

(4.84a)

where

$$\bar{\mathbf{p}}'_i = [F_{x'_i}, F_{y'_i}, F_{z'_i}, M_{x'_i}, M_{y'_i}, M_{z'_i}]^T$$
$$\mathbf{p}'_i = [F_{x'_i}, F_{\hat{y}'_i}, F_{\hat{z}'_i}, M_{\hat{x}'_i}, M_{\hat{y}'_i}, M_{\hat{z}'_i}]^T$$

(4.84b)

$$\mathbf{K}_{11}'^{(e)} = \begin{bmatrix} \dfrac{\hat{D}_a}{l^{(e)}} & 0 & 0 & 0 & 0 & 0 \\[2mm] 0 & \dfrac{\hat{D}_{s_{y'}}}{l^{(e)}} & 0 & 0 & 0 & \dfrac{\hat{D}_{s_{y'}}}{2} \\[2mm] 0 & 0 & \dfrac{\hat{D}_{s_{z'}}}{l^{(e)}} & 0 & -\dfrac{\hat{D}_{s_{z'}}}{2} & 0 \\[2mm] 0 & 0 & 0 & \dfrac{\hat{D}_t}{l^{(e)}} & 0 & 0 \\[2mm] 0 & 0 & -\dfrac{\hat{D}_{s_{z'}}}{2} & 0 & \left(\dfrac{\hat{D}_{s_{z'}}}{4}l^{(e)} + \dfrac{\hat{D}_{b_{y'}}}{l^{(e)}}\right) & 0 \\[3mm] 0 & \dfrac{\hat{D}_{s_{y'}}}{2} & 0 & 0 & 0 & \left(\dfrac{\hat{D}_{s_{y'}}}{4}l^{(e)} + \dfrac{\hat{D}_{b_{z'}}}{l^{(e)}}\right) \end{bmatrix},$$

$$\mathbf{K}_{12}'^{(e)} = \begin{bmatrix} -\dfrac{\hat{D}_a}{l^{(e)}} & 0 & 0 & 0 & 0 & 0 \\[2mm] 0 & -\dfrac{\hat{D}_{s_{y'}}}{l^{(e)}} & 0 & 0 & 0 & \dfrac{\hat{D}_{s_{y'}}}{2} \\[2mm] 0 & 0 & -\dfrac{\hat{D}_{s_{z'}}}{l^{(e)}} & 0 & -\dfrac{\hat{D}_{s_{z'}}}{2} & 0 \\[2mm] 0 & 0 & 0 & -\dfrac{\hat{D}_t}{l^{(e)}} & 0 & 0 \\[2mm] 0 & 0 & \dfrac{\hat{D}_{s_{z'}}}{2} & 0 & \left(\dfrac{\hat{D}_{s_{z'}}}{4}l^{(e)} - \dfrac{\hat{D}_{b_{y'}}}{l^{(e)}}\right) & 0 \\[3mm] 0 & -\dfrac{\hat{D}_{s_{y'}}}{2} & 0 & 0 & 0 & \left(\dfrac{\hat{D}_{s_{y'}}}{4}l^{(e)} - \dfrac{\hat{D}_{b_{z'}}}{l^{(e)}}\right) \end{bmatrix}$$

$$\mathbf{K}_{22}'^{(e)} = \begin{bmatrix} \dfrac{\hat{D}_a}{l^{(e)}} & 0 & 0 & 0 & 0 & 0 \\[2mm] 0 & \dfrac{\hat{D}_{s_{y'}}}{l^{(e)}} & 0 & 0 & 0 & -\dfrac{\hat{D}_{s_{y'}}}{2} \\[2mm] 0 & 0 & \dfrac{\hat{D}_{s_{z'}}}{l^{(e)}} & 0 & \dfrac{\hat{D}_{s_{z'}}}{2} & 0 \\[2mm] 0 & 0 & 0 & \dfrac{\hat{D}_t}{l^{(e)}} & 0 & 0 \\[2mm] 0 & 0 & \dfrac{\hat{D}_{s_{z'}}}{2} & 0 & \left(\dfrac{\hat{D}_{s_{z'}}}{4}l^{(e)} + \hat{D}_{b_{y'}}\right) & 0 \\[3mm] 0 & -\dfrac{\hat{D}_{s_{y'}}}{2} & 0 & 0 & 0 & \left(\dfrac{\hat{D}_{s_{y'}}}{4}l^{(e)} + \dfrac{\hat{D}_{b_{z'}}}{l^{(e)}}\right) \end{bmatrix}$$

$$\mathbf{K}_{21}'^{(e)} = [\mathbf{K}_{12}'^{(e)}]^T$$

Box 4.1 Local stiffness matrices $\mathbf{K}_{ij}'^{(e)}$ for a 2-noded 3D Timoshenko beam element with one point uniform quadrature

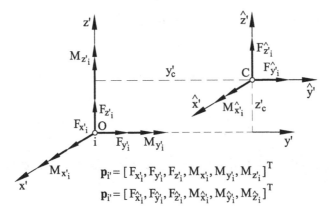

$$\mathbf{p}_{i'} = [\, F_{x_i'}, F_{y_i'}, F_{z_i'}, M_{x_i'}, M_{y_i'}, M_{z_i'} \,]^T$$

$$\mathbf{p}_{i'} = [\, F_{\hat{x}_i'}, F_{\hat{y}_i'}, F_{\hat{z}_i'}, M_{\hat{x}_i'}, M_{\hat{y}_i'}, M_{\hat{z}_i'} \,]^T$$

Fig. 4.17 Components of nodal point force vectors $\mathbf{p}_i'$ and $\bar{\mathbf{p}}_i'$

Eq.(4.84a) is easily deduced if we note that the all the components of $\mathbf{p}_i'$ coincide with those of $\mathbf{p}_i'$, except the torques $M_{x_i'}$ and $M_{\hat{x}_i'}$ that are related by (Figure 4.17)

$$M_{x_i'} = M_{\hat{x}_i'} - z_c' F_{\hat{y}_i'} + y_c' F_{\hat{z}_i'} \tag{4.84c}$$

The transformation of the local element stiffness matrix to global axes is as follows. Matrix $\mathbf{K}_{ij}'^{(e)}$ is first transformed to the x', y', z' system sampled at the neutral point O as

$$\bar{\mathbf{K}}_{ij}'^{(e)} = \mathbf{L}_i^T \mathbf{K}_{ij}'^{(e)} \mathbf{L}_j \quad , \quad i, j = 1, 2 \tag{4.85}$$

For straight elements $\mathbf{L}_i = \mathbf{L}_j$.

Matrix $\bar{\mathbf{K}}_{ij}'^{(e)}$ is finally transformed to the global cartesian system x, y, z as

$$\mathbf{K}_{ij}^{(e)} = \mathbf{T}_i^T \bar{\mathbf{K}}_{ij}'^{(e)} \mathbf{T}_j \quad \text{with} \quad \mathbf{T}_i = [\mathbf{e}_{1_i}, \mathbf{e}_{2_i}, \mathbf{e}_{3_i}] \tag{4.86}$$

where vectors $\mathbf{e}_{1_i}$, $\mathbf{e}_{2_i}$ and $\mathbf{e}_{3_i}$ are defined as explained in Section 4.4.1.

The two transformations are equivalent to using the following modified generalized strain matrix

$$\mathbf{B}_i = \mathbf{B}_i' \mathbf{L}_i \mathbf{T}_i \tag{4.87a}$$

with $\mathbf{B}_i'$ given by Eq.(4.78). The global stiffness matrix is computed as

$$\mathbf{K}_{ij}^{(e)} = \int_{l^{(e)}} \mathbf{B}_i^T \hat{\mathbf{D}}' \mathbf{B}_j \, dx' \tag{4.87b}$$

The transformation of the local equivalent nodal forces to global axes follows a similar steps. First the components of vector $\mathbf{f}'^{(e)}$ of Eq.(4.79) are transformed to the x', y', z' axes by

$$\bar{\mathbf{f}}_i'^{(e)} = \mathbf{L}_i^T \mathbf{f}_i'^{(e)} \tag{4.88}$$

The equivalent nodal forces $\bar{\mathbf{f}}_i^{(e)}$ are then transformed to the global coordinate system by

$$\mathbf{f}_i^{(e)} = \mathbf{T}_i^T \bar{\mathbf{f}}_i'^{(e)} = \mathbf{T}_i^T \mathbf{L}_i^T \mathbf{f}_i'^{(e)} \tag{4.89}$$

4.5 QUASI-EXACT TWO-NODED 3D TIMOSHENKO BEAM ELEMENT

The exact 2-noded Timoshenko beam element of Section 2.9 can be extended to the 3D case. The resulting element yields exact nodal results when applied to the bending analysis of 3D straight plane beams under loads acting on the planes $x'z'$ and $y'z'$. For arbitrary loads (including torques) and curved geometry of the beam (discretized with straight 2-noded elements) the results are not longer "exact", in the sense of perfect agreement with analytical values. However, the element has an excellent behaviour for thick and slender 3D beams and frame structures and its accuracy its typically better (for the same number of elements) than that of the 2-noded 3D Timoshenko beam element of the previous section.

Box 4.2 shows the stiffness matrix of the so-called "quasi-exact" 2-noded 3D beam element.

The equivalent nodal force vector $\mathbf{f}'^{(e)}$ for $f_{\hat{x}'}$, $f_{\hat{y}'}$, $f_{\hat{z}'}$ and $m_{\hat{x}'}$ constant and $m_{y'} = m_{z'} = 0$ is

$$\mathbf{f}'^{(e)} = \left\{ \begin{matrix} \mathbf{f}_1'^{(e)} \\ \mathbf{f}_2'^{(e)} \end{matrix} \right\} \tag{4.90a}$$

$$\mathbf{f}_1'^{(e)} = \left[\frac{l^{(e)}}{2} f_{x'}, f_{\hat{y}_1'}, f_{\hat{z}_1'}, \frac{l^{(e)}}{2} m_{\hat{x}'}, \left(\frac{l^{(e)}}{2} f_{\hat{y}'} - \int_{l^{(e)}} N_1 F_{\hat{y}'} dx' \right), \right.$$
$$\left. \left(\frac{l^{(e)}}{2} f_{\hat{z}'} - \int_{l^{(e)}} N_1 F_{\hat{z}'} dx' \right) \right]^T \tag{4.90b}$$

$$\mathbf{f}_2'^{(e)} = \left[\frac{l^{(e)}}{2} f_{\hat{x}'}, \left(-f_{\hat{y}_1'} + \int_{l^{(e)}} f_{\hat{y}'} dx' \right), \left(-f_{\hat{z}_1'} + \int_{l^{(e)}} f_{\hat{z}'} dx' \right), \frac{l^{(e)}}{2} m_{\hat{x}'}, \right.$$
$$\left. \left(\frac{l^{(e)}}{2} f_{\hat{y}_1'} - \int_{l^{(e)}} N_2 F_{\hat{y}'} dx' \right), \left(\frac{l^{(e)}}{2} f_{\hat{z}_1'} - \int_{l^{(e)}} N_2 F_{\hat{z}'} dx' \right) \right]^T \tag{4.90c}$$

$$\mathbf{K}_{11}^{\prime(e)} = \begin{bmatrix} \dfrac{\hat{D}_a}{l^{(e)}} & 0 & 0 & 0 & 0 & 0 \\ 0 & 12\phi_{y'}\hat{D}_{b_{z'}} & 0 & 0 & 0 & 6\phi_{y'}\hat{D}_{b_{z'}}l^{(e)} \\ 0 & 0 & 12\phi_{z'}\hat{D}_{b_{y'}} & -6\phi_{z'}\hat{D}_{b_{y'}}l^{(e)} & 0 \\ 0 & 0 & 0 & \dfrac{\hat{D}_t}{l^{(e)}} & 0 & 0 \\ 0 & 0 & -6\phi_{z'}\hat{D}_{b_{y'}}l^{(e)} & 0 & (4+\beta_{z'})\hat{D}_{b_{y'}}(l^{(e)})^2 & 0 \\ 0 & 6\phi_{y'}\hat{D}_{b_{z'}}l^{(e)} & 0 & 0 & 0 & (4+\beta_{y'})\hat{D}_{b_{z'}}(l^{(e)})^2 \end{bmatrix},$$

$$\mathbf{K}_{12}^{\prime(e)} = \begin{bmatrix} -\dfrac{\hat{D}_a}{l^{(e)}} & 0 & 0 & 0 & 0 & 0 \\ 0 & -12\phi_{y'}\hat{D}_{b_{z'}} & 0 & 0 & 0 & 6\phi_{y'}\hat{D}_{b_{z'}}l^{(e)} \\ 0 & 0 & -12\phi_{z'}\hat{D}_{b_{y'}} & 0 & -6\phi_{z'}\hat{D}_{b_{y'}}l^{(e)} & 0 \\ 0 & 0 & 0 & -\dfrac{\hat{D}_t}{l^{(e)}} & 0 & 0 \\ 0 & 0 & 6\phi_{z'}\hat{D}_{b_{y'}}l^{(e)} & 0 & (2-\beta_{z'})\hat{D}_{b_{y'}}(l^{(e)})^2 & 0 \\ 0 & -6\phi_{y'}\hat{D}_{b_{z'}}l^{(e)} & 0 & 0 & 0 & (2-\beta_{y'})\hat{D}_{b_{z'}}(l^{(e)})^2 \end{bmatrix}$$

$$\mathbf{K}_{22}^{\prime(e)} = \begin{bmatrix} \dfrac{\hat{D}_a}{l^{(e)}} & 0 & 0 & 0 & 0 & 0 \\ 0 & 12\phi_{y'}\hat{D}_{b_{z'}} & 0 & 0 & 0 & -6\phi_{y'}\hat{D}_{b_{z'}}l^{(e)} \\ 0 & 0 & 12\phi_{z'}\hat{D}_{b_{y'}} & 0 & 6\phi_{z'}\hat{D}_{b_{y'}}l^{(e)} & 0 \\ 0 & 0 & 0 & \dfrac{\hat{D}_t}{l^{(e)}} & 0 & 0 \\ 0 & 0 & 6\phi_{z'}\hat{D}_{b_{y'}}l^{(e)} & 0 & (4+\beta_{z'})\hat{D}_{b_{y'}}(l^{(e)})^2 & 0 \\ 0 & -6\phi_{y'}\hat{D}_{b_{z'}}l^{(e)} & 0 & 0 & 0 & (4+\beta_{y'})\hat{D}_{b_{z'}}(l^{(e)})^2 \end{bmatrix}$$

$$\beta_{y'} = \frac{12\hat{D}_{b_{z'}}}{\hat{D}_{s_{y'}}(l^{(e)})^2} \quad , \quad \beta_{z'} = \frac{12\hat{D}_{b_{y'}}}{\hat{D}_{s_{z'}}(l^{(e)})^2}$$

$$\phi_{y'} = \frac{1}{(1+\beta_{y'})(l^{(e)})^3} \quad , \quad \phi_{z'} = \frac{1}{(1+\beta_{z'})(l^{(e)})^3}$$

$$\mathbf{K}_{21}^{\prime(e)} = [\mathbf{K}_{12}^{\prime(e)}]^T$$

Box 4.2 Quasi-exact 2-noded 3D Timoshenko beam element. Local stiffness matrices

In the derivation of Eqs.(4.90) we have assumed that $f_{x'}$ and $m_{\hat{x}'}$ are constant along the element. Also

$$N_1 = 1 - \frac{x'}{l^{(e)}} \quad , \quad N_2 = \frac{x'}{l^{(e)}} \quad , \quad F_\alpha = \int_0^{x'} f_\alpha dx' \quad , \quad \alpha = \hat{y}', \hat{z}' \quad (4.91)$$

In Eqs.(4.90b,c), $f_{\hat{y}'_1}, f_{\hat{z}'_1}$ are deduced from Eqs.(2.94b) as

$$f_{\hat{y}'_1} = \frac{1}{l^{(e)}(1 + \beta_{y'})} \int_{l^{(e)}} (6N_m + \beta_{y'})F_{\hat{y}'} dx' \tag{4.92a}$$

$$f_{\hat{z}'_1} = \frac{1}{l^{(e)}(1 + \beta_{z'})} \int_{l^{(e)}} (6N_m + \beta_{z'})F_{\hat{z}'} dx' \tag{4.92b}$$

where $F_{\hat{y}'}$ and $F_{\hat{z}'}$ are deduced from Eq.(2.87), $N_m = \left(1 - \frac{x'}{l^{(e)}}\right) \frac{x'}{l^{(e)}}$ and $\beta_{y'}, \beta_{z'}$ are given in Figure 4.2.

The expression of $\mathbf{f}_1'^{(e)}$ and $\mathbf{f}_2'^{(e)}$ for uniformly and triangular distributed loads and point loads can be deduced from Table 2.2.

If shear deformation effects are neglected, then $\beta_{y'} = \beta_{z'} = 0$ and vectors $\mathbf{f}'^{(e)}$ and $\mathbf{f}_2'^{(e)}$ of Eq.(4.90) are

$$\mathbf{f}_1'^{(e)} = \frac{l^{(e)}}{2} \left[f_{x'}, f_{\hat{y}'}, f_{\hat{z}'}, m_{\hat{x}'}, -f_{\hat{z}'}\frac{l^{(e)}}{6}, f_{\hat{y}'}\frac{l^{(e)}}{6} \right]^T \tag{4.93a}$$

$$\mathbf{f}_2'^{(e)} = \frac{l^{(e)}}{2} \left[f_{x'}, f_{\hat{y}'}, f_{\hat{z}'}, m_{\hat{x}'}, f_{\hat{z}'}\frac{l^{(e)}}{6}, -f_{\hat{y}'}\frac{l^{(e)}}{6} \right]^T \tag{4.93b}$$

4.6 CURVED TIMOSHENKO BEAM ELEMENTS

The previous formulation is applicable to moderately curved 3D beams, i.e. when $\frac{a}{R} \ll 1$, where R is the curvature radius of the reference line and a is a characteristic cross section dimension.

The curved geometry is approximated by an isoparametric interpolation using curved n-noded 1D elements [On4]. The tangent vector $\mathbf{e}_1$ is obtained as

$$\mathbf{e}_1 = \frac{1}{|\frac{\partial \mathbf{x}}{\partial s}|} \frac{\partial \mathbf{x}}{\partial s} = \frac{1}{|\sum_{i=1}^n \frac{\partial N_i}{\partial s} x_i|} \sum_{i=1}^n \frac{\partial N_i}{\partial s} x_i \tag{4.94}$$

For end nodes shared by two elements the nodal tangent vector $\mathbf{e}_{1_i}$ is obtained by a simple averaging procedure.

The unit vectors $\mathbf{e}_{2_i}$ and $\mathbf{e}_{3_i}$ are defined nodally along the principal directions y' and z' for each nodal section respectively. The position of vectors $\mathbf{e}_1, \mathbf{e}_2$ and $\mathbf{e}_3$ within the element is obtained by Eq.(4.75).

The element expressions are deduced from those for straight beams simply substituting the derivative $\frac{\partial N_i}{\partial x'}$ by $\frac{\partial N_i}{\partial s}$ and dx' by ds, where s is the curvilinear coordinate. The curvilinear derivatives are computed as

$$\frac{\partial N_i}{\partial s} = \frac{\partial N_i}{\partial \xi} \frac{\partial \xi}{\partial s} = \frac{1}{J} \frac{\partial N_i}{\partial \xi} \tag{4.95}$$

where $J = \frac{\partial s}{\partial \xi}$ is obtained from the isoparametric description using Eq.(4.73) (see also Section 9.8.2). The element integrals are evaluated numerically noting that $ds = Jd\xi$.

The curvilinear formulation of curved beams coincides with that derived in Chapter 9 as a particular case of curved axisymmetric shell elements (Section 9.8). A general formulation for curved 3D beams obtained by degeneration of solid elements is presented in Section 4.12.

4.7 3D EULER-BERNOUILLI BEAMS. SAINT-VENANT THEORY

3D Saint-Venant beam elements following Euler-Bernoulli theory can readily derived from the formulation of the previous section simply imposing that the local rotations $\theta_{y'}$ and $\theta_{z'}$ coincide with the slopes of the neutral axis. With the sign criterion of Figure 4.9,

$$\theta_{z'} = \frac{\partial v'_c}{\partial x'} \quad \text{and} \quad \theta_{y'} = -\frac{\partial w'_c}{\partial x'} \tag{4.96}$$

Substituting these expressions into Eqs.(4.26) gives the strain field as

$$\varepsilon_{x'} = \frac{\partial u'_0}{\partial x'} - z' \frac{\partial^2 w'_c}{\partial x'^2} - y' \frac{\partial^2 v'_c}{\partial x'^2}$$

$$\gamma_{x'y'} = \left[\frac{\partial \omega}{\partial y'} - (z' - z'_c)\right] \frac{\partial \theta_{\hat{x}'}}{\partial x'} \quad ; \quad \gamma_{x'z'} = \left[\frac{\partial \omega}{\partial z'} + (y' - y'_c)\right] \frac{\partial \theta_{\hat{x}'}}{\partial x'} \tag{4.97}$$

i.e., the shear strains are due to torsion only (i.e. a bending state induces no shear strains). The generalized local strain and resultant stress vectors are

$$\hat{\varepsilon}' = \left[\frac{\partial u'_0}{\partial x'}, \frac{\partial^2 w'_c}{\partial x'^2}, \frac{\partial^2 v'_c}{\partial x'^2}, \frac{\partial \theta_{\hat{x}'}}{\partial x'}\right]^T \quad , \quad \hat{\sigma}' = [N, \, M_{y'}, M_{z'}, \, M_{\hat{x}'}]^T \tag{4.98}$$

The new form of matrix $\mathbf{S}_1$ of Eq.(4.28) is

$$\mathbf{S}_1 = \begin{bmatrix} 1 & 0 & 0 & -z' & -y' & 0 \\ 0 & 0 & 0 & 0 & 0 & A \\ 0 & 0 & 0 & 0 & 0 & B \end{bmatrix} \tag{4.99}$$

with $A = \frac{\partial \omega}{\partial y'} - (z' - z'_c)$ and $B = \frac{\partial \omega}{\partial z'} + (y' - y'_c)$. Similar changes are introduced in $\mathbf{S}_2$ and $\mathbf{S}_1$ of Eqs.(4.30) and (4.38b).

The shear forces $Q_{y'}$, $Q_{z'}$ are computed from the shear stress distribution due to bending following classical procedures of Strength of Materials [Ti2,3]. The PVW is given by Eq.(4.72). C^1 continuous finite elements approximations for v_c' and w_c' are required due to the presence of their second derivatives in $\hat{\boldsymbol{\varepsilon}}'$.

The simplest 3D Euler-Bernoulli beam element has two nodes and uses a linear C^0 continuous interpolation for the axial displacement u_0' and the twist rotation $\theta_{x'}$, and a cubic Hermite C^1 continuous approximation for v_c' and w_c'. The displacement interpolation is written as

$$
\mathbf{u}' = \begin{Bmatrix} u_0' \\ v_c' \\ w_c' \\ \theta_{\hat{x}'} \end{Bmatrix} = \sum_{i=1}^{2} \begin{bmatrix} N_i & 0 & 0 & 0 & 0 & 0 \\ 0 & N_i^H & 0 & 0 & 0 & \bar{N}_i^H \\ 0 & 0 & N_i^H & 0 & -\bar{N}_i^H & 0 \\ 0 & 0 & 0 & N_i & 0 & 0 \end{bmatrix} \begin{Bmatrix} \bar{u}_{0_i}' \\ v_{c_i}' \\ w_{c_i}' \\ \theta_{\hat{x}_i'} \\ \theta_{y_i'} \\ \theta_{z_i'} \end{Bmatrix} = \sum_{i=1}^{2} \mathbf{N}_i \mathbf{a}_i'^{(e)}
$$

$$(4.100)$$

where $N_i = \frac{1}{2}(1 + \xi\xi_i)$ and N_i^H and $\bar{N}_i^H$ are the cubic Hermite shape functions (Eq.(1.11a)). The negative sign in $\bar{N}_i^H$ in the third row of $\mathbf{N}_i$ is a consequence of the definition of $\theta_{y'}$ (see Eq.(4.96)). Substituting Eq.(4.100) into (4.98) yields

$$
\hat{\boldsymbol{\varepsilon}}' = \sum_{i=1}^{2} \begin{bmatrix} \dfrac{\partial N_i}{\partial x'} & 0 & 0 & 0 & 0 & 0 \\ 0 & 0 & \dfrac{\partial^2 N_i^H}{\partial x'^2} & 0 & -\dfrac{\partial^2 \bar{N}_i^H}{\partial x'^2} & 0 \\ 0 & \dfrac{\partial^2 N_i^H}{\partial x'^2} & 0 & 0 & 0 & \dfrac{\partial^2 \bar{N}_i^H}{\partial x'^2} \\ 0 & 0 & 0 & \dfrac{\partial N_i}{\partial x'} & 0 & 0 \end{bmatrix} \mathbf{a}_i'^{(e)} = \sum_{i=1}^{2} \mathbf{B}_i' \mathbf{a}_i'
$$

$$(4.101)$$

The local stiffness matrix for the element is given by Eq.(4.79). Explicit integration is possible and the resulting expression coincides with that of classical matrix structural analysis (Box 4.3). Matrix $\mathbf{K}_{ij}'^{(e)}$ can be deduced by neglecting shear deformation effects (i.e. making $\beta_{y'} = \beta_{z'} = 0$) in the expressions of Box 4.2. The transformation to global axes follows precisely the steps explained in Section 4.4.2.

The equivalent nodal force vector is given by Eq.(4.79). Distributed loads induce now nodal bending moments due to the Hermite interpolation for v_c' and w_c', similarly as for Euler-Bernoulli beam elements. Vector $\mathbf{f}_i'^{(e)}$ for uniform distributed loading coincides with Eqs.(4.93).

The formulation of curved Euler-Bernoulli beam elements using a curvilinear description follows the arguments of Section 4.6.

The 2-noded 3D Euler-Bernoulli beam element can be enhanced by using an Hermite interpolation of the geometry. This is a better approximation for curved beams and it is usually combined with a Hermite

$$\mathbf{K}'^{(e)}_{11} = \frac{1}{(l^{(e)})^3} \begin{bmatrix} \hat{D}_a(l^{(e)})^2 & 0 & 0 & 0 & 0 & 0 \\ 0 & 12\hat{D}_{b_{z'}} & 0 & 0 & 0 & 6\hat{D}_{b_{z'}}l^{(e)} \\ 0 & 0 & 12\hat{D}_{b_{y'}} & 0 & -6\hat{D}_{b_{y'}}l^{(e)} & 0 \\ 0 & 0 & 0 & \hat{D}_t(l^{(e)})^2 & 0 & 0 \\ 0 & 0 & -6\hat{D}_{b_{y'}}l^{(e)} & 0 & 4\hat{D}_{b_{y'}}(l^{(e)})^2 & 0 \\ 0 & 6\hat{D}_{b_{z'}}l^{(e)} & 0 & 0 & 0 & 4\hat{D}_{b_{z'}}(l^{(e)})^2 \end{bmatrix},$$

$$\mathbf{K}'^{(e)}_{12} = \frac{1}{(l^{(e)})^3} \begin{bmatrix} -\hat{D}_a(l^{(e)})^2 & 0 & 0 & 0 & 0 & 0 \\ 0 & -12\hat{D}_{b_{z'}} & 0 & 0 & 0 & 6\hat{D}_{b_{z'}}l^{(e)} \\ 0 & 0 & -12\hat{D}_{b_{y'}} & 0 & -6\hat{D}_{b_{y'}}l^{(e)} & 0 \\ 0 & 0 & 0 & -\hat{D}_t(l^{(e)})^2 & 0 & 0 \\ 0 & 0 & 6\hat{D}_{b_{y'}}l^{(e)} & 0 & 2\hat{D}_{b_{y'}}(l^{(e)})^2 & 0 \\ 0 & -6\hat{D}_{b_{z'}}l^{(e)} & 0 & 0 & 0 & 2\hat{D}_{b_{z'}}(l^{(e)})^2 \end{bmatrix}$$

$$\mathbf{K}'^{(e)}_{22} = \frac{1}{(l^{(e)})^3} \begin{bmatrix} \hat{D}_a(l^{(e)})^2 & 0 & 0 & 0 & 0 & 0 \\ 0 & 12\hat{D}_{b_{z'}} & 0 & 0 & 0 & -6\hat{D}_{b_{z'}}l^{(e)} \\ 0 & 0 & 12\hat{D}_{b_{y'}} & 0 & 6\hat{D}_{b_{y'}}l^{(e)} & 0 \\ 0 & 0 & 0 & \hat{D}_t(l^{(e)})^2 & 0 & 0 \\ 0 & 0 & 6\hat{D}_{b_{y'}}l^{(e)} & 0 & 4\hat{D}_{b_{y'}}(l^{(e)})^2 & 0 \\ 0 & -6\hat{D}_{b_{z'}}l^{(e)} & 0 & 0 & 0 & 4\hat{D}_{b_{z'}}(l^{(e)})^2 \end{bmatrix}$$

$$\mathbf{K}'^{(e)}_{21} = [\mathbf{K}'^{(e)}_{12}]^T \quad , \quad \bar{\mathbf{K}}'^{(e)}_{ij} = \mathbf{L}^T \mathbf{K}'_{ij} \mathbf{L} \quad , \quad \mathbf{L} = \mathbf{L}_i = \mathbf{L}_j \ (\text{Eq.}(4.84b))$$

Box 4.3 Local stiffness matrix for a 2-noded 3D Euler-Bernoulli beam element

approximation for the axial displacement. The higher order approximation makes it more difficult for obtaining an explicit form of the element stiffness matrix.

Euler-Bernoulli beam elements are naturally free of shear locking although in their curve form can suffer from membrane locking. The remedy is to use a high order approximation for the axial displacement and uniform reduced integration (Section 9.15).

3D Euler-Bernoulli beam elements coincide in their plane version with the plane arch elements derived in Section 9.9.3.2 as a particular case of thin axisymmetric curved shell elements.

4.8 PLANE GRILLAGE

It is usual to find structures formed by an assembly of beams such that:

- the beams are linked by their gravity centers all placed on the same global plane xy,

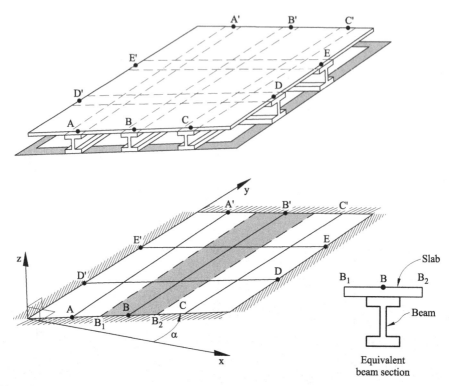

Fig. 4.18 Plane grillage model of a skewed slab-beam assembly. Equivalent section for a beam in the grillage

- the beams are oriented such that plane xy coincides with the symmetry plane $x'y'$ ($z'_c = 0$) of all the beams. The principal axis z' is then parallel to the global z axis.

This type of structure is called a *plane grillage*. Plane grillage models are typically used for analysis of slab-beam assemblies in bridges and building floors, among other applications. Figure 4.18 shows an schematic example of this type of model for a skewed slab-beam bridge.

Figure 4.19 shows the local and global axes for a beam in a grillage. Each beam is subjected to concentrated or distributed forces on the z direction acting on the elastic axis (inducing bending in the plane $x'z'$) and to torques around the $\hat{x}'$ axis. Hence, the kinematic variables satisfy

$$u'_0 = v' = \theta_{z'} = 0 \tag{4.102}$$

The displacement field is expressed as

$$\mathbf{u}' = [w'_c, \theta_{\hat{x}'}, \theta_{y'}]^T \tag{4.103}$$

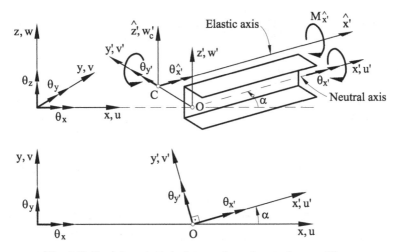

Fig. 4.19 Local and global axes for a beam in a grillage

The local generalized strain vector and the resultant stress vector are

$$\hat{\varepsilon}' = \left[\left(\frac{\partial w'_c}{\partial x} + \theta_{y'} \right), \frac{\partial \theta_{y'}}{\partial y'}, \frac{\partial \theta_{\hat{x}'}}{\partial x'} \right]^T \tag{4.104}$$

$$\hat{\sigma}' = [Q_{z'}, M_{y'}, M_{\hat{x}'}]^T \tag{4.105}$$

The generalized constitutive matrix is

$$\hat{\mathbf{D}}' = \begin{bmatrix} \hat{D}_{s_{z'}} & 0 & \vdots & 0 \\ 0 & \hat{D}_{b_{y'}} & \vdots & 0 \\ \cdots & \cdots & \cdots & \cdots \\ 0 & 0 & \vdots & \hat{D}_t \end{bmatrix} \tag{4.106}$$

The PVW expression is given by Eq.(4.72). The displacement interpolation for a 2-noded grillage element is written by Eq.(4.76) with

$$\mathbf{a}'^{(e)}_i = [w'_{c_i}, \theta_{\hat{x}'_i}, \theta_{y'_i}]^T \tag{4.107}$$

The generalized strain matrix is

$$\mathbf{B}'_i = \begin{bmatrix} \dfrac{\partial N_i}{\partial x'} & 0 & N_i \\ 0 & 0 & \dfrac{\partial N_i}{\partial y'} \\ 0 & \dfrac{\partial N_i}{\partial x'} & 0 \end{bmatrix} \tag{4.108}$$

The local stiffness matrix $\mathbf{K}_{ij}^{\prime(e)}$ for a 2-noded Timoshenko grillage beam element is deduced from Box 4.1 as

$$
\mathbf{K}_{ij}^{\prime(e)} = \begin{bmatrix} \alpha_{ij}\hat{D}_{sz'} & 0 & \beta_i \hat{D}_{sz'} \\ 0 & \alpha_{ij}\hat{D}_t & 0 \\ \beta_j \hat{D}_{sz'} & 0 & \left(\dfrac{\hat{D}_{sz'}}{4} + \alpha_{ij}\hat{D}_{by'} \right) \end{bmatrix} \tag{4.109a}
$$

with

$$
\alpha_{ij} = \frac{(-1)^{i+j}}{l^{(e)}}, \ \beta_i = \frac{(-1)^i}{2} \quad i,j = 1,2 \tag{4.109b}
$$

Box 4.4 shows the local stiffness matrix for the *quasi-exact* 2-noded Timoshenko beam element and the 2-noded Euler-Bernoulli beam element for a plane grillage. These matrices can be deduced from the expressions in Boxes 4.2 and 4.3, respectively.

Quasi-exact 2-noded Timoshenko beam element for plane grillage

$$
\mathbf{K}_{ij}^{\prime(e)} = \begin{bmatrix} 12\phi_{z'}\alpha_{ij}\hat{D}_{b_{y'}} & 0 & 6\phi_{z'}\hat{\beta}_i D_{b_{y'}} l^{(e)} \\ 0 & \dfrac{\alpha_{ij}}{l^{(e)}}\hat{D}_t & 0 \\ 6\phi_{z'}\beta_j \hat{D}_{b_{y'}} l^{(e)} & 0 & (c_j + \phi_{z'}\alpha_{ij})\,\hat{D}_{b_{y'}}(l^{(e)})^2 \end{bmatrix} \quad i,j = 1,2
$$

$$
\phi_{z'} = \frac{1}{(1+\beta_{z'})(l^{(e)})^3} \quad , \quad \beta_{z'} = \frac{12\hat{D}_{b_{y'}}}{\hat{D}_{s_{z'}}(l^{(e)})^2}
$$

2-noded Euler-Bernoulli beam element for plane grillage

$$
\mathbf{K}_{ij}^{\prime(e)} = \frac{1}{(l^{(e)})^3}\begin{bmatrix} 12\alpha_{ij}\hat{D}_{b_{y'}} & 0 & 6\beta_i \hat{D}_{b_{y'}} l^{(e)} \\ 0 & \dfrac{\alpha_{ij}}{l^{(e)}}\hat{D}_t & 0 \\ 6\beta_j \hat{D}_{b_{y'}} l^{(e)} & 0 & c_j \beta_j \hat{D}_{b_{y'}}(l^{(e)})^2 \end{bmatrix} \quad i,j = 1,2
$$

For both cases

$$
\alpha_{ij} = (-1)^{i+j} \quad , \quad \beta_i = (-1)^i \quad , \quad c_j = \begin{cases} 4 & j=1 \\ 2 & j=2 \end{cases}
$$

Box 4.4 Plane grillage. Local stiffness matrix for the quasi-exact 2-noded Timoshenko beam element and the 2-noded Euler-Bernoulli beam element

The global stiffness matrix is given by Eqs.(4.85)–(4.86) with

$$\mathbf{L}_i = \mathbf{L}_j = \begin{bmatrix} 1 & y'_{c_i} & 0 \\ 0 & 1 & 0 \\ 0 & 0 & 1 \end{bmatrix} \quad \text{and} \quad \mathbf{T}_i = \mathbf{T}_j = \begin{bmatrix} 1 & 0 & 0 \\ 0 & C & -S \\ 0 & S & C \end{bmatrix} \tag{4.110}$$

where $C = \cos\alpha$, $S = \sin\alpha$ and α is the angle that the neutral axis (x') forms with the global axis x (Figure 4.18).

The equivalent nodal force vector for distributed loads is computed in global axes as

$$\mathbf{f}_i^{(e)} = \mathbf{T}_i \mathbf{L}_i^T \mathbf{f}_i'^{(e)} \quad \text{with} \quad \mathbf{f}_i'^{(e)} = \int_{l^{(e)}} N_i \begin{Bmatrix} f_{\hat{z}'} \\ m_{\hat{x}'} \\ m_{y'} \end{Bmatrix} dx' \tag{4.111}$$

For uniformly distributed loading, $\mathbf{f}_i'^{(e)} = \frac{l^{(e)}}{2}[f_{\hat{z}'}, m_{\hat{x}'}, m_{y'}]^T$.

Recall that the vertical forces $f_{\hat{z}'}$ and the distributed torque $m_{\hat{x}'}$ act on the elastic axis $\hat{x}'$ (Figure 4.15).

4.9 EXAMPLES OF THE PERFORMANCE OF 3D TIMOSHENKO BEAM ELEMENTS

The first example is the analysis of a circular cantilever rectangular beam with a point load acting on the free edge. Figure 4.20 shows the geometry and the different solutions for the end deflection obtained with the following meshes of 3D Timoshenko beam elements: ten 2-noded (linear) elements and one six-noded (quintic) element. The numerical solution in the later case practically coincides with the analytical one [Ji,Ti3].

The second example is the clamped helycoidal beam shown in Figure 4.21. Self-weight loading is considered. Table 4.2 shows the convergence of the central deflection and maxima and minima of some resultant stresses for different meshes of the following 3D Timoshenko beam elements: linear straigth $(n = 2)$ and quadratic $(n = 3)$, cubic $(n = 4)$, quartic $(n = 5)$ and quintic $(n = 6)$ curved beams. Note the higher accuracy of curved elements for coarse meshes (in particular for predicting the maximum torque).

The computational efficiency of curved element can be enhanced by condensing the internal DOFs prior to the global solution process.

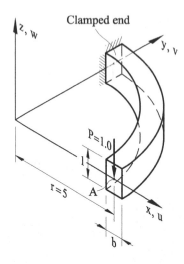

b	10 linear elements	1 quintic element [Ji]	Timoshenko [Ti3]
1	0.11530	0.11589	0.11582
2	0.05010	0.05018	0.05014

Vertical deflection $(-w)$ at point A

$E = 10^6 \, \text{lib/pulg}^2, \quad \nu = 0$

Fig. 4.20 Cantilever beam with point load acting on free edge. Deflection at the free edge for meshes of ten 2-noded and one 6-noded 3D Timoshenko beam elements. Dimensions are in inches and forces in pounds

4.10 THIN-WALLED BEAMS WITH OPEN SECTION

When a thin-walled beam is subjected to an applied torque, shear stresses are generated. In turn, these shear stresses cause out-of-plane deformations of the cross-section called *warping*. Although the magnitude of warping displacements is typically small, they can have an influence on the torsional behaviour of the structure.

Warping effects are particularly relevant when dealing with the non-uniform torsion or a constrained uniform torsion of open sections. In both cases, the twist rate varies along the beam's axis. This contrasts with the Saint-Venant theory studied on Section 4.3 which assumes that the twist rate is constant along the beam's length.

In the following lines we will develop a finite element formulation for analysis of thin-walled beams with open section *under combined axial and flexural effects allowing for warping*. The beam kinematics are assumed to follow Timoshenko beam theory (Chapter 2).

Warping can also appear in slender thin-walled closed sections under a non uniform torque. These effects are less relevant than for open sections and these problems will not be considered here. The interested reader is referred to the literature on the subject [BB,BC,BLD,BT2,OR,Pi,Vl].

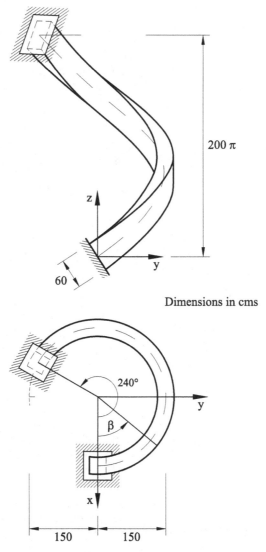

200 π

z

y

60

Dimensions in cms

240°

y

β

x

150   150

Fig. 4.21 Clamped helycoidal beam with square section under self-weight. $E = 2.1 \times 10^6$ Kg/cm^2, $\nu = 0.5$, specific weight $= 2.5$ T/m^3

4.10.1 Geometric description

We consider a straight thin-walled open beam of length L with geometrical and material properties which are independent of the x' coordinate. We will assume that the neutral axis x' and the principal axes y' and z' are known (origin at point O), as well as the position of the shear center C

Variable	No. nodes	Number of elements			
		2	8	16	32
w $(\beta = 120^{\circ})$	2	-0.282	-0.341	-0.360	-0.365
	3	-0.054	-0.367	-0.366	-0.367
	4	-0.286	-0.367	-0.367	-0.367
	5	-0.361	-0.367	-0.367	-0.367
	6	-0.366	-0.367	-0.367	-0.367
N_{max} $(\beta = 240^{\circ})$	2	2.171	2.181	2.171	2.170
	3	2.060	2.147	2.153	2.156
	4	2.138	2.157	2.157	2.158
	5	2.169	2.158	2.158	2.158
	6	2.160	2.158	2.158	2.158
$Q_{\bar{z}max}$ $(\beta = 0^{\circ})$	2	1.501	1.572	1.691	1.601
	3	1.688	1.628	1.617	1.613
	4	1.625	1.613	1.612	1.612
	5	1.599	1.612	1.612	1.612
	6	1.610	1.612	1.612	1.612
T_{max} $(\beta = 240^{\circ})$	2	-0.262	-0.072	0.081	0.162
	3	0.103	0.251	0.256	0.255
	4	0.249	0.257	0.255	0.255
	5	0.249	0.255	0.255	0.255
	6	0.251	0.255	0.255	0.255
$M_{\bar{y}max}$ $(\beta = 120^{\circ})$	2	0.581	0.753	0.801	0.810
	3	0.514	0.798	0.815	0.816
	4	0.763	0.816	0.816	0.816
	5	0.813	0.816	0.816	0.816
	6	0.816	0.816	0.816	0.816
$M_{\bar{y}min}$ $(\beta = 0^{\circ})$	2	-2.132	-2.181	-2.151	-2.110
	3	-1.515	-2.025	-2.058	-2.060
	4	-1.815	-2.059	-2.061	-2.060
	5	-2.027	-2.060	2.060	2.060
	6	-2.060	-2.060	2.060	2.060

Table 4.2 Clamped helycoidal beam with square section under self-weight. Convergence of central deflection and maxima and minima of some resultant stresses using different meshes of 2,3,4,5 and 6-noded 3D Timoshenko beam elements

and the elastic axis local coordinate $(y'_c, 0)$. The thin wall is defined by a middle surface parametrized by the coordinate s with $0 \le s \le L_s$ and by a thickness t which is assumed to be constant for simplicity. The end points at $s = 0$ and $s = L_s$ are D and F, respectively (Figure 4.22).

The position of an arbitrary point p on the middle line within a section is defined in the fixed cartesian system $\mathbf{i}, \mathbf{j}, \mathbf{k}$ by

$$\mathbf{r}_p = y'_p(s)\mathbf{j} + z'_p(s)\mathbf{k} \tag{4.112}$$

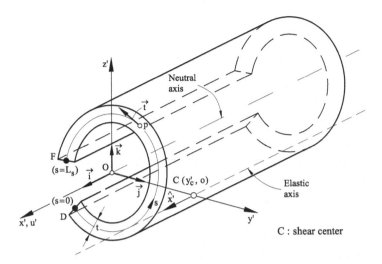

a) Open tube

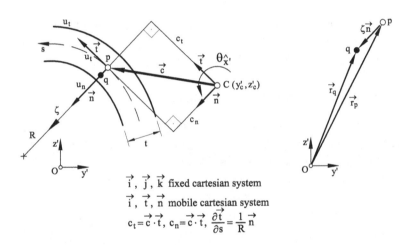

$\vec{i}, \vec{j}, \vec{k}$ fixed cartesian system

$\vec{i}, \vec{t}, \vec{n}$ mobile cartesian system

$c_t = \vec{c} \cdot \vec{t}, \ c_n = \vec{c} \cdot \vec{t}, \ \dfrac{\partial \vec{t}}{\partial s} = \dfrac{1}{R} \vec{n}$

b) Geometric relationships for a general thin open section

Fig. 4.22 Thin-walled tubular beam with open section

and

$$ds = (dy_p'^2 + dz_p'^2)^{1/2} \tag{4.113}$$

The tangent vector to the middle line in p is (in component form)

$$\mathbf{t} = \frac{\partial \mathbf{r}_p}{\partial s} = \frac{\partial y_p'}{\partial s}\mathbf{j} + \frac{\partial z_p'}{\partial s}\mathbf{k} \tag{4.114}$$

The unit normal at p is defined as

$$\mathbf{n} = \mathbf{i} \times \mathbf{t} = n_{y'}\mathbf{j} + n_{z'}\mathbf{k} \tag{4.115a}$$

with

$$n_{y'} = -\frac{\partial z'_p}{\partial s} \quad , n_{z'} = \frac{\partial y'_p}{\partial s} \tag{4.115b}$$

The position of an arbitrary point q across the thickness is defined by the thickness coordinate ζ (Figure 4.22b). In the fixed cartesian system

$$\mathbf{r}_q = \mathbf{r}_p + \zeta \mathbf{n} \quad ; \quad -\frac{t}{2} \leq \zeta \leq \frac{t}{2} \tag{4.116}$$

The coordinates of point q $(y'_q, z'_q) \equiv (y', z')$ can be expressed in the $\mathbf{i}, \mathbf{j}, \mathbf{k}$ system as

$$y'(s,\zeta) = y'_p(s) + \zeta n_{y'}(s) \quad ; \quad z'(s,\zeta) = z'_p(s) + \zeta n_{z'}(s) \tag{4.117}$$

For completeness we define vector $\mathbf{c}$ joining the shear center C and point p on the middle line (Figure 4.22b). The tangent vector $\mathbf{t}$ can be also obtained in terms of vector $\mathbf{c}$ as $\mathbf{t} = \frac{d\mathbf{c}}{ds}$.

4.10.2 Kinematic assumptions. Timoshenko theory

As already mentioned, a torque acting on a thin-walled open section can produce a significant axial displacement due to warping effects. If warping is restrained due to the presence of stiffners or a clampled end, for instance, *the torsion induces an axial stress* $\sigma_{x'}(x', s, \zeta)$ which must be added to the axial stress induced by bending effects.

The displacements of the arbitrary point q *due to torsion* can be expressed as

$$u' = \omega(s,\zeta)\frac{\partial \theta_{\hat{x}'}}{\partial x'} \quad , \quad u_t = -(c_n + \zeta)\theta_{\hat{x}'} \quad , \quad u_n = c_t \theta_{\hat{x}'} \tag{4.118}$$

where, as usual, $\theta_{\hat{x}'}$ is the twist rotation, u', u_t and u_n denote the torsion displacements along the local axes x', t, n, respectively, c_n and c_t are the projections of vector $\mathbf{c}$ along the t and n axes, respectively and ζ is the coordinate of point q along the normal direction (Figure 4.22b).

Eqs.(4.118) clearly show that *the torsion introduces a linear variation of the tangential displacement* u_t *across the wall thickness.*

The relationship between the displacement v', w' and u_t, u_n is

$$v' = u_n n_{y'} - u_t n_{z'} \quad , \quad w' = u_n n_{z'} + u_t n_{y'} \tag{4.119}$$

The total displacements are obtained by adding the displacements due to axial and bending effects in Timoshenko beam theory to the torsion displacements. From Eqs.(4.24), (4.118) and (4.119) we obtain the local displacement vector as

$$\mathbf{u}' = \left\{ \begin{array}{c} u' \\ v' \\ w' \end{array} \right\} = \underbrace{\left\{ \begin{array}{c} u'_0 \\ 0 \\ 0 \end{array} \right\}}_{\text{axial}} + \underbrace{\left\{ \begin{array}{c} z'\theta_{y'} - y'\theta_z \\ v'_c \\ w'_c \end{array} \right\}}_{\text{bending}} + \underbrace{\left\{ \begin{array}{c} \omega \dfrac{\partial \theta_{\hat{x}'}}{\partial x'} \\ -[c_t n_{y'} + (c_n + \zeta) n_{z'}]\theta_{\hat{x}'} \\ [c_t n_{z'} - (c_n + \zeta) n_{y'}]\theta_{\hat{x}'} \end{array} \right\}}_{\text{non-uniform torsion}}$$

$$\tag{4.120}$$

A key difference with Sain-Venant formulation is that the twist rotation $\theta_{\hat{x}'}$ *is not a linear function* and, hence, $\dfrac{\partial^2 \theta_{\hat{x}'}}{\partial x'^2} \neq 0$. Consequently, *torsion originates a non zero axial strain and stress* as shown next.

4.10.3 Warping function and strains and stresses due to torsion

Let us assume an thin-walled open beam under a torque only. The displacement field induced by torsion is given by Eqs.(4.118).

The warping function ω is defined so that in a torsion state the shear strain $\gamma_{x'\zeta}$ is zero over the section and the shear strain $\gamma_{x's}$ is zero at the center line, i.e.

$$\gamma_{x'\zeta} = 0 \quad \forall \quad s \quad \text{and } \zeta \tag{4.121}$$

$$\gamma_{x's} = 0 \quad \forall \quad s \quad \text{for } \zeta = 0 \tag{4.122}$$

Assumption (4.121) and Eq.(4.118) lead to

$$\gamma_{x'\zeta} = \frac{\partial u'}{\partial \zeta} + \frac{\partial u_n}{\partial x'} = \left(\frac{\partial \omega}{\partial \zeta} + c_t \right) \frac{\partial \theta_{\hat{x}'}}{\partial x'} = 0 \tag{4.123}$$

Eq.(4.123) implies that $\frac{\partial \omega}{\partial \zeta} + c_t = 0$. Integrating this along ζ gives

$$\omega(s, \zeta) = g(s) - c_t(s)\zeta \tag{4.124}$$

Assumption (4.122) and Eqs.(4.118) lead to

$$\gamma_{x's}|_{\zeta=0} = \left(\frac{\partial u'}{\partial s} + \frac{\partial u_t}{\partial x'} \right)_{\zeta=0} = \left(\frac{\partial \omega}{\partial s} \Big|_{\zeta=0} - c_n \right) \frac{\partial \theta_{\hat{x}'}}{\partial x'} = 0 \tag{4.125}$$

Hence,

$$\frac{\partial w}{\partial s}\bigg|_{\zeta=0} - c_n = 0 \quad \text{and} \quad w(s,0) = w_s(s) = \int_0^s c_n \, ds \qquad (4.126)$$

Combining Eqs.(4.124) and (4.126) gives

$$g(s) = w_s(s) + w_D \qquad (4.127)$$

where w_D is a constant (typically $w_D = w(0,0)$). The computation of w_s and w_D is detailed in Appendix F.

Function w_s in Eq.(4.127) is called *sectorial area coordinate* (Figure 4.23). Eqs.(4.127) and (4.124) lead to the following expression for the warping function

$$\boxed{w(s,\zeta) = w_s(s) + w_D - c_t(s)\zeta} \qquad (4.128)$$

which gives (noting that $\frac{\partial w_s}{\partial s} = c_n$ from Eq.(4.126))

$$\frac{\partial w}{\partial s} = c_n - \frac{\partial c_t}{\partial s}\zeta \qquad (4.129)$$

The strain field due to torsion is obtained from Eqs.(4.118) and (4.119) as

$$\varepsilon_{x'} = \frac{\partial u'}{\partial x'} = w(s,\zeta)\frac{\partial^2 \theta_{x'}}{\partial x'^2}$$

$$\gamma_{x's} = \frac{\partial u'}{\partial s} + \frac{\partial u_t}{\partial x'} = \left[\frac{\partial w}{\partial s} - (c_n + \zeta)\right]\frac{\partial \theta_{\hat{x}'}}{\partial x'} = -\zeta\left(1 + \frac{\partial c_t}{\partial s}\right)\frac{\partial \theta_{\hat{x}'}}{\partial x'} \qquad (4.130)$$

The expression for $\gamma_{x's}$ can be rewritten using the following relationships

$$\frac{\partial c_t}{\partial s} = \frac{\partial}{\partial s}(\mathbf{c}^T \cdot \mathbf{t}) = \frac{\partial \mathbf{c}^T}{\partial s}\mathbf{t} + \mathbf{c}^T\frac{\partial \mathbf{t}}{\partial s} = \mathbf{t}^T\mathbf{t} + \mathbf{c}^T\frac{\partial \mathbf{t}}{\partial s} = 1 + \frac{c_n}{R} \qquad (4.131)$$

In the derivation of Eq.(4.131) we have used the identities $\mathbf{t} = \frac{\partial \mathbf{c}}{\partial s}$ and $\frac{\partial \mathbf{t}}{\partial s} = \frac{1}{R}\mathbf{n}$ where R is the curvature radius of the curved wall (Figure 4.23) and $\mathbf{c}^T \cdot \mathbf{n} = c_n$.

The *non zero strains due to torsion* are finally expressed as

$$\boxed{\begin{aligned} \varepsilon_{x'} &= w\frac{\partial^2 \theta_{\hat{x}'}}{\partial x'^2} \\ \gamma_{x's} &= -\zeta\left(2 + \frac{c_n}{R}\right)\frac{\partial \theta_{\hat{x}'}}{\partial x'} \end{aligned}} \qquad (4.132)$$

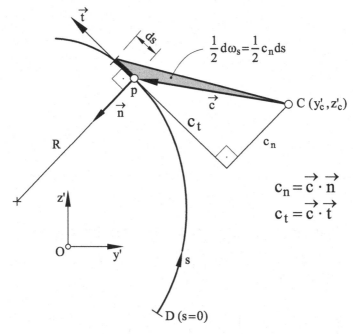

Fig. 4.23 Sectorial area coordinate ω_s

For a section formed by an assembly of straight segments $R = \infty$ and $\gamma_{x's} = -2\zeta \frac{\partial \theta_{\hat{x}'}}{\partial x'}$.

The torsion strains are added to the axial and bending strains as for the Saint-Venant torsion. This is detailed in the next section.

The stresses induced by torsion are deduced from Eqs.(4.132) and (4.1) as

$$\sigma_{x'}^{\omega} = E\varepsilon_{x'} = E\omega \frac{\partial^2 \theta_{\hat{x}'}}{\partial x'^2} \tag{4.133a}$$

$$\tau_{x's} = G\gamma_{x's} = -G\zeta \left(2 + \frac{c_n}{R}\right) \frac{\partial \theta_{\hat{x}'}}{\partial x'} \tag{4.133b}$$

In Eq.(4.133b) we have assumed $G_{y'} = G_{z'} = G$.

The upper-index ω in $\sigma_{x'}^{\omega}$ denotes that this axial stress is due to warping effects. Recall that this stress is zero in Saint-Venant theory. The axial stress $\sigma_{x'}^{\omega}$ varies linearly with the thickness via the ζ-dependance of the warping function ω (Eq.(4.128)). In practice function ω is generally assumed to be constant across the thickness and so is $\sigma_{x'}^{\omega}$ (Section 4.10.7 and [BC,OR,Pi,Ti3]).

Eq.(4.133b) shows that the shear stress $\tau_{x's}$ (or τ_s for short) varies linearly across the wall thickness. *This is a key difference with closed thin-*

walled sections where τ_s is constant across the thickness (Section 4.3.5).

The relationship between the shear strains and the shear stresses in the x', y', z' and x', t, n axes is

$$\gamma_{x'y'} = \gamma_{x's}n_{z'} \quad , \quad \gamma_{x'z'} = -\gamma_{x's}n_{y'}$$
$$\tau_{x'y'} = \tau_{x's}n_{z'} \quad , \quad \tau_{x'z'} = -\tau_{x's}n_{y'} \tag{4.134}$$

Also

$$\tau_{x's} = \tau_{x'y'}n_{z'} - \tau_{x'z'}n_{y'} \quad , \quad \gamma_{x's} = \gamma_{x'y'}n_{z'} - \gamma_{x'z'}n_{y'} \tag{4.135}$$

More details on the computation of the stresses due to torsion in thin-walled open sections are given in Section 4.10.7.

4.10.4 Resultant stresses and generalized constitutive equation

The non zero strains are a sum of the strains induced by the axial, bending and torsion effects, i.e.

$$\boldsymbol{\varepsilon}' = \left\{ \begin{array}{c} \varepsilon_{x'} \\ \gamma_{x'y'} \\ \gamma_{x'z'} \end{array} \right\} = \underbrace{\left\{ \begin{array}{c} \dfrac{\partial u_0'}{\partial x'} \\ 0 \\ 0 \end{array} \right\}}_{\text{axial}} + \underbrace{\left\{ \begin{array}{c} z'\dfrac{\partial \theta_{y'}}{\partial x'} - y'\dfrac{\partial \theta_{z'}}{\partial x'} \\ \dfrac{\partial v_c'}{\partial x'} - \theta_{z'} \\ \dfrac{\partial w_c'}{\partial x'} + \theta_{y'} \end{array} \right\}}_{\text{bending}} + \underbrace{\left\{ \begin{array}{c} \omega\dfrac{\partial^2 \theta_{\hat{x}'}}{\partial x'^2} \\ -\zeta\left(2 + \dfrac{c_n}{R}\right)n_{z'}\dfrac{\partial \theta_{\hat{x}'}}{\partial x'} \\ \zeta\left(2 + \dfrac{c_n}{R}\right)n_{y'}\dfrac{\partial \theta_{\hat{x}'}}{\partial x'} \end{array} \right\}}_{\text{torsion}} \tag{4.136}$$

The stress-strain relationship coincides with Eq.(4.1) with $\mathbf{D}'$ given by Eq.(4.2). The resultant stresses are defined as

$$\hat{\boldsymbol{\sigma}}' = \left\{ \begin{array}{c} N \\ Q_{y'} \\ Q_{z'} \\ M_{y'} \\ M_{z'} \\ M_\omega \\ M_{\hat{x}'} \end{array} \right\} = \iint_A \left\{ \begin{array}{c} \sigma_{x'} \\ \tau_{x'y'} \\ \tau_{x'z'} \\ z'\sigma_{x'} \\ -y'\sigma_{x'} \\ \omega\sigma_{x'} \\ -\zeta\left(2 + \frac{c_n}{R}\right)\tau_{x's} \end{array} \right\} dA \tag{4.137}$$

where M_ω is called the *bimoment* (Figure 4.24) [BC,OR,Ti3].

Substituting the tangential shear stress $\tau_{x's}$ in terms of $\tau_{x'y'}$ and $\tau_{x'z'}$ via Eq.(4.135) and using the stress-strain relationship (Eq.(4.1)) into (4.137) gives

$$\hat{\boldsymbol{\sigma}}' = \hat{\mathbf{D}}'\hat{\boldsymbol{\varepsilon}}' \tag{4.138}$$

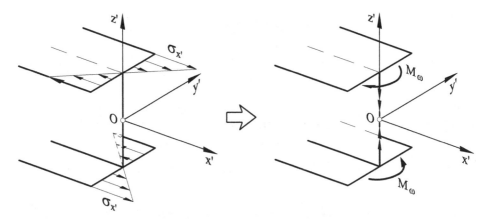

Fig. 4.24 Bimoment M_ω in a thin-walled open section

where $\hat{\varepsilon}'$ is the generalized strain vector and $\hat{\mathbf{D}}'$ is the generalized constitutive matrix respectively given by

$$\hat{\varepsilon}' = \left[\frac{\partial u'_0}{\partial x'}, \left(\frac{\partial v'_c}{\partial x'} - \theta_{z'}\right), \left(\frac{\partial w'_c}{\partial x'} + \theta_{y'}\right), \frac{\partial \theta_{y'}}{\partial x'}, \frac{\partial \theta_{z'}}{\partial x'}, \frac{\partial^2 \theta_{\hat{x}'}}{\partial x'^2}, \frac{\partial \theta_{\hat{x}'}}{\partial x'}\right]^T \quad (4.139a)$$

and

$$\hat{\mathbf{D}}' = \begin{bmatrix} \hat{\mathbf{D}}_a & \mathbf{0} & \mathbf{0} \\ \mathbf{0} & \hat{\mathbf{D}}_f & \mathbf{0} \\ \mathbf{0} & \mathbf{0} & \hat{\mathbf{D}}_t \end{bmatrix} \quad (4.139b)$$

where $\hat{D}_a$ and $\hat{\mathbf{D}}_f$ are given in Eqs.(4.35) and the torsion constitutive
matrix is

$$\hat{\mathbf{D}}_t = \begin{bmatrix} \hat{D}_w & 0 \\ 0 & \hat{D}_t \end{bmatrix} \quad (4.140a)$$

with

$$\hat{D}_w = \iint_A \omega^2 E \, dA \quad , \quad \hat{D}_t = \iint_A \zeta^2 \left(2 \mid \frac{c_n}{R}\right)^2 G \, dA \quad (4.140b)$$

In the expression for $\hat{D}_t$ we have assumed $G_{y'} = G_{z'} = G$.
For homogeneous material

$$\hat{D}_t = GJ \quad \text{and} \quad \hat{D}_w = EI_\omega \quad (4.141)$$

where $J = \iint_A \zeta^2 \left(2 + \frac{c_n^2}{R}\right) dA$ is the *torsional inertia* and $I_\omega = \iint_A \omega^2 dA$
is the *warping inertia modulus*. Figure 4.27 shows the values of J and I_w
for several sections.

For a curved section of uniform thickness

$$J \simeq \frac{1}{3} L_s \frac{t^3}{3} \quad \text{and} \quad \hat{D}_t \simeq \frac{G}{3} \sum_{i=1}^{n} L_s \frac{t^3}{3} \tag{4.142a}$$

where L_s is the length of the middle line (Figure 4.22). These expressions are exact for a circular section [OR,Ti3].

For thin-walled homogeneous open sections formed by n straight segments of thickness t_i and length l_i [OR,Ti3],

$$J = \frac{1}{3} \sum_{i=1}^{n} t_i^3 l_i \quad \text{and} \quad \hat{D}_t = \frac{G}{3} \sum_{i=1}^{n} t_i^3 l_i \tag{4.142b}$$

The maximum shear stress at each segment will occur at the wall edges located at a distance $\pm t/2$ from the midline. Its value is deduced from Eq.(4.133b) for $\zeta = \pm \frac{t}{2}$ and $R = \infty$ and Eq.(4.138) as

$$\tau_s^{\max} = \pm G t \frac{\partial \theta_{\hat{x}'}}{\partial x'} = \pm G t \frac{M_{\hat{x}'}}{\hat{D}_t} \tag{4.143}$$

Clearly the maximum shear stress will be found in the segment featuring the largest thickness.

Example 4.2: Comparison of open and closed thin-walled sections.

The torsional behaviour of thin-walled closed sections is quite different from that of open sections. For closed sections, the shear stress is uniformly distributed through the thickness of the wall (Figure 4.25), whereas a linear distribution through the wall thickness is found in open sections. The torsional stiffness $\hat{D}_t$ is proportional to the square of the enclosed area for a closed section (Eq.(4.52)) in contrast with a thickness cubed proportionality for open sections (Eq.(4.142a) [BC].

Consider, for instance, a thin ring of circular shape and a thin-walled open circular tube, both of identical mean radius R_m and thickness t, as depicted in Figure 4.25. The torsional stiffness of the closed and open sections, denoted $\hat{D}_t^{closed}$ and $\hat{D}_t^{open}$, respectively, are given by Eqs.(4.52) and (4.142a), respectively, as $\hat{D}_t^{closed} = 2\pi G R_m^3 t$ and $\hat{D}_t^{open} = 2\pi G R_m t^3/3$. Their ratio is

$$\frac{\hat{D}_t^{closed}}{\hat{D}_t^{open}} = 3 \left(\frac{R_m}{t} \right)^2$$

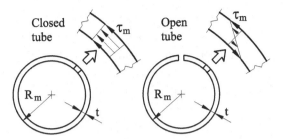

Fig. 4.25 Thin-walled closed and open tubes

If the two sections are subjected to the same torque, $M_{\hat{x}'}$, the maximum shear stresses in the open and closed sections, denoted $\tau_{\max}^{open}$ and $\tau_{\max}^{closed}$, respectively, are given by Eqs.(4.143) and (4.53), respectively, as

$$\tau_{\max}^{open} = \frac{M_{\hat{x}'}t}{\hat{D}_t^{open}} = \frac{3M_{\hat{x}'}}{2\pi R_m t^2} \quad , \quad \tau_{\max}^{closed} = \frac{M_{\hat{x}'}}{2\pi R_m^2 t}$$

Their ratio can then be expressed as

$$\frac{\tau_{\max}^{open}}{\tau_{\max}^{closed}} = 3\left(\frac{R_m}{t}\right)$$

For a typical thin-walled beam with $R_m = 20t$. The torsional stiffness of the closed section will be 1200 times larger than that of the open section. Under the same applied torque, the maximum shear stress in the open section will be 60 times larger than that of the closed section. In other words, the closed section can carry a 60 times larger torque for an equal shear stress level [BC].

4.10.5 Virtual work expression

The PVW expression for a distributed load is

$$\iiint_V \delta\boldsymbol{\varepsilon}'^T \boldsymbol{\sigma}'\, dV - \int_l \delta\mathbf{u}'^T \mathbf{t}' dx' = 0 \qquad (4.144)$$

The internal virtual work can be written after some algebra using Eqs.(4.136)–(4.138) as

$$\iiint_V \delta\boldsymbol{\varepsilon}'^T \boldsymbol{\sigma}'\, dV = \int_L \left(\delta\hat{\varepsilon}'_{x'} N_{x'} + \delta\left(\frac{\partial v'_c}{\partial x'} - \theta_{z'}\right) Q_{y'} + \delta\left(\frac{\partial w'_c}{\partial x'} + \theta_{y'}\right) Q_{z'} \right.$$
$$\left. + \frac{\partial \delta\theta_{y'}}{\partial x'} M_{y'} + \frac{\partial \delta\theta_{z'}}{\partial x'} M_{z'} + \frac{\partial^2 \delta\theta_{\hat{x}'}}{\partial x'^2} M_\omega + \frac{\partial \delta\theta_{\hat{x}'}}{\partial x'} M_{\hat{x}'} \right) dx'$$
$$= \int_L \delta\hat{\boldsymbol{\varepsilon}}'^T \hat{\boldsymbol{\sigma}}'\, dx' \qquad (4.145)$$

where $\hat{\boldsymbol{\sigma}}'$ and $\hat{\boldsymbol{\varepsilon}}'$ are defined in Eqs.(4.137) and (4.139a).

The underlined terms in Eq.(4.145) are due to torsional effects.

The external work for distributed loads is written as

$$\delta W = \int_L \delta \mathbf{u}'^T \mathbf{t}' \, dx' = \int_L \left[\delta \mathbf{u}'^T \mathbf{t}' + \frac{\partial \delta \theta_{\hat{x}'}}{\partial x'} f_\omega \right] dx' \qquad (4.146a)$$

with vectors $\delta \mathbf{u}'$ and $\mathbf{t}'$ defined as in Eq.(4.63) and

$$f_\omega = \iint_A \omega f_{x'} \, dA \qquad (4.146b)$$

The expression of δW for point loads and concentrated moments is

$$\delta W = \sum_i \left[\delta \mathbf{u}_i'^T \mathbf{p}_i' + \delta \left(\frac{\partial \theta_{\hat{x}'}}{\partial x'} \right)_i F_{\omega_i} \right] \qquad (4.147a)$$

where $\mathbf{p}_i'$ is defined in Eq.(4.63) and

$$F_{\omega_i} = \omega F_{x_i'} \qquad (4.147b)$$

Index i in above equations denotes the point along the x' axis where the axial load $F_{x_i'}$ is applied.

4.10.6 Two-noded Timoshenko beam element with thin-walled open section

The formulation of 3D Timoshenko beam elements with thin-walled open section can be derived by superimposing axial effects, flexural effects, with or without shear deformation, and torsion effects. C^1 continuity is required to approximate the twisting rotation $\theta_{\hat{x}'}$, as the second derivatives of $\theta_{\hat{x}'}$ appear in the PVW (Eq.(4.145)). This can be implemented in 2-noded straight elements by choosing $\partial \theta_{\hat{x}'}/\partial x'$ as the 7th nodal DOF and a standard cubic Hermite interpolation for $\theta_{\hat{x}'}$.

The displacement interpolation is written as

$$\mathbf{u}' = \sum_{i=1}^{2} \mathbf{N}_i \mathbf{a}_i' \qquad (4.148a)$$

with

$$\mathbf{u}' = \left[u_0', v_c', w_c', \theta_{\hat{x}'}, \theta_{y'}, \theta_{z'} \right]^T \qquad (4.148b)$$

$$\mathbf{a}_i'^{(e)} = \left[u_{0_i}', v_{c_i}', w_{c_i}', \theta_{\hat{x}_i'}, \theta_{y_i'}, \theta_{z_i'}, \frac{\partial \theta_{\hat{x}_i'}}{\partial x_i'} \right]^T \qquad (4.148c)$$

and

$$\mathbf{N}_i = \begin{bmatrix} N_i & 0 & 0 & 0 & 0 & 0 & 0 \\ 0 & N_i & 0 & 0 & 0 & 0 & 0 \\ 0 & 0 & N_i & 0 & 0 & 0 & 0 \\ 0 & 0 & 0 & N_i^H & 0 & 0 & \bar{N}_i^H \\ 0 & 0 & 0 & 0 & N_i & 0 & 0 \\ 0 & 0 & 0 & 0 & 0 & N_i & 0 \end{bmatrix} \tag{4.148d}$$

where N_i is the standard linear shape function and N_i^H, $\bar{N}_i^H$ are the cubic Hermite shape functions (Eqs.(1.11a)).

The generalized strain-displacement relationship is written as

$$\hat{\varepsilon}' = \sum_{i=1}^{2} \mathbf{B}_i' \mathbf{a}_i' \quad \text{where} \quad \mathbf{B}_i' = \begin{bmatrix} \mathbf{B}_{a_i} \\ \cdots \\ \mathbf{B}_{f_i} \\ \cdots \\ \mathbf{B}_{t_i} \end{bmatrix} \tag{4.149}$$

with $\mathbf{B}_{a_i}$ and $\mathbf{B}_{f_i}$ given by Eq.(4.81b) and

$$\mathbf{B}_{t_i} = \begin{bmatrix} 0 & 0 & 0 & \dfrac{\partial^2 N_i^H}{\partial x_i'^2} & 0 & 0 & \dfrac{\partial^2 \bar{N}_i^H}{\partial x_i'^2} \\ 0 & 0 & 0 & \dfrac{\partial N_i^H}{\partial x_i'} & 0 & 0 & \dfrac{\partial \bar{N}_i^H}{\partial x_i'} \end{bmatrix} \tag{4.150}$$

The element stiffness matrix is obtained as explained in Section 4.4.1. The different stiffness terms coincide with those shown in Box 4.1 with exception of the following terms that complete the 14×14 stiffness matrix for the element (7 DOFs per node)

$$k_{4,4}' = \frac{12\hat{D}_\omega}{(l^{(e)})^3} + \frac{6\hat{D}_t}{5l^{(e)}}, \quad k_{4,7}' = \frac{6\hat{D}_\omega}{(l^{(e)})^2} + \frac{\hat{D}_t}{10}$$

$$k_{4,11}' = -k_{4,4}', \quad k_{4,14}' = k_{4,7}', \quad k_{7,7}' = \frac{4\hat{D}_\omega}{l^{(e)}} + \frac{2l^{(e)}\hat{D}_t}{15} \tag{4.151}$$

$$k_{7,11}' = -k_{4,7}', \quad k_{7,14}' = \frac{2\hat{D}_\omega}{l^{(e)}} - \frac{l^{(e)}\hat{D}_t}{30}, \quad k_{11,11}' = k_{4,4}'$$

$$k_{11,14}' = -k_{4,7}', \quad k_{14,14}' = k_{7,7}'$$

with $k_{ij}' = k_{ji}'$.

An alternative interpolation for $\theta_{\hat{x}'}$ giving "exact" results at the nodes can be chosen using hyperbolic shape functions as described in [BD5].

The transformation of the first six nodal DOFs to the global axes is identical as described in Section 4.4.2. The transformation of $\frac{\partial\theta_{\hat{x}'}}{\partial x'}$ is more cumbersome for folded beams and straight rods with variable section [GP,Gu,Sh]. The degree of nodal compatibility between two elements depends on the satisfaction of the following equation [BD5]

$$\left(\frac{\partial\theta_{\hat{x}'}}{\partial x'}\right)^{(a)}_i = c\left(\frac{\partial\theta_{\hat{x}'}}{\partial x'}\right)^{(b)}_i \tag{4.152}$$

where $c \leq 1$ and indexes a and b denote values at each of the two adjacent elements. Different values for c have been proposed on the basis of local finite element analyses for beam assemblies with sections shaped in H and U [Sh]. An alternative is to assume that the nodal quantities $\frac{\partial\theta_{\hat{x}'}}{\partial x'}$ are discontinuous between elements and that there are as many warping variables at a node as beam elements connected to the node [Ak,BD5].

4.10.7 Computation of stresses due to torsion in thin-walled open sections

The axial and shear stresses induced by torsion in thin-walled open beams can be computed by Eq.(4.133). For straight walls $(R \to \infty)$ and

$$\tau_{x's} = -2\zeta G\frac{\partial\theta_{\hat{x}'}}{\partial x'} \tag{4.153}$$

The maximum values occur at the wall edges $(\zeta = \pm\frac{t}{2})$, i.e.

$$\max|\tau_{x's}| = tG\left|\frac{\partial\theta_{\hat{x}'}}{\partial x'}\right| \tag{4.154}$$

This tangential stress is sometimes referred as Saint-Venant shear stress (denoted hereafter as $\tau^{sv}_{x's}$), meaning that it does not include the shear stress induced by the axial stress due to warping.

The axial stress induced by warping effects can be computed using Eqs.(4.133a) and (4.138) as

$$\sigma^{\omega}_{x'}(s,\zeta) = \omega(s,\zeta)E\frac{\partial^2\theta_{\hat{x}'}}{\partial x'^2} = \omega(s,\zeta)\frac{E}{\hat{D}_\omega}M_\omega \tag{4.155}$$

with $\omega(s,\zeta) = g(s) - c_t(s)\zeta$ (Eq.(4.124)).

If $g(s) \neq 0$, the contribution of the term $c_t\xi$ is typically negligible. Figure 4.26 shows the distribution of $g(s)$ and $\sigma^{\omega}_{x'}(s,0)$ for three sections. The full derivation for one of the sections is presented in Example 4.3.

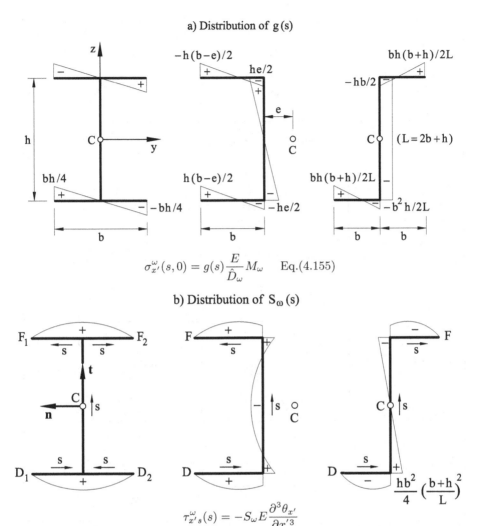

$$\sigma_{x'}^{\omega}(s,0) = g(s)\frac{E}{\hat{D}_{\omega}}M_{\omega} \quad \text{Eq.}(4.155)$$

b) Distribution of $S_\omega(s)$

$$\tau_{x's}^{\omega}(s) = -S_{\omega}E\frac{\partial^3\theta_{x'}}{\partial x'^3}$$

Fig. 4.26 (a) Distribution of $g(s)$, $\sigma_{x'}^{\omega}(s,0)$, $S_\omega(s)$ and (b) $\tau_{x's}^{\omega}(s)$ for three thin-walled open sections

The maximum values of $\sigma_{x'}^{\omega}$ are usually located at the extremities of the section (points D in Figure 4.27) and are independent of ζ. If $g(s) = 0$ these values are placed at the extremity points and at the corners. In general we can write

$$\max|\sigma_{x'}^{\omega}| = \beta_1 E\left|\frac{\partial^2\theta_{\hat{x}'}}{\partial x'^2}\right| \tag{4.156}$$

Figure 4.27 shows the value and position of β_1 (and hence of $\max|\sigma_{x'}^{\omega}|$) for different sections.

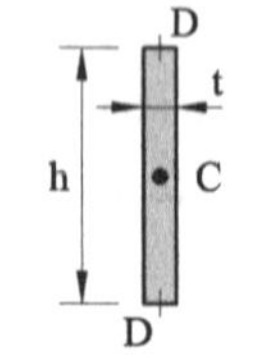 $\omega(s)=0$	$J=\dfrac{1}{3}t^3 h$ $I_\omega=\dfrac{h^3 t^3}{144}=\dfrac{A^3}{144}$	$\beta_1=\dfrac{ht}{4}$ in D, $\zeta=\pm\dfrac{1}{2}$ $\beta_2=0$
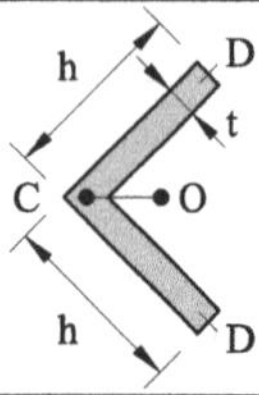 $\omega(s)=0$	$CO=\dfrac{\sqrt{2}}{4}h \quad t\ll h$ $J=\dfrac{t^3}{3}2h$ $I_\omega=\dfrac{t^3 h^3}{18}=\dfrac{A^3}{144}$	$\beta_1=\dfrac{ht}{2}$ in D, $\zeta=\pm\dfrac{1}{2}$ $\beta_2=0$
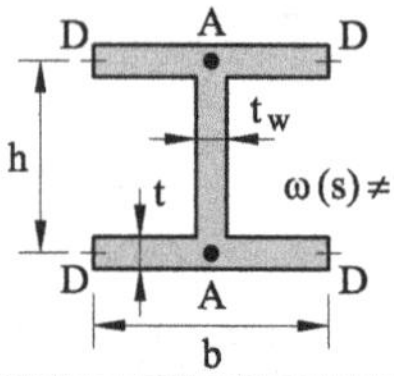 $\omega(s)\neq 0$	$J=\dfrac{1}{3}(2t^3 b+t_w^3 h)$ $I_\omega=h^2\dfrac{t b^3}{24}$	$\beta_1=\dfrac{hb}{4}$ in D, $\forall\zeta$ $\beta_2=\dfrac{hb^2}{16}$ in A, $\forall\zeta$
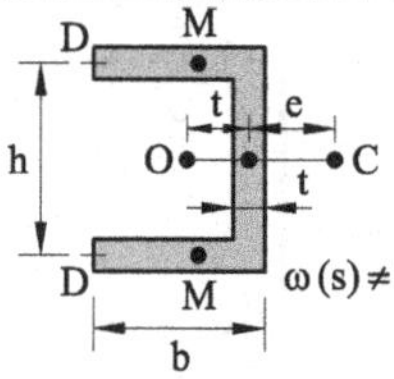 $\omega(s)\neq 0$	$e=\dfrac{3b^2}{h+6b}\ ;\ f=\dfrac{b^2}{h+2b}$ $J=\dfrac{t^3}{3}(h+2b)$ $I_\omega=\dfrac{h^2 b^3 t}{12}\dfrac{2h+3b}{h+6b}$	$\beta_1=\dfrac{hb}{2}\dfrac{h+3b}{h+6b}$ in D, $\forall\zeta$ $\beta_2=\dfrac{hb^2}{4}\left(\dfrac{h+3b}{h+6b}\right)^2$ in M, $\forall\zeta$ $\left(DM=b\dfrac{h+3b}{h+6b}\right)$
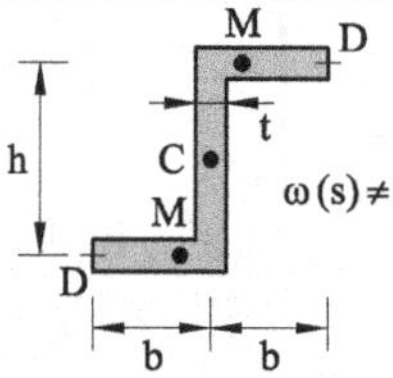 $\omega(s)\neq 0$	$J=\dfrac{t^3}{3}(h+2b)$ $I_\omega=\dfrac{h^2 b^3 t}{12}\dfrac{(b+2h)}{(2b+h)}$	$\beta_1=\dfrac{hb}{2}\dfrac{b+h}{2b+h}$ in D, $\forall\zeta$ $\beta_2=\dfrac{hb^2}{4}\left(\dfrac{b+h}{2b+h}\right)^2$ in M, $\forall\zeta$ $\left(DM=b\dfrac{b+h}{2b+h}\right)$
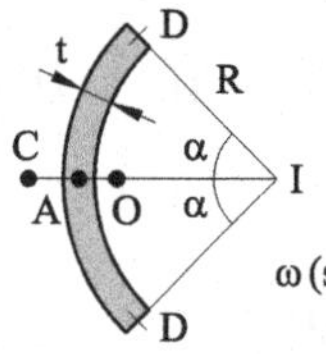 $\omega(s)\neq 0$	$OI=R\dfrac{S}{\alpha}\ ;\ CI=2R\dfrac{S-\alpha C}{\alpha-SC}$ $J=\dfrac{t^3}{3}2R\alpha$ $I_\omega=\dfrac{2tR^5}{3}\left(\alpha^3-6\dfrac{(S-\alpha C)^2}{(\alpha-SC)}\right)$ $S=\sin\alpha\ ;\ C=\cos\alpha$	$e=CA=CI-OI$ $\beta_1=R^2\alpha-ReS$ in D, $\forall\zeta$ $\beta_2=R^2\left(e(1-C)-\dfrac{R\alpha^2}{2}\right)$ in A, $\forall\zeta$

$$\max\left|\sigma_{x'}^\omega\right| = \beta_1 E\left|\frac{\partial^2\theta_{\hat{x}'}}{\partial x'^2}\right| \quad , \quad \max\left|\tau_{x's}^\omega\right| = \beta_2 E\left|\frac{\partial^3\theta_{\hat{x}'}}{\partial x'^3}\right|$$

Fig. 4.27 Torsional inertia (J), warping inertia modulus (I_ω) and value and position of the maximum stresses due to warping for different thin-walled sections

Shear stress due to the warping axial stress $\sigma_{x'}^\omega$

The axial stress $\sigma_{x'}^\omega$ induces an additional shear stress field. The so-called warping shear stress $(\tau_{x's}^\omega)$ can be computed *by integrating the local equilibrium equations a posteriori*. The method follows the arguments used for computing the shear stress distribution in plane beams under bending loads (Section 3.7 and Appendix D). The equilibrium equations are

$$\frac{\partial \sigma_{x'}^\omega}{\partial x'} + \frac{\partial \tau_{x's}^\omega}{\partial s} = 0 \quad \text{with} \quad \tau_{x's}^\omega = 0 \quad \text{at } s = 0, L_s \tag{4.157}$$

Introducing Eq.(4.155) into (4.157) gives

$$\frac{\partial \tau_{x's}^\omega}{\partial s} = -\omega(s,\zeta) E \frac{\partial^3 \theta_{\hat{x}'}}{\partial x'^3} \tag{4.158}$$

The thickness variation of the warping shear stresses is typically neglected. Hence, Eq.(4.158) is rewritten (using Eq.(4.124)) to give

$$\tau_{x's}^\omega(s,\zeta) = -S_\omega(s) E \frac{\partial^3 \theta_{\hat{x}'}}{\partial x'^3} \quad \text{with} \quad S_\omega(s) = \int_0^s g(s)ds \tag{4.159}$$

Consequently, the shear stresses due to warping is zero if $g(s) = 0$. The maximum values are placed at the points where S_ω is a maximum, i.e.

$$\max |\tau_{x's}^\omega| = \beta_2 E |\frac{\partial^3 \theta_{\hat{x}'}}{\partial x'^3}| \tag{4.160}$$

Figure 4.27 shows the value and position of β_2 (and $\max |\tau_{x's}^\omega|$) for different sections.

The total shear stress due to torsion is the sum of the Saint-Venant shear stress (Eq.(4.153)) and the warping shear stress (Eq.(4.159)).

Figure 4.28 shows the distribution of $\tau_{x's}$ in a thin-walled open section.

Example 4.3: Computation of ω, S_ω and I_ω for a double L-shaped section.

- Solution

Let us consider the thin-walled section shown in the next page:

Distribution of $\omega_s(s)$

DA segment: $0 \le s \le b$; $-b \le y \le 0$; $z = -\frac{h}{2}$

$$ds = dy \quad, \quad c_n = -\frac{h}{2} \quad, \quad \omega_s(s) = \int_0^s c_n ds = -\frac{h}{2}s$$

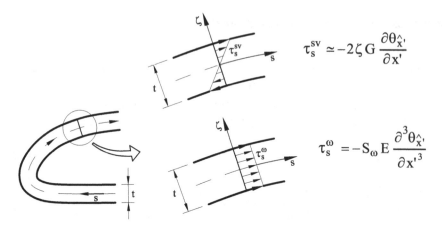

Fig. 4.28 Shear stresses across the thickness at a point of a thin-walled open section. The total shear stress is the sum of the Saint-Venant shear stress and the warping shear stress

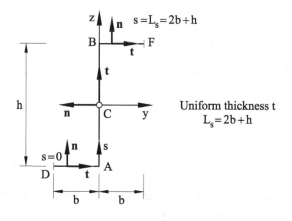

AB segment: $b \leq s \leq b+h$; $y = 0$; $-\frac{h}{2} \leq z \leq \frac{h}{2}$

$$ds = dz \quad , \quad c_n = 0 \quad , \quad \omega_s(s = b) = -\frac{h}{2}b$$

BF segment: $b + h \leq s \leq b + 2h$; $0 \leq y \leq b$; $z = \frac{h}{2}$

$$ds = dy \quad , \quad c_n = \frac{h}{2} \quad , \quad \omega_s(s) = -\frac{h}{2}b + \int_{b+h}^{s} \frac{h}{2}ds = -\frac{h}{2}b + \frac{h}{2}(s - b - h)$$

Mean value: $\omega_m = -\omega_D$

$$\omega_m = \frac{1}{2b + h} \int_0^{2b+h} \omega_s ds = -\frac{bh(b + h)}{2L_s}$$

Sectorial area ω and S_ω:

$$\omega(s,\zeta) = g(s) - c_t(s)\zeta \quad \text{with} \quad g(s) = \omega_s(s) + \omega_D$$

Neglecting the thickness variation, $\omega(s) = g(s)$ and $S_\omega(s,0) = \int_0^s g(s)ds$.

Sectorial inertia modulus: $I_\omega = \int_A \omega^2 dA = \frac{th^2b^3}{12}\left(\frac{b+2h}{2b+h}\right)$

The contribution of the term $-c_t\zeta$ of $\omega(s,\zeta)$ in I_ω is negligible. Its value is $I_\omega = \frac{t^3}{12}\left(\frac{2b^3}{3} + \frac{h^3}{12}\right)$

The figures below shows the distribution of $\omega_s(s), g(s)$ and $S_\omega(s)$ in the double L-shaped section considered.

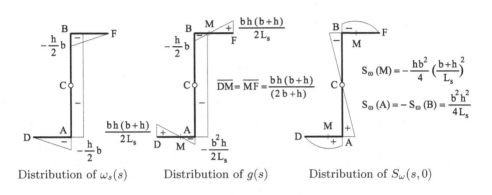

Distribution of $\omega_s(s)$ Distribution of $g(s)$ Distribution of $S_\omega(s,0)$

4.11 THIN-WALLED OPEN TIMOSHENKO BEAM ELEMENTS ACCOUNTING FOR SHEAR STRESSES DUE TO TORSION

4.11.1 Basic equations

The shear stresses induced by torsion can be important in short clamped beams and in open thin-walled beams with composite material. These terms can be taken into account in the theory presented previously following similar arguments to those used for introducing transverse shear deformation effects in classical beam theory. The twist angle (ϕ_ω) is now assumed to be the sum of the change of the twist rotation $\left(\frac{\partial\theta_{\hat{x}'}}{\partial x'}\right)$ and an additional angle (ϕ_s) induced by the shear stresses due to torsion (Figure 4.29), i.e.

$$\phi_\omega = \frac{\partial\theta_{\hat{x}'}}{\partial x'} + \phi_s \tag{4.161}$$

The angle ϕ_s can be interpreted as (minus) the shear deformation γ_t introduced by torsion effects (Figure 4.29). Clearly if $\phi_s = 0$, then $\phi_\omega =$

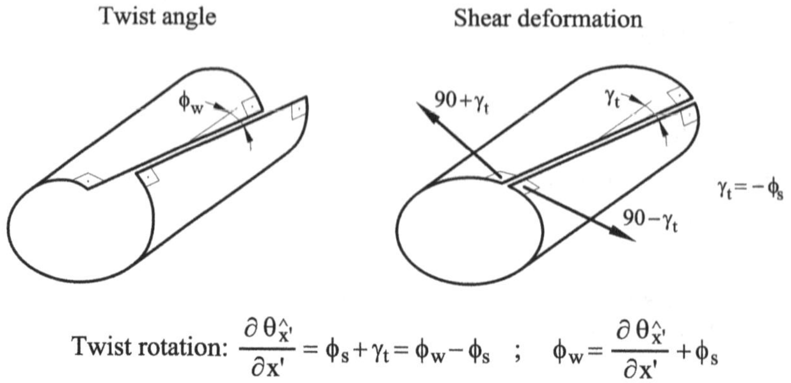

Twist angle   Shear deformation

Twist rotation: $\dfrac{\partial \theta_{\hat{x}'}}{\partial x'} = \phi_s + \gamma_t = \phi_w - \phi_s$; $\phi_w = \dfrac{\partial \theta_{\hat{x}'}}{\partial x'} + \phi_s$

Fig. 4.29 Twist angle (ϕ_ω) and shear deformation due to torsion (γ_t)

$\frac{\partial \theta_{\hat{x}'}}{\partial x'}$ and we recover the classical definition for the twist angle of Eq.(4.23) [Va,VOO].

The displacement field induced by torsion effects is written in the local axes x', t, n in the new theory as

$$u' = \omega \phi_\omega \quad , \quad u_t = -(c_n + \zeta)\theta_{\hat{x}'} \quad , \quad u_n = c_t \theta_{\hat{x}'} \tag{4.162}$$

The only difference with Eq.(4.118) is the definition of the axial displacement. Note that ϕ_ω is now taken as an independent variable.

The *strains induced by torsion* (denoted hereafter $\boldsymbol{\varepsilon}'_t$) are

$$\boldsymbol{\varepsilon}'_t = \left\{ \begin{array}{c} \omega \dfrac{\partial \phi_\omega}{\partial x'} \\[2mm] \dfrac{\partial \omega}{\partial s}\phi_\omega - (c_n + \zeta)\dfrac{\partial \theta_{\hat{x}'}}{\partial x'} \\[2mm] \dfrac{\partial \omega}{\partial \zeta}\phi_\omega + c_t \dfrac{\partial \theta_{\hat{x}'}}{\partial x'} \end{array} \right\} \tag{4.163}$$

Using Eq.(4.129) and assuming that $\frac{\partial \omega}{\partial \zeta} = -c_t$ and $\frac{\partial c_t}{\partial s} = 1$ we find

$$\boldsymbol{\varepsilon}'_t = \left\{ \begin{array}{c} \omega \dfrac{\partial \phi_\omega}{\partial x'} \\[2mm] c_n\phi_s - \zeta\left(\phi_\omega + \dfrac{\partial \theta_{\hat{x}'}}{\partial x'}\right) \\[2mm] c_t\left(\dfrac{\partial \theta_{\hat{x}'}}{\partial x'} - \phi_\omega\right) \end{array} \right\} = \mathbf{S}_t \hat{\boldsymbol{\varepsilon}}'_t \tag{4.164a}$$

with

$$\mathbf{S}_t = \begin{bmatrix} \omega & 0 & 0 \\ 0 & -2\zeta & c_n \\ 0 & 0 & c_t \end{bmatrix} \quad , \quad \hat{\boldsymbol{\varepsilon}}'_t = \left\{ \begin{array}{c} \kappa_w \\ \kappa_{x's} \\ \gamma_t \end{array} \right\} = \left\{ \begin{array}{c} \dfrac{\partial \phi_\omega}{\partial x'} \\[2mm] \frac{1}{2}\left(\phi_\omega + \dfrac{\partial \theta_{\hat{x}'}}{\partial x'}\right) \\[2mm] \dfrac{\partial \theta_{\hat{x}'}}{\partial x'} - \phi_\omega \end{array} \right\} \tag{4.164b}$$

The resultant stresses due to torsion are (accepting that $\frac{c_n}{R} = 0$)

$$\hat{\boldsymbol{\sigma}}'_t = \left\{ \begin{matrix} M_\omega \\ M_{\hat{x}'} \\ M_t \end{matrix} \right\} = \iint_A \left\{ \begin{matrix} \omega\sigma_{x'} \\ -2\zeta\tau_{x's} \\ c_t\tau_{x'\zeta} + c_n\tau_{x's} \end{matrix} \right\} dA = \iint_A [\mathbf{S}_t]^T \boldsymbol{\sigma}'_t dA \qquad (4.165)$$

The constitutive equation is assumed to be of the form

$$\boldsymbol{\sigma}'_t = \left\{ \begin{matrix} \sigma_{x'} \\ \tau_{x's} \\ \tau_{x'\zeta} \end{matrix} \right\} = \begin{bmatrix} E & 0 & 0 \\ 0 & G & 0 \\ 0 & 0 & G \end{bmatrix} \left\{ \begin{matrix} \varepsilon_{x'} \\ \gamma_{x's} \\ \gamma_{x'\zeta} \end{matrix} \right\} = \mathbf{D}'\boldsymbol{\varepsilon}'_t \qquad (4.166)$$

Substituting this equation into (4.165) gives

$$\hat{\boldsymbol{\sigma}}'_t = \hat{\mathbf{D}}_t\hat{\boldsymbol{\varepsilon}}'_t \quad \text{with} \quad \hat{\mathbf{D}}_t = \iint_A [\mathbf{S}_t]^T \mathbf{D}'\mathbf{S}_t dA \qquad (4.167)$$

A simple multiplication gives

$$\hat{\mathbf{D}}_t = \iint_A \begin{bmatrix} E\omega^2 & 0 & 0 \\ 0 & 4\zeta^2 G & -2\zeta c_n G \\ 0 & -2\zeta c_n G & |c|^2 G \end{bmatrix} dA \quad \text{with} \quad |c|^2 = c_n^2 + c_t^2 \qquad (4.168)$$

A comparison of Eqs.(4.140a) and (4.168) shows the terms introduced by shear deformation in the torsion constitutive matrix. For homogeneous material

$$\hat{\mathbf{D}}_t = \begin{bmatrix} \hat{D}_\omega & 0 & 0 \\ 0 & \hat{D}_t & 0 \\ 0 & 0 & \hat{D}_{s_t} \end{bmatrix} \qquad (4.169)$$

where $\hat{D}_\omega$ and $\hat{D}_t$ coincide with the expressions in Eq.(140b) $\left(\text{for } \frac{c_n}{R} = 0\right)$ and $\hat{D}_{s_t} = \iint_A |c|^2 G dA$.

Using Eqs.(4.166), (4.164b) and (4.167) we can deduce that torsion effects contribute the following terms to the PVW

$$\iiint_V \left(\delta\varepsilon_{x'}\sigma_{x'} + \delta\varepsilon_{x's}\tau_{x's} + \delta\varepsilon_{x'\zeta}\tau_{x'\zeta} \right) dA dx' =$$
$$= \int_L \left(\delta\kappa_\omega M_\omega + \delta\gamma_t M_t + \delta\kappa_s M_{\hat{x}'} \right) dx' = \int_L \delta\hat{\boldsymbol{\varepsilon}}'_t \boldsymbol{\sigma}'_t dx' \qquad (4.170)$$

If $\phi_s = 0$ then $\gamma_t = 0$, $\kappa_s = \frac{\partial\theta_{\hat{x}'}}{\partial x'}$, $\kappa_\omega = \frac{\partial^2\theta_{\hat{x}'}}{\partial x'^2}$ and the PVW recovers the expression of Eq.(4.145) for the torsion terms.

We note that only the first derivative of the twist angle $\theta_{\hat{x}'}$ appears in the PVW. This allows us to use a C^0 *continuous interpolation for all the displacement variables.* These variables include the twist angle ϕ_w as an additional degree of freedom.

4.11.2 Finite element discretization

The displacement interpolation for a two-noded beam element is written as

$$\mathbf{u}' = \sum_e N_i \mathbf{a}_i^{'(e)} \quad \text{with} \quad \mathbf{a}_i^{'(e)} = [u'_{0_i}, v'_{c_i}, w'_{c_i}, \theta_{\hat{x}'_i}, \theta_{y'_i}, \theta_{z'_i}, \phi_{w_i}]^T \quad (4.171)$$

In Eq.(4.171) N_i are the standard 1D linear shape functions (Figure 2.4).

The generalized strains are expressed in terms o the nodal DOFs as

$$\hat{\varepsilon}' = \sum_{i=1}^{2} \mathbf{B}'_i \mathbf{a}_i^{'(e)} \quad (4.172a)$$

with

$$\hat{\varepsilon}' = \left[\frac{\partial u'_0}{\partial x'}, \left(\frac{\partial v'_c}{\partial x'} - \theta_{z'} \right), \left(\frac{\partial w'_c}{\partial x'} + \theta_{y'} \right), \frac{\partial \theta_{y'}}{\partial x'}, \frac{\partial \theta_{z'}}{\partial z'}, \frac{\partial \phi_w}{\partial x'}, \right.$$
$$\left. \frac{1}{2} \left(\phi_w + \frac{\partial \theta_{\hat{x}'}}{\partial x'} \right), \left(\frac{\partial \theta_{\hat{x}'}}{\partial x'} - \phi_w \right) \right]^T \quad (4.172b)$$

and

$$\mathbf{B}'_i = \begin{bmatrix} \mathbf{B}_{a_i} \\ \cdots \\ \mathbf{B}_{f_i} \\ \cdots \\ \mathbf{B}_{t_i} \end{bmatrix} \quad (4.172c)$$

where $\mathbf{B}_{a_i}$ and $\mathbf{B}'_{f_i}$ are obtained by extending the expressions in Eq.(4.78) with a column of zeros and $\mathbf{B}'_{t_i}$ are the torsion contributions to the generalized strain matrix given by

$$\mathbf{B}'_{t_i} = \begin{bmatrix} 0\,0\,0 & 0 & 0\,0 & \dfrac{\partial N_i}{\partial x'} \\ 0\,0\,0 & \dfrac{1}{2}\dfrac{\partial N_i}{\partial x'} & 0\,0 & \dfrac{1}{2}N_i \\ 0\,0\,0 & \dfrac{\partial N_i}{\partial x'} & 0\,0 & -N_i \end{bmatrix} \quad (4.173)$$

The element stiffness matrix has the standard form

$$\mathbf{K}_{ij}^{\prime(e)} = \int_{l^{(e)}} \mathbf{B}_i^{\prime T} \hat{\mathbf{D}}' \mathbf{B}_j' \, dx' \quad , \quad i,j = 1,2 \qquad (4.174a)$$

where

$$\hat{\mathbf{D}}' = \begin{bmatrix} \hat{D}_a & 0 & 0 \\ 0 & \hat{\mathbf{D}}_f & 0 \\ 0 & 0 & \hat{\mathbf{D}}_t \end{bmatrix} \qquad (4.174b)$$

Introducing Eqs.(4.172c) and (4.174b) into (4.174a) gives

$$\mathbf{K}_{ij}^{\prime(e)} = \int_{l^{(e)}} \left[\mathbf{B}_{a_i}^T \hat{D}_a \mathbf{B}_{a_j} + \mathbf{B}_{f_i}^T \hat{\mathbf{D}}_f \mathbf{B}_{f_j} + \mathbf{B}_{t_i}^T \hat{\mathbf{D}}_t \mathbf{B}_{t_j} \right] dx' \qquad (4.175)$$

Shear locking induced by flexural and torsion effects can be eliminated in the 2-noded beam element using a one-point reduced quadrature for integrating *all the terms* in the element stiffness matrix.

The same procedure can be followed for deriving three-noded (quadratic) and four-noded (cubic) 3D thin-walled open beam elements accounting for shear deformation effects due to torsion. The performance of the quadratic element improves using a uniform reduced two-point quadrature. The cubic element has an excellent behaviour using a full four-point quadrature [Va,VOS].

Alternative numerical and analytical procedures for analysis of thin-walled open beams accounting for the shear deformation induced by torsion can be found in [BT2,BW2,FM2, Ko2,KP,KSK,Le,LL3,PK,ST].

4.11.3 Examples

Cantilever composite laminated beam under end torque

Figure 4.30 shows a section of a composite laminated cantilever beam of $L = 1000$ mm with double-T section. Each member has seven plies with symmetric orientations $[0,90,0,90,0,90,0]$ with respect to the beam axis. We have considered two different loading cases: (a) a torque of 1 KN×mm acting at the free end, and (b) a vertical load of 1KN also acting at the free end. Both problems have been solved with different meshes of 2-noded linear beam elements and 3-noded quadratic beam elements with full and reduced integration and 4-noded cubic beam elements with full integration. The same problems have been analyzed with the mesh of 550 flat shell DKQ quadrilaterals (Section 8.12.3) also shown in Figure 4.30 which is taken as the reference solution for comparison purposes.

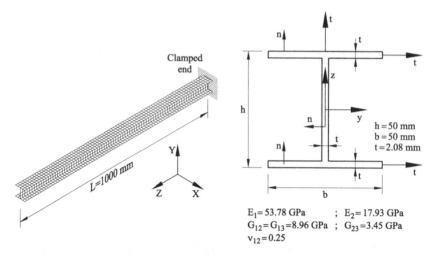

Fig. 4.30 Double-T cantilever beam. Geometry and mesh of 550 DKQ shell quadrilaterals used as a reference solution [Va]

Figure 4.31 displays the ratio between the beam and shell results for the twist rotation of the free end section (for the torque load) and the vertical deflection ratio at the gravity center of the free end section (for the end load) in terms of the beam slenderness ratio $\lambda = L/h$ *for the one element mesh.* The graphs show that:

- The 2-noded linear beam element with full 2 point integration (L-F) locks for slender beams. The one-point uniform reduced quadrature (L-R) eliminates shear locking and yields excellent results for thick and slender beams.
- The 3-noded quadratic beam element with full 3 point integration (Q-F) presents a slight locking behaviour for slender beams. Excellent results are obtained for all cases using the two-point uniform reduced quadrature (Q-R).
- The 4-noded cubic beam element with full 4 point integration (C-F) is locking-free and gives accurate solutions for thick and slender beams.

Figure 4.32 shows the convergence of the free end deflection and twist rotation ratios for a slender beam ($\lambda = 20$) with the number of elements for the linear element (L-R), the quadratic element (Q-F and Q-R) and the cubic element (C-F). All solutions converge fast to the reference values.

We note that $\theta \equiv \theta_{\hat{x}'}$ has been used in Figures 4.31 and 4.32 for simplicity.

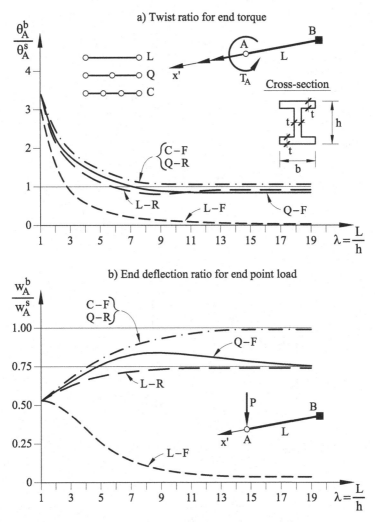

Fig. 4.31 Double-T cantilever beam. Results for twist ratio θ_A^b/θ_A^s under end torque (a) and deflection ratio w_A^b/w_A^s for end point load (b) at end point A in terms of the beam slenderness (λ) for single element meshes of linear (L), quadratic (Q) and cubic (C) beam elements with full (F) and reduced (R) integration. $(\cdot)^b$ beam solution; $(\cdot)^s$ solution with 500 DKQ flat shell elements

U-shaped cantilever and clamped beams under end point loads

The next example is the analysis of a U-shaped composite laminated cantilever beam under two point loads acting at the free end. Figure 4.33 shows the geometry of the beam and the material properties. The problem has been solved with the following three elements:

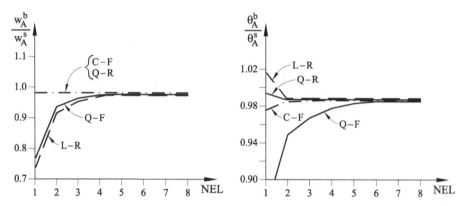

Fig. 4.32 Double-T cantilever slender beam ($\lambda = 20$). Composite material. Convergence of deflection ratio and twist ratio at the beam end with the number of beam elements (NEL). $(\cdot)^b$ and $(\cdot)^s$ denote beam and shell solutions

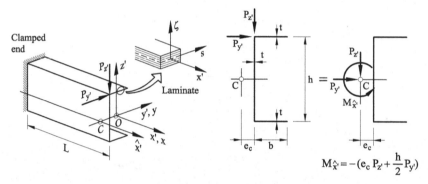

$$M_{\hat{x}'} = -(e_c\,P_{z'} + \frac{h}{2}P_{y'})$$

Fig. 4.33 U-shaped cantilever beam under end point loads. Geometric description and loads. Dimensions: $h = 200$ mm; $b = 60$ mm; $t = 10$ mm; $L = 2000$ mm

- Mesh of twenty 2-noded open thin-walled Timoshenko beam elements accounting for the shear stresses due to torsion with one-point uniform reduced integration (Section 4.11).
- Mesh of twenty 3-noded open thin-walled Euler-Bernoulli beam elements based on Saint-Venant theory (Section 4.7).
- Mesh of 550 9-noded DKQ shell quadrilaterals (Section 8.12.3).

The problem has been solved for the following two different materials and loads:

Homogeneous isotropic material (steel).

$E = 210000$ MPa , $\nu = 0,30$
Loads: $P_{y'} = 25000$ N , $P_{z'} = -25000$ N

Composited laminated material. Laminate with ten plies $[90,0_4]_s$ of glass epoxy matrix with the following properties:

10 plies

$E_1 = 53780$ MPa   , $E_2 = 17930$ MPa , $\nu_{12} = 0,25$

$G_{12} = G_{11} = 8960$ MPa   , $G_{23} = 3450$ MPa

Loads: $P_{y'} = 250$ N   , $P_{z'} = -250$ N

90°
0°
0°
0°
0° — Symmetry axis

	2-noded Euler-Bernoulli/ Saint-Venant beam element (Section 4.7)		2-noded Timoshenko/ beam element (Section 4.10)		DKQ shell element	
	Homogeneous material	Composite material	Homogeneous material	Composite material	Homogeneous material	Composite material
v_c mm	301.86	13.648	301.91	13.662	304.68	13.676
w_c mm	-16.998	-0.766	-17.219	-0.787	-18.667	-0.790
$\theta_{\hat{x}'}$ rad	-0.692	-0.062	-0.545	-0.044	-0.512	-0.045

Table 4.3 U-shaped cantilever under end point loads. Maximum value of lateral and vertical displacements of the shear center (v_c and w_c) and twist rotation ($\theta_{\hat{x}'}$)

Figure 4.34 shows the distribution along the beam of the vertical deflection at the shear center and the twist angle for homogeneous material for the three elements considered. Results are normalized with the maximum value of the DKQ solution. Results for the composite laminated material are practically coincident with those of Figure 4.34.

The 2-noded Timoshenko beam element yields very accurate results. Note the discrepancy in the twist angle results for the 2-noded Euler-Bernoulli beam element based on Saint-Venant theory.

Table 4.3 shows the maximum values for the lateral and vertical dis placements of the shear center and the twist rotation with the three elements considered for the homogeneous and composite laminated sections. The distribution of the horizontal and lateral displacements along the beam is practically coincident for the three elements.

Figure 4.35 and Table 4.4 show a similar set of results for a *clamped* U beam of the same dimensions under eccentric point loads acting at the central section. The beam has been studied for the same two types of homogeneous and composite laminated material. The conclusions are the same as for the cantilever beam.

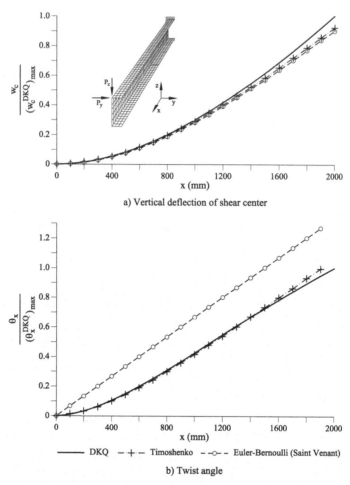

a) Vertical deflection of shear center

b) Twist angle

— DKQ   — + — Timoshenko   – –o– – Euler-Bernoulli (Saint Venant)

Fig. 4.34 U-shaped homogeneous cantilever under end point loads. Distribution of vertical deflection of shear center (a) and twist angle (b). Results for meshes of 20 Timoshenko and 20 Euler-Bernoulli (Saint-Venant) 2-noded beam elements and 550 DKQ shell quadrilaterals (shown in figure). Results are normalized with maximum deflection and twist obtained with the DKQ element mesh

As a general conclusion, the simple two-noded 3D rod element with a single point integration has an excellent behaviour for analysis of thin-walled open beams.

4.12 DEGENERATED 3D BEAM ELEMENTS

3D beam elements can be also derived by imposing the following constrains to solid elements:

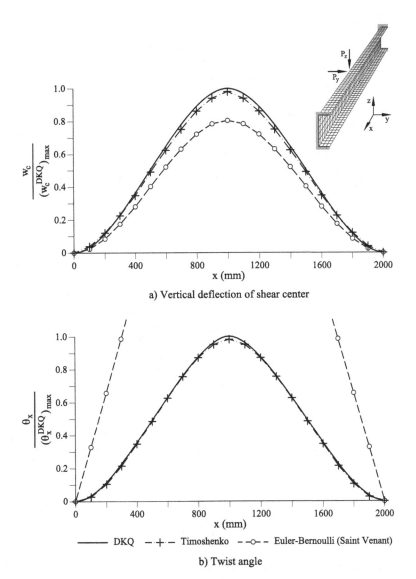

a) Vertical deflection of shear center

b) Twist angle

Fig. 4.35 Clamped homogeneous U-shaped beam under eccentric point loads acting at the center. Distribution of vertical deflection of shear center (a) and twist angle (b) for meshes of 20 Timoshenko and 20 Euler-Bernoulli (Saint-Venant) 2-noded beam elements and 550 DKQ shell quadrilaterals (shown in figure). Results are normalized with the deflection and maximum twist obtained with the DKQ element mesh

1. Linear variation of the displacements over each section (Saint-Venant's plane section assumption),
2. No changes of the section dimensions (which limits the nodal DOFs),

	2-noded Euler-Bernoulli/ (Saint-Venant) beam element (Section 4.7)		2-noded Timoshenko/ beam element (Section 4.10)		DKQ shell element	
	Homogeneous material	Composite material	Homogeneous material	Composite material	Homogeneous material	Composite material
v_c mm	0.472	0.213	0.473	0.217	0.492	0.233
w_c mm	-0.0266	-0.01197	-0.0321	-0.01733	-0.0328	-0.01762
$\theta_{\hat{x}'}$ rad	-0.0173	-0.01560	-0.0051	-0.00281	-0.0052	-0.00292

Table 4.4 Clamped U-shaped beam under eccentric loads acting at central section. Maximum value of lateral (v_c) and vertical (w_c) displacements of the shear center and twist rotation for homogeneous and composite laminated materials

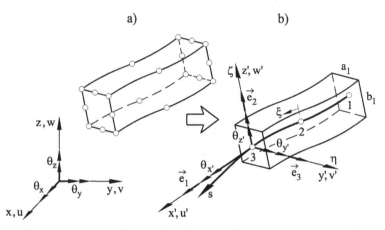

Fig. 4.36 (a) 20-noded quadratic hexahedral element and (b) Degenerated 3-noded quadratic 3D beam element

3. Plane stress assumption in local axes ($\sigma_{y'} = \sigma_{z'} = \gamma_{y'z'} = 0$).

The process is analogous to that explained to derive degenerated shell elements in Chapter 10.

The starting point is a prismatic element. For simplicity only hexahedra will be considered here (Figure 4.36). This limits the formulation to beams with rectangular cross section. Other sections can be modelled by using an equivalent rectangular section with the same area and inertia properties.

4.12.1 Description of geometry and displacement field

The beam reference axis is defined as the line joining the centers of gravity of the sections. A local curvilinear coordinate system x', y', z' is defined at each node on the reference line so that x' is tangent to the beam axis and

y', z' are the principal inertia directions of the section (Figure 4.36). The element geometry is expressed in isoparametric form as

$$\mathbf{x} = \begin{Bmatrix} x \\ y \\ z \end{Bmatrix} = \sum_{i=1}^{n} N_i(\xi) \left(\mathbf{x}_i + \frac{\eta a_i}{2} \mathbf{e}_{2_i} + \frac{\zeta b_i}{2} \mathbf{e}_{3_i} \right) \tag{4.176}$$

where n is the number of elements, $N_i(\xi)$ is the 1D Lagrangian shape function of node i [On4], $\mathbf{x}_i = [x_i, y_i, z_i]^T$ contains the cartesian coordinates of the node, a_i and b_i are the dimensions of the section at the node and η, ζ are the transverse natural coordinates (Figure 4.36).

The displacement field is defined following the assumption of Timoshenko beam theory for the section rotation as

$$\mathbf{u} = \begin{Bmatrix} u \\ v \\ w \end{Bmatrix} = \sum_{i=1}^{n} N_i(\xi) \left[\mathbf{u}_i + \mathbf{T}_i \left(\frac{\eta a_i}{2} \begin{Bmatrix} -\theta_{z_i'} \\ 0 \\ \theta_{x_i'} \end{Bmatrix} + \frac{\zeta b_i}{2} \begin{Bmatrix} \theta_{y_i'} \\ -\theta_{x_i'} \\ 0 \end{Bmatrix} \right) \right] = \sum_{i=1}^{n} \mathbf{N}_i \mathbf{a}'^{(e)}_i \tag{4.177a}$$

where $\mathbf{a}'^{(e)}_i = [u_i, v_i, w_i, \theta_{x_i'}, \theta_{y_i'}, \theta_{z_i'}]^T$, and

$$\mathbf{N}_i = \left[N_i(\xi)\mathbf{I}_3, \mathbf{T}_i \left(\frac{\eta a_i}{2}\mathbf{I}_\eta + \frac{\xi b_i}{2}\mathbf{I}_\zeta \right) \right] \tag{4.177b}$$

with

$$\mathbf{I}_\eta = \begin{bmatrix} 0 & 0 & -1 \\ 0 & 0 & 0 \\ 1 & 0 & 0 \end{bmatrix}, \quad \mathbf{I}_\zeta = \begin{bmatrix} 0 & 1 & 0 \\ -1 & 0 & 0 \\ 0 & 0 & 0 \end{bmatrix}, \quad \mathbf{T}_i = [\mathbf{e}_{1i}, \mathbf{e}_{2i}, \mathbf{e}_{3i}] \tag{4.177c}$$

We note that the components of vectors $\mathbf{e}_{1i}, \mathbf{e}_{2i}$ and $\mathbf{e}_{3i}$ are expressed in the global coordinate system.

Vector $\mathbf{a}'^{(e)}_i$ contains the three *global* displacements of node i: u_i, v_i, w_i and the three *local* nodal rotations: $\theta_{x_i'}, \theta_{y_i'}, \theta_{z_i'}$ (defined in vector form).

4.12.2 Strain field

Taking into account that the displacements are expressed in different axes, the local and global strains at a point are related as

$$\varepsilon' = \begin{Bmatrix} \varepsilon_{x'} \\ \gamma_{x'y'} \\ \gamma_{x'z'} \end{Bmatrix} = \mathbf{S}\,\varepsilon \tag{4.178a}$$

where $\boldsymbol{\varepsilon}$ is the standard strain vector of 3D elasticity [On4,ZTZ]

$$\boldsymbol{\varepsilon} = [\varepsilon_x, \varepsilon_y, \varepsilon_z, \gamma_{xy}, \gamma_{xz}, \gamma_{yz}]^T = \left[\frac{\partial u}{\partial x}, \frac{\partial v}{\partial y}, \frac{\partial w}{\partial z}, \frac{\partial u}{\partial y} + \frac{\partial v}{\partial x}, \frac{\partial u}{\partial z} + \frac{\partial w}{\partial x}, \frac{\partial v}{\partial z} + \frac{\partial w}{\partial y}\right]^T$$

$$(4.178b)$$

and

$$\mathbf{S} = \begin{bmatrix} (e_1^x)^2 & (e_1^y)^2 & (e_1^z)^2 & e_1^x e_1^y & e_1^x e_1^z & e_1^y e_1^z \\ 2e_1^x e_2^x & 2e_1^y e_2^y & 2e_1^z e_2^z & (e_1^y e_2^x + e_1^x e_2^y)(e_1^z e_2^x + e_1^x e_2^z)(e_1^z e_2^y + e_1^y e_2^z) \\ 2e_1^x e_3^x & 2e_1^y e_3^y & 2e_1^z e_3^z & (e_1^y e_3^x + e_1^x e_3^y)(e_1^z e_3^x + e_1^x e_3^z)(e_1^z e_3^y + e_1^y e_3^z) \end{bmatrix}$$

$$(4.178c)$$

where e_1^x, e_1^y, e_1^z are the components in global axes of $\mathbf{e}_1$ at the point where the strains are computed. These components can be obtained by interpolation of the nodal values. The same applies for the components of $\mathbf{e}_2$ and $\mathbf{e}_3$.

The global strain components are obtained in terms of the displacements as follows.

First, the natural derivatives of the global displacements are obtained as

$$\frac{\partial \mathbf{u}}{\partial \xi} = \sum_{i=1}^{n} \frac{\partial N_i}{\partial \xi} \left[\mathbf{u}_i + \mathbf{T}_i \left(\frac{\eta a_i}{2} \left\{ \begin{matrix} -\theta_{z_i'} \\ 0 \\ \theta_{x_i'} \end{matrix} \right\} + \frac{\zeta b_i}{2} \left\{ \begin{matrix} \theta_{y_i'} \\ -\theta_{x_i'} \\ 0 \end{matrix} \right\} \right) \right]$$

$$\frac{\partial \mathbf{u}}{\partial \eta} = \sum_{i=1}^{n} \frac{a_i}{2} N_i \mathbf{T}_i \left\{ \begin{matrix} -\theta_{z_i'} \\ 0 \\ \theta_{x_i'} \end{matrix} \right\} ; \qquad \frac{\partial \mathbf{u}}{\partial \zeta} = \sum_{i=1}^{n} \frac{b_i}{2} N_i \mathbf{T}_i \left\{ \begin{matrix} \theta_{y_i'} \\ -\theta_{x_i'} \\ 0 \end{matrix} \right\} \qquad (4.179)$$

The cartesian derivatives and the natural derivatives of the global displacements are related by the inverse of the 3D jacobian matrix $\mathbf{J}$ as

$$\left[\frac{\partial \mathbf{u}}{\partial x}, \frac{\partial \mathbf{u}}{\partial y}, \frac{\partial \mathbf{u}}{\partial z}\right] = \mathbf{J}^{-1} \left[\frac{\partial \mathbf{u}}{\partial \xi}, \frac{\partial \mathbf{u}}{\partial \eta}, \frac{\partial \mathbf{u}}{\partial \zeta}\right]^T \qquad (4.180a)$$

where

$$\mathbf{J} = \left[\frac{\partial \mathbf{x}}{\partial \xi}, \frac{\partial \mathbf{x}}{\partial \eta}, \frac{\partial \mathbf{x}}{\partial \zeta}\right]^T \quad \text{with} \quad \mathbf{x} = [x, y, z]^T \qquad (4.180b)$$

The elements of the jacobian matrix $\mathbf{J}$ are computed from the isoparametric description (4.176) by

$$
\begin{aligned}
\frac{\partial \mathbf{x}}{\partial \xi} &= \sum_{i=1}^{n} \frac{\partial N_i}{\partial \xi} \left(\mathbf{x}_i + \frac{\eta a_i}{2} \mathbf{e}_{2_i} + \frac{\zeta b_i}{2} \mathbf{e}_{3_i} \right) \\
\frac{\partial \mathbf{x}}{\partial \eta} &= \sum_{i=1}^{n} \frac{a_i}{2} N_i \mathbf{e}_{2_i}; \qquad \frac{\partial \mathbf{x}}{\partial \zeta} = \sum_{i=1}^{n} \frac{b_i}{2} N_i \mathbf{e}_{3_i}
\end{aligned}
\tag{4.181}
$$

The global strain matrix $\mathbf{B}$ is obtained by substituting the displacement interpolation (4.177a) into (4.178b). The full expression is given in Box 4.5. Using this result and Eq.(4.178a) gives the relationship between the local strains and the local displacement as

$$
\boldsymbol{\varepsilon}' = \sum_{i=1}^{n} \bar{\mathbf{B}}' \mathbf{a}_i'^{(e)} = \bar{\mathbf{B}}_i' \mathbf{a}'^{(e)}
\tag{4.182a}
$$

with

$$
\bar{\mathbf{B}}_i' = \left\{ \begin{matrix} \bar{\mathbf{B}}'_{ab_i} \\ \bar{\mathbf{B}}'_{s_i} \end{matrix} \right\} = \mathbf{S} \mathbf{B}_i; \quad \bar{\mathbf{B}}'_{ab_i} = \mathbf{S}_1 \mathbf{B}_i; \quad \bar{\mathbf{B}}'_{s_i} = \mathbf{S}_2 \mathbf{B}_i
\tag{4.182b}
$$

where $\bar{\mathbf{B}}'_{ab_i}$ and $\bar{\mathbf{B}}'_{s_i}$ contain the axial-bending and shear contributions to the local strain matrix and $\mathbf{S}_1$ and $\mathbf{S}_2$ are the first row and the last two rows of matrix $\mathbf{S}$ of Eq.(4.178c), respectively.

The nodal rotations are next expressed in global axes giving the final relationship between the *local strains* and the *global displacements* as

$$
\boldsymbol{\varepsilon}' = \sum_{i=1}^{n} \bar{\mathbf{B}}_i' \mathbf{Q}_i \mathbf{a}_i^{(e)} = \sum_{i=1}^{n} \mathbf{B}_i' \mathbf{a}^{(e)}
\tag{4.183a}
$$

where

$$
\mathbf{B}_i' = \left\{ \begin{matrix} \mathbf{B}'_{ab_i} \\ \mathbf{B}'_{s_i} \end{matrix} \right\} \quad \text{with} \quad \mathbf{B}'_{ab_i} = \bar{\mathbf{B}}'_{mb_i} \mathbf{Q}_i \quad, \quad \mathbf{B}'_{s_i} = \bar{\mathbf{B}}'_{s_i} \mathbf{Q}_i
\tag{4.183b}
$$

with

$$
\mathbf{Q}_i = \begin{bmatrix} \mathbf{I}_3 & \mathbf{0} \\ \mathbf{0} & \mathbf{T}_i \end{bmatrix}
\tag{4.183c}
$$

In above $\mathbf{a}_i^{(e)} = [u_i, v_i, w_i, \theta_{x_i}, \theta_{y_i}, \theta_{z_i}]^T$ and $\mathbf{T}_i$ is the transformation matrix of Eq.(4.177c). Eq.(4.148a) *relates the local strains* and *the global nodal DOFs*.

The constitutive equation is expressed in local axes by Eq.(4.1). This formulation allows us to consider an arbitrary heterogeneous material over the beam section in a straigthforward manner.

$$\boldsymbol{\varepsilon} = [\varepsilon_x, \varepsilon_y, \varepsilon_z, \varepsilon_{xy}, \varepsilon_{xz}, \varepsilon_{yz}]^T = \sum_{i=1}^{n} \mathbf{B}_i \mathbf{a}_i^{\prime(e)} = \mathbf{B} \mathbf{a}^{\prime(e)}$$

$$\mathbf{B} = [\mathbf{B}_1, \mathbf{B}_2, \cdots \cdots, \mathbf{B}_n], \qquad \mathbf{a}_i^{\prime(e)} = [u_i, v_i, w_i, \theta_{x_i'}, \theta_{y_i'}, \theta_{z_i'}]^T$$

$$\mathbf{B}_i = \begin{bmatrix} N_i^1 & 0 & 0 & (\mathbf{G}_i^1)_{11} & (\mathbf{G}_i^1)_{12} & (\mathbf{G}_i^1)_{13} \\ 0 & N_i^2 & 0 & (\mathbf{G}_i^2)_{21} & (\mathbf{G}_i^2)_{22} & (\mathbf{G}_i^1)_{23} \\ 0 & 0 & N_i^3 & (\mathbf{G}_i^3)_{31} & (\mathbf{G}_i^3)_{32} & (\mathbf{G}_i^1)_{33} \\ N_i^2 & N_i^1 & 0 & [(\mathbf{G}_i^2)_{11} + (\mathbf{G}_i^1)_{21}] & [(\mathbf{G}_i^2)_{12} + (\mathbf{G}_i^1)_{22}] & [(\mathbf{G}_i^2)_{13} + (\mathbf{G}_i^1)_{23}] \\ N_i^3 & 0 & N_i^1 & [(\mathbf{G}_i^3)_{11} + (\mathbf{G}_i^1)_{31}] & [(\mathbf{G}_i^3)_{12} + (\mathbf{G}_i^1)_{32}] & [(\mathbf{G}_i^3)_{13} + (\mathbf{G}_i^1)_{33}] \\ 0 & N_i^3 & N_i^2 & [(\mathbf{G}_i^3)_{21} + (\mathbf{G}_i^2)_{31}] & [(\mathbf{G}_i^3)_{22} + (\mathbf{G}_i^2)_{32}] & [(\mathbf{G}_i^3)_{23} + (\mathbf{G}_i^1)_{33}] \end{bmatrix}$$

$$N_i^j = J_{j_1}^{-1} \frac{\partial N_i}{\partial \xi}; \qquad \mathbf{G}_i^j = N_i^j \hat{\mathbf{T}}_i + \frac{a_i}{2} J_{j_2}^{-1} N_i \mathbf{T}_i \mathbf{T}_\eta + \frac{b_i}{2} J_{j_3}^{-1} N_i \mathbf{T}_i \mathbf{T}_\xi$$

$$\hat{\mathbf{T}}_i = \mathbf{T}_i + \frac{\eta a_i}{2} \mathbf{I}_\eta + \frac{\zeta b_i}{2} \mathbf{I}_\xi, \qquad \mathbf{T}_i = [\mathbf{e}_{1_i}, \mathbf{e}_{2_i}, \mathbf{e}_{3_i}]$$

$$\mathbf{I}_\eta = \begin{bmatrix} 0 & 0 & -1 \\ 0 & 0 & 0 \\ 1 & 0 & 0 \end{bmatrix}, \qquad \mathbf{I}_\zeta = \begin{bmatrix} 0 & 1 & 0 \\ -1 & 0 & 0 \\ 0 & 0 & 0 \end{bmatrix}, \qquad \begin{array}{l} J_{ij}^{-1} : \text{term } ij \text{ of the inverse} \\ \text{jacobian matrix} \end{array}$$

Box 4.5. Global strain matrix for a degenerated 3D beam element

4.12.3 Stiffness matrix and equivalent nodal force vector for the element

Substituting Eq.(4.177a) and (4.182a) and the constitutive equation (4.1) into the PVW expression (Eq.(4.62)) gives, after the usual algebra, the element stiffness matrix and the equivalent nodal force vector in *global axes* by

$$\mathbf{K}_{ij}^{(e)} = \iiint_{V^{(e)}} \left[\mathbf{B}_{ab_i}^{\prime T} \mathbf{D}^\prime \mathbf{B}_{ab_j}^\prime + \mathbf{B}_{s_i}^{\prime T} \mathbf{D}^\prime \mathbf{B}_{s_j}^\prime \right] dV = \mathbf{K}_{ab_{ij}}^{(e)} + \mathbf{K}_{s_{ij}}^{(e)}$$

(4.184a)

$$\mathbf{f}_i^{(e)} = \iiint_{V^{(e)}} \mathbf{N}_i^T \mathbf{b} \, dV + \iint_{A^{(e)}} \mathbf{N}_i^T \mathbf{t} \, dA + \mathbf{p}_i^{(e)} \qquad (4.184b)$$

where $V^{(e)}$ is the *volume* of the parent solid element. $\mathbf{K}_{ab}^{(e)}$ and $\mathbf{K}_s^{(e)}$ are the axial-bending and shear contributions to the global stiffness matrices, respectively, $\mathbf{b}$ are the volumetric body forces (self-weight), $\mathbf{t}$ are distributed forces acting on one of the element faces ($\eta = \pm 1$ or $\xi - \pm 1$) and $\mathbf{p}_i^{(e)}$ are nodal point force vectors, respectively. All the load components are defined in global axes as

$$\mathbf{b} = [b_x, b_y, b_z, 0, 0, 0]^T$$
$$\mathbf{t} = [f_x, f_y, f_z, m_x, m_y, m_z]^T \tag{4.185}$$
$$\mathbf{p}_i = [F_{x_i}, F_{y_i}, F_{z_i}, M_{x_i}, M_{y_i}, M_{z_i}]^T$$

If the distributed forces $\mathbf{t}$ act along the beam axis, then the area integral in Eq.(4.184b) is substituted by a line integral over the element length and $\eta = \zeta = 0$ in the expression of N_i of Eq.(4.177b).

The explicit integration over the section for curved elements has some difficulties. This is nevertheless possible following similar procedures as detailed in Chapter 10 for degenerated shell elements. For straight beams the analytical computation of the element stiffness matrix is straightforward. For a 2-noded beam element the stiffness matrix and the equivalent nodal force vector have identical expressions to those derived in Section 4.4 starting from 3D beam theory. In practice, a 3D Gauss quadrature is used for the integration of the element matrices and vectors, i.e.

$$\mathbf{K}_{ij}^{(e)} = \sum_{p=1}^{n_\xi} \sum_{q=1}^{n_\eta} \sum_{r=1}^{n_\zeta} \left(\mathbf{B}_i'^T \, \mathbf{D}' \, \mathbf{B}_j' |\mathbf{J}^{(e)}| \right)_{p,q,r} W_p W_q W_r$$

$$\mathbf{f}_i^{(e)} = \sum_{p=1}^{n_\xi} \sum_{q=1}^{n_\eta} \sum_{r=1}^{n_\zeta} \left(\mathbf{N}_i^T \, \mathbf{b} |\mathbf{J}^{(e)}| \right)_{p,q,r} W_p W_q W_r + \tag{4.186}$$

$$+ \sum_{p=1}^{n_\xi} [\mathbf{N}_i^T \mathbf{t} |\mathbf{J}|]_{P,\eta=0, \, \zeta=0} W_p$$

where n_ξ, n_η n_ζ are the integration points in the directions ξ, η, ζ, respectively and W_p, W_q, W_r are the corresponding weights. For homogeneous material $n_\eta = n_\zeta = 2$ is usually chosen. A higher order quadrature (or even a cell integration) over the cross section is necessary for arbitrary heterogeneous beams. For a laminated section a layer integration suffices.

Shear locking is avoided by using a reduced quadrature for integrating the shear stiffness matrix $\mathbf{K}_s^{(e)}$ along the element length. Typically $n_\xi = 1$ and $n_\xi = 2$ are chosen for the 2 and 3-noded degenerate beam elements, respectively, as for plane beams. For straight elements $\mathbf{K}_{ab}^{(e)}$ a full quadrature is typically used ($\eta_\xi = 2$ and $n_\xi = 3$ are taken for 2 and 3-noded beam elements, respectively). Uniform reduced integration for *all the stiffness* terms in the curved case is recommended to alleviate axial locking (Sections 9.5 and 10.11.1).

This formulation can be adapted to Euler-Bernoulli theory by making $\theta_{z'} = \frac{\partial v'}{\partial x'}$, $\theta_{x'} = -\frac{\partial w'}{\partial x'}$, thus satisfying the normal orthogonality condition. This leads to the vanishing of the shear strains due to bending and introduces the need for a C^1 continuous approximation for the local displacements v' and w', as their second derivatives appear in the expression of the axial strain. A C° continuous interpolation can be still chosen for the axial displacement u'. This can be implemented by defining the displacement interpolation in the local axes followed by the transformation to global displacement components. The simplest 2-noded degenerated Euler-Bernoulli beam element uses a cubic Hermite approximation for v' and w' and a linear field for u'. For homogeneous material, constant section and uniform loading the expression for the stiffness matrix and the equivalent nodal force vector coincide with those expressions derived using classical 3D beam theory (Section 4.7).

4.13 CONCLUDING REMARKS

We have shown the formulation of 3D beam elements adequate for analysis of composite beams using Timoshenko and Euler-Bernoulli theories. Both the Saint-Venant free-torsion theory and the more sophisticated torsion theory allowing for warping effects in thin-walled open sections have been studied in some detail. The simple 2-noded Timoshenko beam element with a single integration point is a useful alternative for practical analysis of all kind of 3D beams.

3D beam elements obtained by degeneration of 3D solid element can be an interesting option for some cases.

The coupling of 3D beam elements with plate/shell elements is straightforward and provides a useful approach for analysis of stiffned shell/plate structures. This topic is studied in Section 10.21.

5

THIN PLATES.
KIRCHHOFF THEORY

5.1 INTRODUCTION

This chapter introduces the study of structures formed by "thin surfaces" such as plates and shells. Plates will be studied in this and the two following chapters. Shell structures formed by assembly of flat plates will be considered in Chapter 8. Axisymmetric shells will be treated in Chapter 9. Finally, the more general case of curved shell structures of arbitrary shape will be studied in Chapter 10.

Plate theory is a simplification of 3D elasticity analogous to that made in Chapters 1–3 for the analysis of beams, which serves as a solid foundation to what follows.

Like for beams, plate theories differ on the assumptions for the rotation of the normal to the middle plane. The classic thin plate theory establishes that the normal remains straight and orthogonal to the middle plane after deformation. Thin plate theory is based on the assumptions formalized by Kirchhoff in 1850 [Ki] and indeed his name is often associated with this theory, through an early version was proposed by Sophie Germain in 1811 [BD7,Re3,TW]. The more advanced thick plate theory proposed by Reissner [Re] and Mindlin [Mi] assumes that normals remain straight, though not necessarily orthogonal to the middle plane after deformation.

This chapter presents the formulation of plate elements following Kirchhoff thin plate theory. Like for Euler-Bernouilli beam elements, Kirchhoff plate elements require C^1 continuity of the deflection field due to the presence of second derivatives of the deflection in the virtual work expression. However, unlike beam elements, Kirchhoff plate elements have serious difficulties for satisfying the continuity requirements between elements. This leads to "non-conforming" elements, some of which can be still applied to practical situations.

E. Oñate, *Structural Analysis with the Finite Element Method. Linear Statics: Volume 2: Beams, Plates and Shells*, Lecture Notes on Numerical Methods in Engineering and Sciences, DOI 10.1007/978-1-4020-8743-1_5,
© International Center for Numerical Methods in Engineering (CIMNE), 2013

A part of this chapter studies rotation-free Kirchhoff plate elements with the deflection as the only nodal variable. These elements are an extension of the rotation-free beam elements studied in Chapter 1 and are competitive for many practical applications. Their formulation combines finite element and finite volume concepts.

Plate elements based on Reissner-Mindlin theory will be studied in the next chapter. These elements are analogous to Timoshenko beam elements and include the effect of shear deformation. This makes them applicable for both thick and thin situations and they require only C^0 continuity for the deflection. Reduced integration and equivalent procedures are necessary to avoid shear locking, like for slender Timoshenko beam elements.

Which of the plate theories is more appropiate? This is the logical question that an unexperienced reader would ask in order not to waste time studying concepts that may be of little use, or even obsolete. In answer, we would say that the study of both Kirchhoff and Reissner-Mindlin theories is highly recommended. Kirchhoff plate elements are available in most commercial codes and are continuously evolving due to the introduction of new concepts, such as the rotation-free formulation [OZ] and the isogeometric theory (Section 10.10 and [CHB]). Reissner-Mindlin plate elements, on the other hand, are attractive thanks to their versatility for analysis of thick and thin plates and the simplicity of their formulation. However, special care should be taken when using Reissner-Mindlin elements to avoid problems like shear locking or spureous mechanisms. In the next chapter we will show that some interesting thin plate elements can be derived by constraining the transverse shear strain to zero in Reissner-Mindlin elements.

5.2 KIRCHHOFF PLATE THEORY

5.2.1 Main assumptions

A plate is defined as a flat solid whose thickness is much smaller than its other dimensions. We assume that the *middle plane* is equidistant from the upper and lower faces. This plane is taken as the reference plane ($z = 0$) for deriving the plane kinematic equations. A plate with homogeneous isotropic material carries lateral loads by bending, like a straight beam (Figure 5.1). Hence the axial straining is zero, the middle plane coincides with the neutral plane and the displacement field can be expressed in terms of the lateral deflection and the rotations of the normal (the so called *bending*

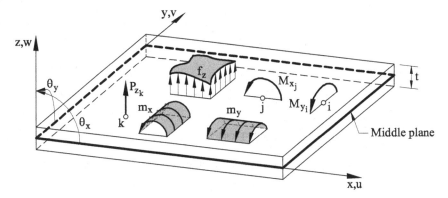

Fig. 5.1 Geometric definition of a plate. Sign convention for displacements, rotations, distributed and point loads and moments

state). If in-plane loading is present or the material is heterogeneous, the axial strains are not zero. This situation is studied in Chapters 7 and 8.

The assumptions of Kirchhoff thin plate theory are the following:

1. In the points belonging to the middle plane ($z = 0$)

$$u = v = 0 \qquad (5.1)$$

In other words, the points on the middle plane only move vertically.
2. The points along a normal to the middle plane have the same vertical displacement (i.e. the thickness does not change during deformation).
3. The normal stress σ_z is negligible (plane stress assumption).
4. A straight line normal to the undeformed middle plane *remains straight* and *normal* to the deformed middle plane (normal orthogonality condition).

Assumptions 1, 2 and 4 allow the displacement field to be defined over the whole plate. Assumption 3 affects the stress-strain relationship, as shown in Section 5.2.3.

5.2.2 Displacement field

From assumptions 1, 2, 4 and Figure 5.2, we deduce

$$\left.\begin{array}{l} u(x,y,z) = -z\theta_x(x,y) \\ v(x,y,z) = -z\theta_y(x,y) \end{array}\right\} \text{ (assumptions 1 and 4)} \qquad (5.2)$$
$$w(x,y,z) = w(x,y) \qquad \text{(assumption 2)}$$

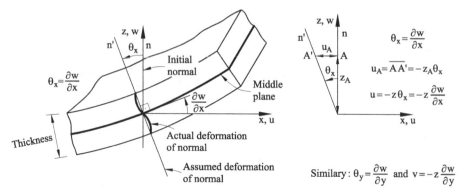

Fig. 5.2 Deformation of the normal vector and in-plane displacement field in a thin plate

where w is the vertical displacement (deflection) of the points on the middle plane and the rotations θ_x and θ_y coincide with the angles followed by the normal vectors contained in the planes xz and yz respectively in their motions (assumption 4). Vector

$$\mathbf{u} = [w, \theta_x, \theta_y]^T \tag{5.3}$$

is the *displacement vector* of a point on the middle plane of the plate. Note that $\mathbf{u}$ contains the deflection and the two rotations.

From assumption 4 and Figure 5.2 we deduce

$$\theta_x = \frac{\partial w}{\partial x} \quad , \quad \text{and similarly} \quad \theta_y = \frac{\partial w}{\partial y} \tag{5.4}$$

i.e. *the rotations of the normal coincide with the slopes of the middle plane* at each point.

The *displacement field* in a plate can thus be expressed as

$$u(x, y, z) = -z\frac{\partial w(x, y)}{\partial x}$$
$$v(x, y, z) = -z\frac{\partial w(x, y)}{\partial y} \tag{5.5}$$
$$w(x, y, z) = w(x, y)$$

The displacement vector is written as

$$\mathbf{u} = \left[w, \frac{\partial w}{\partial x}, \frac{\partial w}{\partial y}\right]^T \tag{5.6}$$

The assumption that the normals remain straight is only an approximation since in practice the normals are distorted as shown in Figure 5.2 and the angles θ_x and θ_y depend on the thickness coordinate. The hypothesis of a straight normal is equivalent to assuming an "average" uniform rotation for the normal, which obviously simplifies the kinematics.

The normal orthogonality condition only holds for thin plates (thickness/average side ratio: $t/L \leq 0.05$). For moderately thick ($0.05 \leq t/L < 0.10$) and very thick ($t/L \geq 0.10$) plates, the distortion of the normal during deformation increases. Reissner-Mindlin theory studied in the next chapter represents a better approximation to the actual deformation of the plate in these cases. If the distortion of the normal is large, as for thick slabs or for particular loading types or boundary conditions, then it is necessary to make use of 3D elasticity theory [On4].

5.2.3 Strain and stress fields and constitutive equation

Using the strain-displacement expressions from 3D elasticity [On4] and Eqs.(5.5) gives

$$
\varepsilon_x = \frac{\partial u}{\partial x} = -z\frac{\partial^2 w}{\partial x^2}
$$

$$
\varepsilon_y = \frac{\partial v}{\partial y} = -z\frac{\partial^2 w}{\partial y^2}; \quad \varepsilon_z = 0
$$

$$
\gamma_{xy} = \frac{\partial u}{\partial y} + \frac{\partial v}{\partial x} = -2z\frac{\partial^2 w}{\partial x \partial y} \tag{5.7}
$$

$$
\gamma_{xz} = \frac{\partial w}{\partial x} + \frac{\partial u}{\partial z} = \frac{\partial w}{\partial x} - \frac{\partial w}{\partial x} = 0
$$

$$
\gamma_{yz} = \frac{\partial w}{\partial y} + \frac{\partial v}{\partial z} = \frac{\partial w}{\partial y} - \frac{\partial w}{\partial y} = 0
$$

Eq.(5.7) shows that the normal orthogonality assumption leads to zero transverse shear strains γ_{xz} and γ_{yz}. Therefore, the transverse shear stresses do not contribute to the deformation work. This does not mean that these stresses are insignificant. They can be computed "a posteriori" using the equilibrium conditions as shown in a next section. Note also that the condition of vanishing normal strain ($\varepsilon_z = 0$) is redundant, as the normal stress σ_z vanishes due to plane stress assumption and, hence, the work performed by the normal strain σ_z is zero (i.e. $\sigma_z\varepsilon_z = 0$).

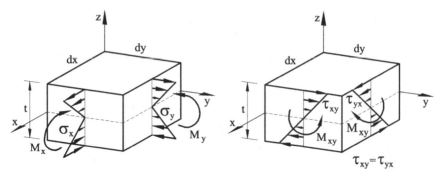

Fig. 5.3 Sign convention for stresses and bending moments

The strain vector containing the three significant strains is

$$\varepsilon = \left[\varepsilon_x, \varepsilon_y, \gamma_{xy}\right]^T = \left[-z\frac{\partial^2 w}{\partial x^2}, -z\frac{\partial^2 w}{\partial y^2}, -2z\frac{\partial^2 w}{\partial x \partial y}\right]^T = \mathbf{S}\hat{\varepsilon}_b \qquad (5.8)$$

with

$$\mathbf{S} = -z\begin{bmatrix}1 & 0 & 0\\0 & 1 & 0\\0 & 0 & 1\end{bmatrix} = -z\mathbf{I}_3 \quad , \quad \hat{\varepsilon}_b = \left[\frac{\partial^2 w}{\partial x^2}, \frac{\partial^2 w}{\partial y^2}, 2\frac{\partial^2 w}{\partial x \partial y}\right]^T \qquad (5.9)$$

where $\hat{\varepsilon}_b$ is the generalized strain vector (or *curvature vector*). Index b in $\hat{\varepsilon}_b$ denotes the bending strains. Matrix $\mathbf{S}$ transforms the curvatures of the middle plane surface into the strains at any point across the thickness.

The strain vector is conjugate to the stress vector

$$\sigma = [\sigma_x, \sigma_y, \tau_{xy}]^T \qquad (5.10)$$

For sign convention, see Figure 5.3.

The constitutive relationship between stresses and strains is written in the standard form as

$$\sigma = \mathbf{D}\varepsilon \qquad (5.11)$$

The constitutive matrix $\mathbf{D}$ is obtained from the general expression of 3D elasticity by introducing the plane stress assumption ($\sigma_z = 0$) and the conditions $\sigma_{xz} = \gamma_{yz} = 0$. The expression of $\mathbf{D}$ coincides with that of *plane stress* theory (Chapter 4 of [On4]) as the significative strains and stresses are the same in both cases. The general expression of $\mathbf{D}$ for an orthotropic material will be derived in Section 6.2.3. In this chapter we will consider *isotropic material* for simplicity. Hence,

$$\mathbf{D} = \frac{E}{1 - \nu^2}\begin{bmatrix}1 & \nu & 0\\\nu & 1 & 0\\0 & 0 & \frac{1-\nu}{2}\end{bmatrix} \qquad (5.12)$$

5.2.4 Bending moments and generalized constitutive matrix

The resultant stress vector (or bending moment vector) is defined as

$$\hat{\sigma}_b = \left\{ \begin{array}{c} M_x \\ M_y \\ M_{xy} \end{array} \right\} = \int_{-\frac{t}{2}}^{+\frac{t}{2}} \mathbf{S}^T \left\{ \begin{array}{c} \sigma_x \\ \sigma_y \\ \tau_{xy} \end{array} \right\} dz = \int_{-\frac{t}{2}}^{+\frac{t}{2}} \mathbf{S}^T \boldsymbol{\sigma} dz \tag{5.13}$$

where M_x and M_y are the bending *moments* produced by the stresses σ_x and σ_y, respectively, and M_{xy} is the *torque* produced by the tangential stress τ_{xy}. The sign convention for the moments is shown in Figure 5.3. The signs of M_x and M_y are consistent with those of θ_x and θ_y, respectively. Also note that $\mathbf{S} \equiv \mathbf{S}^T$, as $\mathbf{S}$ is diagonal (Eq.(5.9)).

Substituting Eqs.(5.11) into (5.13) yields

$$\hat{\sigma}_b = \int_{-\frac{t}{2}}^{+\frac{t}{2}} \mathbf{S}^T \mathbf{D} \boldsymbol{\varepsilon} dz = \int_{-\frac{t}{2}}^{+\frac{t}{2}} \mathbf{S}^T \mathbf{D} \mathbf{S} \hat{\varepsilon}_b dz = \hat{\mathbf{D}}_b \, \hat{\varepsilon}_b \tag{5.14}$$

The generalized bending constitutive matrix $\hat{\mathbf{D}}_b$ in Eq.(5.14) is obtained as

$$\hat{\mathbf{D}}_b = \int_{-\frac{t}{2}}^{+\frac{t}{2}} \mathbf{S}^T \mathbf{D} \mathbf{S} dz = \int_{-\frac{t}{2}}^{+\frac{t}{2}} z^2 \mathbf{D} dz \tag{5.15a}$$

For homogeneous material

$$\hat{\mathbf{D}}_b = \frac{t^3}{12} \mathbf{D} \tag{5.15b}$$

The principal bending moments M_I and M_{II} are the roots of the characteristic polynomial

$$\det[[M] - \lambda \mathbf{I}_2] = 0 \tag{5.16a}$$

where

$$[M] = \begin{bmatrix} M_x & M_{xy} \\ M_{xy} & M_y \end{bmatrix} \quad \text{and} \quad \mathbf{I}_2 = \begin{bmatrix} 1 & 0 \\ 0 & 1 \end{bmatrix} \tag{5.16b}$$

From Eq.(5.16a) we obtain

$$\begin{aligned} M_I &= \frac{M_x + M_y}{2} + \frac{1}{2}[(M_x - M_y)^2 + 4M_{xy}^2]^{1/2} \\ M_{II} &= \frac{M_x + M_y}{2} - \frac{1}{2}[(M_x - M_y)^2 + 4M_{xy}^2]^{1/2} \end{aligned} \tag{5.16c}$$

The sign for the principal bending moments is shown in Figure 5.4. The angle that the principal direction I forms with the x axis is obtained from

$$\text{tg}2\alpha = \frac{2M_{xy}}{M_x - M_y} \tag{5.16d}$$

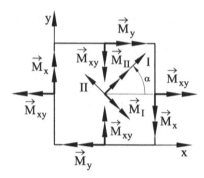

Fig. 5.4 Principal bending moments

5.2.5 Principle of virtual work

The PVW is written as

$$\iiint_V \delta\varepsilon^T \sigma \, dV = \iint_A \delta\mathbf{u}^T \mathbf{t} \, dA + \sum_i \delta\mathbf{u}_i^T \mathbf{p}_i \qquad (5.17a)$$

with

$$\delta\mathbf{u} = \left[\delta w, \delta\left(\frac{\partial w}{\partial x}\right), \delta\left(\frac{\partial w}{\partial y}\right) \right]^T \quad , \quad \delta\mathbf{u}_i = \left[\delta w_i, \delta\left(\frac{\partial w}{\partial x}\right)_i, \delta\left(\frac{\partial w}{\partial y}\right)_i \right]^T$$

$$\mathbf{t} = [f_x, m_x, m_y]^T \quad , \quad \mathbf{p}_i = [P_{z_i}, M_{x_i}, M_{y_i}]^T \qquad (5.17b)$$

In Eqs.(5.17), $\delta\mathbf{u}$ is the virtual displacement vector, $\mathbf{t}$ is the distributed force vector and $\mathbf{p}_i$ is the point force vector. As for the components of the load vectors, f_z is a distributed vertical force, m_x and m_y are distributed bending moments around the x and y axes, respectively, and F_{z_i}, M_{x_i} and M_{y_i} are the external vertical point load and the bending moments acting at point i, respectively (Figure 5.1). Moments are taken as positive if they act anticlockwise in the plane xz or yz (Figure 5.1).

The kinematic and constitutive expressions of the previous section allow us to simplify the virtual strain work δU given by the l. h. s. of Eq.(5.17a) into a surface integral over the plate middle plane (the "reference" geometry) in terms of the bending moments and the virtual curvatures. Making use of Eqs.(5.8) and (5.13) gives

$$\delta U = \iiint_V \delta\varepsilon^T \sigma \, dV = \iiint_V \left[\mathbf{S}\delta\hat{\boldsymbol{\varepsilon}}_b^T \right]^T \sigma \, dV =$$

$$= \iint_A \delta\hat{\boldsymbol{\varepsilon}}_b^T \left[\int_{-\frac{t}{2}}^{+\frac{t}{2}} \mathbf{S}^T \sigma \, dz \right] dA = \iint_A \delta\hat{\boldsymbol{\varepsilon}}_b^T \hat{\boldsymbol{\sigma}}_b \, dA \qquad (5.18)$$

Substituting Eq.(5.18) into (5.17a) gives the expression of the PVW as

$$\iint_A \delta\hat{\varepsilon}_b^T \hat{\sigma}_b dA = \iint_A \delta\mathbf{u}^T \mathbf{t} dA + \sum_i \delta\mathbf{u}_i^T \mathbf{p}_i \qquad (5.19)$$

Consequently, *the plate can be treated as a 2D solid*, since all the variables and integrals in the PVW are functions of the coordinates of the middle plane only.

It is interesting to rewrite the expression of the virtual strain work of Eq.(5.18) as

$$\delta U = \iint_A \left[\frac{\partial^2 w}{\partial x^2} M_x + \frac{\partial^2 w}{\partial y^2} M_y + 2\frac{\partial^2 w}{\partial x \partial y} M_{xy} \right] dA \qquad (5.20)$$

Eq.(5.20) clearly shows that the virtual strain work can be obtained as the integral over the plate area of the work performed by the bending moments over the corresponding virtual curvatures.

The integrand of (5.20) contains second derivatives of the deflection. This requires the continuity of the deflection and its first derivatives; i.e. C^1 continuity requirement (Section 3.8.3 of [On4]).

The C^1 continuity requirement for the deflection field is a particular feature of Kirchhoff plate elements, like it was for Euler-Bernouilli-beam elements (Chapter 1).

5.2.6 Equilibrium equations

The equilibrium equations are of particular interest in Kirchhoff plate theory. Among other things, they allow us to compute the shear forces from the nodal deflections. Also, the equilibrium equations have a simple differential form in terms of the deflection which has been widely used for the analytical (and numerical) solution of thin plate problems [TW].

The equilibrium of external forces, bending moments and shear forces over a differential element of a plate under distributed vertical forces f_z only (i.e. $m_x = m_y = 0$) (Figure 5.5) gives

Equilibrium of vertical forces

$$\sum F_z = o \quad \Rightarrow \quad \left(\frac{\partial Q_x}{\partial x} dx \right) dy + \left(\frac{\partial Q_y}{\partial y} dy \right) dx + f_z dx dy = 0 \quad (5.21a)$$

Dividing by the area differential $(dxdy)$ gives

$$\frac{\partial Q_x}{\partial x} + \frac{\partial Q_y}{\partial y} + f_z = 0 \qquad (5.21b)$$

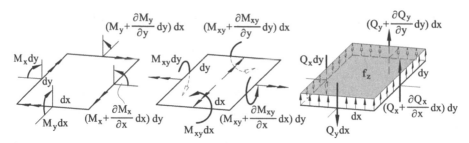

Fig. 5.5 Loads, moments and shear forces in a differential plate element

Equilibrium of moments

$$\sum M_x = 0 \quad \Rightarrow \quad \left(\frac{\partial M_x}{\partial x}dx\right)dy + \left(\frac{\partial M_{xy}}{\partial y}dy\right)dx+$$
$$+ (Q_x dy)\,dx + \left(\frac{\partial Q_y}{\partial y}dy\right)dx\frac{dx}{2} - f_z dxdy\frac{dy}{2} = 0 \tag{5.22a}$$

$$\sum M_y = 0 \quad \Rightarrow \quad \left(\frac{\partial M_y}{\partial y}dy\right)dx + \left(\frac{\partial M_{xy}}{\partial x}dx\right)dy+$$
$$+(Q_y dx)dy + \left(\frac{\partial Q_x}{\partial x}dx\right)dy\frac{dy}{2} - f_z dxdy\frac{dy}{2} = 0 \tag{5.22b}$$

Ignoring second order terms we have after simplification

$$\frac{\partial M_x}{\partial x} + \frac{\partial M_{xy}}{\partial y} + Q_x = 0 \tag{5.23a}$$

$$\frac{\partial M_y}{\partial y} + \frac{\partial M_{xy}}{\partial x} + Q_y = 0 \tag{5.23b}$$

Differentiating Eqs.(5.23a) and (5.23b) with respect to y and x respectively, and substituting the derivatives of the shear forces into Eq.(5.21b) gives

$$\frac{\partial^2 M_x}{\partial x^2} + 2\frac{\partial^2 M_{xy}}{\partial x \partial y} + \frac{\partial^2 M_y}{\partial y^2} - f_z = 0 \tag{5.24}$$

Eq.(5.24) can be rewritten for an isotropic material using (5.14) as

$$\frac{\partial^4 w}{\partial x^4} + 2\frac{\partial^4 w}{\partial x^2 \partial y^2} + \frac{\partial^4 w}{\partial y^4} - \frac{f_z}{D} = 0 \tag{5.25a}$$

with

$$D = \frac{Et^3}{12(1 - \nu^2)} \tag{5.25b}$$

Eq.(5.25a) is a fourth order differential equation relating the deflection to the applied distributed loading and the material properties of the plate.

Substituting Eq.(5.14) into (5.23) gives the expression for the shear forces in terms of the deflection as

$$Q_x = -D\left(\frac{\partial^3 w}{\partial x^3} + \frac{\partial^3 w}{\partial x \partial y^2}\right); \quad Q_y = -D\left(\frac{\partial^3 w}{\partial y^3} + \frac{\partial^3 w}{\partial y \partial x^2}\right) \quad (5.26)$$

The "exact" thickness distribution of the shear stresses can be found in terms of Q_x and Q_y from elasticity theory [Ug,VK,TW]. Accepting a parabolic distribution for the tangential stresses across the plate thickness, similarly as for beams, gives the maximum value of the shear stresses as [TW]

$$(\tau_{xz})_{\max} = \frac{3}{2}\frac{Q_x}{t}; \quad (\tau_{yz})_{\max} = \frac{3}{2}\frac{Q_y}{t} \quad (5.27)$$

5.2.7 The boundary conditions

The boundary conditions which have to be imposed on the problem are:

1. *Fixed boundary* where displacements at restrained points of the boundary are given specified values. These conditions are expressed as

$$w = \bar{w}, \ \theta_n = \bar{\theta}_n \text{ and } \theta_s = \bar{\theta}_s \quad (5.28)$$

Here n and s are directions normal and tangential to the boundary line (Figures 5.6 and 5.7) and $(\bar{\cdot})$ denotes a prescribed value. Note that in Kirchhoff thin plate theory the specification of w along s automatically prescribes θ_s (as $\theta_s = \frac{\partial w}{\partial s}$, Figure 5.7b) but this is not the case for thick plates where w and θ_s are independently prescribed.

A *clamped edge* is a special case of Eq.(5.28) with zero values assigned to the prescribed values. A point support is characterized by $w_i = 0$ (Figure 5.7a).

2. *Traction boundary* where the resultant stresses M_n, M_{ns} and Q_n (Figure 5.6) are given prescribed values

$$M_n = \bar{M}_n \quad ; \quad M_{ns} = \bar{M}_{ns} \quad ; \quad Q_n = \bar{Q}_n \quad (5.29a)$$

The expressions for M_n, M_{ns} and Q_n can be obtained in terms of the bending moment and the shear forces as

$$M_n = M_x n_x^2 + 2M_{xy}n_x n_y + M_y n_y^2$$

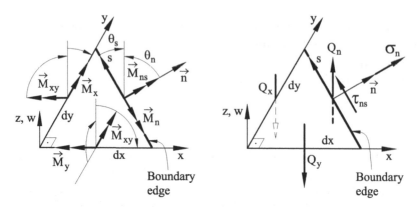

Fig. 5.6 Local rotations, local stresses and resultant stresses at a boundary edge

$$M_{ns} = -M_x n_x n_y + M_{xy}(n_x^2 - n_y^2) + M_y n_x n_y$$
$$Q_n = Q_x n_x + Q_y n_y \tag{5.29b}$$

where n_x, n_y are the components of the unit normal vector $\mathbf{n} = [n_x, n_y]^T$ pointing towards the exterior of the boundary edge (Figure 5.6).

M_n and M_{ns} are the moments at the boundary edge induced by the normal stress σ_n and the tangential stress τ_{ns} at the edge obtained from

$$\begin{Bmatrix} \sigma_n \\ \tau_{ns} \end{Bmatrix} = \begin{bmatrix} n_x^2 & n_y^2 & 2n_x n_y \\ -n_x n_y & n_x n_y & (n_x^2 - n_y^2) \end{bmatrix} \begin{Bmatrix} \sigma_x \\ \sigma_y \\ \tau_{xy} \end{Bmatrix} \tag{5.29c}$$

Eq.(5.29c) can be deduced from the stress transformation given in Eq.(4.12b) of [On4]. For the sign of σ_n and τ_{ns} see Figure 5.6

The bending moments M_n and M_{ns} and the shear force Q_n are work-conjugate to the local rotations θ_n and θ_s and to the deflection w, respectively.

A *free edge* is a special case with zero values assigned to $\bar{M}_n, \bar{M}_{ns}$ and $\bar{\theta}_n$.

3. *Mixed boundary conditions*, where both traction and displacement components can be specified. A typical case is the *simply supported (SS) edge* (Figure 5.7b). Here it is clear that $w = 0$ (and consequently $\theta_s = 0$) and $M_n = 0$. It is less obvious whether $M_{ns} = 0$ needs to be specified. In practice it suffices to prescribe $w = 0$ at the nodes on the SS edge [ZT2]. For a curved SS edge modelled as a polynomial, a unique normal to each node must be specified and used to prescribe $w = \theta_s = 0$ at the node.

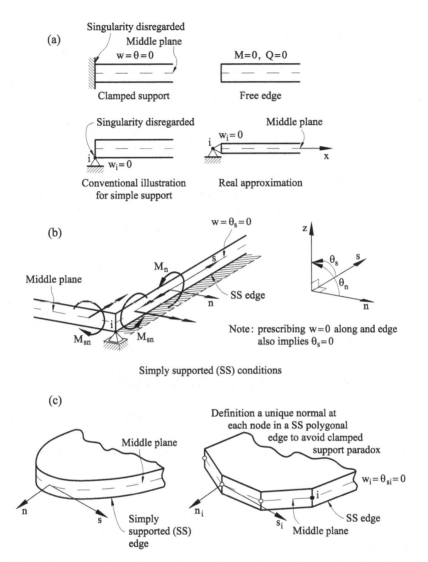

Fig. 5.7 (a) Supported (end) conditions for a plate. Conventionally illustrated simple support and real approximation. (b) Simply supported (SS) conditions. (c) Definition of unique normal and tangent directions in a SS polygonal edge

Otherwise the solution is equivalent, paradoxically, to that of a clamped support [Bab,BS].

The treatment of boundary conditions on edges which are inclined with respect to the cartesian axes is discussed in Section 6.3.5.

The alternatives for integrating the differential equations (5.25a) with the appropriate boundary conditions will not be discussed here. The more popular analytical procedures are based on double Fourier series and the numerical ones on the finite difference method. Both techniques have been widely used to study a large variety of thin isotropic plates, generally of rectangular shape [Ga4,Pan,Sz,TW,Ug,VK].

The analytical study of thin plates with arbitrary geometry, heterogeneous material and complex boundary conditions is difficult. There are also severe limitations in the application of traditional numerical techniques such as the finite difference method.

The finite element method is free of most, if not all, the drawbacks mentioned above, as it can be easily applied to any plate problem, despite the complexity of the geometry or the material properties. It is interesting that it was in the solution of plate bending problems that the FEM gained its popularity in the late 1960's (see references in Chapter 11 of [ZT2]). The generality of the FEM versus the limitations of analytical solutions, or other numerical techniques, like grillage methods, and even the finite difference method, favoured the fast and broad development of the FEM for analysis of thin plates. Some of the more popular thin plate elements are described in the next sections.

5.3 FORMULATION OF THIN PLATE ELEMENTS

The intuitive way to satisfy the C^1 continuity requirement for the deflection is to choose the deflection and the two rotations as the nodal variables, in a similar way to what we do for Euler-Bernouilli beam elements.

Consequently, a thin plate element would have in principle three variables per node: w_i, $\left(\frac{\partial w}{\partial x}\right)_i$ and $\left(\frac{\partial w}{\partial y}\right)_i$, and the total number of variables for a n-noded element would be $3n$. This defines the number of polynomial terms approximating the deflection w within each element.

In general

$$w = \alpha_1 + \alpha_2 x + \alpha_3 y + \alpha_4 x^2 + \alpha_5 xy + ...(\text{up to} \quad 3n \quad \text{terms}) \qquad (5.30)$$

The α_i parameters are obtained by imposing at each node

$$\left.\begin{array}{l} w_i = (w)_i \\[2mm] \theta_{x_i} = \left(\dfrac{\partial w}{\partial x}\right)_i ; \quad \theta_{y_i} = \left(\dfrac{\partial w}{\partial y}\right)_i \end{array}\right\} i = 1, 2, ..., n \qquad (5.31)$$

which gives $3n$ equations.

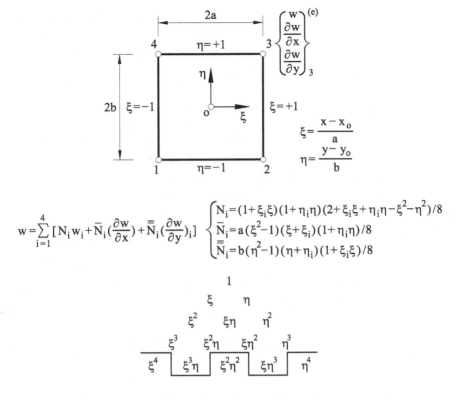

Fig. 5.8 Non-conforming 4-noded MZC thin plate rectangle

The key issue is the selection of the adequate polynomial terms in Eq.(5.30) and, in general, several alternatives are possible. Each one defines a different plate element which performance must be carefully assesed since many of the elements just do not work in practice. In the following sections the pros and cons of some of the more popular rectangular and triangular Kirchhoff thin plate elements are presented.

5.4 RECTANGULAR THIN PLATE ELEMENTS

5.4.1 Non-conforming 4-noded MZC rectangle

The element is shown in Figure 5.8. It has four nodes and hence the number of polynomial terms in Eq.(5.30) is 12. It is however impossible to choose a complete polynomial for describing w, as the complete polynomials of third and quartic order have 10 and 15 terms, respectively (Figure 5.8). Thus, three terms of the quartic polynomial must be omitted. The selection of these terms is not a trivial issue. Melosh [Me3] and Zienkiewicz and Cheung [ZCh,ZCh2] developed a popular 4-noded plate rectangle, denoted hereonwards MZC, on the basis of the following approximation

$$w = \alpha_1 + \alpha_2 x + \alpha_3 y + \alpha_4 x^2 + \alpha_5 xy + \alpha_6 y^2 + \alpha_7 x^3 + \alpha_8 x^2 y +$$
$$+ \alpha_9 xy^2 + \alpha_{10} y^3 + \alpha_{11} x^3 y + \alpha_{12} xy^3 \tag{5.32}$$

Above expression guarantees geometry invariance [On4,ZTZ,ZT2]. Along the sides $x =$constant and $y =$constant the deflection varies as a complete third order polynomial which can be uniquely defined in terms of the two deflections and the two rotations at the end nodes of each side. This ensures the continuity of w between adjacent elements.

The $\alpha_1, \ldots, \alpha_{12}$ parameters are obtained making use of Eq.(5.31). It can be found after some algebra

$$\mathbf{a}^{(e)} = [w_1, \theta_{x_1}, \theta_{y_1}, \cdots, w_4, \theta_{x_4}, \theta_{y_4}] = \mathbf{A} [\alpha_1, \alpha_2, \alpha_3, \cdots, \alpha_{11}, \alpha_{12}]^T \tag{5.33a}$$

with

$$\mathbf{A} = \begin{bmatrix} 1 & x_1 & y_1 & x_1^2 & x_1 y_1 & y_1^2 & x_1^3 & x_1^2 y_1 & x_1 y_1^2 & y_1^3 & x_1^3 y_1 & x_1 y_1^3 \\ 0 & 1 & 0 & 2x_1 & y_1 & 0 & 3x_1^2 & 2x_1 y_1 & y_1^2 & 0 & 3x_1^2 y_1 & y_1^3 \\ 0 & 0 & 1 & 0 & x_1 & 2y_1 & 0 & x_1^2 & 2x_1 y_1 & 3y_1^2 & x_1^3 & 2x_1 y_1^2 \\ & \vdots & & & \vdots & & & \vdots & & & & \\ & \vdots & & & \vdots & & & \vdots & & & & \\ 0 & 0 & 1 & 0 & x_4 & 2y_4 & 0 & x_4^2 & 2x_4 y_4 & 3y_4^2 & x_4^3 & 2x_4 y_4^2 \end{bmatrix} \tag{5.33b}$$

Eq.(5.33a) gives

$$\boldsymbol{\alpha} = \mathbf{A}^{-1} \mathbf{a}^{(e)} \tag{5.34}$$

Combining Eqs.(5.32) and (5.34) yields, finally

$$w = \mathbf{P}^T \boldsymbol{\alpha} = \mathbf{P}^T \mathbf{A}^{-1} \mathbf{a}^{(e)} = \mathbf{N} \mathbf{a}^{(e)} \tag{5.35a}$$

where

$$\mathbf{N} = \mathbf{P}^T \mathbf{A}^{-1} \tag{5.35b}$$

is the shape function matrix with

$$\mathbf{P} = [1, x, y, x^2, xy, y^2, x^3, x^2 y, xy^2, y^3, x^3 y, xy^3]^T \tag{5.35c}$$

This process is quite tedious and costly. Melosh [Me3] derived an explicit form for the shape functions. Eq. (5.32) is rewritten as

$$w = \sum_{i=1}^{4} \left[N_i w_i + \bar{N}_i \left(\frac{\partial w}{\partial x} \right)_i + \bar{\bar{N}}_i \left(\frac{\partial w}{\partial y} \right)_i \right] = \mathbf{N} \mathbf{a}^{(e)} \tag{5.36}$$

where

$$\mathbf{N} = [\mathbf{N}_1, \mathbf{N}_2, \mathbf{N}_3, \mathbf{N}_4] \quad ; \quad \mathbf{a}^{(e)} = \begin{Bmatrix} \mathbf{a}_1^{(e)} \\ \mathbf{a}_2^{(e)} \\ \mathbf{a}_3^{(e)} \\ \mathbf{a}_4^{(e)} \end{Bmatrix} \tag{5.37}$$

and

$$\mathbf{N}_i = [N_i, \bar{N}_i, \bar{\bar{N}}_i] \quad ; \quad \mathbf{a}_i^{(e)} = \left[w_i, \left(\frac{\partial w}{\partial x} \right)_i, \left(\frac{\partial w}{\partial y} \right)_i \right]^T$$

Figure 5.8 shows the analytical form of the shape functions N_i, $\bar{N}_i$ and $\bar{\bar{N}}_i$ in natural coordinates. Similarly, as for the Hermite shape functions of the analogous 2-noded Euler-Bernouilli beam element (Eq.(1.11a)) function N_i corresponding to the deflection takes a unit value at node i, whereas its first derivatives are zero at the node. Conversely, functions $\bar{N}_i$ and $\bar{\bar{N}}_i$ have a zero value and unit slopes in the directions ξ and η, respectively at node i.

The curvature matrix is obtained from Eqs.(5.9) and (5.36) as

$$\hat{\varepsilon}_b = \begin{Bmatrix} \dfrac{\partial^2 w}{\partial x^2} \\ \dfrac{\partial^2 w}{\partial y^2} \\ 2\dfrac{\partial^2 w}{\partial x \partial y} \end{Bmatrix} = \sum_{i=1}^{4} \mathbf{B}_{b_i} \mathbf{a}_i^{(e)} = \mathbf{B}_b \mathbf{a}^{(e)} \tag{5.38}$$

with

$$\mathbf{B}_b = [\mathbf{B}_{b_1}, \mathbf{B}_{b_2}, \mathbf{B}_{b_3}, \mathbf{B}_{b_4}] \quad , \quad \mathbf{B}_{b_i} = \begin{bmatrix} \dfrac{\partial^2 N_i}{\partial x^2} & \dfrac{\partial^2 \bar{N}_i}{\partial x^2} & \dfrac{\partial^2 \bar{\bar{N}}_i}{\partial x^2} \\ \dfrac{\partial^2 N_i}{\partial y^2} & \dfrac{\partial^2 \bar{N}_i}{\partial y^2} & \dfrac{\partial^2 \bar{\bar{N}}_i}{\partial y^2} \\ 2\dfrac{\partial^2 N_i}{\partial x \partial y} & 2\dfrac{\partial^2 \bar{N}_i}{\partial x \partial y} & 2\dfrac{\partial^2 \bar{\bar{N}}_i}{\partial x \partial y} \end{bmatrix} \tag{5.39}$$

The computation of the second derivatives of the shape functions in $\mathbf{B}_{b_i}$ is immediate simply noting that

$$\frac{\partial^2}{\partial x^2} = \frac{1}{a^2}\frac{\partial^2}{\partial \xi^2} \qquad \text{and} \qquad \frac{\partial^2}{\partial y^2} = \frac{1}{b^2}\frac{\partial^2}{\partial \eta^2} \tag{5.40}$$

The bending moment field is expressed in terms of the nodal DOFs by substituting Eq.(5.38) into (5.14). This gives

$$\hat{\boldsymbol{\sigma}}_b = \hat{\mathbf{D}}_b\mathbf{B}_b\mathbf{a}^{(e)} \tag{5.41}$$

It is deduced from Eqs.(5.35c) and (5.39) that the curvature field, and hence the bending moment field, is linear within the MZC element.

Following the usual procedure it is deduced that

$$\delta w = \mathbf{N}\delta\mathbf{a}^{(e)} \qquad \text{and} \qquad \delta\hat{\boldsymbol{\varepsilon}}_b = \mathbf{B}_b\delta\mathbf{a}^{(e)} \tag{5.42}$$

The PVW for a single element can be written as (see Eq.(5.19))

$$\iint_{A^{(e)}} \delta\hat{\boldsymbol{\varepsilon}}_b^T\hat{\boldsymbol{\sigma}}_b dA = \iint_{A^{(e)}} \delta\mathbf{u}^T t dA + \sum_{i=1}^{3}[\delta\mathbf{a}_i^{(e)}]^T\mathbf{q}_i^{(e)} \tag{5.43a}$$

with

$$\delta\mathbf{a}_i^{(e)} = \left[\delta w_i, \delta\left(\frac{\partial w}{\partial x}\right)_i, \delta\left(\frac{\partial w}{\partial y}\right)_i\right]^T \quad , \quad \mathbf{q}_i^{(e)} = [F_{z_i}, M_{x_i}, M_{y_i}]^T \tag{5.43b}$$

where $\mathbf{q}_i^{(e)}$ is the vector of equilibrating nodal forces for node i.

Substituting Eqs.(5.42) and the constitutive equation (5.41) into the PWV, the standard equilibrium equation for the element is found as

$$\mathbf{K}^{(e)}\mathbf{a}^{(e)} - \mathbf{f}^{(e)} = \mathbf{q}^{(e)} \tag{5.44}$$

The element stiffness matrix is

$$\mathbf{K}_{ij}^{(e)} = \iint_{A^{(e)}} \mathbf{B}_{b_i}^T\hat{\mathbf{D}}_b\mathbf{B}_{b_j} dx dy \tag{5.45}$$

The equivalent nodal force vector for a distributed vertical loading and distributed bending moments is

$$\mathbf{f}_i^{(e)} = \begin{Bmatrix} f_{z_i} \\ m_{x_i} \\ m_{y_i} \end{Bmatrix} = \iint_{A^{(e)}} \begin{bmatrix} N_i & N_{i,x} & N_{i,y} \\ \bar{N}_i & \bar{N}_{i,x} & \bar{N}_{i,y} \\ \bar{\bar{N}}_i & \bar{\bar{N}}_{i,x} & \bar{\bar{N}}_{i,y} \end{bmatrix} \begin{Bmatrix} f_z \\ m_x \\ m_y \end{Bmatrix} dx dy \tag{5.46}$$

$$\mathbf{K}^{(e)} = D\,[\mathbf{K}_1^{(e)} + \mathbf{K}_2^{(e)} + \mathbf{K}_3^{(e)} + \mathbf{K}_4^{(e)}] \qquad ; \qquad D = \frac{Et^3}{12(1-\nu^2)}$$

$$\mathbf{K}_1^{(e)} = \frac{b}{6a^3}\begin{bmatrix}
6 \\
6a & 8a^2 \\
0 & 0 & 0 \\
-6 & -6a & 0 & 6 \\
6a & 4a^2 & 0 & -6a & 8a^2 & & & & & & \text{Symmetric} \\
0 & 0 & 0 & 0 & 0 & 0 \\
-3 & -3a & 0 & 3 & -3a & 0 & 6 \\
3a & 2a^2 & 0 & -3a & 4a^2 & 0 & -6a & 8a^2 \\
0 & 0 & 0 & 0 & 0 & 0 & 0 & 0 & 0 \\
3 & 3a & 0 & -3 & 3a & 0 & -6 & 6a & 0 & 6 \\
3a & 4a^2 & 0 & -3a & 2a^2 & 0 & -6a & 4a^2 & 0 & 6a & 8a^2 \\
0 & 0 & 0 & 0 & 0 & 0 & 0 & 0 & 0 & 0 & 0 & 0
\end{bmatrix}$$

$$\mathbf{K}_2^{(e)} = \frac{a}{6b^3}\begin{bmatrix}
6 \\
0 & 0 \\
6b & 0 & 8b^2 \\
3 & 0 & 3b & 6 \\
0 & 0 & 0 & 0 & 0 & & & & & & \text{Symmetric} \\
3b & 0 & 4b^2 & 6b & 0 & 8b^2 \\
-3 & 0 & -3b & -6 & 0 & -6b & 6 \\
0 & 0 & 0 & 0 & 0 & 0 & 0 & 0 \\
3b & 0 & 2b^2 & 6b & 0 & 4b^2 & -6b & 0 & 8b^2 \\
-6 & 0 & -6b & -3 & 0 & -3b & 3 & 0 & -3b & 6 \\
0 & 0 & 0 & 0 & 0 & 0 & 0 & 0 & 0 & 0 & 0 \\
6b & 0 & 4b^2 & 3b & 0 & 2b^2 & -3b & 0 & 4b^2 & -6b & 0 & 8b^2
\end{bmatrix}$$

$$\mathbf{K}_3^{(e)} = \frac{\nu}{2ab}\begin{bmatrix}
1 \\
a & 0 \\
b & 2ab & 0 \\
-1 & 0 & -b & 1 \\
0 & 0 & 0 & -a & 0 & & & & & & \text{Symmetric} \\
-b & 0 & 0 & b & -2ab & 0 \\
1 & 0 & 0 & -1 & a & 0 & 1 \\
0 & 0 & 0 & a & 0 & 0 & -a & 0 \\
0 & 0 & 0 & 0 & 0 & 0 & -b & 2ab & 0 \\
-1 & -a & 0 & 1 & 0 & 0 & -1 & 0 & b & 1 \\
-a & 0 & 0 & 0 & 0 & 0 & 0 & 0 & 0 & a & 0 \\
0 & 0 & 0 & 0 & 0 & 0 & 0 & 0 & 0 & -b & -2ab & 0
\end{bmatrix}$$

$$\mathbf{K}_4^{(e)} = \frac{1-\nu}{30ab}\begin{bmatrix}
21 \\
3a & 8a^2 \\
3b & 0 & 8b^2 \\
-21 & -3a & -3b & 21 \\
3a & -2a^2 & 0 & -3a & 8a^2 & & & & & & \text{Symmetric} \\
-3b & 0 & -8b^2 & 3b & 0 & 8b^2 \\
21 & 3a & 3b & -21 & 3a & -3b & 21 \\
-3a & 2a^2 & 0 & 3a & -8a^2 & 0 & -3a & 8a^2 \\
-3b & 0 & 2b^2 & 3b & 0 & -2b^2 & -3b & 0 & 8b^2 \\
-21 & -3a & -3b & 21 & -3a & 3b & -21 & 3a & 3b & 21 \\
-3a & -2a^2 & 0 & 3a & 2a^2 & 0 & -3a & -2a^2 & 0 & 3a & 8a^2 \\
3b & 0 & -8b^2 & -3b & 0 & 2b^2 & 3b & 0 & -8b^2 & -3b & 0 & 8b^2
\end{bmatrix}$$

Box 5.1. 4-noded MZC thin plate rectangle $(2a \times 2b)$. Stiffness matrix for homogeneous isotropic material

$$\hat{\boldsymbol{\sigma}}_b = \left\{ \begin{array}{c} M_x \\ M_y \\ M_{xy} \end{array} \right\} = \sum_{i=1}^{4} \hat{\mathbf{D}}_b \, \mathbf{B}_{b_i} \, \mathbf{a}_i^{(e)}$$

$$\hat{\mathbf{D}}_b \mathbf{B}_{b_i} = \begin{bmatrix} (\hat{d}_{11} \, N_i^{xx} + \hat{d}_{12} \, N_i^{yy}) & \hat{d}_{12} \, \bar{N}_i^{xx} & \hat{d}_{11} \, \bar{N}_i^{yy} \\ (\hat{d}_{21} \, N_i^{xx} + \hat{d}_{22} \, N_i^{yy}) & \hat{d}_{21} \, \bar{N}_i^{xx} & \hat{d}_{22} \, \bar{\bar{N}}_i^{yy} \\ 2\hat{d}_{33} \, N_i^{xy} & 2\hat{d}_{33} \, \bar{N}_i^{xy} & 2\hat{d}_{33} \, \bar{\bar{N}}_i^{xy} \end{bmatrix}$$

$$N_i^{xx} = -\frac{1}{4a^2}(3\xi_i\xi + 3\xi_i\eta_i\xi\eta) \qquad \bar{N}_i^{xx} = \frac{1}{4a}(3\xi + \xi_i\eta_i\eta + 3\eta_i\xi\eta + \xi_i)$$

$$N_i^{yy} = -\frac{1}{4b^2}(3\eta_i\eta + 3\xi_i\eta_i\xi\eta) \qquad \bar{N}_i^{xy} = \frac{1}{8b}(3\eta_i\xi^2 + 2\xi_i\eta_i\xi - \eta_i)$$

$$N_i^{xy} = \frac{1}{8ab}(4\xi_i\eta_i - 3\xi_i\eta_i\xi^2 - 3\xi_i\eta_i\eta^2) \qquad \bar{\bar{N}}_i^{yy} = \frac{1}{4b}(3\eta + \xi_i\eta_i\xi + 3\xi_i\xi\eta + \eta_i)$$

$$\bar{\bar{N}}_i^{xy} = \frac{1}{8a}(3\xi_i\eta^2 + 2\xi_i\eta_i\eta - \xi)$$

$\hat{d}_{ij}$ term ij of matrix $\hat{\mathbf{D}}_b$

Box 5.2. 4-noded MZC thin plate rectangle. Matrix $\hat{\mathbf{D}}\mathbf{B}_{b_i}$ for computing the bending moments

where f_{z_i}, m_{x_i} and m_{y_i} are the vertical force and the bending moments acting at node i and $N_{i,x} = \frac{\partial N_i}{\partial x}$, etc.

Note that a distributed vertical loading originates nodal bending moments, like for Euler-Bernouilli beam elements. Similarly, distributed bending moments give nodal vertical forces. This is due to the dependence of the deflection field with the nodal rotations (Eq.(5.36)).

The integrals in Eqs.(5.45) and (5.46) can be computed numerically by a Gauss quadrature [On4]. However, the exact integration of the MZC rectangular element stiffness matrix is straightforward and its expression is given in Box 5.1 for homogeneous isotropic material. Box 5.2 shows the explicit product $\hat{\mathbf{D}}_b\mathbf{B}_i$ needed for computing the bending moment field via Eq.(5.41). The equivalent nodal force vector for a uniformly distributed vertical load $f_z = q$ is

$$\mathbf{f}^{(e)} = 4qab \left[\frac{1}{4}, \frac{a}{12}, \frac{b}{12}, \frac{1}{4}, -\frac{a}{12}, \frac{b}{12}, \frac{1}{4}, -\frac{a}{12}, -\frac{b}{12}, \frac{1}{4}, \frac{a}{12}, -\frac{b}{12} \right]^T \quad (5.47a)$$

The expression of $\mathbf{f}_i$ for external forces acting at a node is

$$\mathbf{f}_i = [P_{z_i}, M_{x_i}, M_{y_i}]^T \quad (5.47b)$$

where P_{z_i}, M_{x_i} and M_{y_i} are the vertical point load and the bending moments acting at node i, respectively (Figure 5.1).

The global system of equations $\mathbf{Ka} = \mathbf{f}$ is obtained by assembling the element stiffness matrices and equivalent nodal force vectors in the standard manner. Reactions at the prescribed nodes are computed a posteriori as explained in Section 1.3.4 for beams.

Example 5.1: Derive the term 11 of $\mathbf{K}_{11}^{(e)}$ for the MZC thin plate rectangle.

- Solution

From Eq.(5.45) we deduce

$$[\mathbf{K}_{ij}^{(e)}]_{11} = \iint_{A^{(e)}} [\mathbf{B}_{b_i}^T]_1 \, \hat{\mathbf{D}}_b [\mathbf{B}_{b_j}]_1 \, dxdy$$

where index 1 in $\mathbf{B}_i$ denotes the first row and in $\mathbf{B}_j$ the first column. Using Eqs.(5.39) and (5.40) and the expressions of Figure 5.8 we obtain

$$[\mathbf{K}_{ij}^{(e)}]_{11} = \int_{-1}^{+1} \int_{-1}^{+1} \Big[9 \, \hat{d}_{11} \, b^4 (1 - \eta_i \eta)(1 + \eta_i \eta) \xi_i \xi_j \xi^2 +$$
$$+ 18 d_{12} \, a^2 \, b^2 \left((1 + \eta_i \eta)(1 + \xi_j \eta) \xi_i \eta_j \xi \eta + (1 + \xi_i \xi)(1 + \eta_j \eta) \eta_i \xi_j \xi \eta \right) +$$
$$+ a \, \hat{d}_{22} \, a^4 (1 + \xi_i \xi)(1 + \xi_j \xi) \eta_i \eta_j \eta^2 + \hat{d}_{33} \, a^2 \, b^2 (3\xi^2 + 3\eta^2 - 4) \xi_i \eta_i \xi_j \eta_j \Big] \, a \, b \, d\xi \, d\eta$$

where $\hat{d}_{ij}$ is the term ij of matrix $\hat{\mathbf{D}}_b$ in Eq.(5.15a). Integration gives

$$[\mathbf{K}_{ij}^{(e)}]_{11} = \frac{1}{4a^2 b^2} \Big[3 \, \hat{d}_{11} \, b^4 \xi_i \xi_j (1 + \tfrac{1}{3} \eta_i \eta_j) + 3 \, \hat{d}_{22} \, a^4 \eta_i \eta_j (1 + \tfrac{1}{3} \xi_i \xi_j) +$$
$$+ 2 d_{12} \, a^2 \, b^2 \xi_i \xi_j \eta_i \eta_j + \frac{28}{5} \hat{d}_{33} \, a^2 \, b^2 \xi_i \xi_j \eta_i \eta_j \Big]$$

Particularizing for $i = j = 1$ yields

$$[\mathbf{K}_{11}^{(e)}]_{11} = \hat{d}_{11} \frac{b}{a^3} + \hat{d}_{22} \frac{a}{b^3} + \frac{\hat{d}_{12}}{2 \, ab} + \frac{7}{5} \frac{\hat{d}_{33}}{ab}$$

The same procedure yields the rest of stiffness matrix terms of Box 5.1.

Incompatibility of the normal rotation

It is important to understand that even though the deflection field of Eq.(5.36) guarantees the continuity of w between elements, however *it does not guarantee the continuity of the first derivatives of w (slopes)*, except at the nodes where it is obvious that they are uniquely defined. To

$$w^{(A)} = 0$$

$$w^{(B)} = (\xi^2 - 1)(\xi - 1)(1 + \eta)/8$$

$$\left(\frac{\partial w}{\partial y}\right)^{(A)} = 0$$

$$\left(\frac{\partial w}{\partial y}\right)^{(B)} = \left(\frac{\partial w}{\partial \eta} \frac{\partial \eta}{\partial y}\right)^{(B)} = \frac{l_{25}}{2}(\xi^2 - 1)(\xi - 1)/8$$

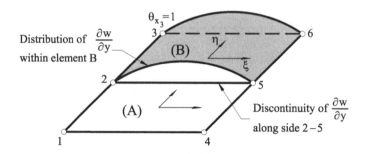

All displacements are zero except $\theta_{x_3} = \left(\frac{\partial w}{\partial x}\right)_3 = 1$

Fig. 5.9 Discontinuity of the slope $\frac{\partial w}{\partial y}$ along a side of a 4-noded rectangular plate

clarify this, let us consider the two elements A and B of Figure 5.9. It is assumed that all the nodal variables are zero with the exception of $\theta_{x_3} = 1$. Since $w = 0$ over element 1, then $\left(\frac{\partial w}{\partial y}\right) = 0$ along side 2-5 shared by both elements. For element B, $w = (\xi^2 - 1)(\xi - 1)(1 + \eta)/8$ and $\left(\frac{\partial w}{\partial y}\right) = \frac{l_{2-3}}{2}(\xi^2 - 1)(\xi - 1)/8$ along side 2-5. The slope $\frac{\partial w}{\partial y}$ is therefore discontinuous along the common side. This indicates that the MZC element is incompatible (non-conforming) [ZT2,Ya].

The discontinuity in the slope orthogonal to a side, termed hereafter the *normal rotation* (Figure 5.10a) implies that the cross derivatives $\frac{\partial^2 w}{\partial x \partial y}$ and $\frac{\partial^2 w}{\partial y \partial x}$ take a different value at the nodes, and this violates one of the basic requirements for the continuity of w. Figure 5.10b shows that $\frac{\partial w}{\partial y}$ along side 1-2 depends on $\left(\frac{\partial w}{\partial y}\right)_1$ and $\left(\frac{\partial w}{\partial y}\right)_2$, whereas $\frac{\partial w}{\partial x}$ along side 2-3 depends on $\left(\frac{\partial w}{\partial x}\right)_2$ and $\left(\frac{\partial w}{\partial x}\right)_3$. Therefore, the derivative $\frac{\partial^2 w}{\partial x \partial y}$ along side 1-2 depends on $\left(\frac{\partial w}{\partial y}\right)_1$. Similarly $\frac{\partial^2 w}{\partial y \partial x}$ along side 2-3 depends on $\left(\frac{\partial w}{\partial x}\right)_3$. Generally $\left(\frac{\partial w}{\partial y}\right)_1$ and $\left(\frac{\partial w}{\partial x}\right)_3$ will take different values and obviously $\left(\frac{\partial^2 w}{\partial x \partial y}\right)_2 \neq \left(\frac{\partial^2 w}{\partial y \partial x}\right)_2$.

It is therefore imposible to guarantee the conformity of the MZC thin plate rectangle simply by taking the deflection and its first derivatives as nodal variables. This, however, does not invalidate the element which

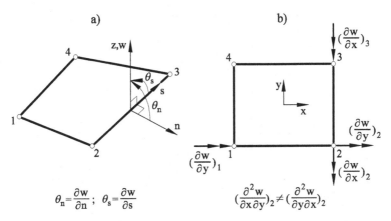

$$\theta_n = \frac{\partial w}{\partial n}; \quad \theta_s = \frac{\partial w}{\partial s} \qquad\qquad (\frac{\partial^2 w}{\partial x \partial y})_2 \neq (\frac{\partial^2 w}{\partial y \partial x})_2$$

Fig. 5.10 (a) Normal (θ_n) and tangential (θ_s) rotations. (b) Discontinuity of $\dfrac{\partial^2 w}{\partial x \partial y}$ at node 2

satisfies the patch test (Section 5.9) [TZSC,Ya,ZTZ]. This ensures the convergence as the mesh is refined.

Unfortunately the patch test is not satisfied for arbitrary quadrilateral shapes as the constant curvature criterion is violated in those situations. This *limits the application of the MZC element plate domains that can be discretized into rectangular plate elements*. However, in these cases it is an accurate element, as shown in the examples presented below.

Henshell *et al.* [HWW] studied the performance of the MZC thin plate element (and also some other plate quadrilaterals of higher order) formulated in curvilinear coordinates and concluded that reasonable accuracy is obtained for arbitrary quadrilateral shapes. The use of curvilinear coordinate for extending the MZC element to parallelogram shapes is discussed in [Da,ZCh,ZCh2,ZT2]. Argyris [Ar2] proposed a different approach for deriving the shape function of a similar 4-noded plate quadrilateral.

Example 5.2: Obtain the deflection and bending moment distributions along the central line of the homogeneous isotropic clamped plate shown in Figure 5.11 under uniformly distributed loading of intensity $f_z = -q$. Use a mesh of 2×2 MZC elements with $\nu = 0.3$.

- Solution

Only a quarter of the plate is analyzed due to the double symmetry of the problem. After eliminating the columns and rows of the prescribed nodal

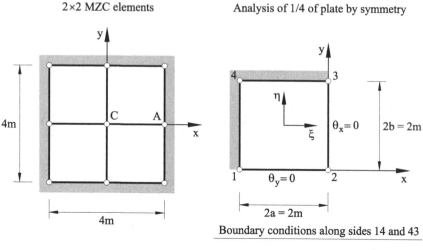

Fig. 5.11 Square plates analyzed with a mesh of 2×2 MZC thin plate rectangular elements

DOFs in the stiffness matrix of the single element considered and using the force vector of Eq.(5.47a), the following equation for the only free nodal variable, i.e. the deflection w_2, is obtained

$$\hat{d}_{11} \left[\frac{b}{a^3} + \frac{a}{b^3} + \frac{\nu}{2ab} + \frac{21(1-\nu)}{30ab} \right] w_2 = -qab$$

Making $\nu = 0.3$ and $a = b = 1$ (Figure 5.11) gives

$$2.64\hat{d}_{11}w_2 = -q$$

and

$$w_2 = -0.378\frac{q}{D} \quad \text{with} \quad \hat{d}_{11} = \frac{Et^3}{12(1-\nu^2)} = D$$

The exact solution for this problem is (for $L = 4$) [TW]

$$w_2 = -0.00126L^4\frac{q}{\hat{d}_{11}} = -0.322\frac{q}{D}$$

The error between the exact and computed deflections is $\approx 17\%$. The deflection field is obtained by substituting w_2 in the expression for w of Figure 5.8 to give

$$w = -0.0472\frac{q}{D}(1+\xi)(1-\eta)(2+\xi-\eta-\xi^2-\eta^2)$$

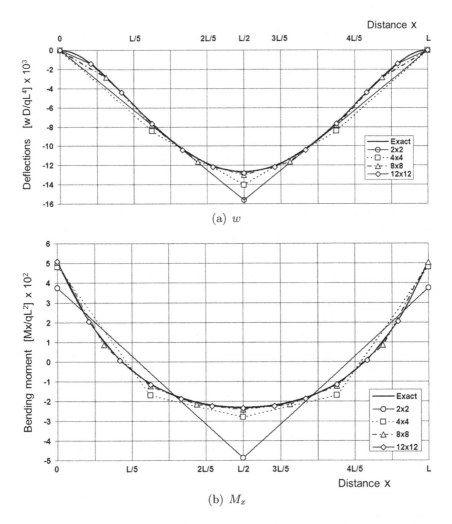

Fig. 5.12 Clamped square plate under uniform loading. Deflection and bending moment (M_x) diagrams along the central line for meshes of 2×2, 4×4, 8×8 and 12×12 MZC elements

Along the central line $(\eta = -1)$

$$w = -0.0944 \frac{q}{D}(1+\xi)(2+\xi-\xi^2)$$

which is a cubic polynomial in ξ as expected. This solution is compared with the exact one [TW] in Figure 5.12a and also with that obtained using meshes of 2×2, 4×4 and 8×8 MZC elements. Note the excellent accuracy of the central deflection value obtained with the simple 2×2 mesh.

The bending moment distribution within the element is obtained from the value of w_2 and the expressions of Box 5.2 (with $\nu = 0.3$, $\hat{d}_{11} = \hat{d}_{22} = D$ and

$\hat{d}_{33} = \frac{1-\nu}{2}D = 0.35D)$ giving

$$M_x = -\frac{3\hat{d}_{11}}{4a^2b^2}[\xi(1-\eta)b^2 - 0.3\eta(1+\xi)a^2]w_2 = -0.284q[\xi - 0.3\eta - 1.3\xi\eta]$$

$$M_y = -\frac{3\hat{d}_{11}}{4a^2b^2}[0.3\xi(1-\eta)b^2 - \eta(1+\xi)a^2]w_2 = -0.284q[0.3\xi - \eta - 1.3\xi\eta]$$

$$M_{xy} = -\frac{\hat{d}_{33}}{4ab}[3\xi^2 + 3\eta^2 - 4]w_2 = -0.034q[3\xi^2 + 3\eta^2 - 4]$$

The distribution of M_x and M_y along side 1-2 ($\eta = -1$) is linear in ξ whereas M_{xy} is quadratic. The linear distribution of M_x along the central line is compared in Figure 5.12b with the exact one [TW] and with that obtained using different meshes. The accuracy of the bending moment solution is poorer than that for the deflection, as expected. However, even for the coarse 4×4 mesh the values of M_x at the center and the clamped edge are a good estimate for design purposes.

The bending moments computed at the 2×2 Gauss points approximate better the exact values. This is further evidence of the benefit of computing the stresses at the "optimal sampling points" ([ZTZ] and Section 6.7 of [On4]).

Table 5.1 shows the percentage error values versus the exact solution of the deflection and the bending moment M_x at the plate center C and at the mid-point of the clamped edge (point A in Figure 5.11).

Figure 5.13 shows the convergence of the distribution of w and M_x along the central line for different meshes. We see that the MZC thin plate element element, although it is incompatible, converges in its rectangular form. Finally, Figure 5.14 shows the contours for w, M_x and M_y in one quarter of plate for a 14×14 mesh.

MCZ mesh	w_C	M_x^C	M_x^A
2×2	23,49%	112,77%	-11,52%
4×4	11,25%	22,17%	-2,63%
8×8	3,08%	4,99%	-0,79%
12×12	1,38%	2,17%	-0,39%
Analytical [TW]	$w_C = -0,00126\frac{qL^4}{D}$	$M_x^C = -2,291\times$ $10^{-2}qL^2$	$M_x^A = -2,24M_x^C$

Table 5.1 Deflection at plate center (C) and bending moment M_x at plate center (C) and the mid-point of clamped edge (A) for different meshes of MZC elements. Numbers show percentage error versus the analytical solution [TW]

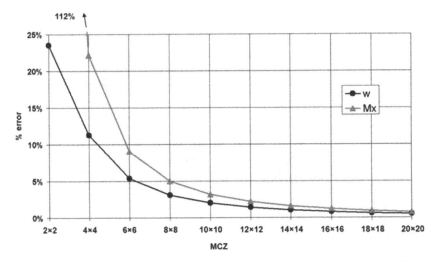

Fig. 5.13 Clamped square plate under uniform loading. Convergence of the error in the deflection w and the bending moment M_x at the central point for different meshes of MZC thin plate rectangles

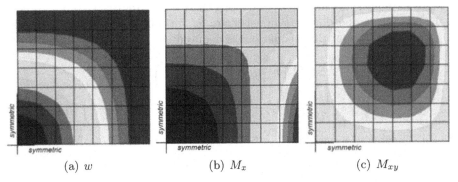

(a) w (b) M_x (c) M_{xy}

Fig. 5.14 Clamped square plate under uniform loading. Contours of vertical deflection w (a) and bending moments M_x (b) and M_{xy} (c) for a mesh of 14×14 MZC rectangles. Results are shown in one quarter of the plate due to symmetry

Example 5.3: Obtain the central deflection for the clamped plate of Figure 5.11 under a central point load $(-P)$. Use a mesh of 2×2 MZC elements.

- Solution

The process is identical to that followed in the previous example. The load acting at node 2 is $-\frac{P}{4}$ due to symmetry. The equation for w_2 is

$$2.64 \hat{d}_{11} w_2 = -\frac{P}{4} \quad \text{and} \quad w_2 = \frac{-P}{10.56 \hat{d}_{11}} = -0.0946 \frac{P}{\hat{d}_{11}}$$

The error with respect to the analytical solution ($w_2 = -0.0896\frac{P}{d_{11}}$ [TW]) is $\approx 5.5\%$. The bending moment distribution is obtained as in the previous example.

Example 5.4: Obtain the central deflection for the square plate of Figure 5.11 with simple supported edges under: a) Uniformly distributed loading $(-q)$, and b) Central point load $(-P)$ using a mesh of 2×2 MZC thin plate rectangular elements.

- Solution

Case a) Uniformly distributed loading

Along the simply supported sides 14 and 43 we just need to prescribe $w = 0$. This automatically implies $\theta_s = 0$ along these edges. Hence the only non-zero DOFs due to the double symmetry are θ_{x_1}, w_2 and θ_{y_3} (Figure 5.11). The resulting system of three equations is deduced as

$$\left(\frac{4b}{3a} + \frac{4(1-\nu)a}{15b}\right)\theta_{x_1} - \left(\frac{b}{a^2} + \frac{(1-\nu)}{10b}\right)w_2 = -\frac{qab^2}{3D}$$

$$-\left(\frac{b}{a^2} + \frac{(1-\nu)}{10a}\right)\theta_{x_1} + \left(\frac{b}{a^3} + \frac{a}{b^3} + \frac{\nu}{2ab} + \frac{7(1-\nu)}{10ab}\right)w_2 +$$

$$+\left(\frac{a}{b^2} + \frac{(1-\nu)}{10a}\right)\theta_{y_3} = -\frac{qab}{\hat{D}}$$

$$\left(\frac{a}{b^2} + \frac{(1-\nu)}{10a}\right)w_2 + \left(\frac{4a}{3b} + \frac{4(1-\nu)b^2}{15ab}\right)\theta_{y_3} = \frac{qab^2}{3D}$$

with $D = \hat{d}_{11}$.

Making $\nu = 0.3$ and $a = b = 1$ and noting that due to symmetry $\theta_{y_3} = -\theta_{x_1}$, the above system can be simplified to

$$1.52\theta_{x_1} - 1.07w_2 = -0.33\frac{q}{D}$$

$$-2.14\theta_{x_1} + 2.64w_2 = -\frac{q}{D}$$

which gives

$$w_2 = -1.25\frac{q}{D} \quad \text{and} \quad \theta_{x_1} = -\theta_{x_3} = -1.08\frac{q}{D}$$

The bending moment distribution can be obtained from Box 5.2.
The value of w_2 agrees reasonably well with the exact one of $w_2 = -1.040\frac{q}{D}$ [TW] (error $\approx 20\%$) despite of the simplicity of the mesh.

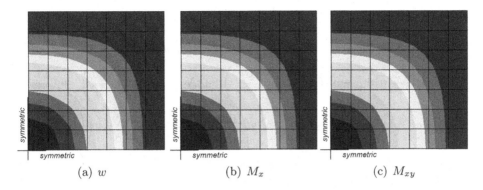

Fig. 5.15 Simple supported square plate. Uniform loading. Contours of vertical deflection (a) and bending moments M_x (b) and M_{xy} (c) for a mesh of 14×14 MZC rectangles. Results shown in one quarter of the plate due to symmetry

Figure 5.15 shows the contours of vertical deflection (a) and bending moments M_x (b) and M_{xy} (c) for a 14×14 mesh.

Case b) Central point load

The system of equations is the same as in the previous example substituting the 1st, 2nd and 3rd right-hand sides for $0, \frac{-P}{4D}$ and 0, respectively. After eliminating θ_{y_3} and making $\nu = 0.3$ and $a = b = 1$ we obtain

$$1.52\theta_{x_1} - 1.07w_2 = 0$$

$$-2.14\theta_{x_1} + 2.64w_2 = 0.33\frac{P}{D}$$

giving

$$w_2 = -0.219\frac{P}{D} \quad \text{and} \quad \theta_{x_1} = -\theta_{y_3} = -0.154\frac{P}{D}$$

The value of w_2 obtained with this very simple mesh is again not far from the analytical value, $w_2 = -0.1856\frac{P}{D}$ [TW] (error $\approx 18\%$).
The bending moment distribution can be obtained from Box 5.2.

5.4.2 12 DOFs plate rectangle proposed by Melosh

A rectangular plate element with the same DOFs as the MZC rectangle was derived almost simultaneously by Melosh [Me,Me2]. The shape functions are obtained by combining the 1D Hermite polynomials used for the 2-noded Euler-Bernouilli beam element (Section 1.3.1) giving the following interpolations for the deflection

$$w = N_1(x)N_1(y)w_1 + N_2(x)N_1(y)w_2 + N_2(x)N_2(y)w_3+$$

$$+N_1(x)N_2(y)w_4 + \bar{N}_1(x)N_1(y)\left(\frac{\partial w}{\partial x}\right)_1 + \bar{N}_2(x)N_1(y)\left(\frac{\partial w}{\partial x}\right)_2 +$$

$$+\bar{N}_2(x)N_2(y)\left(\frac{\partial w}{\partial x}\right)_3 + \bar{N}_1(x)N_2(y)\left(\frac{\partial w}{\partial x}\right)_4 + N_1(x)\bar{N}_1(y)\left(\frac{\partial w}{\partial x}\right)_1 +$$

$$+N_2(x)\bar{N}_1(y)\left(\frac{\partial w}{\partial x}\right)_2 + N_2(x)\bar{N}_2(y)\left(\frac{\partial w}{\partial x}\right)_3 + N_1(x)\bar{N}_2(y)\left(\frac{\partial w}{\partial x}\right)_4$$
$$(5.48)$$

where N_i and $\bar{N}_i$ are the Hermite polynomials of Eq.(1.11a) written in cartesian coordinates and referred to each of the element sides.

This element satisfies the continuity requirements for the normal rotation along the sides. However, the approximation (5.48) does not contain the term $\alpha_6 xy$ and the element can not reproduce a constant torque strain $\left(\frac{\partial^2 w}{\partial x \partial y}\right)$ state and this violates the patch test. This problem does not occur for the MZC element, neither for the BFS one presented in the next section. Consequently, these elements are more reliable for practical purposes.

5.4.3 Conforming BFS plate rectangle

Many authors have attempted to derive conforming Kirchhoff thin plate elements which satisfy the continuity requirements for the normal rotation and the second cross derivative $\frac{\partial^2 w}{\partial x \partial y}$ along the sides. A popular technique is to introduce the cross derivative $\frac{\partial^2 w}{\partial x \partial y}$ as a fourth nodal variable (Figure 5.16). Bogner, Fox and Schmidt [BFS] proposed an element of this type using a 16 term polynomial expansion for w as a product of two complete cubic polynomials in x and y.

An interesting feature of the BFS element is that, as for the Melosh rectangle, the shape functions can be obtained by simple products of 1D cubic Hermite polynomials. The approximation for the deflection is therefore written as

$$w^{(e)} = [\mathbf{N}_w, \mathbf{N}_{\theta_x}, \mathbf{N}_{\theta_y}, \mathbf{N}_\Gamma] \begin{Bmatrix} \mathbf{w} \\ \boldsymbol{\theta}_x \\ \boldsymbol{\theta}_y \\ \boldsymbol{\Gamma} \end{Bmatrix} \qquad (5.49)$$

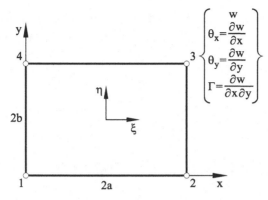

Fig. 5.16 4-noded conforming BFS plate rectangle

with

$$\mathbf{w} = [w_1, \ldots, w_4]^T; \qquad \boldsymbol{\theta}_x = \left[\left(\frac{\partial w}{\partial x}\right)_1, \ldots, \left(\frac{\partial w}{\partial x}\right)_4\right]^T$$

$$\boldsymbol{\theta}_y = \left[\left(\frac{\partial w}{\partial y}\right)_1, \ldots, \left(\frac{\partial w}{\partial y}\right)_4\right]^T; \quad \boldsymbol{\Gamma} = \left[\left(\frac{\partial w}{\partial x \partial y}\right)_1, \ldots, \left(\frac{\partial w}{\partial x \partial y}\right)_4\right]^T \tag{5.50}$$

and

$$\begin{aligned} N_{w_i} &= N_i(\xi), N_i(\eta) &, \quad N_{\theta_{xi}} &= \bar{N}_i(\xi), N_i(\eta) \\ N_{\theta_{yi}} &= N_i(\xi), \bar{N}_i(\eta) &, \quad N_{\Gamma_i} &= \bar{N}_i(\xi), \bar{N}_i(\eta) \end{aligned} \tag{5.51}$$

where N_i and $\bar{N}_i$ are the 1D Hermite shape functions of Eq.(1.11a).

Eqs.(5.50) and (5.51) allow the element stiffness matrix to be derived in a straightforward manner. Details can be found in [WJ,Ya]. The BFS element satisfies the continuity requirements for the normal and cross-derivatives along the sides. The reason is that the normal rotation varies along a side as a cubic polynomial uniquely defined by four parameters, i.e. one rotation and the cross derivative at each of the two end nodes. The element is therefore conforming and it satisfies the patch test. The BFS rectangle is more accurate than the MZC one [WJ]. This is not due to the conformity of the former, but to the fact that the BFS rectangle involves a higher approximation with more nodal variables.

Unfortunately, the practical use of the BFS element is limited to rectangular shapes only. The continuity requirements for the second cross derivatives at the nodes in arbitrary quadrilaterals requires all the second derivatives of w with respect to the different side directions meeting at each node to be defined as nodal variable. Obviously this is quite difficult to generalize for practical purposes.

A development of the BFS element to include the continuity of higher derivatives was outlined in [Sp].

The derivation of conforming plate quadrilaterals starting from triangular plate elements will be studied in Section 5.6.

5.5 TRIANGULAR THIN PLATE ELEMENTS

Triangular plate elements are of interest for the analysis of plates with irregular shapes. Their formulation, however, has the same difficulties for satisfying conformity as the rectangular elements previously studied. Some of the more popular non-conforming and conforming Kirchhoff plate triangles are presented next.

5.5.1 Non-conforming thin plate triangles

Let us consider first the 3-noded triangle. The obvious choice of nodal variables gives a total of nine DOFs (w_i, $\left(\frac{\partial w}{\partial x}\right)_i$ and $\left(\frac{\partial w}{\partial y}\right)_i$ at each node). A complete cubic polynomial has ten terms and, hence, a problem arises when choosing the term to be dropped out. A number of authors have proposed different elements on the basis of the term omitted. Unfortunately all of them require substantial manipulation for ensuring conformity.

Adini and Clough [AC] omitted the xy term in the cubic expansion, i.e.

$$w(x,y) = a_1 + a_2x + a_3y + a_4x^2 + a_6y^2 + a_7x^3 + a_8x^2y + a_9xy^2 + a_{10}y^3 \quad (5.52)$$

This simple criterion yields a poor element which is unable to reproduce constant torsion curvature $\left(\frac{\partial^2 w}{\partial x \partial y}\right)$ states. In addition, the element does not satisfy the C^1 continuity requirement.

Tocher and Kapur [TK] grouped the terms a_8 and a_9 of the cubic polynomial as

$$w(x,y) = a_1 + a_2x + a_3y + a_4x^2 + a_5xy + a_6y^2 +$$
$$+a_7x^3 + a_8(x^2y + xy^2) + a_9y^3 \quad (5.53)$$

This element does not respect the continuity of the normal rotation along the sides. Also, matrix $\mathbf{A}$ of Eq.(5.33b) becomes singular when the sides of the triangle are parallel to the x, y axes [TK,To].

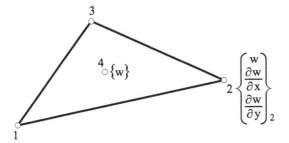

Fig. 5.17 4-noded plate triangle with 10 degrees of freedom [HK]

Harvey and Kelsey [HK] obtained a complete cubic deflection field by adding a fourth central node with a single deflection variable (Figure 5.17). The internal DOF can be eliminated by static condensation. The resulting element does not satisfy the continuity requirement for the normal rotation and it has poor convergence. The performance of this element can be substantially improved by imposing the continuity requirement using Lagrange multipliers. Harvey and Kelsey [HK] showed that the enhanced element satisfies the patch test and it converges monotonically to the exact solution. Further details can be found in [Ya].

Bazeley *et al.* [BCIZ] developed a 3-noded plate triangle with 9 DOFs. The element was subsequently modified by Cheung, King and Zienkiewicz [CKZ] (termed hereafter CKZ element). The starting point is an incomplete cubic expansion of the deflection using area coordinates as

$$
w = a_1 L_1 + a_2 L_2 + a_3 L_3 + a_4 \left(L_1^2 L_2 + \frac{L_1 L_2 L_3}{2} \right) + a_5 \left(L_2^2 L_1 + \frac{L_1 L_2 L_3}{2} \right)
$$
$$
+ a_6 \left(L_2^2 L_3 + \frac{L_1 L_2 L_3}{2} \right) + a_7 \left(L_3^2 L_2 + \frac{L_1 L_2 L_3}{2} \right) +
$$
$$
+ a_8 \left(L_3^2 L_1 + \frac{L_1 L_2 L_3}{2} \right) + a_9 \left(L_1^2 L_3 + \frac{L_1 L_2 L_3}{2} \right) \tag{5.54}
$$

The bracketed terms guarantee the reproduction of an arbitrary curvature field (including that of constant curvature) for a zero value of the nodal deflections.

Following a similar procedure as for the MZC rectangle, Eq.(5.54) can be written in the form

$$
w = \sum_{i=1}^{3} \left(N_i w_i + \bar{N}_i \left(\frac{\partial w}{\partial x} \right)_i + \bar{\bar{N}}_i \left(\frac{\partial w}{\partial y} \right)_i \right) \tag{5.55}
$$

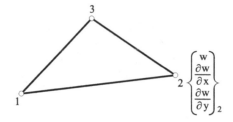

Shape functions :

$$N_1 = L_1 + L_1^2 L_2 + L_1^3 L_3 - L_1 L_2^2 - L_1 L_3^2$$
$$\bar{N}_1 = c_3(L_1^2 L_2 + L) - c_2(L_3 L_1^2 + L)$$
$$\bar{\bar{N}}_1 = b_3(L_1^2 L_2 + L) - b_2(L_3 L_1^2 + L)$$
$$N_2 = L_2 + L_2^2 L_3 + L_2^2 L_1 - L_2 L_3^2 - L_2 L_1^2$$
$$\bar{N}_2 = c_1(L_2^2 L_3 + L) - c_3(L_1 L_2^2 + L)$$

$$\bar{\bar{N}}_2 = b_1(L_2^2 L_3 + L) - b_3(L_1 L_2^2 + L)$$
$$N_3 = L_3 + L_3^2 L_1 + L_3^2 L_2 - L_3 L_1^2 - L_3 L_2^2$$
$$\bar{N}_3 = c_2(L_3^2 L_1 + L) - c_1(L_2 L_3^2 + L)$$
$$\bar{\bar{N}}_3 = b_2(L_3^2 L_1 + L) - b_1(L_2 L_3^2 + L)$$

$$L = \frac{L_1 L_2 L_3}{2} \qquad b_i = y_j - y_m \qquad c_i = x_m - x_j$$

Fig. 5.18 Shape functions for the 3-noded CKZ plate triangle [CKZ]

The shape functions N_i , $\bar{N}_i$ and $\bar{\bar{N}}_i$ are given in Figure 5.18. The stiffness matrix for this element can be found in [CKZ].

The CKZ triangle violates the continuity requirement for the normal rotation and hence is non conforming. However, it converges in a monotonic manner and this has contributed to its popularity [CKZ].

Bazeley *et al.* [BCIZ] proposed a correction to the deflection field leading to a linear distribution of the normal rotation along the sides. This modification does not improve the CKZ element substantially and the performance of the original form is sometimes superior.

Different authors have tried to enhance the behaviour of the CKZ triangle so that it passes the patch test [BN,FB2,KA]. A simple proposal was due to Specht [Sp] who achieved conformity by adding 4th degree terms to the cubic expansion (5.54) as

$$w = a_1 L_1 + a_2 L_2 + a_3 L_3 + a_4 L_1 L_2 + a_5 L_2 L_3 + a_6 L_1 L_3 +$$

$$+ a_7 \left[L_1^2 L_2 + \frac{L}{2}(3(1-\gamma_3))L_1 - (1+3\gamma_3)L_2 + (1+3\gamma_3)L_3 \right] +$$

$$+ a_8 \left[L_2^2 L_3 + \frac{L}{2}(3(1-\gamma_1))L_2 - (1+3\gamma_1)L_3 + (1+3\gamma_1)L_1 \right] +$$

$$+ a_9 \left[L_3^2 L_1 + \frac{L}{2}(3(1-\gamma_2))L_3 - (1+3\gamma_2)L_1 + (1+3\gamma_2)L_2 \right] \quad (5.56)$$

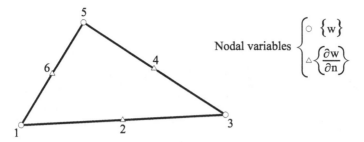

Fig. 5.19 Morley plate triangle of constant curvature

with $L = L_1 L_2 L_3$, $\gamma_1 = \frac{l_3^2 - l_2^2}{l_1^2}$; $\gamma_2 = \frac{l_1^2 - l_3^2}{l_2^2}$; $\gamma_3 = \frac{l_2^2 - l_1^2}{l_3^2}$ and l_1, l_2 and l_3 are the element sides. The element passes all patch tests and performs excellently [Sp,TZSC,ZT2].

Morley 6 DOFs constant curvature triangle

Morley proposed a simple non-conforming 6-noded triangle with just 6 DOFs [Mo,Mo2]. The element uses a complete quadratic expansion of the deflection in terms of the three corner deflection values and the three normal rotations at the mid-sides (Figure 5.19). The Morley triangle has constant curvature and bending moment fields. It satisfies the patch test and converges despite the violation of the C^1 continuity requirement, which demands a cubic deflection field. The element stiffness can be explicitly obtained by

$$\mathbf{K}^{(e)} = A^{(e)} \mathbf{B}^T \mathbf{D}_b \mathbf{B} \tag{5.57}$$

with

$$\mathbf{B} = \frac{1}{A^{(e)}} [\mathbf{G}_1, \mathbf{G}_2] \tag{5.58a}$$

$$\mathbf{G}_1 = - \begin{bmatrix} C_4 S_4 - C_6 S_6 & C_5 S_5 - C_4 S_4 & C_6 S_6 - C_5 S_5 \\ -C_4 S_4 + C_6 S_6 & -C_5 S_5 + C_4 S_4 & -C_6 S_6 + C_5 S_5 \\ -C_4^2 + S_4^2 + C_6^2 - S_6^2 & -C_5^2 + S_5^2 + C_4^2 - S_4^2 & -C_6^2 + S_6^2 + C_5^2 S_5^2 \end{bmatrix} \tag{5.58b}$$

$$\mathbf{G}_2 = \left[\mathbf{G}_2^4, \mathbf{G}_2^5, \mathbf{G}_2^6 \right] \quad , \quad \mathbf{G}_2^k = - \begin{bmatrix} C_k^2 l_k \\ S_k^2 l_k \\ 2 S_k C_k l_k \end{bmatrix} \quad , \quad k = 4, 5, 6 \tag{5.58c}$$

where $C_k = y_{ji}/l_k$, $S_k = -x_{ji}/l_k$, $x_{ji} = x_j - x_i$, $y_{ji} = y_j - y_i$ and $l_k = (x_{ji}^2 + y_{ji}^2)^{1/2}$ is the lengh of side k. The constant bending moment field is given by

$$\hat{\sigma}_b = \mathbf{D}_b \mathbf{B} \mathbf{a}^{(e)} \qquad (5.59a)$$

with

$$\mathbf{a}^{(e)} = \left[w_1, w_2, w_3, \left(\frac{\partial w}{\partial n}\right)_4, \left(\frac{\partial w}{\partial n}\right)_5, \left(\frac{\partial w}{\partial n}\right)_6 \right]^T \qquad (5.59b)$$

The equivalent nodal force vector due to a uniformly distributed loading $f_z = q$ is

$$\mathbf{f}^{(e)} = q\frac{A^{(e)}}{3}[1, 1, 1, 0, 0, 0]^T \qquad (5.60)$$

Nodal point loads are assumed to act at the corner nodes only. Full details on the derivation of the Morley triangle can be found in [Wo].

The Morley triangle is so far the simplest Kirchhoff plate triangle involving deflections and rotations as variables. Its simplicity is comparable to that of the constant strain triangle for plane elasticity problems. Despite its slow convergence, the Morley triangle enjoys big popularity for analysis of plates and shells. A thin plate triangle with identical features as the Morley triangle can be derived starting from the Reissner-Mindlin TLLL triangle (Section 6.8.3) using a Discrete Kirchhoff approach (see Section 6.8.3).

5.5.2 Conforming thin plate triangles

Satisfying the conformity requirements in triangles is a challenging task. The technique of using the curvatures as additional nodal variables is cumbersome and it also makes the extension of the elements for shell analysis difficult. A more successful alternative is to guarantee the continuity of the normal rotation along the sides using additional mid-side variables. Some of these elements are described next.

A conforming plate triangle emerges as a modification of the CKZ element previously described. The shape functions of Eq.(5.54) define a quadratic variation of the normal rotation along each side which can not be uniquely described by the two end values. A solution to this problem is adding three additional mid-side variables which coincide with the normal rotation to each side (Figure 5.20) [ZT]. This suffices to define a complete cubic variation of $\frac{\partial w}{\partial n}$ along each side and conformity is thus satisfied.

Clough and Tocher [CT] developed another conforming triangle starting from an idea of Hsieh in correspondence with Clough [Ya] (denoted here as HCT element). The shape functions are obtained by dividing the element into three inner triangular subdomains as shown in Figure 5.21a.

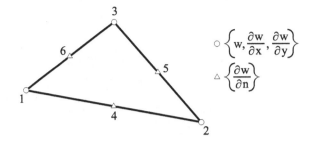

Fig. 5.20 Conforming 12 DOFs triangle

a)   HCT element

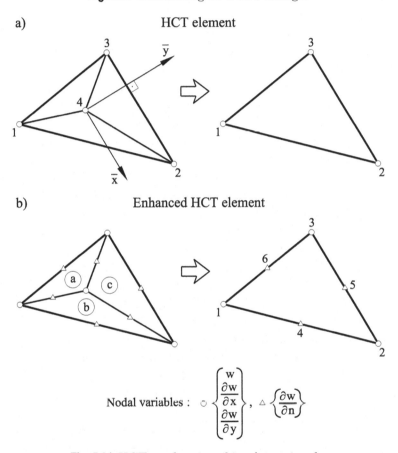

b)   Enhanced HCT element

Nodal variables : $\circ \begin{Bmatrix} w \\ \dfrac{\partial w}{\partial x} \\ \dfrac{\partial w}{\partial y} \end{Bmatrix}$, $\triangle \left\{ \dfrac{\partial w}{\partial n} \right\}$

Fig. 5.21 HCT conforming thin plate triangles

A nine-term incomplete cubic expansion is written in the local axes $\bar{x}$, $\bar{y}$ for every triangular subdomain $4ij$ with $\bar{y}$ chosen orthogonal to the side ij. Thus, for the triangle 423 (Figure 5.21) we write

$$w_A = C_1 + C_2\bar{x} + C_3\bar{y} + C_4\bar{x}^2 + C_5\bar{y}^2 + C_6\bar{x}\bar{y} + C_7\bar{x}^3 + C_8\bar{x}\bar{y}^2 + C_9\bar{y}^3 \quad (5.61)$$

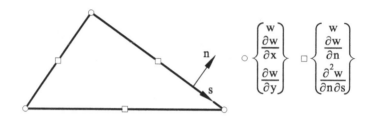

Fig. 5.22 18 DOFs conforming plate triangle proposed by Irons [Ir]

with $\bar{y}$ being orthogonal to side 23. Similar expressions are used for triangles 412 and 431. The omission of the term $\bar{x}^2\bar{y}$ in Eq.(5.61) guarantees that the normal rotation varies linearly along the external sides, whereas the deflection varies quadratically. The local stiffness matrices for each triangular subdomain are transformed to global axes for assembly purposes. The three DOFs of the central node are eliminated by imposing continuity of the normal rotation at the mid-point of the inner sides (three conditions). Further details can be found in [CMPW,CT,Ya]. The HCT element is conforming and it has 9 DOFs as the CKZ element. However, it has a slightly stiffer behaviour.

The performance of the HCT element can be enhanced by starting from three triangular subdomains where, in addition to the standard corner variables, a normal rotation variable is introduced at each mid-side point. This defines a quadratic variation of the rotation along the sides. After eliminating the internal variables a 12 DOFs plate triangle similar to the modified CKZ element is obtained (Figure 5.18b) [Ga2].

A drawback of elements with mid-side normal rotations as variables is that they involve a different number of DOFs per node. To overcome this problem Irons [Ir] proposed a 18 DOFs quartic triangle where the deflection and the curvature $\frac{\partial^2 w}{\partial n \partial s}$ are added as mid-side variables (Figure 5.22).

Other authors have proposed different 3-noded conforming plate triangles based on cubic and quartic expansions of the normal rotation $\frac{\partial w}{\partial n}$ along the sides. Cowper *et al.* [CKLO] proposed a 18 DOFs plate triangle with $w, \frac{\partial w}{\partial x}, \frac{\partial w}{\partial y}, \frac{\partial^2 w}{\partial x^2}, \frac{\partial^2 w}{\partial y^2}, \frac{\partial^2 w}{\partial x \partial y}$ at each node (Figure 5.23a). The shape functions omit three terms of a complete quintic polynomial (which has 21 terms) preserving a cubic variation of $\frac{\partial w}{\partial n}$ along the three sides.

This element can be enhanced by adding three mid-side nodes with the normal rotation as variable [AFS,Be,Ir] (Figure 5.23b). The shape

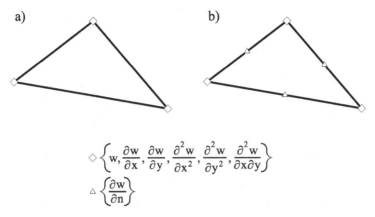

Fig. 5.23 18 and 21 DOFs conforming plate triangles

functions are now complete quintic polynomials (21 terms) and $\frac{\partial w}{\partial n}$ has a quartic variation along the sides.

More information on these elements can be found in [Ya,ZT2] in addition to the previously quoted references. Despite their accuracy, the practical acceptance of thin plate triangles with curvatures as nodal variables has been limited. The main reason for this is the intrinsic difficulties for their extension to shell analysis.

5.6 CONFORMING THIN PLATE QUADRILATERALS OBTAINED FROM TRIANGLES

One of the first conforming plate quadrilaterals derived from triangles is due to Fraejis de Veubeke [FdV,FdV2]. The element was later developed by Sander [San]. The starting point is the splitting of the quadrilateral into four inner triangles as shown in Figure 5.24a. A complete 10-term cubic polynomial is used to approximate the deflection within each subdomain and, thus, the total number of initial variables is 40. After eliminating the internal variables the DOFs are reduced to 16, i.e. the three standard corner variables and the normal rotations at the mid-side points. The element is conforming and it can be distorted to arbitrary quadrilateral shapes. The mid-side normal rotations are eliminated by imposing a linear variation of $\frac{\partial w}{\partial n}$ along each side. This, however, does not improve the performance of the element [FdV2].

A second thin plate quadrilateral was developed by Clough and Felippa [CF] almost at the same time as the Fraejis de Veubeke element described

a) Conforming quadrilateral of Fraeijs de Veubeke [FdV2]

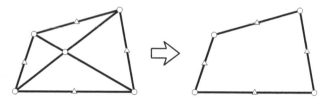

b) Conforming quadrilateral of Clough and Felippa [CF]

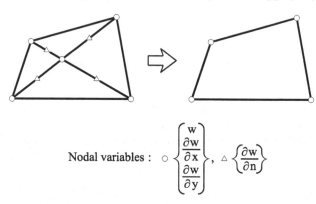

Nodal variables : $\circ \left\{ \begin{matrix} w \\ \dfrac{\partial w}{\partial x} \\ \dfrac{\partial w}{\partial y} \end{matrix} \right\}, \; \triangle \left\{ \dfrac{\partial w}{\partial n} \right\}$

Fig. 5.24 Conforming plate quadrilaterals of Fraejis de Veubeke [FdV2] (a) and Clough and Felippa [CF] (b)

above. The triangular subdivision is very similar in both cases, as shown in Figure 5.24b. A complete cubic expansion is again used for the deflection within each inner triangle. All internal DOFs are eliminated after assembly by simple static condensation and also by imposing the continuity of the normal rotation at the internal mid-side nodes. The resulting conforming quadrilateral has only 12 DOFs and is quite accurate.

5.7 CONFORMING THIN PLATE ELEMENTS DERIVED FROM REISSNER-MINDLIN FORMULATION

Conforming thin plate elements can also be derived by "degeneration" of the C^o continuous Reissner-Mindlin plate elements to be studied in the next chapter. The process consists in constraining the transverse shear strains to take a zero value at a discrete number of points of the original Reissner-Mindlin element. Consequently, the "effective" shear energy is zero over the element. These elements are termed DK (for Discrete Kirchhoff) and they perform like Kirchhoff thin plate elements. However, *they only require C^o continuity for the displacement field* and this guarantees compatibility.

This technique can be considered as a generalization of that used in Chapter 2 for deriving Euler-Benouilli beam elements from Timoshenko elements (Section 2.8.4 and Examples 2.5 and 2.6).

The formulation of DK plate elements is described in the next chapter when dealing with Reissner–Mindlin thick plate theory. Nevertheless, we should keep in mind that they are another class of thin plate elements.

5.8 ROTATION-FREE THIN PLATE TRIANGLES

Rotation-free plate elements can be derived by extending the concepts explained in Section 1.4 for rotation-free Euler-Bernouilli beam elements.

The idea of using the deflection as the only nodal variable for plate bending analysis was originally exploited by finite difference (FD) practitioners [Ug]. The obvious difficulties in FD techniques are the treatment of boundary conditions and the problems when dealing with non-orthogonal or unstructured grids.

Several authors have proposed plate and shell finite elements with displacements as the nodal variables. Nay and Utku [NU] derived a rotation-free 3-noded thin plate triangle using a least square quadratic approximation to describe the deflection field within the patch surrounding a node in terms of the deflections of the patch nodes. The element stiffness matrix was obtained by the standard minimum potential energy approach. Later Barnes [Bar] proposed a method for deriving a 3-noded plate triangle with the nodal deflections as the only DOFs based on the computation of the curvatures in terms of the normal rotations at the mid-side points determined from the nodal deflections of adjacent elements. This method was exploited by Hampshire *et al.* [HTC] assuming that the elements are hinged together at their common boundaries, the bending stiffness being represented by torsional springs resisting the rotations about the hinge lines. Phaal and Calladine [PC2,3] presented a similar class of rotation-free triangles for plate and shell analysis. Yang *et al.* [YJS+] derived a family of triangular elements of this type for sheet stamping analysis based on so called bending energy augmented membrane approach which basically reproduces the hinge bending stiffness procedure described in [HTC]. Brunet and Sabourin [BS5,6,SB2] used a different method to compute the constant curvature field within each triangle in terms of the six nodal displacements of a macro-element (quadrilateral). The triangle was successfully applied to non-linear shell analysis using an explicit dynamic approach. Rio *et al.* [RTL] have used the concept of side hinge bending stiffness for deriving a thin shell triangle of "translational" kind for explicit dynamic analysis of sheet stamping problems.

Oñate and Cervera [OC2] obtained a competititve and simple three node plate triangle with the deflection as the only nodal DOF by blending ideas from finite element and finite volume methods [IO,OCZ,ZO]. This work was generalized and extended by Oñate and Zarate [OZ] who used for the first time the name *rotation-free element* and derived two families of rotation-free plate and shell triangles by combining a standard linear finite element approximation over 3-noded triangles with cell centred (CC) and cell vertex (CV) finite volume schemes. These elements were extended to linear and non linear analysis of shells by Flores and Oñate [FO2,3,4] and Oñate *et al.* [OCM,OF,OFN]. Details of this two families of rotation-free plate triangles are given in the next sections.

Rotation-free thin plate and shell quadrilaterals have been derived by Savourin *et al.* [SCB2] and Flores and Estrada [FE]. These elements require some stabilization to avoid spurious deformation modes and are not so popular as the rotation-free triangles.

Other rotation-free plate and shell elements based on the upgrading of membrane theory to shells [LWB], isogeometric formulations [BBHH, KBLW] and Bezier interpolations over triangular patches [UO] have been recently proposed. Indeed, this topic seems to continue attracting much interest in the computational mechanics community.

5.8.1 Formulation of rotation-free triangles by a combined finite element and finite volume method

Let us consider an arbitrary discretization of the plate into standard 3-noded triangles. The curvature and the bending moments are described by constant fields within appropriate *non-overlapping control domains* (also termed "control volumes" in the finite volume literature [IO,OCZ,ZO]) covering the whole plate as

$$\hat{\boldsymbol{\varepsilon}}_b = \hat{\boldsymbol{\varepsilon}}_b^p \quad , \quad \hat{\boldsymbol{\sigma}}_b = \hat{\boldsymbol{\sigma}}_b^p \tag{5.62}$$

where $(\cdot)^p$ denotes constant values for the p-th control domain.

Two modalities of control domains are considered: a) that formed by a single triangular element (Figure 5.25a), and b) the control domain formed by one third of the areas of the elements surrounding a node (Figure

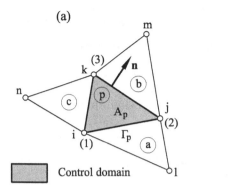 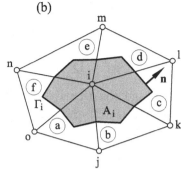

Fig. 5.25 Cell-centred (a) and cell-vertex (b) schemes. BPT and BPN triangles. Numbers in brackets denote local node numbers. Element sides are defined by opposite local node numbers, i.e. side 1 has nodes 2 and 3, etc.

5.25b). In the finite volume literature the two options are termed "cell-centred" (CC) and "cell-vertex" (CV) schemes, respectively [OCZ,ZO].

In the CC scheme each control domain coincides with a standard 3-noded finite element triangle. In the CV scheme a control domain is contributed by different elements, as shown in Figure 5.25b.

We identify the "patch of elements" associated with a control domain. In the CC scheme the patch is always formed by four elements (except in elements sharing a boundary segment). In the CV scheme the number of elements in the patch can vary.

In a CC scheme the chosen variables (i.e. the curvatures and bending moments) are "sampled" at the center of the cells discretizing the analysis domain (i.e. the 3-noded triangles). In a CV scheme the variables are sampled at the mesh nodes.

Let us now integrate the constitutive equation (5.14) and the curvature-deflection relationship (5.9) over each control domain as

$$\iint_{A_p} (\hat{\boldsymbol{\sigma}}_b - \hat{\mathbf{D}}_b\hat{\boldsymbol{\varepsilon}}_b)dA = 0 \tag{5.63}$$

$$\iint_{A_p} (\hat{\boldsymbol{\varepsilon}}_b - \mathbf{L}w)dA = 0 \tag{5.64}$$

where A_p is pth control domain area and the curvature operator $\mathbf{L}$ is

$$\mathbf{L} = \left[\frac{\partial^2}{\partial x^2}, \frac{\partial^2}{\partial y^2}, 2\frac{\partial^2}{\partial x\partial y}\right]^T \tag{5.65}$$

Eqs.(5.63) and (5.64) express the satisfaction of the constitutive and curvature-deflection equations over a control domain in a mean sense.

Introducing into Eq.(5.63) the assumed constant bending moment and curvature fields within each control domain gives

$$\hat{\boldsymbol{\sigma}}_b^p = \hat{\mathbf{D}}_b^p \hat{\boldsymbol{\varepsilon}}_b^p \ , \quad \text{where} \quad \hat{\mathbf{D}}_b^p = \frac{1}{A_p} \iint_{A_p} \hat{\mathbf{D}}_b dA \tag{5.66}$$

is the average constitutive matrix over a control domain.

From Eqs.(5.62) and (5.64) we also obtain

$$\hat{\boldsymbol{\varepsilon}}_b^p = \frac{1}{A_p} \iint_{A_p} \mathbf{L} w \, dA \tag{5.67}$$

A simple integration by parts of the r.h.s. of Eq.(5.67) leads to

$$\boxed{\hat{\boldsymbol{\varepsilon}}_b^p = \frac{1}{A_p} \int_{\Gamma_p} \mathbf{T} \nabla w d\Gamma} \tag{5.68}$$

where

$$\nabla = \left[\frac{\partial}{\partial x}, \frac{\partial}{\partial y} \right]^T \ , \quad \mathbf{T} = \begin{bmatrix} n_x & 0 & n_y \\ 0 & n_y & n_x \end{bmatrix}^T \tag{5.69a}$$

and n_x, n_y are the components of the outward unit normal $\mathbf{n}$ to the boundary Γ_p of the pth control domain (Figure 5.25). For the ith side of p with length l_i^p

$$\mathbf{n}_i = \frac{1}{l_i^p}[-b_i^p, -c_i^p] \ , \quad \mathbf{T}_i = \frac{1}{l_i^p} \begin{bmatrix} b_i^p & 0 & c_i^p \\ 0 & c_i^p & b_i^p \end{bmatrix}^T \tag{5.69b}$$

$$b_i^p = y_j^p - y_k^p \ , \quad c_i^p = x_k^p - x_j^p \quad i,j = 1,2,3 \tag{5.69c}$$

where parameters b_i^p and c_i^p are obtained by cyclic permutation of the indexes i, j, k.

Note that the element sides are defined by the opposite local node number, i.e. side 1 is defined by nodes 2,3, etc. (Figure 5.26).

Eq.(5.68) defines the average curvatures for each control volume in terms of the deflection gradients along its boundaries. The transformation of the area integral of Eq.(5.67) into the line integral of Eq.(5.68) is typical of finite volume methods [IO,OCZ,ZO].

The line integral in Eq.(5.68) poses a difficulty when the deflection gradient is discontinuous at the control domain boundary and some smoothing procedure is required. This issue is discussed below.

The PVW for the case of a distributed vertical load (Eq.(5.19)) is written as (noting that $\delta\hat{\boldsymbol{\varepsilon}}_b = \mathbf{L}\delta w$)

$$\sum_p \iint_{A_p} (\mathbf{L}\delta w)^T \hat{\boldsymbol{\sigma}}_b^p dA - \iint_A \delta w f_z dA = 0 \tag{5.70}$$

The sum in the first term of Eq.(5.70) extends over all the control domains in the mesh.

Integrating by parts the first integral in Eq.(5.70) and recalling that the bending moments are constant within each control domain, gives

$$\sum_p \left(\int_{\Gamma_p} [\mathbf{T}\boldsymbol{\nabla}\delta w]^T d\Gamma \right) \hat{\boldsymbol{\sigma}}_b^p - \iint_A \delta w f_z dA = 0 \tag{5.71}$$

Substituting Eqs.(5.66) and (5.68) into (5.71) yields finally

$$\sum_p \left\{ \left(\int_{\Gamma_p} [\mathbf{T}\boldsymbol{\nabla}\delta w]^T d\Gamma \right) \frac{1}{A_p} \hat{\mathbf{D}}_b^p \left(\int_{\Gamma_p} \mathbf{T}\boldsymbol{\nabla} w \, d\Gamma \right) \right\} - \iint_A \delta w f_z dA = 0 \tag{5.72}$$

Eq.(5.72) is the basis for deriving the final set of equilibrium equations, after the appropriate discretization of the deflection field.

An alternative derivation of Eq.(5.72) using a Hu-Washizu variational principle and a mixed formulation can be found in [OZ].

Derivation of the discretized equations

The deflection field is interpolated linearly within each triangular element in terms of the nodal values as in the standard FEM, i.e.

$$w = \sum_{i=1}^3 N_i w_i = \mathbf{N}^{(e)} \mathbf{w}^{(e)} \tag{5.73}$$

with $\mathbf{N}^{(e)} = [N_1, N_2, N_3]$ and $\mathbf{w}^{(e)} = [w_1, w_2, w_3]^T$. In Eq.(5.73) N_i are the standard linear shape functions for the 3-noded triangle. Substituting Eq.(5.73) into (5.68) gives

$$\hat{\boldsymbol{\varepsilon}}_b^p = \frac{1}{A_p} \int_{\Gamma_p} \mathbf{T}\boldsymbol{\nabla}\mathbf{N}^{(e)} \mathbf{w}^{(e)} = \mathbf{B}^p \bar{\mathbf{w}}^p \tag{5.74}$$

where vector $\bar{\mathbf{w}}^p$ lists the deflections of the nodes linked to the p-th control domain and $\mathbf{B}^p$ is the curvature matrix relating the constant curvature field within each control domain and the nodal deflections $\bar{\mathbf{w}}^p$. Matrix $\mathbf{B}^p$ is different for the CC and CV schemes.

Substituting Eq.(5.73) into (5.72) and using (5.74) gives the final system of equilibrium equations as

$$\mathbf{Kw} = \mathbf{f} \tag{5.75}$$

where vector $\mathbf{w}$ contains the nodal deflections of all mesh nodes. The global stiffness matrix $\mathbf{K}$ is obtained by assembling in the usual manner the stiffness contributions from the different control domains given by

$$\mathbf{K}^p = [\mathbf{B}^p]^T \hat{\mathbf{D}}_b^p \mathbf{B}^p A_p \tag{5.76}$$

The components of the nodal force vector $\mathbf{f}$ in Eq.(5.75) are obtained as for standard C° linear finite element triangles [On4], i. e.

Point loads

$$f_i = P_{z_i} \tag{5.77}$$

where P_{z_i} is the vertical point load acting on the i-th node

Distributed loading

$$f_i^{(e)} = \iint_{A^{(e)}} N_i f_z(x) dA \tag{5.78}$$

The global nodal force component f_i is obtained by assembling the element contributions $f_i^{(e)}$. For a constant distributed load $f_z = q$

$$f_i = \sum_e \frac{qA^{(e)}}{3} \tag{5.79}$$

where the sum extends over all the triangles sharing the i-th node, and $A^{(e)}$ is the area of element e.

The previous equations are particularized next for CC and CV schemes.

5.8.2 Cell-centred patch. BPT rotation-free plate triangle

In CC patches the control domain coincides with an individual element and the evaluation of the constant curvature field of Eq.(5.68) can be simply written as

$$\hat{\varepsilon}_b^p = \frac{1}{A_p} \sum_{j=1}^3 l_j^p \mathbf{T}_j^p (\nabla w)_j^p = \mathbf{C}_p (\overline{\nabla w})^p \tag{5.80a}$$

with

$$\mathbf{C}_p = [\mathbf{C}_1^p, \mathbf{C}_2^p, \mathbf{C}_3^p] \quad , \quad \mathbf{C}_i^p = \frac{1}{l_i^p} \begin{bmatrix} b_i^p & 0 \\ 0 & c_i^p \\ c_i^p & 0 \end{bmatrix} \quad , \quad (\overline{\nabla w})^p = \left\{ \begin{array}{c} (\nabla w)_1^p \\ (\nabla w)_2^p \\ (\nabla w)_3^p \end{array} \right\} \tag{5.80b}$$

In Eq.(5.80a) the sum extends over the three sides of element p coinciding with the pth control domain, $\mathbf{T}_j^p$ is the transformation matrix of

Eq.(5.69) for the side j of element p, $(\nabla w)_j^p$ is the deflection gradient at the side mid-point, x_j^p, y_j^p are the coordinates of node j of element p.

The evaluation of the deflection gradient $(\nabla w)_i^p$ at the element sides in Eq.(5.80a) poses a difficulty as these gradients are discontinuous across elements for linear interpolations of w. Oñate and Cervera [OC2] overcome this problem by computing the deflection gradient at the triangle sides as the average values of the gradients contributed by the two elements sharing the side. This gives

$$(\overline{\nabla w})^p = \mathbf{M}_p(\widehat{\nabla w})^p \tag{5.81a}$$

with

$$\mathbf{M}_p = \frac{1}{2}\begin{bmatrix} \mathbf{I}_2 & \mathbf{0}_2 & \mathbf{I}_2 & \mathbf{0}_2 \\ \mathbf{I}_2 & \mathbf{0}_2 & \mathbf{0}_2 & \mathbf{I}_2 \\ \mathbf{I}_2 & \mathbf{I}_2 & \mathbf{0}_2 & \mathbf{0}_2 \end{bmatrix} \quad, \quad (\widehat{\nabla w})^p = \begin{Bmatrix} (\nabla w)^p \\ (\nabla w)^a \\ (\nabla w)^b \\ (\nabla w)^c \end{Bmatrix} \quad, \quad \mathbf{I}_2 = \begin{bmatrix} 1 & 0 \\ 0 & 1 \end{bmatrix} \quad, \quad \mathbf{0}_2 = \begin{bmatrix} 0 & 0 \\ 0 & 0 \end{bmatrix}$$

$$\tag{5.81b}$$

In Eq.(5.81b) $(\nabla w)^k$ is the (constant) deflection gradient at element k.

Substituting the linear interpolation of the deflection (Eq.(5.73)) into (5.81a) gives

$$(\overline{\nabla w})^p = \mathbf{M}_p\mathbf{G}_p\bar{\mathbf{w}}^p \tag{5.82}$$

with

$$\mathbf{G}_p = \begin{bmatrix} \bar{b}_1^p & \bar{b}_2^p & \bar{b}_3^p & 0 & 0 & 0 \\ \bar{c}_1^p & \bar{c}_2^p & \bar{c}_3^p & 0 & 0 & 0 \\ \bar{b}_3^a & \bar{b}_2^a & 0 & \bar{b}_1^a & 0 & 0 \\ \bar{c}_3^a & \bar{c}_2^a & 0 & \bar{c}_1^a & 0 & 0 \\ 0 & \bar{b}_3^b & \bar{b}_2^b & 0 & \bar{b}_1^b & 0 \\ 0 & \bar{c}_3^b & \bar{c}_2^b & 0 & \bar{c}_1^b & 0 \\ \bar{b}_2^c & 0 & \bar{b}_3^c & 0 & 0 & \bar{b}_1^c \\ \bar{c}_2^c & 0 & \bar{c}_3^c & 0 & 0 & \bar{c}_1^c \end{bmatrix} \quad, \quad \bar{\mathbf{w}}^p = \begin{Bmatrix} w_i \\ w_j \\ w_k \\ w_l \\ w_m \\ w_n \end{Bmatrix} \tag{5.83}$$

where $\bar{b}_i^p = \frac{b_i^p}{2A^p}$, $\bar{c}_i^p = \frac{c_i^p}{2A^p}$ with b_i^p, c_i^p given in Eq.(5.69c).

Substituting Eq.(5.82) into (5.80a) gives finally

$$\hat{\boldsymbol{\varepsilon}}_b^p = \mathbf{C}_p\mathbf{M}_p\mathbf{G}_p\bar{\mathbf{w}}^p = \mathbf{B}^p\bar{\mathbf{w}}^p \tag{5.84a}$$

with the curvature matrix $\mathbf{B}^p$ for element p given by

$$\mathbf{B}^p = \mathbf{C}_p\mathbf{M}_p\mathbf{G}_p \tag{5.84b}$$

$\mathbf{B}^p$ is a 3×6 matrix relating the curvatures with the deflections at the six nodes of the four element patch contributing to the control domain. Consequently, $\mathbf{K}^p$ is a 6×6 stiffness matrix.

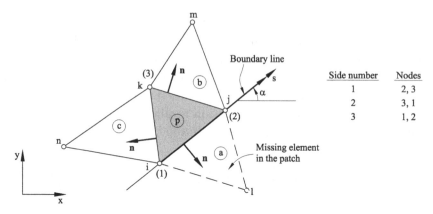

Fig. 5.26 Basic plate triangle (BPT) next to a boundary line. Numbers in brackets denote local node numbers. Definition of element sides

The resulting plate element is termed BPT (for Basic Plate Triangle). The element can be viewed as a standard finite element plate triangle with one DOF per node and a wider bandwidth, as each element is linked to its neighbours through Eq.(5.84a).

The BPT element can be extended to shell analysis leading to a simple and accurate rotation-free shell triangle (Section 8.13).

5.8.2.1 Boundary conditions for the BPT element

A difference between the BPT element and standard plate elements is that the conditions on the prescribed rotations must be imposed when building the curvature matrix $\mathbf{B}_p$.

Free edge

A BPT element with a side along a free boundary edge has one of the elements contributing to the patch missing. The contribution of this element is therefore omitted in matrix $\mathbf{M}^p$ of Eq.(5.81b) when performing the average of the deflection gradients. Thus, if side 3 corresponding to nodes i, j lies on a free boundary (Figure 5.26), matrix $\mathbf{M}^p$ is modified as

$$\mathbf{M}_p = \frac{1}{2} \begin{bmatrix} \mathbf{I}_2 & \mathbf{0}_2 & \mathbf{I}_2 & \mathbf{0}_2 \\ \mathbf{I}_2 & \mathbf{0}_2 & \mathbf{0}_2 & \mathbf{I}_2 \\ 2\mathbf{I}_2 & \mathbf{0}_2 & \mathbf{0}_2 & \mathbf{0}_2 \end{bmatrix} \tag{5.85}$$

while matrix $\mathbf{G}_p$ of Eq.(5.83) can remain unaltered. Clearly, the deflection at node 1 in the missing element should be prescribed to zero.

Additional conditions must be imposed on boundary edges where the rotations and/or the deflections are constrained as explained next.

Simply supported edge $\left(w = \dfrac{\partial w}{\partial s} = 0\right)$

The condition $\frac{\partial w}{\partial s} = 0$, where s is the boundary direction, is imposed by prescribing $w = 0$ at the boundary nodes. The "missing" element in the patch at the boundary edge is treated as described above for the free edge.

Clamped edge $(w = \nabla w = 0)$

The condition $w = 0$ at clamped edges is satisfied by making the corresponding nodal deflections equal to zero when solving the global system of equations, as it is standard in the FEM.

The condition of zero rotations is imposed by disregarding the contributions from the clamped edges when computing the sum along the element sides in Eq.(5.80a). For instance, if side ij is clamped, matrix $\mathbf{M}_p$ of Eq.(5.81b) is modified as

$$\mathbf{M}_p = \frac{1}{2} \begin{bmatrix} \mathbf{I}_2 & \mathbf{0}_2 & \mathbf{I}_2 & \mathbf{0}_2 \\ \mathbf{I}_2 & \mathbf{0}_2 & \mathbf{0}_2 & \mathbf{I}_2 \\ \mathbf{0}_2 & \mathbf{0}_2 & \mathbf{0}_2 & \mathbf{0}_2 \end{bmatrix} \tag{5.86}$$

Symmetry line $\left(\dfrac{\partial w}{\partial n} = 0\right)$

The condition of zero normal rotation is imposed by neglecting the contribution from the prescribed rotation term at the symmetry line when computing Eq.(5.80a).

Let us assume that side 3 of element p with nodes i, j is a symmetry axis (Figure 5.26). The deflection gradient at that side is expressed in term of the tangential and normal rotations as

$$(\nabla w)_3^p = \begin{bmatrix} C_\alpha & S_\alpha \\ S_\alpha & -C_\alpha \end{bmatrix} \left\{ \begin{array}{c} \dfrac{\partial w}{\partial s} \\[2mm] \dfrac{\partial w}{\partial n} \end{array} \right\}_3^p \tag{5.87}$$

where s and n are the directions along the side and normal to the side, respectively, $C_\alpha = \cos \alpha$ and $S_\alpha = \sin \alpha$ and α is the angle that the side forms with the x axis (Figure 5.26).

The condition of zero normal rotation at the side is now introduced in Eq.(5.87), i.e.

$$(\nabla w)_3^p = \begin{bmatrix} C_\alpha & S_\alpha \\ S_\alpha & C_\alpha \end{bmatrix} \left\{ \begin{array}{c} \frac{w_j - w_i}{l_3^p} \\ 0 \end{array} \right\} = \frac{1}{l_3^p} \begin{bmatrix} -C_\alpha & C_\alpha \\ -S_\alpha & S_\alpha \end{bmatrix} \left\{ \begin{array}{c} w_i \\ w_j \end{array} \right\} \tag{5.88}$$

Matrix $\mathbf{M}_p$ is modified to account for the contribution of the element adjacent to the side lying on the symmetry axis only as

$$\mathbf{M}_p = \frac{1}{2} \begin{bmatrix} \mathbf{I}_2 & \mathbf{0}_2 & \mathbf{I}_2 & \mathbf{0}_2 \\ \mathbf{I}_2 & \mathbf{0}_2 & \mathbf{0}_2 & \mathbf{I}_2 \\ \mathbf{0}_2 & 2\mathbf{I}_2 & \mathbf{0}_2 & \mathbf{0}_2 \end{bmatrix} \tag{5.89}$$

Finally, the third and four rows of matrix $\mathbf{G}_p$ are modified as

$$\frac{1}{l_3^p} \begin{bmatrix} -C_\alpha & C_\alpha & 0 & 0 & 0 & 0 \\ -S_\alpha & S_\alpha & 0 & 0 & 0 & 0 \end{bmatrix} \tag{5.90}$$

5.8.3 Cell-vertex patch. BPN rotation-free plate triangle

A different class of rotation-free plate triangles can be derived starting from the "cell vertex" (CV) scheme (Figure 5.25b) [OZ]. The advantage is that the deflection gradient is now continuous along the control domain boundary. This allows the constant curvature vector to be computed directly over the control domain as

$$\hat{\varepsilon}_b^i = \frac{1}{A_i} \int_{\Gamma_i} \mathbf{T}\nabla\mathbf{N}_i w_i d\Gamma = \mathbf{B}_i \bar{\mathbf{w}}_i \tag{5.91}$$

where $\mathbf{N}_i$ contains the contributions of the shape functions from all the elements participating to the i-th nodal control domain. Eq.(5.91) can be rewritten taking into account that the deflection gradients are constant within each element, as

$$\hat{\varepsilon}_b^i = \frac{1}{A_i} \sum_{j=1}^{n_i} \frac{l_j}{2} \mathbf{T}_j \nabla\mathbf{N}^{(j)} \mathbf{w}^{(j)} = \mathbf{B}_i \bar{\mathbf{w}}^i \tag{5.92}$$

where the sum extends over the n_i elements contributing to the i-th control domain (for instance $n_i = 5$ in Figure 5.27), l_j is the external side of element j, $\mathbf{T}_j$ is the transformation matrix of Eq.(5.69) for side l_j, superindex j refers to element values and $A_i = \frac{1}{3} \sum_{k=1}^{n_i} A^{(k)}$ where $A^{(k)}$ is the area of element k.

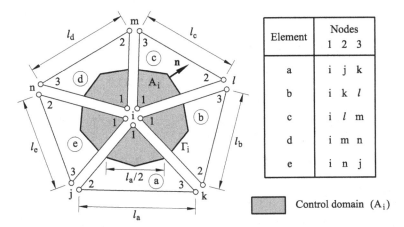

Element	Nodes 1 2 3
a	i j k
b	i k l
c	i l m
d	i m n
e	i n j

▨ Control domain (A_i)

Fig. 5.27 BPN element. Typical CV control domain and numbering of nodes

Vector $\bar{\mathbf{w}}_i$ in Eq.(5.92) lists all the patch nodes. For instance in Figure 5.27, $\bar{\mathbf{w}}_i = [w_i, w_j, w_k, w_l, w_m, w_n]^T$.

The computation of the curvature matrix $\mathbf{B}_i$ for the patch is not so straightforward as its size depends on the number of nodes in the patch contributing to a nodal control domain. Typically,

$$
\underset{3 \times n}{\mathbf{B}_i} = \overset{1 \quad 2 \quad \dots \quad n_p}{\left[\mathbf{B}^i, \mathbf{B}^a, \dots, \mathbf{B}^r \right]}
\tag{5.93}
$$

where n_p is the number of nodes in the patch (i.e. $n_p = 6$ for the patch of Figure 5.27) and upper indexes $i, a, \dots r$ refer to global node numbers. Box 5.3 shows matrix $\mathbf{B}_i$ for the control domain of Figure 5.27.

Note that $\mathbf{B}_i$ is the *global curvature matrix* for the central i-th node. The stiffness matrix for the i-th control domain is obtained as

$$
\mathbf{K}_i = \mathbf{B}_i^T \hat{\mathbf{D}}_b^i \mathbf{B}_i A_i
\tag{5.94}
$$

where $\hat{\mathbf{D}}_b^i$ is the average constitutive matrix for the i-th control domain. The global stiffness matrix is assembled from the nodal stiffness matrices $\mathbf{K}_i$ for the different control domains. The process is analogous to that followed for assembling the stiffness matrices for the CVB rotation-free beam element (Section 1.4.2). An element stiffness matrix can be also found by combining the nodal stiffness matrix of the three nodes as described for the CVB element. However, the direct assembly of the nodal stiffness matrix is recommended in practice.

$$\mathbf{B}_i = [\mathbf{B}^i, \mathbf{B}^j, \mathbf{B}^k, \mathbf{B}^l, \mathbf{B}^m, \mathbf{B}^n]$$

$$\mathbf{B}^i = \frac{1}{2A_i}[l_a\mathbf{T}_a\mathbf{G}_1^{(a)} + l_b\mathbf{T}_b\mathbf{G}_1^{(b)} + l_c\mathbf{T}_c\mathbf{G}_1^{(c)} + l_d\mathbf{T}_d\mathbf{G}_1^{(d)} + l_e\mathbf{T}_e\mathbf{G}_1^{(e)}]$$

$$\mathbf{B}^j = \frac{1}{2A_i}[l_a\mathbf{T}_a\mathbf{G}_2^{(a)} + l_e\mathbf{T}_e\mathbf{G}_3^{(e)}] \quad , \quad \mathbf{B}^k = \frac{1}{2A_i}[l_a\mathbf{T}_a\mathbf{G}_3^{(a)} + l_b\mathbf{T}_b\mathbf{G}_2^{(b)}]$$

$$\mathbf{B}^l = \frac{1}{2A_i}[l_b\mathbf{T}_b\mathbf{G}_3^{(b)} + l_c\mathbf{T}_c\mathbf{G}_2^{(c)}] \quad , \quad \mathbf{B}^m = \frac{1}{2A_i}[l_c\mathbf{T}_c\mathbf{G}_3^{(c)} + l_d\mathbf{T}_d\mathbf{G}_2^{(d)}]$$

$$\mathbf{B}^n = \frac{1}{2A_i}[l_d\mathbf{T}_d\mathbf{G}_3^{(d)} + l_e\mathbf{T}_e\mathbf{G}_2^{(e)}]$$

$$\mathbf{G}_i^{(k)} = \boldsymbol{\nabla} N_i^{(k)} = \frac{1}{2A^{(k)}} \begin{Bmatrix} b_i \\ c_i \end{Bmatrix}^{(k)} , \quad b_i^{(k)} = y_j^{(k)} - y_k^{(k)}, \quad c_i^{(k)} = x_k^{(k)} - x_j^{(k)}$$

Box 5.3 Curvature matrix for the BPN control domain of Figure 5.27

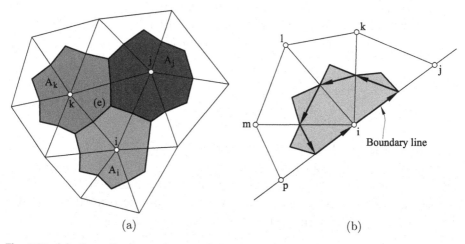

(a)   (b)

Fig. 5.28 (a) Contribution of control domains (in grey) to a rotation-free BPN triangle in the cell-vertex scheme. (b) Control domain sharing a boundary line. Arrows show the integration patch for computing the curvature matrix

The nodal force vector $\mathbf{f}_i$ is computed as explained in Section 5.8.1.

This plate element is termed BPN (for Basic Plate Nodal patch) [OZ]. The term "element" is somehow ambiguous as the BPN element combines a standard finite element interpolation over triangles with non-standard integration regions (the control domains) (Figure 5.28a).

The boundary conditions in the BPN element are imposed following the lines explained for the BPT element. The procedure is simpler as the deflection gradient is continuous on the control domain boundaries which lay within the elements. The conditions on the nodal deflections are imposed by prescribing w_i at the equations solution level. The conditions on the rotations, however, are imposed when building the curvature matrix. The integration patch for computing the curvature matrix should be adequately defined in this case. An example is shown in Figure 5.28b.

Clamped edges and symmetry lines. Zero rotations at clamped edges and symmetry lines are imposed by eliminating the contributions from these rotation terms in the sum of Eq.(5.92).

Simple supported edges. The condition $\frac{\partial w}{\partial s} = 0$ along an edge direction is accounted for by making zero the deflection at the edge nodes.

Free edges. No special treatment for the rotations is required at free edges. The performance of the element can be improved by prescribing the edge bending moments M_n and M_{sn} to a zero value. This can be simply done by making the appropriate rows in the constitutive matrix $\hat{\mathbf{D}}_b^p$ equal to zero at the patches containing a free edge. For free edges which are not parallel to one of the cartesian axes, a transformation of the constitutive equation to edge axes is necessary. This procedure can also be applied for prescribing the condition $M_n = 0$ at simply supported edges [OZ].

Similarly as for the CVB beam element (Section 1.4.3), the performance of the BPN element is superior to the BPT one for regular meshes. However, its accuracy slightly deteriorates for non-structured meshes and particularly when the control domains involve less than six nodes, since the evaluation of the curvatures is not as accurate in these cases. Also, its extension to shells is not so straightforward as for the BPT element.

5.9 PATCH TESTS FOR KIRCHHOFF PLATE ELEMENTS

The three modalities of patch tests A, B and C explained in Section 6.10 of [On4] are applicable to plate elements. Patch test A consists in prescribing a displacement field at all nodes of a patch of elements and checking that the equilibrium conditions are satisfied. In patch test B the displacements of the nodes at the boundary of the patch are prescribed. Then the values of the displacement at the internal nodes are computed and compared to the exact ones. Patch test C consists in assembling the matrix system for the whole patch and finding the solution after fixing the minimum number of DOFs necessary to eliminate rigid body motion. The computed solution is then compared to the exact one. Satisfaction of patch tests A and B is a necessary conditions for convergence of the elements, while patch test C assesses the stability of the solution and provides a necessary and sufficient condition for convergence [On4,ZTZ].

A simple patch test of type B can be applied in thin plate elements in order to verify the good representation of rigid body displacements and the absence of spurious modes. The following displacement field is imposed at the boundary nodes

$$w = c - ax - by \tag{5.95}$$

where a, b and c are arbitrary numbers.

After solving the system of equations the internal DOFs must comply with Eq.(5.95) and the curvatures must be zero at each point in the patch.

A similar type B patch test can be devised for verifying the capability of the element for reproducing a constant curvature field. The test is based on imposing to the patch boundary nodes the quadratic displacement field

$$w(x, y) = \frac{1}{2}(ax^2 + by^2 + cxy) \tag{5.96}$$

where again a, b and c are arbitrary numbers. The numerical solution for the deflection at internal nodes must be in accordance with Eq.(5.96). Also, the curvature field must be constant at each point within the element giving $\hat{\varepsilon}_b = [a, b, c]^T$.

For compatible Kirchhoff plate elements (i.e. those satisfying the C^1 continuity requirement) the patch tests are not theoretically needed. The tests however are useful for verifying the absence of programming errors.

5.10 COMPARISON OF KIRCHHOFF PLATE ELEMENTS

Some of the thin plate elements presented in this chapter are compared in the analysis of a clamped square plate under a central point load (Figure 5.29) and a uniformly distributed load (Figure 5.30). Only a quarter of the plate is discretized due to symmetry. Figures 5.29a and 5.30a show the error in the central deflection obtained with different meshes of rectangular elements. The non-conforming MZC element converges from above, thus giving an upper bound for the correct solution. This is so for most non-conforming elements. All the conforming elements considered converge from stiffer solutions in a fast and monotonic manner.

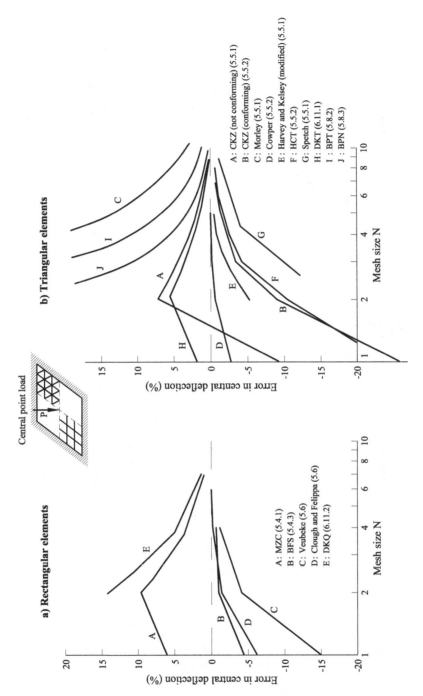

Fig. 5.29 Comparison of different rectangular and triangular Kirchhoff plate elements in the analysis of a clamped square plate under a central point load

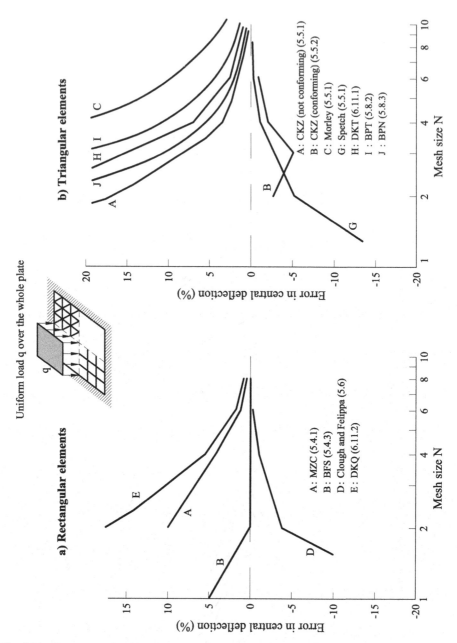

Fig. 5.30 Comparison of different rectangular and triangular Kirchhoff plate elements in the analysis of a clamped square plate under a uniformly distributed loading

The same type of analysis now using triangular elements is shown in Figures 5.29b and 5.30b. The solution for the Morley element is poor for coarse meshes as it involves fewer DOFs. Both the CKZ (non-conforming) and the Morley triangle elements converge from "the flexible side" due to their non-conformity. Conversely, the conforming triangles considered, i.e. Cooper *et al.* [CKLO], modified CKZ [ZT], HCT [CT] and Spetch [Sp] converge monotonically from the stiff side. The best results for coarse meshes were obtained with the first of these three elements. The modified 10 DOFs triangle of Harvey and Kelsey [HK] also converges to the exact solution. Very poor results (not shown in the figure) are found if the original version of this element is used. Note the monotonic convergence from the "flexible" side of the BST and BSN rotation-free triangles which *involve approximately one third of the DOFs* required for the rest of the elements tested. The good performance of the Discrete-Kirchhoff thin plate triangle (DKT) and quadrilateral (DKQ) described in Section 6.11 is also shown. The performance of all the element tested is very similar for simple supported plates.

More examples on the comparison of Kirchhoff plate elements can be found in [CMPW,Raz,SD,Ya,ZT2].

Abel and Desai [AD] assessed the "computational efficiency" of different Kirchhoff plate elements. This was defined as the computational effort required to obtain a prescribed accuracy and was measured in terms of the number of algebraic operations to solve the global equation system. The conclusion was that lower order elements are more efficient than higher order ones. Paradoxically, Rossow and Chen [RCh] arrived at the opposite conclusion simply by modifying the efficiency criterion and including the effect of the boundary conditions. The issue is therefore still open for discussion.

The advantage of higher order plate elements is their better ability to approximate complex bending moment fields over larger regions. Lower order elements are typically less accurate and require finer meshes. However, they usually include "physical" nodal variables (i.e. deflection and rotations) which avoids the cumbersome interpretation of the equivalent nodal force terms associated to higher order nodal variables, such as curvatures. Lower order triangular elements such as the Morley triangle and the BST and BSN rotation-free triangles are particularly attractive for practical applications and adaptive mesh refinement analysis.

5.11 CONCLUDING REMARKS

Kirchhoff thin plate theory requires C^1 continuity for the deflection field. This results in a higher complexity for deriving conforming plate elements that satisfy the C^1 continuity requirement.

Different conforming and non-conforming thin plate triangles and quadrilaterals have been studied and their behaviour has been compared in simple but illustrative examples of application.

A conclusion is that although conformity is not an essential requirement for the convergence of an element, it does guarantee a good performance for arbitrary geometries, and mainly for quadrilateral elements. Thus, any of the conforming quadrilaterals and triangles studied can be used with full confidence in practice. We note the simplicity of the MZC rectangle and the non-conforming Morley triangle which are two candidates for practical purposes.

The BPT and BPN rotation-free triangles have the deflection as the only nodal variable and their performance has been found to be very good for solving practical plate and shell problems [FO2,3,4,OCM,OF,OFN].

Kirchhoff plate elements based on mixed and hybrid formulations for which the displacements and the bending moments are simultaneously interpolated have not been studied in this chapter. Some successful elements of this type can be found in [ZT2]. A different family of plate elements is based on an explicit representation of the free strain states. This approach is called "natural formulation" in the thin plate and shell elements derived by Argyris et al. [AHM,AHMS,APAK,ATO,ATPA] and "free formulation" in the thin and thick elements proposed by Bergan et al. [BN,BW,FM]

In the next chapter we will derive an interesting family of Discrete-Kirchhoff thin plate elements starting from a "thick" (Reissner-Mindlin) plate formulation.

6

THICK/THIN PLATES.
REISSNER-MINDLIN THEORY

6.1 INTRODUCTION

Kirchhoff plate elements studied in the previous chapter are restricted to thin plate situations only (thickness/average side ≤ 0.10). Also the C^1 continuity requirement for Kirchhoff elements poses severe difficulties for deriving a conforming deflection field. These problems can be overcome by using the plate formulation due to Reissner [Re] and Mindlin [Mi] presented in this chapter.

The so called Reissner-Mindlin plate theory assumes that the normals to the plate do not remain orthogonal to the mid–plane after deformation, thus allowing for transverse shear deformation effects. This assumption is analogous to that made for the rotation of the transverse cross section in Timoshenko beam theory (Chapter 2). This allows us to use C^o continuous elements. Unfortunately, some difficulties arise when Reissner-Mindlin elements are used for thin plate situations due to the excessive influence of the transverse shear deformation terms. The "shear locking" defect is analogous to that found when Timoshenko beam elements are applied to slender beams. Elimination of shear locking is possible via reduced integration, linked interpolations or assumed transverse shear strain fields, similarly as described for beams in Chapter 2.

Reissner-Mindlin plate elements can be taken as the starting point for deriving C^o continuous thin plate elements by adequately constraining the transverse shear deformation to be zero at selected element points. Some of the so called Discrete-Kirchhoff (DK) plate elements are described.

A method for extending the basic rotation-free plate triangle described in Section 5.8.2 to accounting for shear deformation effects is also outlined in the last part of the chapter.

E. Oñate, *Structural Analysis with the Finite Element Method. Linear Statics:*
Volume 2: Beams, Plates and Shells, Lecture Notes on Numerical Methods
in Engineering and Sciences, DOI 10.1007/978-1-4020-8743-1_6,
© International Center for Numerical Methods in Engineering (CIMNE), 2013

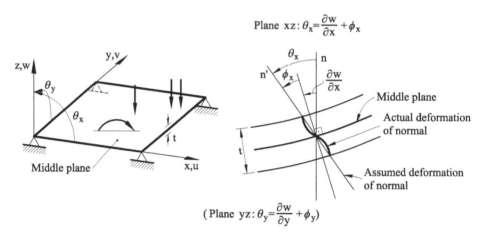

Fig. 6.1 Reissner-Mindlin plate theory. Sign convention for the displacements and the rotations of the normal. For loading types see Figure 4.1.

Reissner-Mindlin plate theory is very adequate for studying composite laminate plates for which shear deformation effects are important. The derivation of finite elements for this type of structures is presented in the next chapter.

Reissner-Mindlin plate theory can be readily extended to shell analysis. The study of this and the following chapters is therefore recommended as an introduction to the chapters dealing with shell structures.

The simplicity of Reissner-Mindlin plate elements and their versatility for analysis of thick and thin plates with homogeneous and composite material have contributed to their popularity for practical applications (see references in chapters on plate analysis in [CMPW,Cr,HO,ZT2]).

6.2 REISSNER-MINDLIN PLATE THEORY

Reissner-Mindlin plate bending theory shares the first three assumptions of Kirchhoff plate theory (Section 5.2.1). The fourth assumption on the rotation of the normal is different and reads as follows:

4) A straight line normal to the undeformed middle plane *remains straight* but *not necessarily orthogonal* to the middle plane after deformation (Figure 6.1).

The reader will recognize the analogy of this assumption with that for the rotation of the cross section in Timoshenko beams (Section 2.2). There are in fact many common features between both plate and beam theories.

6.2.1 Displacement field

The 3D displacement field is expressed in terms of the middle plane kinematic variables w, θ_x and θ_y as

$$
\begin{aligned}
u(x, y, z) &= -z\theta_x(x, y) \\
v(x, y, z) &= -z\theta_y(x, y) \\
w(x, y, z) &= w(x, y)
\end{aligned}
\tag{6.1}
$$

where θ_x and θ_y are the angles defining the rotation of the normal vector. Eq.(6.1) is identical to (5.2) for Kirchhoff theory. Here again the middle plane is taken as the reference plane $(z = 0)$. The displacement vector is

$$
\mathbf{u} = [w, \theta_x, \theta_y]^T
\tag{6.2}
$$

Assumption 4 of previous page allows us to express the rotation of the normal on the plane xz as (Figure 6.1)

$$
\theta_x = \frac{\partial w}{\partial x} + \phi_x
\tag{6.3}
$$

Similarly, for the plane yz

$$
\theta_y = \frac{\partial w}{\partial y} + \phi_y
\tag{6.4}
$$

The rotation of the normal in each of the two vertical planes xz and yz is obtained as the sum of two terms: 1) the adequate slope of the plate middle plane, and 2) an additional rotation ϕ resulting from the lack of orthogonality of the normal with the middle plane after deformation (Figure 6.1). Consequently, the rotations θ_x and θ_y can not be computed in terms of the deflection only and, therefore, are treated as *independent variables*. This is a substantial difference between Reissner–Mindlin and Kirchhoff plate theories.

The assumption of straight normals is an approximation of the true plate kinematics as in reality the plate normals are distorted during deformation. Clearly this effect is more important for thick plates. The angles θ_x and θ_y can be interpreted as the rotations of the straight line representing the "average" deformation of the normal (Figure 6.1).

6.2.2 Strain and stress fields

Substituting the displacement field (6.1) into the expression for the strains in a 3D solid (see Eq.(8.3) of [On4]) gives

$$
\varepsilon_x = \frac{\partial u}{\partial x} = -z\frac{\partial \theta_x}{\partial x}
$$

$$\varepsilon_y = \frac{\partial v}{\partial y} = -z\frac{\partial \theta_y}{\partial y}$$

$$\varepsilon_z = \frac{\partial w}{\partial z} = 0$$

$$\gamma_{xy} = \frac{\partial u}{\partial y} + \frac{\partial v}{\partial x} = -z\left(\frac{\partial \theta_x}{\partial y} + \frac{\partial \theta_y}{\partial x}\right) \tag{6.5}$$

$$\gamma_{xz} = \frac{\partial u}{\partial z} + \frac{\partial w}{\partial x} = -\theta_x + \frac{\partial w}{\partial x} = -\phi_x$$

$$\gamma_{yz} = \frac{\partial v}{\partial z} + \frac{\partial w}{\partial y} = -\theta_y + \frac{\partial w}{\partial y} = -\phi_y$$

The non-orthogonality of the normal vector induces *non zero transverse shear strains* γ_{xz} and γ_{yz}, which coincide (in absolute value) with the rotations ϕ_x and ϕ_y. These strains are independent of the coordinate z.

Making zero the transverse shear strains we recover the normal orthogonality condition of Kirchhoff plate theory as $\theta_x = \frac{\partial w}{\partial x}$ and $\theta_y = \frac{\partial w}{\partial y}$.

The same comments on the irrelevance of the normal strain ε_z made in Section 5.2.3 apply in this case.

The strain vector containing the significant strains is therefore

$$\varepsilon = \begin{Bmatrix} \varepsilon_x \\ \varepsilon_y \\ \gamma_{xy} \\ \cdots \\ \gamma_{xz} \\ \gamma_{yz} \end{Bmatrix} = \begin{Bmatrix} -z\dfrac{\partial \theta_x}{\partial x} \\[2mm] -z\dfrac{\partial \theta_y}{\partial y} \\[2mm] -z\left(\dfrac{\partial \theta_x}{\partial y} + \dfrac{\partial \theta_y}{\partial x}\right) \\[1mm] \cdots\cdots\cdots \\[1mm] \dfrac{\partial w}{\partial x} - \theta_x \\[2mm] \dfrac{\partial w}{\partial y} - \theta_y \end{Bmatrix} = \begin{Bmatrix} \varepsilon_b \\ \cdots \\ \varepsilon_s \end{Bmatrix} \tag{6.6}$$

where vectors ε_b and ε_s contain the bending and transverse shear strains, respectively. The strain vector of Eq.(6.6) can be expressed as

$$\varepsilon = \mathbf{S}\hat{\varepsilon} \tag{6.7}$$

where

$$\hat{\varepsilon} = \begin{Bmatrix} \hat{\varepsilon}_b \\ \hat{\varepsilon}_s \end{Bmatrix} \tag{6.8}$$

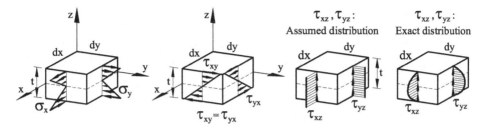

Fig. 6.2 Sign convention for the in-plane stresses σ_x, σ_y, τ_{xy} and the transverse shear stresses τ_{xz} and τ_{yz}

is the *generalized strain vector* and

$$\hat{\varepsilon}_b = \left\{ \begin{array}{c} \dfrac{\partial \theta_x}{\partial x} \\[2mm] \dfrac{\partial \theta_y}{\partial y} \\[2mm] \left(\dfrac{\partial \theta_x}{\partial y} + \dfrac{\partial \theta_y}{\partial x} \right) \end{array} \right\} \quad \text{and} \quad \hat{\varepsilon}_s = \left\{ \begin{array}{c} \dfrac{\partial w}{\partial x} - \theta_x \\[2mm] \dfrac{\partial w}{\partial y} - \theta_y \end{array} \right\} \tag{6.9}$$

are the generalized bending strain vector and the transverse shear strain vector, respectively. The strain transformation matrix $\mathbf{S}$ in Eq.(6.7) is

$$\mathbf{S} = \begin{bmatrix} -z & 0 & 0 & 0 & 0 \\ 0 & -z & 0 & 0 & 0 \\ 0 & 0 & -z & 0 & 0 \\ 0 & 0 & 0 & 1 & 0 \\ 0 & 0 & 0 & 0 & 1 \end{bmatrix} \tag{6.10}$$

The stress vector is

$$\sigma = \left\{ \begin{array}{c} \sigma_x \\ \sigma_y \\ \tau_{xy} \\ \cdots\cdots \\ \tau_{xz} \\ \tau_{yz} \end{array} \right\} = \left\{ \begin{array}{c} \sigma_b \\ \cdots\cdots \\ \sigma_s \end{array} \right\} \tag{6.11}$$

where σ_b and σ_s are the stresses due to pure bending and transverse shear effects, respectively. As usual in plate theory σ_z has been excluded from Eq.(6.11) due to the plane stress assumption ($\sigma_z = 0$). The sign convention for the stresses is shown in Figure 6.2.

6.2.3 Stress–strain relationship

We will consider in the following the material to be *homogeneous* and isothermal conditions only. This is consistent with the assumption of plate bending theory, as no axial strains or axial forces are introduced during the plate deformation. Axial effects appear for composite laminated plates or when thermal effects are taken into account. These situations are treated in the next chapter.

Starting from the constitutive equation of 3D elasticity (Eq.(8.5) of [On4]) and using the plane stress assumption ($\sigma_z = 0$), we can find a relationship between the non–zero stresses and strains. For an *orthotropic* material with orthotropy axes 1,2,3 with $3 = z$ and satisfying the condition of plane anisotropy (i.e. the plane 1,2 is a plane of material symmetry [BD4]) we can write

$$\sigma_I = \begin{Bmatrix} \sigma_1 \\ \sigma_2 \end{Bmatrix} = \left[\begin{array}{c:c} \mathbf{D}_1 & \mathbf{0} \\ \hdashline \mathbf{0} & \mathbf{D}_2 \end{array} \right] \begin{Bmatrix} \varepsilon_1 \\ \varepsilon_2 \end{Bmatrix} = \mathbf{D}_I \varepsilon_I \tag{6.12}$$

where

$$\sigma_1 = [\sigma_1, \ \sigma_2, \ \tau_{12}]^T, \quad \sigma_2 = [\tau_{13}, \ \tau_{23}]^T$$
$$\varepsilon_1 = [\varepsilon_1, \ \varepsilon_2, \ \gamma_{12}]^T, \quad \varepsilon_2 = [\gamma_{13}, \ \gamma_{23}]^T \tag{6.13}$$

are the bending and transverse shear stresses and strains in the principal orthotropy axes and

$$\mathbf{D}_1 = \frac{1}{1 - \nu_{12}\nu_{21}} \begin{bmatrix} E_1 & \nu_{21}E_1 & 0 \\ \nu_{12}E_1 & E_2 & 0 \\ 0 & 0 & (1 - \nu_{12}\nu_{21})G_{12} \end{bmatrix}, \quad \mathbf{D}_2 = \begin{bmatrix} G_{13} & 0 \\ 0 & G_{23} \end{bmatrix} \tag{6.14}$$

with $\nu_{12}E_2 = \nu_{21}E.5_1$. If isotropy exists in the direction 1 (i.e. on the plane $2 - z$, Figure 6.3) as for fiber composites with fibers in the direction 1 covered by a matrix, then $G_{13} = G_{12}$.

The following relationships hold [BC,BD4,CMPW]

$$\varepsilon_1 = \mathbf{T}_1 \varepsilon_b, \quad \varepsilon_2 = \mathbf{T}_2 \varepsilon_s$$
$$\sigma_b = \mathbf{T}_1^T \sigma_1, \quad \sigma_s = \mathbf{T}_2^T \sigma_2 \tag{6.15a}$$

with

$$\mathbf{T}_1 = \begin{bmatrix} C^2 & S^2 & CS \\ S^2 & C^2 & -CS \\ -2CS & 2CS & C^2 - S^2 \end{bmatrix}, \quad \mathbf{T}_2 = \begin{bmatrix} C & S \\ -S & C \end{bmatrix} \tag{6.15b}$$

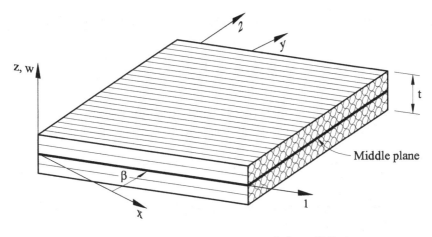

1, 2, z   Orthotropy axes

x, y, z   Global axes

Fig. 6.3 Orthotropic material in the plane 1-2 and isotropic in the plane $2 - z$

where $C = \cos\beta$, $S = \sin\beta$ and β is the angle between the axes 1 and x (Figure 6.3) and ε_b, ε_s are defined in Eq.(6.6).

The inverse relationship can be obtained simply by replacing β by $-\beta$ in Eqs.(6.15) [BC], i.e.

$$\varepsilon_b = \bar{\mathbf{T}}_1 \varepsilon_1 \quad , \quad \varepsilon_s = \bar{\mathbf{T}}_2 \varepsilon_2$$
$$\boldsymbol{\sigma}_1 = \bar{\mathbf{T}}_1^T \boldsymbol{\sigma}_b \quad , \quad \boldsymbol{\sigma}_2 = \bar{\mathbf{T}}_2^T \boldsymbol{\sigma}_s \tag{6.16a}$$

with

$$\bar{\mathbf{T}}_1 = \begin{bmatrix} C^2 & S^2 & -CS \\ S^2 & C^2 & CS \\ 2CS & -2CS & C^2 - S^2 \end{bmatrix} \quad , \quad \bar{\mathbf{T}}_2 = \begin{bmatrix} C & -S \\ S & C \end{bmatrix} \tag{6.16b}$$

Combining Eqs.(6.12) and (6.15) gives the constitutive equations for the bending and transverse shear stresses in global axes as

$$\boldsymbol{\sigma}_b = \mathbf{D}_b \varepsilon_b \quad , \quad \boldsymbol{\sigma}_s = \bar{\mathbf{D}}_s \varepsilon_s \tag{6.17a}$$

or

$$\boldsymbol{\sigma} = \begin{Bmatrix} \boldsymbol{\sigma}_b \\ \boldsymbol{\sigma}_s \end{Bmatrix} = \mathbf{D} \begin{Bmatrix} \varepsilon_b \\ \varepsilon_s \end{Bmatrix} = \mathbf{D}\varepsilon \quad \text{with} \quad \mathbf{D} = \begin{bmatrix} \mathbf{D}_b & \mathbf{0} \\ \mathbf{0} & \bar{\mathbf{D}}_s \end{bmatrix} \tag{6.17b}$$

and

$$\mathbf{D}_b = \mathbf{T}_1^T \mathbf{D}_1 \mathbf{T}_1, \quad \bar{\mathbf{D}}_s = \mathbf{T}_2^T \mathbf{D}_2 \mathbf{T}_2 \tag{6.17c}$$

The case of isotropic material is simply recovered for $\beta = 0$, giving

$$\mathbf{D}_b = \frac{E}{1 - \nu^2} \begin{bmatrix} 1 & \nu & 0 \\ \nu & 1 & 0 \\ 0 & 0 & \frac{1-\nu}{2} \end{bmatrix}, \quad \bar{\mathbf{D}}_s = G \begin{bmatrix} 1 & 0 \\ 0 & 1 \end{bmatrix} \tag{6.18}$$

Substituting Eq.(6.7) into (6.17) gives the relationship between the stresses at a point across the thickness and the generalized strains as

$$\boldsymbol{\sigma} = \begin{Bmatrix} \boldsymbol{\sigma}_b \\ \boldsymbol{\sigma}_s \end{Bmatrix} = \mathbf{DS}\hat{\boldsymbol{\varepsilon}} = \mathbf{D} \begin{Bmatrix} -z\hat{\boldsymbol{\varepsilon}}_b \\ \hat{\boldsymbol{\varepsilon}}_s \end{Bmatrix} \tag{6.19}$$

Eq.(6.19) shows that the distribution of the "in-plane" bending stresses σ_x, σ_y and τ_{xy} varies linearly with z, while the transverse shear stresses τ_{xz} and τ_{yz} are constant across the thickness (Figure 6.2).

The "exact" thickness distribution of the transverse shear stresses obtained from 3D elasticity theory is not uniform and they vanish at the upper and lower plate surfaces. For isotropic material the distribution is parabolic (Figure 6.2) [TW]. This problem, which also appeared for Timoshenko beam theory (Section 2.2.3), can be overcome by scaling the internal work associated to the transverse shear stresses so that it coincides with the exact internal work obtained from 3D elasticity theory. Thus, the total strain work is computed exactly, although the transverse shear stresses have not a correct thickness distribution locally.

In practice this implies modifying the shear constitutive relationship of Eq.(6.17a) as

$$\boldsymbol{\sigma}_s = \mathbf{D}_s \boldsymbol{\varepsilon}_s \quad \text{with} \quad \mathbf{D}_s = \begin{bmatrix} k_{11}\bar{D}_{s_{11}} & k_{12}\bar{D}_{s_{12}} \\ k_{12}\bar{D}_{s_{12}} & k_{22}\bar{D}_{s_{22}} \end{bmatrix} \tag{6.20}$$

where $\mathbf{D}_{s_{ij}}$ are the components of $\bar{\mathbf{D}}_s$ of Eq.(6.17c) and k_{ij} are shear correction parameters. Their computation follows a procedure analogous to that explained for Timoshenko beams. For an isotropic plate $k_{12} = 0$ and $k_{11} = k_{22} = 5/6$, as for rectangular beams (Section 2.2.3.1). The computation of the shear correction parameters for composite laminated beams is presented in Section 7.3 and in Annex D.

6.2.4 Resultant stresses and generalized constitutive matrix

The vector of resultant stresses at a point of the plate middle plane is

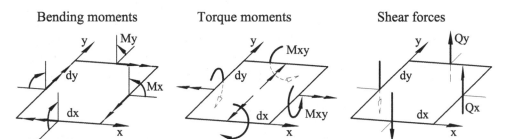

Fig. 6.4 Sign convention for the resultant stresses in a plate

defined as

$$
\hat{\sigma} = \left\{ \begin{array}{c} \hat{\sigma}_b \\ \cdots \cdots \\ \hat{\sigma}_s \end{array} \right\} = \left\{ \begin{array}{c} M_x \\ M_y \\ M_{xy} \\ \cdots \cdots \\ Q_x \\ Q_y \end{array} \right\} = \int_{-\frac{t}{2}}^{+\frac{t}{2}} \left\{ \begin{array}{c} -z\sigma_x \\ -z\sigma_y \\ -z\tau_{xy} \\ \cdots \cdots \\ \tau_{xz} \\ \tau_{yz} \end{array} \right\} dz =
$$

$$
= \int_{-\frac{t}{2}}^{+\frac{t}{2}} \left\{ \begin{array}{c} -z\boldsymbol{\sigma}_b \\ \cdots \cdots \\ \boldsymbol{\sigma}_s \end{array} \right\} dz = \int_{-\frac{t}{2}}^{+\frac{t}{2}} \mathbf{S}^T \boldsymbol{\sigma}\, dz \tag{6.21}
$$

where $\mathbf{S}$ is the transformation matrix of Eq.(6.10) and vectors $\hat{\sigma}_b$ and $\hat{\sigma}_s$ contain the *moments* and the *shear forces*, respectively. For sign convention see Figure 6.4.

Eq.(6.21) can be modified using Eqs.(6.17) and (6.7) as

$$
\hat{\sigma} = \left\{ \begin{array}{c} \hat{\sigma}_b \\ \cdots \\ \hat{\sigma}_s \end{array} \right\} = \int_{-\frac{t}{2}}^{+\frac{t}{2}} \mathbf{S}^T \mathbf{D}\boldsymbol{\varepsilon}\, dz = \int_{-\frac{t}{2}}^{+\frac{t}{2}} \mathbf{S}^T \mathbf{D}\mathbf{S}\hat{\boldsymbol{\varepsilon}}\, dz = \int_{-\frac{t}{2}}^{+\frac{t}{2}} \left\{ \begin{array}{c} z^2 \mathbf{D}_b \hat{\boldsymbol{\varepsilon}}_b \\ \cdots \cdots \\ \mathbf{D}_s \hat{\boldsymbol{\varepsilon}}_s \end{array} \right\} dz = \hat{\mathbf{D}}\hat{\boldsymbol{\varepsilon}}
$$
$$\tag{6.22}$$

where $\hat{\varepsilon}$ is the generalized strain vector of Eq.(6.9) and

$$
\hat{\mathbf{D}} = \int_{-\frac{t}{2}}^{+\frac{t}{2}} \mathbf{S}^T \mathbf{D}\mathbf{S}\, dz = \int_{-\frac{t}{2}}^{+\frac{t}{2}} \begin{bmatrix} z^2 \mathbf{D}_b & \mathbf{0} \\ \mathbf{0} & \mathbf{D}_s \end{bmatrix} dz = \begin{bmatrix} \hat{\mathbf{D}}_b & \mathbf{0} \\ \mathbf{0} & \hat{\mathbf{D}}_s \end{bmatrix} \tag{6.23}
$$

is the *generalized constitutive matrix* with $\mathbf{D}_b$ and $\mathbf{D}_s$ given in Eqs.(6.17c) and (6.20), respectively.

For isotropic material

$$
\hat{\mathbf{D}}_b = \frac{t^3}{12}\mathbf{D}_b \quad \text{and} \quad \hat{\mathbf{D}}_s = t\mathbf{D}_s = \frac{5}{6}tG\mathbf{I}_2 \tag{6.24}
$$

where $\mathbf{D}_b$ is given by Eq.(6.18) and $\mathbf{I}_2$ is the 2×2 unit matrix.

Eq.(6.22) relates the resultant stresses and the generalized strains at any point of the plate surface. The stresses across the thickness can be recovered from the generalized strains combining Eqs.(6.7) and (6.17) as

$$\boldsymbol{\sigma} = \mathbf{DS}\hat{\boldsymbol{\varepsilon}} \tag{6.25}$$

Eq.(6.25) gives an inaccurate constant distribution of the transverse shear stresses across the thickness. Enhanced values can be obtained by assuming a parabolic distribution of these stresses and computing their maximum value at $z = 0$ from the shear forces using Eq.(5.27).

6.2.5 Virtual work principle

Let us consider a plate loaded by a vertical distributed loads $\mathbf{t}$ and point loads $\mathbf{p}_i$. The virtual work expression is written as

$$\iiint_V \delta\boldsymbol{\varepsilon}^T \boldsymbol{\sigma} dV = \iint_A \delta\mathbf{u}^T \mathbf{t} \, dA + \sum_i \delta\mathbf{u}_i^T \mathbf{p}_i \tag{6.26a}$$

where

$$\delta\mathbf{u} = [\delta w, \delta\theta_x, \delta\theta_y]^T \quad , \quad \mathbf{t} = [f_z, m_x, m_y]^T \quad , \quad \mathbf{p}_i = [P_{z_i}, M_{x_i}, M_{y_i}]^T \tag{6.26b}$$

and $\delta\mathbf{u}$ are the virtual displacement vector, $\delta\boldsymbol{\varepsilon}$ is the virtual strain vector, f_z, m_x and m_y are the vertical distributed load and the distributed moments acting on the xz and yz planes, respectively. P_{z_i}, M_{x_i} and M_{y_i} are the vertical point load and the concentrated bending moments acting at a point i, respectively (Figure 5.1).

The virtual internal work over the plate domain can be expressed in terms of the middle plane variables (i.e. resultant stresses and generalized strains) using Eqs.(6.7) and (6.21) as follows

$$\iiint_V \delta\boldsymbol{\varepsilon}^T \boldsymbol{\sigma} dV = \iint_A \delta\hat{\boldsymbol{\varepsilon}}^T \left[\int_{-t/2}^{t/2} \mathbf{S}^T \boldsymbol{\sigma} dz \right] dA = \iint_A \delta\hat{\boldsymbol{\varepsilon}}^T \hat{\boldsymbol{\sigma}} dA \tag{6.27}$$

The PVW is finally written in terms of integrals over the plate surface as

$$\iint_A \delta\hat{\boldsymbol{\varepsilon}}^T \hat{\boldsymbol{\sigma}}^T dA = \iint_A \delta\mathbf{u}^T \mathbf{t} \, dA + \sum_i \delta\mathbf{u}_i^T \mathbf{p}_i \tag{6.28a}$$

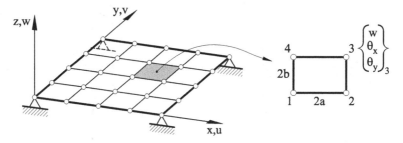

Fig. 6.5 Discretization of a plate using 4-noded Reissner-Mindlin rectangles

Above integrals contain displacement derivatives up to first order only. This allows us to use C^o continuous elements which are simpler than Kirchhoff plate elements.

Sometimes it is interesting to split the virtual internal work in terms of the bending and transverse shear contributions using Eqs.(6.8) and (6.21). The PVW is then written as

$$\iint_A [\delta\hat{\boldsymbol{\varepsilon}}_b^T \hat{\boldsymbol{\sigma}}_b + \delta\hat{\boldsymbol{\varepsilon}}_s^T \hat{\boldsymbol{\sigma}}_s]dA = \iint_A \mathbf{u}^T \mathbf{t}\, dA + \sum_i \delta\mathbf{u}_i^T \mathbf{p}_i \qquad (6.28b)$$

6.3 FINITE ELEMENT FORMULATION

6.3.1 Discretization of the displacement field

The plate middle plane is discretized into a mesh of n node elements; for instance Figure 6.5 shows a discretization in 4-noded plate rectangles. The deflection and the two rotations *are independent variables* and the displacement field is interpolated in the standard C^o form; i.e.

$$\mathbf{u} = \begin{Bmatrix} w \\ \theta_x \\ \theta_y \end{Bmatrix} = \sum_{i=1}^{n} \begin{Bmatrix} N_i w_i \\ N_i \theta_{x_i} \\ N_i \theta_{y_i} \end{Bmatrix} = \begin{bmatrix} N_1 & 0 & 0 & \vdots & & \vdots & N_n & 0 & 0 \\ 0 & N_1 & 0 & \vdots & \cdots & \vdots & 0 & N_n & 0 \\ 0 & 0 & N_1 & \vdots & & \vdots & 0 & 0 & N_n \end{bmatrix} \begin{Bmatrix} w_1 \\ \theta_{x_1} \\ \theta_{y_1} \\ \cdots \\ \vdots \\ \cdots \\ w_n \\ \theta_{x_n} \\ \theta_{y_n} \end{Bmatrix} =$$

$$= [\mathbf{N}_1, \mathbf{N}_2, \dots, \mathbf{N}_n] \begin{Bmatrix} \mathbf{a}_1^{(e)} \\ \vdots \\ \mathbf{a}_n^{(e)} \end{Bmatrix} = \mathbf{N}\mathbf{a}^{(e)} \qquad (6.29)$$

where

$$\mathbf{N} = [\mathbf{N}_1, \mathbf{N}_2, \ldots, \mathbf{N}_n] \quad , \quad \mathbf{a}^{(e)} = \left\{ \begin{array}{c} \mathbf{a}_1^{(e)} \\ \mathbf{a}_2^{(e)} \\ \vdots \\ \mathbf{a}_n^{(e)} \end{array} \right\} \tag{6.30a}$$

and

$$\mathbf{N}_i = \begin{bmatrix} N_i & 0 & 0 \\ 0 & N_i & 0 \\ 0 & 0 & N_i \end{bmatrix} \quad , \quad \mathbf{a}_i^{(e)} = \left\{ \begin{array}{c} w_i \\ \theta_{x_i} \\ \theta_{y_i} \end{array} \right\} \tag{6.30b}$$

are the shape function matrix and the displacement vector for the element and a node i, respectively.

6.3.2 Discretization of the generalized strains and resultant stress fields

The generalized strains are expressed in terms of the nodal displacements (using Eqs.(6.8), (6.9) and (6.29)) as

$$\hat{\boldsymbol{\varepsilon}} = \left\{ \begin{array}{c} \hat{\boldsymbol{\varepsilon}}_b \\ \cdots \\ \hat{\boldsymbol{\varepsilon}}_s \end{array} \right\} = \left\{ \begin{array}{c} \dfrac{\partial \theta_x}{\partial x} \\[4pt] \dfrac{\partial \theta_y}{\partial y} \\[4pt] \left(\dfrac{\partial \theta_x}{\partial y} + \dfrac{\partial \theta_y}{\partial x} \right) \\ \cdots\cdots\cdots \\ \dfrac{\partial w}{\partial x} - \theta_x \\[4pt] \dfrac{\partial w}{\partial y} - \theta_y \end{array} \right\} = \sum_{i=1}^{n} \left\{ \begin{array}{c} \dfrac{\partial N_i}{\partial x} \theta_{x_i} \\[4pt] \dfrac{\partial N_i}{\partial y} \theta_{y_i} \\[4pt] \left(\dfrac{\partial N_i}{\partial y} \theta_{x_i} + \dfrac{\partial N_i}{\partial x} \theta_{y_i} \right) \\ \cdots\cdots\cdots\cdots \\ \dfrac{\partial N_i}{\partial x} w_i - N_i \theta_{x_i} \\[4pt] \dfrac{\partial N_i}{\partial y} w_i - N_i \theta_{y_i} \end{array} \right\} =$$

$$= \sum_{i=1}^{n} \left\{ \begin{array}{c} \mathbf{B}_{b_i} \\ \mathbf{B}_{s_i} \end{array} \right\} \mathbf{a}_i^{(e)} = [\mathbf{B}_1, \ldots, \mathbf{B}_n] \left\{ \begin{array}{c} \mathbf{a}_1^{(e)} \\ \vdots \\ \mathbf{a}_n^{(e)} \end{array} \right\} = \mathbf{B}\mathbf{a}^{(e)} \tag{6.31}$$

where $\mathbf{B}$ and $\mathbf{B}_i$ are the generalized strain matrices for the element and a node i, respectively. From Eq.(6.31) we deduce

$$\mathbf{B}_i = \left\{ \begin{array}{c} \mathbf{B}_{b_i} \\ \cdots \\ \mathbf{B}_{s_i} \end{array} \right\} \quad \text{with} \quad \mathbf{B}_{b_i} = \begin{bmatrix} 0 & \dfrac{\partial N_i}{\partial x} & 0 \\[6pt] 0 & 0 & \dfrac{\partial N_i}{\partial y} \\[6pt] 0 & \dfrac{\partial N_i}{\partial y} & \dfrac{\partial N_i}{\partial x} \end{bmatrix} \quad , \quad \mathbf{B}_{s_i} = \begin{bmatrix} \dfrac{\partial N_i}{\partial x} & -N_i & 0 \\[6pt] \dfrac{\partial N_i}{\partial y} & 0 & -N_i \end{bmatrix}$$

$$\tag{6.32}$$

$\mathbf{B}_{b_i}$ and $\mathbf{B}_{s_i}$ are the bending and transverse shear strain matrices associated to the ith node, respectively.

The resultant stresses are expressed in terms of the nodal displacements using Eqs.(6.22) and (6.31) as

$$\hat{\boldsymbol{\sigma}} = \hat{\mathbf{D}}\mathbf{B}\mathbf{a}^{(e)} \quad \text{and} \quad \hat{\boldsymbol{\sigma}}_b = \hat{\mathbf{D}}_b\mathbf{B}_b\mathbf{a}^{(e)} \quad , \quad \hat{\boldsymbol{\sigma}}_s = \hat{\mathbf{D}}_s\mathbf{B}_s\mathbf{a}^{(e)} \tag{6.33a}$$

with

$$\mathbf{B}_b = [\mathbf{B}_{b_1}, \mathbf{B}_{b_2}, \cdots, \mathbf{B}_{b_n}] \quad , \quad \mathbf{B}_s = [\mathbf{B}_{s_1}, \mathbf{B}_{s_2}, \cdots, \mathbf{B}_{s_n}] \tag{6.33b}$$

6.3.3 Derivation of the equilibrium equations for the element

The PVW for a single element under distributed loads reads (Eq.(6.28a))

$$\iint_{A^{(e)}} \delta\hat{\boldsymbol{\varepsilon}}^T\hat{\boldsymbol{\sigma}}\,dA = \iint_{A^{(e)}} \delta\mathbf{u}^T\mathbf{t}\,dA + [\delta\mathbf{a}^{(e)}]^T\mathbf{q}^{(e)} \tag{6.34}$$

where, as usual, $\delta\mathbf{a}^{(e)}$ is the virtual nodal displacement vector and the last term of the r.h.s. is the virtual work of the equilibrating nodal forces $\mathbf{q}^{(e)}$, with

$$\delta\mathbf{a}^{(e)} = \begin{Bmatrix} \delta\mathbf{a}_1^{(e)} \\ \vdots \\ \delta\mathbf{a}_n^{(e)} \end{Bmatrix}, \quad \delta\mathbf{a}_i^{(e)} = \begin{Bmatrix} \delta w_i \\ \delta\theta_{x_i} \\ \delta\theta_{y_i} \end{Bmatrix}, \quad \mathbf{q}^{(e)} = \begin{Bmatrix} \mathbf{q}_1^{(e)} \\ \vdots \\ \mathbf{q}_n^{(e)} \end{Bmatrix}, \quad \mathbf{q}_i^{(e)} = \begin{Bmatrix} F_{z_i} \\ M_{x_i} \\ M_{y_i} \end{Bmatrix} \tag{6.35}$$

Substituting Eq.(6.33a) into (6.34) and using Eqs.(6.29) and (6.31) yields the standard discrete equilibrium expression for the element

$$\mathbf{K}^{(e)}\mathbf{a}^{(e)} - \mathbf{f}^{(e)} = \mathbf{q}^{(e)} \tag{6.36}$$

where

$$\mathbf{K}_{ij}^{(e)} = \iint_{A^{(e)}} \mathbf{B}_i^T\hat{\mathbf{D}}\mathbf{B}_j\,dA \tag{6.37}$$

$$\mathbf{f}_i^{(e)} = \iint_{A^{(e)}} \mathbf{N}_i[f_z, m_x, m_y]^T\,dA \tag{6.38}$$

are the element stiffness matrix connecting nodes i and j and the equivalent nodal force vector due to a distributed vertical load f_z and distributed bending moments m_x and m_y.

The global system of equations $\mathbf{Ka} = \mathbf{f}$ is obtained by assembling the element contributions to the global stiffness matrix $\mathbf{K}$ and the global equivalent nodal force vector $\mathbf{f}$ in the usual manner.

The element stiffness matrix can be split into the bending and transverse shear contributions using Eqs.(6.23) and (6.32) as follows.

$$
\mathbf{K}_{ij}^{(e)} = \iint_{A^{(e)}} [\mathbf{B}_{b_i}^T, \mathbf{B}_{s_i}^T]^T \hat{\mathbf{D}} \left\{ \begin{matrix} \mathbf{B}_{b_j} \\ \mathbf{B}_{s_j} \end{matrix} \right\} dA =
$$
$$
= \int_{A^{(e)}} \left(\mathbf{B}_{b_i}^T \hat{\mathbf{D}}_b \mathbf{B}_{b_j} + \mathbf{B}_{s_i}^T \hat{\mathbf{D}}_s \mathbf{B}_{s_j} \right) dA = \mathbf{K}_{b_{ij}}^{(e)} + \mathbf{K}_{s_{ij}}^{(e)} \qquad (6.39a)
$$

where

$$
\mathbf{K}_{b_{ij}}^{(e)} = \iint_{A^{(e)}} \mathbf{B}_{b_i}^T \hat{\mathbf{D}}_b \mathbf{B}_{b_j} dA \quad ; \quad \mathbf{K}_{s_{ij}}^{(e)} = \iint_{A^{(e)}} \mathbf{B}_{s_i}^T \hat{\mathbf{D}}_s \mathbf{B}_{s_j} dA \qquad (6.39b)
$$

Above splitting provides a more economical way for computing the element stiffness matrix. It also helps explaining the behaviour of the element for thin plate situations.

Differently to Kirchhoff plate elements, vertical loads and bending moments contribute to the terms of $\mathbf{f}_i^{(e)}$ in an uncoupled manner, i.e. vertical loads do not introduce bending moment components in $\mathbf{f}_i^{(e)}$. This is due to the independent interpolation for the deflection and the rotations.

Self-weight is treated as a vertical distributed loading. For the gravity g acting in opposite direction to the global z axis, $f_z = -\rho g t$ and

$$
\mathbf{f}_i^{(e)} = - \iint_{A^{(e)}} \mathbf{N}_i \rho g t [1, 0, 0]^T dA \qquad (6.40a)
$$

where ρ and t are the material density and the plate thickness, respectively. Finally, the equivalent nodal force vector due to an external vertical point load P_{z_j} and concentrated bending moments M_{x_j}, M_{y_j} acting at a node with global number j is

$$
\mathbf{p}_j = [P_{z_j}, M_{x_j}, M_{y_j}]^T \qquad (6.40b)
$$

As usual concentrated forces acting at nodes are assembled directly into the global vector $\mathbf{f}$.

The reactions at the prescribed nodes are computed a posteriori once the nodal displacements have been found as explained in Section 1.3.4.

6.3.4 Numerical integration

The integrals appearing in the element stiffness matrix and the equivalent nodal force vector are typically computed by a Gauss quadrature.

From Eq.(6.39a) we deduce

$$\mathbf{K}_{ij}^{(e)} = \sum_{p=1}^{n_{b_p}} \sum_{q=1}^{n_{b_q}} \left[\mathbf{B}_{b_i}^T \hat{\mathbf{D}}_b \mathbf{B}_{b_j} |\mathbf{J}^e|\right]_{p,q} W_p W_q + \sum_{p=1}^{n_{s_p}} \sum_{q=1}^{n_{s_q}} \left[\mathbf{B}_{s_i}^T, \hat{\mathbf{D}}_s \mathbf{B}_{b_j} |\mathbf{J}^e|\right]_{p,q} W_p W_q$$

(6.41)

where $|\mathbf{J}^e|$ is the Jacobian determinant [On4], (n_{b_p}, n_{b_q}) and (n_{s_p}, n_{s_q}) are the integration points for the bending and transverse shear stiffness matrices, respectively and W_p, W_q are the corresponding weights. Eq.(6.11) allows us to use a selective integration rule for the bending and transverse shear stiffness matrices.

The Gauss quadrature for the equivalent nodal force vector of Eq.(6.38) is

$$\mathbf{f}_i^{(e)} = \sum_{p=1}^{n_p} \sum_{q=1}^{n_q} \left[N_i[f_z, m_x, m_y]^T |\mathbf{J}^e|\right]_{p,q} W_p W_q$$

(6.42)

6.3.5 The boundary conditions

The standard boundary conditions are:

Point support: $w_i = 0$

Symmetry axis (of geometry and loading): $\theta_n = 0$, where n is the orthogonal direction to the symmetry axis.

Clamped (CL) side: $w = \theta_x = \theta_y = 0$

Simply supported (SS) side:

- Hard support: $w = \theta_s = 0$
- Soft support: $w = 0$

where s is the direction of the side. Recall that in Kirchhoff theory constraining of w along a direction s automatically specifies $\theta_s = 0$ (Section 5.2.7). This is not the case for Reissner–Mindlin theory where both w and θ_s have to be independently prescribed.

Figure 6.6 shows graphically the different boundary conditions. A free edge requires no specific constraint, as usual.

In plates with corners, the soft support condition introduces a boundary layer of order t adjacent to the prescribed side for the resultant stresses Q_s, Q_n and M_{ns} where n is the orthogonal direction to the side. The hard

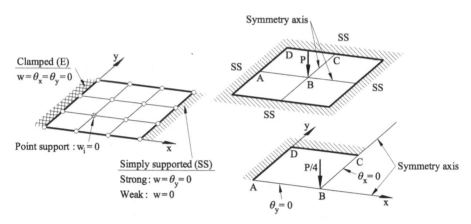

Fig. 6.6 Some boundary conditions in Reissner–Mindlin plates

a) SS (hard support)   b) SS (soft support)

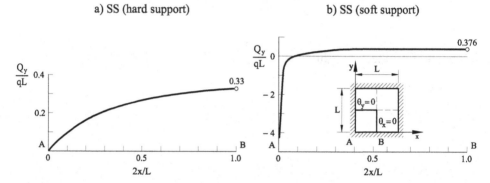

Fig. 6.7 Simply supported (SS) square plate under uniform distributed loading q. Distribution of shear force Q_y along the support line AB for hard support (a) and soft support (b) conditions. Solution obtained with an adapted mesh of 10×10 QLQL elements in a quarter of the plate and $t/L = 0.02$

support condition is recommended in those cases, as capturing the boundary layer requires a very fine discretization near the side. Figure 6.7 shows a situation of this kind for a simply supported square plate under uniformly distributed loading $(f_z = q)$ and $\frac{t}{L} = 0.02$. The distribution of the shear force Q_y along the support line AB is plotted for the hard support $(w = \theta_x = 0)$ and soft support $(w = 0)$ conditions. Note the strong variation of Q_y in the vicinity of the corner A for the soft support assumption whereas the value at the center is quite insensible to the boundary conditions chosen. The analysis was performed with a mesh of 10×10 QLQL elements in a quarter of plate due to symmetry (Section 6.7.4). The mesh

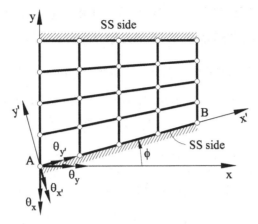

Fig. 6.8 Boundary conditions in a plate with inclined simply supported (SS) side

density was higher near the plate edges in order to capture the boundary layer [BD5].

The condition $\theta_s = 0$ along an inclined axis requires transforming the cartesian rotations to the boundary axes.

For instance, the boundary conditions on the simply supported inclined side AB in the plate of Figure 6.8 are $w' = \theta_{x'} = 0$ (note the definition of the rotation vectors in the figure). The displacement transformation for all nodes belonging to the inclined boundary line AB is written as

$$\mathbf{a}_i^{(e)} = [w_i, \theta_{x_i}, \theta_{y_i}]^T = \mathbf{L}_i[w_i, \theta_{x_i'}, \theta_{y_i'}]^T = \mathbf{L}_i\mathbf{a}_i'^{(e)} \tag{6.43a}$$

where

$$\mathbf{L}_i = \begin{bmatrix} 1 & 0 & 0 \\ 0 & \cos\phi_i & -\sin\phi_i \\ 0 & \sin\phi_i & \cos\phi_i \end{bmatrix} \tag{6.43b}$$

Clearly in the plate of Figure 6.8 $\phi_i = \phi$ for all nodes on the boundary AB.

The stiffness matrix of the element adjacent to the boundary line AB is transformed as (Section 9.2 of [On4])

$$\bar{\mathbf{K}}_{ij} = \hat{\mathbf{L}}_i^T \mathbf{K}_{ij} \hat{\mathbf{L}}_j \tag{6.44}$$

with

$$\hat{\mathbf{L}}_i = \begin{cases} \mathbf{L}_i, & \text{if node } i \text{ lays on the inclined boundary} \\ \mathbf{I}_3, & \text{if node } i \text{ belongs to the interior domain} \end{cases}$$

where $\mathbf{I}_3$ is the 3×3 unit matrix.

6.4 PERFORMANCE OF REISSNER–MINDLIN PLATE ELEMENTS FOR THIN PLATE ANALYSIS

6.4.1 Locking. Reduced integration. Constraint index

Reissner–Mindlin plate elements suffer from the same drawbacks as Timoshenko beam elements, i.e. the numerical solution "locks" for thin plate situations.

This defect can be observed following the same procedure as for beams (Section 2.5). Let us consider an isotropic plate of constant thickness under nodal point loads. The global equilibrium equations are written as

$$(\mathbf{K}_b + \mathbf{K}_s)\mathbf{a} = \mathbf{f} \tag{6.45}$$

where $\mathbf{K}_b$ and $\mathbf{K}_s$ assemble the individual bending and shear stiffness contributions for each element. Since the material properties and the thickness are constant we can rewrite Eq.(6.45) as

$$\left(\frac{Et^3}{12(1 - \nu^2)} \bar{\mathbf{K}}_b + Gt\bar{\mathbf{K}}_s \right) \mathbf{a} = \mathbf{f} \tag{6.46}$$

The "exact" thin Kirchhoff plate solution, termed hereafter $\mathbf{a}_k$, is inversely proportional to $\frac{Et^3}{12(1-\nu^2)}$ [TW]. Dividing Eq.(6.46) by this expression gives

$$\left(\bar{\mathbf{K}}_b + \frac{1}{\beta} \bar{\mathbf{K}}_s \right) \mathbf{a} = \frac{12(1 - \nu^2)}{Et^3} \mathbf{f} = O(\mathbf{a}_k) \tag{6.47a}$$

with

$$\beta = \frac{Et^2}{12(1 - \nu^2)G} \tag{6.47b}$$

The r.h.s. of Eq.(6.47a) is *of the order of magnitude* of the exact thin plate solution. Clearly, as $t \to 0$, then $\beta \to 0$ and $\frac{1}{\beta} \to \infty$, i.e. the transverse shear terms in Eq.(6.47a) dominate the solution as the plate is thinner. The influence of the bending terms is negligible for very thin plates and Eq.(6.47a) tends to

$$\frac{1}{\beta} \bar{\mathbf{K}}_s \mathbf{a} = O(\mathbf{a}_k) \quad \text{and} \quad \bar{\mathbf{K}}_s \mathbf{a} = \beta O(\mathbf{a}_k) \to \mathbf{0} \tag{6.48}$$

Hence, for very thin plates the solution is infinitely stiffer than the "exact" thin plate solution. From Eq.(6.48) we deduce that the only possibility for obtaining a solution different from $\mathbf{a} = 0$ is that $\mathbf{K}_s$ be singular.

Singularity of $\mathbf{K}_s$ can be achieved by reduced integration, similarly as for Timoshenko beam elements [ZT2,ZTZ]. Eq.(2.50) can be used to check the singularity of $\mathbf{K}_s$ for each mesh "a priori".

To clarify concepts the following terminology will be used hereonwards:

Full integration: exact integration for $\mathbf{K}_b^{(e)}$ and $\mathbf{K}_s^{(e)}$

Selective integration: exact integration for $\mathbf{K}_b^{(e)}$ and reduced integration for $\mathbf{K}_s^{(e)}$

Reduced integration: reduced integration for both $\mathbf{K}_b^{(e)}$ and $\mathbf{K}_s^{(e)}$

Exact integration is only possible for simple geometrical element shapes, i.e. rectangles or straight-sided triangles, as for 2D solid elements. The term full integration hereonwards therefore refers to the Gauss quadrature that yields an exact integration for the element in its rectangular or straight-sided triangular form [On4].

The singularity of $\mathbf{K}_s$ can be anticipated by evaluating the *constraint index* (CI) of each element. The CI is obtained by applying Eq.(2.50) for a single quadrilateral plate element with two adjacent edges clamped and the other two edges free. This gives (for $s = 2$)

$$\text{CI} = \text{Free DOFs} - 2 * NGP \tag{6.49}$$

where NGP is the number of Gauss points chosen for integrating $\mathbf{K}_s^{(e)}$.

Values of CI ≥ 4 ensure the singularity of $\mathbf{K}_s$ for any mesh [BD5,ZT]. Values of CI < 0 indicate that $\mathbf{K}_s^{(e)}$ is not singular and invalidate the element. A value of CI close to zero indicates that the element is not reliable and that some situations where singularity is not satisfied can be found (leading to locking). Figure 6.9 shows the CI for some plate elements. The CI is the same for reduced and selective integration as it only depends on the quadrature for $\mathbf{K}_s^{(e)}$.

Alternative indexes based on the rank of $\mathbf{K}_s^{(e)}$ for assessing the tendency of the element to lock have been proposed [BD5]. The same conclusions regarding the merits using a reduced quadrature for $\mathbf{K}_s^{(e)}$ were found.

6.4.2 Mechanisms induced by reduced integration

An element has a mechanism when it can adopt a deformed shape compatible with the boundary conditions, without consuming internal work (strain energy). For this reason mechanisms are also called zero energy

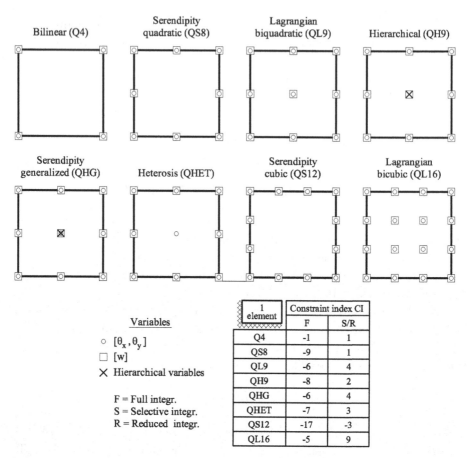

Fig. 6.9 Some Reissner-Mindlin plate quadrilaterals. Constraint index (CI) for full (F), selective (S) and reduced (R) integration

modes. An individual element, free of external constraints, has the standard rigid body mechanisms of translation and rotation which disappear by prescribing the boundary conditions. Thus, a beam element has two mechanisms (the vertical deflection and the rotation) which vanish when the beam is simply supported at two points, or clamped at one end. Similarly, a plate element has three mechanisms: the deflection and two rotations. (Figure 6.10). These mechanisms can be identified by computing the zero eigenvalues of the stiffness matrix for an unconstrained element [CMPW,Cr,ZT2]. Each zero eigenvalue corresponds to a mechanism whose shape is given by the corresponding eigenmode. This procedure can be applied to any element assembly.

Rigid body displacements   Elimination of mechanisms

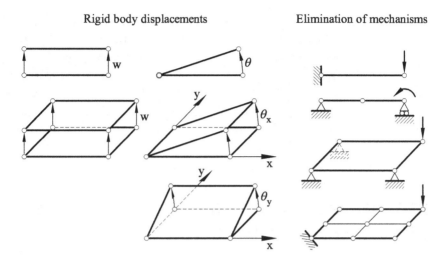

Fig. 6.10 Rigid body mechanisms for beam and plate elements

Reduced integration can induce additional zero eigenvalues in the element stiffness matrix and hence originate new mechanisms in addition to the rigid body motions. These new mechanisms can or can not propagate themselves within a mesh. This depends on their compatibility with adjacent elements and with the boundary conditions. Consequently, the singularity of $\mathbf{K}_s^{(e)}$ must always be verified together with the existence of spurious mechanisms in the global element stiffness matrix $\mathbf{K}^{(e)}$ and on their capacity to propagate within a mesh.

6.4.3 Requirements for the ideal plate element

On the basis of the preceedings arguments the ideal plate element should fulfil the following requirements:

a) It should be applicable to both thick and thin situations.
b) It should not have other mechanisms than the standard rigid body modes.
c) It should satisfy the usual invariance and convergence conditions.
d) It should give accurate solutions for the deflection, the rotations, the bending moments and the shear forces.
e) It should be insensitive to geometry distorsions.
f) Its formulation should not be based on user's dependent adjustement parameters.
g) It should be easy to implement and use in a computer program.

The more popular Reissner–Mindlin plate elements satisfying most of these conditions are presented in the next sections. In order to facilitate their study, the elements have been classified in three groups: i) those based on *selective/reduced integration* techniques, ii) those based on *assumed transverse shear strain fields*, and iii) those based on *linked interpolations*. Recent research experiences favour the second, and to some extent the third, approaches. However, didactic and historical reasons justify the overview of the elements based on selective and reduced quadratures which have enjoyed popularity in last decades.

6.5 REISSNER-MINDLIN PLATE QUADRILATERALS BASED ON SELECTIVE/REDUCED INTEGRATION

All elements presented in this section are based on the strict application of the concepts explained in Section 6.4.1, i.e. the singularity of matrix $\mathbf{K}_s^{(e)}$ is sought by using a lower order quadrature.

6.5.1 Four-noded plate quadrilateral (Q4)

The bilinear 4-noded plate quadrilateral was initially developed using selective integration by Hughes *et al.* [HTK] and Pugh *et al.* [PHZ]. This element is contemporary of the 8 and 9-noded elements of next sections.

The Q4 element has the standard bilinear shape functions of the 4-noded quadrilateral (Section 4.4.1 of [On4]). Full integration for rectangular shapes requires a 2×2 Gauss quadrature for $\mathbf{K}_b^{(e)}$ and $\mathbf{K}_s^{(e)}$ (Table 6.1). Figure 6.9 shows that the CI is negative in this case. The one point quadrature for $\mathbf{K}_s^{(e)}$ makes this matrix singular and gives CI = 1, far from the optimum value of CI ≥ 4. The quadratures shown in Table 6.1 are applicable for both rectangular and quadrilateral shapes.

Reduced integration leads to seven zero energy modes (Table 6.2). After substracting the three rigid body modes the element still has the four spurious mechanisms shown in Figure 6.11 which invalidate the quadrature. Belytschko *et al.* [BT,BTL] derived a procedure to stabilize these mechanisms. This allows using the Q4 element with the simple one point reduced quadrature.

Selective integration (Table 6.1) brings down the number of spurious mechanisms to the first two of Figure 6.11. The second one can not propagate in a mesh, but the first one can lead to problems. An example is a square plate simply supported at the four corners where the stiffness

Element	Quadrature	Quadrature for integration of $\mathbf{K}_b^{(e)}$	$\mathbf{K}_s^{(e)}$
Q4	F	2×2	2×2
	S	2×2	1×1
	R	1×1	1×1
QS8,QL9	F	3×3	3×3
QH9,QHG	S	3×3	2×2
QHET	R	2×2	2×2
	F	4×4	4×4
QS12,	S	4×4	3×3
QL16	R	3×3	3×3

Table 6.1 Full (F), selective (S) and reduced (R) quadratures for various Reissner-Mindlin plate quadrilaterals

Element	Number of zero eigenvalues (mechanisms) for $\mathbf{K}^{(e)}$ in an isolated element		
	Full integration	Selective integration	Reduced integration
Q4	3 (0)	5 (2)	7 (4)
QS8	3 (0)	3 (0)	4 (1)
QL9	3 (0)	4 (1)	7 (4)
QH9,QHG	3 (0)	3 (0)	4 (1)
QHET	3 (0)	3 (0)	6 (3)
QS12	3 (0)	3 (0)	3 (0)
QL16	3 (0)	4 (1)	7 (4)

Table 6.2 Mechanisms in quadrilateral Reissner-Mindlin plate elements induced by different quadratures. Numbers within brackets denote the number of spurious mechanisms

matrix is singular. These mechanisms can be stabilized, yielding a safe element [BT,BTL]. The element performs well with selective integration for a clamped plate, as shown in Figure 6.10.

The resultant stresses must be computed at the 2×2 Gauss points, while the central point is optimal for evaluating the shear forces [HTK].

McNeal [Ma2] proposed an interesting version of this element using a modified shear modulus similarly as done for Timoshenko beams in Section 2.6. This idea was applied to higher order plate elements by Tessler and Hughes [TH] (Section 6.7.6).

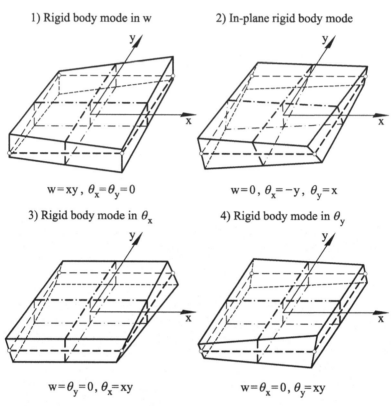

1) Rigid body mode in w

$$w=xy, \; \theta_x=\theta_y=0$$

2) In-plane rigid body mode

$$w=0, \; \theta_x=-y, \; \theta_y=x$$

3) Rigid body mode in θ_x

$$w=\theta_y=0, \; \theta_x=xy$$

4) Rigid body mode in θ_y

$$w=\theta_x=0, \; \theta_y=xy$$

Fig. 6.11 Q4 plate element. Spurious mechanisms induced by reduced (1-4) and selective (1 and 2) integration

6.5.2 Eight-noded Serendipity plate quadrilateral (QS8)

The QS8 8-noded Serendipity Reissner-Mindlin plate quadrilateral (Figure 6.9) has an irregular behaviour for thick and thin plates. Full integration requires a 3×3 Gauss quadrature for both $\mathbf{K}_b^{(e)}$ and $\mathbf{K}_s^{(e)}$ (Table 6.1) and CI is negative in this case. The 2×2 reduced quadrature for $\mathbf{K}_s^{(e)}$ gives an insufficient value of CI= 1 (Figure 6.9).

Reduced integration introduces a spurious zero eigenvalue in $\mathbf{K}^{(e)}$ (Table 6.2). The associated mechanism *can not propagate* in a mesh and the element is "safe" for practical purposes. Unfortunately, reduced integration does not suffice to ensure the singularity of $\mathbf{K}_s^{(e)}$ and the element locks for some problems. An example is the clamped plate of Figure 6.12.

In conclusion, the QS8 plate element should be used with precaution. It is free from spurious mechanisms and performs well for thick and moder-

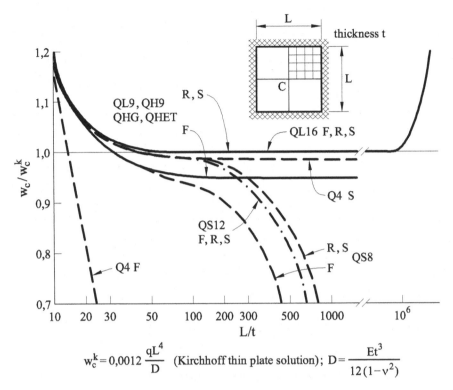

$$w_c^k = 0{,}0012 \, \frac{qL^4}{D} \quad \text{(Kirchhoff thin plate solution)}; \quad D = \frac{Et^3}{12(1-v^2)}$$

Fig. 6.12 Clamped square plate under uniform load. Central deflection versus L/t for different quadrilateral elements: Bilinear (Q4), Serendipity quadratic and cubic (QS8 and QS12), Lagrangian biquadratic and cubic (QL9 and QL16), Heterosis (QHET) and Hierarchical (QH9, QHG) with full (F), selective (S) and reduced (R) integration. A quarter of the plate is analyzed with a 4×4 mesh due to symmetry

ately thin plates. Its behaviour for very thin plates depends on the boundary conditions.

6.5.3 Nine-noded Lagrange plate quadrilateral (QL9)

The QL9 9-noded Lagrange plate quadrilateral (Figure 6.9) has an opposite behaviour to the QS8 element. A 3×3 full integration yields a negative CI. The 2×2 reduced quadrature for $\mathbf{K}_s^{(e)}$ raises the CI to 4, thus anticipating a good behaviour for thin situations (Figure 6.9). This is shown in the clamped plate of Figure 6.12 where excellent results are obtained for a wide range of thicknesses. The solution deteriorates for $\frac{L}{t} \geq 10^6$ due to round–off errors when solving the equations. This deficiency can be avoided by using a fictitious thickness when computing $\mathbf{K}_s^{(e)}$

such that $(\frac{L}{t})^2 \geq 10^{p/2}$, where p is the number of decimal digits which the computer can store, while the bending stiffness is computed with the correct thickness. In this manner, the shear strain is computed incorrectly when its effect is so small that it does not influence the numerical solution [CMPW,Fr,Fr2,HTK].

Reduced integration excites four spurious mechanisms (Table 6.2), while selective integration induces just one mechanism. This mechanism can propagate in a mesh for some boundary conditions and thus the element is not reliable. Figure 6.13 shows an example of two square plates under a corner point load with the minimum boundary constraints to avoid rigid body movements. The QL9 element with reduced integration yields such poor results that can not even be graphically plotted. Selective integration induces an oscillatory solution due to the propagable mechanism (Figure 6.13b). The QS8 element yields the correct solution (for moderate thickness), as well as the 9-noded Hierarchical and Heterosis quadrilaterals presented in the next sections.

6.5.4 Nine-noded hierarchical plate quadrilateral (QH9)

The behaviour of the QS8 and QL9 quadrilaterals evidences the need for an element which incorporates their best features, i.e. a plate quadrilateral free of spurious mechanisms and valid for all range of thicknesses.

Cook [CPMW] modified the QS8 element by adding a central node with a bubble function for the deflection only (Figure 6.9). The displacement field is defined as

$$\theta_x = \sum_{i=1}^{8} N_i \theta_{x_i}, \quad \theta_y = \sum_{i=1}^{8} N_i \theta_{y_i} \qquad (6.50a)$$

$$w = \left(\sum_{i=1}^{8} N_i w_i \right) + \bar{N}_9 \bar{w}_9 \qquad (6.50b)$$

where N_i are the standard shape functions for the QS8 element and $\bar{N}_9 = (1 - \xi^2)(1 - \eta^2)$. The hierarchical variable $\bar{w}_9$ is the difference between the nodal deflections obtained with the QS8 and QL9 elements. For the isoparametric description the N_i shape functions are used.

Reduced integration (Table 6.1) gives CI $= 2$, which places the element at an intermediate position between the QS8 and QL9. Reduced integration excites one spurious mechanism (Table 6.2). The mechanism can be eliminated by multiplying the diagonal stiffness coefficient corresponding

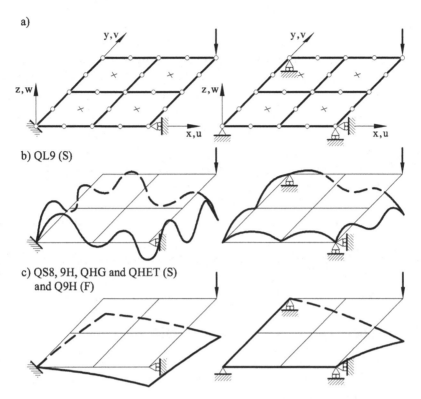

Fig. 6.13 Analysis of two square plates under corner point load with minimum boundary constraints: (a) Geometry, b.c. and loading. (b) and (c) Deformed shapes obtained using QL9, QS8, QH9, QHG and QHET elements with full (F) and selective (S) integration

to $\bar{w}_9$ by $(1+e)$ where e is a small number ($e = 10^{-2}t/L$ is recommended). This is equivalent to adding a small "spring" to the hierarchical node. Selective integration eliminates the internal mechanism (Table 6.2) and leads to good results for thick and thin plates (Figures 6.12 and 6.13).

6.5.5 A generalization of the 9-noded hierarchical plate quadrilateral (QHG)

Oñate *et al.* [OHG] generalized and improved the QH9 element by including the two rotations as hierarchical variables. The displacement field is

$$\left\{ \begin{array}{c} w \\ \theta_x \\ \theta_y \end{array} \right\} = \left(\sum_{i=1}^{8} N_i \left\{ \begin{array}{c} w_i \\ \theta_{x_i} \\ \theta_{y_i} \end{array} \right\} \right) + \bar{N}_9 \left\{ \begin{array}{c} \bar{w}_9 \\ \theta_{x_9} \\ \theta_{y_9} \end{array} \right\} \tag{6.51}$$

where N_i and $\bar{N}_9$ have the same meaning as in Eq.(6.50).

This allows us to retain three variables per node. Also, the value of CI with the reduced 2×2 quadrature increases to 4 (Figure 6.9).

The diagonal coefficients of $\mathbf{K}^{(e)}$ corresponding to the three hierarchical variables are multiplied by $(1 + e_1)$, $(1 + e_2)$ and $(1 + e_3)$, respectively. By choosing $e_1 = e_2 = e_3 = 0$ the element performs as the QL9 one. Making e_1, e_2 and e_3 very large is equivalent to eliminating the hierarchical variables and the behaviour of the QS8 element is reproduced. Finally, making e_1 and e_2 very large reproduces the QH9 element.

Reduced integration introduces one spurious mechanism (Table 6.2). This can be eliminated by adjusting the spring parameters ($e_1 = e_2 = e_3 = 0.004$ is recommended in [OHG]) or by selective integration. Figures 6.12 and 6.13 show the good behaviour of the QHG element with selective integration.

The choice of different spring parameters for the hierarchical variables allows us to derive plate elements with features laying between the QS8 and QL9. An example is the element presented in the next section.

6.5.6 Nine-noded heterosis plate quadrilateral (QHET)

Hughes and Cohen [HC] proposed a 9-noded plate quadrilateral termed Heterosis (hybrid wich inherits the best properties of its parents) with the following interpolation

$$\theta_x = \sum_{i=1}^{9} N_i^L \theta_{xi}; \quad \theta_y = \sum_{i=1}^{9} N_i^L \theta_{yi} \quad y \quad w = \sum_{i=1}^{8} N_i^S w_i \quad (6.52)$$

where N_i^L and N_i^S are the shape functions of the 9-noded Lagrangian and the 8-noded Serendipity quadrilaterals, respectively (Appendix I). The Serendipity shape functions are used to interpolate the element geometry.

The element (termed here QHET) has an acceptable value of CI = 3 (Figure 6.9). Selective integration eliminates the spurious mechanism typical of the QL9 element (Table 6.2). This preserves the good performance for thin plate analysis as shown in Figures 6.12 and 6.13.

The QHET element is a particular case of the QHG of previous section by making $e_1 = e_2 = 0$ and e_3 equal to a large number [OHG].

An inconvenient of the QHET element is that, similarly as for the QH9, it has a different number of DOFs at the central node. Also it does not satisfy the patch test for irregular shapes [Cr].

6.5.7 Higher order Reissner–Mindlin plate quadrilaterals with 12 and 16 nodes

Higher order Serendipity and Lagrange Reissner–Mindlin plate quadrilaterals present a behaviour analogous to the QS8 and QL9 ones. We consider here the cubic 12-noded Serendipity (QS12) and the 16-noded Lagrange (QL16) quadrilaterals (Figure 6.9 and Table 6.1). Selective integration for the QS12 element eliminates all spurious mechanisms, similarly as for the QS8 one (Table 6.2). However, its behaviour for thin plate analysis is poor as indicated by the negative value of CI $= -3$ (Figure 6.9). The QL16 element has CI $= 9$ with reduced and selective integration and it behaves well for thin plates. Unfortunately, both quadratures induce a spurious mechanism, similarly as for the Q9L element (Table 6.2). Figure 6.12 shows the performance of the QS12 and QL16 elements for a clamped plate under uniform loading. The merit of selective/reduced integration versus the full quadrature is not relevant for these elements [PHZ,ZT2].

6.6 REISSNER-MINDLIN PLATE ELEMENTS BASED ON ASSUMED TRANSVERSE SHEAR STRAIN FIELDS

6.6.1 Basic concepts

A thin plate element must satisfy Kirchhoff condition of zero transverse shear strains. From Eq.(6.31) we can write

$$\hat{\boldsymbol{\varepsilon}}_s = \mathbf{B}_s \mathbf{a}^{(e)} = \boldsymbol{\alpha}_1(w_i,\boldsymbol{\theta}_i) + \boldsymbol{\alpha}_2(w_i,\boldsymbol{\theta}_i)\xi + \boldsymbol{\alpha}_3(w_i,\boldsymbol{\theta}_i)\eta + \cdots\cdots$$
$$\cdots\cdots + \boldsymbol{\alpha}_n(w_i,\boldsymbol{\theta}_i)\xi^p\eta^q = 0$$

$$(6.53)$$

The fulfilment of Eq.(6.53) implies

$$\boldsymbol{\alpha}_j(w_i,\boldsymbol{\theta}_i) = 0; \qquad j = 1,n \qquad (6.54)$$

Eq.(6.54) imposes a linear constraint between the nodal deflections and the rotations which can be interpreted from a physical standpoint. Elements which are able to satisfy Eq.(6.54) can reproduce the thin plate condition without locking.

However, in many elements the α_j's are a function of the nodal rotations only. The condition $\boldsymbol{\alpha}_j(\boldsymbol{\theta}_i) = 0$ is then too restrictive (and sometimes even non physical!) and this leads to locking.

Example. The 4-noded plate rectangle

As an example let us consider the simple 4-noded plate rectangle of Figure 6.14. Using the standard bilinear shape functions, $N_i = \frac{1}{4}(1+\xi\xi_i)(1+\eta\eta_i)$ (Appendix I) the shear strain γ_{xz} is given by

$$
\begin{aligned}
\gamma_{xz} &= \frac{\partial w}{\partial x} - \theta_x = \sum_{i=1}^{4}\left[(\frac{\xi_i}{4a}w_i - \frac{1}{4}\theta_{x_i}) + (\frac{\xi_i\eta_i}{4a}w_i - \frac{\eta_i}{4}\theta_{x_i})\eta + \right.\\
&\left. + (\frac{\xi_i}{4}\theta_{x_i})\xi - (\frac{\xi_i\eta_i}{4}\theta_{x_i})\xi\eta \right] = \alpha_1(w_i, \theta_{x_i}) + \\
&+ \alpha_2(w_i, \theta_{x_i})\eta + \alpha_3(\theta_{x_i})\xi + \alpha_4(\theta_{x_i})\xi\eta
\end{aligned}
\tag{6.55}
$$

The Kirchhoff constraint $\gamma_{xz} = 0$, implies $\alpha_1 = \alpha_2 = \alpha_3 = \alpha_4 = 0$. The conditions on α_1 and α_2 are physically posible and they impose a relationship between the average rotation θ_x over the element and the nodal deflections. However, the element is unable to satisfy naturally the conditions $\alpha_3 = 0$ and $\alpha_4 = 0$, unless $\theta_{x_i} = 0$ (which leads to $w_i = 0$ and, hence, to locking). Identical conclusions are found for γ_{yz} simply by interchanging ξ by η and θ_{x_i} by θ_{y_i}. Note the analogy with the performance of the 2-noded Timoshenko beam element (Section 2.8.4).

We deduce from above that a simple way to avoid locking is to evaluate the transverse shear strains at points where the undesirable terms $\alpha_j(\theta_i)$ vanish. For the 4-noded rectangle the terms α_3 and α_4 are zero if γ_{xz} and γ_{yz} are sampled along the lines $\xi = 0$ and $\eta = 0$, respectively. The resulting expansion for $\hat{\boldsymbol{\varepsilon}}_s$ is

$$
\hat{\boldsymbol{\varepsilon}}_s = \begin{Bmatrix} \gamma_{xz} \\ \gamma_{yz} \end{Bmatrix} = \begin{Bmatrix} \alpha_1(w_i, \theta_{x_i}) + \alpha_2(w_i, \theta_{y_i})\eta \\ \bar{\alpha}_1(w_i, \theta_{x_i}) + \bar{\alpha}_2(w_i, \theta_{y_i})\xi \end{Bmatrix} = \bar{\mathbf{B}}_s\, \mathbf{a}^{(e)}
\tag{6.56}
$$

This transverse shear strain field can satisfy naturally the thin plate condition. *The substitute transverse shear strain matrix* $\bar{\mathbf{B}}_s$ *is used instead of the original matrix* $\mathbf{B}_s$ *for computing the shear stiffness matrix in Eq.(6.39b)*, i.e.

$$
\mathbf{K}_s^{(e)} = \iint_{A^{(e)}} \bar{\mathbf{B}}_s^T \hat{\mathbf{D}}_s \bar{\mathbf{B}}_s\, dA
\tag{6.57}
$$

The new shear stiffness matrix can now be integrated "exactly" (using a 2×2 quadrature.

The $\bar{\mathbf{B}}_s$ matrix was already introduced in Section 2.4.1 for the Timoshenko beam element.

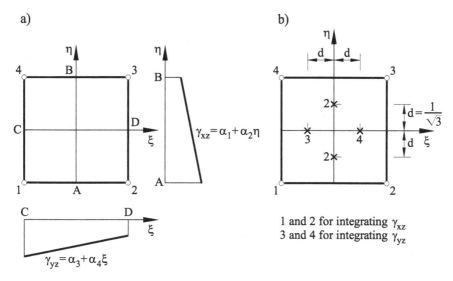

Fig. 6.14 Four-noded plate quadrilateral. a) Assumed transverse shear strain field. b) Reduced quadrature for integrating the shear terms in the original $\mathbf{K}_s^{(e)}$ matrix

The "assumed" transverse shear strain field is displayed in Figure 6.14. The element is identical to that proposed by Bathe and Dvorkin [BD] (termed here QLLL) and is studied in some detail in Section 6.7.

For rectangles the shear stiffness matrix of Eq.(6.57) is identical to that obtained with the original $\mathbf{B}_s$ matrix and the special reduced quadrature shown in Figure 6.14b. The reasons for this coincidence are:

1. Sampling the original shear strain field at the quadrature points shown in the figure leads precisely to the assumed field of Eq.(6.56).
2. The quadrature points along the lines $\xi = 0$ and $\eta = 0$ integrate exactly the quadratic terms in η and ξ of the original shear stiffness matrix, respectively. This leads to the same stiffness matrix of Eq.(6.57) using 2×2 integration.

Above arguments show us the coincidence between the assumed shear strain method and an "ad hoc" reduced quadrature for $\mathbf{K}_s^{(e)}$ in the standard stiffness matrix. Unfortunately, the "special" reduced quadrature is not so simple to identify in other cases and the general procedure described next is recommended in practice.

The general method is based in imposing "a priori" a transverse shear strain field which fulfils condition (6.54), thus allowing the vanishing of $\hat{\varepsilon}_s$

in the thin limit. The assumed transverse shear strain interpolation is

$$\hat{\varepsilon}_s = \sum_{k=1}^{m} N_{\gamma_k} \, \gamma_k = \mathbf{N}_\gamma \, \gamma^{(e)} \tag{6.58}$$

where $\gamma_k^{(e)}$ are the transverse shear strain values at m points within the element and N_{γ_k} are the shear interpolating functions. Combining Eqs.(6.53) and (6.58) gives

$$\hat{\varepsilon}_s = \sum_{k=1}^{m} N_{\gamma_k} \, \mathbf{B}_{s_k} \, \mathbf{a}_k = \bar{\mathbf{B}}_s \, \mathbf{a}^{(e)} \tag{6.59}$$

Eq.(6.59) can be written in the form (6.53) and this ensures a locking-free element, as long as the conditions described in the next section are satisfied.

6.6.2 Selection of the transverse shear strain field

The assumed transverse shear strain field must satisfy certain conditions. The starting point is a three field mixed formulation where the deflection, the rotations and the transverse shear strains are interpolated independently as

$$w = \mathbf{N}_w \, \mathbf{w}^{(e)}; \quad \boldsymbol{\theta} = \mathbf{N}_\theta \, \boldsymbol{\theta}^{(e)}; \quad \hat{\varepsilon}_s = \mathbf{N}_\gamma \, \gamma^{(e)} \tag{6.60}$$

The conditions which must satisfy the three fields are [ZL,ZQTN,ZT2]

$$\boxed{\begin{aligned} n_\theta \; + \; n_w &\geq n_\gamma \\ n_\gamma &\geq n_w \end{aligned}} \tag{6.61}$$

where n_w, n_θ y n_γ are the number of variables involved in the interpolation of the deflection, the rotations and the transverse shear strains, respectively (after eliminating the prescribed values). The proof of Eq.(6.61) is given in Appendix G. It is important to point out that the rotation field in Eq.(6.60) has to be C^o *continuous*, whereas a *discontinuous* interpolation can be used for the deflection and the transverse shear strain fields.

Note also that the conditions (6.61) are identical to (2.79) for Timoshenko beam elements.

Conditions (6.61) apply for each element, or any patch of elements, as a *necessary* (although not always sufficient) requirement for the stability of the solution [BFS3,ZL,ZQTN,ZTPO,ZT2]. The convergence must be verified via the patch test. It is interesting that the condition $n_w + n_\theta > n_\gamma$ is analogous to the singularity condition for $\mathbf{K}_s$ (Appendix G). This shows the link between reduced integration and assumed transverse shear strain techniques [MH]. A systematic way for deriving $\bar{\mathbf{B}}_s$ is presented in the next section.

6.6.3 Derivation of the substitute transverse shear strain matrix

We consider an isoparametric Reissner–Mindlin plate element with n nodes for which the deflection, the rotation and the transverse shear strains are interpolated independently according to Eq.(6.60). We also assume that this interpolation satisfies conditions (6.61).

Step 1

The transverse shear strains are interpolated in the natural coordinate system ξ, η as

$$
\gamma' = \left\{ \begin{array}{c} \gamma_\xi \\ \gamma_\eta \end{array} \right\} = \left[\begin{array}{ccccccc|cccccc} 1 & \xi & \eta & \xi\eta & \cdots & \xi^p\eta^q & 0 & 0 & 0 & \cdots & & 0 \\ 0 & 0 & 0 & 0 & \cdots & 0 & 1 & \xi & \eta & \cdots & & \xi^r\eta^s \end{array} \right] \left\{ \begin{array}{c} \alpha_1 \\ \alpha_2 \\ \vdots \\ \alpha_{n\gamma} \end{array} \right\} = \mathbf{A}\boldsymbol{\alpha}
$$

(6.62)

where n_γ is the number of sampling points defining the polynomial expansion for γ_ξ and γ_η within the element. To simplify the notation, the generalized transverse shear strain vector $\hat{\boldsymbol{\varepsilon}}_s$ is denoted hereonwards $\boldsymbol{\gamma}$.

The transverse shear strains in the cartesian system are obtained as

$$
\gamma = \left\{ \begin{array}{c} \gamma_{xz} \\ \gamma_{yz} \end{array} \right\} = \mathbf{J}^{-1}\gamma'
$$

(6.63a)

where $\mathbf{J}$ is the 2D Jacobian matrix [Hu,On4,ZTZ]

$$
\mathbf{J} = \left[\begin{array}{cc} \dfrac{\partial x}{\partial \xi} & \dfrac{\partial y}{\partial \xi} \\ \dfrac{\partial x}{\partial \eta} & \dfrac{\partial y}{\partial \eta} \end{array} \right]
$$

(6.63b)

Step 2

The transverse shear strain along a natural direction $\bar{\xi}_i$ is defined as

$$
\gamma_{\bar{\xi}_i} = \cos \beta_i \gamma_\xi + \sin \beta_i \gamma_\eta
$$

(6.64)

where β_i is the angle that the $\bar{\xi}_i$ direction forms with the natural ξ axis. The $\bar{\xi}_i$ direction along the element sides is taken so that it follows the increasing numbering of the corner nodes (Figure 6.16).

The shear strains $\gamma_{\bar{\xi}_i}$ are sampled at each of the n_γ points placed along the $\bar{\xi}_i$ directions. From Eq.(6.64)

$$\gamma_{\bar{\xi}} = \mathbf{T}(\beta_i)\hat{\gamma}' \tag{6.65}$$

where

$$\gamma_{\bar{\xi}} = [\gamma_{\bar{\xi}}^1, \gamma_{\bar{\xi}}^2, \cdots, \gamma_{\bar{\xi}}^{n_\gamma}]^T \tag{6.66a}$$

$$\hat{\gamma}' = [\gamma_{\xi}^1, \gamma_{\eta}^1, \gamma_{\xi}^2, \gamma_{\eta}^1, \cdots, \gamma_{\xi}^{n_\gamma}, \gamma_{\eta}^{n_\gamma}]^T \tag{6.66b}$$

where $(\cdot)^1$, $(\cdot)^2$ etc. denote values at each samping point.

From Eq.(6.62) it is found

$$\hat{\gamma}' = \begin{bmatrix} \mathbf{A}^1 \\ \vdots \\ \mathbf{A}^{n_\gamma} \end{bmatrix} \alpha = \hat{\mathbf{A}}(\xi_i, \eta_i)\alpha \tag{6.67}$$

Substituting Eq.(6.67) into (6.65) gives

$$\gamma_{\bar{\xi}} = \mathbf{P}(\xi_i, \ \eta_i, \ \beta_i)\alpha \tag{6.68}$$

where $\mathbf{P} = \mathbf{T}\hat{\mathbf{A}}$ is a $n_\gamma \times n_\gamma$ matrix.

Vector α is obtained as

$$\alpha = \mathbf{P}^{-1}\gamma_{\bar{\xi}} \tag{6.69}$$

Step 3

Combining Eqs.(6.62), (6.65) and (6.69) gives

$$\gamma' = \mathbf{A}\,\mathbf{P}^{-1}\,\mathbf{T}\,\hat{\gamma}' \tag{6.70}$$

In many cases the relationship between γ' and $\hat{\gamma}'$ can be written from simple observation of the assumed transverse shear strain field. This allows writing directly the matrix resulting from the product $\mathbf{A}\mathbf{P}^{-1}\mathbf{T}$.

Step 4

The relationship between the cartesian and the natural transverse shear strains at each sampling point is

$$\hat{\gamma}_i' = \begin{Bmatrix} \gamma_{\xi}^i \\ \gamma_{\eta}^i \end{Bmatrix} = \mathbf{J}^i \begin{Bmatrix} \gamma_{xz}^i \\ \gamma_{yz}^i \end{Bmatrix} = \mathbf{J}^i\,\hat{\gamma}^i \tag{6.71}$$

where $\mathbf{J}^i$ is the Jacobian matrix at the ith sampling point. Thus,

$$\hat{\boldsymbol{\gamma}}' = \begin{bmatrix} \mathbf{J}^1 & & \mathbf{0} \\ & \ddots & \\ \mathbf{0} & & \mathbf{J}^{n_\gamma} \end{bmatrix} \left\{ \begin{array}{c} \hat{\boldsymbol{\gamma}}^1 \\ \vdots \\ \hat{\boldsymbol{\gamma}}^{n_\gamma} \end{array} \right\} = \mathbf{C}\,\hat{\boldsymbol{\gamma}} \tag{6.72}$$

Substituting Eqs.(6.70) and (6.72) into (6.63a) gives

$$\boldsymbol{\gamma} = \mathbf{J}^{-1}\mathbf{A}\mathbf{P}^{-1}\mathbf{T}\mathbf{C}\hat{\boldsymbol{\gamma}} = \mathbf{N}_\gamma\hat{\boldsymbol{\gamma}} \tag{6.73}$$

where

$$\mathbf{N}_\gamma = \mathbf{J}^{-1}\mathbf{A}\mathbf{P}\mathbf{T}^{-1}\mathbf{C} \tag{6.74}$$

are the shape functions which interpolate the cartesian transverse shear strains in terms of their values at the sampling points.

Step 5

The general relationship between the cartesian transverse shear strains and the nodal displacements for each element can be written using a weighted residual procedure [OZST,ZT2,ZTZ] as

$$\iint_{A^{(e)}} W_k[\boldsymbol{\gamma} - (\nabla w - \boldsymbol{\theta})]dA = 0 \tag{6.75}$$

where $\nabla = \left[\frac{\partial}{\partial x}, \frac{\partial}{\partial y}\right]^T$ and W_k are arbitrarily chosen weighting functions. Eq.(6.75) imposes the satisfaction of the equality $\boldsymbol{\gamma} = \nabla w - \boldsymbol{\theta}$ in a mean integral sense over the element.

Substituting Eqs.(6.73) and (6.31) into (6.75) and chosing a Galerkin weighting , i.e. $W_k = N_{\gamma_k}$ [ZTZ] gives

$$\left[\iint_{A^{(e)}} \mathbf{N}_\gamma^T\mathbf{N}_\gamma d\Lambda\right]\hat{\boldsymbol{\gamma}} = \left[\iint_{A^{(e)}} \mathbf{N}_\gamma^T\mathbf{B}_s dA\right]\mathbf{a}^{(e)} \tag{6.76}$$

where $\mathbf{B}_s$ is the standard transverse shear strain matrix of Eq.(6.32). Eq.(6.76) is used to obtain the shear strains at the sampling points for each element as

$$\hat{\boldsymbol{\gamma}} = \hat{\mathbf{B}}_s\mathbf{a}^{(e)} \tag{6.77}$$

with

$$\hat{\mathbf{B}}_s = \left[\iint_{A^{(e)}} \mathbf{N}_\gamma^T\mathbf{N}_\gamma dA\right]^{-1} \iint_{A^{(e)}} \mathbf{N}_\gamma^T\mathbf{B}_s d \tag{6.78}$$

A simpler procedure is to use point collocation in Eq.(6.75). This implies choosing $W_i = \delta_i$ where δ_i is the Dirac delta at the ith sampling point $(i = 1, \cdots n_\gamma)$. This gives

$$\hat{\mathbf{B}}_s = \left\{ \begin{array}{c} \mathbf{B}_s^1 \\ \vdots \\ \mathbf{B}_s^{n_\gamma} \end{array} \right\} \tag{6.79}$$

where $\mathbf{B}_s^i$ is the value of the original transverse shear strain matrix at the ith sampling point.

Step 6

Combining steps 1–5 yields finally

$$\boldsymbol{\gamma} = \mathbf{J}^{-1} \mathbf{A} \mathbf{P}^{-1} \mathbf{TC} \hat{\mathbf{B}}_s \mathbf{a}^{(e)} = \bar{\mathbf{B}}_s \mathbf{a}^{(e)} \tag{6.80}$$

where $\bar{\mathbf{B}}_s$ is the substitutive transverse shear strain matrix given by

$$\boxed{\bar{\mathbf{B}}_s = \mathbf{J}^{-1} \mathbf{A} \mathbf{P}^{-1} \mathbf{T} \mathbf{C} \hat{\mathbf{B}}_s} \tag{6.81}$$

The weighting procedure described in Step 5 is not the only alternative to derive a relationship between the transverse shear strains and the nodal displacements. One could for instance require integrals such as

$$\int_{\Gamma^{(e)}} W_k \left[\gamma_{\bar{\xi}} - \left(\frac{\partial w}{\partial \bar{\xi}} - \theta_{\bar{\xi}} \right) \right] d\Gamma \tag{6.82}$$

to vanish on a segment of the element boundary $\Gamma^{(e)}$. This allows us to obtain directly a relationship of the form

$$\gamma_{\bar{\xi}} = [TCB]\mathbf{a}^{(e)} \tag{6.83}$$

and the substitute shear strain matrix is given by [OTZ,OZST,ZT2]

$$\bar{\mathbf{B}}_s = \mathbf{J}^{-1} \mathbf{A} \mathbf{P}^{-1}[TCB] \tag{6.84}$$

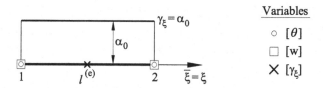

Variables
○ $[\theta]$
□ $[w]$
✕ $[\gamma_\xi]$

Fig. 6.15 2-noded Timoshenko beam element

Example 6.1: Obtain the substitutive transverse shear strain matrix for the 2-noded Timoshenko beam element using a constant shear strain field.

- Solution

The element is shown in Figure 6.15. In this case

$$\gamma_\xi = (\alpha_0 + \alpha_1\xi)_{\xi=0} = \alpha_0, \quad \mathbf{A} = [1]$$

The natural direction $\bar{\xi}$ coincides with ξ and therefore

$$\gamma_{\bar{\xi}} = \gamma_\xi; \quad \beta = 0 \quad \text{and} \quad \mathbf{P} = [1], \quad \mathbf{T} = [1]$$

Also

$$(\gamma_\xi)_{\xi=0} = \frac{l^{(e)}}{2}(\gamma_x)_{\xi=0} \quad \text{and} \quad \mathbf{J} = \left[\frac{l^{(e)}}{2}\right], \quad \mathbf{C} = \left[\frac{l^{(e)}}{2}\right]$$

$$(\gamma_x)_{\xi=0} = \frac{1}{l^{(e)}}\left[-1, -\frac{l^{(e)}}{2}, 1, -\frac{l^{(e)}}{2}\right] \quad \mathbf{a}^{(e)} = \hat{\mathbf{B}}_s\, \mathbf{a}^{(e)}$$

(6.85)

Finally from Eq.(6.81)

$$\bar{\mathbf{B}}_s = \mathbf{J}^{-1}\, \mathbf{A}\, \mathbf{P}^{-1}\, \mathbf{T}\, \mathbf{C}\, \hat{\mathbf{B}}_s = \frac{1}{l^{(e)}}\left[-1, -\frac{l^{(e)}}{2}, 1, -\frac{l^{(e)}}{2}\right]$$

Note that $\bar{\mathbf{B}}_s$ coincides with the original $\mathbf{B}_s$ matrix (Eq.(2.15)) evaluated at the element mid–point. This is another evidence of the analogy between the assumed transverse shear strain procedure and reduced integration (see also Section 2.8.4 and Example 2.8).

6.7 REISSNER–MINDLIN PLATE QUADRILATERALS BASED ON ASSUMED TRANSVERSE SHEAR STRAIN FIELDS

6.7.1 4-noded plate quadrilateral with linear shear field (QLLL)

This popular element was initially developed by Bathe and Dvorkin [BD,DB]. Its formulation can be considered a particularization of the procedures based on auxiliary transverse shear modes proposed by Mac Neal [Ma,Ma2] and Hughes et al. [HTT]. Donea and Lamain [DL] and Oñate et al. [OTZ,OZBT] derived the element using assumed strain concepts. In the following lines we will derive the element following the methodology described in Section 6.6.3.

The starting point is the standard 4-noded Q4 element of Section 6.5.1 with a bilinear interpolation of the deflections and the rotations. We define an assumed transverse shear strain field in the natural system ξ, η as (Figure 6.14a)

$$\gamma_\xi = \alpha_1 + \alpha_2\eta$$

$$\gamma_\eta = \alpha_3 + \alpha_4\xi; \quad \text{i.e.} \quad \mathbf{A} = \begin{bmatrix} 1 & \eta & 0 & 0 \\ 0 & 0 & 1 & \xi \end{bmatrix} \quad (6.86)$$

The α_i parameters are found by sampling the natural transverse shear strains $\gamma_{\bar\xi}$ at the four mid-side points shown in Figure 6.16, with

$$\gamma_{\bar\xi_i} = (\alpha_1 + \alpha_2\eta)\cos\beta_i + (\alpha_3 + \alpha_4\xi)\sin\beta_i; \qquad i = 1,4 \quad (6.87)$$

where β_i is the angle between the $\bar\xi_i$ and ξ axis. A simple operation gives

$$\underbrace{\begin{bmatrix} 1 & -1 & 0 & 0 \\ 0 & 0 & 1 & 1 \\ -1 & -1 & 0 & 0 \\ 0 & 0 & 1 & -1 \end{bmatrix}}_{\mathbf{P}} \begin{Bmatrix} \alpha_1 \\ \alpha_2 \\ \alpha_3 \\ \alpha_4 \end{Bmatrix} = \begin{Bmatrix} \gamma_{\bar\xi}^1 \\ \gamma_{\bar\xi}^2 \\ \gamma_{\bar\xi}^3 \\ \gamma_{\bar\xi}^4 \end{Bmatrix} \quad \text{and} \quad \mathbf{P}^{-1} = \frac{1}{2}\begin{bmatrix} 1 & 0 & -1 & 0 \\ -1 & 0 & -1 & 0 \\ 0 & 1 & 0 & 1 \\ 0 & 1 & 0 & -1 \end{bmatrix} \quad (6.88)$$

The strains $\gamma_{\bar\xi}^i$ are related to $\gamma_\xi^i, \gamma_\eta^i$ by

$$\begin{Bmatrix} \gamma_{\bar\xi}^1 \\ \gamma_{\bar\xi}^2 \\ \gamma_{\bar\xi}^3 \\ \gamma_{\bar\xi}^4 \end{Bmatrix} = \begin{bmatrix} 1 & 0 & & & \mathbf{0} \\ & 0 & 1 & & \\ & & -1 & 0 & \\ \mathbf{0} & & & 0 & 1 \end{bmatrix} \begin{Bmatrix} \gamma_\xi^1 \\ \gamma_\eta^1 \\ \vdots \\ \gamma_\xi^4 \\ \gamma_\eta^4 \end{Bmatrix} = \mathbf{T}\hat\gamma' \quad (6.89)$$

The cartesian transverse shear strains at the sampling points are related to the natural transverse shear strains by

$$\hat\gamma' = \begin{bmatrix} \mathbf{J}^1 & & & \mathbf{0} \\ & \mathbf{J}^2 & & \\ & & \mathbf{J}^3 & \\ \mathbf{0} & & & \mathbf{J}^4 \end{bmatrix} \begin{Bmatrix} \hat\gamma^1 \\ \vdots \\ \hat\gamma^4 \end{Bmatrix} = \mathbf{C}\,\hat\gamma \quad ; \quad \hat\gamma^i = \begin{Bmatrix} \gamma_{xz}^i \\ \gamma_{yz}^i \end{Bmatrix} \quad (6.90)$$

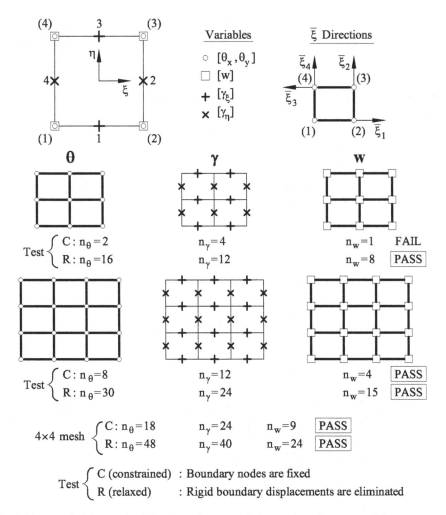

Fig. 6.16 4-noded Reissner-Mindlin plate quadrilateral with assumed linear transverse shear strain field (QLLL). Numbers within brackets denote node numbers

The relationship between the cartesian transverse shear strains at the four sampling points and the nodal displacements is

$$\hat{\gamma} = \begin{Bmatrix} \mathbf{B}_s^1 \\ \mathbf{B}_s^2 \\ \mathbf{B}_s^3 \\ \mathbf{B}_s^4 \end{Bmatrix} \mathbf{a}^{(e)} = \hat{\mathbf{B}}_s\, \mathbf{a}^{(e)} \tag{6.91}$$

The substitute transverse shear strain matrix is obtained by Eq.(6.81).

This element has been given different names in the literature [ZT2]. Here it is termed QLLL (for **Q**uadrilateral, bi**L**inear deflection, bi**L**inear rotations and **L**inear transverse shear strain fields). The QLLL element satisfies conditions (6.61) for meshes of more than 2×2 elements (Figure 6.16) and it is considered robust for practical applications. Computation of the stiffness matrix requires a full 2×2 quadrature for *all* terms, and this preserves the element from spurious mechanisms. The bending moments and shear forces are constant along each natural direction. Hence fine meshes are required for certain applications. The extension for shell analysis is studied in Chapter 9.

The product $\mathbf{A}\mathbf{P}^{-1}\mathbf{T}$ in Eq.(6.70) is

$$\mathbf{A}\,\mathbf{P}^{-1}\,\mathbf{T} = \frac{1}{2}\begin{bmatrix} (1-\eta)\,0\,0 & 0 & | & (1+\eta)\,0\,0 & 0 \\ 0 & 0\,0\,(1+\xi) & | & 0 & 0\,0\,(1-\xi) \end{bmatrix} \qquad (6.92)$$

This matrix could have been anticipated by writting directly the assumed transverse shear strain field as

$$\begin{aligned} \gamma_\xi &= \frac{1}{2}(1-\eta)\gamma_\xi^1 + \frac{1}{2}(1+\eta)\gamma_\xi^3 \\ \gamma_\eta &= \frac{1}{2}(1+\xi)\gamma_\eta^2 + \frac{1}{2}(1-\xi)\gamma_\eta^4 \end{aligned} \qquad (6.93)$$

6.7.2 8-noded Serendipity plate quadrilateral with assumed quadratic transverse shear strain field (QQQQ-S)

The starting point is the QS8 plate quadrilateral of Section 6.5.2. The fulfilment of conditions (6.61) requires an assumed transverse shear strain field containing ten terms. The simplest choice is

$$\begin{aligned} \gamma_\xi &= \alpha_1 + \alpha_2\xi + \alpha_3\eta + \alpha_4\xi\eta + \alpha_5\eta^2 \\ \gamma_\eta &= \alpha_6 + \alpha_7\xi + \alpha_8\eta + \alpha_9\xi\eta + \alpha_{10}\xi^2 \end{aligned} \qquad (6.94)$$

Figure 6.17 shows the sampling points for the transverse shear strains γ_ξ and γ_η (two points along each side and the central point). The $\bar{\xi}$ directions coincide with those defined for the QLLL element (Figure 6.16).

The derivation of the substitute transverse shear strain matrix can be simplified if we define the assumed field for γ_ξ as

$$\gamma_\xi = \frac{1}{4}\left[(1+\frac{1}{a}\xi)\gamma_\xi^1 + (1-\frac{1}{a}\xi)\gamma_\xi^2\right](1+\eta) +$$

$$+ \frac{1}{4}\left[(1+\frac{1}{a}\xi)\gamma_\xi^4 + (1-\frac{1}{a}\xi)\gamma_\xi^5\right](1-\eta) + (1-\eta^2)\gamma_\xi^3 \quad (6.95)$$

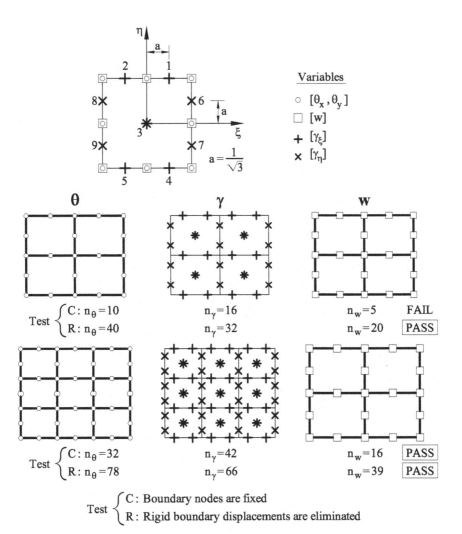

Fig. 6.17 8-noded Reissner-Mindlin plate quadrilateral with an assumed quadratic transverse shear strain field (QQQQ-S)

with $a = \frac{1}{\sqrt{3}}$. A similar interpolation can be written for γ_η simply interchanging ξ by η and the points 1,2,4,5 by 6,7,8 and 9, respectively in Eq.(6.94). This yields directly the product $\mathbf{A}\,\mathbf{P}^{-1}\,\mathbf{T}$ in Eq.(6.70) as

$$
\mathbf{A}\,\mathbf{P}^{-1}\,\mathbf{T} =
\begin{bmatrix}
\dfrac{A\eta_1}{4} & 0 & \dfrac{B\eta_1}{4} & 0 & \eta_3 & 0 & \dfrac{A\eta_2}{4} & 0 & \dfrac{B\eta_2}{4} & 0 & \underset{1\times 8}{\mathbf{0}} \\[2mm]
\underset{1\times 8}{\mathbf{0}} & 0 & \dfrac{\bar{A}\xi_1}{4} & 0 & \dfrac{\bar{B}\xi_1}{4} & 0 & \xi_3 & 0 & \dfrac{\bar{A}\xi_2}{4} & 0 & \dfrac{\bar{B}\xi_2}{4}
\end{bmatrix}
\tag{6.96}
$$

with

$$A = 1 + \sqrt{3}\xi \ , \ B = 1 - \sqrt{3}\xi \ , \ \bar{A} = 1 + \sqrt{3}\eta \ , \ \bar{B} = 1 - \sqrt{3}\eta$$
$$s_1 = 1 + s \ , \ s_2 = 1 - s \ , \ s_3 = 1 - s^2 \quad \text{with } s = \xi, \eta \tag{6.97}$$

The following expressions for $\mathbf{C}$ and $\hat{\mathbf{B}}_s$ are necessary for computing the sustitutive transverse shear strain matrix $\bar{\mathbf{B}}_s$ via Eq.(6.81)

$$\mathbf{C} = \begin{bmatrix} \mathbf{J}_1 & & & \mathbf{0} \\ & \mathbf{J}_2 & & \\ & & \ddots & \\ \mathbf{0} & & & \mathbf{J}_9 \end{bmatrix} , \quad \hat{\mathbf{B}}_s = \begin{Bmatrix} \mathbf{B}_s^1 \\ \mathbf{B}_s^2 \\ \vdots \\ \mathbf{B}_s^9 \end{Bmatrix} \tag{6.98}$$

The global stiffness matrix is integrated with a full 3×3 quadrature which eliminates any spurious mechanisms [DL,HH2,HH4].

This element is termed QQQQ-S (for **S**erendipity plate **Q**uadrilateral with **Q**uadratic deflection, **Q**uadratic rotations and **Q**uadratic transverse shear strain fields). Conditions (6.61) are satisfied for meshes of more than 2×2 elements (Figure 6.17).

6.7.3 9-noded Lagrange plate quadrilateral with assumed quadratic transverse shear strain field (QQQQ-L)

The starting point is the QL9 element of Section 6.5.3. Satisfaction of condition (6.61) requires the following assumed transverse shear strain field containing 12 terms

$$\gamma_\xi = \alpha_1 + \alpha_2\xi + \alpha_3\eta + \alpha_4\xi\eta + \alpha_5\eta^2 + \alpha_6\xi\eta^2$$
$$\gamma_\eta = \alpha_7 + \alpha_8\xi + \alpha_9\eta + \alpha_{10}\xi\eta + \alpha_{11}\xi^2 + \alpha_{12}\eta\xi^2 \tag{6.99}$$

Figure 6.18 displays the sampling points for the transverse shear strains. The $\bar{\xi}$ directions coincide again with those shown in Figure 6.16.

Direct observation gives the γ_ξ field as

$$\gamma_\xi = \frac{1}{4}\left[(1+\frac{1}{a}\xi)\gamma_\xi^1 + (1-\frac{1}{a}\xi)\gamma_\xi^2\right]\eta(1+\eta) +$$
$$+ \frac{1}{2}\left[(1+\frac{1}{a}\xi)\gamma_\xi^3 + (1-\frac{1}{a}\xi)\gamma_\xi^4\right](1-\eta^2) +$$
$$+ \frac{1}{4}\left[(1+\frac{1}{a}\xi)\gamma_\xi^5 + (1-\frac{1}{a}\xi)\gamma_\xi^6\right]\eta(\eta-1) \tag{6.100}$$

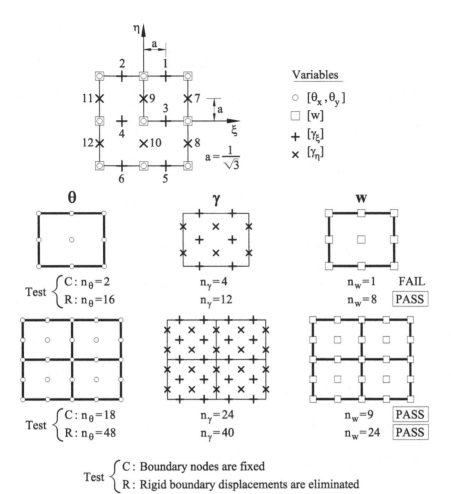

Fig. 6.18 9-noded Lagrangian plate quadrilateral with assumed quadratic transverse shear strains (QQQQ-L)

with $a = \frac{1}{\sqrt{3}}$. The interpolation for γ_η is obtained simply by interchanging ξ by η and points $1, \ldots, 6$ by $7, \ldots, 12$, respectively in Eq.(6.99). The product $\mathbf{A} \, \mathbf{P}^{-1} \, \mathbf{T}$ in Eq.(6.70) can be written directly as

$$
\mathbf{A} \, \mathbf{P}^{-1} \, \mathbf{T} = \begin{bmatrix} \frac{A\eta_1}{4} & 0 & \frac{B\eta_1}{4} & 0 & \frac{A\eta_2}{2} & 0 & \frac{B\eta_2}{2} & 0 & \frac{A\eta_3}{4} & 0 & \frac{B\eta_3}{4} & 0 & \mathbf{0}_{1\times 12} \\[2mm] \mathbf{0}_{1\times 12} & 0 & \frac{\bar{A}\xi_1}{4} & 0 & \frac{\bar{B}\xi_1}{4} & 0 & \frac{\bar{A}\xi_2}{4} & 0 & \frac{\bar{B}\xi_2}{4} & 0 & \frac{\bar{A}\xi_3}{4} & 0 & \frac{\bar{B}\xi_3}{4} \end{bmatrix}
$$

$$(6.101)$$

where $A, B, \bar{A}$ and $\bar{B}$ coincide with the values given in Eq.(6.97) and

$$s_1 = s(1 + s) \ , \ s_2 = 1 - s^2 \ , \ s_3 = s(1 - s) \text{ with } s = \xi, \eta \qquad (6.102)$$

The $\mathbf{C}$ and $\hat{\mathbf{B}}_s$ matrices for computing the substitute transverse shear strain matrix are

$$\mathbf{C} = \begin{bmatrix} \mathbf{J}_1 & & & \mathbf{0} \\ & \mathbf{J}_2 & & \\ & & \ddots & \\ \mathbf{0} & & & \mathbf{J}_{12} \end{bmatrix}, \quad \hat{\mathbf{B}}_s = \begin{Bmatrix} \mathbf{B}_s^1 \\ \mathbf{B}_s^2 \\ \vdots \\ \mathbf{B}_s^{12} \end{Bmatrix} \qquad (6.103)$$

A full 3×3 quadrature is used for integrating *all* the stiffness matrix terms and this eliminates the spurious mechanisms. The element satisfies conditions (6.61) for meshes with more than one element (Figure 6.18).

The element is termed QQQQ-L (for **L**agrange plate **Q**uadrilateral with **Q**uadratic deflection, **Q**uadratic rotations and **Q**uadratic trasverse shear strain fields). This element was simultaneously developed by Hinton and Huang [HH3,HH4] and Bathe and Dvorkin [BD2]. Donea y Lamain [DL] presented a different version of the element based in the direct derivation of expressions (6.100) in cartesian coordinates. The performance of the QQQQ-L element is superior to the QQQQ-S.

6.7.4 Sixteen DOFs plate quadrilateral (QLQL)

The interpolating fields are the following.

1. The deflection is bi-linearly interpolated as

$$w = \sum_{i=1}^{4} N_i w_i \qquad (6.104)$$

where N_i are the bilinear shape functions of the Q4 element.

2. An incomplete quadratic interpolation is chosen for the rotations as

$$\boldsymbol{\theta} = \sum_{i=1}^{4} N_i \boldsymbol{\theta}_i + f(\xi)(1 - \eta)\mathbf{e}_{12}\Delta\theta_{s_5} + f(\eta)(1 + \xi)\mathbf{e}_{23}\Delta\theta_{s_6} +$$
$$+ f(\xi)(1 + \eta)\mathbf{e}_{23}\Delta\theta_{s_7} + f(\eta)(1 - \xi)\mathbf{e}_{14}\Delta\theta_{s_8} \qquad (6.105)$$

where $f(\xi) = 1 - \xi^2$ and $f(\eta) = 1 - \eta^2$. In above $\Delta\theta_{S_i}$ is a hierarchical tangential mid-side rotation and $\mathbf{e}_{ij}$ are unit vectors along the side directions (Figure 6.19).

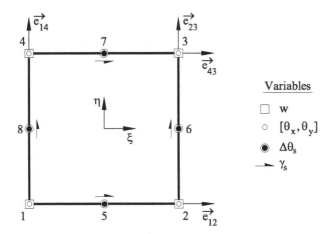

Fig. 6.19 Sixteen DOFs QLQL Reissner-Mindlin plate quadrilateral, (w linear, $\boldsymbol{\theta}$, cuadratic and $\boldsymbol{\gamma}$ linear)

3. The transverse shear strains are assumed to vary linearly as in Eq.(6.86). The substitute transverse shear strain matrix is found as explained for the QLLL element (Section 6.7.1). A 2×2 quadrature is used for all the terms of the stiffness matrix.

The element is termed QLQL (for **Q**uadrilateral, bi**L**inear deflection, **Q**uadratic rotations and **L**inear transverse shear strain fields) and it satisfies conditions (6.61) for all meshes.

The hierarchical side rotations can be eliminated by imposing the condition of zero transverse shear strains along the sides. This leads to a 4-noded Discrete Kirchhoff thin plate quadrilateral identical to that proposed by Batoz and Ben Tahar [BBt] (Section 6.11.2). A procedure for eliminating the side rotations while still preserving some shear strain energy within the element is described in [BBt,BD5].

6.7.5 4-noded plate quadrilateral of Tessler-Hughes

Tessler and Hughes [TH] derived a plate quadrilateral with a quadratic interpolation for the deflection, a linear one for the rotations and a constant transverse shear strain along the sides. The element is free from spurious mechanisms and applicable to both thick and thin plates.

The formulation shares some similarities with the Heterosis and Hierarchical plate elements (Sections 6.5.4–6.5.6). The starting point is the standard 8-noded Serendipity interpolation for the deflection, while the

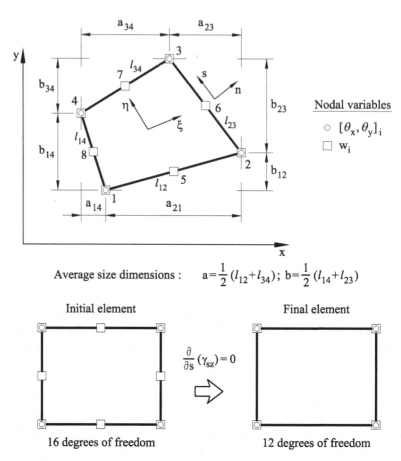

Average size dimensions : $a=\dfrac{1}{2}\,(l_{12}+l_{34});\ b=\dfrac{1}{2}\,(l_{14}+l_{23})$

Initial element   Final element

$$\frac{\partial}{\partial s}\,(\gamma_{sz})=0$$

16 degrees of freedom   12 degrees of freedom

Fig. 6.20 Four-noded plate quadrilateral of Tessler and Hughes [TH]

rotations are bilinearly interpolated using the corner node values (Figure 6.20). Thus

$$\mathbf{u}=\begin{Bmatrix} w \\ \theta_x \\ \theta_y \end{Bmatrix}=\begin{bmatrix} \bar{\mathbf{N}}(\xi^2,\eta^2) & 0 & 0 \\ 0 & \mathbf{P}(\xi,\eta) & 0 \\ 0 & 0 & \mathbf{P}(\xi,\eta) \end{bmatrix}\begin{Bmatrix} \mathbf{w} \\ \boldsymbol{\theta}_x \\ \boldsymbol{\theta}_y \end{Bmatrix}^{(e)}=\hat{\mathbf{N}}\mathbf{a}^{(e)} \qquad (6.106)$$

where $\bar{\mathbf{N}}(\xi^2,\eta^2)=[N_1,N_2,\ldots,N_8]$ are the quadratic shape functions of the QS8 element and $\mathbf{P}(\xi,\eta)=[P_1,P_2,P_3,P_4]$ are the bilinear shape functions of the QL4 element. The nodal displacement vector is written as

$$\mathbf{a}^{(e)}_{16\times 1}=[w_1,\cdots,w_8,\theta_{x_1},\cdots,\theta_{x_4},\theta_{y_1},\cdots,\theta_{y_4}]^T \qquad (6.107)$$

The 4-noded configuration is reached after eliminating the mid-side deflections w_5, w_6, w_7 and w_8 by imposing a constant transverse shear strain γ_{sz} along the sides, where s is the size direction (Figure 6.20). This ensures continuity of the tangential transverse shear strain across the sides of adjacent elements. The final displacement interpolation is written as

$$\mathbf{u} = \mathbf{N}\hat{\mathbf{a}}^{(e)} \tag{6.108}$$

where

$$\hat{\mathbf{a}}^{(e)}_{12 \times 1} = [w_1, \cdots w_4, \theta_{x1}, \cdots \theta_{x4}, \theta_{y1}, \cdots, \theta_{y4}]^T$$

$$\mathbf{N} = \begin{bmatrix} \mathbf{P}(\xi, \eta) & \mathbf{N}_x(\xi^2, \eta^2) & \mathbf{N}_y(\xi^2, \eta^2) \\ \mathbf{0} & \mathbf{P}(\xi, \eta) & \mathbf{0} \\ \mathbf{0} & \mathbf{0} & \mathbf{P}(\xi, \eta) \end{bmatrix} \tag{6.109}$$

with $\mathbf{N}_x = [N_{x1}, N_{x2}, N_{x3}, N_{x4}]$, $\mathbf{N}_y = [N_{y1}, N_{y2}, N_{y3}, N_{y4}]$ and

$$N_{x1} = \frac{1}{8}[-b_{12}N_5 - b_{14}N_8]; \qquad N_{x2} = \frac{1}{8}[b_{12}N_5 - b_{23}N_6]$$

$$N_{x3} = \frac{1}{8}[b_{23}N_6 + b_{34}N_7]; \qquad N_{x4} = \frac{1}{8}[-b_{34}N_7 + b_{14}N_8]$$

$$N_{y1} = \frac{1}{8}[-a_{12}N_5 + a_{14}N_8]; \qquad N_{y2} = \frac{1}{8}[a_{12}N_5 + a_{23}N_6] \tag{6.110}$$

$$N_{y3} = \frac{1}{8}[a_{34}N_7 - a_{23}N_6]; \qquad N_{y4} = \frac{1}{8}[-a_{34}N_7 - a_{14}N_8]$$

where a_{ij} and b_{ij} are given in Figure 6.20.

Note that the interpolation for the deflection involves also the rotations. This can be seen as a class of "linked interpolation" (Section 6.10).

The geometry is defined in sub-parametric manner using the corner nodes coordinates. The bending and transverse shear strain matrices are

$$\mathbf{B}_b = \begin{bmatrix} 0 & \frac{\partial}{\partial x}\mathbf{P} & 0 \\ 0 & 0 & \frac{\partial}{\partial y}\mathbf{P} \\ 0 & -\frac{\partial}{\partial y}\mathbf{P} & -\frac{\partial}{\partial x}\mathbf{P} \end{bmatrix}, \quad \mathbf{B}_s = \begin{bmatrix} \frac{\partial}{\partial x}\mathbf{P} & \left(\frac{\partial}{\partial x}\mathbf{N}_x - \mathbf{P}\right) & \frac{\partial}{\partial x}\mathbf{P} \\ \frac{\partial}{\partial y}\mathbf{P} & \frac{\partial}{\partial y}\mathbf{P} & \left(\frac{\partial}{\partial y}\mathbf{N}_y - \mathbf{P}\right) \end{bmatrix}$$

$$\tag{6.111}$$

The equivalent nodal force vector for distributed loads is

$$\mathbf{f}^{(e)} = \iint_{A^{(e)}} \left(f_z \begin{Bmatrix} \mathbf{P}^T \\ \mathbf{N}_x^T \\ \mathbf{N}_y^T \end{Bmatrix} + \begin{Bmatrix} 0 \\ m_x\mathbf{P}^T \\ m_y\mathbf{P}^T \end{Bmatrix} \right) dA \tag{6.112}$$

Note that bending moment components are introduced in $\mathbf{f}^{(e)}$ by the distributed load f_z.

A 2×2 reduced quadrature eliminates spurious mechanisms and avoids locking. However, the behaviour of the element is somehow stiff for very thin plates.

An enhanced element can be derived by modifying the shear correction parameter k as we did by using a substitute shear modulus for Timoshenko beams (Section 2.6). The substitute shear correction parameter k^* is found by comparing the analytical solution for a simple supported rectangular plate with that obtained with a single element in a quarter of the plate (using symmetry conditions) giving [TH]

$$k^* = \frac{C_b\psi}{1 + C_b\psi} k \qquad \text{with} \qquad \psi = \frac{w_s}{w_b} \qquad (6.113)$$

where w_b and w_s are the bending and shear contributions to the analytical solution for the central deflection and C_b is a number close to unity deduced from numerical experiments. Good results were reported in [TH] using $C_b = 0.9$ and the value of ψ obtained from the analysis of a rectangular plate $(a \times b)$ under sine loading giving

$$\psi = \frac{\pi^2}{6(1-\nu)k} \left(\frac{t}{a}\right)^2 \left[1 + \frac{a^2}{b^2}\right] \qquad (6.114)$$

For arbitrary shape elements a and b are taken as the average side lengths (Figure 6.20).

Similar substitute shear correction parameters for Reissner-Mindlin plate elements have been suggested [Fr,Fr2,FY,Ma,Ma2,SV,TH2]. A simple one was proposed by Carpenter *et al.* [CBS] as

$$k^* = \frac{c}{1+c} k \qquad \text{with} \qquad c = \frac{E\,t^2}{2\alpha G(1-\nu^2)A^{(e)}} \qquad (6.115)$$

6.7.6 8-noded plate quadrilateral proposed by Crisfield

Crisfield [Cr3] developed an 8-noded thick quadrilateral following similar arguments as for the 4-noded Tessler-Hughes element of the previous section. The starting point is the 27 DOFs quadrilateral of Figure 6.21. The standard quadratic 8-noded Serendipity interpolation is used for the rotations and a quadratic 9-noded Lagrange field is used for the deflections.

The four mid-side deflections are eliminated by imposing a constant tangential shear strains along the sides. The central deflection is eliminated by imposing an "effective" constant shear strain field along the two diagonals (Example 2.8).

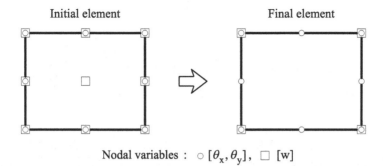

Nodal variables :   ○ $[\theta_x, \theta_y]$,   □ [w]

Fig. 6.21 8-noded plate quadrilateral of Crisfield [Cr3]

The final element has 20 DOFs as shown in Figure 6.21. A full 3×3 quadrature is required for all the stiffness terms. This element has similar features as the QLQL of Section 6.7.4 but has more DOFs.

6.7.7 Higher order 12 and 16-noded plate quadrilaterals with assumed transverse shear strain fields

Hinton and Huang [HH2,HH4] developed 12 and 16-noded plate quadri-laterals using assumed transverse shear strain fields with 16 and 24 parameters, respectively. The elements perform well for thick and thin plates but contain an excessive number of variables for practical purposes.

6.8 REISSNER-MINDLIN PLATE TRIANGLES

Most Reissner-Mindlin plate triangles have been developed using the assumed transverse shear strain methodology. Some popular elements are presented next.

6.8.1 6-noded quadratic triangle with assumed linear shear strains (TQQL)

Zienkiewicz *et al.* [ZTPO] developed a 6-noded plate triangle with a standard quadratic interpolation for the deflection and the rotations and the following assumed linear transverse shear strain field

$$
\begin{aligned}
\gamma_\xi &= \alpha_1 + \alpha_2\xi + \alpha_3\eta \\
\gamma_\eta &= \alpha_4 + \alpha_5\xi + \alpha_6\eta
\end{aligned}, \quad \text{i.e} \quad \mathbf{A} = \begin{bmatrix} 1 & \xi & \eta & 0 & 0 & 0 \\ 0 & 0 & 0 & 1 & \xi & \eta \end{bmatrix}
\tag{6.116}
$$

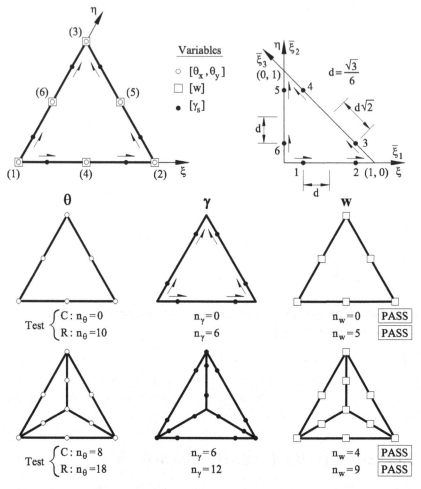

Fig. 6.22 Eighteen DOFs TQQL Reissner-Mindlin plate triangle (w Quadratic, θ Quadratic, γ, Linear). Figures within brackets denote node numbers

Figure 6.22 shows the position of the six shear sampling points and the $\bar{\xi}_i$ directions. Following the procedure of Section 6.6.3 we find

$$
\mathbf{P} = \begin{bmatrix}
1 & \xi_1 & \eta_1 & 0 & 0 & 0 \\
1 & \xi_2 & \eta_2 & 0 & 0 & 0 \\
-a & -a\xi_3 & -a\eta_3 & a & a\xi_3 & a\eta_3 \\
-a & -a\xi_4 & -a\eta_4 & a & a\xi_4 & a\eta_4 \\
0 & 0 & 0 & 1 & \xi_5 & \eta_5 \\
0 & 0 & 0 & 1 & \xi_6 & \eta_6
\end{bmatrix}
; \quad
\mathbf{T} = \begin{bmatrix}
1 & 0 & & & & \mathbf{0} \\
& 1 & 0 & & & \\
& & -a & a & & \\
& & & -a & a & \\
& & & & 0 & 1 \\
\mathbf{0} & & & & & 0 & 1
\end{bmatrix}
$$

$$C = \begin{bmatrix} \mathbf{J}^1 & & & 0 \\ & \mathbf{J}^2 & & \\ & & \ddots & \\ 0 & & & \mathbf{J}^6 \end{bmatrix}, \quad \hat{\mathbf{B}}_s = \begin{Bmatrix} \mathbf{B}^1 \\ \mathbf{B}^2 \\ \vdots \\ \mathbf{B}^6 \end{Bmatrix}, \quad a = \frac{\sqrt{2}}{2} \quad (6.117)$$

Matrix $\bar{\mathbf{B}}_s$ is computed by Eq.(6.81). A full 3-point Gauss quadrature is used for the numerical integration of $\mathbf{K}_s^{(e)}$.

This element was termed TRI-6R in [ZTPO] and T6D6 in [ZT2]. In our notation we call it TQQL (for **T**riangle, **Q**uadratic deflection, **Q**uadratic rotation and **L**inear transverse shear strain fields). The element satisfies conditions (6.61) for all meshes (Figure 6.22).

The TQQL element performes well although sometimes is too flexible [ZTPO,ZT2]. Improvements are found by imposing a linear variation for the normal rotation θ_n along the three sides as

$$\theta_{n_i} - \frac{1}{2}(\theta_{n_{i-2}} + \theta_{n_{i+2}}) = 0 \quad (6.118)$$

where $i = 4, 5, 6$ are the mid-side nodes (Figure 6.22).

Eq.(6.118) can be used for eliminating the normal rotation at the mid-side nodes [OC,ZTPO]. Conditions (6.61) still hold in this case. A plate triangle based on this concept is presented next.

6.8.2 Quadratic/Linear Reissner–Mindlin plate triangle (TLQL)

Zienkiewicz *et al.* [ZTPO], Papadopoulus and Taylor [PT] and Oñate *et al.* [OC,OTZ,OZBT] developed very similar enhanced versions of the TQQL element of previous section. The basic assumptions are

1. The deflection varies linearly as:

$$w = \sum_{i=1}^{3} N_i w_i \quad (6.119)$$

2. A incomplete quadratic interpolation is used for the rotations:

$$\boldsymbol{\theta} = \sum_{i=1}^{3} N_i \boldsymbol{\theta}_i + 4N_1 N_2 \mathbf{e}_{12} \Delta \theta_{s_4} + 4N_2 N_3 \mathbf{e}_{23} \Delta \theta_{s_5} + 4N_1 N_3 \mathbf{e}_{13} \Delta \theta_{s_6}$$

$$(6.120)$$

3. The transverse shear strains vary linearly within the element in terms of the three shear strain γ_s at the element mid-side points. This defines a constant shear strain along each side.

In above expressions N_i are the standard linear shape functions for the 3-noded triangle, $\Delta\theta_{s_i}$ is a hierarchical tangential rotation at the mid-side points and $\mathbf{e}_{ij}$ are unit vectors along the side directions (Figure 6.23). Eq.(6.120) defines a linear variation for the normal rotation along the sides, while the tangential rotation varies quadratically. This is equivalent to imposing condition (6.118) explicitly. The vector of nodal variables is

$$\mathbf{a}^{(e)} = [w_1, \theta_{x_1}, \theta_{y_1}, w_2, \theta_{x_2}, \theta_{y_2}, w_3, \theta_{x_3}, \theta_{y_3}, \Delta\theta_{s_1}, \Delta\theta_{s_1}, \Delta\theta_{s_3}]^T$$
$$(6.121)$$

This element was termed DRM in [ZTPO] and T3T3 in [ZT]. Here it is called TLQL (for **T**riangle, **L**inear deflection, **Q**uadratic rotations and **L**inear transverse shear strain fields). The TLQL element satisfies conditions (6.61), for all meshes (Figure 6.23). It can be considered the triangular counterpart of the QLQL quadrilateral of Section 6.7.4. Matrix $\bar{\mathbf{B}}_s$ can be derived by constraining the quadratic shear strain field of the TQQL element. A more direct procedure is to write the assumed linear shear strain field in the natural coordinate system as [OZBT]

$$\gamma' = \left\{ \begin{matrix} \gamma_\xi \\ \gamma_\eta \end{matrix} \right\} = \begin{bmatrix} 1 - \eta & -\sqrt{2}\eta & \eta \\ \xi & \sqrt{2}\xi & 1 - \xi \end{bmatrix} \left\{ \begin{matrix} \gamma_{\bar{\xi}}^{12} \\ \gamma_{\bar{\xi}}^{23} \\ \gamma_{\bar{\xi}}^{13} \end{matrix} \right\} = [A^{-1}PT]\gamma_{\bar{\xi}} \qquad (6.122)$$

where $\gamma_{\bar{\xi}}^{ij} = [\gamma_{\bar{\xi}}^{12}, \gamma_{\bar{\xi}}^{23}, \gamma_{\bar{\xi}}^{13}]^T$ contains the local shear strains at the element mid-side points. The signs in Eq.(6.122) correspond to the side coordinates $\bar{\xi}_1$, $\bar{\xi}_2$ and $\bar{\xi}_3$ as defined in Figure 6.23.

The relationship between the shear strains and the nodal displacements is obtained by imposing the condition $\gamma_{\bar{\xi}}^{ij} = \left(\frac{\partial w}{\partial \bar{\xi}} - \theta_s \right)$ to be satisfied along each side in a weighted residual form, similarly as explained in step 5 of Section 6.6.3. Thus, we write

$$\int_{l^{ij}} W_k \left[\gamma_{\bar{\xi}}^{ij} - \frac{\partial w}{\partial \bar{\xi}} + \theta_s \right] d\bar{\xi} = 0 \qquad (6.123)$$

where l^{ij} is the length of the side ij.

Choosing a Galerkin weighting with $W_k = 1$ [ZTZ] and assuming that $\gamma_{\bar{\xi}}^{ij}$ is constant along each side leads, after substituting Eqs.(6.119) and

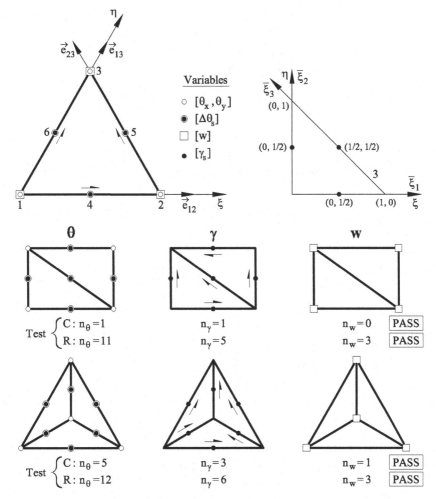

Fig. 6.23 Twelve DOFs TLQL Reissner-Mindlin plate triangle (w **L**inear, Q **Q**uadratic, $\boldsymbol{\gamma}$ **L**inear)

(6.120) into (6.123), to the following equation for each side

$$\gamma_{\bar{\xi}}^{ij} = \frac{1}{l_{\bar{\xi}}^{ij}} \int_{l_{\bar{\xi}}} \left(\frac{\partial w}{\partial \bar{\xi}} - \theta_{\bar{\xi}} \right) d\bar{\xi} = \frac{1}{l_{\bar{\xi}}^{ij}}(w_j - w_i) - \frac{l^{ij}}{2l_{\bar{\xi}}^{ij}}\mathbf{e}_{ij}^T(\boldsymbol{\theta}_i + \boldsymbol{\theta}_j) - \frac{2}{3}\Delta\theta_{s_k}\frac{l^{ij}}{l_{\bar{\xi}}^{ij}}$$

$$(6.124)$$

where $k = 3 + i$, $l_{\bar{\xi}}^{12} = l_{\bar{\xi}}^{13} = 1$ and $l_{\bar{\xi}}^{23} = \sqrt{2}$. Vectors $\mathbf{e}_{ij}$ are defined so that i and j coincide with the corner nodes with smaller and larger global numbers, respectively (Figure 6.23). From Eq.(6.124) we can write

$$\boldsymbol{\gamma}_{\bar{\xi}} = [TCB]\mathbf{a}^{(e)} \qquad\qquad (6.125)$$

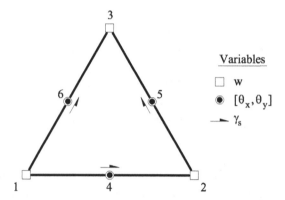

Fig. 6.24 Nine DOFs TLLL Reissner-Mindlin plate triangle (w linear, $\boldsymbol{\theta}$ quadratic, γ linear) [OZF]

where

$$[TCB] = \begin{bmatrix} -1 & \frac{x_{12}}{2} & \frac{y_{12}}{2} & 1 & \frac{x_{12}}{2} & \frac{y_{12}}{2} & 0 & 0 & 0 & -\frac{2}{3}l^{12} & 0 & 0 \\ 0 & 0 & 0 & -\frac{1}{\sqrt{2}} & \frac{x_{23}}{\sqrt{2}} & \frac{y_{23}}{\sqrt{2}} & \frac{1}{\sqrt{2}} & \frac{x_{23}}{\sqrt{2}} & \frac{\bar{y}_{23}}{\sqrt{2}} & 0 & -\frac{\sqrt{2}}{3}l^{23} & 0 \\ 1 & -\frac{\bar{C}_{13}}{2} & -\frac{\bar{S}_{13}}{2} & 0 & 0 & 0 & -1 & -\frac{\bar{C}_{13}}{2} & -\frac{\bar{S}_{13}}{2} & 0 & 0 & -\frac{2}{3}l^{13} \end{bmatrix}$$

$$(6.126)$$

with $x_{ij} = x_i - x_j$ and $y_{ij} = y_i - y_j$. The substitute transverse shear strain matrix is

$$\bar{\mathbf{B}}_s = \mathbf{J}^{-1}[A^{-1}PT][TCB] \tag{6.127}$$

where $[A^{-1}PT]$ is deduced from Eq.(6.122). A full 3 point quadrature is used and this prevents the element from having spurious mechanisms.

The TLQL element performs well and examples can be found in [OZBT] and in Section 6.11. The element yields linear bending moments and constant transverse shear distributions along the sides which satisfy the equilibrium conditions (5.23). This contributes to its good behaviour.

The three hierarchical tangential rotations $\Delta\theta_{s_k}$ at the mid-side points can be eliminated by imposing a zero value of the transverse shear strain along the sides. This leads to the nine DOFs Discrete Kirchhoff thin plate triangle (DKT) described in Section 6.11.1.

6.8.3 Linear plate triangle with nine DOFs (TLLL)

Oñate *et al.* [OZF] proposed a low order Reissner-Mindlin plate triangle based on the following fields (Figure 6.24)

1. The deflection is linearly interpolated in terms of the three corner values by Eq.(6.119)
2. The rotations are linearly interpolated in terms of mid-side values by

$$\boldsymbol{\theta} = \sum_{i=4}^{6} N_i^\theta \boldsymbol{\theta}_i; \quad \boldsymbol{\theta}_i = [\theta_{x_i}, \ \theta_{y_i}]^T \qquad (6.128)$$

where

$$N_4^\theta = 1 - 2\eta, \ N_s^\theta = 2\xi + 2\eta - 1, \ N_6^\theta = 1 - 2\xi \qquad (6.129a)$$

For convenience, we define the displacement vector as

$$\mathbf{a}^{(e)} = [w_1, \ w_2, \ w_3, \ \theta_{x_4}, \ \theta_{y_4}, \ \theta_{x_5}, \ \theta_{y_5}, \ \theta_{x_6}, \ \theta_{y_6}]^T \qquad (6.129b)$$

Eq.(6.128) defines an incompatible rotation field with interelemental compatibility satisfied at the mid-side nodes only. The good performance of the element is ensured as the patch test is satisfied [OFZ].

3. A linear assumed transverse shear strain field is assumed identical to that of Eq.(6.122) for the TLQL element. The derivation of $\bar{\mathbf{B}}_s$ follows the arguments of the preceeding section with matrix $[A^{-1}PT]$ deduced from Eq.(6.122) and matrix $[TCB]$ given by

$$[TCB] = \begin{bmatrix} -1 & 1 & 0 & x_{12} & y_{12} & 0 & 0 & 0 & 0 \\ 0 & \frac{1}{\sqrt{2}} & \frac{1}{\sqrt{2}} & 0 & 0 & \frac{x_{23}}{\sqrt{2}} & \frac{x_{23}}{\sqrt{2}} & 0 & 0 \\ -1 & 0 & 1 & 0 & 0 & 0 & 0 & x_{13} & y_{13} \end{bmatrix} \qquad (6.130)$$

A full 3 point quadrature is used.

The TLLL element (for **T**riangle and **L**inear interpolation for the deflection, the rotations and the transverse shear fields) satisfies conditions (6.61) and it is free from spurious mechanisms. Examples of its good behaviour can be found in Section 6.14.1 and [OFZ].

Flores and Oñate [FO2] have enhanced the behaviour of the TLLL element for plate and shell analysis by using a one-point reduced integration for the shear stiffness matrix. A stabilized method is used for eliminating the spureous energy modes induced by reduced integration.

The tangential side rotations can be eliminated by constraining the mid-side shear strains to be zero. This yields a six DOFs Discrete Kirchhoff thin plate triangle with identical features as the Morley triangle (Section 5.5.1) [On4].

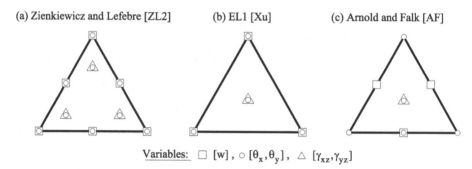

(a) Zienkiewicz and Lefebre [ZL2]   (b) EL1 [Xu]   (c) Arnold and Falk [AF]

Variables: □ [w] , ○ [θ_x,θ_y] ,   △ [γ_{xz},γ_{yz}]

Fig. 6.25 Reissner-Mindlin plate triangles based on assumed transverse shear strain fields

6.9 MORE PLATE TRIANGLES BASED ON ASSUMED SHEAR STRAIN FIELDS

Zienkiewicz and Lefebre [ZL2] developed a quadratic triangle with a linear interpolation for the transverse shear strains. Three hierarchical quartic bubbles are added to the standard quadratic field giving a total of 24 DOFs (Figure 6.25a). Good performance is obtained by integrating $\mathbf{K}_b$ and $\mathbf{K}_s$ with seven and four point quadratures, respectively.

A linear triangle with an additional cubic bubble to the linear rotational interpolation and a constant shear field was developed by Xu [Xu] (Figure 6.25b). The original 9 DOFs element satisfies Eqs.(6.61) but it locks for very thin plates. The so-called EL1 element was enhanced using the concept of linked interpolation described in the next section [XZZ].

Arnold and Falk [AF] proposed a 9 DOFs plate triangle with an incomplete cubic interpolation for the rotations, a linear *discontinuous* deflection field and a constant transverse shear strain field (Figure 6.25c). The element performs reasonably well after eliminating the internal rotations [AF,Di]. However, the representation of the deflection field requires an adequate smoothing to avoid non-physical discontinuities.

Belytschko *et al.* [BSC] developed a 3-noded plate triangle by splitting the strain energy into bending and transverse shear modes. The bending mode is defined so that the bending energy coincides with that of an equivalent Kirchhoff mode with zero transverse shear strains. The rest of the element strain energy is associated to the transverse shear mode.

Hughes and Taylor [HT] proposed a 3-noded plate triangle based on a degeneration of a 4-noded quadrilateral with a linear transverse shear field. This element only works correctly with a reduced one point quadrature.

The stiffness matrix is then identical to that of the triangle proposed by Belytschko *et al.* [BSC].

The popularity of many of these elements has been limited by the difficulties for their extension to shell analysis. A review of Reissner-Mindlin plate elements can be found in [BD5,On3,Tu,ZT2].

6.10 REISSNER-MINDLIN PLATE ELEMENTS BASED ON LINKED INTERPOLATIONS

A different class of Reissner-Mindlin plate elements can be derived by using an interpolation for the deflection field of one order higher than for the rotations. This favours satisfying the thin plate conditions $\theta_x = \frac{\partial w}{\partial x}$ and $\theta_y = \frac{\partial w}{\partial y}$ in the thin limit. The concept is similar to that studied for Timoshenko beams in Section 2.8.2.

An effective procedure of this type is via "linked interpolations", where the deflection field is enriched with additional higher order polynomial terms involving the nodal rotations (Section 2.8.3).

For the 3-noded linked triangle and the 4-noded linked quadrilateral we require that

1. The deflection field along a side is defined by the rotations at the end side nodes only, in order to guarantee C° continuity;
2. The rotation terms must introduce quadratic expansions into the deflection field;
3. The transverse shear strains are constant along the sides.

The following interpolation satisfies above conditions

$$w = \sum_{j=1}^{n} N_i w_i + \frac{1}{8} \sum_{k=1}^{n_s} \overline{N}_i l_{ij}(\theta_{s_i} - \theta_{s_j}) \qquad (6.131a)$$

$$\boldsymbol{\theta} = \sum_{j=1}^{n} N_i \boldsymbol{\theta}_i \qquad (6.131b)$$

where N_i are the shape functions of the original element (with $n = 3$ and $n = 4$ for the linear triangle and the bilinear quadrilateral, respectively), n_s is the number of nodes along the element sides, l_{ij} is the length of side ij, θ_{s_i} and θ_{s_j} are the rotations along the tangential directions to the k-th side with nodes i and j and $\overline{N}_i$ are quadratic shape functions vanishing at the corner nodes.

For 3-noded triangles,

$$\overline{\mathbf{N}} = [\overline{N}_1, \overline{N}_2, \overline{N}_3] = 4[L_1 L_2, L_2 L_3, L_3 L_1] \qquad (6.132a)$$

For 4-noded quadrilaterals,

$$\overline{\mathbf{N}} = [\overline{N}_1, \overline{N}_2, \overline{N}_3, \overline{N}_3] = \frac{1}{2} \left[(1 - \xi^2)(1 - \eta), (1 - \xi)(1 - \eta^2), \right.$$
$$\left. (1 - \xi^2)(1 + \eta), (1 - \xi)(1 + \eta^2) \right] \qquad (6.132b)$$

The tangential rotations θ_{s_i} are expressed in terms of the cartesian components at each node. For side ij

$$\theta_{s_i} = \theta_{x_i} \cos \phi_{ij} + \theta_{y_i} \sin \phi_{ij} \qquad (6.133)$$

where ϕ_{ij} is the angle that side ij forms with the x axis. Substituting Eq.(6.133) into (6.131a) yields the deflection field in terms of the deflection and the two cartesian rotations at the corner nodes.

For a 4-noded rectangle Eq.(6.131a) takes the following form for side 12 with $\eta = -1$

$$w = \frac{1}{2}(1 - \xi)w_1 + \frac{1}{2}(1 + \xi)w_2 + (1 - \xi^2)\frac{l_{12}}{8}(\theta_{s_1} - \theta_{s_2}) \qquad (6.134)$$

Eqs.(6.134) and (6.131b) guarantee a constant shear distribution along the side. Assuming a rectangular shape with $s = \xi = x$ gives for side 12

$$\gamma_{xz} = \frac{\partial w}{\partial x} - \theta_x = \frac{w_2 - w_1}{l_{12}} - \frac{\theta_{x_1} + \theta_{x_2}}{2} \qquad (6.135)$$

and $\gamma_{yz} = 0$. A similar result is obtained for the three other sides. Note the analogy with the process of Section 2.8.3 for Timoshenko beams.

The linked interpolation introduces bending moments in the equivalent nodal force vector for a distributed loading, similarly as for linked beam elements (Section 2.8.3).

Linear and quadratic plate triangles based on linked interpolations have been derived by Tessler [Te], Tessler and Hughes [TH], Xu *et al.* [XZZ], Lynn *et al.* [GL,LD], Aurichio and Lodavina [AL] and Taylor *et al.* [AT2,PT,TA]. Linked plate quadrilaterals have been proposed by Crisfield [Cr], Auricchio and Taylor [AT] and Zienkiewicz *et al.* [ZXZ+].

A different strategy for deriving Reissner-Mindlin plate elements based on a sort of linked interpolation is by starting from the analytical solution for a Timosnhenko beam [ZT2]. Different interesting shear-locking free 3-noded triangles and 4-noded quadrilaterals of this kind for thick/thin plate analysis have been proposed [ChC4,ChC5,Ib,SLC,SCLL,ZK].

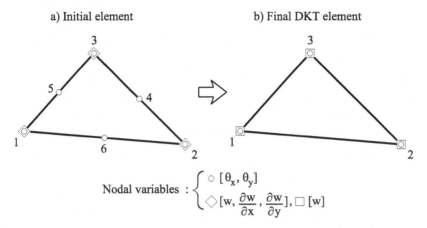

Fig. 6.26 3-noded DKT plate element

6.11 DISCRETE–KIRCHHOFF PLATE ELEMENTS

A family of thin plate elements can be derived by imposing the Kirchhoff constraints ($\gamma_{xz} = \gamma_{yz} = 0$) at selected points within a Reissner-Mindlin element so that the transverse shear strain energy is effectively zero.

The so-called Discrete-Kirchhoff (DK) plate elements were originally proposed by Wempner *et at.* [WOK], Stricklin *et al.* [SHTG] and Dhatt [Dh,Dh2] as early as in 1968-70 as a way for overcoming the C^1 continuity limitations of Kirchhoff plate theory. Several DK plate and shell elements were subsequently developed [BD5,6,BRI,Cr,Cr3,DMM], the most successful ones being the 3-noded DK triangle (DKT) detailed in the next section and the "semi-loof" shell element [IA,Ir2] (Section 7.12.5). The derivation of DK elements can be viewed as a particular class of assumed transverse shear strain techniques, leading to the vanishing of the transverse shear strain energy over the element. A state of the art on DK plate elements is presented in [BD3].

6.11.1 3-noded DK plate triangle (DKT)

The DKT element was initially developed by Striklin *et al.* [SHTG] and subsequently modified by Dhatt [Dh,Dh2] and Batoz *et al.* [Bat,BBH,BD5] who analyzed its performance extensively.

The starting point is the 6-noded Reissner-Mindlin triangle of Figure 6.26a under the following constraints:

1. The rotations θ_x and θ_y vary quadraticaly over the element (12 DOFs).

2. The deflection varies as a cubic Hermite polynomial along each side ij in terms of w_i, $\left(\frac{\partial w}{\partial s}\right)_i$, w_j, $\frac{\partial w}{\partial s_j}$. After transformation, this gives a total of 9 DOFs (w, $\frac{\partial w}{\partial x}$ and $\frac{\partial w}{\partial y}$ at each corner node).
3. A linear variation for the normal rotation θ_n is imposed along each side.
4. The conditions of zero transverse shear strain are imposed:
   a) at the corner nodes ($\gamma_{xz} = \gamma_{yz} = 0$)
   b) at the mid-side nodes ($\gamma_{sz} = \frac{\partial w}{\partial s} + \theta_s = 0$).
5. Only the contribution of the bending terms is taken into account for computing the element siffness matrix, i.e. $\mathbf{K}^{(e)} = \mathbf{K}_b^{(e)}$.

Conditions 3 and 4 impose twelve constraints which allow us to eliminate the slopes $\frac{\partial w}{\partial x}$ and $\frac{\partial w}{\partial y}$ at the corner nodes and the rotations θ_n and θ_s at the mid-side nodes. The rotation field is finally expressed in terms of the standard nine nodal DOFs as

$$\begin{Bmatrix} \theta_x \\ \theta_y \end{Bmatrix} = \begin{bmatrix} N_{x1}, N_{x2}, N_{x3}, N_{x4}, N_{x5}, N_{x6}, N_{x7}, N_{x8}, N_{x9} \\ N_{y1}, N_{y2}, N_{y3}, N_{y4}, N_{y5}, N_{y6}, N_{y7}, N_{y8}, N_{y9} \end{bmatrix} \mathbf{a}^{(e)} \qquad (6.136)$$

where

$$\mathbf{a}^{(e)} = [w_1, \theta_{x1}, \theta_{y1}, w_2, \theta_{x2}, \theta_{y2}, w_3, \theta_{x3}, \theta_{y3},]^T \qquad (6.137)$$

is the nodal displacement vector. The shape functions N_{x_i} and N_{y_i} are shown in Box 6.1.

Eq.(6.136) allows us to obtain the bending strain matrix $\mathbf{B}_b$ from which the element stiffness matrix can be exactly computed using a 3 point quadrature. The bending moments are sampled at the quadrature points.

Batoz [Bat,BD5] derived an explicit form for the stiffness matrix of the DKT element. First the matrix is evaluated in the local axes $\bar{x}$, $\bar{y}$ of Figure 6.27

$$\bar{\mathbf{K}}^{(e)} = \frac{1}{A^{(e)}} \mathbf{Q} \, \mathbf{S} \qquad (6.138)$$

Matrices $\mathbf{Q}$ and $\mathbf{S}$ are shown in Figure 6.28. The global stiffness matrix is obtained by

$$\mathbf{K}_{ij}^{(e)} = \mathbf{T}^T \, \bar{\mathbf{K}}_{ij}^{(e)} \, \mathbf{T} \quad \text{with} \quad \mathbf{T} = \begin{bmatrix} 1 & 0 & 0 \\ 0 & -\sin\alpha & \cos\alpha \\ 0 & -\cos\alpha & -\sin\alpha \end{bmatrix} \qquad (6.139)$$

The bending moments are obtained at any element point by

$$\hat{\sigma}_b = \frac{1}{2A^{(e)}} \hat{\mathbf{D}}_b \, \mathbf{N}_3 \, \mathbf{S} \, \mathbf{a}^{(e)} \qquad (6.140)$$

$$\left\{ \begin{matrix} \theta_x \\ \theta_y \end{matrix} \right\} = \sum_{i=1}^{3} \begin{bmatrix} N_{x_i} & N_{x_{i+1}} & N_{x_{i+2}} \\ N_{y_i} & N_{y_{i+1}} & N_{y_{i+2}} \end{bmatrix} \left\{ \begin{matrix} w_i \\ \theta_{x_i} \\ \theta_{y_i} \end{matrix} \right\}$$

$$N_{x_1} = 1.5(a_6 N_6 - a_5 N_5) \quad ; \quad N_{x_2} = b_5 N_5 + b_6 N_6 \qquad N_{x_3} = N_1 - c_5 N_5 - c_6 N_6$$
$$N_{x_4} = 1.5(a_4 N_4 - a_6 N_6) \quad ; \quad N_{x_5} = b_6 N_6 + b_4 N_4 \qquad N_{x_6} = N_2 - c_4 N_4 - c_6 N_6$$
$$N_{x_7} = 1.5(a_5 N_5 - a_4 N_4) \quad ; \quad N_{x_8} = b_4 N_4 + b_4 N_5 \qquad N_{x_9} = N_3 - c_4 N_4 - c_5 N_5$$
$$N_{y_1} = 1.5(d_6 N_6 - d_5 N_5) \quad ; \quad N_{y_2} = -N_1 + e_5 N_5 + e_6 N_6 \quad ; \quad N_{y_3} = -N_{x_2}$$
$$N_{y_4} = 1.5(d_4 N_4 - d_6 N_6) \quad ; \quad N_{y_5} = -N_2 + e_4 N_4 + e_6 N_6 \quad ; \quad N_{y_6} = -N_{x_5}$$
$$N_{y_7} = 1.5(d_5 N_5 - d_4 N_4) \quad ; \quad N_{y_8} = -N_3 + e_4 N_4 + e_5 N_5 \quad ; \quad N_{y_9} = -N_{x_8}$$

$$a_k = -\frac{x_{ij}}{l_{ij}^2} \quad ; \quad b_k = \frac{3}{4 l_{ij}^2} x_{ij} y_{ij} \qquad c_k = \left(\frac{1}{4} x_{ij}^2 - \frac{1}{2} y_{ij}^2 \right) / l_{ij}^2$$

$$d_k = -\frac{y_{ij}}{l_{ij}^2} \quad ; \quad l_k = \left(\frac{1}{4} y_{ij}^2 - \frac{1}{2} x_{ij}^2 \right) / l_{ij}^2 \quad ; \quad l_{ij}^2 = (x_{ij}^2 + y_{ij}^2)$$

$x_{ij} = x_i - x_j$, $y_{ij} = y_i - y_j$, $k = 4, 5, 6$ for sides $ij = 23, 31, 12$.
$N_i =$ shape functions of the 6-noded quadratic triangle (Appendix I)

Box 6.1 Shape functions for the DKT element

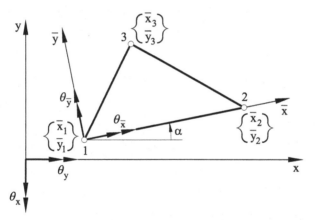

Fig. 6.27 Sign convention for rotations and local axes for computing the stiffness matrix for the DKT plate element. Note the vectorial definition of rotations

where $\mathbf{N}_3$ is the linear shape function matrix for the 3-noded triangle [Bat,BD5]. Figure 6.28 shows the explicit form for the local stiffness matrix of the DKT element.

Alternative derivation of the DKT element

We present an alternative procedure for deriving a DKT element with identical features to that of the previous section. The starting point is the TLQL element of Section 6.8.2. The transverse shear strain is made zero

$$\bar{\mathbf{K}}^{(e)} = \frac{1}{2\mathbf{A}^{(e)}}\,\mathbf{Q}\,\mathbf{S}$$

$$\mathbf{Q} = \frac{1}{24}\begin{bmatrix} (\hat{d}_{11}\ \mathbf{S}_{11}^T + \hat{d}_{12}\ \mathbf{S}_{21}^T)\mathbf{R} & | & (\hat{d}_{12}\ \mathbf{S}_{11}^T + \hat{d}_{22}\ \mathbf{S}_{21}^T)\mathbf{R} & | & \hat{d}_{33}\ \mathbf{S}_{31}^T\mathbf{R} \\ (\hat{d}_{11}\ \mathbf{S}_{12}^T + \hat{d}_{12}\ \mathbf{S}_{22}^T)\mathbf{R} & | & (\hat{d}_{12}\ \mathbf{S}_{12}^T + \hat{d}_{22}\ \mathbf{S}_{22}^T)\mathbf{R} & | & \hat{d}_{33}\ \mathbf{S}_{32}^T\mathbf{R} \\ (\hat{d}_{11}\ \mathbf{S}_{13}^T + \hat{d}_{22}\ \mathbf{S}_{23}^T)\mathbf{R} & | & (\hat{d}_{12}\ \mathbf{S}_{13}^T + \hat{d}_{12}\ \mathbf{S}_{23}^T)\mathbf{R} & | & \hat{d}_{33}\ \mathbf{S}_{33}^T\mathbf{R} \end{bmatrix} \quad ; \quad \mathbf{R} = \begin{bmatrix} 2 & 1 & 1 \\ 1 & 2 & 1 \\ 1 & 1 & 2 \end{bmatrix}$$

$$\mathbf{S} = \left[\begin{array}{ccc|ccc|ccc}
6\bar{y}_3a_6 & 0 & -4\bar{y}_3 & -6\bar{y}_3a_6 & 0 & -2\bar{y}_3 & 0 & 0 & 0 \\
-6\bar{y}_3a_6 & 0 & 2\bar{y}_3 & 6\bar{y}_3a_6 & 0 & 4\bar{y}_3 & 0 & 0 & 0 \\
\bar{y}_3p_5 & -\bar{y}_3q_5 & \bar{y}_3(2-r_5) & 6\bar{y}_3a_4 & 4\bar{y}_3b_4 & \bar{y}_3(r_4-2) & -\bar{y}_3(6a_4+p_5) & \bar{y}_3(4b_4-q_5) & \bar{y}_3(r_4-r_5) \\
\cdots & \cdots & \cdots & \cdots & \cdots & \cdots & \cdots & \cdots & \cdots \\
-\bar{x}_2t_5 & \bar{x}_{23}+\bar{x}_2r_5 & -\bar{x}_2q_5 & 0 & \bar{x}_3 & 0 & \bar{x}_2t_5 & \bar{x}_2(r_5-1) & \bar{x}_2q_5 \\
0 & \bar{x}_{23} & 0 & 6\bar{x}_2d_4 & \bar{x}_3+\bar{x}_2r_4 & -4\bar{x}_2b_4 & 6\bar{x}_2d_4 & \bar{x}_2(r_4-1) & -4\bar{x}_2b_4 \\
\bar{x}_{23}t_5 & \bar{x}_{23}(1-r_5) & \bar{x}_{23}q_5 & -6\bar{x}_3d_4 & \bar{x}_3(1-r_4) & 4\bar{x}_3b_4 & \begin{pmatrix}-\bar{x}_{23}t_5\\+6\bar{x}_3d_4\end{pmatrix} & \begin{pmatrix}-\bar{x}_{23}r_5-\bar{x}_3r_4\\-\bar{x}_2\end{pmatrix} & 4\bar{x}_3b_4+\bar{x}_{23}q_5 \\
\cdots & \cdots & \cdots & \cdots & \cdots & \cdots & \cdots & \cdots & \cdots \\
\begin{pmatrix}-6\bar{x}_3a_6\\-\bar{x}_2p_5\end{pmatrix} & (\bar{x}_2q_5+y_3) & \begin{pmatrix}-4\bar{x}_{23}\\+\bar{x}_2r_5\end{pmatrix} & 6\bar{x}_3a_6 & -\bar{y}_3 & 2\bar{x}_3 & \bar{x}_2p_5 & \bar{x}_2q_5 & (r_5-2)\bar{x}_2 \\
-6\bar{x}_{23}a_6 & \bar{y}_3 & 2\bar{x}_{23} & \begin{pmatrix}6\bar{x}_{23}a_6\\+6\bar{x}_2a_4\end{pmatrix} & -(\bar{y}_3+4\bar{x}_2b_4) & (-4\bar{x}_3+\bar{x}_2r_4) & -6\bar{x}_2a_4 & 4\bar{x}_2b_4 & (r_4-2)\bar{x}_2 \\
\begin{pmatrix}\bar{x}_{23}p_5\\+\bar{y}_3t_5\end{pmatrix} & \begin{pmatrix}-\bar{x}_{23}p_5\\+(1-r_5)\bar{y}_3\end{pmatrix} & \begin{pmatrix}(2-r_5)\bar{x}_{23}\\+\bar{y}_3q_5\end{pmatrix} & \begin{pmatrix}6\bar{x}_3q_4\\-6\bar{y}_3d_4\end{pmatrix} & \begin{pmatrix}(r_4-1)\bar{y}_3\\-4\bar{x}_3b_4\end{pmatrix} & \begin{pmatrix}(2-r_4)\bar{x}_3\\-4\bar{y}_3b_4\end{pmatrix} & \begin{pmatrix}-\bar{x}_{23}p_5\\+6\bar{x}_3a_4\\-(6d_4+t_5)\bar{y}_3\end{pmatrix} & \begin{pmatrix}-\bar{x}_{23}q_5\\-4\bar{x}_3b_4\\+(r_4-r_5)\bar{y}_3\end{pmatrix} & \begin{pmatrix}\bar{x}_{23}r_5\\-\bar{x}_3r_4\\+4\bar{x}_2\\+(q_5-4b_4)\bar{y}_3\end{pmatrix}
\end{array}\right]$$

$p_5 = -6x_3/l_{31}^2$, $t_5 = -6y_3/l_{31}^2$, $q_5 = 3x_3y_3/l_{31}^2$, $r_4 = 3y_{23}^2/l_{23}^2$, $r_5 = 3y_{31}^2/l_{31}^2$
$\bar{x}_i, \bar{y}_i$ = local coordinates of node i ; $\bar{x}_{ij}, a_4, a_6, d_4, b_4$ as in Box 6.1 changing $\bar{x}_i$, $\bar{y}_i$ for x_i, y_i.

Fig. 6.28 Local stiffness matrix for the DKT element

over the element by constraining the tangential transverse shear strains along the sides to a zero value. This allows us to eliminating the hierarchical tangential rotations at the mid-side nodes as

$$\gamma_\xi^{ij} = 0 \Rightarrow \Delta\theta_{s_k} = \frac{3}{2l^{ij}}(w_j - w_i) - \frac{3}{4}e_{ij}(\boldsymbol{\theta}_i + \boldsymbol{\theta}_j) \tag{6.141}$$

Substituting Eq.(6.141) into (6.120) gives the new rotation field as

$$\boldsymbol{\theta} = \sum_{i=1}^{3} \bar{\mathbf{N}}_i \mathbf{a}_i^{(e)} \quad ; \quad \mathbf{a}_i^{(e)} = \left\{ \begin{array}{c} w_i \\ \theta_{x_i} \\ \theta_{y_i} \end{array} \right\} \tag{6.142}$$

with

$$\bar{\mathbf{N}}_1 = \left[-\left(\frac{6L_1L_2}{l^{12}}\mathbf{e}_{12} + \frac{6L_1L_3}{l^{13}}\mathbf{e}_{23} \right), \ (L_1 - 3L_1L_2 - 3L_1L_3)\,\mathbf{I}_2 \right]$$

$$\bar{\mathbf{N}}_2 = \left[\left(\frac{6L_1L_2}{l^{12}}\mathbf{e}_{12} - \frac{6L_2L_3}{l^{23}}\mathbf{e}_{23} \right), \ (L_2 - 3L_1L_2 - 3L_2L_3)\,\mathbf{I}_2 \right]$$

$$\bar{\mathbf{N}}_3 = \left[\left(\frac{6L_2L_3}{l^{23}}\mathbf{e}_{23} + \frac{6L_1L_3}{l^{13}}\mathbf{e}_{13} \right), \ (L_3 - 3L_2L_3 - 3L_1L_3)\,\mathbf{I}_2 \right] \tag{6.143}$$

where $\mathbf{I}_2$ is the 2×2 unit matrix.

The bending strain matrix $\mathbf{B}_b$ is obtained by using the shape functions $\bar{N}_i$ instead of N_i in Eq.(6.32). The bending stiffness matrix is computed with a 3 point quadrature.

Figures 6.27 and 6.28 show the performance of the DKT element for analysis of a clamped plate under point and uniform loads.

6.11.2 DK plate quadrilaterals

The first DK quadrilaterals where developed by Dhatt and Venkatasubby [DV] and Baldwin *et. al.* [BRI]. Later Irons proposed several DK plate and shell quadrilaterals [IA], among which the semi-loof [Ir2] has enjoied big popularity (Section 7.12.5). Lyons [Ly], Crisfield [Cr,Cr2] and Batoz and Ben Tahar [BBt] have also derived DK plate quadrilaterals and same of these are displayed in Figure 6.29.

Derivation of a 4-noded DKQ element

The QLQL quadrilateral of Section 6.7.4 is starting point for deriving a 12 DOFs DK quadrilateral. The arguments follow as explained in the previous section for deriving a 3-noded DK triangle from the TLQL element. The

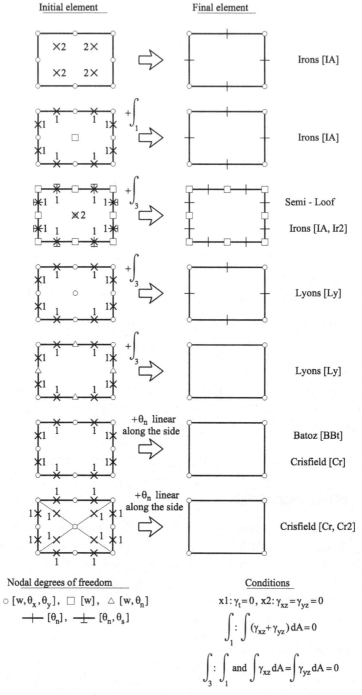

Fig. 6.29 Some DK plate quadrilaterals

constraint of vanishing transverse shear strain at the element sides (Figure 6.19) allows us to eliminate the four hierarchical tangential side rotations by an expression identical to Eq.(6.141). The final rotation field can be found substituting Eq.(6.141) into (6.105) to give

$$\boldsymbol{\theta} = \sum_{i=1}^{4} \bar{\mathbf{N}}_i \mathbf{a}_i \quad \text{with} \quad \mathbf{a}_i = [w_i, \, \theta_{x_i}, \, \theta_{y_i}]^T \tag{6.144}$$

and $\bar{\mathbf{N}}_i = [\bar{\mathbf{N}}_1, \bar{\mathbf{N}}_2, \bar{\mathbf{N}}_3, \bar{\mathbf{N}}_4]$ with

$$\bar{\mathbf{N}}_1 = \left[\left(-\tfrac{3}{4l^{12}} f(\xi)(1-\eta)\mathbf{e}_{12} - \tfrac{3}{4l^{14}} f(\eta)(1-\xi)\mathbf{e}_{14} \right), \left(N_1 - \tfrac{3}{8} f(\xi)(1-\eta) - \tfrac{3}{8} f(\eta)(1-\xi) \right) \mathbf{I}_2 \right]$$

$$\bar{\mathbf{N}}_2 = \left[\left(\tfrac{3}{4l^{12}} f(\xi)(1-\eta)\mathbf{e}_{12} - \tfrac{3}{4l^{23}} f(\eta)(1+\xi)\mathbf{e}_{23} \right), \left(N_2 - \tfrac{3}{8} f(\xi)(1-\eta) - \tfrac{3}{8} f(\eta)(1+\xi) \right) \mathbf{I}_2 \right]$$

$$\bar{\mathbf{N}}_3 = \left[\left(\tfrac{3}{4l^{43}} f(\xi)(1+\eta)\mathbf{e}_{43} + \tfrac{3}{4l^{23}} f(\eta)(1+\xi)\mathbf{e}_{23} \right), \left(N_3 - \tfrac{3}{8} f(\xi)(1+\eta) - \tfrac{3}{8} f(\eta)(1+\xi) \right) \mathbf{I}_2 \right]$$

$$\bar{\mathbf{N}}_4 = \left[\left(\tfrac{3}{4l^{14}} f(\eta)(1-\xi)\mathbf{e}_{14} - \tfrac{3}{4l^{43}} f(\eta)(1+\xi)\mathbf{e}_{43} \right), \left(N_4 - \tfrac{3}{8} f(\eta)(1-\xi) - \tfrac{3}{8} f(\xi)(1+\eta) \right) \mathbf{I}_2 \right]$$
$$\tag{6.145}$$

where $f(\xi) = 1 - \xi^2$, $f(\eta) = 1 - \eta^2$ and $\mathbf{I}_2$ is the 2×2 unit matrix.

Matrix $\mathbf{B}_{b_i}$ is obtained using $\bar{N}_i$ instead of N_i in Eq.(6.32). The bending stiffness matrix is computed with a 2×2 quadrature.

The element is basically identical to that derived by Batoz and Ben Tahar [BBt]. Examples of its good performance can be found in [BBt,BD5].

6.12 DK ELEMENTS ACCOUNTING FOR SHEAR DEFORMATION EFFECTS: DST ELEMENT

Batoz and Lardeur [BL] presented an extension of the DKT element accounting for shear deformation effects. The starting point is the displacement field of the DKT element (Section 6.11.1). The fourth constraints on the transverse shear strains are modified as

4a) At the corner nodes 1, 2, 3

$$\frac{\partial w}{\partial x} - \theta_x = \gamma_{xz} \quad , \quad \frac{\partial w}{\partial y} - \theta_y = \gamma_{yz} \tag{6.146a}$$

4b) At the mid-side points 4, 5, 6

$$\left(\frac{\partial w}{\partial s} \right)_k + (\theta_s) = \gamma_{sz} = -S_k \gamma_{xz} + C_k \gamma_{yz} \quad k = 4, 5, 6 \tag{6.146b}$$

with $S_k = \sin\phi_{ij}$, $C_k = \cos\phi_{ij}$ where ϕ_{ij} is the angle between the normal to the side ij and the x axis.

The transverse shear strains γ_{xz} and γ_{yz} are obtained from the equilibrium equations (Eqs.(5.23)) as

$$
\begin{aligned}
\gamma_{xz} &= \frac{Q_x}{\hat{D}_s} = -\frac{1}{\hat{D}_s}\left(\frac{\partial M_x}{\partial x} + \frac{\partial M_{xy}}{\partial y}\right)\\
\gamma_{yz} &= \frac{Q_y}{\hat{D}_s} = -\frac{1}{\hat{D}_s}\left(\frac{\partial M_y}{\partial y} + \frac{\partial M_{xy}}{\partial x}\right)
\end{aligned}
\tag{6.147}
$$

Substituting the bending moments in terms of the rotations into Eq.(6.147) via Eqs.(6.22) and (6.9) gives

$$
\begin{aligned}
\bar{\gamma}_{xz} &= -\frac{\hat{D}_b}{\hat{D}_s}\left[\frac{\partial^2\theta_x}{\partial x^2} + \nu\frac{\partial\theta_y}{\partial x\partial y} + \frac{1-\nu}{2}\left(\frac{\partial^2\theta_x}{\partial y^2} + \frac{\partial\theta_y}{\partial x\partial y}\right)\right]\\
\bar{\gamma}_{yz} &= -\frac{\hat{D}_b}{\hat{D}_s}\left[\frac{\partial^2\theta_y}{\partial y^2} + \nu\frac{\partial\theta_x}{\partial x\partial y} + \frac{1-\nu}{2}\left(\frac{\partial\theta_x}{\partial x\partial y} + \frac{\partial\theta_y}{\partial x^2}\right)\right]
\end{aligned}
\tag{6.148}
$$

In the above $\hat{D}_b = \frac{Et^3}{12(1-\nu^2)}$ and $\hat{D}_s = kGt$.

The nine conditions introduced by Eqs.(6.146), plus the three constraints imposed by prescribing a linear variation of θ_n along the sides, allow us to eliminating the same twelve DOFs, similarly as for the DKT element. The rotation field is finally expressed as

$$
\boldsymbol{\theta} = \mathbf{N}_\theta \mathbf{a}^{(e)} \quad,\quad \mathbf{a}^{(e)} = \left\{\begin{array}{c}\mathbf{a}_1^{(e)}\\\mathbf{a}_2^{(e)}\\\mathbf{a}_3^{(e)}\end{array}\right\} \quad,\quad \mathbf{a}_i^{(e)} = \left\{\begin{array}{c}w_i\\\theta_{x_i}\\\theta_{y_i}\end{array}\right\}
\tag{6.149}
$$

The bending and transverse shear matrices are deduced as

$$
\hat{\boldsymbol{\varepsilon}}_b = \hat{\mathbf{B}}_b \mathbf{a}^{(e)} \quad \text{and} \quad \hat{\boldsymbol{\varepsilon}}_s = \hat{\mathbf{B}}_s \mathbf{a}^{(e)}
\tag{6.150}
$$

Matrices $\mathbf{N}_\theta$, $\hat{\mathbf{B}}_b$ and $\hat{\mathbf{B}}_s$ can be found in [BD5,BL]. The stiffness matrix is obtained by adding the bending and transverse shear contributions, using $\hat{\mathbf{B}}_b$ and $\hat{\mathbf{B}}_s$ instead of $\mathbf{B}_b$ and $\mathbf{B}_s$ in Eqs. (6.39).

The DST element behaves well if the plate is moderately thin and transverse shear deformation effects are not very important. Its extension to shell analysis is possible but is not so simple [BD5].

Batoz and Katili [BK]derived an enhanced version of the DST element using constant bending modes plus incompatible energy orthogonal higher order bending modes. Katili [Ka2,3] proposed other simple triangular and quadrilateral elements for analysis of thick and thin plates obtained as an extension of the DKT and DKQ elements of previous sections.

6.13 PATCH TESTS FOR REISSNER-MINDLIN PLATE ELEMENTS

The absence of spurious modes in the element can be verified by computing the number of zero eigenvalues in excess of three in the stiffness matrix of a single unconstraint element. The procedure is the same as for assessing the mechanisms in plate elements induced by reduced and selective integration quadratures (Section 6.4.2). The absence of spurious modes and the good representation of rigid body motions can be assessed by imposing the following displacement field to the nodes laying on the boundary of a patch (patch test of type B, Section 4.9 and [On4])

$$w = c + ax + by \quad , \quad \theta_x = a \quad , \quad \theta_y = b \tag{6.151}$$

where a, b and c are arbitrary numbers. The numerical solution for the displacements at the internal node must agree with Eq.(6.151). Also the curvatures and the transverse shear strains must be zero everywhere in the patch.

A patch test can be devised for assessing the capability of Reissner-Mindlin plate elements to reproduce a pure bending state. The following displacement field is imposed to the boundary nodes of a type B patch

$$w = \frac{1}{2}(ax^2 + by^2 + cxy) \quad , \quad \theta_x = ax + \frac{1}{2}cxy \quad , \quad \theta_y = by + \frac{1}{2}cx \tag{6.152}$$

where a, b and c are arbitrary numbers. The solution for the internal nodal displacements must comply with Eq.(6.152). Also a constant curvature field $\hat{\varepsilon}_b = [a, b, c]^T$ and a zero transverse shear strain field $\hat{\varepsilon}_s = [0, 0]^T$ must be obtained everywhere in the patch.

A similar but less severe patch test is based in imposing the following displacement field

$$w = 0 \quad \text{at all patch nodes}$$

$$\theta_x = ax + \frac{1}{2}cy \, , \quad \theta_y = by + \frac{1}{2}cx \tag{6.153}$$

and $\alpha = 0$ or very small in order to make $\mathbf{K}_s$ a null matrix at each element. The numerical solution must satisfy the same conditions as for the previous test. These two tests are also applicable to DK plate elements.

The following displacement field can be imposed to the boundary nodes of a type B patch in order to verify the element ability to reproduce a constant transverse shear strain field and a zero curvature field

$$w = \frac{1}{2}(ax + by) \quad , \quad \theta_x = \frac{-a}{2} \quad , \quad \theta_y = \frac{-b}{2} \qquad (6.154)$$

Naturally, the displacements found at the internal nodes must be in accordance with Eqs.(6.154).

The stability of the element can be assessed via a patch test C (Section 4.9 of [On4]), i.e. by computing the rank of the stiffness matrix in element patches with the minimum number of DOFs prescribed. This test when applied to a single element is equivalent to the zero eigenvalue check mentioned above.

We note finally that the "count" inequalities of Eq.(6.61) are necessary conditions to be satisfied by plate elements based on assumed transverse shear strain fields. The kinematic patch tests described above are however mandatory to ascertain the good behaviour of the element.

6.14 EXAMPLES

6.14.1 Performance of some plate elements based on assumed transverse shear strain fields

The performance of some of the plate elements described in this chapter is studied first for the analysis of *a square plate* under a uniform load and a central point load. The analysis is performed for simple supported (SS) edges with hard support ($w = \theta_s = 0$) and fully clamped edges. The material properties are $E = 10.92$ and $\nu = 0.3$ (units in the International System). The problem is solved for two values of the a/t ratio of $\frac{a}{t} = 10$ (thick) and 100 (thin), where a and t are the plate side length and thickness, respectively. The elements studied are the QLLL (Section 6.7.1), QLQL (Section 6.7.4), TLQL (Section 6.8.2) and TLLL (Section 6.8.3). All problems were solved with uniform meshes in a quarter of the plate due to symmetry.

Results are compared with analytical and series values for the thin and thick cases when available [SG,SR,TW] or, alternatively, with FEM results for the deflection at the center of the middle plane obtained using a mesh of $40 \times 40 \times 6$ 8-noded hexahedra [On4] in a quarter of plate. *For the point load case the analytical value of the deflection under the load given by thick plate theory is infinity.* Hence, results for thick plates are compared at the mid-point along a central line in this case.

Figure 6.30 shows the convergence of the normalized values of the central and mid-side deflection for the QLLL and QLQL elements under point and uniform loads. Both elements converge well to the reference solutions for thick and thin situations.

Figure 6.31 shows similar results for the TLQL element for the two mesh orientations shown. Results are sensitive to the mesh orientation. This does not preclude the convergence of the element.

Figure 6.32 shows results for the same SS squate plate obtained with the TLLL element for the SS (hard) case under uniform loading. The results plotted are the central deflection and the central bending moment. The performance is good for thick and thin situations and results are less sensitive to the mesh orientation than for the TLQL element.

Figure 6.33 shows results for a *moderately thick clamped circular plate* using the TLQL, QLLL and QLQL elements. Graphics show the convergence of the deflection and the bending moment at the center toward the analytical solution (Appendix H and [BD5,TW]) and the distribution of the radial bending moment and the radial shear force along a radial line. All elements behave well. Note the accuracy of the QLLL element.

Table 6.3 shows results for the deflection and the bending moment at the center for thick and thin SS and clamped circular plates obtained with the TLLL element. The performance of this simple element is remarkably good with less than 10% error versus the reference solution for meshes of 64 elements in all cases.

Note that in all cases the deflection for thick plates converges to larger values than those given by thin plate theory. The reason is the increased capacity of a Reissner-Mindlin plate to deform under external loads due to the shear deformation terms in the PVW.

Table 6.4 shows the convergence of the central deflection at the two free corners in *cantilever skew plates* under uniform load and different skew angles. Results obtained with the TLLL element are shown as well as those obtained with the DRM triangle [ZTPO,ZT2] (an equivalent of the TLQL of Section 6.8.2) and the EL1 triangle of Figure 6.25b for comparison purposes. The performance of the TLLL element is again noticeable.

More evidence of the good performance of the QLLL element can be found in [BD,OZST]. Results for the TLQL, QLQL and QQQQ elements are presented in [OZST]. The good behaviour of the TLLL element in its original and enhanced versions is studied in [FO,OZF].

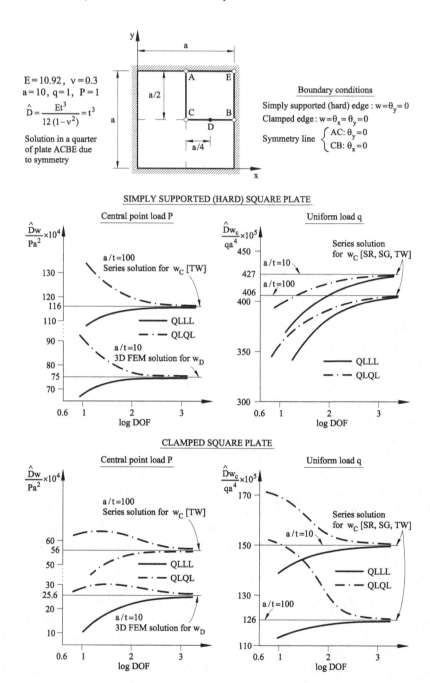

Fig. 6.30 QLLL and QLQL elements. Square plate with SS (hard) and clamped edges under central point load and uniform load. Convergence of the deflection at the plate center C (uniform load and point load for $a/t = 100$) and at a mid-side point D along a center line (point load for $a/t = 10$)

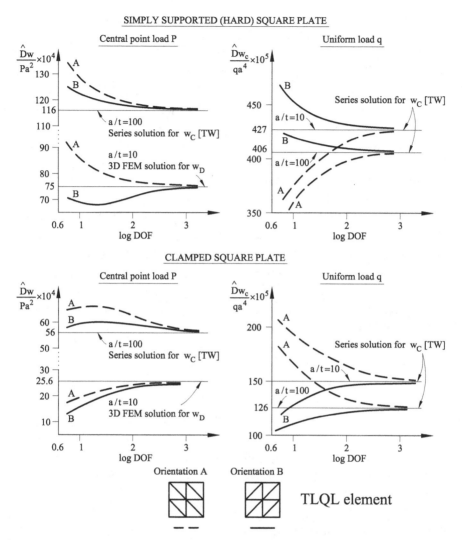

Fig. 6.31 TLQL element. Square plate with SS (hard) and clamped edges under central point load and uniform load. Convergence of the deflection at the plate center C (uniform load and point load for $a/t = 100$) and at a mid-side point D along a center line (point load for $a/t = 10$) for two different thicknesses and two mesh orientations

6.14.2 Simple supported plate under uniform load. Adaptive solution

The next example is the analysis of a square plate with soft SS conditions $(w = 0)$ and uniform loading using adaptive mesh refinement (AMR) following the two AMR strategies explained in [OB] and in Section 9.9.4 of [On4]. Figure 6.34 shows the geometry and material properties of the

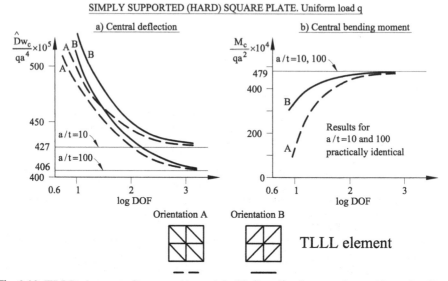

Fig. 6.32 TLLL element. Square plate with SS (hard) edges under uniform loading. Convergence of central deflection and central bending moment for two different thicknesses and two mesh orientations

		$R/t = 10$		$R/t = 100$	
Elem.	DOF	$w_c \times 10^2$	M_c^*	$w_c \times 10^5$	M_c^*

CIRCULAR PLATES. TLLL ELEMENT

Clamped circular plate

Elem.	DOF	$w_c \times 10^2$	M_c^*	$w_c \times 10^5$	M_c^*
4	12	3.7668	6.9599	3.6950	6.9606
16	54	2.2352	7.7235	2.1657	7.7431
64	220	1.7882	7.9751	1.7186	8.0146
144	498	1.7023	8.0243	1.6326	8.0726
225	780	1.6774	8.0391	1.6077	8.0904
Anal. (App. H)		1.6339	8.1250	1.5625	8.1250

SS circular plate (soft, $w = 0$)

Elem.	DOF	$w_c \times 10^2$	M_c^*	$w_c \times 10^5$	M_c^*
4	17	7.2815	1.6400	7.2096	1.6402
16	62	6.7248	1.9427	6.6553	1.9449
64	236	6.5191	2.0279	6.4495	2.0319
144	522	6.4763	2.0437	6.4066	2.0487
225	810	6.4637	2.0484	6.3939	2.0536
Anal. (App. H)		6.4416	2.0625	6.3702	2.0625

*Clamped plate: $M_c^* = M_c$; SS plate: $M_c^* = M_c \times 10$

Table 6.3 TLLL element. Convergence of central deflection and central bending moment ($M_{x_c} = M_{y_c} = M_c$) for thick and thin situations in clamped and SS (soft) circular plates under uniform load. Material properties as in Figure 6.33

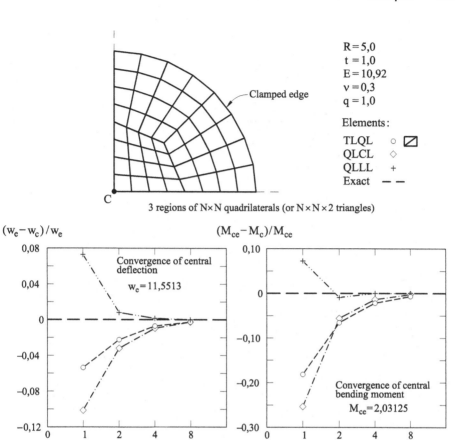

Fig. 6.33 Clamped circular plate ($t/R = 0.20$) under uniform load analyzed with TLQL, QLLL and QLQL elements. Convergence of central deflection w_c and central bending moment ($M_{x_c} = M_{y_c} = M_c$) towards exact values w_e and M_{ce}. Distribution of radial bending moment M_r and radial shear force Q_r along a radius. Exact results taken from Appendix H

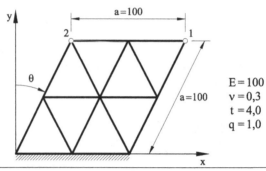

CANTILEVER SKEW PLATES. TLLL ELEMENT							
		20°		40°		60°	
Mesh	DOF	$\bar{w}_1$	$\bar{w}_2$	$\bar{w}_1$	$\bar{w}_2$	$\bar{w}_1$	$\bar{w}_2$
2×2	14	3.0093	2.6744	2.5112	1.3772	2.1821	0.4959
4×4	41	1.9701	1.7478	1.7478	0.8321	1.3882	0.2940
8×8	137	1.6032	1.1611	1.3950	0.6349	1.0800	0.2111
16×16	497	1.4802	1.0741	1.2610	0.5724	0.9521	0.1781
32×32	1889	1.4442	1.0517	1.2159	0.5554	0.9030	0.1672
TLQL/DRM [Xu,ZT2]	416	1.4269	1.0436	1.1789	0.5456	0.8435	0.1553
EL1 [Xu]	472	1.4237	1.0421	1.1722	0.5441	0.8314	0.1538

Table 6.4 TLLL element. Cantilever skew plates under uniform loading $f_z = q$. Convergence of the normalized deflection ($\bar{w}_i = Et^3 w_i / qa^4$) at the two free corners for different skew angles θ

plate and the initial unstructured mesh of 68 TLQL elements. The weak support conditions yield a zero torque M_{xy} along the supporting sides. This originates a boundary layer for the torque M_{xy} in the vecinity of these sides [OCK]. The percentage of admissible global error in the energy norm is $\eta = 5\%$. We have taken $p = 1$ and $d = 2$ in the expression for the mesh refinement parameter $\beta^{(e)}$ for each AMR strategy (Section 9.9.4 of [On4]). The computed value of the global error parameter for the initial mesh is $\xi_g = 2.7074$. The aim is to reduce this parameter to a value close to unity by adaptive mesh refinement.

Figure 6.34 shows the sequence of the refined meshes obtained using two mesh optimality criteria: one based on the equidistribution of the global error (Section 8.9.4.1 of [On4]) and the based in the equidistribution of the error density in the mesh (Section 8.9.4.2 of [On4]). The number of elements and the value of ξ_g for each mesh are also shown. The criterion based on the equidistribution of the error density leads to meshes with a

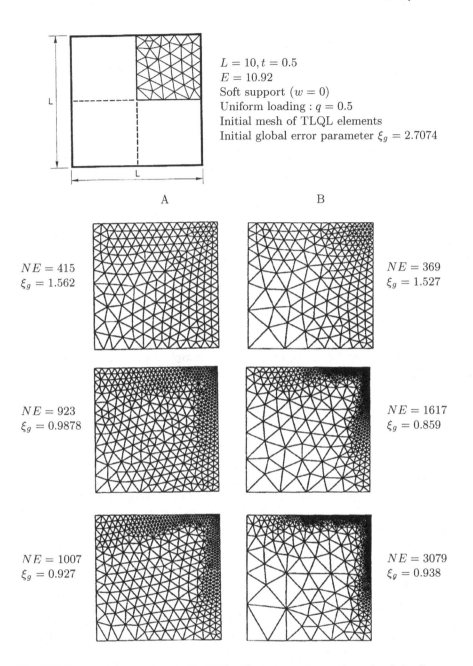

Fig. 6.34 Symmetric quadrant of a SS (soft) plate under uniform loading. Sequence of meshes obtained with mesh adaption strategies based on: Uniform distribution of the global error (column A); Uniform distribution of the error density (column B); Target global error: $\eta = 5\%$; NE = number of elements [On4,OCK]

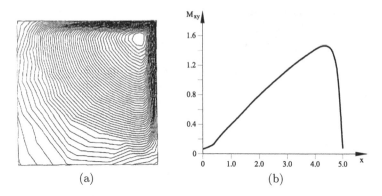

Fig. 6.35 Symmetric quadrant (5.0×5.0) of a SS (soft) plate under uniform loading. a) Isolines for the torque M_{xy}; b) Distribution of M_{xy} along the line $y = 2.5$. Results obtained with the mesh of 3079 TLQL elements of Figure 6.34

larger number of elements. Its advantage, however, is that it captures very well the boundary layer for the torque M_{xy} (Figure 6.35).

6.14.3 Effect of shear deformation in a plate simply supported at three edges under a line load acting on the free edge

The plate is displayed in Figure 6.36, where the geometry, the boundary conditions and the uniformly distributed loading along the edge is also shown. The edge loading induces high transverse shear stresses and strains near the edge, which play an important role in total deflection value for thick situations. This example was analyzed in [On] using 20 Reissner-Mindlin linear strip elements (Chapter 11) and in [Bl] using a mesh of 10×10 QLLL plate elements. In both cases half of the plate was studied due to symmetry, with a finer mesh in the vicinity of the free edge. Both experimental [AR] and analytical results (based on Kirchhoff thin plate theory) [TW] are available.

Figure 6.36a shows the distribution of the maximum σ_x stress in the center of the free edge for different line load intensities ($\frac{d}{a}$) and two values of the ratio $\frac{t}{a} = \frac{1}{20}$ and $\frac{1}{4}$. The Kirchhoff analytical solution is independent of the thickness and it differs in excess from the Reissner-Mindlin finite element solution as the load concentrates and the thickness increases. This difference raises to 60% for a thick plate ($t/a = 1/4$).

Figures 6.36b and c show the distribution of σ_x on the plate surface along several lines parallel to the loaded edge for a load intensity of $\frac{d}{a} = 0.01$ and two thickness ratios of $\frac{t}{a} = \frac{1}{20}$ and $\frac{1}{4}$, respectively. At

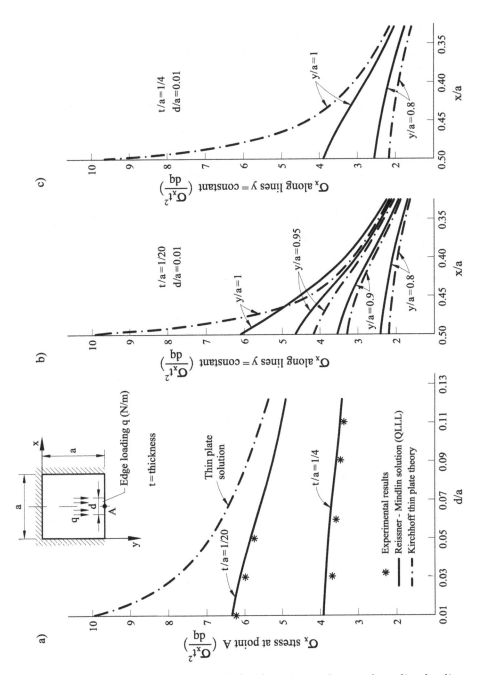

Fig. 6.36 Square plate simply supported (soft) at three edges under a line loading acting on the free edge. Distribution of σ_x using Kirchhoff thin plate theory (analytical) and 10×10 QLLL elements

the mid-point of the free edge $\left(\frac{y}{a} = 1, \frac{x}{a} = 0.5\right)$ the thin plate (Kirchhoff) solution yields a much larger value than the thick (Reissner-Mindlin) one, as expected.

The differences between the thin and thick plate solutions diminish as we move away from the edge and from the plate center. It is interesting that Reissner-Mindlin results are larger than the Kirchhoff values in these regions.

This example clearly shows the importance of accounting for the transverse shear deformation effects in certain situations.

6.15 EXTENDED ROTATION-FREE PLATE TRIANGLE WITH SHEAR DEFORMATION EFFECTS

The rotation-free basic plate triangle (BPT) of Section 5.8.2 can be extended to account for shear deformation effects. The enhanced element is termed BPT+1. The method is similar to that used to introduce shear deformation in the rotation-free CCB beam element (Section 2.10).

The curvatures in the BPT+1 element are expressed in terms of the deflection and the shear angles ϕ_x, ϕ_y substituting Eqs.(6.3) and (6.4) into (6.9) as

$$
\hat{\boldsymbol{\varepsilon}}_b = \left\{ \begin{array}{c} \dfrac{\partial^2 w}{\partial x^2} + \dfrac{\partial \phi}{\partial x} \\[2mm] \dfrac{\partial^2 w}{\partial y^2} + \dfrac{\partial \phi_y}{\partial y} \\[2mm] 2\dfrac{\partial^2 w}{\partial x \partial y} + \left(\dfrac{\partial \phi_x}{\partial y} + \dfrac{\partial \phi_y}{\partial x} \right) \end{array} \right\} = \boldsymbol{\kappa}_w + \boldsymbol{\kappa}_\phi \qquad (6.155)
$$

where

$$
\boldsymbol{\kappa}_w = \left[\frac{\partial^2 w}{\partial x^2}, \frac{\partial^2 w}{\partial y^2}, 2\frac{\partial^2 w}{\partial x \partial y} \right]^T \quad , \quad \boldsymbol{\kappa}_\phi = \left[\frac{\partial \phi_x}{\partial x}, \frac{\partial \phi_y}{\partial y}, \frac{\partial \phi_y}{\partial x} + \frac{\partial \phi_x}{\partial y} \right]^T
$$
$$(6.156)$$

are termed the geometrical curvature and the transverse shear curvature, respectively.

The transverse shear strains are expressed in terms of ϕ_x and ϕ_y as

$$
\hat{\boldsymbol{\varepsilon}}_s = \left[\frac{\partial w}{\partial x} - \theta_x, \frac{\partial w}{\partial y} - \theta_y \right]^T = -[\phi_x, \phi_y]^T = -\boldsymbol{\phi} \qquad (6.157)
$$

The PVW is written substituting Eqs.(6.155) and (6.157) into (6.26a) (assuming a distributed vertical load f_z to act only). This gives

$$\iint_A [[\delta\boldsymbol{\kappa}_w + \delta\boldsymbol{\kappa}_\phi]^T \hat{\boldsymbol{\sigma}}_b - \delta\boldsymbol{\phi}^T \hat{\boldsymbol{\sigma}}_s]dA - \iint_A \delta w f_z dA = 0 \qquad (6.158)$$

Eq.(6.158) is split in the following two independent equations

$$\iint_A \delta\boldsymbol{\kappa}_w^T \hat{\boldsymbol{\sigma}}_b dA - \iint_A \delta w f_z dA = 0 \qquad (6.159)$$

$$\iint_A [\delta\boldsymbol{\kappa}_\phi^T \hat{\boldsymbol{\sigma}}_b - \delta\boldsymbol{\phi}^T \hat{\boldsymbol{\sigma}}_s]dA = 0 \qquad (6.160)$$

The plate is discretized into 3-noded triangles. The deflection is linearly interpolated in terms of the nodal values within the pth triangle as

$$w = \sum_{i=1}^{i} N_i w_i = \mathbf{N}_w^p \mathbf{w}^p \quad , \quad \mathbf{N}_w^p = [N_1, N_2, N_3] \quad , \quad \mathbf{w}^p = [w_1, w_2, w_3]^T$$
$$(6.161)$$

In Eq.(6.161) N_i is the shape function for the 3-noded linear C° triangle with area A^i, i.e.

$$N_i = \frac{1}{2A^i}(a_i + b_{ix} + c_{iy}) \quad \text{with} \quad a_i = x_j y_k - x_i y_j \,,\ b_i = y_j - y_k \,,\ c_i = x_k - x_j$$
$$(6.162)$$

We assume an average value for the geometrical curvature field $\boldsymbol{\kappa}_w^p$ over a 3-noded triangle following the method described in Section 5.8.1, i.e.,

$$\boldsymbol{\kappa}_w = \boldsymbol{\kappa}_w^p = \frac{1}{A_p}\left[\iint_{A^p} \boldsymbol{\kappa}_w dA\right] = \frac{1}{A_p}\int_{\Gamma_p} \mathbf{T}\boldsymbol{\nabla}w d\Gamma \qquad (6.163)$$

where $\mathbf{T}$ contains the components of the outward unit normal to the boundary Γ_p of element p (Eq.(5.69a) and Figure 6.37).

Substituting the approximation (6.161) into (6.163) and following the procedure described in Section 5.8.2 gives

$$\boldsymbol{\kappa}_w^p = \mathbf{B}_w \bar{\mathbf{w}}^p \qquad (6.164)$$

where $\mathbf{B}_w$ coincides with matrix $\mathbf{B}^p$ of Eq.(5.84b) and $\bar{\mathbf{w}}^p$ is defined in Eq.(5.83).

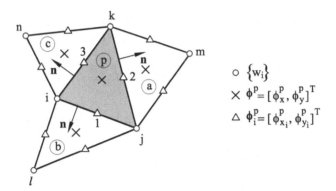

Fig. 6.37 Four-element patch for the pth BPT+1 triangle (ijk). Nodal deflections (w_i), transverse shear strains for each element in the patch (ϕ^p) and at the mid-side points of element $p(\phi_i^p)$

The transverse shear curvature within the central triangle p is computed as follows. First the shear angles are assumed to be constant within each triangle, i.e. for the pth triangle (Figure 6.37)

$$\phi = \phi^p \quad \text{with} \quad \phi^p = \left[\phi_x^p, \phi_y^p,\right]^T \tag{6.165}$$

An average value of the transverse shear curvature over the pth triangle is assumed as $\kappa_\phi = \kappa_\phi^p$ where

$$\kappa_\phi^p = \frac{1}{A_p} \iint_{A_p} \kappa_\phi dA = \frac{1}{A_p} \int_{\Gamma^p} \mathbf{T}\phi d\Gamma = \frac{1}{A_p} \sum_{j=i}^{3} l_j^p \mathbf{T}_j^p \phi_j^p = \mathbf{C}_p \hat{\phi}^p \tag{6.166}$$

where $\mathbf{C}_p$ is given in Eq.(5.80b), l_j^p are the element sides and

$$\hat{\phi}^p = \left\{ \begin{array}{c} \phi_1^p \\ \phi_2^p \\ \phi_3^p \end{array} \right\} \tag{6.167}$$

where ϕ_i^p are the values of the shear angles at the mid-side points of element p (Figure 6.37). As the shear angles are discontinuous at the element sides, vector ϕ^p is expressed in terms of the constant shear angles for the four elements in the patch via a simple averaging procedure as

$$\phi^p = \mathbf{M}_p \bar{\phi}^p \quad \text{with} \quad \bar{\phi}^p = \left\{ \begin{array}{c} \phi^p \\ \phi^a \\ \phi^b \\ \phi^c \end{array} \right\} \tag{6.168}$$

where $\mathbf{M}_p$ is given by Eq.(5.81b).

Substituting Eq.(6.168) into (6.166) gives

$$\boldsymbol{\kappa}_\phi^p = \mathbf{C}^p \mathbf{M}_p \bar{\boldsymbol{\phi}}^p = \mathbf{B}_\phi \bar{\boldsymbol{\phi}}^p \quad \text{with} \quad \mathbf{B}_\phi = \mathbf{C}_p \mathbf{M}_p \qquad (6.169)$$

Substituting Eqs.(6.169) and (6.164) into (6.155) gives

$$\hat{\boldsymbol{\varepsilon}}_b^p = \mathbf{B}_w \bar{\mathbf{w}}^p + \mathbf{B}_\phi \bar{\boldsymbol{\phi}}^p \qquad (6.170)$$

We assume a constant bending moment field $\hat{\boldsymbol{\sigma}}_b^p$ over the element expressed in terms of the nodal deflections and the nodal shear angles as

$$\hat{\boldsymbol{\sigma}}_b^p = \hat{\mathbf{D}}_b^p \hat{\boldsymbol{\varepsilon}}_b^p = \hat{\mathbf{D}}_b^p [\mathbf{B}_w \bar{\mathbf{w}}^p + \mathbf{B}_\phi \bar{\boldsymbol{\phi}}^p] \qquad (6.171)$$

where $\hat{\mathbf{D}}_b^p = \frac{1}{A_p} \iint_{A_p} \hat{\mathbf{D}}_b dA$ is an average bending constitutive matrix.

The constitutive equation for the shear forces is written as

$$\hat{\boldsymbol{\sigma}}_s = \hat{\mathbf{D}}_s \hat{\boldsymbol{\varepsilon}}_s = -\hat{\mathbf{D}}_s \boldsymbol{\phi} = -\hat{\mathbf{D}}_s \mathbf{M}_p \bar{\boldsymbol{\phi}}^p \qquad (6.172)$$

Substituting the assumed constant curvature fields $\boldsymbol{\kappa}_w^p$ and $\boldsymbol{\kappa}_\phi^p$ (via Eqs.(6.164) and (6.169)) in the PVW (Eqs.(6.159) and (6.160)) gives

$$\sum_p \left\{ [\delta \bar{\mathbf{w}}^p]^T \mathbf{B}_w^T \hat{\boldsymbol{\sigma}}_b^p A_p - \iint_{A_p} \delta w f_z dA \right\} = 0$$

$$\sum_p \left\{ [\delta \bar{\boldsymbol{\phi}}^p]^T \mathbf{B}_\phi^T \hat{\boldsymbol{\sigma}}_b^p A_p - \int_{A_p} [\delta \bar{\boldsymbol{\phi}}^p]^T \hat{\boldsymbol{\sigma}}_s dA \right\} = 0 \qquad (6.173)$$

Substituting Eqs.(6.171) and (6.172) into (6.174) gives

$$\sum_p [\delta \bar{\mathbf{w}}^p]^T \left\{ [\mathbf{B}_w^T \hat{\mathbf{D}}_b \mathbf{B}_w \bar{\mathbf{w}}^p + \mathbf{B}_w^T \hat{\mathbf{D}}_b \mathbf{B}_\phi \bar{\boldsymbol{\phi}}^p] A_p - \iint_{A_p} \bar{\mathbf{N}}^p \bar{\mathbf{f}} dA \right\} = 0$$

$$\sum_p [\delta \bar{\boldsymbol{\phi}}^p]^T \left\{ \mathbf{B}_\phi^T \hat{\mathbf{D}}_b \mathbf{B}_w A_p \bar{\mathbf{w}}^p + \left(\mathbf{B}_\phi^T \hat{\mathbf{D}}_b \mathbf{B}_\phi A_p + \iint_{A_p} \mathbf{M}_p^T \hat{\mathbf{D}}_s \mathbf{M}_p dA \right) \bar{\boldsymbol{\phi}}^p \right\} = 0 \qquad (6.174)$$

The sums in Eqs.(6.173) and (6.174) extend over all the triangles in the mesh and

$$\bar{\mathbf{N}}^p = [N_1, N_2, N_3, 0, 0, 0]^T \quad , \quad \bar{\mathbf{f}} = f_z [1, 1, 1, 0, 0, 0]^T \qquad (6.175)$$

Simplification of the virtual DOFs yields the system of equations

$$\mathbf{K}_w \mathbf{w} + \mathbf{K}_{w\phi} \boldsymbol{\phi} = \mathbf{f}_w$$

$$\mathbf{K}_{w\phi}^T \mathbf{w} + \mathbf{K}_\phi \boldsymbol{\phi} = 0 \qquad (6.176)$$

with

$$\mathbf{w} = [w_1, w_2, \cdots, w_N]^T \quad , \quad \boldsymbol{\phi} = [(\boldsymbol{\phi}^1)^T, (\boldsymbol{\phi}^2)^T, \cdots, (\boldsymbol{\phi}^{n_e})^T]^T \quad (6.177)$$

where N is the total number of nodes and n_e is the total number of elements in the mesh.

The global matrices and vectors are assembled from the element contributions in the usual manner with

$$\mathbf{K}_w^p = \mathbf{B}_w^T \hat{\mathbf{D}}_b^p \mathbf{B}_w A^p \quad , \quad \mathbf{K}_{w\phi}^p = \mathbf{B}_w^T \hat{\mathbf{D}}_b^p \mathbf{B}_\phi A^p$$

$$\mathbf{K}_\phi^p = \left[\mathbf{B}_\phi^T \hat{\mathbf{D}}_b^p \mathbf{B}_\phi + \mathbf{M}_p^T \hat{\mathbf{D}}_s \mathbf{M}_p \right] A^p \tag{6.178}$$

In the derivation of the second term in $\mathbf{K}_\phi^p$ we have assumed a constant transverse shear constitutive matrix $\hat{\mathbf{D}}_s$ over the element.

For a uniformly distributed load $f_z = q$ and

$$\mathbf{f}_w^p = \frac{q A^p}{3} [1, 1, 1, 0, 0, 0]^T \tag{6.179}$$

Note the similarity of Eqs.(6.176) with Eq.(2.125) obtained for an analogous rotation-free beam element.

6.15.1 Iterative solution scheme

The following iterative scheme is recommended for solving Eqs.(6.176).

Step 1. Compute the nodal deflections $\mathbf{w}^1$

$$\mathbf{K}_w \mathbf{w}^1 = \mathbf{f}_w \rightarrow \mathbf{w}^1 \quad \text{(Kirchhoff thin plate solution)} \tag{6.180}$$

Step 2. Compute $\boldsymbol{\phi}^i$, $i \geq 1$

$$\mathbf{K}_\phi \boldsymbol{\phi}^i = \mathbf{f}_\phi - \mathbf{K}_{w\phi}^T \mathbf{w}^i \quad \rightarrow \boldsymbol{\phi}^i \tag{6.181}$$

Step 3. Compute $\mathbf{w}^i$, $i > 1$

$$\mathbf{K}_w \mathbf{w}^i = \mathbf{f}_w - \mathbf{K}_{w\phi} \boldsymbol{\phi}^{i-1} \quad \rightarrow \bar{\mathbf{w}}^i \tag{6.182}$$

Return to step 2.

Convergence of the above iterative scheme is quite fast (2–4 iterations), even for thick plates [OZ2,3].

6.15.2 Boundary conditions

Free edge

A BPT+1 element with a side along a free boundary edge has one of the triangles belonging to the patch missing. This is taken into account by ignoring the contributions of this triangle when computing $\mathbf{B}_w$ and $\mathbf{B}_\phi$. This can be implemented by a simple modification of matrix $\mathbf{M}_p$ in Eq.(6.168), similarly as described for the BPT element in Section 5.8.2.1.

Simple supported (SS) edge

Soft (SS) condition $\left(w = 0, \frac{\partial w}{\partial s}\right) = 0$: The two edge deflections w_i and w_j are prescribed to a zero value when solving the system of equations.

Hard (SS) condition $(w = 0, \theta_s = 0)$. The condition $w_i = 0$ at the SS nodes is imposed when solving the global system of equations.

The condition $\theta_s = 0$ is introduced using a penalty method as follows. The expression for the virtual work is extended by adding to Eq.(6.158) the term $\alpha l_i^p \delta\theta_s\theta_s$, where αl_i^p is the length of the SS side and α is a large number that plays the role of a penalty parameter. The tangential rotation is expressed in terms of the nodal deflection and shear angle variables as

$$\theta_s = \theta_x C_\alpha + \theta_y S_\alpha = \left(\frac{\partial w}{\partial x} + \phi_x\right) C_\alpha + \left(\frac{\partial w}{\partial y} + \phi_y\right) S_\alpha$$

$$= \frac{1}{2A_p} \sum_{i=1}^{3}(b_i C_\alpha + c_i S_\alpha)w_i + \phi_x^p C_\alpha + \phi_y^p S_\alpha \qquad (6.183)$$

where b_i, c_i are given in Eq.(6.162), $C_\alpha = \cos\alpha$, $S_\alpha = \sin\alpha$ and α is the angle that the SS side forms with the x axis. Eq.(6.183) is grouped as

$$\theta_s = \mathbf{P}_w \bar{\mathbf{w}}^p + \mathbf{P}_\phi \bar{\phi}^p \qquad (6.184)$$

with
$$\mathbf{P}_w = [P_1, P_2, P_3, 0, 0, 0] \quad , \quad P_i = \frac{1}{2A_p}(b_i C_\alpha + c_i S_\alpha)$$

$$\mathbf{P}_\phi = [C_\alpha, S_\alpha, 0, 0, 0, 0, 0, 0] \qquad (6.185)$$

Introducing the approximation (6.184) in the penalty term gives

$$\alpha l_i^p \delta\theta_s\theta_s = \alpha l_i^p [\mathbf{P}_w^T(\delta\bar{\mathbf{w}}^p)^T + \mathbf{P}_\phi^T(\delta\bar{\phi}^p)^T][\mathbf{P}_w \bar{\mathbf{w}}^p + \mathbf{P}_\phi \bar{\phi}^p] \qquad (6.186)$$

The contribution of Eq.(6.186) to the element matrices is

$$\bar{\mathbf{K}}_w^p = \mathbf{K}_w^p + \alpha l_i^p \mathbf{P}_w^T \mathbf{P}_w \quad , \quad \bar{\mathbf{K}}_{w\phi}^p = \mathbf{K}_{w\phi}^p + \alpha l_i^p \mathbf{P}_\phi^T \mathbf{P}_\phi$$

$$\bar{\mathbf{K}}_\phi^p = \mathbf{K}_\phi^p + \alpha l_i^p \mathbf{P}_\phi^T \mathbf{P}_\phi \qquad (6.187)$$

where $\mathbf{K}_w^p$, $\mathbf{K}_{w\phi}^p$ and $\mathbf{K}_\phi^p$ are given in Eq.(6.178).

Clearly as the parameter α increases, the condition $\theta_s = 0$ is better satisfied. In practice a value $\alpha = 10^5 Et^3$ suffices to obtain accurate results.

Clamped edge ($\theta = 0$)

The contribution of a clamped edge to the curvatures κ_w^p and κ_ϕ^p is neglected. This implies modifying matrix $\mathbf{M}_p$ as explained in Section 5.8.2.1. In addition, the edge deflections are prescribed to a zero value.

Symmetry edge ($\theta_n = 0$)

The condition $\theta_n = 0$ at a symmetry edge is imposed via a penalty approach similar as for the SS (hard) case. The PVW is enhanced with the term

$$\alpha l_i^p \delta\theta_n \theta_n = \alpha l_i^p [\mathbf{P}_w^T (\delta\bar{\mathbf{w}}^p)^T + \mathbf{P}_\phi^T (\delta\bar{\phi}^p)^T][\mathbf{P}_w \bar{\mathbf{w}}^p + \mathbf{P}_\phi \bar{\phi}^p] \qquad (6.188)$$

where $\mathbf{P}_w$ and $\mathbf{P}_\phi$ are given in Eq.(6.185) with $P_i = \frac{1}{2A_p}[b_i S_\alpha - c_i C_\alpha]$. The element matrices are modified as in Eq.(6.187).

6.15.3 Examples of performance of the BPT+1 element

The efficiency and accuracy of the BPT+1 element has been tested in the analysis of simply supported (SS, soft) square plates of side L and circular plates of diameter $2L$ under uniformly distributed loading and clamped or square plates under a central point load. The study was performed for different thicknesses ranging from to $t/L = 10^{-3}$ (very thin plate) to $t/L = 0.1$ (thick plate) and several uniform meshes with increasing number of elements *in all the plate surface*.

Results of the study using the iterative scheme of Section 6.15.1 are presented in Figures 6.38–6.40. Each figure shows:

(a) The convergence of the vertical deflection and the shear angles with the number of iterations measured as

$$L_2^w = \left[\sum_{j=1}^N \frac{(w_j^i - w_j^{i-1})^2}{(w_j^i)^2}\right]^{1/2} \quad , \quad L_2^\phi = \left[\sum_{j=1}^{n_e} \frac{[\phi_j^i - \phi_j^{i-1}]^T(\phi_j^i - \phi_j^{i-1})}{[\phi_j^i]^T\phi_j^i}\right]^{1/2}$$
$$(6.189)$$

where N is the number of nodes in the mesh and an upper index denotes the iteration number. A value of $w_j^0 = 0$ and $\boldsymbol{\phi}_j^0 = \mathbf{0}$ has been taken. The iterative scheme stops when $L_2^w < 10^{-3}$.

(b) The ratio between the normalized central deflection and a reference solution for four values of $t/L = 10^{-3}, 10^{-2}, 5 \times 10^{-2}$ and 10^{-1} for each of the meshes considered. For cases when an analytical solution is not available we have used as the reference solution the 3D FEM results for the deflection at the center of the midle plane using a mesh of $40 \times 40 \times 6$ 8-noded hexahedra [On4] in a quarter of plate.

For the clamped plate under central point load the analytical value for the deflection under the load given by thick plate theory is infinity (See Section 6.16). Hence, results for the deflection for thick plates ($t/L = 0.05$ and $t/L = 0.10$) are compared at the mid-poind D along a central line in this case (Figure 6.39).

(c) The distribution of the bending moment M_x and the shear force Q_x along the central line for the thick case ($t/L = 0.10$) for each of the five meshes considered. The isovalues of M_x and Q_x over a quarter of the plate are also shown for the finer mesh.

The following conclusions are drawn from the examples:

- The BPT+ element reproduces accurately the expected results for the deflection field for thin and thick plates.
- A converged solution for the deflection field was obtained in a maximum of four iterations for all cases considered. The convergence of the shear angles is slightly slower than for the deflection field.
- The distribution of the bending moments and the shear forces was good and in accordance with the expected results for the thick case.
- For thin plates the distribution of bending moments is also very good. The distribution of the shear forces deteriorates slightly if computed via Eq.(6.172) as shear angles tend to zero as the plate is thinner. It is more appropriate in theses cases to compute the shear forces from the bending moments via Eq.(5.23).
- Similar good results were obtained for the "hard" SS condition obtained by prescribing $\bar{\phi}_{s_i} = 0$ at the support nodes (Table 6.5) [OZ2].

A similar 3-noded plate triangle element (called BPT+) was derived in [OZ2] using a continuous linear interpolation for the shear angles within each element. The BPT+1 element has a slighly superior behaviour for capturing shear force jumps [OZ2,3].

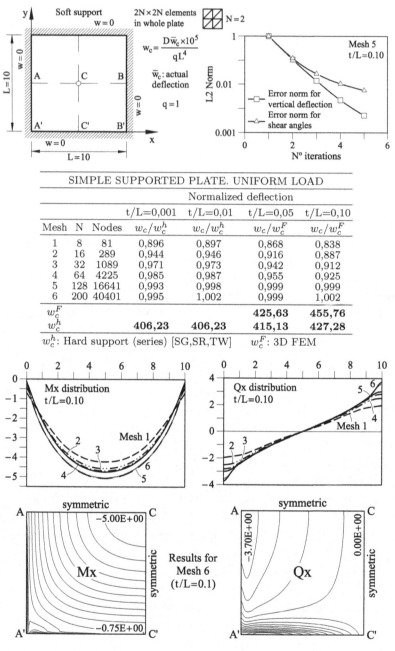

Fig. 6.38 BPT+1 element. SS (Soft) square plate under uniform load ($q = 1$). Table shows convergence of normalized central deflection w_c for different thicknesses. Upper curves show convergence of vertical deflection and shear angles for a thick plate with number of iterations. Lower diagrams show the distribution of M_x and Q_x along the central line and their contours

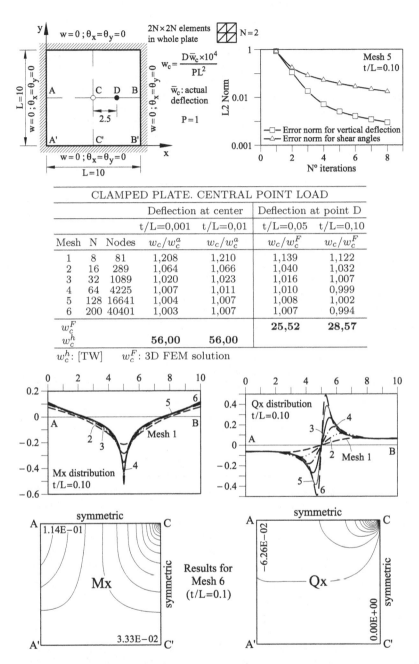

Fig. 6.39 BPT+1 element. Clampled square plate under central point load ($P = 1$). Table shows convergence of normalized central deflection w_c for different thicknesses. Upper curves show convergence of vertical deflection and shear angles for a thick plate with number of iterations. Lower diagrams show the distribution of M_x and Q_x along the central line and their contours

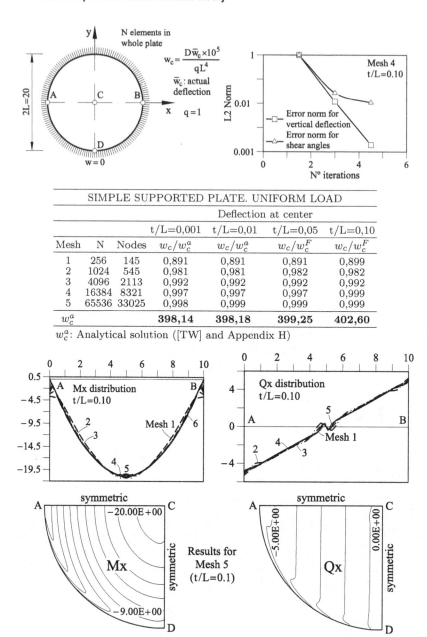

Fig. 6.40 BPT+1 element. SS (soft) circular plate under uniform load ($q = 1$). Table shows convergence of normalized central deflection w_c for different thicknesses. Upper curves show convergence of vertical deflection and shear angles for a thick plate with the number of iterations. Lower diagrams show the distribution of M_x and Q_x along the central line and their contours

SS (hard) square thick plate. Uniform load, $t/L = 0.10$					
	Mesh 1	Mesh 2	Mesh 3	Mesh 4	Mesh 5	Mesh 6
w_c	381,73	404,59	416,31	421,98	424,65	425,59
w_c/w_c^h	0,893	0,947	0,974	0,988	0,994	0,996

w_c^h (Series solution): 427.28 [SG,SR,TW]
Convergence achieved in a maximum of 4 iterations for each mesh

Table 6.5 BPT+1 element. SS square thick plate (hard support) under uniform load. Normalized central deflection values for $t/L = 0.10$

Mesh,	$t/L = 10^{-1}$; $w_c \times 10^{-2}$		$t/L = 10^{-3}$; $w_c \times 10^{-7}$	
M	hard support	soft support	hard support	soft support
2	4.2626	4.6085	4.0389	4.2397
4	4.2720	4.5629	4.0607	4.1297
8	4.2727	4.5883	4.0637	4.0928
16	4.2728	4.6077	4.0643	4.0773
32	4.2728	4.6144	4.0644	4.0700
Series [SG,SR,TW]	4.2728		4.0624	

Table 6.6 BPT+1 element. Normalized center displacement (w_c) for SS plate under uniform load for two t/L ratios; $E = 10.92$, $\nu = 0.3$, $L = 10$, $f_z = 1$.

6.16 LIMITATIONS OF THIN PLATE THEORY

Kirchhoff plate element have obviously limitations to reproduce the behaviour of thick plates. Reissner-Mindlin plate elements are superior as they are applicable to thick and thin situation. Table 6.6 taken from [ZT2] shows some results obtained for SS plates with hard and soft support conditions under uniform load for two different side length/thickness ratios. Numerical results were obtained with the 4-noded QLLL element (Section 6.7.1) but the conclusions hold for any Reissner-Mindlin plate element. The effect of the hard and soft SS conditions is clearly shown in the table. The central deflection for the thick plate always converges to *larger values* than those given by thin plate theory. Also for the SS case the soft support condition yields larger deflection values, as expected. These differences can be more pronounced in different plate configurations.

Figure 6.41 shows results for a simple supported rombic plate for $L/t = 100$ and 1000 analyzed with the triangular element of Zienkiewicz and Lefebre (Figure 6.25). There is an exact solution for this problem using Kirchhoff plate theory [Mo4]. Results for the "thicker" case $(L/t = 100)$

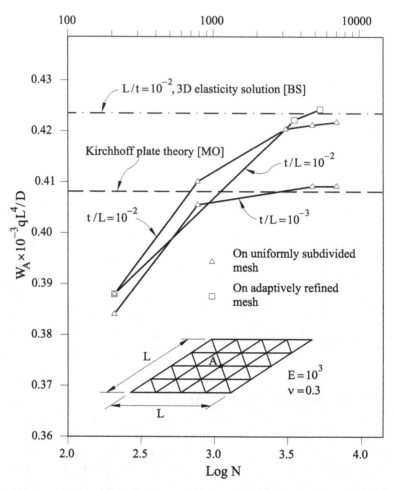

Fig. 6.41 Skew rombic plate (30°) with soft SS conditions under uniform load. Convergence of central deflection with the number of DOFs. The problem was solved with the triangle of Zienkiewicz and Lefebre [ZL2] (Section 6.9)

converge to a central deflection value which is 4% larger than the exact thin plate solution.

Babŭska and Scapolla [BS] studied this problem using 3D elasticity theory with "soft" support conditions which approximate better the physical problem. 3D results for $L/t = 100$ are close to the thick plate solution. This confirms the superiority of Reissner-Mindlin theory for this problem.

We note finally an important difference between thick and thin plates when point loads are involved. In the thin case the deflection w remains finite under the load. However, as mentioned earlier, transverse shear deformation effects *lead to an infinite displacement under the load for thick plates* (as indeed 3D elasticity theory predicts). In finite element approximations one always predicts a finite displacement at point load locations with the magnitude increasing without limit as a mesh is refined near the point load. Thus, *it is meaningless to compare the deflections at point load locations*. It is recommended to compare the total strain energy for such situations [ZT2].

6.17 CONCLUSIONS

Reissner-Mindlin plate theory is the basis for systematically deriving C^o continuous plate elements which include transverse shear strain effects. Practically, the only drawback of Reissner-Mindlin plate elements is the appearance of shear locking for thin plate situations. The reduced integration of the transverse shear stiffness terms is a simple and efficient procedure for eliminating shear locking, although it can introduce spurious mechanisms which can pollute the solution in some cases. The assumed transverse shear strain technique is a more consistent approach for designing robust locking-free plate elements and some of these elements have been presented in detail.

We note the good performance of the low order QLLL quadrilateral and the TLLL triangle based on assumed transverse shear strain fields.

Reissner-Mindlin plate elements are also the starting point for deriving Discrete-Kirchhoff thin plate elements, while keeping all the features of the C^o continuous formulation.

The merits of the Reissner-Mindlin plate formulation will show clearer when dealing with composite plates and shells in the subsequent chapters.

7

COMPOSITE LAMINATED PLATES

7.1 INTRODUCTION

High-performance and lightweight characteristics of composite materials have motivated a wide range of applications of these materials in aeronautic and space vehicles and in naval, automotive and civil structures, among other fields [Bar2,BC,Mar,TH4,Ts,Wh]. The development of cost-effective and reliable composite laminated structures requires advanced stress analysis and failure prediction methods.

Differently from the plate bending theories for homogeneous material studied in previous chapters, the middle plane points in a composite laminated plate *can move in the plane direction* and this originates in-plane elongations and axial forces. This changes the standard bending mode of a homogeneous plate to a mixed mode where bending and axial effects are coupled. Axial effects will be also called *membrane* effects in the following.

The study of composite laminated plates in this chapter is introductory to that of shells to be treated in the following chapters. Composite laminated structures typically combine flat and curved surfaces meeting at different angles and shell theory is needed for their study.

By assuring that the plate surface is flat we can analyze in detail many interesting features of the kinematics, the constitutive relationships and the discretized equations for composite laminated plates. Understanding these concepts will easy the way for studying shell structures.

The standard first-order theory presented in the next section follows precisely the assumption of Reissner-Mindlin plate theory studied in the previous chapter. Kirchhoff thin plate theory is also applicable to composite laminated plates and it is the basis of the so-called "classical laminated plate theory (CLPT)" [Red2]. However, Reissner-Mindlin theory is generally more precise, as the effect of transverse shear deformation is important

E. Oñate, *Structural Analysis with the Finite Element Method. Linear Statics:*
Volume 2: Beams, Plates and Shells, Lecture Notes on Numerical Methods
in Engineering and Sciences, DOI 10.1007/978-1-4020-8743-1_7,
© International Center for Numerical Methods in Engineering (CIMNE), 2013

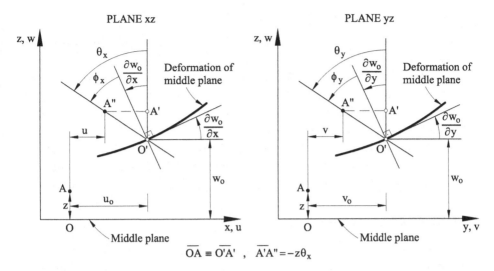

Fig. 7.1 Displacements and rotations in a plate with bending and in-plane effects

in these structures [Wh2]. The case is similar to the analysis of laminated beams using Timoshenko theory (Chapter 3).

Higher order composite laminate plate theories, such as the layer-wise and refined zigzag theories, are studied in the second part of the chapter and the procedures to develop accurate plate finite elements are described. The chapter concludes with an overview of failure theories in composite laminated plates.

7.2 BASIC THEORY

7.2.1 Displacement field

The displacement field of Reissner-Mindlin plate theory Eq.(6.1) is extended by introducing the horizontal (in-plane) displacements $u_0(x, y)$ and $v_0(x, y)$ as (Figure 7.1)

$$\begin{aligned}
u(x, y, z) &= u_0(x, y) - z\theta_x(x, y) \\
v(x, y, z) &= v_0(x, y) - z\theta_y(x, y) \\
w(x, y, z) &= w_0(x, y)
\end{aligned} \tag{7.1}$$

where $(\cdot)_0$ denotes the displacements of the middle plane. For convenience, this plane is taken as the reference plane $(z = 0)$ for defining the plate kinematics, as in the previous two chapters.

More sophisticated laminated plate theories can be derived by assuming higher order expansions for the displacement field in the thickness direction, or even a linear distribution of the kinematic variables within each layer [BOM,Red,Red2]. A selection of these theories is presented in Sections 7.6–7.8. A simplification of Eqs.(7.1) can also be introduced by accepting that the normals remain orthogonal to the middle plane after deformation. This yields the classical laminated plate (CLT) theory following Kirchhoff assumptions [Bar2,Red2,TH]. Reissner-Mindlin plate theory is a good compromise between CLT and higher order theories for practical analysis of composite laminated plates and shells.

7.2.2 Strain and generalized strain vectors

The strain vector is obtained from the expressions of 3D elasticity theory (Chapter 8 of [On4]), neglecting the thickness strain ε_z, as it does not contribute to the internal work due to the plane stress assumption ($\sigma_z = 0$). Substituting Eqs.(7.1) into the strain expressions of 3D elasticity gives

$$
\boldsymbol{\varepsilon} = \left\{ \begin{array}{c} \varepsilon_x \\ \varepsilon_y \\ \gamma_{xy} \\ \cdots \\ \gamma_{xz} \\ \gamma_{yz} \end{array} \right\} = \left\{ \begin{array}{c} \dfrac{\partial u}{\partial x} \\[4pt] \dfrac{\partial v}{\partial y} \\[4pt] \dfrac{\partial u}{\partial y} + \dfrac{\partial v}{\partial x} \\ \cdots \\ \dfrac{\partial u}{\partial z} + \dfrac{\partial w}{\partial x} \\[4pt] \dfrac{\partial v}{\partial z} + \dfrac{\partial w}{\partial y} \end{array} \right\} = \left\{ \begin{array}{c} \dfrac{\partial u_0}{\partial x} \\[4pt] \dfrac{\partial v_0}{\partial y} \\[4pt] \dfrac{\partial u_0}{\partial y} + \dfrac{\partial v_0}{\partial x} \\ \cdots \\ 0 \\ 0 \end{array} \right\} + \left\{ \begin{array}{c} -z\dfrac{\partial \theta_x}{\partial x} \\[4pt] -z\dfrac{\partial \theta_y}{\partial y} \\[4pt] -z\left(\dfrac{\partial \theta_x}{\partial y} + \dfrac{\partial \theta_y}{\partial x}\right) \\ \cdots \\ \dfrac{\partial w_0}{\partial x} - \theta_x \\[4pt] \dfrac{\partial w_0}{\partial y} - \theta_y \end{array} \right\} =
$$

$$
= \left\{ \begin{array}{c} \boldsymbol{\varepsilon}_m \\ \mathbf{0} \end{array} \right\} + \left\{ \begin{array}{c} -z\hat{\boldsymbol{\varepsilon}}_b \\ \hat{\boldsymbol{\varepsilon}}_s \end{array} \right\} = \mathbf{S}\hat{\boldsymbol{\varepsilon}} \tag{7.2}
$$

where

$$
\hat{\boldsymbol{\varepsilon}}_m = \left[\dfrac{\partial u_0}{\partial x}, \dfrac{\partial v_0}{\partial y}, \left(\dfrac{\partial u_0}{\partial y} + \dfrac{\partial v_0}{\partial x}\right) \right]^T
$$

$$
\hat{\boldsymbol{\varepsilon}} = \left\{ \begin{array}{c} \hat{\boldsymbol{\varepsilon}}_m \\ \hat{\boldsymbol{\varepsilon}}_b \\ \hat{\boldsymbol{\varepsilon}}_s \end{array} \right\} \quad \text{with} \quad \hat{\boldsymbol{\varepsilon}}_b = \left[\dfrac{\partial \theta_x}{\partial x}, \dfrac{\partial \theta_y}{\partial y}, \left(\dfrac{\partial \theta_x}{\partial y} + \dfrac{\partial \theta_y}{\partial x}\right) \right]^T \tag{7.3}
$$

$$
\hat{\boldsymbol{\varepsilon}}_s = \left[\dfrac{\partial w_0}{\partial x} - \theta_x, \dfrac{\partial w_0}{\partial y} - \theta_y \right]^T
$$

are the generalized (resultant) strain vectors due to membrane, bending and transverse shear deformation effects, respectively and

$$\mathbf{S} = \begin{bmatrix} \mathbf{I}_3 & -z\mathbf{I}_3 & \mathbf{0}_2 \\ \mathbf{0}_3^T & \mathbf{0}_2 & \mathbf{I}_2 \end{bmatrix} \quad \text{with} \quad \mathbf{0}_2 = \begin{bmatrix} 0 & 0 \\ 0 & 0 \end{bmatrix}, \quad \mathbf{0}_3 = \begin{bmatrix} 0 & 0 \\ 0 & 0 \\ 0 & 0 \end{bmatrix} \quad (7.4)$$

and $\mathbf{I}_n$ is the $n \times n$ unit matrix.

From Eq.(7.2) we deduce the following useful expressions

$$\boldsymbol{\varepsilon} = \left\{ \begin{matrix} \boldsymbol{\varepsilon}_p \\ \boldsymbol{\varepsilon}_s \end{matrix} \right\} \tag{7.5a}$$

with

$$\boldsymbol{\varepsilon}_p = \left\{ \begin{matrix} \varepsilon_x \\ \varepsilon_y \\ \gamma_{xy} \end{matrix} \right\} = \hat{\boldsymbol{\varepsilon}}_m - z\hat{\boldsymbol{\varepsilon}}_b \quad , \quad \boldsymbol{\varepsilon}_s = \left\{ \begin{matrix} \gamma_{xz} \\ \gamma_{yz} \end{matrix} \right\} = \hat{\boldsymbol{\varepsilon}}_s \tag{7.5b}$$

where vectors $\boldsymbol{\varepsilon}_p$ and $\boldsymbol{\varepsilon}_s$ respectively contain the in-plane strains and the transverse shear strains at a point in the thickness direction. The transformation of the generalized strains $\hat{\boldsymbol{\varepsilon}}$ to the actual strains $\boldsymbol{\varepsilon}$ at a point is performed via matrix $\mathbf{S}$. This transformation is typical for beams, plates and shells.

7.2.3 Stress-strain relationship

Let us consider a composite laminated plate formed by a piling of n_l orthotropic layers (also called "laminas" or "plies") with orthotropy axes L, T, z and isotropy in the L axis (i.e. in the plane Tz). The L axis defines the direction of the longitudinal fibers which are embedded in a matrix of polymeric or metallic material (Figure 7.2). We also assume that

- each layer k is defined by the planes $z = z_k$ and $z = z_{k+1}$ with $z_k \leq z \leq z_{k+1}$,
- the orthotropy directions L and T can vary for each layer and are represented by the angle β_i between the global axis x and the directions L_i of the ith layer (Figure 7.2),
- each layer satisfies the plane stress assumption ($\sigma_z = 0$) and the plane anisotropy condition (the z axis is the orthotropy axis for all the layers),
- the displacement field is continuous between the layers and satisfies Eq.(7.1).

The names *layer*, *lamina* and *ply* will be indistinguishable used in the following sections.

Note that the orthotropy axes were denoted as $1, 2, z$ in Figure 6.3 for convenience.

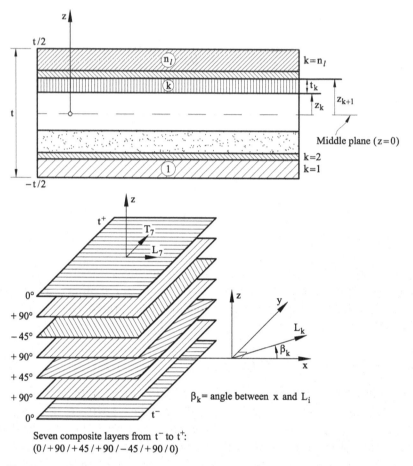

Fig. 7.2 Definition of layers (laminas) in a composite laminated plate

The above assumptions allow us to write the relationships between the in-plane stresses σ_x, σ_y, τ_{xy} and the transverse shear strains τ_{xz}, τ_{yz} with their conjugate strains *for each layer k* as

$$\sigma_p = \begin{Bmatrix} \sigma_x \\ \sigma_y \\ \tau_{xy} \end{Bmatrix} = \mathbf{D}_p \begin{Bmatrix} \varepsilon_x \\ \varepsilon_y \\ \gamma_{xy} \end{Bmatrix} + \begin{Bmatrix} \sigma_x^0 \\ \sigma_y^0 \\ \tau_{xy}^0 \end{Bmatrix} = \mathbf{D}_p \varepsilon_p + \sigma_p^0 \qquad (7.6a)$$

$$\sigma_s = \begin{Bmatrix} \tau_{xz} \\ \tau_{yz} \end{Bmatrix} = \mathbf{D}_s \begin{Bmatrix} \gamma_{xz} \\ \gamma_{yz} \end{Bmatrix} + \begin{Bmatrix} \tau_{xz}^0 \\ \tau_{yz}^0 \end{Bmatrix} = \mathbf{D}_s \varepsilon_s + \sigma_s^0 \qquad (7.6b)$$

or

$$\sigma = \begin{Bmatrix} \sigma_p \\ \sigma_s \end{Bmatrix} = \begin{bmatrix} \mathbf{D}_p & \mathbf{0} \\ \mathbf{0} & \mathbf{D}_s \end{bmatrix} \begin{Bmatrix} \varepsilon_p \\ \varepsilon_s \end{Bmatrix} + \begin{Bmatrix} \sigma_p^0 \\ \sigma_s^0 \end{Bmatrix} = \mathbf{D}\varepsilon + \sigma^0 \qquad (7.7)$$

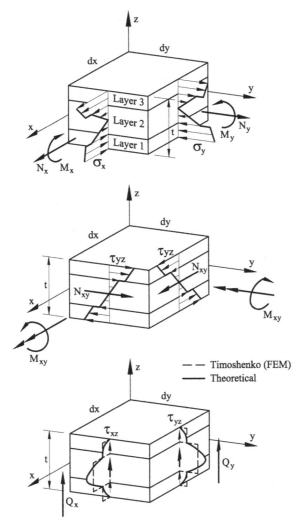

Fig. 7.3 Stresses and resultant stresses in 3-layered (symmetric) plate element

Figure 7.3 shows the sign convention for the stresses.

In Eqs.(7.6) $\boldsymbol{\sigma}_p^0$ and $\boldsymbol{\sigma}_s^0$ are the vectors of initial in-plane and transverse shear stress vectors, respectively. Initial stresses due to initial thermal strains have been included in vectors $\boldsymbol{\sigma}_p^0$ and $\boldsymbol{\sigma}_s^0$ for convenience (see Eqs.(7.14)).

The constitutive matrices $\mathbf{D}_p$ and $\mathbf{D}_s$ are symmetrical and their terms are a function of five independent material parameters and the angle β_k. These matrices can be obtained as follows.

The constitutive relationships in the orthotropy axes L, T, z are

$$\boldsymbol{\sigma}_1 = \mathbf{D}_1 \boldsymbol{\varepsilon}_1 + \boldsymbol{\sigma}_1^0 \quad , \quad \boldsymbol{\sigma}_2 = \mathbf{D}_2 \boldsymbol{\varepsilon}_2 + + \boldsymbol{\sigma}_2^0 \qquad (7.8a)$$

or

$$\boldsymbol{\sigma}_I = \left\{ \begin{matrix} \boldsymbol{\sigma}_1 \\ \boldsymbol{\sigma}_2 \end{matrix} \right\} = \begin{bmatrix} \mathbf{D}_1 & \mathbf{0} \\ \mathbf{0} & \mathbf{D}_2 \end{bmatrix} \left\{ \begin{matrix} \boldsymbol{\varepsilon}_1 \\ \boldsymbol{\varepsilon}_2 \end{matrix} \right\} + \left\{ \begin{matrix} \boldsymbol{\sigma}_1^0 \\ \boldsymbol{\sigma}_2^0 \end{matrix} \right\} = \mathbf{D}_I \boldsymbol{\varepsilon}_I + \boldsymbol{\sigma}_I^0 \qquad (7.8b)$$

$$\begin{aligned}
\boldsymbol{\sigma}_1 &= [\sigma_L, \sigma_T, \tau_{LT}]^T \quad , & \boldsymbol{\sigma}_1^0 &= [\sigma_L^0, \sigma_T^0, \tau_{LT}^0]^T \\
\boldsymbol{\sigma}_2 &= [\tau_{Lz}, \tau_{Tz}]^T \quad , & \boldsymbol{\sigma}_2^0 &= [\tau_{Lz}^0, \tau_{Tz}^0]^T \\
\boldsymbol{\varepsilon}_1 &= [\varepsilon_L, \varepsilon_T, \gamma_{LT}]^T \quad , & \boldsymbol{\varepsilon}_2 &= [\gamma_{Lz}, \gamma_{Tz}]^T
\end{aligned} \qquad (7.9)$$

$$\mathbf{D}_1 = \begin{bmatrix} D_{LL} & D_{LT} & 0 \\ & D_{TT} & 0 \\ \text{Sym.} & & G_{LT} \end{bmatrix} \quad , \quad \mathbf{D}_2 = \begin{bmatrix} G_{Lz} & 0 \\ 0 & G_{Tz} \end{bmatrix}$$

with

$$\begin{aligned}
D_{LL} &= \frac{E_L}{a} \quad , \quad D_{TT} = \frac{E_T}{a} \quad , \quad a = 1 - \nu_{LT}\nu_{TL} \\
D_{LT} &= \frac{E_T \nu_{LT}}{a} = \frac{E_L \nu_{TL}}{a}
\end{aligned} \qquad (7.10)$$

The five independent coefficients can be either

$$D_{LL}, D_{LT}, D_{TT}, G_{Lz} = G_{LT}, G_{Tz} \qquad (7.11a)$$

or

$$E_L, E_T, \nu_{LT} \left(\text{or } \nu_{TL} = \frac{E_T}{E_L}\nu_{LT} \right)$$

$$G_{Lz} = G_{LT}, G_{Tz} \left(\text{or } \nu_{Tz} \text{ and } G_{Tz} = \frac{E_T}{2(1 + \nu_{Tz})} \right) \qquad (7.11b)$$

The constitutive parameters in matrices $\mathbf{D}_1$ and $\mathbf{D}_2$ can be measured experimentally, as described in Example 7.1.

The relationship between matrices $\mathbf{D}_p$ and $\mathbf{D}_s$ of Eqs.(7.6) and $\mathbf{D}_1$ and $\mathbf{D}_2$ is obtained by (Section 6.2.4)

$$\underset{3 \times 3}{\mathbf{D}_p = \mathbf{T}_1^T \mathbf{D}_1 \mathbf{T}_1} \quad , \quad \underset{2 \times 2}{\mathbf{D}_s = \mathbf{T}_2^T \mathbf{D}_2 \mathbf{T}_2} \qquad (7.12)$$

The following relationships hold

$$\begin{aligned}
\boldsymbol{\varepsilon}_1 = \mathbf{T}_1 \boldsymbol{\varepsilon} \quad , \quad \boldsymbol{\sigma}_p = \mathbf{T}_1^T \boldsymbol{\sigma}_1 \quad , \quad \boldsymbol{\varepsilon}_2 = \mathbf{T}_2 \boldsymbol{\varepsilon}_s \quad , \quad \boldsymbol{\sigma}_s = \mathbf{T}_2^T \boldsymbol{\sigma}_2 \\
\boldsymbol{\sigma}_p^0 = \mathbf{T}_1^T \boldsymbol{\sigma}_1^0 \quad , \quad \boldsymbol{\sigma}_s^0 = \mathbf{T}_2^T \boldsymbol{\sigma}_2^0
\end{aligned} \qquad (7.13)$$

where $\mathbf{T}_1$ and $\mathbf{T}_2$ are the transformation matrices of Eq.(6.15b) with $C = \cos\beta_k$ and $S = \sin\beta_k$ and β_k is defined as shown in Figure 7.2.

It is easy to verify that both $\mathbf{D}_p$ and $\mathbf{D}_s$ are symmetrical.

The in-plane stresses $\boldsymbol{\sigma}_p$ are *discontinuous* in the thickness direction when the material properties vary between layers. This also occurred for composite plane beams (Figure 3.5).

For initial stresses due to thermal effects we can obtain

$$\boldsymbol{\sigma}_p^0 = -\mathbf{T}_1^T\mathbf{D}_1\left[\alpha_L\Delta T, \alpha_T\Delta T, 0\right]^T \quad , \quad \boldsymbol{\sigma}_s^0 = \mathbf{0} \qquad (7.14a)$$

where α_L and α_T are the thermal expansion coefficients in the L and T directions, respectively and ΔT is the temperature increment. For homogeneous isotropic material

$$\boldsymbol{\sigma}_p^0 = -\frac{E\alpha\Delta T}{(1-\nu^2)}\left[1, 1, 0\right] \quad , \quad \boldsymbol{\sigma}_s^0 = \mathbf{0} \qquad (7.14b)$$

The examples presented next describe the experiments needed for characterizing the properties of a composite laminated material [BC].

Example 7.1: Experimental measurement of constitutive parameters for a composite laminated material.

Consider a thin sheet of composite material made of unidirectional fibers embedded in a matrix. The constitutive parameters can be measured experimentally by simple tests where the composite is subjected to a known stress field. For this purpose it is more convenient to write the constitutive laws in compliance form as

$$\boldsymbol{\varepsilon}_1 = \begin{Bmatrix} \varepsilon_L \\ \varepsilon_T \\ \gamma_{LT} \end{Bmatrix} = \begin{bmatrix} \frac{1}{E_L} & -\frac{\nu_{LT}}{E_T} & 0 \\ -\frac{\nu_{TL}}{E_L} & \frac{1}{E_T} & 0 \\ 0 & 0 & \frac{1}{G_{LT}} \end{bmatrix} \begin{Bmatrix} \sigma_L \\ \sigma_T \\ \tau_{LT} \end{Bmatrix} = \mathbf{C}_1\boldsymbol{\sigma}_1 \qquad (7.15a)$$

$$\boldsymbol{\varepsilon}_2 = \begin{Bmatrix} \gamma_{Lz} \\ \gamma_{Tz} \end{Bmatrix} = \begin{bmatrix} \frac{1}{G_{Lz}} & 0 \\ 0 & \frac{1}{G_{Tz}} \end{bmatrix} \begin{Bmatrix} \tau_{LT} \\ \tau_{Tz} \end{Bmatrix} = \mathbf{C}_2\boldsymbol{\sigma}_2 \qquad (7.15b)$$

It is easy to verify that $\mathbf{D}_1 = \mathbf{C}_1^{-1}$ and $\mathbf{D}_2 = \mathbf{C}_2^{-1}$. $\mathbf{D}_1$ and $\mathbf{D}_2$ are given in Eq.(7.9)

We consider first the measurement of the constitutive parameters in $\mathbf{D}_1$ by performing simple tests with the applied stresses acting on the $L - T$ plane.

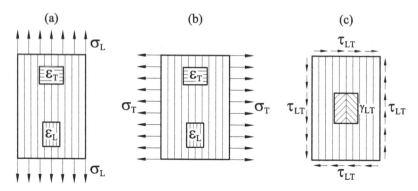

Fig. 7.4 Three simple tests for the determination of the in-plane constitutive parameters in a composite [BC]

In the first test the composite is subjected to a prescribed stress in the fiber direction only σ_L, i.e. $\sigma_T = \tau_{Lz} = 0$ (Figure 7.7a).

The first equation of Eq.(7.15a) now reduces to $\varepsilon_L = \frac{\sigma_L}{E_L}$. The strain in the fiber direction, ε_L, can be measured by means of a strain gauge and the Young modulus in the fiber direction is then computed as $E_L = \frac{\sigma_L}{\varepsilon_L}$.

The strain in the direction transverse to the fiber can also be similarly measured. The Poisons's ratio ν_{LT} is deduced from the second equation of (7.15a) as $\nu_{LT} = -E_L \frac{\varepsilon_T}{\sigma_L}$.

Consider now a second similar test where the composite is subjected to a know stress in the direction transverse to the fiber σ_T, i.e. $\sigma_L = \tau_{Lz} = 0$ (Figure 7.4b). Using the same approach as before, a measurement of the transverse strain ε_T will yield the Young modulus in the direction transverse to the fiber E_T. An additional measurement of the strain in the fiber direction, ε_L, will yield ν_{TL}. The symmetry of the compliance matrix can be verified experimentally by checking that the measured quantities satisfy $\frac{\nu_{LT}}{E_L} = \frac{\nu_{TL}}{E_T}$ (within the expected measurement errors).

Finally in the last test the composite is subjected to a known shear stress ν_{LT} only, i.e. $\sigma_L = \sigma_T = 0$. A measurement of the shear strain γ_{LT} then allows the evaluation of the shear modulus from the third equation of (7.15b) as $G_{LT} = \frac{\tau_{LT}}{\gamma_{LT}}$. This completes the definition of the compliance matrix $\mathbf{C}_1$.

The constitutive matrix $\mathbf{D}_1$ is obtained by inverting $\mathbf{C}_1$. Table 7.1 lists the constitutive parameters for different composite materials, as well as the volume fraction of fibers in the composite and the density of the material. Other material properties for composites are given in Appendix A and in [BC,Mar,PP4,Ts].

As for the determination of the parameters in the transverse shear constitutive matrix $\mathbf{D}_2$ (or $\mathbf{C}_2$) we note that $G_{Lz} = G_{LT}$. The G_{Tz} parameter can be found by subjecting the composite to a prescribed transverse shear stress field as shown in Figure 7.5. The measurement of the shear strain γ_{Tz} yields

Material	V_f	E_L [GPa]	E_T [GPa]	ν_{LT}	G_{LT} [GPa]	Density [kg/m³]
Graphite/Epoxy (T300/5208)	0.70	180	10	0.28	7.0	1600
Graphite/Epoxy (AS/3501)	0.66	138	9	0.30	7.0	1600
Boron/Epoxy (T300/5208)	0.50	204	18	0.23	5.6	2000
Scotchply (1002)	0.45	39	8	0.26	4.0	1800
Kevlar 49	0.60	76	5.5	0.34	2.3	1460

Table 7.1 Material parameters for the in-plane constitutive matrix $\mathbf{D}_1$ for different composite materials [BC]

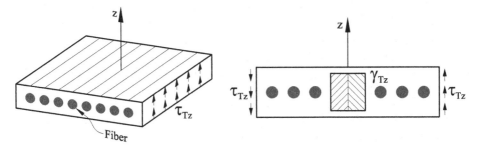

Fig. 7.5 Shear test for determining the transverse shear modulus G_{Tz} [BC]

the sought transverse shear modulus as $G_{Tz} = \frac{\tau_{Tz}}{\gamma_{Tz}}$. This type of test is more difficult to perform that the in-plane tests previously described.

Example 7.2: Effect of the fiber orientation in the constitutive matrix for a composite [BC].

Let us study in some detail the effect of the fiber orientation in the constitutive matrix for a composite made of Graphite/Epoxy T300/5028. For the sake of conciseness we will consider the in-plane constitutive matrix $\mathbf{D}_p$ only. The properties of this material are listed in Table 7.1 of previous example. The terms of $\mathbf{D}_p$ can be expressed as

$$\mathbf{D}_p = \begin{bmatrix} D_{11}^p & D_{12}^p & D_{13}^p \\ D_{12}^p & D_{22}^p & D_{23}^p \\ D_{13}^p & D_{23}^p & D_{33}^p \end{bmatrix} \tag{7.16}$$

Figure 7.6 shows the stiffness components D_{11}^p and D_{22}^p as a function of the lamina angle, β. Note the rapid decline of the stiffness coefficient, D_{11}^p, when

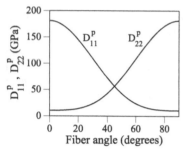
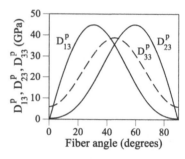

Fig. 7.6 Generalized stiffness coefficients, D_{11}^p, D_{22}^p, D_{13}^p, D_{23}^p and D_{33}^p for a Graphite/Epoxy T300/5028 composite, as a function of the fiber angle β [BC]

the lamina angle moves away from zero degrees. This sharp decline is due to the high directionality of the lamina stiffness properties. The shearing stiffness component, D_{33}^p, shown in Figure 7.6, drastically increases when the lamina angle is 45 degrees. This can be explained as follows: a state of pure shear is equivalent to stresses in tension and compression acting at 45 and 135 degree angles, respectively. Theses stresses are now aligned with the fiber direction, which presents very high stiffness.

The coupling stiffness terms, D_{13}^p and D_{23}^p, do not vanish and their variation in terms of the lamina angle β is shown in Figure 7.6. These terms show a coupling between extension and shearing of the lamina. In contrast, the constitutive matrix $\mathbf{D}_1$, expressed in the fiber aligned triad, has vanishing terms in the corresponding entries. Indeed, when the loading is applied along the fiber direction, which is the intersection of two planes of symmetry, the response of the system must be symmetric, precluding extension-shear coupling. When the loading is no longer aligned with the intersection of the two planes of symmetry, a coupled response of the lamina is intuitively expected.

___

7.2.4 Resultant stresses and generalized constitutive matrix

The resultant stress vectors are defined as

Membrane forces

$$\hat{\boldsymbol{\sigma}}_m = \begin{Bmatrix} N_x \\ N_y \\ N_{xy} \end{Bmatrix} = \int_{-t/2}^{t/2} \boldsymbol{\sigma}_p \, dz \qquad (7.17a)$$

Bending moments

$$\hat{\boldsymbol{\sigma}}_b = \begin{Bmatrix} M_x \\ M_y \\ M_{xy} \end{Bmatrix} = -\int_{-t/2}^{t/2} z \boldsymbol{\sigma}_p \, dz \qquad (7.17b)$$

Transverse shear forces

$$\hat{\boldsymbol{\sigma}}_s = \begin{Bmatrix} Q_x \\ Q_y \end{Bmatrix} = \int_{-t/2}^{t/2} \boldsymbol{\sigma}_s \, dz \qquad (7.17c)$$

where $-t/2$ and $t/2$ are the z coordinates of the plate's upper and lower surfaces, respectively (Figure 7.2). The sign of the resultant stresses is shown in Figure 7.3.

Substituting Eqs.(7.6) into Eqs.(7.15–7.17) gives

$$\begin{aligned}
\hat{\boldsymbol{\sigma}}_m &= \hat{\mathbf{D}}_m \hat{\boldsymbol{\varepsilon}}_m + \hat{\mathbf{D}}_{mb} \hat{\boldsymbol{\varepsilon}}_b + \hat{\boldsymbol{\sigma}}_m^0 \\
\hat{\boldsymbol{\sigma}}_b &= \hat{\mathbf{D}}_{m_b} \hat{\boldsymbol{\varepsilon}}_m + \hat{\mathbf{D}}_b \hat{\boldsymbol{\varepsilon}}_b + \hat{\boldsymbol{\sigma}}_b^0 \\
\hat{\boldsymbol{\sigma}}_s &= \hat{\mathbf{D}}_s \hat{\boldsymbol{\varepsilon}}_s + \hat{\boldsymbol{\sigma}}_s^0
\end{aligned} \qquad (7.18)$$

where

$$\hat{\boldsymbol{\sigma}}_m^0 = \int_{-t/2}^{t/2} \boldsymbol{\sigma}_p^0 dz \,, \qquad \hat{\boldsymbol{\sigma}}_b^0 = -\int_{-t/2}^{t/2} z \boldsymbol{\sigma}_p^0 dz \,, \qquad \hat{\boldsymbol{\sigma}}_s^0 = \int_{-t/2}^{t/2} \boldsymbol{\sigma}_s^0 dz$$

$$\hat{\mathbf{D}}_m = \int_{-t/2}^{t/2} \mathbf{D}_p \, dz \,, \qquad \hat{\mathbf{D}}_{mb} = \int_{-t/2}^{t/2} z \mathbf{D}_p \, dz$$

$$\hat{\mathbf{D}}_b = \int_{-t/2}^{t/2} z^2 \mathbf{D}_p \, dz \,, \qquad \hat{\mathbf{D}}_s = \begin{bmatrix} k_{11} \bar{D}_{s_{11}} & k_{12} \bar{D}_{s_{12}} \\ \text{Sym.} & k_{22} \bar{D}_{s_{22}} \end{bmatrix} \quad \text{with } \bar{D}_{s_{ij}} = \int_{-t/2}^{t/2} D_{s_{ij}} \, dz$$

$$(7.19)$$

Eq.(7.18) can be written in compact form as

$$\hat{\boldsymbol{\sigma}} = \hat{\mathbf{D}} \hat{\boldsymbol{\varepsilon}} + \hat{\boldsymbol{\sigma}}^0 \qquad (7.20a)$$

with

$$\hat{\boldsymbol{\sigma}} = \begin{Bmatrix} \hat{\boldsymbol{\sigma}}_m \\ \hat{\boldsymbol{\sigma}}_b \\ \hat{\boldsymbol{\sigma}}_s \end{Bmatrix} \,, \qquad \hat{\boldsymbol{\sigma}}^0 = \begin{Bmatrix} \hat{\boldsymbol{\sigma}}_m^0 \\ \hat{\boldsymbol{\sigma}}_b^0 \\ \hat{\boldsymbol{\sigma}}_s^0 \end{Bmatrix} \,, \qquad \hat{\mathbf{D}} = \begin{bmatrix} \hat{\mathbf{D}}_m & \hat{\mathbf{D}}_{mb} & \mathbf{0}_3 \\ \hat{\mathbf{D}}_{mb} & \hat{\mathbf{D}}_b & \mathbf{0}_3 \\ \mathbf{0}_2 & \mathbf{0}_2 & \hat{\mathbf{D}}_s \end{bmatrix} \qquad (7.20b)$$

where $\mathbf{0}_n$ is a $n \times n$ zero matrix.

For composite plates where x and y are orthotropy axes for all the layers, matrix $\hat{\mathbf{D}}_s$ is diagonal and it has two transverse shear correction parameters k_{11} and k_{22} only [BD5]. For an isotropic plate $k_{12} = 0$ and $k_{11} = k_{22} = 5/6$ as for rectangular beams (Section 3.5).

The computation of the k_{ij} parameters is described in the next section. Eqs.(7.17) defining the resultant stresses can be grouped as

$$\hat{\boldsymbol{\sigma}} = \int_{-t/2}^{t/2} \mathbf{S}^T \boldsymbol{\sigma} dz \qquad (7.21a)$$

where $\mathbf{S}$ is defined in Eq.(7.4). Substituting Eq.(7.7) into (7.21a) gives

$$\hat{\sigma} = \int_{-t/2}^{t/2} \left[\mathbf{S}^T \mathbf{D} \mathbf{S} \hat{\varepsilon} + \mathbf{S}^T \sigma_0 \right] dz = \hat{\mathbf{D}} \hat{\varepsilon} + \hat{\sigma}^0 \qquad (7.21b)$$

with

$$\hat{\mathbf{D}} = \int_{-t/2}^{t/2} \mathbf{S}^T \mathbf{D} \mathbf{S} dz \quad , \quad \hat{\sigma}_0 = \int_{-t/2}^{t/2} \mathbf{S}^T \sigma^0 dz \qquad (7.21c)$$

Eqs.(7.21c) define the generalized constitutive matrix $\hat{\mathbf{D}}$ and the initial resultant stress vector $\hat{\sigma}_0$ in compact form.

For a laminate with n_l orthotropic layers and homogeneous material within each layer (Figure 7.2) we can write

$$\hat{\mathbf{D}}_m = \sum_{k=1}^{n_l} t_k \mathbf{D}_{p_k} \quad ; \quad \hat{\mathbf{D}}_{mb} = -\sum_{k=1}^{n_l} t_k \bar{z}_k \mathbf{D}_{p_k} \quad ; \quad \hat{\mathbf{D}}_b = \sum_{k=1}^{n_l} \frac{1}{3} [z_{k+1}^3 - z_k^3] \mathbf{D}_{p_k}$$

$$(7.22)$$

where $t_k = z_{k+1} - z_k$, $\bar{z}_k = \frac{1}{2}(z_{k+1} + z_k)$ and $\mathbf{D}_{p_k}$ is the in-plane constitutive matrix for the kth layer.

For homogeneous material, or a material whose properties are symmetrical with respect to the middle plane ($z = 0$), matrix $\hat{\mathbf{D}}_{mb} = 0$. Plane xy then coincides with the neutral plane and the membrane and bending effects are uncoupled. This means that in-plane forces acting on the neutral plane do not produce any curvature and, conversely, bending moments do not produce any membrane strain. However, it must be pointed out that symmetry is not the only way to supress in-plane/flexural coupling; several types of non-symmetrical uncoupled laminates exist, such as [60/30/30/0/60/30] for example [KV].

The constitutive equations (7.18) simplify for the uncoupled case to

$$\hat{\sigma}_m = \hat{\mathbf{D}}_m \hat{\varepsilon}_m + \hat{\sigma}_m^0 \quad , \quad \hat{\sigma}_b = \hat{\mathbf{D}}_b \hat{\varepsilon}_b + \hat{\sigma}_b^0 \quad , \quad \hat{\sigma}_s = \hat{\mathbf{D}}_s \hat{\varepsilon}_s + \hat{\sigma}_s^0 \qquad (7.23)$$

For homogeneous isotropic material

$$\hat{\mathbf{D}}_m = t \mathbf{D}_p \quad , \quad \hat{\mathbf{D}}_b = \frac{t^3}{12} \mathbf{D}_p \quad , \quad \hat{\mathbf{D}}_s = k \bar{\mathbf{D}}_s \quad \text{with } k = 5/6 \qquad (7.24a)$$

with

$$\mathbf{D}_p = \frac{E}{1 - \nu^2} \begin{bmatrix} 1 & \nu & 0 \\ \nu & 1 & 0 \\ 0 & 0 & \frac{1-\nu}{2} \end{bmatrix} \quad , \quad \bar{\mathbf{D}}_s = Gt \begin{bmatrix} 1 & 0 \\ 0 & 1 \end{bmatrix} \qquad (7.24b)$$

The stresses within each layer can be directly computed in terms of the resultant stresses via Eqs.(7.7), (7.2) and (7.20a) as

$$\sigma = \left\{ \begin{matrix} \sigma_p \\ \sigma_s \end{matrix} \right\} = \mathbf{DS\hat{D}}^{-1}(\hat{\sigma} - \hat{\sigma}^0) + \sigma^0 \tag{7.25}$$

The bending moments and the membrane and shear forces have unit of moment and force *per unit width* of the plate, respectively.

7.3 COMPUTATION OF TRANSVERSE SHEAR CORRECTION PARAMETERS

We present first a method for computing the transverse shear correction parameters k_{ij} of Eq.(7.19) based on considerations of static equilibrium and energetic equivalences. The aim is that the transverse shear stiffness of the plate model corresponds as much as possible with that deduced from 3D elasticity. An alternative and simpler procedure based on assuming cylindrical bending is also described.

In the following we will accept that the bending and membrane effects are uncoupled (i.e. $\hat{\mathbf{D}}_{mb} = \mathbf{0}$) and there are no initial stress effects. Matrix $\hat{\mathbf{D}}_s$ is defined so that the shear strain energy density obtained for an exact 3D distribution of the transverse shear stresses τ_{xz} and τ_{yz} (denoted as U_1) is identical to the shear energy associated to the Reissner-Mindlin plate model (denoted as U_2).

For a 3D solid

$$U_1 = \frac{1}{2} \int_{-t/2}^{t/2} \sigma_s^T \mathbf{D}_s^{-1} \sigma_s \, dz \tag{7.26}$$

Reissner-Mindlin theory gives

$$U_2 = \frac{1}{2} \hat{\sigma}_s \hat{\mathbf{D}}_s^{-1} \hat{\sigma}_s \tag{7.27}$$

Equaling U_1 and U_2 yields the expression for $\hat{\mathbf{D}}_s$ and, consequently, the transverse shear correction parameters.

We focus next on the derivation of an expression for U_1 in terms of the resultant stresses.

The transverse shear stresses $\sigma_s = [\tau_{xz}, \tau_{yz}]^T$ are obtained in terms of the in-plane stresses $\sigma_p = [\sigma_x, \sigma_y, \tau_{yz}]^T$ from the equilibrium equations of 3D elasticity (assuming zero body forces, Appendix B)

$$\frac{\partial \sigma_x}{\partial x} + \frac{\partial \tau_{xy}}{\partial y} + \frac{\partial \tau_{xz}}{\partial z} = 0$$
$$\frac{\partial \sigma_y}{\partial y} + \frac{\partial \tau_{xy}}{\partial x} + \frac{\partial \tau_{yz}}{\partial z} = 0 \tag{7.28a}$$

from where we deduce the transverse shear stresses

$$\sigma_s = \begin{Bmatrix} \tau_{xz} \\ \tau_{yz} \end{Bmatrix} = -\int_{-t/2}^{z} \begin{Bmatrix} \dfrac{\partial \sigma_x}{\partial x} + \dfrac{\partial \tau_{xy}}{\partial y} \\ \dfrac{\partial \sigma_y}{\partial y} + \dfrac{\partial \tau_{xy}}{\partial x} \end{Bmatrix} dz \tag{7.28b}$$

with $\tau_{xz} = \tau_{yz} = 0$ for $z = -t/2$ and $t/2$.

The in-plane stresses σ_p are next expressed in terms of the resultant stresses. From Eq.(7.25) we deduce (for $\mathbf{D}_{mb} = \mathbf{0}$ and $\sigma^0 = 0$)

$$\sigma_p = [\sigma_x, \sigma_y, \tau_{xy}]^T = \mathbf{D}_p[\hat{\mathbf{D}}_m^{-1}\hat{\sigma}_m - z\hat{\mathbf{D}}_b^{-1}\hat{\sigma}_b] \tag{7.29a}$$

Accepting a pure bending state ($\hat{\sigma}_m = \mathbf{0}$) we have

$$\sigma_p = -z\mathbf{A}(z)\hat{\sigma}_b \quad \text{with} \quad \underset{3\times3}{\mathbf{A}}(z) = \mathbf{D}_p(z)\hat{\mathbf{D}}_b^{-1} \tag{7.29b}$$

Substituting Eq.(7.29b) into (7.28b) gives after some algebra

$$\sigma_s = \mathbf{D}_{s_1}\hat{\sigma}_s + \mathbf{D}_{s_2}\boldsymbol{\lambda} \tag{7.30}$$

with

$$\mathbf{D}_{s_1} = \int_{-t/2}^{z} \frac{z}{2} \begin{bmatrix} A_{11} + A_{33}, & A_{13} + A_{32} \\ A_{31} + A_{23}, & A_{22} + A_{33} \end{bmatrix} dz$$

$$\mathbf{D}_{s_2} = \int_{-t/2}^{z} \frac{z}{2} \begin{bmatrix} A_{11} - A_{33}, & A_{13} - A_{32}, & 2A_{12}, & 2A_{31} \\ A_{31} - A_{23}, & A_{33} - A_{22}, & 2A_{32}, & 2A_{21} \end{bmatrix} dz \tag{7.31}$$

where A_{ij} are the components of $\mathbf{A}$ of Eq.(7.29b) and

$$\boldsymbol{\lambda} = \left[\frac{\partial M_x}{\partial x} - \frac{\partial M_{xy}}{\partial y}, \frac{\partial M_{xy}}{\partial x} - \frac{\partial M_y}{\partial y}, \frac{\partial M_y}{\partial x}, \frac{\partial M_x}{\partial y} \right]^T \tag{7.32}$$

In the derivation of Eq.(7.30) we have used the relationships between shear forces and bending moments of Eqs.(5.23).

Substituting Eq.(7.30) into the expression of U_1 (Eq.(7.26)) yields (neglecting the contribution of the terms involving $\boldsymbol{\lambda}$)

$$U_1 \simeq \frac{1}{2}\hat{\sigma}_s^T \mathbf{H}_s \hat{\sigma}_s \tag{7.33}$$

with

$$\mathbf{H}_s = \int_{-t/2}^{t/2} \mathbf{D}_{s_1}^T \mathbf{D}_s^{-1} \mathbf{D}_{s_1} \, dz \tag{7.34}$$

Equaling U_1 (Eq.(7.33)) and U_2 (Eq.(7.27)) gives the generalized transverse shear strain constitutive matrix as

$$\hat{\mathbf{D}}_s = \mathbf{H}_s^{-1} = \hat{\mathbf{H}}_s \qquad (7.35)$$

Expression (7.35) identifies the transverse shear correction parameters for a composite laminate plate as

$$k_{11} = \frac{\hat{H}_{s_{11}}}{\bar{D}_{s_{11}}} \quad , \quad k_{12} = \frac{\hat{H}_{s_{12}}}{\bar{D}_{s_{12}}} \quad , \quad k_{22} = \frac{\hat{H}_{s_{22}}}{\bar{D}_{s_{22}}} \qquad (7.36)$$

where $\hat{H}_{s_{ij}}$ are the components of $\hat{\mathbf{H}}_s = \mathbf{H}_s^{-1}$ and $\bar{D}_{s_{ij}}$ are defined in Eq.(7.19).

For an *homogeneous plate* of thickness t

$$\mathbf{D}_{s_1} = \frac{6}{t^3}\left(\frac{t^2}{4} - z^2\right)\mathbf{I}_2 \quad , \quad k_{11} = k_{12} = k_{22} = \frac{5}{6} \qquad (7.37a)$$

For an *homogeneous isotropic plate* $(G_{L_z} = G_{T_z} = G)$, $k_{12} = 0$ and

$$\hat{\mathbf{D}}_s = \frac{5}{6}tG\mathbf{I}_2 \qquad (7.37b)$$

The expression of $\hat{\mathbf{D}}_s$ of Eq.(7.37b) coincides with that of Eq.(7.24a) for homogeneous isotropic plates, as expected. Substituting $\mathbf{D}_{s_1}$ of Eq.(7.37a) into (7.30) and neglecting the terms involving λ yields the standard parabolic distribution for the transverse shear stresses in the thickness direction (see also Eq.(3.24)).

The validity of the choice $\hat{\mathbf{D}}_s = \hat{\mathbf{H}}_s^{-1}$ (Eq.(7.35)) can be verified *a posteriori* when an estimate of the solution is available (typically obtained using the displacement and stress fields from plane stress theory) as follows. First the transverse shear field is obtained from Eq.(7.30), then the correct expression for U_1 is found from Eqs.(7.26) and (7.30) as

$$U_1 = \frac{1}{2}[\hat{\sigma}_s^T, \lambda^T]\begin{bmatrix} \mathbf{C}_{11} & \mathbf{C}_{12} \\ \mathbf{C}_{12}^T & \mathbf{C}_{22} \end{bmatrix}\begin{Bmatrix} \hat{\sigma}_s \\ \lambda \end{Bmatrix} \qquad (7.38)$$

with

$$\mathbf{C}_{11} = \int_{-t/2}^{t/2} \mathbf{D}_{s_1}^T \mathbf{D}_s^{-1} \mathbf{D}_{s_1} dz \quad , \quad \mathbf{C}_{12} = \int_{-t/2}^{t/2} \mathbf{D}_{s_1}^T \mathbf{D}_s^{-1} \mathbf{D}_{s_2} dz \qquad (7.39)$$

and

$$\mathbf{C}_{22} = \int_{-t/2}^{t/2} \mathbf{D}_{s_2}^T \mathbf{D}_s^{-1} \mathbf{D}_{s_2} dz \qquad (7.40)$$

where $(\mathbf{D}_{s_1}, \mathbf{D}_{s_2})$ and $\mathbf{D}_s$ are given in Eqs.(7.31) and (7.12), respectively. The final step is to compare U_1 and U_2 (Eq.(7.27)) and this completes the verification process.

Computation of shear correction parameters assuming cylindrical bending

Assuming cylindrical bending in the in-plane directions (i.e. $\tau_{xy} = 0$), the equilibrium equations are simply (Appendix B)

$$
\frac{\partial \sigma_x}{\partial x} + \frac{\partial \tau_{xz}}{\partial z} = 0
$$
$$
\frac{\partial \sigma_y}{\partial y} + \frac{\partial \tau_{yz}}{\partial z} = 0
$$

(7.41)

Integrating Eqs.(7.41) in the thickness direction gives

$$
\tau_{xz} = -\int_{-t/2}^{z} \frac{\partial \sigma_x}{\partial x} dz \quad , \quad \tau_{yz} = -\int_{-t/2}^{z} \frac{\partial \sigma_y}{\partial y} dz
$$

(7.42)

Assuming that the axial forces are negligible and that $\mathbf{D}_p$ and $\hat{\mathbf{D}}_b$ are diagonal matrices, Eqs.(7.29a) yield

$$
\sigma_x = -z \frac{D_{p11}}{\hat{D}_{b11}} M_x \quad , \quad \sigma_y = -z \frac{D_{p22}}{\hat{D}_{b22}} M_y
$$

(7.43)

where (D_{p11}, D_{p22}) and $(\hat{D}_{b11}, \hat{D}_{b22})$ are diagonal elements of $\mathbf{D}_p$ and $\hat{\mathbf{D}}_b$, respectively. For isotropic material, $D_{p11} = D_{p22} = E$ and $\hat{D}_{b11} = \hat{D}_{b22} = \frac{Et^3}{12(1-\nu^2)}$.

Substituting Eqs.(7.43) into (7.42) gives

$$
\tau_{xz} = \frac{-Q_x}{\hat{D}_{b11}} g_1(z) \quad , \quad \tau_{yz} = \frac{-Q_y}{\hat{D}_{b22}} g_2(z)
$$

(7.44)

with

$$
g_1(z) = \int_{-t/2}^{z} z D_{p11} \, dz \quad , \quad g_2(z) = \int_{-t/2}^{z} z D_{p22} \, dz
$$

(7.45)

In the derivation of Eqs.(7.44) we have used $Q_x = -\frac{\partial M_x}{\partial x}$, $Q_y = -\frac{\partial M_y}{\partial y}$ deduced from Eqs.(5.23), accepting that $M_{xy} = 0$.

The strain energy for each transverse shear stress component is deduced from Eq.(7.25) (assuming that $\mathbf{D}_s$ is a diagonal matrix) as

$$
U_{xz} = \frac{1}{2} \int_{-t/2}^{t/2} \frac{\tau_{xz}^2}{G_{xz}} dz = \frac{Q_x^2}{2\hat{D}_{b11}^2} \int_{-t/2}^{t/2} \frac{g_1^2(z)}{G_{xz}} dz
$$
$$
U_{yz} = \frac{1}{2} \int_{-t/2}^{t/2} \frac{\tau_{yz}^2}{G_{yz}} dz = \frac{Q_y^2}{2\hat{D}_{b22}^2} \int_{-t/2}^{t/2} \frac{g_2^2(z)}{G_{yz}} dz
$$

(7.46)

Let us now assume a constant distribution of the transverse shear stresses in the thickness direction. The internal energy for each shear stress is

$$\bar{U}_{xz} = \frac{1}{2} \int_{-t/2}^{t/2} \tau_{xz}\gamma_{xz}dz = \frac{1}{2}\frac{Q_x^2}{\hat{D}_{s_{11}}}$$

$$\bar{U}_{yz} = \frac{1}{2} \int_{-t/2}^{t/2} \tau_{yz}\gamma_{yz}dz = \frac{1}{2}\frac{Q_y^2}{\hat{D}_{s_{22}}}$$

(7.47)

with $\hat{D}_{s_{11}} = k_{11}\int_{-t/2}^{t/2} G_{xz}dz$ and $\hat{D}_{s_{22}} = k_{22}\int_{-t/2}^{t/2} G_{yz}dz$. For isotropic material, $\hat{D}_{s_{11}} = \hat{D}_{s_{22}} = ktG$.

Equalling Eqs.(7.46) and (7.47) we finally deduce

$$k_{11} = \frac{\bar{U}_{xz}}{U_{xz}} = (\hat{D}_{b_{11}}^2) \left[\bar{G}_{xz} \int_{-t/2}^{t/2} \frac{g_1^2(z)}{G_{xz}} dz \right]^{-1}$$

$$k_{22} = \frac{\bar{U}_{yz}}{U_{yz}} = (\hat{D}_{b_{22}}^2) \left[\bar{G}_{yz} \int_{-t/2}^{t/2} \frac{g_2^2(z)}{G_{yz}} dz \right]^{-1}$$

(7.48a)

where

$$(\bar{G}_{xz}, \bar{G}_{yz}) = \int_{-t/2}^{t/2} (G_{xz}, G_{yz})dz$$

(7.48b)

Eqs.(7.45) are analogous to (3.26) for composite laminated beams. Once again, for homogeneous material $k_{11} = k_{22} = k = \frac{5}{6}$.

Eqs.(7.48) can also be found by assuming that the constitutive matrices $\mathbf{D}_p$ (Eqs.(7.12)) are proportional from one layer to another [BD5].

7.4 PRINCIPLE OF VIRTUAL WORK

The PVW is written in terms of the generalized strains, the resultant stresses and the external distributed loads $\mathbf{t}$ as

$$\iiint_V \delta\boldsymbol{\varepsilon}^T \boldsymbol{\sigma} \, dV = \iint_A \delta\mathbf{u}^T \mathbf{t} \, dA$$

(7.49)

where V and A are the volume and the mid-plane surface of the plate, respectively, $\delta\mathbf{u} = [\delta u_0, \delta v_0, \delta w_0, \delta\theta_x, \delta\theta_y]^T$ and $\mathbf{t} = [f_x, f_y, f_z, m_x, m_y]^T$. In vector $\mathbf{t}$, f_x and f_y are in-plane distributed loads acting in the direction of the x and y axes, respectively and the rest of loads are defined in Figure 5.1.

The integral in the l.h.s. of Eq.(7.49) represents the virtual internal work. This expression is written in terms of the work of the resultant stresses on the virtual generalized strains over the plate middle surface as follows. From Eq.(7.2) we obtain

$$\delta\boldsymbol{\varepsilon} = \mathbf{S}\delta\hat{\boldsymbol{\varepsilon}} \tag{7.50}$$

Substituting Eq.(7.50) into the l.h.s. of Eq.(7.49) and using Eq.(7.21a) gives

$$\iiint_V \delta\boldsymbol{\varepsilon}^T \boldsymbol{\sigma}\, dV = \iiint_V \delta\hat{\boldsymbol{\varepsilon}}^T \mathbf{S}\boldsymbol{\sigma}\, dV = \iint_A \delta\hat{\boldsymbol{\varepsilon}}^T \hat{\boldsymbol{\sigma}}\, dA \tag{7.51}$$

The PVW is therefore written in terms of integrals over the plate midplane as

$$\iint_A \delta\hat{\boldsymbol{\varepsilon}}^T \hat{\boldsymbol{\sigma}}\, dA = \iint_A \delta\mathbf{u}^T \mathbf{t}\, dA \tag{7.52}$$

The effect of the initial resultant stresses in the PVW can be taken into account by substituting Eq.(7.21b) in the left-hand side of Eq.(7.52) giving

$$\iint_A \delta\hat{\boldsymbol{\varepsilon}}^T \hat{\mathbf{D}}\hat{\boldsymbol{\varepsilon}}\, dA + \iint_A \delta\hat{\boldsymbol{\varepsilon}}^T \hat{\boldsymbol{\sigma}}^0 dA = \iint_A \delta\mathbf{u}^T \mathbf{t}\, dA \tag{7.53}$$

The displacement derivatives in the integrals of Eq.(7.53) are of first order, as usual for Reissner-Mindlin plate theory. Therefore, a C° continuous finite element interpolation can be used for all the displacement variables.

7.5 COMPOSITE LAMINATED PLATE ELEMENTS

7.5.1 Displacement interpolation

Let us consider a plate discretized into n-noded finite elements of triangular or quadrilateral shape. The displacements and rotations are interpolated within each element as

$$\mathbf{u} = \begin{Bmatrix} u_0 \\ v_0 \\ w_0 \\ \theta_x \\ \theta_y \end{Bmatrix} = \sum_{i=1}^{n} \mathbf{N}_i \mathbf{a}_i^{(e)} = [\mathbf{N}_1, \mathbf{N}_2, \cdots, \mathbf{N}_n] \begin{Bmatrix} \mathbf{a}_1^{(e)} \\ \mathbf{a}_2^{(e)} \\ \vdots \\ \mathbf{a}_n^{(e)} \end{Bmatrix} = \mathbf{N}\mathbf{a}^{(e)} \tag{7.54a}$$

with

$$
\mathbf{N}_i = \begin{bmatrix} N_i & 0 & 0 & 0 & 0 \\ 0 & N_i & 0 & 0 & 0 \\ 0 & 0 & N_i & 0 & 0 \\ 0 & 0 & 0 & N_i & 0 \\ 0 & 0 & 0 & 0 & N_i \end{bmatrix} \; ; \quad \mathbf{a}_i^{(e)} = [u_{0_i}, v_{0_i}, w_{0_i}, \theta_{x_i}, \theta_{y_i}]^T \qquad (7.54\text{b})
$$

where $N_i(\xi, \eta)$ is the C° continuous shape function of node i.

7.5.2 Generalized strain matrices

Substituting Eqs.(7.54a) into the expression for $\hat{\boldsymbol{\varepsilon}}$ of (7.3) yields

$$
\hat{\boldsymbol{\varepsilon}} = \left\{ \begin{array}{c} \hat{\boldsymbol{\varepsilon}}_m \\ \cdots \\ \hat{\boldsymbol{\varepsilon}}_b \\ \cdots \\ \hat{\boldsymbol{\varepsilon}}_s \end{array} \right\} = \left\{ \begin{array}{c} \dfrac{\partial u_0}{\partial x} \\[6pt] \dfrac{\partial v_0}{\partial y} \\[6pt] \dfrac{\partial u_0}{\partial y} + \dfrac{\partial v_0}{\partial x} \\[6pt] \cdots \\ \dfrac{\partial \theta_x}{\partial x} \\[6pt] \dfrac{\partial \theta_y}{\partial y} \\[6pt] \left(\dfrac{\partial \theta_x}{\partial y} + \dfrac{\partial \theta_y}{\partial x} \right) \\[8pt] \cdots \\ \dfrac{\partial w_0}{\partial x} - \theta_x \\[6pt] \dfrac{\partial w_0}{\partial y} - \theta_y \end{array} \right\} = \sum_{i=1}^n \left\{ \begin{array}{c} \dfrac{\partial N_i}{\partial x} u_{0_i} \\[6pt] \dfrac{\partial N_i}{\partial y} v_{0_i} \\[6pt] \left(\dfrac{\partial N_i}{\partial y} u_{0_i} + \dfrac{\partial N_i}{\partial x} v_{0_i} \right) \\[8pt] \cdots \\ \dfrac{\partial N_i}{\partial x} \theta_{x_i} \\[6pt] \dfrac{\partial N_i}{\partial y} \theta_{y_i} \\[6pt] \left(\dfrac{\partial N_i}{\partial y} \theta_{x_i} + \dfrac{\partial N_i}{\partial x} \theta_{y_i} \right) \\[8pt] \cdots \\ \dfrac{\partial N_i}{\partial x} w_{0_i} - N_i \theta_{x_i} \\[6pt] \dfrac{\partial N_i}{\partial y} w_{0_i} - N_i \theta_{y_i} \end{array} \right\} =
$$

$$
= [\mathbf{B}_1, \mathbf{B}_2, \cdots, \mathbf{B}_n] \left\{ \begin{array}{c} \mathbf{a}_1^{(e)} \\ \mathbf{a}_2^{(e)} \\ \vdots \\ \mathbf{a}_n^{(e)} \end{array} \right\} = \mathbf{B} \mathbf{a}^{(e)} \qquad (7.55)
$$

where $\mathbf{B}_i$ is the generalized strain matrix for the ith node. This matrix can be split into its membrane, bending and transverse shear contributions as

$$\mathbf{B}_i = \begin{Bmatrix} \mathbf{B}_{m_i} \\ \mathbf{B}_{b_i} \\ \mathbf{B}_{s_i} \end{Bmatrix} \quad \text{with} \quad \mathbf{B}_{m_i} = \begin{bmatrix} \dfrac{\partial N_i}{\partial x} & 0 & 0 & 0 & 0 \\[2mm] 0 & \dfrac{\partial N_i}{\partial y} & 0 & 0 & 0 \\[2mm] \dfrac{\partial N_i}{\partial y} & \dfrac{\partial N_i}{\partial x} & 0 & 0 & 0 \end{bmatrix}$$

$$\mathbf{B}_{b_i} = \begin{bmatrix} 0 & 0 & 0 & -\dfrac{\partial N_i}{\partial x} & 0 \\[2mm] 0 & 0 & 0 & 0 & -\dfrac{\partial N_i}{\partial y} \\[2mm] 0 & 0 & 0 & -\dfrac{\partial N_i}{\partial y} & -\dfrac{\partial N_i}{\partial x} \end{bmatrix} \quad , \quad \mathbf{B}_{s_i} = \begin{bmatrix} 0 & 0 & \dfrac{\partial N_i}{\partial x} & -N_i & 0 \\[2mm] 0 & 0 & \dfrac{\partial N_i}{\partial y} & 0 & -N_i \end{bmatrix}$$

(7.56)

7.5.3 Stiffness matrix and equivalent nodal force vector

The discretized equilibrium equations relating the nodal forces and the nodal displacements are obtained by substituting the discretization equations into the PVW in the standard manner. The stiffness matrix and the equivalent nodal force for an element are

$$\mathbf{K}_{ij}^{(e)} = \iint_{A^{(e)}} \mathbf{B}_i^T \hat{\mathbf{D}} \mathbf{B}_j \, dA \tag{7.57}$$

$$\mathbf{f}_i^{(e)} = \iint_{A^{(e)}} N_i t \, dA - \iint_{A^{(e)}} \mathbf{B}_i^T \hat{\boldsymbol{\sigma}}^0 \, dA \tag{7.58}$$

Matrix $\mathbf{K}_{ij}^{(e)}$ can be obtained by adding the contributions from the stiffness matrices due to membrane, bending, transverse shear and membrane-bending coupling effects. These matrices are written in compact form as

$$\mathbf{K}_{a_{ij}}^{(e)} = \iint_{A^{(e)}} \mathbf{B}_{a_i}^T \hat{\mathbf{D}}_a \mathbf{B}_{a_j} \, dA \quad , \quad a = m, b, s$$

$$\mathbf{K}_{mb_{ij}}^{(e)} = \iint_{A^{(e)}} \left[\mathbf{B}_{m_i}^T \hat{\mathbf{D}}_{mb} \mathbf{B}_{b_j} + \mathbf{B}_{b_i}^T \hat{\mathbf{D}}_{mb} \mathbf{B}_{m_j} \right] dA$$

(7.59)

All integrals are computed using a Gauss quadrature [On4]. Clearly, $\mathbf{K}_{mb}^{(e)}$ vanishes for the cases when $\hat{\mathbf{D}}_{mb} = 0$ (Section 7.2.4).

Composite laminated plate elements based on Reissner-Mindlin theory suffer from shear locking, similarly as explained for homogeneous plates.

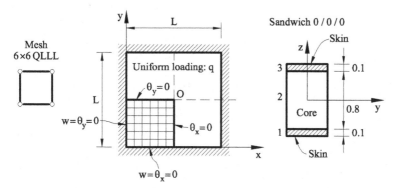

Mesh
6×6 QLLL

$w=\theta_y=0$

$w=\theta_x=0$

Orthotropic material

Skin: $E_L = 3.4156$, $E_T = 1.793$, $G_{LT} = 1$, $G_{LZ} = 0.608$, $G_{TZ} = 1.015$, $\nu_{LT} = \nu_{TL} = 0.44$

Core: E and G modules of skin divided by C

Results: $\bar{w} = w\dfrac{G_{LT}(\text{core})}{qt}$, $\bar{\sigma}_x = \dfrac{\sigma_x}{q}$

Fig. 7.7 Simply supported (hard) sandwich square plate

Shear locking can be eliminated using any of the procedures detailed in the previous chapter. Indeed any of the locking-free triangular and rectangular plate elements can be used for analysis of composite laminated plates.

For instance, for the 4-noded QLLL element based on an assumed transverse shear strain approach (Section 6.7.1) matrix $\mathbf{B}_s$ should be replaced by the substitute shear matrix $\bar{\mathbf{B}}_s$ of Eq.(6.81).

The coupling between membrane and bending stiffness matrices at element can induce the so-called *membrane locking*. This defect, of less importance than shear locking in composite Reissner-Mindlin plates, can be eliminated using similar techniques as for shear locking. Membrane locking is more important in curved shells. More details are given in Sections (8.11), (9.5.2), (9.15), (10.11.1) and (10.15).

7.5.4 Simply supported sandwich square plate under uniform loading

We present some results for the analysis of a simply supported (SS) three-layered sandwich square plate (hard support conditions: $w = \theta_s = 0$) under uniformly distributed loading (Figure 7.7). The two skin layers and the core are formed by orthotropic material with the same axes of orthotropy. The E and G values of the core are C times weaker than those of the skin. The analysis has been performed for values of $C = 1$, 10 and

C	Model	$\bar{w}$ $(z=0)$	$\bar{\sigma}_x$ $(z=0.4)$	$\bar{\sigma}_x$ $(z=-0.4)$	$\bar{\sigma}_x$ $(z=0.5)$
	QLLL 6 × 6	180.05	29.10	29.10	36.37
$C=1$	DST 6 × 6 [LB]	180.88	27.48	27.48	34.34
$k_{11}=k_{22}=0.8333$	Heterosis 4 × 4 [OF2]	183.99	28.98	28.98	36.22
	3D elasticity [Sr]	181.05	28.45	28.45	35.94
	CLPT [Sr]	168.38	28.88	28.88	36.10
	QLLL 6 × 6	39.50	5.40	54.37	67.90
$C=10$	DST 6 × 6	41.92	4.71	47.06	58.82
$k_{11}=k_{22}=0.3521$	Heterosis 4 × 4	41.92	4.87	48.73	65.23
	3D elasticity	41.91	4.86	48.61	65.08
	CLPT	31.24	5.36	53.56	66.95
	QLLL 6 × 6	16.25	1.18	59.04	73.80
$C=50$	DST 6 × 6	16.65	1.06	53.05	66.32
$k_{11}=k_{22}=0.0938$	Heterosis 4 × 4	16.85	0.93	46.65	58.31
	3D elasticity	16.75	0.74	37.15	66.90
	CLPT	6.76	1.16	57.97	72.46

Table 7.2 Displacement and stresses at the center of a SS sandwich square plate

50 using a mesh of 6 × 6 QLLL elements in a quarter of the plate due to symmetry. This problem was also solved by Owen and Figueiras [OF2] using the Heterosis quadrilateral (Section 6.5.6) and by Lardeur and Batoz [LB] using the DST element (Section 6.12). An analytical solution using 3D elasticity theory was reported by Srinivas [Sr]. Table 7.2 shows the shear correction factors for each case (note that k rapidly decreases with the value of C) and the normalized values of the deflection w and the σ_x stress at the plate center for the different numerical and analytical solutions considered. Results of classical laminate plate theory (CLPT) based on Kirchhoff thin plate assumptions are also presented to show the importance of accounting for transverse shear deformation effects. A good correlation between all results is observed. Note that for $C=50$ the central displacement is 2.5 times larger than the CLPT solution.

Satisfactory results are also obtained for the σ_x stress. For $C=50$ the 3D distribution of σ_x shown in Figure 7.8a indicates that the assumption of constant transverse shear strain is not valid as the warping of the section is non linear.

Figure 7.8b shows the thickness distribution of $\frac{\tau_{xz}}{q}$ for $C=1$, 10 and 50. The value of C greatly influences the distribution of τ_{xz}. For $C=50$, τ_{xz} is almost linear in the skin layers and constant in the core [LB].

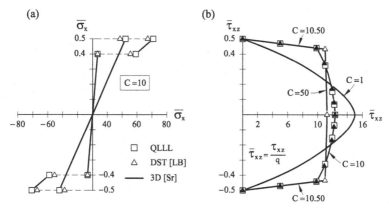

Fig. 7.8 SS (hard) sandwich square plate. Distribution of $\bar{\sigma}_x(C = 10)$ and $\bar{\tau}_{xz}(C = 1, 10, 50)$ at the plate center

7.6 HIGHER ORDER COMPOSITE LAMINATED PLATE ELEMENTS

A number of higher order theories for analysis of composite laminate plate and shell structures have been developed in an effort to obtain accurate results for stress analysis and failure prediction. Indeed, the use of a 3D finite element model is prohibitive for laminates with hundred of layers and this has motivated the search for refined finite element models based on extension of the plate/shell kinematics. This topic was already touched upon in Chapter 3 when dealing with composite beams.

Higher order plate theories are based on representing the displacement field across the thickness by quadratic or cubic polynomials or trigonometric expressions in terms of the thickness coordinate [ChP,LL,Red]. These theories typically avoid the need for the transverse shear correction parameters, at the expense of using more than five kinematic variables at each node. Two relatively simple and effective higher order theories for composite laminated plates are the layer-wise theory [Red,Red2] that assumes a linear variation of the in-plane displacement within each layer, and the refined zigzag theory (RZT) [GTD,TDG2] for which the number of kinematic variables are independent of the number of layers.

In the following sections we present the formulation of a TLQL triangular plate element based on the layer-wise theory and a QLLL quadrilateral plate element based on the RF theory.

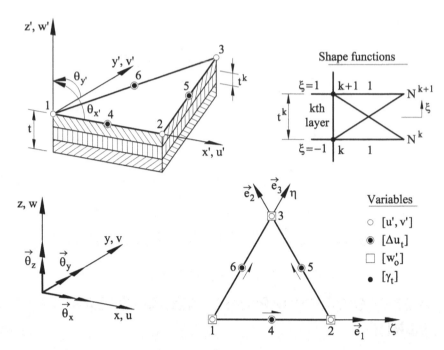

Fig. 7.9 TLQL plate element with layer-wise approximation

7.7 LAYER-WISE TLQL TRIANGULAR PLATE ELEMENT

A number of plate/shell elements based on the *layer-wise* theory of composite laminates have been developed [Red2]. Here we present the layer-wise extension of the TLQL element of Section 6.8.2 proposed by Botello *et al.* [BOM]. Figure 7.9 shows the element with the thickness discretized in layers $k = 1, 2, \cdots n_l$. Each layer k is defined by the interfaces $k, k+1$.

7.7.1 Displacement field

A linear variation for the *in-plane* displacements is assumed within each layer as

$$\begin{Bmatrix} u \\ v \end{Bmatrix} = \sum_{i=1}^{3} N_i(\xi, \eta) \left\{ N^k(\zeta) \begin{Bmatrix} u_i^k \\ v_i^k \end{Bmatrix} + N^{k+1}(\zeta) \begin{Bmatrix} u_i^{k+1} \\ v_i^{k+1} \end{Bmatrix} \right\} +$$

$$+ \sum_{i=4}^{6} N_i(\xi, \eta) e_{i-3}[N^k(\zeta)\Delta u_{t_i}^k + N^{k+1}(\zeta)\Delta u_{t_i}^{k+1}] \qquad (7.60)$$

where u_i^k, v_i^k are the in-plane displacement of the kth interface, $\Delta u_{t_i}^k$ ($i = 4, 5, 6$) are the nodal displacement increments at the mid-side nodes for the kth interface in the direction of the unit vectors $\mathbf{e}_{i-3}$.

The *vertical displacement* is assumed to be constant across the thickness and it is interpolated in terms of the corner node values as

$$w = \sum_{i=1}^{3} N_i(\xi, \eta) w_i \tag{7.61}$$

In Eq.(7.61), the linear shape functions $N_i(\xi, \eta)$ are given by Eqs.(6.162) and $N^k(\zeta) = \frac{1-\zeta}{2}, N^{k+1}(\zeta) = \frac{1+\zeta}{2}$.

The transverse shear strain field is assumed to be linear within the element as described in Section 7.8.2.

Indeed, for a single layer the element coincides with the TLQL plate triangle of Section 6.8.2.

7.7.2 Generalized strain matrices

Eqs.(7.60) and (7.61) together with the assumed linear transverse shear strain field allow us to write for every layer the relationship between the generalized strains and the displacements as

$$\varepsilon_b^k = \mathbf{B}_b \mathbf{a}^k \quad , \quad \varepsilon_s^k = \mathbf{B}_s \mathbf{a}^k \tag{7.62a}$$

where

$$\varepsilon_b = \left[\frac{\partial u}{\partial x}, \frac{\partial v}{\partial y}, \frac{\partial u}{\partial y} + \frac{\partial v}{\partial x} \right]^T \quad , \quad \varepsilon_s = \left[\frac{\partial w}{\partial x} + \frac{\partial u}{\partial z}, \frac{\partial w}{\partial y} + \frac{\partial v}{\partial z} \right]^T \tag{7.62b}$$

$$\mathbf{a}^k = \left[[\bar{\mathbf{a}}^k]^T, [\bar{\mathbf{a}}^{k+1}]^T, w_1, w_2, w_3 \right]^T$$

and

$$\bar{\mathbf{a}}^k = [u_1^k, v_1^k, u_2^k, v_2^k, u_3^k, v_3^k, \Delta u_{t_4}^k, \Delta u_{t_5}^k, \Delta u_{t_6}^k]^T \tag{7.62c}$$

Matrices $\mathbf{B}_b$ and $\mathbf{B}_s$ are

$$\mathbf{B}_b = [\mathbf{B}_b^k, \mathbf{B}_b^{k+1}, \mathbf{B}_{b_i}^0] \quad , \quad \mathbf{B}_b^k = [\mathbf{B}_{b_1}^k, \mathbf{B}_{b_2}^k, \cdots, \mathbf{B}_{b_6}^k] \tag{7.63}$$

with

$$\mathbf{B}_{b_i}^k = \begin{bmatrix} \dfrac{\partial N_i}{\partial x} & 0 \\[2mm] 0 & \dfrac{\partial N_i}{\partial y} \\[2mm] \dfrac{\partial N_i}{\partial y} N^k & \dfrac{\partial N_i}{\partial x} N^k \end{bmatrix} \quad i = 1, 2, 3 \tag{7.64a}$$

$$\mathbf{B}_{b_i}^k = \mathbf{B}_{b_{i-3}}^k \mathbf{e}_{i-3} \quad i = 4, 5, 6 \tag{7.64b}$$

$$\mathbf{B}_s = \mathbf{J}^{-1} \mathbf{P}[\mathbf{B}_s^k, \mathbf{B}_s^{k+1}, \mathbf{B}_w] , \quad \mathbf{B}_s^k = \mathbf{B}_s , \quad \mathbf{B}_s^{k+1} = -\mathbf{B}_s^k \tag{7.64c}$$

$$\mathbf{B}_s^k = \begin{bmatrix} a_{12} & b_{12} & a_{12} & b_{12} & 0 & 0 & c_{12} & 0 & 0 \\ 0 & 0 & a_{23} & b_{23} & a_{23} & b_{23} & 0 & c_{23} & 0 \\ a_{31} & b_{31} & 0 & 0 & a_{32} & b_{32} & 0 & 0 & c_{32} \end{bmatrix} , \quad \mathbf{B}_w = \begin{bmatrix} -1 & 1 & 0 \\ 0 & -1 & 1 \\ -1 & 0 & 1 \end{bmatrix} \tag{7.64d}$$

$$\mathbf{P} = \begin{bmatrix} 1 - \eta & -\eta & \eta \\ \xi & -\xi & 1 - \xi \end{bmatrix} , \quad a_{ij} = -\frac{C_i l^{ij}}{2t^k} , \quad b_{ij} = -\frac{S_i l^{ij}}{2t^k} , \quad c_{ij} = -\frac{2l^{ij}}{3t^k} \tag{7.65}$$

$\mathbf{J}$ is the jacobian matrix, C_i, S_i are the components of the unit side vector $\mathbf{e}_i = [C_i, S_i]^T$, l^{ij} is the length of side ij and t^k is the thickness of the kth layer.

7.7.3 Element stiffness matrix

The element stiffness matrix is given by

$$\mathbf{K}^{(e)} = \iint_{A^{(e)}} \left(\sum_{k=1}^{n_l} \int_{t^k} \mathbf{B}^T \hat{\mathbf{D}}^k \mathbf{B} dz \right) dA \tag{7.66}$$

where $\mathbf{B} = \begin{Bmatrix} \mathbf{B}_b \\ \mathbf{B}_s \end{Bmatrix}$ and $\hat{\mathbf{D}}^k$ is the constitutive matrix for the kth layer given by Eq.(7.20). Typically a simple one point integration is used for the thickness integration within each layer, while a full three point quadrature is chosen for the area integration over the element plane, as for the TLQL element (Section 6.8.2).

The stiffness matrix can be assembled over the thickness follows the general rule as for 1D elements. The global system of equations can be written in the form

$$
\begin{bmatrix}
\mathbf{K}_{11}^1 & \mathbf{K}_{12}^1 & \mathbf{0} & \mathbf{0} & \cdots & \mathbf{K}_3^1 \\
\mathbf{K}_{21}^1 & (\mathbf{K}_{22}^1 + \mathbf{K}_{11}^2) & \mathbf{K}_{12}^2 & \mathbf{0} & \cdots & (\mathbf{K}_{23}^1 + \mathbf{K}_{23}^3) \\
\mathbf{0} & \mathbf{K}_{21}^2 & (\mathbf{K}_{22}^2 + \mathbf{K}_{11}^3) & \mathbf{K}_{12}^3 & \cdots & (\mathbf{K}_{13}^1 + \mathbf{K}_{23}^3) \\
& & \ddots & & & \vdots \\
\text{Symm.} & & & \mathbf{K}_{22}^{n_l} & & \mathbf{K}_{23}^{n_l} \\
& & & & & \mathbf{K}_w
\end{bmatrix}
\begin{Bmatrix}
\mathbf{a}^1 \\ \mathbf{a}^2 \\ \mathbf{a}^3 \\ \vdots \\ \mathbf{a}^{n_l+1} \\ \mathbf{a}_w
\end{Bmatrix}
=
\begin{Bmatrix}
\mathbf{f}^1 \\ \mathbf{f}^2 \\ \mathbf{f}^3 \\ \vdots \\ \mathbf{f}^{n_l+1} \\ \mathbf{f}_w
\end{Bmatrix}
\tag{7.67}
$$

where $\mathbf{K}_{ij}^k$ is the stiffness matrix linking the interfaces i and j for the kth layer, $\mathbf{K}_w$ assembles the stiffness contributions affecting the vertical deflection DOFs $\mathbf{a}_w$, $\mathbf{a}^k$ are the in-plane nodal displacements for the kth interface and $\mathbf{a}_w = [w_1, w_2, \cdots, w_n]^T$ where n is the total number of corner nodes with vertical deflection DOFs.

The form of Eq.(7.67) allows one the elimination of the interface variables $\mathbf{a}^1, \mathbf{a}^2 \cdots \mathbf{a}^{n_l}$ using a substructuring technique. From the first row of Eq.(7.67) it follows

$$
\mathbf{a}^1 = [\mathbf{K}_{11}^1]^{-1}[\mathbf{f}_1^1 - \mathbf{K}_{12}^1 \mathbf{a}^2 - \mathbf{K}_{13}^1 \mathbf{a}_w]
\tag{7.68}
$$

and the new equation system can be written in terms of $\mathbf{a}^2, \mathbf{a}^3, \cdots \mathbf{a}_w$ only. Again it is possible to eliminate $\mathbf{a}^2$ (the in-plane variables for the second interface) in a similar way. The procedure is repeated for every layer so that the final condensed system of equations contains only the in-plane displacement for the top layer $\mathbf{a}^{n_l+1}$ and the nodal vertical displacements $\mathbf{a}_w$, i.e.

$$
\begin{bmatrix}
\bar{\mathbf{K}}_{11} & \bar{\mathbf{K}}_{12} \\
\bar{\mathbf{K}}_{21} & \bar{\mathbf{K}}_{22}
\end{bmatrix}
\begin{Bmatrix}
\mathbf{a}^{n_l+1} \\
\mathbf{a}_w
\end{Bmatrix}
=
\begin{Bmatrix}
\mathbf{f}^{n_l+1} \\
\mathbf{f}_w
\end{Bmatrix}
\tag{7.69}
$$

where $(\bar{\cdot})$ denote the modified stiffness matrices and equivalent nodal force vectors. Once the displacement $\mathbf{a}^{n_l+1}$ and $\mathbf{a}_w$ have been found. Eq.(7.68) can be used to obtain the in-plane displacement at each interface. This substructuring technique was initially suggested by Owen and Li [OL,OL2] and later used in [BOM] for linear and linear analysis of composite laminated plates and shells under static and dynamic loads.

We present next an application of the layer-wise TLQL element to the analysis of a SS composite laminated square plate.

7.7.4 Simple supported multilayered square plates under sinusoidal load

We study a graphite-epoxy simply supported (hard) plate. The geometry and material properties are shown in Figure 7.10. We consider five plies

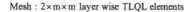

Mesh : 2×m×m layer wise TLQL elements

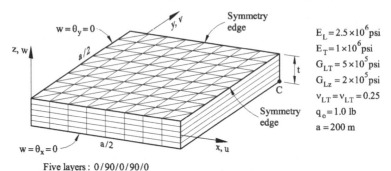

$w = \theta_y = 0$

Symmetry edge

$w = \theta_x = 0$

a/2

Symmetry edge

$E_L = 2.5 \times 10^6 \, \text{psi}$
$E_T = 1 \times 10^6 \, \text{psi}$
$G_{LT} = 5 \times 10^5 \, \text{psi}$
$G_{Lz} = 2 \times 10^5 \, \text{psi}$
$\nu_{LT} = \nu_{LT} = 0.25$
$q_o = 1.0 \, \text{lb}$
$a = 200 \, \text{m}$

Five layers : 0/90/0/90/0

Fig. 7.10 Simply supported (hard) square laminated plate (five graphite-epoxy laminates) geometry and material properties

with orientations $0°/90°/0°/90°/0°$ and thickness $t/6/t/4/t/6/t/4/t/6$. The distributed load over the plate is given by $q = q_0 \sin \frac{\pi x}{a} \sin \frac{\pi y}{a}$.

The problem was solved using the layer-wise TLQL element described in the previous section. The thickness DOFs were eliminated using the condensation technique there explained. Only a quarter of plate was analyzed due to symmetry. The discretization is shown in Figure 7.10. Table 7.3 shows results for the vertical deflection at the plate center and the stresses at some characteristic points for different side length/thickness ratios (a/t). Results are compared with those obtained by Stavsky [Sta2] using classical laminated plate theory. Note the difference in the results when the thickness increases. The discrepance is due to the effect of transverse shear deformation which is important for thick plates.

7.8 COMPOSITE LAMINATED PLATE ELEMENTS BASED ON THE REFINED ZIGZAG THEORY

As mentioned in Section 3.16, zigzag theories aim to reproducing a zigzag-like distribution for the in-plane displacements through the laminate thickness with discontinuous derivatives in the thickness direction along the lamina interfaces, *while ensuring a fixed number of kinematic variables regardless the number of material layers*. Indeed this can be achieved by using the layer-wise theory and the substructuring technique described in the previous section. However, this has a computational cost that zigzag theory aims to reduce. See Section 3.16 for a review of zigzag theories.

MESH	a/h	$\bar{w}_c$ $(a/2,a/2)$	$\bar{\sigma}_x$ $(a/2,a/2,\pm h/2)$	$\bar{\sigma}_y$ $(a/2,a/2,\pm h/2)$	$\bar{\tau}_{xz}$ $(0,a/2,0)$	$\bar{\tau}_{yz}$ $(a/2,0,0)$
$m=2$	4	5.0097	$\pm.585$	$\pm.397$	.160	.171
$m=4$	4	5.6431	$\pm.625$	$\pm.455$	.202	.212
$m=8$	4	5.7582	$\pm.625$	$\pm.473$	.21775	.226
CLPT	4	4.291	$\pm.651$	$\pm.626$	N.A.	.233
$m=2$	20	1.0226	$\pm.4575$	$\pm.340$	.170	.151
$m=4$	20	1.1624	$\pm.5225$	$\pm.373$	.228	.191
$m=8$	20	1.1946	$\pm.535$	$\pm.378$	.250	.207
CLPT	20	1.145	$\pm.539$	$\pm.380$	.268	.212
$m=2$	100	.8439	$\pm.468$	$\pm.318$	.178	.148
$m=4$	100	.971693	$\pm.528$	$\pm.396$	.259	.186
$m=8$	100	1.001176	$\pm.539$	$\pm.362$	.272	.210
CLPT	100	1.0	$\pm.539$	$\pm.359$	.272	.205

$$\bar{w}_c = \frac{\pi^4 Q w_c}{12 s^4 t q_0}, \ (\bar{\sigma}_x, \bar{\sigma}_y, \bar{\tau}_{xz}, \bar{\tau}_{yz}) = \frac{1}{q_0 s^2}(\sigma_x, \sigma_y, \tau_{xz}, \tau_{yz})$$

$$s = \frac{a}{t}, \ Q = 4G_{LT} + [E_L + E_T(1 + 2\nu_{LT})](1 - \nu_{LT}\nu_{TZ})^{-1}$$

Table 7.3 Deflection and stresses at the center of SS graphite-epoxy laminated plate (5 laminates) under sinusoidal distributed load. Results for different meshes of layer wise TLQL elements and classical laminate plate theory (CLPT) [Sta2]

Following the arguments explained in Section 3.16 for laminated beams, the displacements in zigzag plate theory are expressed as

$$u^k(x, y, z) = u_0(x, y) - z\theta_x(x, y) + \bar{u}^k(x, y, z)$$
$$v^k(x, y, z) = v_0(x, y) - z\theta_y(x, y) + \bar{v}^k(x, y, z) \quad (7.70a)$$
$$w(x, y, z) = w_0(x, y)$$

where $\bar{u}^k$ and $\bar{v}^k$ are the *zigzag interfacial displacements* defined as

$$\bar{u}^k(x, y, z) = \phi_x^k(z)\psi_x(x, y)$$
$$\bar{v}^k(x, y, z) = \phi_y^k(z)\psi_y(x, y) \quad (7.70b)$$

Superscript k denotes quantities between the k and $k+1$ laminae. Thus the kth lamina thickness coordinate is defined in the range $z \in [z_{k-1}, z_k]$, $k = 1, n_l$ where n_l is the number of layers. The ϕ_x^k, ϕ_y^k functions represent the through-the-thickness piecewise-linear zigzag functions, associated to non homogeneous plates, to be defined later, while ψ_x, ψ_y are the spatial amplitude of the zigzag displacement. These two amplitudes together with the five standard kinematic variables $(u_0, v_0, w_0, \theta_x, \theta_y)$ are the problem unknowns.

The zigzag interfacial displacements $\bar{u}^k$ and $\bar{v}^k$ may be regarded as corrections to the in-plane displacements of standard Reissner-Mindlin plate theory (Eq.(7.1)) due to the laminate heterogeneity.

Di Sciuva [DiS] developed a "linear zigzag model" for plate bending problems that employs only five kinematic variables namely u, v, w, ψ_x and ψ_y. The zigzag functions in Di Sciuva's theory are determined by enforcing the transverse shear stresses to be continuous along the adjacent lamina interfaces and, in addition, by requiring that the zigzag functions vanish in a single "fixed layer" that is selected a priori. Di Sciuva's theory has the following characteristics: (a) the transverse shear stresses are uniform through the thickness and they correspond to those in the "fixed layer"; (b) the laminate transverse shear stiffness is governed by the transverse shear moduli of the "fixed layer" alone; (c) the transverse shear strains and stresses erroneously vanish along fully clamped edges; and (d) the integral of the transverse shear stress across the laminate thickness does not correspond to the shear force obtained from the plate equilibrium equations.

To remove the flaws associated with Di Sciuva's zigzag model, Tessler et al. [TDG2] introduced a *refined* zigzag theory (RZT) for laminated-composite plates in which: (1) a novel zigzag function is used to produce non-vanishing zigzag displacements in every lamina, thus removing the shear stiffness bias associated with the "fixed layer" approach, and (2) the equilibrium of transverse shear stresses along adjacent lamina interfaces is fulfilled only in an average sense. The resulting theory overcomes must of the aforementioned flaws of the previous zigzag theories and has been shown to demonstrate consistently superior results [GTD,TDG2,3]. 3 and 6-noded C° continuous RZT-based plate triangles with anisoparametric shape functions have been recently proposed [VGM+].

The RZT theory for composite laminated beams was detailed in Section 3.17. In the following sections we describe this theory for plates.

7.8.1 Refined zigzag theory

7.8.1.1 Definition of the zigzag functions

Tessler et al. [TDG2] proposed the following linear and C° continuous zigzag functions within each layer of a composite laminated plate

$$
\begin{aligned}
\phi_x^k &= \frac{1}{2}(1 - \zeta^k)\bar{\phi}_x^{k-1} + \frac{1}{2}(1 + \zeta^k)\bar{\phi}_x^k \\
\phi_y^k &= \frac{1}{2}(1 - \zeta^k)\bar{\phi}_y^{k-1} + \frac{1}{2}(1 + \zeta^k)\bar{\phi}_y^k
\end{aligned}
\tag{7.71}
$$

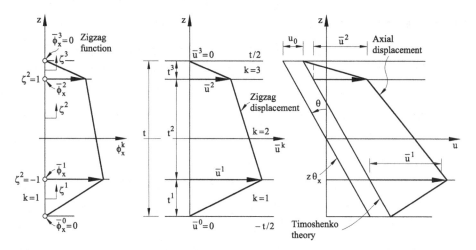

Fig. 7.11 Notation for a three-layered laminate and zigzag function in the xz plane. Same applies for ϕ_y^k by interchanging x for y and u by v

where $\bar{\phi}_x^k, \bar{\phi}_y^k$ and $\bar{\phi}_x^{k-1}, \bar{\phi}_y^{k-1}$ are the values of the zigzag functions at the k and $k-1$ interface, respectively, and

$$\zeta^k = \frac{2(z - z_{k-1})}{t^k} \quad , \quad k = 1, \cdots, n_l \tag{7.72}$$

with the first layer begining at $z_0 = -t/2$, the last n_lth layer ending at $z_{n_l} = t/2$ and the kth layer ending at $z_k = z_{k-1} + t^k$ where t^k is the layer thickness. Figure 7.11 shows the graphic representation of the zigzag function across the thickness of a three-layered laminate.

Its very important to note that the zigzag function values (and hence the interfacial displacements) at the bottom and top plate surface are set herein to vanish identically, i.e.,

$$\bar{u}^0 = \bar{u}^{n_l} = \bar{v}^0 = \bar{v}^{n_l} = 0 \tag{7.73}$$

The linear form of the zigzag function allows us to define the following constant functions within each layer

$$\beta^k = [\beta_x^k, \beta_y^k]^T = \left[\frac{\partial \phi_x^k}{\partial z}, \frac{\partial \phi_y^k}{\partial z}\right]^T = \frac{1}{t^k}[(\bar{\phi}_x^k - \bar{\phi}_x^{k-1}), (\phi_y^k - \phi_y^{k-1})]^T \tag{7.74}$$

It is simple to show that the through the thickness integration of functions β_α^k vanish identically, i.e.

$$\int_{-t/2}^{t/2} \beta_\alpha^k dz = 0 \tag{7.75}$$

7.8.1.2 Strain and stress fields

The in-plane and transverse shear strains within each layer consistent with Eqs.(7.70a) are

$$
\begin{aligned}
\varepsilon_x^k &= \frac{\partial u_0}{\partial x} - z\frac{\partial \theta_x}{\partial x} + \phi_x^k\frac{\partial \psi_x}{\partial x} \\
\varepsilon_y^k &= \frac{\partial v_0}{\partial x} - z\frac{\partial \theta_y}{\partial y} + \phi_y^k\frac{\partial \psi_y}{\partial y} \\
\gamma_{xy}^k &= \frac{\partial u_0}{\partial y} + \frac{\partial v_0}{\partial x} - z\left(\frac{\partial \theta_x}{\partial y} + \frac{\partial \theta_y}{\partial x}\right) + \phi_x^k\frac{\partial \psi_x}{\partial y} + \phi_y^k\frac{\partial \psi_y}{\partial x}
\end{aligned}
\tag{7.76}
$$

The transverse shear strains can be written as

$$
\gamma_{\alpha z}^k = \gamma_\alpha + \beta_\alpha^k \psi_\alpha \quad, \quad \alpha = x, y
\tag{7.77a}
$$

where

$$
\gamma_\alpha = \frac{\partial w_0}{\partial \alpha} - \theta_y \quad, \quad \beta_\alpha^k = \frac{\partial \phi_\alpha^k}{\partial z} \quad, \quad \alpha = x, y
\tag{7.77b}
$$

Integrating the expression for $\gamma_{\alpha z}^k$ of Eq.(7.77b) across the laminate thickness and noting Eq.(7.75) gives

$$
\begin{Bmatrix} \gamma_x \\ \gamma_y \end{Bmatrix} = \frac{1}{t}\int_{-t/2}^{t/2} \begin{Bmatrix} \gamma_{xz} \\ \gamma_{yz} \end{Bmatrix}^k dz
\tag{7.78}
$$

i.e., γ_x, γ_y represent the average transverse shear strains, coinciding with the standard transverse shear strains of Reissner-Mindlin theory (Eq.(7.3)). Eq.(7.78) also shows that the zigzag amplitude variables do not contribute to the average transverse shear strains.

The strains at each layer can be expressed in terms of generalized strains by

$$
\varepsilon_p^k = \begin{Bmatrix} \varepsilon_x^k \\ \varepsilon_y^k \\ \gamma_{xy}^k \end{Bmatrix} = \mathbf{S}_p^k \hat{\varepsilon}_p \quad \text{with } \hat{\varepsilon}_p = \begin{Bmatrix} \hat{\varepsilon}_m \\ \hat{\varepsilon}_b \end{Bmatrix}
\tag{7.79a}
$$

$$
\varepsilon_s^k = \begin{Bmatrix} \gamma_{xz}^k \\ \gamma_{yz}^k \end{Bmatrix} = \mathbf{S}_t^k \hat{\varepsilon}_t
\tag{7.79b}
$$

where

$$
\hat{\varepsilon}_m = \left[\frac{\partial u_0}{\partial x}, \frac{\partial v_0}{\partial y}, \frac{\partial u_0}{\partial y} + \frac{\partial v_0}{\partial x}\right]^T
$$

$$\hat{\varepsilon}_b = \left[\frac{\partial \theta_x}{\partial x}, \frac{\partial \theta_y}{\partial y}, \left(\frac{\partial \theta_x}{\partial y} + \frac{\partial \theta_y}{\partial x}\right), \frac{\partial \psi_x}{\partial x}, \frac{\partial \psi_x}{\partial y}, \frac{\partial \psi_y}{\partial x}, \frac{\partial \psi_y}{\partial y}\right]^T$$
(7.80)

$$\hat{\varepsilon}_t = \left[\left(\frac{\partial w}{\partial x} - \theta_x\right), \left(\frac{\partial w}{\partial y} - \theta_y\right), \psi_x, \psi_y\right]^T$$

$$\mathbf{S}_p^k = [\mathbf{I}_3, \mathbf{S}_\phi^k] \quad ; \quad \mathbf{S}_\phi^k = [-z\mathbf{I}_3, [\phi^k]] \quad ; \quad [\phi^k] = \begin{bmatrix} \phi_x^k & 0 & 0 & 0 \\ 0 & \phi_y^k & 0 & 0 \\ 0 & 0 & \phi_x^k & \phi_y^k \end{bmatrix}$$
(7.81)

$$\mathbf{S}_t^k = [\mathbf{I}_2, [\beta^k]] \quad ; \quad [\beta^k] = \begin{bmatrix} \beta_x^k & 0 \\ 0 & \beta_y^k \end{bmatrix}$$

where $\hat{\varepsilon}_m$, $\hat{\varepsilon}_b$ and $\hat{\varepsilon}_t$ are generalized membrane bending and transverse shear strain vectors. Note that $\hat{\varepsilon}_m$ coincides wit the expression of Eq.(7.3), while $\hat{\varepsilon}_b$ and $\hat{\varepsilon}_t$ contain contributions from functions ψ_x and ψ_y.

The generalized Hooke's law for the kth orthotropic lamina, whose principal material directions are arbitrary with respect to the middle place reference coordinates x, y is written as (disregarding initial stresses)

$$\sigma_p^k = \begin{Bmatrix} \sigma_x \\ \sigma_y \\ \tau_{xy} \end{Bmatrix}^k = \mathbf{D}_p^k \begin{Bmatrix} \varepsilon_x \\ \varepsilon_y \\ \gamma_{xy} \end{Bmatrix}^k = \mathbf{D}_p^k \varepsilon_p^k = \mathbf{D}_p^k \mathbf{S}_p^k \hat{\varepsilon}_p$$
(7.82)

$$\sigma_s^k = \begin{Bmatrix} \tau_{xy} \\ \tau_{zz} \end{Bmatrix}^k = \mathbf{D}_s^k \begin{Bmatrix} \gamma_{xz} \\ \gamma_{yz} \end{Bmatrix}^k = \mathbf{D}_s^k \varepsilon_s^k = \mathbf{D}_s^k \mathbf{S}_t^k \hat{\varepsilon}_t$$

where the constitutive matrices $\mathbf{D}_p^k$ and $\mathbf{D}_s^k$ for the kth lamina are obtained as explained in Section 7.2.3.

7.8.1.3 Computation of the zigzag function

The β_α^k functions are determined by first casting the transverse shear strains $\gamma_{\alpha z}^k$ (Eq.(7.77a)) in terms of transverse shear strain measures $\eta_\alpha = \gamma_\alpha - \psi_\alpha$ ($\alpha = x, y$) and the zigzag amplitude function as

$$\begin{Bmatrix} \gamma_{xz} \\ \gamma_{yz} \end{Bmatrix}^k = \begin{Bmatrix} \eta_x \\ \eta_y \end{Bmatrix} + \begin{bmatrix} (1 + \beta_x^k) & 0 \\ 0 & (1 + \beta_y^k) \end{bmatrix} \begin{Bmatrix} \psi_x \\ \psi_y \end{Bmatrix}$$
(7.83)

The η_α strain measures are set to vanish explicitly in Di Sciva's theory [DiS] and enforced to vanish by way of penalty constraints in Averill's theory [Av,AY]. Tessler et $al.$ [TDG2,3] do not impose such constraints on these strain measures. Alternatively, they express the constitutive equation for the transverse shear stresses in the following form

$$
\boldsymbol{\sigma}_s^k = \left\{ \begin{matrix} \tau_{xz} \\ \tau_{yz} \end{matrix} \right\}^k = \mathbf{D}_s^k \left[\left\{ \begin{matrix} \eta_x \\ \eta_y \end{matrix} \right\} + \left[\begin{matrix} (1+\beta_x^k) & 0 \\ 0 & (1+\beta_y^k) \end{matrix} \right] \left\{ \begin{matrix} \psi_x \\ \psi_y \end{matrix} \right\} \right]
\tag{7.84}
$$

where $\eta_\alpha = \gamma_\alpha - \psi_\alpha$, $\alpha = x, y$ is a difference function.

In Eq.(7.84) the tangential stress vector associated with the η_α strain measure is independent of the zigzag functions. The second vector includes the coefficients $D_{s11}^k (1 + \beta_x^k)$ and $D_{s22}^k (1 + \beta_y^k)$ that are dependent on the zigzag functions through β_α^k. These coefficients are set to be constant quantities denoted as G_α ($\alpha = x, y$), thus imposing constraint conditions on the distribution of the zigzag functions. These constraints yield the following expression for β_α^k

$$
\left\{ \begin{matrix} \beta_x^k \\ \beta_y^k \end{matrix} \right\} = \left\{ \begin{matrix} \dfrac{G_x}{D_{s11}^k} - 1 \\ \dfrac{G_y}{D_{s22}^k} - 1 \end{matrix} \right\}
\tag{7.85}
$$

The G_x and G_y constant material parameters are obtained by integrating Eq.(7.85) through the laminate thickness and using the property that the thickness integration of the β_α^k functions is zero. The result is

$$
\left\{ \begin{matrix} G_x \\ G_y \end{matrix} \right\} = \left\{ \begin{matrix} \left(\dfrac{1}{t} \displaystyle\sum_{k=1}^{n_l} \dfrac{t^k}{D_{s11}^k} \right)^{-1} \\ \left(\dfrac{1}{t} \displaystyle\sum_{k=1}^{n_l} \dfrac{t^k}{D_{s22}^k} \right)^{-1} \end{matrix} \right\}
\tag{7.86}
$$

Substituting Eq.(7.74) into (7.75) gives the following recursion relation for the zigzag function values at the layer interfaces

$$
\left\{ \begin{matrix} \bar{\phi}_x^k \\ \bar{\phi}_y^k \end{matrix} \right\} = \sum_{i=1}^{k} \left\{ \begin{matrix} t^i \beta_x^i \\ t^i \beta_y^i \end{matrix} \right\} , \quad k = 1, \cdots, n_l
\tag{7.87}
$$

Substituting the expressions for $\bar{\phi}_x^k, \bar{\phi}_y^k$ of Eq.(7.87) into Eq.(7.71) gives an explicit form of the zigzag functions in terms of the β_α^k functions as

$$
\left\{ \begin{matrix} \phi_x^k \\ \phi_y^k \end{matrix} \right\} = \frac{t^k}{2} (\zeta^k - 1) \left\{ \begin{matrix} \beta_x^k \\ \beta_y^k \end{matrix} \right\} + \sum_{i=1}^{k} t^i \left\{ \begin{matrix} \beta_x^i \\ \beta_y^i \end{matrix} \right\}
\tag{7.88}
$$

For homogeneous plates the ratios $\frac{G_x}{D^k_{s_{11}}}$ and $\frac{G_y}{D^k_{s_{22}}}$ in Eq.(7.85) are unit-valued and the β^k_α functions are zero. This automatically leads to the vanishing of the zigzag function (Eq.(7.87)) and the interfacial displacements (Eqs.(7.70b). Consequently, the kinematic and constitutive equations coincide with those of Reissner-Mindlin flat shell theory (Section 7.2).

7.8.1.4 Resultant constitutive equations

The stresses are defined as

$$
\hat{\sigma} = \left\{ \begin{matrix} \hat{\sigma}_p \\ \hat{\sigma}_t \end{matrix} \right\} = \left\{ \begin{matrix} \hat{\sigma}_m \\ \hat{\sigma}_b \\ \cdots \\ \hat{\sigma}_t \end{matrix} \right\} = \int_{-t/2}^{t/2} \left\{ \begin{matrix} [\mathbf{S}^k_p]^T \, \sigma^k_p \\ \cdots \quad \cdots \\ [\mathbf{S}^k_t]^T \, \sigma^k_s \end{matrix} \right\} dz \tag{7.89}
$$

where $\hat{\sigma}$ is the resultant stress vector. The axial resultant stresses $\hat{\sigma}_m$ coincide with those of Eq.(7.17a) while the bending and transverse shear resultant stress vectors are

$$
\hat{\sigma}_b = [M_x, M_y, M_{xy}, M^\phi_x, M^\phi_{xy}, M^\phi_y, M^\phi_{yx}]^T
$$
$$
\hat{\sigma}_t = [Q_x, Q_y, Q^\phi_x, Q^\phi_y]^T \tag{7.90}
$$

where $(M^\phi_x, M^\phi_{xy}, M^\phi_y, M^\phi_{yx})$ and (Q^ϕ_x, Q^ϕ_y) are pseudo-bending moments and pseudo-shear forces introduced by the RZT that have not a physical meaning.

The resultant constitutive equations are obtained by substituting Eqs.(7.82) into (7.89), i.e.

$$
\hat{\sigma} = \int_{-t/2}^{t/2} \left\{ \begin{matrix} [\mathbf{S}^k_p]^T \, \mathbf{D}^k_p \, \mathbf{S}^k_p \, \hat{\varepsilon}_p \\ [\mathbf{S}^k_t]^T \, \mathbf{D}^k_s \, \mathbf{S}^k_t \, \hat{\varepsilon}_s \end{matrix} \right\} dt = \begin{bmatrix} \hat{\mathbf{D}}_p & \mathbf{0} \\ \mathbf{0} & \hat{\mathbf{D}}_t \end{bmatrix} \left\{ \begin{matrix} \hat{\varepsilon}_p \\ \hat{\varepsilon}_t \end{matrix} \right\} = \hat{\mathbf{D}}\hat{\varepsilon} \tag{7.91}
$$

with

$$
\hat{\mathbf{D}}_p = \int_{-t/2}^{t/2} [\mathbf{S}^k_p]^T \mathbf{D}^k_p \mathbf{S}^k_p \quad , \quad \hat{\mathbf{D}}_t = \int_{-t/2}^{t/2} [\mathbf{S}^k_t]^T \mathbf{D}^k_s \mathbf{S}^k_t \, dt \tag{7.92}
$$

where $\mathbf{D}_p$ and $\mathbf{D}_t$ are given in Eqs.(7.12) and $\mathbf{S}^k_p$ and $\mathbf{S}^k_t$ in Eqs.(7.81).

Substituting the expression for $\hat{\varepsilon}_p$ of Eqs.(7.79a) yields after small algebra an expanded form of the generalized constitutive equation as

$$
\hat{\sigma} = \left\{ \begin{matrix} \hat{\sigma}_m \\ \hat{\sigma}_b \\ \hat{\sigma}_t \end{matrix} \right\} = \begin{bmatrix} \hat{\mathbf{D}}_m & \hat{\mathbf{D}}_{mb} & \mathbf{0} \\ \hat{\mathbf{D}}^T_{mb} & \hat{\mathbf{D}}_b & \mathbf{0} \\ \mathbf{0} & \mathbf{0} & \hat{\mathbf{D}}_t \end{bmatrix} \left\{ \begin{matrix} \hat{\varepsilon}_m \\ \hat{\varepsilon}_b \\ \hat{\varepsilon}_t \end{matrix} \right\} = \hat{\mathbf{D}}\hat{\varepsilon} \tag{7.93}
$$

where $\hat{\mathbf{D}}_m$ coincides with the expression of Eq.(7.19),

$$\hat{\mathbf{D}}_b = \int_{-t/2}^{t/2} [\mathbf{S}_\phi^k]^T \mathbf{D}_p^k \mathbf{S}_\phi^k \, dz \quad \text{and} \quad \hat{\mathbf{D}}_{mb} = \int_{-t/2}^{t/2} \mathbf{D}_p P k \mathbf{S}_\phi^k \, dz \qquad (7.94)$$

where $\mathbf{S}_\phi^k$ is given in Eqs.(7.81).

7.8.1.5 Principle of virtual work

The expression of the internal virtual work (l.h.s. of Eq.(7.49)) is written in terms of the new generalized strains and resultant stresses using Eq.(7.93) as

$$\iiint_V \left\{ [\delta \boldsymbol{\varepsilon}_p^k]^T \boldsymbol{\sigma}_p^k + [\delta \boldsymbol{\varepsilon}_s^k]^T \boldsymbol{\sigma}_s^k \right\} dV = \iiint_V \left\{ [\delta \boldsymbol{\varepsilon}_p^k]^T \mathbf{D}_p^k [\mathbf{S}_p^k] \hat{\boldsymbol{\varepsilon}}_p + [\delta \boldsymbol{\varepsilon}_s^k]^T \mathbf{D}_t^k \mathbf{S}_t^k \hat{\boldsymbol{\varepsilon}}_t \right\}$$

$$= \iint_A \left[\delta \hat{\boldsymbol{\varepsilon}}_p^T \hat{\mathbf{D}}_p \hat{\boldsymbol{\varepsilon}}_p + \delta \hat{\boldsymbol{\varepsilon}}_t \hat{\mathbf{D}}_t \hat{\boldsymbol{\varepsilon}}_t \right] dA = \iint_A \delta \hat{\boldsymbol{\varepsilon}}^T \hat{\boldsymbol{\sigma}} dA \tag{7.95}$$

The PVW is therefore finally expressed by Eq.(7.52)

7.8.1.6 Element matrices

Composite laminated plate elements based on the RFT can be derived by adding functions ψ_x and ψ_y to the standard five kinematic variables $(u_0, v_0, w, \theta_x, \theta_y)$. A C° interpolation is used for all the kinematic variables as it is usual is Reissner-Mindlin elements, i.e.

$$\mathbf{u} = [u_0, v_0, w, \theta_x, \theta_y, \psi_x, \psi_y]^T = \sum_{i=1}^n N_i \mathbf{a}_i^{(e)} \tag{7.96}$$

where N_i are the element shape functions and

$$\mathbf{a}_i = [u_{0_i}, v_{0_i}, w, \theta_{x_i}, \theta_{y_i}, \psi_{x_i}, \psi_{y_i}]^T \tag{7.97}$$

Substituting Eq.(7.96) into the expression for the generalized strains in Eqs.(7.80) yields

$$\hat{\boldsymbol{\varepsilon}}_m = \mathbf{B}_m \mathbf{a}^{(e)} \quad , \quad \hat{\boldsymbol{\varepsilon}}_b = \mathbf{B}_b \mathbf{a}^{(e)} \quad , \quad \hat{\boldsymbol{\varepsilon}}_t = \mathbf{B}_t \mathbf{a}^{(e)} \tag{7.98}$$

with the generalized strain matrices given by

$$
\mathbf{B}_{m_i} = \begin{bmatrix} \dfrac{\partial N_i}{\partial x} & 0 & 0\,0\,0\,0\,0 \\[2mm] 0 & \dfrac{\partial N_i}{\partial y} & 0\,0\,0\,0\,0 \\[2mm] \dfrac{\partial N_i}{\partial y} & \dfrac{\partial N_i}{\partial x} & 0\,0\,0\,0\,0 \end{bmatrix} \quad , \quad
\mathbf{B}_{b_i} = \begin{bmatrix} 0\,0\,0 & \dfrac{\partial N_i}{\partial x} & 0 & 0 & 0 \\[2mm] 0\,0\,0 & 0 & \dfrac{\partial N_i}{\partial y} & 0 & 0 \\[2mm] 0\,0\,0 & \dfrac{\partial N_i}{\partial y} & \dfrac{\partial N_i}{\partial x} & 0 & 0 \\[2mm] 0\,0\,0 & 0 & 0 & \dfrac{\partial N_i}{\partial x} & 0 \\[2mm] 0\,0\,0 & 0 & 0 & \dfrac{\partial N_i}{\partial y} & 0 \\[2mm] 0\,0\,0 & 0 & 0 & 0 & \dfrac{\partial N_i}{\partial x} \\[2mm] 0\,0\,0 & 0 & 0 & 0 & \dfrac{\partial N_i}{\partial y} \end{bmatrix}
$$

$$
\mathbf{B}_{t_i} = \begin{bmatrix} 0 & 0 & \dfrac{\partial N_i}{\partial x} & -N_i & 0 & 0 & 0 \\[2mm] 0 & 0 & \dfrac{\partial N_i}{\partial y} & -N_i & 0 & 0 & 0 \\[1mm] \cdots\cdots & \cdots & \cdots\cdots & \cdots\cdots\cdots \\[1mm] 0 & 0 & 0 & 0 & 0 & N_i & 0 \\[2mm] 0 & 0 & 0 & 0 & 0 & 0 & N_i \end{bmatrix} = \left\{ \begin{array}{c} \mathbf{B}_{s_i} \\ \cdots \\ \mathbf{B}_{\psi_i} \end{array} \right\} \tag{7.99}
$$

The stiffness matrix is obtained by Eq.(7.59) with matrix $\mathbf{K}_t^{(e)}$ substituting $\mathbf{K}_s^{(e)}$ given by

$$
\mathbf{K}_t^{(e)} = \iint_{A^e} \mathbf{B}_t^T \hat{\mathbf{D}}_t \mathbf{B}_t \, dA \tag{7.100}
$$

The membrane stiffness matrix $\mathbf{K}_m^{(e)}$ is integrated with a full quadrature while the bending and transverse shear matrices are integrated with the same quadratures as explained for Reissner-Mindlin plate elements in Chapter 5. Shear (and membrane) locking in RTT element can be avoided using selective/reduced integration, assumed transverse shear strain techniques or linked interpolations.

For example, if an assumed transverse shear technique is used, matrix $\mathbf{B}_{s_i}$ of Eq.(7.92) should be substituted by

$$
\mathbf{B}_{s_i} = [\bar{\mathbf{B}}_{s_i}, \mathbf{0}_2] \tag{7.101}
$$

where $\bar{\mathbf{B}}_{s_i}$ is the substitute transverse shear strain matrix and $\mathbf{0}_2$ is a 2×2 null matrix.

The equivalent nodal force vector for a distributed load f_z is

$$\mathbf{f}^{(e)} = \iint_{A^e} N_i f_z [0, 0, 1, 0, 0, 0, 0]^T dA \qquad (7.102)$$

The global stiffness equations are assembled in the standard manner. The boundary conditions for the new nodal variables ψ_{x_i}, ψ_{y_i} are the same as for the rotations, i.e.

$$\begin{aligned}
\psi_{x_i} = \psi_{y_i} = 0 & \quad \text{at a clamped edge} \\
\psi_{\alpha_i} = 0 & \quad \text{if } \alpha \text{ is a symmetry axis } (\alpha = x, y) \\
\psi_{y_i} = 0 & \quad \text{if } x \text{ is a SS edge (likewise for } \psi_{x_i})
\end{aligned} \qquad (7.103)$$

7.8.2 QLRZ plate element

Oñate *et al.* [OEO,OEO2] have extended the QLLL plate element of Section 6.7.1 based on a linear assumed transverse shear strain field using the RZT. Matrices $\mathbf{B}_{m_i}$, $\mathbf{B}_{b_i}$ and $\mathbf{B}_{\psi_i}$ are computed using bi-linear shape functions while the substitute transverse shear strain matrix $\bar{\mathbf{B}}_{s_i}$ of Eq.(6.7.1) replaces matrix $\mathbf{B}_{s_i}$ in Eq.(7.99) as shown in Eq.(7.101). All the terms in the stiffness matrix and the equivalent nodal force vector for the element are computed using a 2×2 quadrature. Some applications of the so-called QLRZ element to the analysis of composite laminated plates are shown in the next section. For further details see [EOO].

7.8.2.1 Convergence study for the QLRZ plate element

The convergence of the QLRZ plate element is studied in the analysis of simply supported (SS) and clamped square plates of three-layered composite laminated material under uniformly distributed loading. The side length/thickness ratio is $\lambda = \frac{a}{t} = 40$. Three different materials and thickness distributions for the three layers are shown in Table 7.4. Each layer is assumed to behave isotropically. The "reference" solution obtained with a mesh of $10 \times 10 \times 9$ 20-noded hexahedra with 13497 DOFs [EOO] (Figure 7.12).

Tables 7.5 and 7.6 show the error in the vertical deflection at the plate center, and in the σ_x stress and the zigzag amplitude function ψ_x at the upper surface of point E for the three composite materials considered .

The analysis of the results for the SS plate (Table 7.5) shows that errors for the three variables chosen are less than 2.5% for the 8×8 mesh and

		Layer 1 (top)	Layer 2	Layer 3 (bottom)
	t [m]	0,0333	0,0333	0,0333
Composite A	E [MPa]	0,219	0,219E-1	0,44
	ν	0,25	0,25	0,25
	t [m]	0,0333	0,0333	0,0333
Composite B	E [MPa]	0,219	0,219E-2	0,219
	ν	0,25	0,25	0,25
	t [m]	0,01	0.08	0.01
Composite C	E [MPa]	0,219	0,725E-3	0,73E-1
	ν	0,25	0,25	0,25

Table 7.4 3-layered composite plates used for convergence analysis

Simple supported plate - Relative error (%) Convergence analysis								
Mesh of	w_c			σ_x at E			ψ_x at E	
QLRZ elements	A	B	C	A	B	C	A	B	C
2×2	2,69	19,36	25,83	26,98	32,89	33,24	-9,11	41,06	51,92
4×4	0,68	6,50	10,14	4,86	7,70	9,05	-3,99	8,95	13,67
8×8	0,25	1,54	2,22	-0,30	-0,79	0,44	-0,71	-0,40	-1,84
16×16	0,15	0,38	0,35	-1,55	-3,04	-1,92	0,07	-0,45	-1,44
32×32	0,12	0,12	-0,02	-1,86	-3,49	-2,07	0,00	0,00	0,00

Table 7.5 SS square plate under uniformly distributed load ($\lambda = 40$). Relative error for the deflection at the plate center w_c and for σ_x and ψ_x at point E

Clamped plate - Relative error (%) Convergence analysis								
Mesh of	w_c			σ_x at E			ψ_x at E	
QLRZ elements	A	B	C	A	B	C	A	B	C
2×2	11,71	50,28	60,99	99,99	100	100	26,13	80,09	86,48
4×4	4,65	30,16	43,47	20,86	44,14	45,53	-6,28	43,34	54,80
8×8	1,60	12,32	22,44	2,90	14,35	17,24	-1,47	13,68	18,58
16×16	0,29	3,67	9,25	-1,21	-0,40	-1,15	-0,30	2,58	2,22
32×32	-0,14	0,69	2,85	-2,22	-4,70	-4,62	0,00	0,00	0,00

Table 7.6 Clamped square plate under uniformly distributed load ($\lambda = 40$). Relative error for the central vertical deflection w_c and for σ_x and ψ_x at point A

the three composite materials. For the clamped plate (Table 7.6) errors are less than 10% for the 16×16 mesh in all cases. Convergence is always

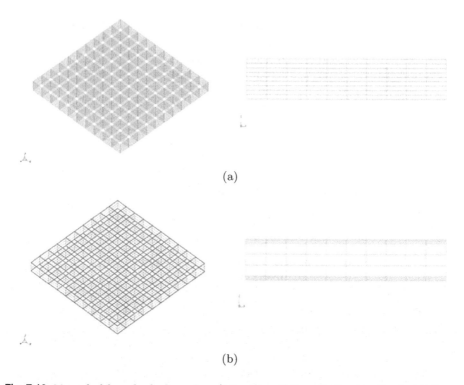

(a)

(b)

Fig. 7.12 20-noded hexahedral meshes $(10 \times 10 \times 9)$ for: a) Composite A; b) Composite C

slower for the more heterogeneous material (material C). It is remarkable that errors for material A do not exceed 3% for the 8×8 mesh for all the problems analyzed.

Figure 7.13 shows the convergence curves for the vertical deflection for the SS and clamped cases.

7.8.2.2 Simply supported of square and circular multilayered plates

We study next a 4-layered square SS plate and a 9-layered circular SS plate under uniformly distributed loading. The layer distribution and the material properties of each layer for each case are shown in Table 7.7.

The numerical solution for the square plate has been obtained with two meshes of 8×8 and 16×16 QLRZ elements as well as with a mesh of 16×16 standard QLLL composite plate elements derived as explained in Section 7.5. Results are compared with a "reference" 3D solution obtained using a mesh of $10 \times 10 \times 12$ 20-noded hexahedra (Figure 7.15a).

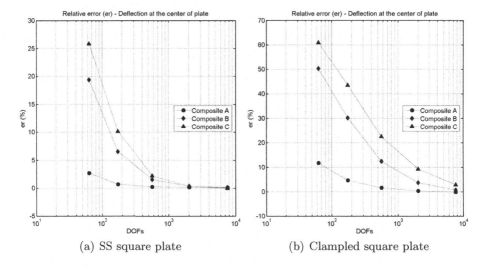

Fig. 7.13 Convergence curves for the central deflection in a square plate under uniform loading. a) SS plate. b) Clamped plate

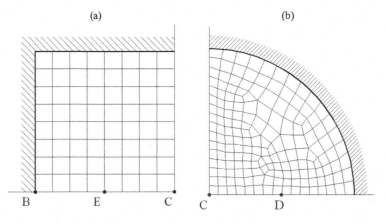

Fig. 7.14 QLRZ and QLLL element meshes for the analysis of square and circular SS composite laminated plates under uniformly distributed load. Only one quarter of the plate is discretized due to symmetry. a) 8×8 mesh. b) 168 elements

Figure 7.16a shows the distribution of the deflection along the central line. Note the accuracy of the QLRZ results and the erroneous results obtained with the QLLL element.

Figures 7.16b,c,d respectively show the thickness distribution of the horizontal displacement u and the stress σ_x at the plate center and the transverse shear stress τ_{xz} at three points on the middle line. The zigzag distribution of the axial displacement is accurately captured with the

Composite laminate plates		
Composite	**Materials**	t^k/t
C1 (4-layers)	(A/B/C/D)	(0,1/0,3/0,5/0,1)
C2 (9-layers)	(A/C/A/C/B/C/A/C/A)	(0,1/0,1/0,1/0,1/0,2/0,1/0,1/0,1/0,1)

Material	E_1	E_2	E_3	ν_{12}	ν_{13}	ν_{23}	G_{12}	G_{13}	G_{23}
A	15790	958	958	0,32	0,32	0,49	593	593	323
B	19,15	19,15	191,5	6,58E-4	6,43E-8	6,43E-8	42,3E-7	36,51	124,8
C	10,4			0,30			4		
D	10410			0,31			3973		

Table 7.7 Material properties. 4 and 9-layered plates. Units for E and G in MPa

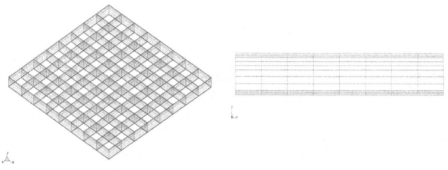

(a) 4-layered square SS plate. Material C1

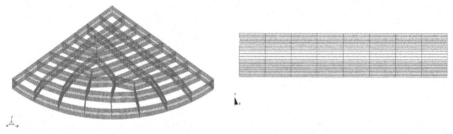

(b) 9-layered circular SS plate. Material C2

Fig. 7.15 Meshes of 20-noded hexahedra for analysis of SS 4-layered square plate and 9-layered circular plate under uniformly distributed loading. A quarter of the plate is analyzed due to symmetry

QLRZ element while the QLLL results yield a wrong linear distribution. Very accurate results are also obtained with the QLRZ element for the thickness distribution of the axial stress σ_x and the transverse shear stress

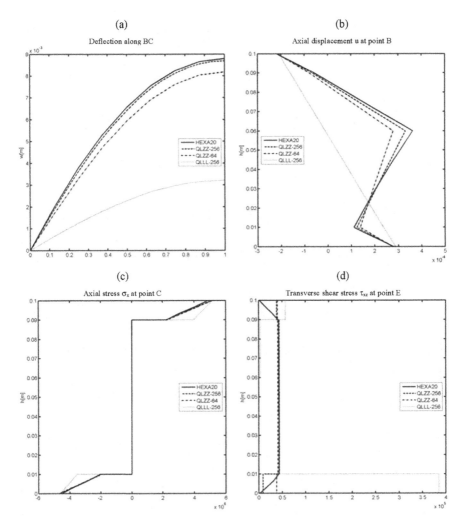

Fig. 7.16 SS square plate under uniformly distributed loading. Material C1. a) Distribution of vertical deflection along central line BC. b) Thickness distribution of the in-plane displacement u at boundary point B. c) of the axial stress σ_x at central point C and d) of the transverse shear stress τ_{xz} at point E

τ_{xz}. The QLLL results are reasonably good for σ_x but are far from the correct value for τ_{xz}.

Figures 7.17 display a similar set of results for the circular plate. The mesh used for the 3D analysis is shown in Figure 7.15b. The conclusion is identical to the square plate case: the QLRZ element reproduces the complex kinematics of the composite laminate and yields an accurate distribution of the stresses across the thickness. Conversely, the QLLL element

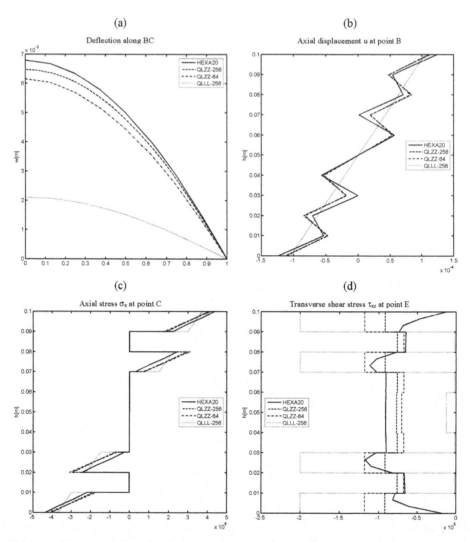

Fig. 7.17 SS circular plate under uniformly distributed loading. Material C2. a) Distribution of vertical deflection along central line BC. b) Thickness distribution of the in-plane displacement u at point B. c) of the axial stress σ_x at central point C. d) Thickness distribution of the transverse shear stress τ_{xz} at point E

gives wrong results for the deflection curve and an inaccurate thickness distribution for the axial and transverse shear stresses.

More evidence of the good behaviour of the QLRZ element for analysis of composite laminated plates can be found in [EOO].

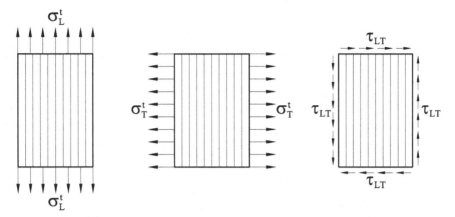

Fig. 7.18 Experimental tests for estimating the ultimate stresses in a lamina

7.9 FAILURE THEORIES

Failure at a point in a solid can be identified when the maximum principal stress reaches a limit value, typically called the ultimate stress or the failure stress. This criterion is typically used for detecting the onset of failure in fragile material (concrete, glass, ceramics, etc.). More sophisticated failure criteria are based on verifying the limit bound for an appropriate stress invariant [ZT2].

In the following section we present some basic concepts on the ultimate stress (also called the limit strength) of a lamina in a composite under different loading conditions.

7.9.1 Ultimate stress of a lamina under simple loading conditions

The ultimate stress of a lamina (i.e. a layer) made of composite material formed by unidirectional fibers embedded in a matrix can be experimentally determined as follows [BC]. Assuming that the lamina is under a plane stress state we apply a single tensile stress along the fiber direction (σ_L^t) to an isolated lamina (Figure 7.18a). The applied stress is increased until the material fails for a stress level $\sigma_{L_u}^t$. The same test can be repeated for a compressive stress σ_L^c giving $\sigma_{L_u}^c$ as the *absolute value* of the ultimate compressive stress at failure. Typically $\sigma_{L_u}^t$ and $\sigma_{L_u}^c$ will be different.

In a second test (Figure 7.18b) the lamina is subjected to a single tensile stress σ_T^t applied in the direction transverse to the fiber. The stress level that corresponds to failure in the lamina (i.e. in the matrix material) is denoted $\sigma_{T_u}^t$. The absolute value of the compressive stress applied in the direction transverse the fiber yielding material failure is denoted $\sigma_{T_u}^c$.

In the third test shown in Figure 7.18c the lamina is subjected to a shear stress τ_{LT}. The failure level for this stress is termed τ_u for simplicity.

These tests are conceptually simple but quite difficult to perform in practice. The ends of the test specimens must be reinforced in the tensile test to prevent premature failure near the grips. Also the specimen must be long enough to eliminate end effects. Buckling of the specimen should be prevented in the compressive tests by providing lateral support to the test sample. The shear test in a flat rectangular specimen is also complex. Tubular specimens can be used at a greater cost. Table 7.8 lists the typical failure stress level for lamina made of different materials [BC].

Lamina material	$\sigma_{L_u}^t$	$\sigma_{L_u}^c$	$\sigma_{T_u}^t$	$\sigma_{T_u}^c$	τ_u
Graphite/Epoxy (T300/5208)	1500	1500	40	240	68
Graphite/Epoxy (AS/3501)	1450	1450	52	205	93
Boron/Epoxy (T300/5208)	1260	2500	61	202	67
E-glass/Epoxy (Scotchply 1002)	1060	60	31	118	72
Arand (Kevlar 49)/Epoxy	1400	235	12	53	34

Table 7.8 Typical ultimate stresses for lamina made of different composite materials. Stress values in MPa

7.9.2 Failure stress of a lamina under combined loading conditions

In practical applications, a lamina might be subjected to several stress components simultaneously. Figure 7.19 shows the so-called failure stress envelope in the two-dimensional (2D) stress space. The intersection with the coordinate axes corresponds to the failure stress levels $\sigma_{L_u}^t$, $\sigma_{L_u}^c$, $\sigma_{T_u}^t$, $\sigma_{T_u}^c$ defined in the previous section. Points laying on the failure envelope denote the values of the stresses acting along the fiber and transverse directions for which the lamina will fail. For instance, point A found by the intersection of the failure envelope with the 45 degree line corresponds to the case for which failure occurs for equal tensile stresses acting along the fiber and the transverse direction, i.e. $\sigma_{L_u}^t = \sigma_{T_u}^t$.

All stress states within the failure envelope correspond to stress levels the lamina material can sustain without failure, whereas the stress states on and outside the failure envelope result in failure.

The tangential stress τ_{LT} has not been taken into account in the 2D failure envelope of Figure 7.19. If this stress is taken into account (as it is the case in practical design situations) the failure envelope becomes a

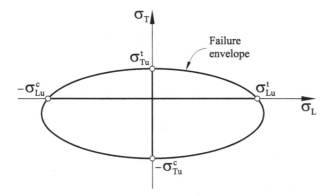

Fig. 7.19 Failure stress envelope in two-dimensional stress space

3D surface which is a function of the three stresses acting on the lamina coordinate system σ_L, σ_T and τ_{LT}.

The failure envelope could be obtained experimentally by performing a large number of tests with various combination of applied stress components σ_L, σ_T and τ_{LT}. This approach is non practical as it requires many tests to determine the failure envelope. A more practical approach is to define the failure envelope in terms of the five failure stress levels $\sigma^t_{L_u}$, $\sigma^c_{L_u}$, $\sigma^t_{T_u}$, $\sigma^c_{T_u}$ and τ_u. This can be achieved by mean of a failure criterion that predicts failure (i.e. the point laying on the failure envelope) under combined loads. Many different failure criteria have been proposed [Red2]. However, their agreement with experimental results for the failure stresses is not fully satisfactory. In the next section we briefly present a failure criterion widely used for design of composite structures.

Note finally, that failure of the matrix due to a transverse load does not decrease much the ability of the lamina to continue carrying out load in the fiber direction. However, if fiber failure occurs then the load carrying capability of the lamina is completely lost. This means that in a composite material the failure mode is as important as the failure stress [BC].

7.9.3 The Tsai-Wu failure criterion

The Tsai-Wu failure criterion states that failure occurs in a lamina when the combined applied stresses (expressed in the fiber aligned triad) satisfy the following equality

$$F_{LL}\sigma^2_L + 2F_{LT}\sigma_L\sigma_T + F_{TT}\sigma^2_T + F_S\tau^2_{LT} + F_L\sigma_L + F_T\sigma_T = 1 \quad (7.104)$$

The coefficient multiplying the stresses in Eq.(7.104) are determined experimentally, as follows. The five coefficients F_{LL}, F_{TT}, F_s and F_T are computed by applying the three tests explained in Section 7.9.1 where a single stress component is applied for each test. For instance, if the stress σ_L is applied, at failure in tension and in compression the satisfaction of Eq.(7.104) implies

$$F_{LL}(\sigma_{L_u}^t)^2 + F_L\sigma_{L_u}^t = 1 \quad \text{and} \quad F_{LL}(\sigma_{L_u}^c)^2 - F_L\sigma_{L_u}^c = 1 \qquad (7.105)$$

The second test involves the stress component σ_T only and it yields

$$F_{TT}(\sigma_{T_u}^t)^2 + F_T\sigma_{T_u}^t = 1 \quad \text{and} \quad F_{TT}(\sigma_{T_u}^c)^2 - F_T\sigma_{T_u}^c = 1 \qquad (7.106)$$

Finally, the last test involves τ_{LT} only and implies $F_S\tau_u^2 = 1$. These five equations can be solved for the five coefficient sought. Introducing these in Eq.(7.104) and rearranging the terms, the following expression is found

$$\bar{\sigma}_L^2 + 2\bar{F}_{LT}\bar{\sigma}_L\bar{\sigma}_T + \bar{\sigma}_T^2 + \bar{\tau}_{LT}^2 + \bar{F}_1\bar{\sigma}_L + \bar{F}_2\bar{\sigma}_T = 1 \qquad (7.107)$$

where the following non-dimensional stress components are defined

$$\bar{\sigma}_L = \frac{\sigma_L}{(\sigma_{L_u}^t \sigma_{L_u}^c)^{1/2}} \quad , \quad \bar{\sigma}_T = \frac{\sigma_T}{(\sigma_{T_u}^t \sigma_{T_u}^c)^{1/2}} \quad , \quad \bar{\tau}_{LT} = \frac{\tau_{LT}}{\tau_u} \qquad (7.108)$$

as well as the following non-dimensional coefficients

$$\bar{F}_1 = \frac{\sigma_{L_u}^c - \sigma_{L_u}^t}{(\sigma_{L_u}^t \sigma_{L_u}^c)^{1/2}} \quad , \quad \bar{F}_2 = \frac{\sigma_{T_u}^c - \sigma_{T_u}^t}{(\sigma_{T_u}^t \sigma_{T_u}^c)^{1/2}} \qquad (7.109)$$

Coefficient $\bar{F}_{LT}$ is yet undetermined. Clearly, this could be found by an additional test where both σ_L and σ_T are applied simultaneously (i.e. a biaxial stress state). This test is quite difficult to perform and, therefore, $\bar{F}_{LT}$ is often selected by fitting the prediction of the failure criterion to experimental data. The best fit has been found for $\bar{F}_{LT} = -1/2$ and the final expression for the Tsai-Wu criterion is [BC]

$$\bar{\sigma}_L^2 - \bar{\sigma}_L\bar{\sigma}_T + \bar{\sigma}_T^2 + \bar{\tau}_{LT}^2 + \bar{F}_1\bar{\sigma}_L + \bar{F}_2\bar{\sigma}_T = 1 \qquad (7.110)$$

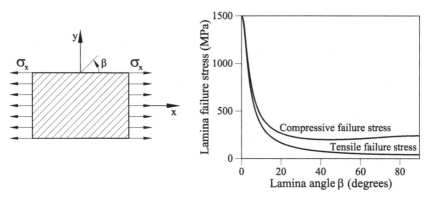

Fig. 7.20 (a) Failure test for a lamina at an angle β. (b) Variation of the tensile and compressive failure loads with lamina angle β

Example 7.3: Tsai-Wu failure criterion for uniaxial stress.

In this example taken from [BC] we apply the Tsai-Wu criterion to a simple test where a single stress component σ_x is applied to a lamina with fibers running at an angle β (Figure 7.20). The stress rotation formula (Eqs.(6.16)) yields the applied stresses in the fiber aligned triad as $\sigma_L = \sigma_x \cos^2 \beta$, $\sigma_T = \sigma_x \sin^2 \beta$ and $\tau_{LT} = -\sigma_x \sin \beta \cos \beta$.

The level of applied stress that correspond to failure satisfies the failure criterion (7.110), i.e.

$$\sigma_x^2 \left[\frac{C^4}{A^2} - \frac{S^2 C^2}{AB} + \frac{S^4}{B^2} + \frac{S^2 C^2}{\tau_u^2} \right] + \sigma_y \left[\frac{C^2}{A} + \frac{S^2}{B} \right] = 1 \qquad (7.111)$$

where $A = (\sigma_{L_u}^t \sigma_{L_u}^c)^{1/2}$, $B = (\sigma_{T_u}^t \sigma_{T_u}^c)^{1/2}$, $C = \sin \beta$ and $C = \cos \beta$.

The two solutions of the above second order equation yield the failure stress in tension and compression. Figure 7.20 shows the absolute value of these failure stresses as a function of the lamina angle β for the Graphite/Epoxy material (T300/5208) whose properties are given in Table 7.8. Note the sudden drop in the failure stress as the lamina angle moves away from zero degrees.

7.9.4 The reserve factor

The reserve factor, R, is defined as the number by which the applied stress can be multiplied to reach failure, i.e.

$$\sigma_{L_u}^t = R \sigma_L^t \quad , \quad \sigma_{L_u}^c = R \sigma_L^c \quad , \text{etc} \qquad (7.112)$$

From this definition it follows that

- $R = 1$ means that the applied stress strictly induces lamina failure.
- $R > 1$ means that the applied stress is below the failure level, i.e. the stress level is safe. For instance $R = 3$ would mean that the applied stress can be tripled before lamina failure occurs.
- $R < 1$ means that the applied stress is above the failure stress.

Assuming that failure can be predicted by the Tsai-Wu criterion and substituting in Eq.(7.110), the stress component by R times their value, the failure condition (7.110) can be written as [BC]

$$(\bar{\sigma}_L^2 - \bar{\sigma}_L\bar{\sigma}_T + \bar{\sigma}_T^2 + \bar{\tau}_{LT})R^2 + (\bar{F}_1\bar{\sigma}_L + \bar{F}_2\bar{\sigma}_T)R = 1 \qquad (7.113)$$

The positive root for R in Eq.(7.113) indicates the failure stress level, while the negative root gives the failure stress level when the sign of the applied stresses is reversed.

In general, the modules of these two roots are different since the failure stress levels in tension and in compression are not equal.

7.9.5 Some conclusions on failure criteria of laminates

Practical failure criteria of laminates are typically based on the classical laminated plate theory and typically assume elastic constant ply properties. A relevant design criterion is the Tsai-Wu criterion because it includes some interaction between stresses and different strengths in tension or compression and residual and hygrothermal stresses can be incorporated without difficulties. It also involves simple algebraic calculations and thus is easier to apply than the criteria involving inequalities.

The performance of a given criterion always has to be validated using some well-founded mechanical experiments on materials and structures. Moreover, the applicability of a criterion is usually a function of many external (loading history, environmental conditions) and internal (rheological properties of the matrix) parameters. In particular, the post first ply failure prediction requires particular assumptions on the damage process. Failure criteria that can be easily used in structural design are usually not directly related to failure modes and their couplings. Hence, they only have to be regarded as empirical and oversimplified ways to predict occurrence of some internal damage, i.e. of the irreversible and unacceptable loss of the load bearing capabilities of a laminate [PP3,Red2].

A more accurate prediction of failure in composite laminates is possible using non linear finite element methods that incorporate sophisticated material model for characterizing onset of play failure and its evolution in the laminate as well as the effect of ply delamination [OO].

7.10 MODELING OF DELAMINATION VIA ZIGZAG THEORY

Delamination effects in composite laminated plates due to interlamina failure can be effectively reproduced with the refined zigzag theory (RZT) theory presented in Section 7.8.1. Similarly, to beams, the delamination model simply implies introducing a very thin "interface layer" between adjacent material layers. Delamination is produced when the material properties of the "interface layer" are reduced to almost a zero value in comparison to those of the adjacent layers due to interlamina failure. The RZT can take into account the reduction of the overall plate stiffness due to the failure of the interface layer. Moreover, the refined zigzag model can accurately represent the jump in the axial displacement field across the interphase layer, as well as the change in the axial and tangential stress distribution over the plate section.

Figures 7.21 and 7.22 show an example of the capabilities of the QLRZ element of Section 7.8.2 to model delamination. The problem represents the analysis of a SS plate under uniformly distributed loading. The plate has a 9 layers of composite material which properties are shown in Table 7.7 (material C2). Delamination is modelled by progressively reducing the value of the shear modulus at the fourth layer (starting from the bottom), as shown in Table 7.9. The fourth layer is therefore assumed to act as an interface layer that induces delamination between the third and fifth layers.

Shear modules values for layer 4					
Model No.	G(MPa)	Model No.	G(MPa)	Model No.	G(MPa)
1	4000	4	0,4	7	0.0004
2	40	5	0,04		
3	4	6	0,004		

Table 7.9 Shear modulus for the delamination layer in square SS plate

Figure 7.21 shows results for the central deflection value in terms of the properties of the interface layer. Note that the deflection increases one order of magnitude versus the non-delaminated case. It is also interesting that the deflection does not change after the material properties of the interface layer are reduced beyond 6 orders of magnitude. Results agree

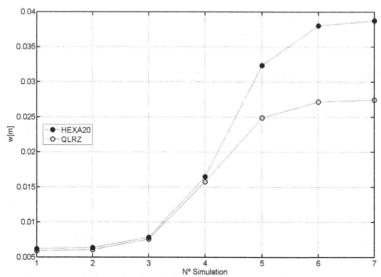

Fig. 7.21 Delamination study in SS square plate under uniformly distributed loading. Values of central deflection for each of the material models studied

reasonably well with those obtained with the 3D model of Figure 7.12b introducing a similar reduction in the material properties and an *ad hoc* interface layer. The discrepancy for the smaller values of the shear modulus for the interface layer (G_5, G_6 and G_7) is due to the intrinsic higher stiffness of the plate bending model versus the 3D model.

The jump of the axial displacement across the thickness at the interface layer during delamination can be clearly seen in Figure 7.22. Again note that the displacement jump remains stationary after a reduction of the material properties in the interphase layer of 8 orders of magnitude. Results agree well with the 3D solution also shown in the figure .

The example shows clearly the capability of the QLRZ element to model a complex phenomena such as delamination *without introducing additional kinematic variables*. More evidence of the good behaviour of the QLRZ element for delamination studies can be found in [OEO2].

7.11 EDGE DELAMINATION IN COMPOSITE LAMINATES

The simplest theory for determining the in-plane elastic response of a laminated composite assumes that a state of plane stress exists for symmetric laminates under in-plane traction, wherein the interlaminar stress components vanish. This assumption is accurate for interior regions far

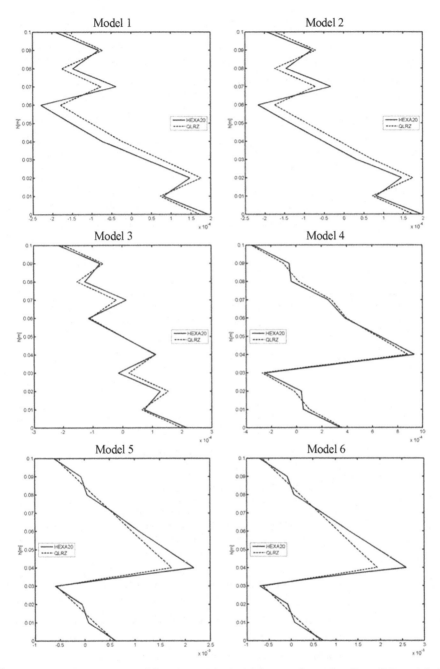

Fig. 7.22 Delamination in SS square plate under uniform loading. Material C2 of Table 7.7. Thickness distribution of in-plane displacement u at point B for models 1–6. Results are compared with a 3D solution (HEXA20)

from geometric discontinuities, such as free edges. However, there exists a boundary layer, near the laminate free edges, where the stress state is three-dimensional. There is a stress transfer between plies through the action of interlaminar stresses that are substantially increased in that region [On2]. The width of this boundary layer is a function of the elastic properties of the laminate, its fibre orientations and the laminate geometry. A simple rule proposed by Pipes and Pagano [PP1] for hard-polymer-matrix composites is that the thickness of the boundary layer is approximately equal to the laminate thickness.

The primary consequences of the laminate boundary layer are delamination induced failures that initiate within this region and distortions due to the presence of interlaminar stress components (Figure 7.23).

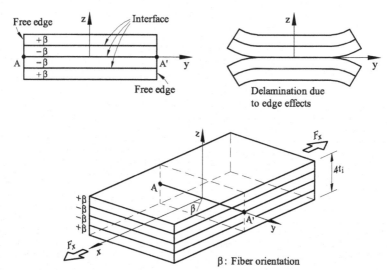

Fig. 7.23 Schematic representation of the deformed cross section of a laminate caused by delamination due to edge effects [On4]

The laminate stacking sequence may have a significant influence upon the values and distribution of the interlaminar stresses within the boundary layer causing delamination. Pagano and Pipes [PP2] analyzed the stacking sequence in a boron-epoxy angel ply laminate under uniaxial tension with built-in thermal stresses assumed. Numerical results show that the interlaminar tensile stress (which is the most probable cause of delamination failure) is highest in a $[\pm 15/\pm 45]_s$ laminate becoming much lower in a $[15/\pm 45/-15]_s$ orientation. On the other hand, a $[\pm 45/\pm 15]_s$ orientation gives the highest normal compressive stresses, and a $[45/\pm 15/-45]_s$ orientation is optimum because both interlaminar shear and normal stresses are kept to a minimum. Similar results were observed experimentally by Foye and Baker [FB].

Kim and Soni [KS2] performed investigations towards the suppression of free-edge delamination by reduction of the predominant interlaminar stress components. Their conclusions are that the mixing of S-glass and carbon reinforced plies reduced the interlaminar stress values. The amount of this reduction varies depending upon the laminate type, including the orientation and the location of the S-glass plies. As a general rule, laminate resistance to delamination is enhanced by hybridization.

Soni and Kim [SK] studied the suppression of free-edge delamination by introducing an adhesive layer at appropriate interfaces of high interlaminar stresses. This procedure leads to significant higher ultimate strength of laminates.

7.12 CONCLUDING REMARKS

The study of composite laminated plates with the FEM requires introducing in-plane (membrane) effects into the standard plate bending theories. In this chapter we have derived a family of composite laminated plate elements based on extensions of classical Reissner-Mindlin plate theory. These elements account for transverse shear deformation effects which are important in composite laminated plates and, therefore, in general are more accurate than elements derived via the traditional Kirchhoff thin plate theory.

Higher order formulations based on the layer-wise and the refined zigzag theory (RFT) have also been explained. We note the simplicity and accuracy of the RFT for reproducing the complex distribution of the in-plane displacements and the stresses across the thickness in a laminated plate. The RZF has also possibilities for modeling delamination effects and, as such, it is indeed a very promising procedure for linear and non-linear analysis of composite laminated plates with the FEM.

8

ANALYSIS OF SHELLS WITH FLAT ELEMENTS

8.1 INTRODUCTION

A shell can be seen, in essence, as the extension of a plate to a non-planar surface. The non-coplanarity introduces axial (membrane) forces in addition to flexural (bending and shear) forces, thus providing a higher overall structural strength.

Shell-type structures are common in many engineering constructions such as roofs, domes, bridges, containment walls, water and oil tanks and silos, as well as in airplane and spacecraft fuselages, ship hulls, automobile bodies, mechanical parts, etc.

The way in which a shell supports external loads by the combined action of axial and flexural effects is similar to that of arches and frame structures. Thus, while a beam and a plate typically resist the external forces by flexural effects only, frames, arches and shells offer a higher resistance to load due to the coupled action of axial and flexural forces. A good structural knowledge of frames and arches is therefore helpful to understand the behaviour of shells. Figure 8.1 shows a simple scheme of the axial forces acting on a plane frame and on a folded shell formed by assembly of two plates.

It is important to understand that the coupling of membrane and flexural effects can also occur in flat shells made of composite laminated material. This case was studied in the previous chapter when dealing with composite laminated plates.

Shells are typically classified by the shape of their middle surface. In this book we will study shells with arbitrary shape (Chapters 8 and 10), axisymmetric shells (Chapter 9) and prismatic shells (Chapter 11).

E. Oñate, *Structural Analysis with the Finite Element Method. Linear Statics: Volume 2: Beams, Plates and Shells*, Lecture Notes on Numerical Methods in Engineering and Sciences, DOI 10.1007/978-1-4020-8743-1_8,

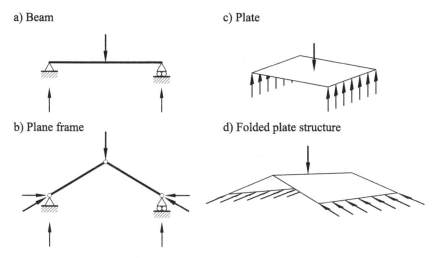

Fig. 8.1 Axial forces in frames and folded plate structures

The governing equations of a curved shell (equilibrium and kinematic equations, etc.) are quite complex due to the curvature of the middle surface [Fl,Kr,Ni,No2,TW,Vl2]. A way of overcoming this problem is considering the shell as formed by a number of folded plates (Figures 8.2 and 8.3). This is precisely the approach to be followed in this chapter.

The chapter stars with the formulation of flat shell elements as a direct extension of the Reissner-Mindlin thick plate theory studied in Chapter 5. It will be shown that in many cases the element stiffness matrix can be formed by assembling the flexural and axial contributions of the corresponding plate and plane stress elements, similarly as for straight members in frames. The second part of the chapter deals with flat shell elements following Kirchhoff thin folded plate theory. The derivation of rotation-free thin shell triangles as an extension of the rotation-free plate elements of Section 4.8 is also presented. The chapter concludes with a description of some higher order theories for modelling composite laminated shells.

8.2 FLAT SHELL THEORY

A "shell element" combines a flexural (bending and shear) behaviour and an "in-plane" (membrane) one. The membrane state induces axial forces contained in the shell middle surface. If the shell element is flat then the flexural and in-plane states are typically decoupled at the element level, an exception being the case of composite laminated shells. This

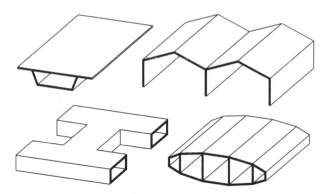

Fig. 8.2 Examples of folded plate structures

a)   b)

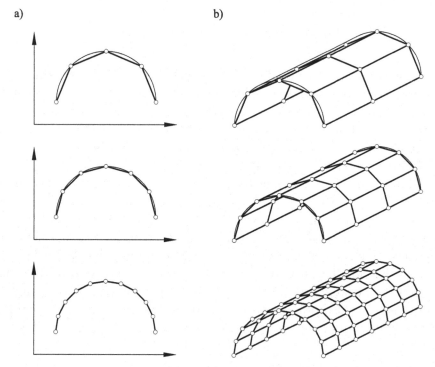

Fig. 8.3 a) Discretization of a curved arch in segments. b) Discretization of a cylindrical surface in flat rectangular elements

decoupling extends to the element stiffness matrix which is formed by a simple superposition of the flexural and membrane contributions. The full flexural-membrane coupling appears when flat elements meeting at different angles are assembled in the global stiffness matrix.

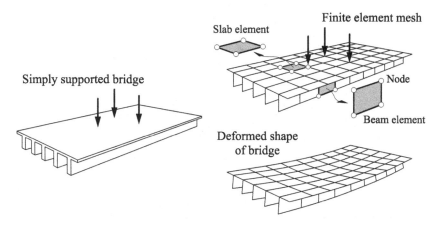

Fig. 8.4 Discretization of slab-beam bridge in rectangular flat shell elements

Flat elements are "natural" for folded plate structures such as bridges, plane roofs and some mechanical parts (Figure 8.2) and can also be used to discretize a curved shell as shown in Figure 8.3. Figure 8.4 shows the discretization of a slab-beam bridge in rectangular flat shell elements

Consequently, flat elements provide a general procedure for analysis of shells of arbitrary shape. The simplicity of their formulation versus the more complex curved shell elements (Chapter 10) makes flat shell elements a good and popular option for practical purposes.

The formulation of flat shell elements follows the lines of the previous chapters. First the basic kinematic, constitutive and equilibrium (virtual work) equations will be derived for an individual element. Then the global assembly process will be studied. The formulation based on Reissner-Mindlin theory (adequate for thin and thick situations) will be presented first. Thin flat shell elements based on Kirchhoff theory will be studied in the second half of the chapter.

8.3 REISSNER-MINDLIN FLAT SHELL THEORY

8.3.1 Displacement field

Let us a consider the rectangular shell domain of Figure 8.5 defined in a global coordinate system x, y, z. As in plates, the middle plane is taken as the reference surface for the kinematic description. A local system x', y', z' is defined where z' is the normal to the middle plane and x', y' are two arbitrary orthogonal directions contained in it. These directions will be as-

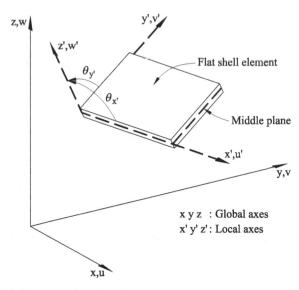

Fig. 8.5 Rectangular flat shell domain. Local and global axes

sumed to coincide with two adjacent sides of the rectangular shell domain, for simplicity. A more general definition for the local axes will be introduced in a later section. The deformation of the domain points referred to the local coordinate system will be considered next.

We will assume that Reissner-Mindlin assumptions for the normal rotation hold (Section 6.2). Accordingly, the displacements of a point A along the normal direction OA are expressed as (Figure 8.6)

$$
\begin{aligned}
u'(x', y', z') &= u'_0(x', y') - z'\theta_{x'}(x', y') \\
v'(x', y', z') &= v'_0(x', y') - z'\theta_{y'}(x', y') \\
w'(x', y', z') &= w'_0(x', y')
\end{aligned}
\tag{8.1}
$$

where u'_0, v'_0 and w'_0 are the displacements of point O over the middle plane along the local directions x', y' and z', respectively; $\theta_{x'}$ and $\theta_{y'}$ are the rotations of the normal OA contained in the local planes $x'z'$ and $y'z'$, respectively, and $z' = \overline{OA}$. The local displacement vector of point A is

$$
\mathbf{u}' = \left[u'_0, v'_0, w'_0, \theta_{x'}, \theta_{y'}\right]^T
\tag{8.2}
$$

In addition to the "bending" displacements $(w'_0, \theta_{x'}, \theta_{y'})$, the middle plane points have in-plane displacements (u'_0, v'_0). This in-plane motion introduces membrane strains and axial forces.

The above displacement field is analogous to that used for composite laminated plates (Section 7.2) where the different material properties

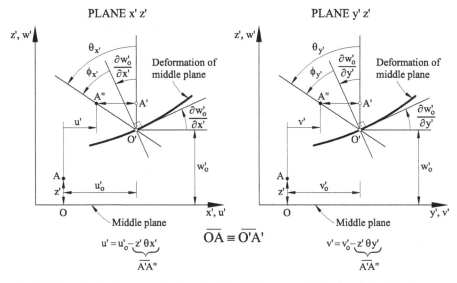

Fig. 8.6 Local displacements in a flat shell element. Reissner-Mindlin theory

across the thickness induced in-plane displacements in addition to the bending modes. *The kinematics of a flat shell element are in fact identical to those of a composite laminate plate, if the local coordinate system in the shell is made coincident with the global system in the plate.*

8.3.2 Strain field

As for plates, the normal strain $\varepsilon_{z'}$ does not play any role in the internal work due to the plane stress assumption ($\sigma_{z'} = 0$). The relevant strains are written in the *local axes* using Eqs.(7.3) as

$$
\varepsilon' = \left\{ \begin{array}{c} \varepsilon_{x'} \\ \varepsilon_{y'} \\ \gamma_{x'y'} \\ \cdots \\ \gamma_{x'z'} \\ \gamma_{y'z'} \end{array} \right\} = \left\{ \begin{array}{c} \dfrac{\partial u'}{\partial x'} \\[2mm] \dfrac{\partial v'}{\partial y'} \\[2mm] \dfrac{\partial u'}{\partial y'} + \dfrac{\partial v'}{\partial x'} \\[2mm] \cdots \\[2mm] \dfrac{\partial u'}{\partial z'} + \dfrac{\partial w'}{\partial x'} \\[2mm] \dfrac{\partial v'}{\partial z'} + \dfrac{\partial w'}{\partial y'} \end{array} \right\} = \left\{ \begin{array}{c} \dfrac{\partial u'_0}{\partial x'} \\[2mm] \dfrac{\partial v'_0}{\partial y'} \\[2mm] \dfrac{\partial u'_0}{\partial y'} + \dfrac{\partial v'_0}{\partial x'} \\[2mm] \cdots \\[2mm] 0 \\[2mm] 0 \end{array} \right\} + \left\{ \begin{array}{c} -z'\dfrac{\partial \theta_{x'}}{\partial x'} \\[2mm] -z'\dfrac{\partial \theta_{y'}}{\partial y'} \\[2mm] -z'\left(\dfrac{\partial \theta_{x'}}{\partial y'} + \dfrac{\partial \theta_{y'}}{\partial x'}\right) \\[2mm] \cdots \\[2mm] \dfrac{\partial w'_0}{\partial x'} - \theta_{x'} \\[2mm] \dfrac{\partial w'_0}{\partial y'} - \theta_{y'} \end{array} \right\}
$$

(8.3)

or

$$\varepsilon' = \left\{ \begin{array}{c} \varepsilon'_p \\ \cdots \\ \varepsilon'_s \end{array} \right\} = \left\{ \begin{array}{c} \hat{\varepsilon}'_m \\ \cdots \\ 0 \end{array} \right\} + \left\{ \begin{array}{c} -z'\hat{\varepsilon}'_b \\ \cdots \\ \hat{\varepsilon}'_s \end{array} \right\} \tag{8.4}$$

i.e.

$$\begin{array}{c} \varepsilon'_p = \hat{\varepsilon}'_m - z'\hat{\varepsilon}'_b \\ \varepsilon'_s = \hat{\varepsilon}'_s \end{array} \tag{8.5}$$

Vectors ε'_p and ε'_s contain in-plane strains due to membrane-bending effects ($\varepsilon_{x'}$, $\varepsilon_{y'}$, $\gamma_{x'y'}$) and transverse shear strains ($\tau_{x'z'}$, $\tau_{y'z'}$), respectively and

$$\hat{\varepsilon}'_m = \left[\frac{\partial u'_0}{\partial x'}, \frac{\partial v'_0}{\partial y'}, \left(\frac{\partial u'_0}{\partial y'} + \frac{\partial v'_0}{\partial x'} \right) \right]^T \tag{8.6a}$$

$$\hat{\varepsilon}'_b = \left[\frac{\partial \theta_{x'}}{\partial x'}, \frac{\partial \theta_{y'}}{\partial y'}, \left(\frac{\partial \theta_{x'}}{\partial y'} + \frac{\partial \theta_{y'}}{\partial x'} \right) \right]^T \tag{8.6b}$$

$$\hat{\varepsilon}'_s = \left[\frac{\partial w'_0}{\partial x'} - \theta_{x'}, \frac{\partial w'_0}{\partial y'} - \theta_{y'} \right]^T = [-\phi_{x'}, -\phi_{y'}]^T \tag{8.6c}$$

are *generalized local strain vectors* due to membrane (elongations), bending (curvatures) and transverse shear effects, respectively. As in plates, the transverse shear strains $\gamma_{x'z'}$ and $\gamma_{y'z'}$ represent (with opposite sign) the transverse shear rotations $\phi_{x'}$ and $\phi_{y'}$, respectively (Eq.(8.6c)).

Eq.(8.4) shows that the total strains at a point are the sum of the membrane (axial) and flexural (bending and shear) contributions.

Eq.(8.4) can be rewritten as

$$\varepsilon' = S\hat{\varepsilon}' \tag{8.7}$$

where

$$\hat{\varepsilon}' = \left\{ \begin{array}{c} \hat{\varepsilon}'_m \\ \hat{\varepsilon}'_b \\ \hat{\varepsilon}'_s \end{array} \right\} \quad \text{and} \quad S = \begin{bmatrix} 1 & 0 & 0 & -z' & 0 & 0 & 0 & 0 \\ 0 & 1 & 0 & 0 & -z' & 0 & 0 & 0 \\ 0 & 0 & 1 & 0 & 0 & -z' & 0 & 0 \\ 0 & 0 & 0 & 0 & 0 & 0 & 1 & 0 \\ 0 & 0 & 0 & 0 & 0 & 0 & 0 & 1 \end{bmatrix} \tag{8.8}$$

are the generalized local strain vector and the transformation matrix relating 3D strains and generalized strains.

8.3.3 Stress field. Constitutive relationship

The stress-strain relationship of 3D elasticity is written in local axes in order to introduce the plane stress condition $(\sigma_{z'} = 0)$. The following relationship between the significant local stresses and strains is obtained after eliminating $\varepsilon_{z'}$,

$$
\boldsymbol{\sigma}' = \left\{ \begin{array}{c} \sigma_{x'} \\ \sigma_{y'} \\ \tau_{x'y'} \\ \cdots \\ \tau_{x'z'} \\ \tau_{y'z'} \end{array} \right\} = \left\{ \begin{array}{c} \boldsymbol{\sigma}'_p \\ \cdots \\ \boldsymbol{\sigma}'_s \end{array} \right\} = \left[\begin{array}{ccc} \mathbf{D}'_p & \vdots & \mathbf{0} \\ \cdots & \cdots & \cdots \\ \mathbf{0} & \vdots & \mathbf{D}'_s \end{array} \right] \left\{ \begin{array}{c} \varepsilon_{x'} \\ \varepsilon_{y'} \\ \gamma_{x'y'} \\ \cdots \\ \gamma_{x'z'} \\ \gamma_{y'z'} \end{array} \right\} + \left\{ \begin{array}{c} \sigma_{x'}^0 \\ \sigma_{y'}^0 \\ \tau_{x'y'}^0 \\ \cdots \\ \tau_{x'z'}^0 \\ \tau_{y'z'}^0 \end{array} \right\} = \mathbf{D}'\boldsymbol{\varepsilon}' + \boldsymbol{\sigma}'^0
$$

$$\tag{8.9a}$$

with

$$
\boldsymbol{\sigma}'^0 = \left\{ \begin{array}{c} \boldsymbol{\sigma}_p'^0 \\ \boldsymbol{\sigma}_s'^0 \end{array} \right\} \quad , \quad \boldsymbol{\sigma}_p'^0 = [\sigma_{x'}^0, \sigma_{y'}^0, \tau_{x'y'}^0]^T \quad , \quad \boldsymbol{\sigma}_s'^0 = [\tau_{x'z'}^0, \tau_{y'z'}^0]^T \tag{8.9b}
$$

where $\boldsymbol{\sigma}'_p$ and $\boldsymbol{\sigma}'_s$ contain the in-plane stresses $(\sigma_{x'}, \sigma_{y'}, \tau_{x'y'})$ and the transverse shear stresses $(\tau_{x'z'}, \tau_{y'z'})$ in local axes, respectively and $\boldsymbol{\sigma}'^0$ is an initial stress vector. For isotropic material

$$
\mathbf{D}'_p = \frac{E}{1-\nu^2} \begin{bmatrix} 1 & \nu & 0 \\ \nu & 1 & 0 \\ 0 & 0 & \frac{1-\nu}{2} \end{bmatrix} \quad , \quad \mathbf{D}'_s = G \begin{bmatrix} 1 & 0 \\ 0 & 1 \end{bmatrix} \tag{8.10}
$$

Figure 8.7 shows the distribution of the stresses across the shell thickness for homogeneous material.

Let us consider now a composite laminated shell formed by a number of orthotropic layers with orthotropy axes L, T, z' with $\vec{z}' = \vec{n}$ and satisfying the plane anisotropy conditions (Figure 8.8). The constitutive equation in the orthotropy axes L, T, z' for each layer is written as

$$
\boldsymbol{\sigma}_I = \mathbf{D}_I \boldsymbol{\varepsilon}_I \tag{8.11}
$$

where vectors $\boldsymbol{\sigma}_I$, $\boldsymbol{\varepsilon}_I$ and matrix $\mathbf{D}_I$ are given by Eqs.(7.8)-(7.11).

The constitutive matrices in local axes x', y', z' are found as explained in Section 7.2.3 for composite laminated plates. The result is

$$
\mathbf{D}'_p = \mathbf{T}_1^T \mathbf{D}_1 \mathbf{T}_1, \quad \mathbf{D}'_s = \mathbf{T}_2^T \mathbf{D}_2 \mathbf{T}_2 \tag{8.12}
$$

The constitutive matrices $\mathbf{D}_1$ and $\mathbf{D}_2$ are given in Eq.(7.9) and the transformation matrices $\mathbf{T}_1$ and $\mathbf{T}_2$ in Eq.(6.15b).

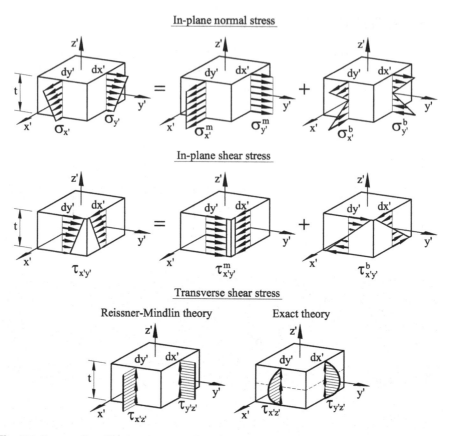

Fig. 8.7 Stress distribution across the shell thickness for homogeneous material. Membrane $(\cdot)^m$ and bending $(\cdot)^b$ contributions to the in-plane stresses

The initial stress vector $\boldsymbol{\sigma}'^0$ of Eq.(8.9a) induced by a temperature increment is

$$\boldsymbol{\sigma}'^0 = \left\{ \begin{array}{c} \boldsymbol{\sigma}_p'^0 \\ \cdots \\ 0 \\ 0 \end{array} \right\} \quad \text{with} \quad \boldsymbol{\sigma}_p'^0 = -\mathbf{T}_1^T \mathbf{D}_1 [\alpha_L \Delta T, \ \alpha_T \Delta T, \ 0]^T \qquad (8.13)$$

where α_L and α_T are the thermal expansion coefficients in the orthotropy directions L and T, respectively, and ΔT is the temperature increment. For isotropic material $\beta = 0$, $\alpha_L = \alpha_T = \alpha$ and

$$\boldsymbol{\sigma}_p'^0 = -\frac{E\alpha\Delta T}{1 - \nu^2} [1, 1, 0]^T \qquad (8.14)$$

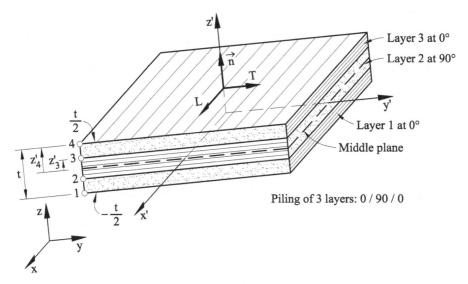

Fig. 8.8 Composite laminated flat shell element. Piling of three layers at $0/90/0$

Substituting Eqs.(8.5) into (8.9a) gives the relationship between the local stresses and the generalized local strains at a point as

$$\boldsymbol{\sigma}' = \begin{Bmatrix} \boldsymbol{\sigma}'_p \\ \boldsymbol{\sigma}'_s \end{Bmatrix} = \mathbf{D}' \begin{bmatrix} \hat{\boldsymbol{\varepsilon}}'_m - z'\hat{\boldsymbol{\varepsilon}}'_b \\ \hat{\boldsymbol{\varepsilon}}'_s \end{bmatrix} + \boldsymbol{\sigma}'^0 = \mathbf{D}'\mathbf{S}\hat{\boldsymbol{\varepsilon}}' + \boldsymbol{\sigma}'^0 \tag{8.15}$$

Eq.(8.14) shows that the stresses $\sigma_{x'}$, $\sigma_{y'}$ and $\tau_{x'y'}$ vary linearly across the shell thickness and they are not necessarily zero for $z' = 0$. This is analogous to the distribution of normal stresses in beam cross sections under axial and bending forces, i.e. the total stress is obtained as the sum of a uniform stress field due to the axial forces and a symmetric linear bending stress field. This is shown in Figure 8.7 where superindices m and b denote the membrane and bending contributions to the stress field.

Eq.(8.15) evidences that the tangential stresses $\tau_{x'z'}$ and $\tau_{y'z'}$ are constant across the thickness. Hence, the transverse shear moduli must be properly modified, specially for composite laminated material [Co1]. The correct distribution of the tangential stresses can be computed "a posteriori" using the transverse shear forces, similarly as for plates. The exact distribution of the tangential stresses is parabolic for homogeneous material (Figure 8.7).

For composite laminated shells the distribution of stresses across the thickness follows the pattern of Figure 7.3.

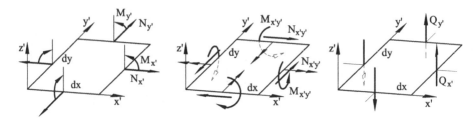

Fig. 8.9 Sign convention for resultant stresses in a flat shell element

8.3.4 Resultant stresses and generalized constitutive matrix

The resultant stress vector at a point of the shell middle plane is

$$
\hat{\sigma}' = \left\{ \begin{array}{c} \hat{\sigma}'_m \\ \cdots \\ \hat{\sigma}'_b \\ \cdots \\ \hat{\sigma}'_s \end{array} \right\} = \left\{ \begin{array}{c} N_{x'} \\ N_{y'} \\ N_{x'y'} \\ \cdots \\ M_{x'} \\ M_{y'} \\ M_{x'y'} \\ \cdots \\ Q_{x'} \\ Q_{y'} \end{array} \right\} = \int_{-\frac{t}{2}}^{\frac{t}{2}} \left\{ \begin{array}{c} \sigma'_p \\ \cdots \\ z' \sigma'_p \\ \cdots \\ \sigma'_s \end{array} \right\} dz' = \int_{-\frac{t}{2}}^{\frac{t}{2}} \mathbf{S}^T \sigma' \, dz' \qquad (8.16)
$$

where $\hat{\sigma}'_m$, $\hat{\sigma}'_b$ and $\hat{\sigma}'_s$ are resultant stress vectors corresponding to membrane, bending and transverse shear effects and t is the shell thickness. The definition of axial forces, bending moments and transverse shear forces coincides with that for composite laminated plates (Section 7.2.4 and Figures 7.3 and 8.9). The resultant stresses have units of moment or force per unit width of the shell surface.

Introducing Eq.(8.9a) into (8.16) and using Eq.(8.7) gives the relationship between resultant stresses and local generalized strains (including initial stresses) as

$$
\hat{\sigma}' = \left\{ \begin{array}{c} \hat{\sigma}'_m \\ \cdots \\ \hat{\sigma}'_b \\ \cdots \\ \hat{\sigma}'_s \end{array} \right\} = \int_{-\frac{t}{2}}^{\frac{t}{2}} \mathbf{S}^T (\mathbf{D}' \varepsilon' + \sigma'^0) dz' = \hat{\mathbf{D}}' \hat{\varepsilon}' + \hat{\sigma}'^0 \qquad (8.17)
$$

where the generalized constitutive matrix is

$$\hat{\mathbf{D}}' = \int_{-\frac{t}{2}}^{\frac{t}{2}} \mathbf{S}^T \mathbf{D}' \mathbf{S} dz' = \int_{-\frac{t}{2}}^{\frac{t}{2}} \begin{bmatrix} \mathbf{D}'_p & -z' \mathbf{D}'_p & 0 \\ -z' \mathbf{D}'_p & z'^2 \mathbf{D}'_p & 0 \\ 0 & 0 & \mathbf{D}'_s \end{bmatrix} dz' = \begin{bmatrix} \hat{\mathbf{D}}'_m & \hat{\mathbf{D}}'_{mb} & 0 \\ \hat{\mathbf{D}}'_{mb} & \hat{\mathbf{D}}'_b & 0 \\ 0 & 0 & \hat{\mathbf{D}}'_s \end{bmatrix}$$

$$\hat{\mathbf{D}}'_m = \int_{-\frac{t}{2}}^{\frac{t}{2}} \mathbf{D}'_p dz' \quad ; \quad \hat{\mathbf{D}}'_{mb} = -\int_{-\frac{t}{2}}^{\frac{t}{2}} z' \mathbf{D}'_p dz'$$

$$\hat{\mathbf{D}}'_b = \int_{-\frac{t}{2}}^{\frac{t}{2}} z'^2 \mathbf{D}'_p dz' \quad ; \quad \hat{\mathbf{D}}'_s = \begin{bmatrix} k_{11} \bar{D}'_{s_{11}} & k_{12} \bar{D}'_{s_{12}} \\ \text{Sym.} & \bar{k}_{22} \bar{D}'_{s_{22}} \end{bmatrix} \tag{8.18a}$$

with

$$\bar{D}_{s_{ij}} = \int_{-\frac{t}{2}}^{\frac{t}{2}} D'_{s_{ij}} dz' \tag{8.18b}$$

and

$$\hat{\boldsymbol{\sigma}}'^0 = \begin{Bmatrix} \hat{\boldsymbol{\sigma}}'^0_m \\ \hat{\boldsymbol{\sigma}}'^0_b \\ \hat{\boldsymbol{\sigma}}'^0_s \end{Bmatrix} = \int_{-\frac{t}{2}}^{\frac{t}{2}} \mathbf{S}^T \boldsymbol{\sigma}'^0 dz' \tag{8.19a}$$

or

$$\hat{\boldsymbol{\sigma}}'^0_m = \int_{-\frac{t}{2}}^{\frac{t}{2}} \boldsymbol{\sigma}'^0_p dz' \quad , \quad \hat{\boldsymbol{\sigma}}'^0_b = \int_{-\frac{t}{2}}^{\frac{t}{2}} \boldsymbol{\sigma}'^0_p dz' \quad , \quad \hat{\boldsymbol{\sigma}}'^0_s = \int_{-\frac{t}{2}}^{\frac{t}{2}} z' \boldsymbol{\sigma}'^0_s dz' \tag{8.19b}$$

In above $\hat{\mathbf{D}}'_m$, $\hat{\mathbf{D}}'_b$ and $\hat{\mathbf{D}}'_s$ are the membrane, bending and transverse shear constitutive matrices, respectively; $\hat{\mathbf{D}}'_{mb}$ is the *membrane-bending coupling* constitutive matrix and $\hat{\boldsymbol{\sigma}}'^0$ is the initial resultant stress vector deduced from the initial stress field. All matrices are symmetrical.

The computation of the shear correction parameters is performed as explained for composite laminated plates in Section 7.3. For isotropic material $k_{11} = k_{22} = 5/6$ and $k_{12} = 0$.

An arbitrary initial stress field induces axial forces as well as bending moments in $\hat{\boldsymbol{\sigma}}'^0$. For internal thermal stresses the temperature increment is typically defined in both shell faces and a linear distribution is accepted across the thickness direction. This simplifies the computation of $\hat{\boldsymbol{\sigma}}'^0$.

For a composite shell formed by n_l orthotropic layers with constant material properties within each layer, matrices $\hat{\mathbf{D}}'_m$, $\hat{\mathbf{D}}'_b$ and $\hat{\mathbf{D}}'_{mb}$ can be computed by

$$\hat{\mathbf{D}}'_m = \sum_{i=1}^{n_l} \mathbf{D}'_{m_i} \Delta z'_i \quad , \quad \hat{\mathbf{D}}'_b = \frac{1}{3} \sum_{i=1}^{n_l} \mathbf{D}'_{m_i} (z'^3_{i+1} - z'^3_i)$$

$$\hat{\mathbf{D}}'_{mb} = -\frac{1}{2} \sum_{i=1}^{n_l} \mathbf{D}'_{m_i} (z^2_{i+1} - z^2_i) \tag{8.20}$$

where $\Delta z_i' = z_{i+1}' - z_i'$ is the layer thickness (Figure 8.8).

This formulation can also be applied for introducing the effect of steel bars in reinforced concrete shells.

Indeed, the position of *a neutral plane* can be found. Taking the neutral surface as the reference surface leads to the decoupling of the bending and membrane effects at each point. Finding the neutral plane for heterogeneous materials is a tedious task, and, hence, the middle plane is usually chosen as the reference surface.

If the material properties are homogeneous, or symmetric with respect to the middle plane $\hat{\mathbf{D}}'_{mb} = 0$, the *middle plane is a neutral plane* and each local resultant stress vector can be computed from the corresponding generalized local strains in a decoupled manner as

$$\hat{\boldsymbol{\sigma}}'_m = \hat{\mathbf{D}}'_m \hat{\boldsymbol{\varepsilon}}'_m + \hat{\boldsymbol{\sigma}}'^0_m \quad ; \quad \hat{\boldsymbol{\sigma}}'_b = \hat{\mathbf{D}}'_b \hat{\boldsymbol{\varepsilon}}'_b + \hat{\boldsymbol{\sigma}}'^0_b \quad ; \quad \hat{\boldsymbol{\sigma}}'_s = \hat{\mathbf{D}}'_s \hat{\boldsymbol{\varepsilon}}'_s + \hat{\boldsymbol{\sigma}}'^0_s \quad (8.21)$$

For *homogeneous material* we obtain the standard expressions

$$\hat{\mathbf{D}}'_m = t\mathbf{D}'_p; \qquad \hat{\mathbf{D}}'_b = \frac{t^3}{12}\mathbf{D}'_p \quad \text{and} \quad \hat{\mathbf{D}}'_s = \frac{5}{6}t\mathbf{D}'_s \qquad (8.22)$$

8.3.5 Principle of Virtual Work

The PVW for a flat shell is

$$\iiint_V \delta\boldsymbol{\varepsilon}'^T \boldsymbol{\sigma}' dV = \iint_A \delta\mathbf{u}'^T \mathbf{t}' dA + \sum_i \delta\mathbf{u}_i'^T \mathbf{p}_i' \qquad (8.23)$$

where V and A are the shell volume and the area of the shell surface, respectively,

$$\mathbf{t}' = \left[f_{x'}, f_{y'}, f_{z'}, m_{x'}, m_{y'}\right]^T \qquad (8.24)$$

is the surface load vector, $f_{x'}$, $f_{y'}$, $f_{z'}$ are distributed loads acting on the shell surface in the local directions x', y', z', respectively; $m_{x'}$, $m_{y'}$ are distributed moments contained in the planes $x'z'$ and $y'z'$, respectively, and

$$\mathbf{p}_i' = [P_{x_i'}, P_{y_i'}, P_{z_i'}, M_{x_i'}, M_{y_i'}]^T \qquad (8.25)$$

are concentrated loads and moments. Substituting Eqs.(8.4) and (8.9a) into the l.h.s. of (8.23) gives (neglecting the initial strains)

$$\iiint_V \delta\boldsymbol{\varepsilon}'^T \boldsymbol{\sigma}' dV = \iiint_V \delta\left[\hat{\boldsymbol{\varepsilon}}'^T_m - z'\hat{\boldsymbol{\varepsilon}}'^T_b, \hat{\boldsymbol{\varepsilon}}'^T_s\right]\begin{Bmatrix}\hat{\boldsymbol{\sigma}}'_p \\ \hat{\boldsymbol{\sigma}}'_s\end{Bmatrix}dV =$$

$$= \iiint_V \left(\delta\hat{\varepsilon}_m'^T \boldsymbol{\sigma}_p' - z' \delta\hat{\varepsilon}_b'^T \boldsymbol{\sigma}_p' + \delta\hat{\varepsilon}_s'^T \boldsymbol{\sigma}_c' \right) dV =$$

$$= \iint_A \left[\delta\hat{\varepsilon}_m'^T \underbrace{\left(\int_{-\frac{t}{2}}^{+\frac{t}{2}} \boldsymbol{\sigma}_p' dz' \right)}_{\hat{\boldsymbol{\sigma}}_m'} + \delta\hat{\varepsilon}_b'^T \underbrace{\left(\int_{-\frac{t}{2}}^{+\frac{t}{2}} -z' \boldsymbol{\sigma}_p' dz' \right)}_{\hat{\boldsymbol{\sigma}}_b'} + \right.$$

$$\left. + \; \delta\hat{\varepsilon}_c'^T \underbrace{\left(\int_{-\frac{t}{2}}^{+\frac{t}{2}} \boldsymbol{\sigma}_s' dz' \right)}_{\hat{\boldsymbol{\sigma}}_s'} \right] dA =$$

$$= \iint_A (\delta\hat{\varepsilon}_m'^T \hat{\boldsymbol{\sigma}}_m' + \delta\hat{\varepsilon}_b'^T \hat{\boldsymbol{\sigma}}_b' + \delta\hat{\varepsilon}_s'^T \hat{\boldsymbol{\sigma}}_s') dA = \iint_A \delta\hat{\varepsilon}'^T \hat{\boldsymbol{\sigma}}' dA \qquad (8.26)$$

i.e. *the internal virtual work is obtained as the sum of the membrane, bending and transverse shear* contributions. Also the integration domain reduces from three to two dimensions, similarly as for plates.

The PVW can therefore be finally written as

$$\iint_A \delta\hat{\varepsilon}'^T \hat{\boldsymbol{\sigma}}' dA = \iint_A \delta\mathbf{u}'^T \mathbf{t}' dA + \sum_i \delta\mathbf{u}_i'^T \mathbf{p}_i' \qquad (8.27)$$

Note that all the derivatives in the integrals in Eq.(8.27) are of first order, which allows us to use C^0 continuous elements.

8.4 REISSNER-MINDLIN FLAT SHELL ELEMENTS

8.4.1 Discretization of the displacement field

Let us consider the shell surface discretized into C^0 isoparametric flat shell elements with n nodes. Figure 8.10 shows a discretization into 8-noded rectangles. Each element is contained in the local plane x', y'. The definition of this plane and the advantages of choosing a particular element will be discussed later. The local displacements are interpolated as

$$\mathbf{u}' = \sum_{i=1}^n \mathbf{N}_i \mathbf{a}_i'^{(e)} = [\mathbf{N}_1, \mathbf{N}_2, \cdots, \mathbf{N}_n] \left\{ \begin{array}{c} \mathbf{a}_1'^{(e)} \\ \mathbf{a}_2'^{(e)} \\ \vdots \\ \mathbf{a}_n'^{(e)} \end{array} \right\} = \mathbf{N}\mathbf{a}'^{(e)} \qquad (8.28a)$$

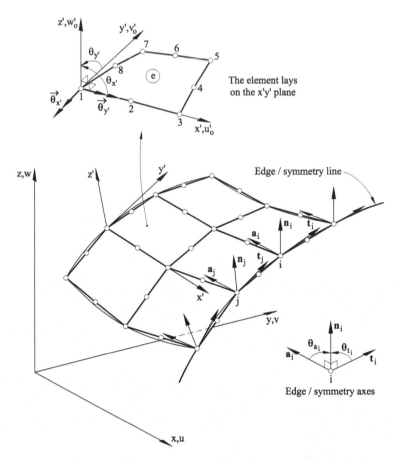

Fig. 8.10 Shell discretized with 8-noded flat rectangles. Local coordinate system for the element (x', y', z'). Local axes at an edge node i (x_i, t_i, n_i)

where

$$\mathbf{N}_i = \begin{bmatrix} N_i & 0 & 0 & 0 & 0 \\ 0 & N_i & 0 & 0 & 0 \\ 0 & 0 & N_i & 0 & 0 \\ 0 & 0 & 0 & N_i & 0 \\ 0 & 0 & 0 & 0 & N_i \end{bmatrix} \quad ; \quad \mathbf{a}_i^{\prime(e)} = \begin{bmatrix} u_{0_i}', v_{0_i}', w_{0_i}', \theta_{x_i'}, \theta_{y_i'} \end{bmatrix}^T \quad (8.28\mathrm{b})$$

are the shape function matrix and the local displacement vector of a node i, respectively. This vector contains the in-plane displacements u_{0_i}' and v_{0_i}', the lateral displacement w_{0_i}' and the local rotations $\theta_{x_i'}$ and $\theta_{y_i'}$.

8.4.2 Discretization of the generalized strain field

Substituting Eq.(8.28a) into the expression for the generalized local strain vector (8.8) gives (using Eqs.(8.6))

$$
\hat{\varepsilon}' = \left\{ \begin{array}{c} \hat{\varepsilon}'_m \\ \cdots \\ \hat{\varepsilon}'_b \\ \cdots \\ \hat{\varepsilon}'_s \end{array} \right\} = \left\{ \begin{array}{c} \dfrac{\partial u'_0}{\partial x'} \\[2mm] \dfrac{\partial v'_0}{\partial y'} \\[2mm] \dfrac{\partial u'_0}{\partial y'} + \dfrac{\partial v'_0}{\partial x'} \\[2mm] \cdots \\ \dfrac{\partial \theta_{x'}}{\partial x'} \\[2mm] \dfrac{\partial \theta_{y'}}{\partial y'} \\[2mm] \dfrac{\partial \theta_{x'}}{\partial y'} + \dfrac{\partial \theta_{y'}}{\partial x'} \\[2mm] \cdots \\ \dfrac{\partial w'_0}{\partial x'} - \theta_{x'} \\[2mm] \dfrac{\partial w'_0}{\partial y'} - \theta_{y'} \end{array} \right\} = \sum_{i=1}^{n} \left\{ \begin{array}{c} \dfrac{\partial N_i}{\partial x'} u'_{o_i} \\[2mm] \dfrac{\partial N_i}{\partial y'} v'_{o_i} \\[2mm] \dfrac{\partial N_i}{\partial y'} u'_{o_i} + \dfrac{\partial N_i}{\partial x'} v'_{o_i} \\[2mm] \cdots \\ \dfrac{\partial N_i}{\partial x'} \theta_{x'_i} \\[2mm] \dfrac{\partial N_i}{\partial y'} \theta_{y'_i} \\[2mm] \dfrac{\partial N_i}{\partial y'} \theta_{x'_i} + \dfrac{\partial N_i}{\partial x'} \theta_{y'_i} \\[2mm] \cdots \\ \dfrac{\partial N_i}{\partial x'} w'_{o_i} - N_i \theta_{x'_i} \\[2mm] \dfrac{\partial N_i}{\partial y'} w'_{o_i} - N_i \theta_{y'_i} \end{array} \right\} =
$$

$$
= \sum_{i=1}^{n} \mathbf{B}'_i \mathbf{a}'^{(e)} = \left[\mathbf{B}'_1, \mathbf{B}'_2, \cdots, \mathbf{B}'_n \right] \left\{ \begin{array}{c} \mathbf{a}_1'^{(e)} \\ \mathbf{a}_2'^{(e)} \\ \vdots \\ \mathbf{a}_n'^{(e)} \end{array} \right\} = \mathbf{B}' \mathbf{a}'^{(e)} \qquad (8.29)
$$

where $\mathbf{B}'$ and $\mathbf{B}'_i$ are the local generalized strain matrices for the element and a node i, respectively. The later can be written as

$$
\mathbf{B}'_i = \left\{ \begin{array}{c} \mathbf{B}'_{m_i} \\ \mathbf{B}'_{b_i} \\ \mathbf{B}'_{s_i} \end{array} \right\} \qquad (8.30)
$$

where $\mathbf{B}'_{m_i}$, $\mathbf{B}'_{b_i}$ and $\mathbf{B}'_{s_i}$ are respectively the membrane, bending and transverse shear strain matrices of a node given by

$$
\mathbf{B}'_{m_i} = \begin{bmatrix} \dfrac{\partial N_i}{\partial x'} & 0 & 0 & 0 & 0 \\[2ex] 0 & \dfrac{\partial N_i}{\partial y'} & 0 & 0 & 0 \\[2ex] \dfrac{\partial N_i}{\partial y'} & \dfrac{\partial N_i}{\partial x'} & 0 & 0 & 0 \end{bmatrix} \quad , \quad \mathbf{B}'_{b_i} = \begin{bmatrix} 0 & 0 & 0 & \dfrac{\partial N_i}{\partial x'} & 0 \\[2ex] 0 & 0 & 0 & 0 & \dfrac{\partial N_i}{\partial y'} \\[2ex] 0 & 0 & 0 & \dfrac{\partial N_i}{\partial y'} & \dfrac{\partial N_i}{\partial x'} \end{bmatrix}
$$

$$
\mathbf{B}'_{s_i} = \begin{bmatrix} 0 & 0 & \dfrac{\partial N_i}{\partial x'} & -N_i & 0 \\[2ex] 0 & 0 & \dfrac{\partial N_i}{\partial y'} & 0 & -N_i \end{bmatrix} \tag{8.31}
$$

8.4.3 Derivation of the element stiffness equations

The PVW (Eq.(8.27)) applied to a single element reads

$$
\iint_{A^{(e)}} \delta \hat{\boldsymbol{\varepsilon}}'^T \hat{\boldsymbol{\sigma}}' dA = \iint_{A^{(e)}} \delta \mathbf{u}'^T \mathbf{t}' dA + \left[\delta \mathbf{a}'^{(e)} \right]^T \mathbf{q}'^{(e)} \tag{8.32a}
$$

In above $\mathbf{q}'^{(e)}$ is the equilibrating nodal force vector with

$$
\mathbf{q}_i'^{(e)} = \left[F_{x_i'}, F_{y_i'}, F_{z_i'}, M_{x_i'}, M_{y_i'} \right]^T \tag{8.32b}
$$

where $F_{x_i'}$, $F_{y_i'}$ and $F_{z_i'}$ are nodal equilibrating point forces acting in the local directions x', y', z', respectively, and $M_{x_i'}$, $M_{z_i'}$ are nodal couples contained in the planes $x'z'$ and $y'z'$, respectively.

Substituting the constitutive equation (8.17) into (8.32a) gives

$$
\iint_{A^{(e)}} \delta \hat{\boldsymbol{\varepsilon}}'^T \hat{\mathbf{D}}' \hat{\boldsymbol{\varepsilon}}' dA + \iint_{A^{(e)}} \delta \hat{\boldsymbol{\varepsilon}}'^T \hat{\boldsymbol{\sigma}}'^o dA - \iint_{A^{(e)}} \delta \mathbf{u}'^T \mathbf{t}' dA = \left[\delta \mathbf{a}'^{(e)} \right]^T \mathbf{q}'^{(e)} \tag{8.33}
$$

Introducing the finite element discretization (Eqs.(8.28a) and (8.29)) into (8.33) and following the standard process, the equilibrium equations for the element are obtained as

$$
\mathbf{q}'^{(e)} = \mathbf{K}'^{(e)} \mathbf{a}'^{(e)} - \mathbf{f}'^{(e)} \tag{8.34}
$$

where the stiffness matrix and the equivalent nodal force vector for the element in local axes are

$$
\mathbf{K}_{ij}'^{(e)} = \iint_{A^{(e)}} \mathbf{B}_i'^T \hat{\mathbf{D}}' \mathbf{B}_j' dA \tag{8.35a}
$$

$$\mathbf{f}_i^{\prime(e)} = [f_{x_i'}, f_{y_i'}, f_{z_i'}, m_{x_i'}, m_{y_i'}]^T = \iint_{A^{(e)}} \mathbf{N}_i^T \mathbf{t}' dA - \iint_{A^{(e)}} \mathbf{B}_i^{\prime T} \hat{\boldsymbol{\sigma}}'^0 dA$$

$$(8.35b)$$

In Eq.(8.35b) only the effect of surface loads and initial stresses has been considered. Other load types can be included in a straightforward manner. For instance, the self-weight is equivalent to a uniformly distributed vertical load of intensity $-\rho t$, where ρ is the specific weight and t the thickness. In practice, it is convenient to write the components of the equivalent nodal force vector directly in global axes. This is explained in the next section when dealing with the global assembly process.

The expression of $\mathbf{K}_{ij}^{\prime(e)}$ of Eq.(8.35a) is rewritten using Eqs.(8.18a) and (8.30) as follows

$$\mathbf{K}_{ij}^{\prime(e)} = \iint_{A^{(e)}} \left[\mathbf{B}_{m_i}^{\prime T}, \mathbf{B}_{b_i}^{\prime T}, \mathbf{B}_{s_i}^{\prime T}\right] \begin{bmatrix} \mathbf{D}_m' & \hat{\mathbf{D}}_{mb}' & 0 \\ \hat{\mathbf{D}}_{mb}' & \hat{\mathbf{D}}_b' & 0 \\ 0 & 0 & \hat{\mathbf{D}}_s' \end{bmatrix} \begin{Bmatrix} \mathbf{B}_{m_j}' \\ \mathbf{B}_{b_j}' \\ \mathbf{B}_{s_j}' \end{Bmatrix} dA =$$

$$= \mathbf{K}_{m_{ij}}^{\prime(e)} + \mathbf{K}_{b_{ij}}^{\prime(e)} + \mathbf{K}_{s_{ij}}^{\prime(e)} + \mathbf{K}_{mb_{ij}}^{\prime(e)} + \left[\mathbf{K}_{mb_{ij}}^{\prime(e)}\right]^T \qquad (8.36)$$

where

$$\mathbf{K}_{m_{ij}}^{\prime(e)} = \iint_{A^{(e)}} \mathbf{B}_{m_i}^{\prime T} \hat{\mathbf{D}}_m' \mathbf{B}_{m_j}' dA \quad ; \quad \mathbf{K}_{b_{ij}}^{\prime(e)} = \iint_{A^{(e)}} \mathbf{B}_{b_i}^{\prime T} \hat{\mathbf{D}}_b' \mathbf{B}_{b_j}' dA$$

$$(8.37)$$

$$\mathbf{K}_{s_{ij}}^{\prime(e)} = \iint_{A^{(e)}} \mathbf{B}_{s_i}^{\prime T} \hat{\mathbf{D}}_s' \mathbf{B}_{s_j}' dA \quad ; \quad \mathbf{K}_{mb_{ij}}^{\prime(e)} = \iint_{A^{(e)}} \mathbf{B}_{m_i}^{\prime T} \hat{\mathbf{D}}_{mb}' \mathbf{B}_{b_j}' dA$$

are the local stiffness matrices corresponding to membrane, bending, transverse shear and membrane-bending coupling effects, respectively. If $\hat{\mathbf{D}}_{mb}'$ is zero, the terms of $\mathbf{K}_{mb}^{\prime(e)}$ vanish and the local stiffness matrix for the element is obtained as the *sum of the membrane, bending and transverse shear contributions.*

The local element stiffness matrix for the element can be directly expressed in this case as

$$\mathbf{K}_{ij}^{\prime(e)} = \begin{bmatrix} (\mathbf{K}_{PS}^{\prime(e)})_{ij} & \vdots & \mathbf{0} \\ {\scriptstyle 2\times2} & \vdots & {\scriptstyle 2\times3} \\ \cdots\cdots\cdots & \vdots & \cdots\cdots\cdots \\ \mathbf{0} & \vdots & \\ {\scriptstyle 3\times2} & \vdots & (\mathbf{K}_{PB}^{\prime(e)})_{ij} \\ & & {\scriptstyle 3\times3} \end{bmatrix} \begin{matrix} u' \\ v' \\ w' \\ \theta_{x'} \\ \theta_{y'} \end{matrix} \qquad (8.38)$$

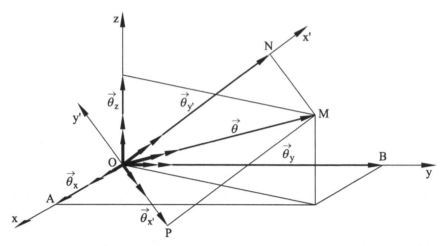

Fig. 8.11 Transformation of the two local rotations $(\theta_{x'}, \theta_{y'})$ to the three global rotations $(\theta_x, \theta_y, \theta_z)$

where $\mathbf{K}_{PS}^{(e)}$ and $\mathbf{K}_{PB}^{(e)}$ are the element stiffness matrices corresponding to the *plane stress* and *plate bending* problem given by Eqs.(6.61) of [On4] and (6.37), respectively.

In conclusion, when the coupling membrane-bending effects can be neglected, the local stiffness matrix for a flat shell element can be readily obtained by combining the plane stress and plate bending stiffness matrices as shown in Eq.(8.38). At the local level the membrane stiffness (plane stress) equilibrates the in-plane forces, and the bending stiffness balances the out of plane forces. The membrane-bending coupling appears when the element stiffness matrices are assembled into the global stiffness matrix, as studied in the next section.

8.5 ASSEMBLY OF THE STIFFNESS EQUATIONS

The global equilibrium equations are obtained as usual on a node by node basis by establishing the equilibrium of all the nodal forces meeting at each node. This requires that these forces are defined in the same global coordinate system. Hence, a transformation of the nodal displacements and forces prior to the assembly process is mandatory, as for bar structures (Chapter 1 of [On4] and [Li]). This is somehow more complicated in shells as the transformation of the two local rotations $\theta_{x'_i}$ and $\theta_{y'_i}$ to the global axes introduces a third global rotation θ_{z_i} (Figure 8.11). The same occurs

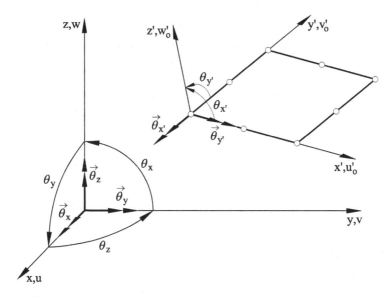

Fig. 8.12 Sign convention for the local and global rotations

with the transformation of the bending moments which introduces a third nodal bending moment M_{z_i}.

The local and global displacements and forces are related by the following transformations

$$\mathbf{a}_i'^{(e)} = \mathbf{L}_i^{(e)} \mathbf{a}_i^{(e)} \quad , \quad \mathbf{f}_i'^{(e)} = \mathbf{L}_i^{(e)} \mathbf{f}_i^{(e)} \tag{8.39}$$

where

$$\mathbf{a}_i^{(e)} = [u_{o_i}, v_{o_i}, w_{o_i}, \theta_{x_i}, \theta_{y_i}, \theta_{z_i}]^T$$
$$\mathbf{f}_i^{(e)} = [f_{x_i}, f_{y_i}, f_{z_i}, m_{x_i}, m_{y_i}, m_{z_i}]^T \tag{8.40}$$

are the global displacement vector and the global load vector of a node, respectively, including the third rotation and the third bending moment, as mentioned above. Note that the global rotations and the bending moments are defined now *in vector form*, i.e. $\overrightarrow{\theta}_x$ is the rotation vector defined by the axial axis x, etc. (Figure 8.12).

As the element is flat, the transformation matrix $\mathbf{L}_i^{(e)}$ is constant for all the element nodes and it has the following expression

$$\mathbf{L}_i^{(e)} = \begin{bmatrix} \boldsymbol{\lambda}_{3\times3}^{(e)} & \mathbf{0} \\ \mathbf{0} & \hat{\boldsymbol{\lambda}}_{2\times3}^{(e)} \end{bmatrix} \quad , \quad \boldsymbol{\lambda}^{(e)} = \begin{bmatrix} \lambda_{xx'} & \lambda_{xx'} & \lambda_{x'z} \\ \lambda_{y'x} & \lambda_{y'y} & \lambda_{y'z} \\ \lambda_{z'x} & \lambda_{z'y} & \lambda_{z'z} \end{bmatrix}^{(e)} \tag{8.41}$$

with $\lambda_{x'x}$ being the cosine of the angle formed by axes x' and x, etc. Keeping in mind the different sign conventions for the local and global rotations, the rotation transformation matrix is

$$\hat{\boldsymbol{\lambda}}^{(e)} = \begin{bmatrix} -\lambda_{y'x} & -\lambda_{y'y} & -\lambda_{y'z} \\ \lambda_{x'x} & \lambda_{x'y} & \lambda_{x'z} \end{bmatrix}^{(e)} \tag{8.42}$$

We deduce from Eq.(8.39)

$$\mathbf{a}'^{(e)} = \mathbf{T}^{(e)}\mathbf{a}^{(e)} \quad , \quad \mathbf{f}'^{(e)} = \mathbf{T}^{(e)}\mathbf{f}^{(e)} \tag{8.43}$$

where

$$\underset{5n \times 6n}{\mathbf{T}^{(e)}} = \begin{matrix} & 1 & 2 \cdots n \\ \begin{bmatrix} \mathbf{L}_1^{(e)} & & \\ & \ddots & \\ & & \mathbf{L}_n^{(e)} \end{bmatrix} & \begin{matrix} 1 \\ 2 \\ \vdots \\ n \end{matrix} \end{matrix} \tag{8.44}$$

is the transformation matrix for the element. As the element is flat $\mathbf{L}_1^{(e)} = \mathbf{L}_2^{(e)} = \ldots = \mathbf{L}_n^{(e)}$.

Combining Eqs.(8.34) and (8.43) gives finally

$$\mathbf{q}^{(e)} = \left[\mathbf{T}^{(e)}\right]^T \mathbf{q}'^{(e)} = \left[\mathbf{T}^{(e)}\right]^T \left[\mathbf{K}'^{(e)}\mathbf{a}'^{(e)} - \mathbf{f}'^{(e)}\right] =$$

$$= \left[\mathbf{T}^{(e)}\right]^T \mathbf{K}'^{(e)}\mathbf{T}^{(e)}\mathbf{a}^{(e)} - \left[\mathbf{T}^{(e)}\right]^T \bar{\mathbf{f}}'^{(e)} = \mathbf{K}^{(e)}\mathbf{a}^{(e)} - \mathbf{f}^{(e)} \tag{8.45}$$

which is the new equilibrium equation for the element, where the displacements and forces are referred to the global axes. In above

$$\mathbf{K}^{(e)} = \left[\mathbf{T}^{(e)}\right]^T \mathbf{K}'^{(e)}\mathbf{T}^{(e)} \quad ; \quad \mathbf{f}^{(e)} = [\mathbf{T}]^T \mathbf{f}'^{(e)} \tag{8.46}$$

are the element stiffness matrix and the equivalent nodal force vector *in global axes*.

External point loads P_i acting directly at a node i are added to the global equivalent nodal force vector $\mathbf{f}$ in the standard manner.

The triple matrix product in Eq.(8.46) is not necessary in practice. Combining Eqs.(8.35a) and (8.46) gives

$$\mathbf{K}_{ij}^{(e)} = \left[\mathbf{L}_i^{(e)}\right]^T \left[\iint_{A^{(e)}} \mathbf{B}_i''^T \hat{\mathbf{D}}' \mathbf{B}_j'\right] \mathbf{L}_j^{(e)} = \iint_{A^{(e)}} \mathbf{B}_i^T \hat{\mathbf{D}}' \mathbf{B}_j dA \tag{8.47}$$

where
$$\mathbf{B}_i = \mathbf{B}_i' \mathbf{L}_i^{(e)} \tag{8.48}$$

Note that $\mathbf{L}_i^{(e)} = \mathbf{L}_j^{(e)}$ can be used in Eq.(8.47), taking advantage from the flat geometry of the element.

The key results sought in shell analysis are the nodal displacements in global axes and the local resultant stresses. The later allow us to evaluate the resistant capacity of the structure and to design the necessary steel reinforcement. A relationship between the local resultant stresses and the global nodal displacements is obtained combining Eqs.(8.17), (8.29), (8.39) and (8.48) (neglecting the initial strains) as

$$\hat{\sigma}' = \hat{\mathbf{D}}'\hat{\varepsilon}' = \hat{\mathbf{D}}' \sum_{i=1}^{n} \mathbf{B}_i' \mathbf{a}_i'^{(e)} = \hat{\mathbf{D}}' \sum_{i=1}^{n} \mathbf{B}_i' \mathbf{L}_i^{(e)} \mathbf{a}_i^{(e)} = \hat{\mathbf{D}}' \sum_{i=1}^{n} \mathbf{B}_i \mathbf{a}_i = \hat{\mathbf{D}}' \mathbf{B} \mathbf{a}^{(e)} \tag{8.49}$$

Matrix $\mathbf{B}_i$ therefore has a double utility; it reduces the matrix operations to obtain the global stiffness matrix for the element, and it can also be used for the direct computation of the local resultant stresses.

The transformation (8.49) can be written separately for each of the membrane, bending and transverse shear strain matrices giving

$$\mathbf{B}_{m_i} = \mathbf{B}_{m_i}' \mathbf{L}_i^{(e)}, \quad \mathbf{B}_{b_i} = \mathbf{B}_{b_i}' \mathbf{L}_i^{(e)} \quad \text{and} \quad \mathbf{B}_{s_i} = \mathbf{B}_{s_i}' \mathbf{L}_i^{(e)} \tag{8.50}$$

The global stiffness matrices can be therefore computed by an expression identical to (8.37) simply substituting $\mathbf{B}_m'$, $\mathbf{B}_b'$ and $\mathbf{B}_s'$, by $\mathbf{B}_m$, $\mathbf{B}_b$ and $\mathbf{B}_s$, respectively.

8.6 NUMERICAL INTEGRATION OF THE STIFFNESS MATRIX AND THE EQUIVALENT NODAL FORCE VECTOR

Both the global stiffness matrix $\mathbf{K}^{(e)}$ and the equivalent nodal force vector are evaluated using numerical integration. A first step is to define the nodal coordinates in the local axes $x'y'z'$. This can be performed by a simple coordinate transformation analogous to Eq.(8.39) for the nodal displacements as

$$\mathbf{x}' = [x', y', z']^T = \boldsymbol{\lambda}^{(e)}(\mathbf{x} - \mathbf{x}_0) \tag{8.51}$$

where $\boldsymbol{\lambda}^{(e)}$ is the transformation matrix of Eq.(8.42) and $\mathbf{x}_0$ are the coordinates of the origin of the local system. During the computation of the element stiffness matrix the coordinates appear in the Jacobian evaluation only, via the terms $\dfrac{\partial x'}{\partial \xi}$, $\dfrac{\partial y'}{\partial \eta}$ etc. Hence, since $\boldsymbol{\lambda}^{(e)}$ is constant within the element and $\dfrac{\partial}{\partial \xi}\mathbf{x}_0 = \dfrac{\partial}{\partial \eta}\mathbf{x}_0 = \mathbf{0}$, the simpler transformation $\mathbf{x}' = \boldsymbol{\lambda}^{(e)}\mathbf{x}$ can be used with identical results. This explains why the element stiffness matrix is independent of the origin of the coordinate system [Li].

The element stiffness matrix can be directly computed in global axes using a Gauss quadrature by

$$
\mathbf{K}_{ij}^{(e)} = \sum_{p_m=1}^{n_{pm}} \sum_{q_m=1}^{n_{qm}} (\mathbf{I}_m^{(e)})_{p_m,q_m} W_{p_m} W_{q_m} + \sum_{p_b=1}^{n_{pb}} \sum_{q_b=1}^{n_{qb}} (\mathbf{I}_b^{(e)})_{p_b,q_b} W_{p_b} W_{q_b} +
$$

$$
+ \sum_{p_s=1}^{n_{ps}} \sum_{q_s=1}^{n_{qs}} (\mathbf{I}_s^{(e)})_{p_s,q_s} W_{p_s} W_{q_s} + \sum_{p_{mb}=1}^{n_{pmb}} \sum_{q_{mb}=1}^{n_{qmb}} (\mathbf{I}_{mb}^{(e)})_{p_{mb},q_{mb}} W_{p_{mb}} W_{q_{mb}}
$$

$$(8.52\mathrm{a})$$

where

$$
\mathbf{I}_a^{(e)} = \mathbf{B}_{a_i}^T \hat{\mathbf{D}}_a' \mathbf{B}_{a_j} \left| \mathbf{J}^{(e)} \right| \qquad a = m, b, s \tag{8.52b}
$$

$$
\mathbf{I}_{mb}^{(e)} = \left[\mathbf{B}_{m_i}^T \hat{\mathbf{D}}_{mb}' \mathbf{B}_{b_j} + \mathbf{B}_{b_i}^T \hat{\mathbf{D}}_{mb}' \mathbf{B}_{m_j} \right] |J^{(e)}| \tag{8.52c}
$$

In Eq.(8.52a), n_{p_a}, W_{p_a} and n_{q_a}, W_{q_a} are the number of integration points and the corresponding weights along each natural directions ξ and η, respectively. Subscripts m, b, s, mb denote membrane, bending, transverse shear and membrane-bending coupling contributions, as usual. Eq.(8.52a) allows us to use different quadrature rules for each of these terms. This is useful to avoid shear (and membrane) locking.

If the number of integration points is large, then the evaluation of $\mathbf{B}_{a_i} (= \mathbf{B}_{a_i}' \mathbf{L}_i^{(e)})$ at each integration point in Eq.(8.52a) can be costly. A more economical option is to compute the local stiffness matrix first and then transform this to the global axes using Eq.(8.47). A disadvantage of this option is the need to repeat the transformations (8.49) to compute "a posteriori" the local resultant stresses at each Gauss point (which are the "optimal" sampling point for the stresses in most cases) [On4]. Which option is the best one depends on the element type and the quadrature chosen. For linear and quadratic elements with 2×2 and 3×3 quadratures, performing first the transformations (8.50) and then directly computing the global stiffness matrix is more advantageous.

The equivalent nodal force vector (in global coordinates) of Eq.(8.35b) is also computed numerically using a Gauss quadrature as

$$\mathbf{f}_i^{(e)} = \sum_{p=1}^{n_p} \sum_{q=1}^{n_q} (\mathbf{I}_f)_{p,q} W_p W_q \quad \text{with} \quad \mathbf{I}_f = (\mathbf{N}_i^T \mathbf{t} - \mathbf{B}^T \hat{\boldsymbol{\sigma}}_0') |\mathbf{J}|^{(e)} \qquad (8.53)$$

Note that in the expression of $\mathbf{I}_f$ of Eq.(8.53) the surface loads are expressed in the global coordinate system.

8.7 BOUNDARY CONDITIONS

Standard boundary conditions in shells are the following.

Point support: $u_{0_i} = v_{0_i} = w_{0_i} = 0$

Clampled edge: $(\mathbf{a}_i = 0)$. All DOFs at nodes laying on a clampled edge are prescribed to a zero value.

Simple supported (SS) edge

- Soft SS edge: $u_{0_i} = v_{0_i} = w_{0_i} = 0$
- Hard SS edge: $u_{0_i} = v_{0_i} = w_{0_i} = \theta_{t_i} = 0$, where t is the tangential direction at the ith edge node (Figures 8.10 and 8.13).

The definition of the tangential direction at an edge node implies the following steps. First the *average unit normal* at a node is computed $(\mathbf{n}_i)$. The unit tangential vector $\mathbf{t}_i$ is defined as orthogonal to $\mathbf{n}_i$ and contained in the plane formed by the two edge sides sharing node i. The *edge coordinate system* at a node $(\mathbf{t}_i, \mathbf{a}_i, \mathbf{n}_i)$ is completed by defining vector $\mathbf{a}_i$ as $\mathbf{a}_i = \mathbf{n}_i \wedge \mathbf{t}_i$.

Symmetry edge: $\theta_{a_i} = 0$, where $\mathbf{a}_i$ is the normal vector to the symmetry plane at node i.

The edge system at a symmetry node is obtained as described for a SS node.

Prescribing the edge rotations θ_{t_i} and θ_{a_i} implies first the definition of the edge coordinate system at each edge node and then transforming the

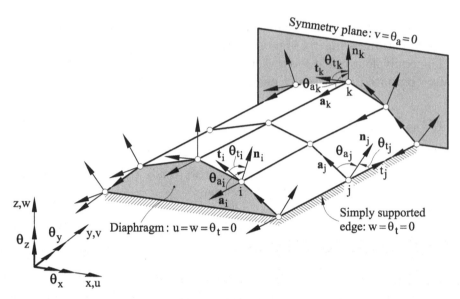

Fig. 8.13 Shallow cylindrical shell discretized with 4-noded rectangles. Schematic representation of boundary conditions at nodes laying on a SS edge, the edges of a rigid diaphragm and a symmetry plane

local nodal rotations to the edge rotations (Figures 8.10 and 8.13). The transformation is [On4]

$$
\boldsymbol{\theta}_i' = \begin{Bmatrix} \theta_{x_i'} \\ \theta_{y_i'} \end{Bmatrix} = \underbrace{\begin{bmatrix} C_{x's} & C_{x't} \\ C_{y's} & C_{y't} \end{bmatrix}}_{\hat{\boldsymbol{\lambda}}_i^{(e)}} \begin{Bmatrix} \theta_{t_i} \\ \theta_{a_i} \end{Bmatrix} \tag{8.54}
$$

where $C_{x's}$ is the cosine of the angle between the x' axis and the t axis, etc. Matrix $\hat{\boldsymbol{\lambda}}_i^{(e)}$ of Eq.(8.54) substitutes the rotation transformation matrix $\hat{\boldsymbol{\lambda}}^{(e)}$ in Eq.(8.42). Note that $\hat{\boldsymbol{\lambda}}_i^{(e)}$ may now vary for each boundary node.

The definition of the local and edge rotations follows the same angular criterium. The nodal DOFs after the local-global transformation are $\mathbf{a}_i^{(e)} = [u_{0_i}, v_{0_i}, w_{0_i}, \theta_{t_i}, \theta_{a_i}]^T$.

Figure 8.14 shows the difference between the local and edge axes at a node on a SS edge.

If all the elements sharing a boundary node lay on the same plane (i.e. the node is *coplanar* (Section 8.9)), then the edge axes t, a, n can be made coincident with the local axes x', y', z'. The nodal variables are $u_{0_i}, v_{0_i}, w_{0_i}, \theta_{x_i'}, \theta_{y_i'}$. The SS (hard) boundary condition is simply prescribed by making $\theta_{x_i'} = 0$, while the symmetry edge condition implies

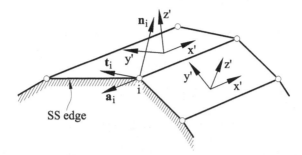

Fig. 8.14 Edge and local coordinate systems at a node on a SS edge

making $\theta_{y'_i} = 0$. An example of this situation is node j laying on a SS edge in Figure 8.13.

8.8 DEFINITION OF THE LOCAL AXES

A good definition of the local axes is essential for identifying the local resultant stresses easily.

The local axis x' can take any arbitrary direction within the element. The selection of x' influences the definition of the local coordinate system x', y', z' and the transformation matrix $\mathbf{T}^{(e)}$. The solution is not unique and several alternatives exist. Some options are presented below.

8.8.1 Definition of local axes from an element side

Vector x' is defined as the direction of one of the element sides. This process is equally valid for triangular and quadrilateral elements.

For the elements shown in Figure 8.15 vector $\mathbf{V}^{(e)}_{x'}$ is computed using the coordinates of two nodes i and j along a side as

$$\mathbf{V}^{(e)}_{x'} = \left\{ \begin{array}{c} x_j - x_i \\ y_j - y_i \\ z_j - z_i \end{array} \right\}^{(e)} = \left\{ \begin{array}{c} x_{ij} \\ y_{ij} \\ z_{ij} \end{array} \right\}^{(e)} \tag{8.55}$$

The unit vector is

$$\mathbf{v}^{(e)}_{x'} = \left\{ \begin{array}{c} \lambda_{x'x} \\ \lambda_{x'y} \\ \lambda_{x'z} \end{array} \right\}^{(e)} = \frac{1}{l^{(e)}_{ij}} \left\{ \begin{array}{c} x_{ij} \\ y_{ij} \\ z_{ij} \end{array} \right\} \tag{8.56}$$

where $l^{(e)}_{ij} = \sqrt{(x^2_{ij} + y^2_{ij} + z^2_{ij})^{(e)}}$ is the length of side ij.

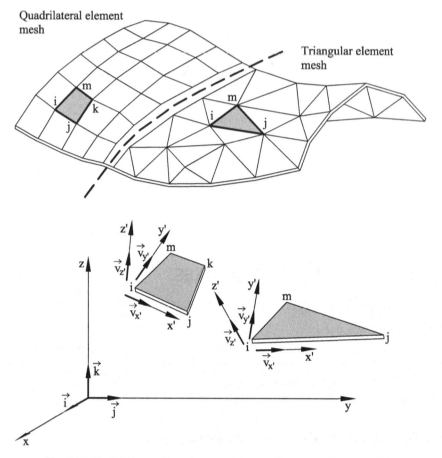

Fig. 8.15 Definition of local axes starting from an element side

The direction cosines of the z' axis are obtained by the cross product of any two sides, i.e.

$$\mathbf{v}_{z'}^{(e)} = \left\{ \begin{array}{c} \lambda_{z'x} \\ \lambda_{z'y} \\ \lambda_{z'z} \end{array} \right\}^{(e)} = \frac{1}{|\mathbf{V}_{ij}^{(e)} \wedge \mathbf{V}_{im}^{(e)}|}(\mathbf{V}_{ij}^{(e)} \wedge \mathbf{V}_{im}^{(e)}) = \frac{1}{d_{z'}^{(e)}} \left\{ \begin{array}{c} y_{ij}z_{im} - z_{ij}y_{im} \\ x_{im}z_{ij} - z_{im}x_{ij} \\ x_{ij}y_{im} - y_{ij}x_{im} \end{array} \right\}^{(e)}$$

(8.57a)

and

$$d_{z'}^{(e)} = \sqrt{[(y_{ij}z_{im} - z_{ij}y_{im})^2 + (x_{im}z_{ij} - z_{im}x_{ij})^2 + (x_{ij}y_{im} - y_{ij}x_{im})^2]^{(e)}}$$

For a triangle $d_{z'}^{(e)}$ is twice its area and this simplifies the computations. The direction cosines of the y' axis are obtained by the cross product

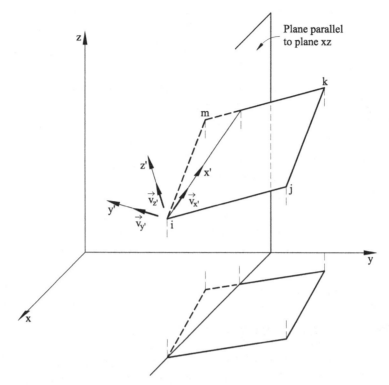

Fig. 8.16 Definition of the local x' axis by intersecting the element with a plane parallel to the global plane xz

of the unit vectors in the z' and x' directions:

$$
\mathbf{v}_{y'}^{(e)} = \left\{ \begin{matrix} \lambda_{y'x} \\ \lambda_{y'y} \\ \lambda_{y'z} \end{matrix} \right\}^{(e)} = \mathbf{v}_{z'}^{(e)} \wedge \mathbf{v}_{x'}^{(e)} = \left\{ \begin{matrix} \lambda_{z'y}\lambda_{x'z} - \lambda_{x'y}\lambda_{z'z} \\ \lambda_{x'x}\lambda_{z'z} - \lambda_{z'x}\lambda_{x'z} \\ \lambda_{z'x}\lambda_{x'y} - \lambda_{z'y}\lambda_{x'x} \end{matrix} \right\}^{(e)} \quad (8.57b)
$$

8.8.2 Definition of local axes by intersection with a coordinate plane

A useful alternative is to define x' as the intersection of the element plane with a plane parallel to one of the global coordinate planes xz or yz.

For instance, the x' axis can be defined by intersecting the element with a plane parallel to the xz plane as shown in Figure 8.16. The projection of x' along the y axis is then zero and,

$$
\mathbf{v}_{x'}^{(e)} = \left\{ \begin{matrix} \lambda_{x'x} \\ 0 \\ \lambda_{x'z} \end{matrix} \right\}^{(e)} \quad (8.58)
$$

As the length of this vector is unity, then

$$(\lambda_{x'x}^{(e)})^2 + (\lambda_{x'z}^{(e)})^2 = 1 \tag{8.59}$$

The second necessary equation comes from the condition that the scalar product of the unit vectors $\mathbf{v}_{z'}^{(e)}$ and $\mathbf{v}_{z'}^{(e)}$ is zero, i.e.

$$\lambda_{x'x}^{(e)}\lambda_{z'x}^{(e)} + \lambda_{x'z}^{(e)}\lambda_{z'z}^{(e)} = 0 \tag{8.60}$$

From Eqs.(8.59) and (8.60) we have

$$\lambda_{x'x}^{(e)} = \frac{1}{\sqrt{1 + (\frac{\lambda_{z'z}^{(e)}}{\lambda_{z'x}^{(e)}})^2}} \quad \text{and} \quad \lambda_{x'z}^{(e)} = \frac{1}{\sqrt{1 + (\frac{\lambda_{z'x}^{(e)}}{\lambda_{z'z}^{(e)}})^2}} \tag{8.61}$$

Vector $\mathbf{v}_{y'}^{(e)}$ is finally obtained by the cross product of $\mathbf{v}_{z'}^{(e)}$ and $\mathbf{v}_{x'}^{(e)}$.

8.8.3 Definition of a local axis parallel to a global one

A mesh of rectangular or triangular elements in a prismatic shell can always be generated so that one of the element sides is always parallel to a global axis (For instance the x axis in Figure 8.17). This side defines the local direction x'. From Figure 8.17 we deduce

$$\mathbf{v}_{x'}^{(e)} = [1, 0, 0]^T \tag{8.62}$$

Vector $\mathbf{v}_{y'}^{(e)}$ is contained in the yz plane and it can be defined from the coordinates of two nodes ij along a side orthogonal to x', i.e.

$$\mathbf{v}_{y'}^{(e)} = \frac{1}{l_{ij}^{(e)}} \left\{ \begin{matrix} 0 \\ y_{ij} \\ z_{ij} \end{matrix} \right\}^{(e)} \quad \text{with} \quad l_{ij}^{(e)} = \sqrt{(y_{ij}^{(e)})^2 + (z_{ij}^{(e)})^2} \tag{8.63}$$

Finally, the normal vector $\mathbf{v}_{z'}^{(e)}$ is obtained by

$$\mathbf{v}_{z'}^{(e)} = \mathbf{v}_{x'}^{(e)} \wedge \mathbf{v}_{y'}^{(e)} = \frac{1}{l_{ij}^{(e)}} \left[0, -z_{ij}^{(e)}, y_{ij}^{(e)} \right]^T \tag{8.64}$$

This technique depends on the shell shape and the element chosen. The method explained in Section 8.8.2 is the more general procedure.

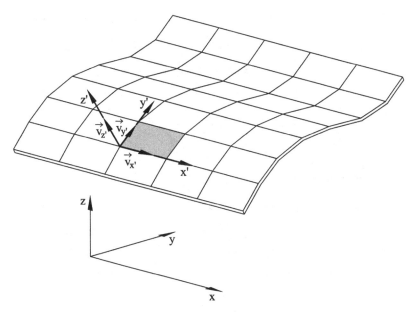

Fig. 8.17 Definition of the local x' axis as parallel to the global x axis

8.9 COPLANAR NODES. TECHNIQUES FOR AVOIDING SINGULARITY

A node is termed *coplanar* when all the elements meeting at the node lay in the same plane. This situation is typical in folded plate structures (Figure 8.18). A local coordinate system can be chosen at a coplanar node so that the local nodal rotations $\theta_{x'_i}$, $\theta_{y'_i}$ are uniquely defined for all the adjacent elements. If the equilibrium equations at the node are assembled in such a local system, six equations are obtained, the last of which (corresponding to the $\theta_{z'}$ direction) is simply $\theta_{z'} = 0$ and the element stiffness matrix is *singular*. If the assembly is performed in global axes, then the six resulting equilibrium equations at the node appear to be correct, although they are equally singular. Singularity means that one of the three equations expressing equilibrium of bending moments at a node is linearly dependent on the other two. The detection "a priori" of this singularity is, in general, not easy. Some alternatives to avoid it are described below.

Singularity can also occur in the so called "*quasi-coplanar*" situation. This is typical in smooth shells when a mesh of flat elements is used. As the mesh is refined the elements meeting at a node tend to lay in the same

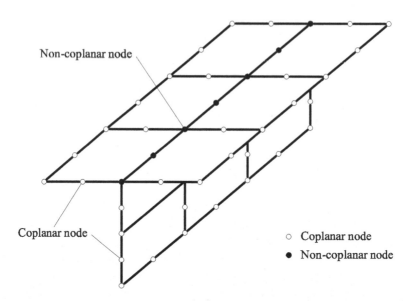

Non-coplanar node

Coplanar node

○ Coplanar node
● Non-coplanar node

Fig. 8.18 Example of coplanar and non coplanar nodes

tangent plane and the singularity explained above arises. This problem is discussed in a next section.

8.9.1 Selective assembly in local axes

The simplest alternative to avoid singularity is to assemble the rotational equations at coplanar (or quasi-coplanar) nodes in the same local nodal coordinate system. The nodal displacement vector for a coplanar node is

$$\mathbf{a}_i^{(e)} = \left[u_{o_i}, v_{o_i}, w_{o_i}, \theta_{x_i'}, \theta_{y_i'}\right]^T \tag{8.65}$$

and the nodal transformation matrix $\mathbf{L}_i^{(e)}$ is

$$\mathbf{L}_i^{(e)} = \begin{bmatrix} \boldsymbol{\lambda}_i^{(e)} & \mathbf{0} \\ \mathbf{0} & \mathbf{I}_2 \end{bmatrix} \quad \text{with} \quad \mathbf{I}_2 = \begin{bmatrix} 0 & 0 \\ 0 & 0 \end{bmatrix} \tag{8.66}$$

For non-coplanar nodes the assembly is performed in the global system in the usual manner, as explained in Section 8.5.

This procedure leads to a different number of DOFs per node (five DOFs in coplanar nodes and six DOFs in non-coplanar ones). This does not pose a serious problem for most FEM codes.

Keeping the local definition of rotations at coplanar nodes also simplifies the treatment of boundary conditions along inclined boundaries where the use of local rotations is mandatory.

The detection of a coplanar node requires verifying the angle between the normal directions for all the elements meeting at the node. This process is repeated for all nodes. When the angle between two normal directions exceeds a prescribed value (say 5°) the node is marked as coplanar. The limit angle should not be too small so that quasi-coplanar situations in smooth shells can be easily identified.

8.9.2 Global assembly with six DOFs using an artificial rotational stiffness

A procedure to keep six DOFs at all nodes is by inserting an arbitrary coefficient $K_{\theta_{z'}}$ in the diagonal term of the local stiffness matrix as

$$
\bar{\mathbf{K}}_{ij}^{\prime(e)} = \begin{bmatrix} \mathbf{K}_{ij}^{(e)} & \mathbf{0} \\ \underset{5\times5}{} & \\ \mathbf{0} & K_{\theta_{z'}} \end{bmatrix} \tag{8.67a}
$$

The local displacement vector is now

$$
\mathbf{a}^{\prime(e)} = \left[u_{o_i}', v_{o_i}', w_{o_i}', \theta_{x_i'}, \theta_{y_i'}, \theta_{z_i'} \right]^T \tag{8.67b}
$$

The sixth equilibrium equations for a coplanar node written in the local axes x', y', z' is

$$
(K_{\theta_{z'}})\theta_{z'} = 0 \tag{8.68}
$$

which gives $\theta_{z_i'} = 0$ and avoids the singularity.

The new local stiffness matrix is *transformed to global axes* in the standard way and the resulting global equations are not singular. Numerical results have proved to be good and quite insensitive to the values of the parameter $K_{\theta_{z'}}$, which is typically chosen of the order of $EtA^{(e)}$ [Ka,ZCh3, ZT2]. The explanation is that the stiffness equations corresponding to $\theta_{z_i'}$ are uncoupled from the rest. The new rotation does not affect the computation of the resultant stresses either.

This procedure can be enhanced so that the computation of the local stiffness matrix is not necessary. Eq.(8.67a) can be rewritten as

$$
\bar{\mathbf{K}}_{ij}^{\prime(e)} = \begin{bmatrix} \mathbf{K}_{ij}^{\prime(e)} & \mathbf{0} \\ \mathbf{0} & 0 \end{bmatrix} + \begin{bmatrix} \mathbf{0} & \mathbf{0} \\ \mathbf{0} & K_{\theta_{z'}} \end{bmatrix} = {}^1\mathbf{K}_{ij}^{\prime(e)} + {}^2\mathbf{K}_{ij}^{\prime(e)} \tag{8.69}
$$

The new local equation requires modifying the transformation matrix of Eq.(8.42), as

$$
\hat{\lambda}^{(e)} = \begin{bmatrix} -\lambda_{y'x} & -\lambda_{y'y} & -\lambda_{y'z} \\ \lambda_{x'x} & \lambda_{x'y} & \lambda_{x'z} \\ \lambda_{z'x} & \lambda_{z'y} & \lambda_{z'z} \end{bmatrix}^{(e)}
\tag{8.70}
$$

The global stiffness matrix is obtained using Eq.(8.47) as

$$
\underset{6\times6}{\bar{\mathbf{K}}_{ij}^{(e)}} = \left[\mathbf{L}_j^{(e)}\right]^T \underset{6\times6}{\bar{\mathbf{K}}_{ij}^{\prime(e)}} \underset{6\times6}{\mathbf{L}_i^{(e)}} = \left[\mathbf{L}_i^{(e)}\right]^T \left[{}^1\mathbf{K}_{ij}^{\prime(e)} + {}^2\mathbf{K}_{ij}^{\prime(e)}\right]\mathbf{L}_j^{(e)}
\tag{8.71}
$$

It is easy to verify that

$$
\left[\mathbf{L}_i^{(e)}\right]^T {}^1\mathbf{K}_{ij}^{\prime(e)}\mathbf{L}_j^{(e)} = \mathbf{K}_{ij}^{(e)}
\tag{8.72}
$$

where $\mathbf{K}_{ij}^{(e)}$ is the global stiffness matrix given by Eq.(8.47). Similarly,

$$
\left[\mathbf{L}_i^{(e)}\right]^T {}^2\mathbf{K}_{ij}^{\prime(e)}\mathbf{L}_j^{(e)} = {}^2\mathbf{K}_{ij}^{(e)}
\tag{8.73}
$$

with

$$
\underset{6\times6}{{}^2\mathbf{K}_{ij}^{(e)}} = K_{\theta_{z'}}\begin{bmatrix} \mathbf{0} & \mathbf{0} \\ \mathbf{0} & \hat{\lambda}_z^{(e)} \end{bmatrix} \quad \text{and} \quad \hat{\lambda}_z^{(e)} = \begin{bmatrix} (\lambda_{z'x'})^2 & \lambda_{z'x}\lambda_{z'y} & \lambda_{z'x}\lambda_{z'z} \\ \lambda_{z'y}\lambda_{z'x} & (\lambda_{z'y})^2 & \lambda_{z'y}\lambda_{z'z} \\ \lambda_{z'z}\lambda_{z'x} & \lambda_{z'z}\lambda_{z'y} & (\lambda_{z'z})^2 \end{bmatrix}^{(e)}
\tag{8.74}
$$

The global element stiffness matrix is finally obtained as

$$
\bar{\mathbf{K}}_{ij}^{(e)} = \mathbf{K}_{ij}^{(e)} + {}^2\mathbf{K}_{ij}^{(e)}
\tag{8.75}
$$

The computational process is as follows:

a) the global stiffness matrix $\mathbf{K}_{ij}^{(e)}$ given by Eq.(8.47), is computed first *for all elements* (wether they contain coplanar nodes or not), and

b) for elements containing coplanar nodes, matrix ${}^2\mathbf{K}_{ij}^{(e)}$ is added to the previous global stiffness matrix.

8.9.3 Drilling degrees of freedom

An alternative technique is to modify the formulation so that the in-plane rotational parameters arise naturally and have a real physical significance. The $\theta_{z'}$ rotation introduced in this way is called a *drilling* DOF, on account of its action to the shell surface.

A simple way for incorporating this drilling effect is by introducing a set of rotational stiffness coefficients such that the overall equilibrium is not disturbed in local coordinates. This can be accomplished by adding to the virtual work expression for each element the term

$$\iint_{A^{(e)}} \alpha_n E t^n \left(\delta \theta_{z'} - \delta \bar{\theta}_{z'} \right) \left(\theta_{z'} - \bar{\theta}_{z'} \right) dA \tag{8.76}$$

where α_n is a fictitious elastic parameter, n is an arbitrary number, $\theta_{z'}$ is the drilling rotation and $\bar{\theta}_{z'}$ is a mean element rotation which allows the element to satisfy equilibrium in an average sense. For a 3-noded triangle, the elimination of $\bar{\theta}_{z'}$ leads to the following relationship between the external moments and the nodal drilling rotations [ZT]

$$\begin{Bmatrix} M_{z'_i} \\ M_{z'_j} \\ M_{z'_k} \end{Bmatrix} = \alpha_n E t^n A^{(e)} \begin{bmatrix} 1 & -0.5 & -0.5 \\ -0.5 & 1 & -0.5 \\ -0.5 & -0.5 & 1 \end{bmatrix} \begin{Bmatrix} \theta_{z'_i} \\ \theta_{z'_j} \\ \theta_{z'_k} \end{Bmatrix} \tag{8.77}$$

This procedure is essentially identical to the addition of rotational stiffness coefficients proposed in [ZPK] with $n = 1$. Examples of application for $n = 3$ (i.e. with the in-plane rotational stiffness proportional to the cubic bending terms) can be found in [CW,GSW,MH,ZT2]. Numerical experiments indicate that mesh refinement reduces the influence of the elastic parameter α_n which can take very small values $(10^{-2} - 10^{-3})$ without affecting the results.

A similar method is based on defining the average drilling rotation $\bar{\theta}_{z'}$ in Eq.(8.76) as a mechanical rotation in terms of the in-plane displacement gradients (Figure 4.2 of [On4]) as

$$\bar{\theta}_{z'} = \frac{1}{2} \left(\frac{\partial v'}{\partial x'} - \frac{\partial u'}{\partial y'} \right) \tag{8.78}$$

Drilling DOF's are also of interest for enhancing the in-plane behaviour of the element. The reason is the following. Many flat shell elements incorporate bending approximations of higher order than the membrane ones. A typical example is the 3-noded plane stress triangle when combined with any of the higher order bending elements of previous chapters. The accuracy of the linear plane stress triangle is relatively poor and, consequently, the membrane error terms dominate the shell behaviour. This problem can be overcome by increasing the interpolation order of the membrane field, i.e. by using a quadratic 6-noded approximation for the in-plane displacement field. An alternative, however, is the introduction of drilling DOFs.

Willam and Scordelis [WS] used the drilling rotation technique for studying in-plane bending of lateral girders in cellular bridges. An early application of the drilling rotation for plane-stress analysis was reported by Reissner [Re2] and extended to the FEM by Hughes and Brezzi [HB2]. The main difficulty is that, although the definition of $\bar{\theta}_{z'}$ of Eq.(8.78) is invariant with respect to the reference coordinate system, there is not a unique relationship between $\bar{\theta}_{z'}$ and the rotations in adjacent element sides, and this violates the C^0 continuity requirement and the patch test in some cases [IA].

The drilling rotations lead to a displacement interpolation along the element sides involving the in-plane rotations also. The procedure can be interpreted as a class of *linked interpolation* analogous to that used for beam and plate elements (Sections 2.8.3 and 6.10) [ZT2].

Several authors have developed successful elements based on the drilling rotation satisfying the patch test. Membrane triangles of this type have been proposed by Allman [Al,Al2,Al3], Carpenter *et al.* [CSB2] and Cook [Co3,Co4].

Bergan and Felippa [BF,BF2] developed an accurate membrane element incorporating drilling DOFs starting from the so called *free formulation* [BH2,Ny]. Elements deriving from this formulation are designed so as to reproduce arbitrary rigid body movements or constant strain fields when interacting with adjacent elements. This condition is called the "individual element test". The resulting elements do not require C^0 continuity and can incorporate the drilling rotation in a straightforward manner.

Other triangular and quadrilateral membrane and shell elements incorporating drilling rotations have been reported in [AHMS,EKE,HMH,IA2, ITW,Je,SP,SWC]. An interesting element is the TRIC facet shell triangle based on the so-called natural formulation [AHM,Ar] for analysis of thin/thick isotropic and composite shells [ATO,ATPA,APAK].

Flat shell element with drilling DOFs typically perform well for folded plate structures or for shells where membrane effects are dominant. Conversely, they are prone to membrane locking for problems where bending effects are important [BD6,CSB2,GB,Ny]. This is due to the excessive influence of the membrane stiffness induced by the drilling rotation $\theta_{z'}$ versus the stiffness associated to the rotations $\theta_{x'}$ and $\theta_{y'}$, via the local-global roation transformation (membrane stiffness terms are proportional to t, while the bending ones to t^3). These problems evidence the difficulties for finding a good shell element that performs equally well for flat and curved shells [BD6,Co5,IA]. At this point we note the excellent performance of

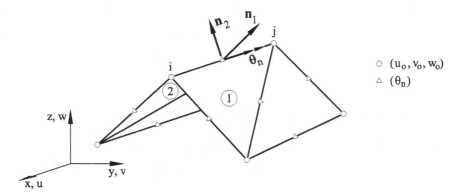

Fig. 8.19 Normal rotation DOF at mid-side nodes. $\mathbf{n}_1$ and $\mathbf{n}_2$ are normal vectors to elements 1 and 2. The normal rotation vector has the direction of the common side ij

the EBST rotation-free shell triangle described in Section 8.13.2 for both membrane and bending dominating shell problems.

8.9.4 Flat shell elements with mid-side normal rotations

Many of the assembly difficulties for flat shell element disappear if the displacements are defined at the nodes, whereas the rotation field is defined in terms of the normal slope at the element midsides. As the normal rotation vector has the direction of the side, clearly full compatibility is achieved for adjacent (coplanar or non-coplanar) elements sharing the side (Figure 8.19). Any transformation is then unnecessary and no additional rotational DOFs are required for assembly purposes.

Plate elements of this kind where studied in Chapters 5 and 6. A popular shell triangle is the extension of the Morley thin plate triangle (Figure 8.19) developed by Dawe [Da]. A more sophisticated shell element of this kind was derived by Irons [IA,Ir2] under the name of *semi-loof*. This element is quite accurate and it will be studied when dealing with degenerated shell elements in Chapter 10 (Section 10.16.1).

8.9.5 Quasi-coplanar nodes in smooth shells

A typical problem in the analysis of smooth shells with flat elements is that the degree of coplanarity depends on the size mesh. Thus, for coarse meshes the standard check on the normal direction at nodes will identify an artificially large number of non-coplanar nodes. This problem disappears

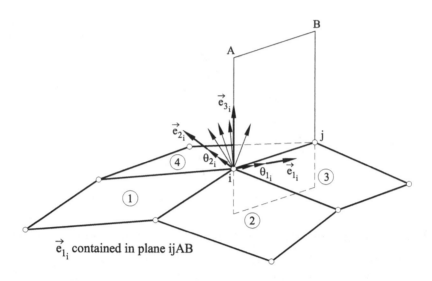

Fig. 8.20 Local axes for five DOFs assembly in quasi-coplanar nodes

when the mesh is refined, as in the limit all nodes are coplanar for a smooth surface.

This ambiguous situation can be overcome by assembling the rotational stiffness equations in a new nodal coordinate system $\mathbf{e}_{1_i}$, $\mathbf{e}_{2_i}$, $\mathbf{e}_{3_i}$ where $\mathbf{e}_{3_i}$ is a vector in the average direction of all the normals meeting at the node (Figure 8.20), $\mathbf{e}_{1_i}$ is orthogonal to $\mathbf{e}_{3_i}$ and is contained in the plane formed by one of the sides ij and $\mathbf{e}_{3_i}$, and $\mathbf{e}_{2_i} = \mathbf{e}_{3_i} \times \mathbf{e}_{1_i}$ [HB]. This allows us to keep the following five DOFs at each node

$$\mathbf{a}_i = [u_{0_i}, v_{0_i}, w_{0_i}, \theta_{1_i}\theta_{2_i}]^T \tag{8.79}$$

where the nodal displacements are defined in global axes and θ_{1_i} and θ_{2_i} are the rotations with axial directions along $\mathbf{e}_{1_i}$ and $\mathbf{e}_{2_i}$, respectively. These rotations are expressed in terms of the original local rotations $\theta_{x'_i}$, $\theta_{y'_i}$ as follows. First, θ_{1_i} and θ_{2_i} are transformed to *global axes* as

$$\boldsymbol{\theta}_i = \bar{\boldsymbol{\lambda}}_i^T \bar{\boldsymbol{\theta}}_i \quad \text{with} \quad \boldsymbol{\theta}_i = \left\{ \begin{matrix} \theta_{x_i} \\ \theta_{y_i} \\ \theta_{z_i} \end{matrix} \right\} \; , \; \bar{\boldsymbol{\theta}}_i = \left\{ \begin{matrix} \theta_{1_i} \\ \theta_{2_i} \end{matrix} \right\} \; , \; \underset{2\times 3}{\bar{\boldsymbol{\lambda}}_i} = \begin{bmatrix} \mathbf{e}_{1_i}^T \\ \mathbf{e}_{2_i}^T \end{bmatrix} \tag{8.80}$$

The sought expression is found using Eqs.(8.39) and (8.41) as

$$\underset{2\times 1}{\boldsymbol{\theta}'_i} = \left\{ \begin{matrix} \theta_{x'_i} \\ \theta_{y'_i} \end{matrix} \right\} = \hat{\boldsymbol{\lambda}}^{(e)} \bar{\boldsymbol{\lambda}}_i^T \bar{\boldsymbol{\theta}}_i = \underset{2\times 1}{\hat{\bar{\boldsymbol{\lambda}}}_i^{(e)}} \bar{\boldsymbol{\theta}}_i \tag{8.81}$$

Matrix $\hat{\bar{\boldsymbol{\lambda}}}^{(e)} = \hat{\boldsymbol{\lambda}}_i^{(e)}\overline{\boldsymbol{\lambda}}_i^T$ substitutes now $\hat{\boldsymbol{\lambda}}^{(e)}$ in Eq.(8.41). Note that this matrix changes for each node.

Carpenter *et al.* [CSB,CSB2] have proposed an alternative transformation technique for the nodal rotation that preserves the rigid body rotation of the element while keeping 5 DOFs per node.

8.10 CHOICE OF REISSNER-MINDLIN FLAT SHELL ELEMENTS

Flat shell elements can be formulated by adequately combining plane stress (membrane) and bending elements and many options are possible. Naturally, the accuracy flat shell elements very much depends on the merits of the membrane and bending approximations chosen. Some popular Reissner-Mindlin flat shell elements are:

a) Four-noded Q4 flat shell quadrilateral obtained by combining the 4-noded plane stress quadrilateral of Section 6.4.1 of [On4] and the Q4 plate element with selective integration of Section 6.5.1.
b) Four-noded QLLL flat shell quadrilateral combining the 4-noded plane stress quadrilateral and the QLLL plate element of Section 6.7.1 [On3].
c) QS8 and QL9 flat shell quadrilaterals combining the 8 and 9-noded plane stress quadrilaterals (Chapter 6 of [On4]) and the QS8 and QL9 plate elements of Sections 6.5.2 and 6.5.3, respectively.
d) QQQQ-S, QQQQ-L and QLQL flat shell quadrilateral obtained by combining the 8 and 9-noded plane stress quadrilaterals with the corresponding plate elements (Sections 6.7.2–6.7.4).
e) TLQL and TLLL flat shell triangles combining the 3-noded linear plane stress triangle (Chapter 5 of [On4]) and the corresponding plate elements of Sections 6.8.2 and 6.8.3 [On3].
f) TQQL flat shell triangle obtained by combining the 6-noded quadratic plane stress triangle (Chapter 6 of [On4]) and the TQQL plate element (Section 6.8.1).

The membrane behaviour of all these elements can be enhanced by using reduced integration for the in-plane tangential terms, or else by introducing incompatible modes or an assumed linear in-plane strain field (see Sections 5.4.2.2–4 of [On4]). This also helps to eliminating membrane locking as explained in the next section.

Table 8.1 shows the displacement interpolations for some of the flat shell elements mentioned above and the number of DOFs for the "smooth"

Aproximation \ Element	u', v', w'	$\theta_{x'}, \theta_{y'}$	$\gamma_{x'z'}, \gamma_{y'z'}$	numbers of DOFs
Q4	bilinear	bilinear	–	20
QS8	biquadratic	biquadratic	–	40
QL9	biquadratic	biquadratic	–	45
QLLL	bilinear	bilinear	linear	20
QLQL	bilinear	quadratic	linear	24
QQQQ-L	biquadratic	biquadratic	quadratic	45
QQQQ-S	biquadratic	biquadratic	quadratic	40
TLQL	linear	quadratic	linear	18
TQQL	quadratic	quadratic	linear	30
TLLL	linear	linear	linear	15

Table 8.1 Interpolations for Reissner-Mindlin flat shell elements based on the standard displacement formulation and the assumed transverse shear strain approach

shell case. An additional rotational DOF per node should be added for kinked or branching shells. The elements are termed after the name of the "parent" plate element, for convenience. Examples showing the behaviour of some of these elements are presented in Section 8.13.

8.11 SHEAR AND MEMBRANE LOCKING

Let us consider the equilibrium equations for a flat shell element written in local axes. Assuming constant thickness Eq.(8.34) can be rewritten as

$$\left[t\mathbf{K}_m'^{(e)} + t^3\mathbf{K}_b'^{(e)} + t^2\bar{\mathbf{K}}_{mb}'^{(e)} + t\mathbf{K}_s'^{(e)}\right]\mathbf{a}'^{(e)} - \mathbf{f}'^{(e)} = \mathbf{q}'^{(e)} \qquad (8.82)$$

where the thickness has been taken out from the matrices.

Eq.(8.82) shows that the influence of the thickness is of the same order for both the membrane and transverse shear matrices. For clarify let us rewrite Eq.(8.82) neglecting the coupling membrane-bending matrix as

$$\left[t^3\mathbf{K}_b'^{(e)} + t\left(\mathbf{K}_s'^{(e)} + \mathbf{K}_m'^{(e)}\right)\right]\mathbf{a}'^{(e)} - \mathbf{f}'^{(e)} = \mathbf{q}'^{(e)} \qquad (8.83)$$

Eq.(8.83) is expressed in local axes and therefore only the bending and transverse shear terms are coupled. Consequently, if the shell degenerates into a flat plate, the bending and membrane displacements can be obtained in a decoupled manner as

$$\left[t^3\mathbf{K}_b' + t\mathbf{K}_s'\right]\mathbf{a}_b' = \mathbf{f}_b' \qquad (8.84)$$

$$t\mathbf{K}_m'\mathbf{a}_m' = \mathbf{f}_m' \qquad (8.85)$$

where $\mathbf{f}'_b$ and $\mathbf{f}'_m$ are the equivalent nodal force vectors due to bending and in-plane loads and

$$\mathbf{a}'_{b_i} = \left[w'_{o_i}, \theta_{x'_i}, \theta_{y'_i}\right]^T, \qquad \mathbf{a}_{m_i} = \left[u'_{o_i}, v'_{o_i}\right]^T \qquad (8.86)$$

The solution of Eq.(8.85) will suffer from shear locking, similarly as explained for plates in Section 6.4.1. However, no problem exists in finding the membrane solution by solving Eq.(8.86).

The global equilibrium equation can be written after assembly as

$$\left[t^3 \mathbf{K}_b + t(\mathbf{K}_s + \mathbf{K}_m)\right] \mathbf{a} = \mathbf{f} \qquad (8.87)$$

where the membrane and flexural terms are now coupled. This equation suffers from the same defect as Eqs.(8.85), i.e. the sum of membrane and transverse shear terms will have an excessive influence for thin situations. Eq.(8.87) degenerates in the thin limit case to

$$(\mathbf{K}_s + \mathbf{K}_m)\mathbf{a} = \mathbf{K}_{sm}\mathbf{a} = 0 \qquad (8.88)$$

which requires the singularity of $\mathbf{K}_{sm} = \mathbf{K}_s + \mathbf{K}_m$ for a non trivial solution. The rule (5.68) indicates that a reduced quadrature is required for $\mathbf{K}_{sm}$ to be singular. This implies that both $\mathbf{K}_m^{(e)}$ and $\mathbf{K}_s^{(e)}$ must be underintegrated to avoid *shear and membrane locking*.

Above explanation shows that membrane locking very much depends on the degree of coupling between the flexural and membrane terms, whereas shear locking is *intrinsic* to the Reissner-Mindlin plate formulation. For shells where bending effects are dominant, membrane locking is of little importance and only shear locking must be accounted for. However for problems with high flexural-membrane coupling, membrane locking can perturb the solution. This occurs for composite shells, or for curved shell elements where coupling between the bending and membrane stiffness terms appears at the element level (Sections 9.15 and 10.11).

Membrane locking can also be interpreted as the incapacity of the element to reproduce a pure bending strain field without introducing spurious membrane strains. Shell elements free of membrane locking must therefore be able to reproduce the condition $\hat{\boldsymbol{\varepsilon}}'_m = 0$ under pure bending loads. This condition is analogous to that of $\hat{\boldsymbol{\varepsilon}}'_s = 0$, for the thin limit in shear locking-free plate elements. These conditions can be satisfied by choosing the adequate reduced integration rule or by using assumed strain fields.

Flat shell elements are less prone to membrane locking, as they typically satisfy individually the condition $\hat{\boldsymbol{\varepsilon}}'_m = 0$ under pure bending. This

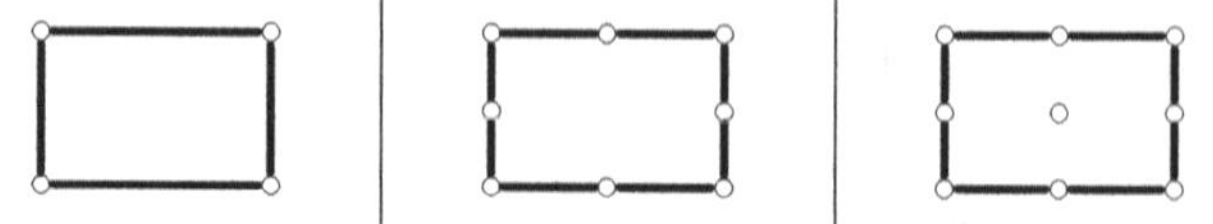

	$K_b^{(e)}$	$K_s^{(e)}$	$K_m^{(e)}$	$K_b^{(e)}$	$K_s^{(e)}$	$K_m^{(e)}$	$K_b^{(e)}$	$K_s^{(e)}$	$K_m^{(e)}$
Gauss-Legendre quadrature	2×2	$1*$	1	3×3	2×2	2×2	3×3	2×2	2×2

For $K_{mb}^{(e)}$ the same quadrature as for $K_b^{(e)}$ is recommended

* It contains one propagable mechanism (Section 5.5.1)

Fig. 8.21 Recommended quadratures for some flat shell quadrilaterals

is because the flexural and membrane stiffness terms are decoupled at element level for homogeneous or symmetric material properties. This is not generally so for shells with composite materials or for curved shell elements which are more sensitive to membrane locking.

Reduced/selective integration is the simplest procedure to avoid membrane and shear locking. Figure 8.21 shows the quadratures recommended for the 4, 8 and 9-noded quadrilaterals. Different checks on the singularity rule (2.50) for these elements are shown in Figure 8.22. The conclusions are similar to the plate bending case: the 4 and 9-noded quadrilaterals with reduced integration for $\mathbf{K}_m^{(e)}$ and $\mathbf{K}_s^{(e)}$ satisfy the singularity rule, whereas the 8-noded quadrilateral fails in some cases and is not recommended for thin shell analysis. The 4 and 9-noded quadrilateral shell element with uniform reduced quadrature and spurious mode control are good candidates for practical applications [BLOL,BT].

Shear and membrane locking can be consistently avoided using *assumed transverse shear and membrane strain fields*. The procedure is identical to reduced integration in some cases, as for beams and plates. The assumed strain approach given is detailed in Chapter 10 where some locking-free curved shell elements are presented. These elements are also applicable for flat situations.

8.12 THIN FLAT SHELL ELEMENTS

Thin flat shell elements are based on Kirchhoff plate theory (Chapter 4). The methodology follows the steps for the Reissner-Mindlin elements studied in previous sections. The relevant expressions are given next.

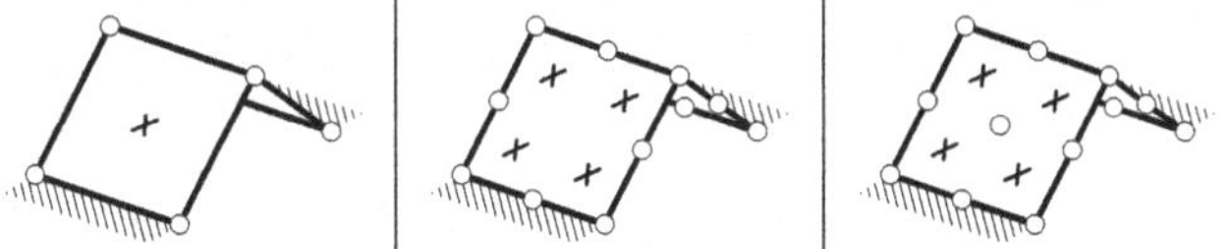

	4 nodes	8 nodes	9 nodes
	NGP = 2	NGP = 8	NGP = 8
SS (hard) FDOF : I :	16 $16-2\times5=6>0$	44 $44-8\times5=4>0$	54 $54-8\times5=14>0$
Clamped edge FDOF : I :	12 $12-2\times5=2>0$	38 $38-8\times5=\boxed{-2<0}$ FAIL	48 $48-8\times5=8>0$

$$I = FDOF - NGP * NMS$$

Fig. 8.22 Singulary tests for 4, 8 and 9-noded flat shell quadrilaterals using reduced integration for $\mathbf{K}_m^{(e)}$ and $\mathbf{K}_s^{(e)}$. FDOF= free DOFs, NGP= No. of Gauss points, NMS= No. of membrane and transverse shear strain components (=5)

8.12.1 Kinematic, constitutive and equilibrium equations

Kirchhoff thin shell theory assumes that the normal rotations $\theta_{x'}$ and $\theta_{y'}$ coincide with the mid-plane slopes $\frac{\partial w'}{\partial x'}$ and $\frac{\partial w'}{\partial y'}$, respectively (Figure 8.6). Introducing this assumption in the displacement field of Eq.(8.1) gives

$$u' = u'_0 - z'\frac{\partial w'}{\partial x'} \quad , \quad v' = v'_0 - z'\frac{\partial w'}{\partial y'} \quad , \quad w' = w'_0 \qquad (8.89)$$

It is easy to verify that the transverse shear strains $\gamma_{x'z'}$ and $\gamma_{y'z'}$ are zero. The local strain vector is simply

$$\varepsilon' = \hat{\varepsilon}'_m - z'\hat{\varepsilon}'_b \qquad (8.90)$$

The membrane strain vector $\hat{\varepsilon}'_m$ coincides with Eq.(8.6a) whereas the bending strain vector is

$$\hat{\varepsilon}'_b = \left[\frac{\partial^2 w'_0}{\partial x'^2}, \frac{\partial^2 w'_0}{\partial y'^2}, 2\frac{\partial^2 w'_0}{\partial x'\partial y'}\right]^T \qquad (8.91)$$

The stress-strain relationship is deduced from Eqs.(8.9a) and (8.4) (for simplicity the initial stresses will be neglected hereafter) as

$$\sigma' = \sigma'_b = \mathbf{D}'_p(\hat{\varepsilon}'_m - z'\hat{\varepsilon}'_b) \qquad (8.92)$$

where the bending constitutive matrix $\mathbf{D}'_p$ is given by Eq.(8.12).

The resultant stress vector contains axial forces and bending moments only. Following a process identical to that of Section 8.3.4, the relationship between resultant stresses and generalized local strains is found as

$$\hat{\sigma}' = \begin{Bmatrix} \hat{\sigma}'_m \\ \hat{\sigma}'_b \end{Bmatrix} = \begin{bmatrix} \hat{\mathbf{D}}'_m & \hat{\mathbf{D}}'_{mb} \\ \hat{\mathbf{D}}'_{mb} & \hat{\mathbf{D}}'_b \end{bmatrix} \begin{Bmatrix} \hat{\varepsilon}'_m \\ \hat{\varepsilon}'_b \end{Bmatrix} = \hat{\mathbf{D}}'\hat{\varepsilon}' \qquad (8.93)$$

where $\hat{\mathbf{D}}'_m$, $\hat{\mathbf{D}}'_b$ and $\hat{\mathbf{D}}'_{mb}$ are given by Eq.(8.18a). Again $\hat{\mathbf{D}}'_{mb} = 0$ for homogeneous or symmetric material properties with respect to the middle plane. This decouples the membrane and bending effects at element level.

The PVW is obtained by neglecting the transverse shear terms in Eq.(8.26); i.e.

$$\iint_A (\delta\hat{\varepsilon}'^T_m \hat{\sigma}'_m + \delta\hat{\varepsilon}'^T_b \hat{\sigma}'_b)dA = \iint_A \delta\mathbf{u}'^T \mathbf{t}' dA + \sum_i \delta\mathbf{u}'^T_i \mathbf{p}'_i \qquad (8.94)$$

8.12.2 Derivation of thin flat shell element matrices

The second derivatives of the transverse displacement w'_0 in the PVW introduce the need for C^1 continuity for the lateral deflection field, similarly as for thin plates. On the other hand, the interpolation of the in-plane displacements u'_0 and v'_0 requires C^0 continuity only. The different continuity requirements for the local displacement components is a drawback of thin flat shell elements [BD6,YSL,ZT2].

The displacement interpolation field can be chosen by considering the plane stress and Kirchhoff plate bending elements studied in Chapter 5 of [On4] and Chapter 4 of this volume, respectively. We will assume for simplicity, that both fields are defined by the same number of nodes. For example, the 3-noded constant strain triangle (CST) (Section 5.3 of [On4]) can be used to define a linear field for the in-plane displacements while an incompatible field can be chosen for w'_0 (such as the CKZ plate element of Section 5.5.1). Another possibility is to combine a bilinear field for u'_0 and v'_0 (using the 4-noded plane stress rectangle) and a cubic field for w'_0 (i.e. the MZC plate element of Section 5.4.1). The local displacement field can be written in both cases as

$$\mathbf{u}' = \sum_{i=1}^{n} \mathbf{N}_i \mathbf{a}'^{(e)}_i = [\mathbf{N}_1, \mathbf{N}_2, \cdots, \mathbf{N}_n] \begin{Bmatrix} \mathbf{a}'^{(e)}_1 \\ \mathbf{a}'^{(e)}_2 \\ \vdots \\ \mathbf{a}'^{(e)}_n \end{Bmatrix} = \mathbf{N}\mathbf{a}'^{(e)} \qquad (8.95)$$

where

$$\mathbf{u}' = \left[u'_0, v'_0, w'_0,\right]^T \quad ; \quad \mathbf{a}'^{(e)}_i = \left[u'_{0_i}, v'_{0_i}, w'_{0_i}, \left(\frac{\partial w'_0}{\partial x'}\right)_i, \left(\frac{\partial w'_0}{\partial y'}\right)_i\right]^T \quad (8.96)$$

and

$$\mathbf{N}_i = \begin{bmatrix} N_i & 0 & \vdots & 0 & 0 & 0 \\ 0 & N_i & \vdots & 0 & 0 & 0 \\ \cdots & \cdots & \vdots & \cdots & \cdots & \cdots \\ 0 & 0 & \vdots & P_i & \bar{P}_i & \bar{\bar{P}}_i \end{bmatrix} = \begin{bmatrix} \mathbf{N}_i^m & \vdots & \mathbf{0} \\ \cdots\cdots & \vdots & \cdots\cdots \\ \mathbf{0} & \vdots & \mathbf{N}_i^b \end{bmatrix} \quad (8.97)$$

In Eq.(8.97) N_i are the C^0 continuous shape functions for the membrane displacements and P_i $\bar{P}_i$ and $\bar{\bar{P}}_i$ are the shape functions expressing the transverse deflection w' in terms of the nodal deflections w'_{0_i} and the slopes $\left(\frac{\partial w'_0}{\partial x'}\right)_i$ and $\left(\frac{\partial w'_0}{\partial y'}\right)_i$ (functions $P_i, \bar{P}_i$ and $\bar{\bar{P}}_i$ coincide with $N_i, \bar{N}_i$ and $\bar{\bar{N}}_i$ in Eq.(5.36) (MZC element) or (5.55) (KZ element).

Substituting Eq.(8.91) into the expression for $\hat{\boldsymbol{\varepsilon}}'$ of Eq.(8.93) gives the generalized local strain matrix as

$$\mathbf{B}' = \left[\mathbf{B}'_1, \mathbf{B}'_2, \cdots, \mathbf{B}'_n\right] \quad \text{with} \quad \mathbf{B}'_i = \begin{Bmatrix} \mathbf{B}'_{m_i} \\ \mathbf{B}'_{b_i} \end{Bmatrix} \quad (8.98)$$

where the membrane matrix $\mathbf{B}'_{mi}$ is identical to (8.31) and

$$\mathbf{B}'_{b_i} = \begin{bmatrix} 0 & 0 & \dfrac{\partial^2 P_i}{\partial x'^2} & \dfrac{\partial^2 \bar{P}_i}{\partial x'^2} & \dfrac{\partial^2 \bar{\bar{P}}_i}{\partial x'^2} \\ 0 & 0 & \dfrac{\partial^2 P_i}{\partial y'^2} & \dfrac{\partial^2 \bar{P}_i}{\partial y'^2} & \dfrac{\partial^2 \bar{\bar{P}}_i}{\partial y'^2} \\ 0 & 0 & 2\dfrac{\partial^2 P_i}{\partial x'\partial y'} & 2\dfrac{\partial^2 \bar{P}_i}{\partial x'\partial y'} & 2\dfrac{\partial^2 \bar{\bar{P}}_i}{\partial x'y'} \end{bmatrix} \quad (8.99)$$

Note that $\mathbf{B}'_{b_i}$ is a simple extension of the bending strain matrix for thin plates (Eq.(5.39)).

Following a process similar to that of Section 8.4.3 yields the local element stiffness matrix

$$\mathbf{K}'^{(e)}_{ij} = \mathbf{K}'^{(e)}_{m_{ij}} + \mathbf{K}'^{(e)}_{b_{ij}} + \mathbf{K}'^{(e)}_{mb_{ij}} + \left[\mathbf{K}'^{(e)}_{mb_{ij}}\right]^T \quad (8.100)$$

The expression of above matrices is identical to Eq.(8.37). For $\mathbf{K}'_{mb} = 0$ the element stiffness matrix can be formed by combining the plane stress and bending stiffness matrices as shown in Eq.(8.38). The global assembly process follows precisely the transformations of Section 8.5.

8.12.3 Selection of thin flat shell elements

The more popular options are:

- 4-noded plane stress rectangle of Section 5.3.1 of [On4] (both in the standard and enhanced forms) combined with the non conforming plate MZC rectangle of Section 5.4.1. This flat shell element was first used in references [ZCh2,3]. Arbitrary quadrilateral forms of the element are not recommended as the patch test is not satisfied.

- The 4-noded plane stress rectangle can be combined with any of the 4-noded compatible plate quadrilaterals of Sections 5.4.2 and 5.6. Arbitrary quadrilateral shapes are now possible.

- The simple 3-noded plane stress triangle (Section 4.3 of [On4]) has been successfully combined with the CKZ incompatible plate triangle with nine DOFs (Section 5.5.1) [Pa,ZPK].

- A Morley flat shell triangle can be derived by combining the 3-noded plane stress triangle with the Morley plate triangle of Section 8.5.1. The convergence of this simple shell triangle is quite poor [Mo,Mo2,ZT2].

- The 6-noded plane stress triangle can be combined with the 12 DOFs HCT conforming plate triangle of Section 5.5.2. This element was introduced by Razzaque [Raz].

- A possibility is combining plane stress triangles and quadrilaterals with the Discrete Kirchhoff (DK) plate elements of Section 6.11. This alternative has been exploited by Batoz *et al.* [B10]. A simple and accurate shell triangle results from combining the 3-noded plane stress triangle and the DKT element of Section 6.11.1. Another good element of this type is obtained by combining the 4-noded plane stress rectangle and the DKQ plate element of Section 6.11.2.

- The rotation-free BPT and BPN plate triangles can be combined with the 3-noded plane stress triangle to yield two interesting rotation-free shell triangles. Details are given in Section 8.13.

- The drilling rotation technique (Section 8.9.3) can be applied to enhance the performance of any of the thin flat shell elements mentioned above [Ta,ZT2].

Examples of other flat shell elements can be found in [ZT2].

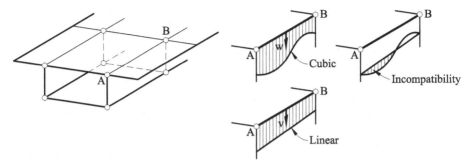

Fig. 8.23 Displacement incompatibility in 4-noded and MZC flat shell rectangle

8.12.4 Incompatibility between membrane and bending fields

Most thin flat shell elements present a displacement incompatibility along common sides in non-coplanar situations. Let us consider, for instance, the box girder bridge of Figure 8.23, analyzed with 4-noded MZC shell rectangles of the type described in the previous section. The in-plane displacements vary linearly along the sides, whereas the transverse displacement is cubic. This leads to a displacement incompatibility along a side belonging to a fold. This defficiency yields overstiff results for coarse meshes and it is corrected by mesh refinement. An alternative is to use special higher order plane stress elements with the same approximation as for the bending deflection. These elements have been developed for cellular bridges by Lim *et al.* [LKM] and Willam and Scordelis [WS].

8.13 BST AND BSN ROTATION-FREE THIN FLAT SHELL TRIANGLES

Rotation-free shell elements that incorporate the three displacements as the only nodal DOFs are very attractive for practical purposes.

We describe next the extension of the rotation-free BPT and BPN thin plate triangles (Section 5.8) to shells. The so-called BST and BSN rotation-free shell triangles were originally derived by Oñate and Zarate [OZ]. The background of these elements was presented in Section 5.8.

8.13.1 BST rotation-free shell triangle

Figure 8.24 shows the patch of four shell triangles typical of the cell-centred (CC) finite volume scheme [OCZ,ZO4]. As usual in the CC scheme

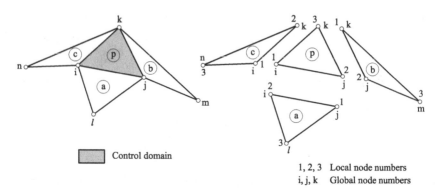

Fig. 8.24 BST element. Control domain and four-element patch

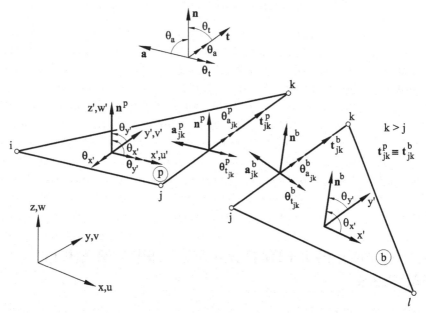

Fig. 8.25 BST element. Definition of global, local and side coordinate systems

the control domain coincides with an individual element. Also in Figure 8.24 the local and global node numbering chosen is shown.

Figure 8.25 shows the local element axes x', y', z' where x' is parallel to side 1–2 (or $i-j$) and in the direction of increasing local node numbers, z' is a direction orthogonal to the element and y' is obtained by cross product of vectors along z' and x'. A side coordinate system is defined with side unit vectors $\mathbf{t}$, $\mathbf{a}$ and $\mathbf{n}$. Vector $\mathbf{t}$ is aligned along the side following the

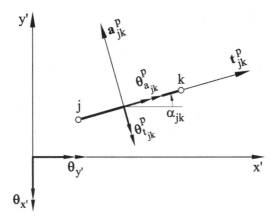

Fig. 8.26 BST element. Transformation from side to local rotations

direction of increasing global node numbers, $\mathbf{n}$ is the normal vector parallel to the z' local axis and $\mathbf{a} = \mathbf{n} \wedge \mathbf{t}$.

The local rotations $\theta_{x'}, \theta_{y'}$ along each side are expressed in terms of the so called tangential and normal side rotations θ_t and θ_a by the following transformation (Figure 8.26)

$$
\boldsymbol{\theta}'^p = \left\{ \begin{array}{c} \theta_{x'}^p \\ \theta_{y'}^p \end{array} \right\}^{(e)} = \begin{bmatrix} c_{jk}^p & -s_{jk}^p \\ s_{jk}^p & c_{jk}^p \end{bmatrix}^{(e)} \left\{ \begin{array}{c} \theta_{t_{jk}}^p \\ \theta_{a_{jk}}^p \end{array} \right\}^{(e)} = \hat{\mathbf{T}}_{jk} \hat{\boldsymbol{\theta}}'^p_{jk} \tag{8.101}
$$

where $\theta_{t_{jk}}^p$ and $\theta_{a_{jk}}^p$ are the tangential and normal rotations along the side ij of element p, $\theta_{x'} = \frac{\partial w'}{\partial x'}$, $\theta_{y'} = \frac{\partial w'}{\partial y'}$ and c_{jk}^p, s_{jk}^p are the components of the side unit vector $\mathbf{t}_{jk}^p$, i.e. $\mathbf{t}_{jk}^p = [c_{jk}^p, s_{jk}^p]^T$. The sign for the rotations is shown in Figures 8.25 and 8.26.

The definition of the curvatures follows the lines given for the BPT element. The local curvatures over the control domain p of area A_p formed by the triangle ijk are given by the line integral (see Eq.(5.68))

$$
\hat{\boldsymbol{\varepsilon}}_b'^p = \frac{1}{A_p} \int_{\Gamma_p} \mathbf{T} \boldsymbol{\nabla}' w' d\Gamma \tag{8.102}
$$

with

$$
\hat{\boldsymbol{\varepsilon}}_b'^p = \left[\frac{\partial^2 w'}{\partial x'^2}, \frac{\partial^2 w'}{\partial y'^2}, 2\frac{\partial^2 w'}{\partial x'\partial y'} \right]^T , \quad \mathbf{T} = \begin{bmatrix} t_{x'} & 0 & t_{y'} \\ 0 & t_{y'} & t_{x'} \end{bmatrix}^T \text{ and } \boldsymbol{\nabla}' = \left[\frac{\partial}{\partial x'}, \frac{\partial}{\partial y'} \right]^T \tag{8.103}
$$

where $t_{x'}, t_{y'}$ are the components of vector $\mathbf{t}$ in the x', y' coordinate system.

Recalling that $\boldsymbol{\theta}' = \boldsymbol{\nabla}'w'$ and using Eq.(8.101), Eq.(8.103) expressing the curvature over the triangle ijk can be written as

$$\hat{\varepsilon}_b'^p = \frac{1}{A_p}[\mathbf{T}_{ij}^p\hat{\mathbf{T}}_{ij}^p\hat{\theta}_{ij}'^p l_{ij} + \mathbf{T}_{jk}^p\hat{\mathbf{T}}_{jk}^p\hat{\theta}_{jk}'^p l_{jk} + \mathbf{T}_{ki}^p\hat{\mathbf{T}}_{ki}^p\hat{\theta}_{ki}'^p l_{ki}] \qquad (8.104)$$

In the derivation of Eq.(8.104) we have taken advantage that the local rotations are constant along each element side.

The tangential side rotations can be expressed in terms of the local deflections along the sides. For instance, for side jk (Figures 8.25 and 8.26)

$$\theta_{t_{jk}}^p = \frac{w_k'^p - w_j'^p}{l_{jk}} \quad \text{for } k > j \qquad (8.105)$$

where l_{jk} is the length of side jk.

Equation (8.105) introduces an approximation as the tangential rotation vectors of adjacent elements sharing a side are not parallel. Therefore the tangential rotations are discontinuous along element sides; hence

$$\theta_{t_{jk}}^p = \frac{w_k'^p - w_j'^p}{l_{jk}} \neq \frac{w_k'^b - w_j'^b}{l_{jk}} = \theta_{t_{jk}}^b \qquad (8.106)$$

This error has little relevance in practice and it vanishes for smooth shells as the mesh is refined. For quasi-coplanar sides $w_k'^p \simeq w_k'^b$, $w_j'^p \simeq w_j'^b$ and, hence, $\theta_{t_{jk}}^p \simeq \theta_{t_{jk}}^b$.

A continuous tangential side rotation $\theta_{t_{jk}}$ can be ensured if defined as the average of the tangential side rotations contributed by the two elements sharing the side [OZ]. The form of Eq.(8.105) is however chosen in the derivations presented hereafter.

Vector $\mathbf{t}_{jk}$ is the same for the two elements sharing a side (i.e. $\mathbf{t}_{jk}^p = \mathbf{t}_{jk}^b$ in Figure 8.26). A *continuous* normal rotation is enforced by defining an average normal rotation along side jk as

$$\theta_{a_{jk}}^p = \frac{1}{2}(\theta_{a_{jk}}^p + \theta_{a_{jk}}^b) \qquad (8.107)$$

This average rotation is expressed in terms of the normal deflections using the inverse of Eq.(8.101) and the fact that $\boldsymbol{\theta}' = \boldsymbol{\nabla}'w'$ as

$$\theta_{a_{jk}}^p = \frac{1}{2}\left(\lambda_{jk}^p(\boldsymbol{\nabla}'w')_{jk}^p + \lambda_{jk}^b(\boldsymbol{\nabla}'w')_{jk}^b\right) \qquad (8.108)$$

where

$$\boldsymbol{\lambda}_{jk}^p = [-s_{jk}^p, c_{jk}^p] \tag{8.109}$$

Substituting Eqs.(8.105) and (8.108) for the three element sides into (8.104) and choosing a linear interpolation for the displacements within each triangle, the curvatures for the central triangle p in Figure 8.24 are expressed in terms of the normal deflections of the nodes in the four-element patch as

$$\hat{\boldsymbol{\varepsilon}}_b^{\prime p} = \mathbf{S}^p \boldsymbol{w}_p^{\prime} \tag{8.110}$$

where

$$\mathbf{S}^p = [\mathbf{S}_{ij}^p, \mathbf{S}_{jk}^p, \mathbf{S}_{ki}^p] \tag{8.111}$$

$$\boldsymbol{w}_p^{\prime} = [w_i^{\prime p}, w_j^{\prime p}, w_k^{\prime p}, w_j^{\prime a}, w_i^{\prime a}, w_l^{\prime a}, w_k^{\prime b}, w_j^{\prime b}, w_m^{\prime b}, w_i^{\prime c}, w_k^{\prime c}, w_n^{\prime c}]^T \tag{8.112}$$

The local curvature matrices $\mathbf{S}_{ij}^p$ are given in Box 8.1. The ordering of the components of vector $\boldsymbol{w}_p^{\prime}$ depends on the convention chosen for the local and global node numbers for the four-element patch (Figure 8.24).

$$\boxed{\hat{\boldsymbol{\varepsilon}}_b^{\prime p} = \mathbf{S}^p \boldsymbol{w}_p^{\prime}}$$

$$\boldsymbol{w}_p^{\prime} = [w_i^{\prime p}, w_j^{\prime p}, w_k^{\prime p}, w_j^{\prime a}, w_i^{\prime a}, w_l^{\prime a}, w_k^{\prime b}, w_j^{\prime b}, w_m^{\prime b}, w_i^{\prime c}, w_k^{\prime c}, w_n^{\prime c}]^T$$

$$\mathbf{S}^p = [\mathbf{S}_{ij}^p, \mathbf{S}_{jk}^p, \mathbf{S}_{ki}^p]; \qquad \mathbf{S}_{ij}^p = \frac{l_{ij}}{A_p} \mathbf{T}_{ij}^p \hat{\mathbf{T}}_{ij}^p \mathbf{A}_{ij}^p$$

$$\mathbf{A}_{ij}^p = \begin{bmatrix} \alpha/l_{ij} & \beta/l_{ij} & 0 & 0 & 0 & 0 \\ \gamma_{iji}^p & \gamma_{ijj}^p & \gamma_{ijk}^p & \gamma_{ijj}^{(a)} & \gamma_{iji}^{(a)} & \gamma_{ijl}^{(a)} \end{bmatrix} \mathbf{0}_3 \ \mathbf{0}_3 \Bigg]; \qquad \begin{array}{l} \alpha = -1, \ \beta = 1, \ j > i \\ \alpha = 1, \quad \beta = -1, \ j < i \end{array}$$

$$\mathbf{A}_{jk}^p = \begin{bmatrix} 0 & \alpha/l_{jk} & \beta/l_{jk} & & 0 & 0 & 0 & \\ \gamma_{jki}^p & \gamma_{jkj}^p & \gamma_{jkk}^p & \mathbf{0}_3 & \gamma_{jkk}^b & \gamma_{jkj}^b & \gamma_{jkm}^b & \mathbf{0}_3 \end{bmatrix}; \qquad \begin{array}{l} \alpha = -1, \ \beta = 1, \ k > j \\ \alpha = 1, \quad \beta = -1, \ k < j \end{array}$$

$$\mathbf{A}_{ki}^p = \begin{bmatrix} \beta/l_{ki} & 0 & \alpha/l_{ki} & & 0 & 0 & 0 \\ \gamma_{kii}^p & \gamma_{kij}^p & \gamma_{kik}^p & \mathbf{0}_3 \ \mathbf{0}_3 & \gamma_{kii}^{(c)} & \gamma_{kik}^{(c)} & \gamma_{kin}^{(c)} \end{bmatrix}; \qquad \begin{array}{l} \alpha = 1, \quad \beta = -1, \ k > i \\ \alpha = -1, \ \beta = 1, \ k < i \end{array}$$

$$\gamma_{ijk}^p = \tfrac{1}{2}\boldsymbol{\lambda}_{ij}^p \boldsymbol{\nabla} N_k^p, \quad \boldsymbol{\lambda}_{ij}^p = [-s_{ij}^p, \ c_{ij}^p], \quad \mathbf{0}_3 = \begin{bmatrix} 0 & 0 & 0 \\ 0 & 0 & 0 \end{bmatrix}$$

$$\boldsymbol{\nabla} N_k^p = \begin{bmatrix} \dfrac{\partial N_k}{\partial x'} \\ \dfrac{\partial N_k}{\partial y'} \end{bmatrix}^p = \frac{1}{2A_p} \begin{Bmatrix} b_i \\ c_i \end{Bmatrix}^p; \quad b_i^p = y_j^{\prime p} - y_k^{\prime p}; \quad c_i^p = x_k^{\prime p} - x_j^{\prime p}$$

Box 8.1 BST element. Local curvature matrix for the control domain of Figure 8.24

The normal nodal deflections for each element are related to the global nodal displacements by the following transformation

$$w_i^{\prime p} = \mathbf{C}_i^p \mathbf{u}_i \quad \text{with} \quad \mathbf{C}_i^p = [c_{z'x}^p, c_{z'y}^p, c_{z'z}^p], \quad \mathbf{u}_i = [u_i, v_i, w_i]^T \quad (8.113)$$

where $e = p, a, b, c$ and $c_{z'x}^p$ is the cosine of the angle between the local z' axis of element p and the global x axis, etc.

Substituting Eq.(8.113) into (8.110) gives finally

$$\hat{\varepsilon}_b^{\prime p} = \mathbf{B}_b^p \mathbf{a}_p \quad \text{with} \quad \mathbf{B}_b^p = \mathbf{S}_p \mathbf{C}_p \quad , \quad \mathbf{a}_p = \begin{Bmatrix} \mathbf{u}_i \\ \mathbf{u}_j \\ \mathbf{u}_k \\ \mathbf{u}_l \\ \mathbf{u}_m \\ \mathbf{u}_n \end{Bmatrix} \quad (8.114)$$

In above $\mathbf{B}_b$ is the curvature matrix of the p-th triangle, $\mathbf{a}_p$ contains the 18 nodal displacements of the six nodes belonging to the four-element patch associated to the p-th triangle and $\mathbf{C}_p$ is the transformation matrix relating the 12 components of the normal deflection vector $\mathbf{w}_p'$ (Eq.(8.112)) and the 18 global displacements of vector $\mathbf{a}_p$. Recall that a triangle coincides with a standard triangle for the BST element.

The bending stiffness matrix for the p-th control domain is obtained by

$$\mathbf{K}_b^p = [\mathbf{B}_b^p]^T \hat{\mathbf{D}}_b^{\prime p} \mathbf{B}_b^p A_p \quad (8.115)$$

where $\hat{\mathbf{D}}_b^{\prime p}$ is the average bending constitutive matrix over the pth triangle obtained as

$$\hat{\mathbf{D}}_b^{\prime p} = \frac{1}{A^p} \iint_{A^p} \hat{\mathbf{D}}_b' dA \quad (8.116)$$

with $\hat{\mathbf{D}}_b'$ given by Eqs.(8.10) and (8.12) for homogeneous and composite materials, respectively.

BST element. Membrane stiffness matrix

The membrane stiffness matrix can be obtained from the expressions for the 3-noded plane stress triangle (Section 5.3 of [On4]). The local membrane strains are defined within each triangle in terms of the local nodal displacements as

$$\hat{\varepsilon}_m' = \sum_{i=1}^{3} \mathbf{B}_{m_i}^{\prime p} \mathbf{u}_i^{\prime p} = \mathbf{B}_m^{\prime p} \mathbf{a}_m^{\prime p} \quad (8.117a)$$

where

$$\hat{\varepsilon}'_m = \left[\frac{\partial u'}{\partial x'}, \frac{\partial v'}{\partial y'}, \frac{\partial u'}{\partial y'} + \frac{\partial v'}{\partial x'}\right]^T \qquad (8.117b)$$

$$\mathbf{B}'^p_{m_i} = \begin{bmatrix} \dfrac{\partial N^p_i}{\partial x'} & 0 \\[2mm] 0 & \dfrac{\partial N^p_i}{\partial y'} \\[2mm] \dfrac{\partial N^p_i}{\partial y'} & \dfrac{\partial N^p_i}{\partial x'} \end{bmatrix} = \frac{1}{2A^p}\begin{bmatrix} b_i & 0 \\ 0 & c_i \\ c_i & b_i \end{bmatrix}^p \quad, \quad \mathbf{a}'^p_m = \begin{Bmatrix} \mathbf{u}'^p_i \\ \mathbf{u}'^p_j \\ \mathbf{u}'^p_k \end{Bmatrix} \text{ and } \mathbf{u}'^p_i = [u'^p_i, v'^p_i]^T$$

$$(8.118)$$

In above u'^p_i and v'^p_i are the local in plane displacements (Figure 8.25) and b^p_i, c^p_i are defined in Box 8.1.

The membrane strains within the pth triangle are expressed in terms of the 18 global nodal displacements of the four-element patch as

$$\hat{\varepsilon}'^p_m = \mathbf{B}'^p_m \mathbf{L}_p \mathbf{a}_p = \mathbf{B}^p_m \mathbf{a}_p \quad \text{where} \quad \mathbf{B}^p_m = \mathbf{B}'^p_m \mathbf{L}_p \qquad (8.119)$$

The transformation matrix $\mathbf{L}_p$ is

$$\mathbf{L}_p = \begin{bmatrix} \mathbf{L}^p & \mathbf{0} & \mathbf{0} \\ \mathbf{0} & \mathbf{L}^p & \mathbf{0} \ \bar{\mathbf{0}} \\ \mathbf{0} & \mathbf{0} & \mathbf{L}^p \end{bmatrix} \text{ with } \mathbf{L}^p = \begin{bmatrix} c_{x'x} & c_{x'y} & c_{x'z} \\ c_{y'x} & c_{y'y} & c_{y'z} \end{bmatrix}^p \qquad (8.120)$$

where $\mathbf{0}$ and $\bar{\mathbf{0}}$ are 2×3 and 6×9 null matrices, respectively.

The membrane stiffness matrix associated to the p-th triangle is finally obtained as

$$\mathbf{K}^p_m = A^p [\mathbf{B}^p_m]^T \hat{\mathbf{D}}'^p_m \mathbf{B}^p_m \qquad (8.121)$$

Full stiffness matrix and equivalent nodal force vector

The stiffness matrix for the BST element is obtained by adding the membrane and bending contributions, i.e.

$$\mathbf{K}^p = \mathbf{K}^p_b + \mathbf{K}^p_m \qquad (8.122)$$

where $\mathbf{K}^p_b$ and $\mathbf{K}^p_m$ are given by Eqs.(8.115) and (8.121), respectively.

The dimensions of $\mathbf{K}^p$ are 18×18 as this matrix links the eighteen global displacements of the six nodes in the four-element patch associated to the BST element. The assembly of the stiffness matrices $\mathbf{K}^p$ into the global equation system follows the standard procedure, i.e. a *control domain is treated as a macro-triangular element with six nodes* [OZ].

The equivalent nodal force vector is obtained as for standard C^0 shell triangular elements, i.e. the contribution of a uniformly distributed load over an element is split into three equal parts among the three element nodes. Nodal point loads are directly assigned to a node, as usual.

Boundary conditions for the BST element

The procedure for prescribing the boundary conditions follows the lines explained for the BPT plate triangle. The process is quite straightforward as the side rotations are expressed in terms of the normal and tangential values. This allows to treat naturally all the boundary conditions found in practice.

The conditions for the normal rotations are introduced when building the curvature matrix, whereas the conditions for the nodal displacements and the tangential rotations are prescribed at the solution equation level.

Clamped edge $(\mathbf{u}_i = \mathbf{u}_j = \mathbf{0}; \theta_{a_{ij}} = \theta_{t_{ij}} = 0)$. The condition $\mathbf{u}_i = \mathbf{u}_j = \mathbf{0}$ is prescribed when solving the global system of equations. The condition $\theta_{t_{ij}} = 0$ is automatically satisfied by prescribing the side displacements to a zero value.

The condition $\theta_{a_{ij}} = 0$ is imposed by making zero the second row of matrix $\mathbf{A}_{ij}^p$ (Box 8.1) as this naturally enforces the condition of zero normal side rotations in Eq.(8.108). The control domain in this case has the element adjacent to the boundary side missing (Figure 5.26).

Simply supported edge $(\mathbf{u}_i = \mathbf{u}_j = \mathbf{0}; \theta_{t_{ij}} = 0)$. This boundary condition is imposed by prescribing $\mathbf{u}_i = \mathbf{u}_j = \mathbf{0}$ at the equation solution level.

Symmetry edge $(\theta_{a_{ij}} = 0)$. The condition of zero normal side rotation is imposed by making zero the second row of matrix $\mathbf{A}_{ij}^p$ as described above.

Free edge. Matrix $\mathbf{A}_{ij}^p$ is modified by ignoring the contribution from the missing adjacent element to the boundary side ij. This is implemented by making $\gamma_{ijj}^a = \gamma_{iji}^a = \gamma_{ijl}^a = 0$ and changing the $1/2$ in the definition of γ_{ijk}^p to a unit value (see Figure 5.26 and Box 8.1).

Flores and Oñate [FO2,OF] have proposed to prescribe the natural boundary condition of zero normal curvature at simply supported and free edges. This can be done by adequately modifying the bending moment-curvature relationship at the edge. The resulting element, called LBST, has a slight better behaviour than the BST element [FO2].

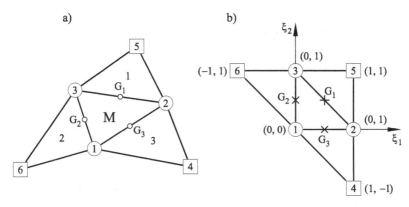

Fig. 8.27 EBST element. Patch of four 3-noded triangles: (a) actual geometry and (b) geometry in natural coordinates

8.13.2 Enhanced BST rotation-free shell triangle

Flores and Oñate [FO3,OF] have proposed an enhancement for the BST element using a non-standard quadratic interpolation for the geometry of the shell mid-surface over patches of four elements as

$$\mathbf{x} = \sum_{i=1}^{6} N_i(\xi, \eta)(\mathbf{x}_i^0 + \mathbf{u}_i) \tag{8.123a}$$

where

$$\mathbf{x} = [x, y, z]^T \quad, \quad \mathbf{x}_i^0 = [x_i^0, y_i^0, z_i^0]^T \quad, \quad \mathbf{u}_i = [u_i, v_i, w_i]^T \tag{8.123b}$$

and

$$N_1 = \xi_3 + \xi_1\xi_2 \quad, \quad N_4 = \frac{\xi_2}{2}(\xi_2 - 1)$$

$$N_2 = \xi_1 + \xi_2\xi_3 \quad, \quad N_5 = \frac{\xi_3}{2}(\xi_3 - 1) \tag{8.123c}$$

$$N_3 = \xi_2 + \xi_1\xi_3 \quad, \quad N_6 = \frac{\xi_1}{2}(\xi_1 - 1)$$

with $\xi_3 = 1 - \xi_1 - \xi_2$. Figure 8.27 shows the definition of the natural coordinates ξ_1, ξ_2 in the normalized space.

In Eq.(8.123a) $\mathbf{x}_i^0$ is the coordinate vector of the shell nodes in the undeformed position and $\mathbf{u}_i$ is the displacement vector of node i. Eq.(8.123a) is the basis for computing the gradient vector, the normal vector and the curvature tensor at any point in the patch in terms of the nodal displacements of the patch nodes. The gradient vector varies linearly over the patch and its value at the three Gauss points G_i located at the mid-side positions in the central element M depends only on the nodes pertaining

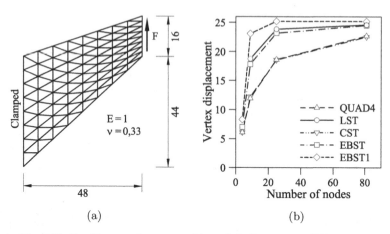

Fig. 8.28 Cook's membrane problem (a) Geometry (b) Results

to the two elements adjacent to the corresponding side (Figure 8.27). This is a key difference with the BST element where the normal rotation at the mid-side points is computed as the average of the values at the two adjacent elements (Eq.(8.107)). The curvature field is constant over the patch and can be obtained from the gradients at the three Gauss points.

The equivalent nodal force vector is computed as for the BST element. The explicit expression for the membrane and bending matrices of the so-called EBST element can be found in [FO2].

A simplified and yet very effective version of the EBST element can be obtained by using a reduced one point quadrature for computing all the element integrals. This element is termed EBST1. This only affects the membrane stiffness matrix and the element performs very well for membrane and bending dominating shell problems. Both the EBST and EBST1 elements are free of spurious energy modes.

The good behaviour of the EBST1 element for membrane dominant problems is shown in Figure 6.28 for the analysis of a thick tapered cantilever beam (the so-called Cook's membrane test problem). Figure 8.28 shows results for the upper vertex displacement for different meshes of EBST and EBST1 elements as well as for the standard 3-noded constant strain (plane stress) triangle (CST), the 6-noded linear strain triangle (LST) and the 4-noded bi-linear quadrilateral (QUAD4). The accuracy of the EBST1 element for coarse meshes is remarkable. Examples of the excellent behaviour of the EBST and EBST1 elements for linear and nonlinear analysis of shells are given in [FO3,4,OF]. A selection of linear examples is presented in Section 8.17.1.

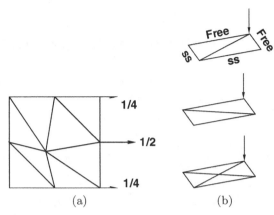

Fig. 8.29 Patch test for uniform tensile stress (a) and torsion (b)

Patch tests

The three elements considered (BST, EBST and EBST1) satisfy the membrane patch test defined in Figure 8.29a. A uniform axial tensile stress is obtained in all cases.

The element bending formulation does not allow to apply external bending moments (there are not rotational DOFs). Hence it is not possible to analyse a patch of elements under loads leading to a uniform bending moment. A uniform torsion can be considered if a point load is applied at the corner of a rectangular plate with two consecutive free sides and two simple supported sides. Figure 8.29b shows three patches leading to correct results both in displacements and stresses. All three patches are structured meshes. When the central node in the third patch is shifted from the center, the results obtained with the EBST and EBST1 elements are not correct. This however does not seems to preclude the excellent performance of these elements, as proved in the examples analyzed in [FO2] and Figures 8.34–8.36. The BST element gives correct results in all torsion patch tests if natural boundary conditions are imposed in the formulation [FO2]. If this is not the case, incorrect results are obtained even with structured meshes.

8.13.3 Extension of the BST and EBST elements for kinked and branching shells

Flores and Oñate [FO3] have extended the capabilities of the BST and EBST elements for the analysis of kinked and branching shells. The computation of the curvature tensor is first redefined in terms of the angle change between the normals of the adjacent element sharing a common side. This allows to deal with arbitrary large angles between adjacent elements at a branch and also to treat kinked surfaces. A relative stiffness between element is introduced to account for changing of material properties of element at a branch or a kink. Details of the formulation and examples of application can be found in [FO3].

8.13.4 Extended EBST elements with transverse shear deformation

Both the BST and EBST elements of previous sections can be extended to account for transverse shear deformation effects. The procedure follows the lines explained for the BPT plate triangle in Section 6.15. Two elements of this kind have been recently derived by Zarate and Oñate [ZO2]. The EBST+ element has five DOFs at ecah node: the 3 displacements and the two shear angles. A linear interpolation is used for all the DOFs.

The EBST+1 element has the standard displacement DOFs at the nodes plus two additional DOFs per element which represent the transverse shear angles. For both elements the bending and membrane stiffness matrices are computed as for the EBST elements and the only difference is the computation of the transverse shear stiffness contribution. Details of the derivation and good performance of the EBST+ and EBST+1 rotation-free shell elements can be found in [ZO2].

8.13.5 Basic shell nodal patch element (BSN)

The rotation-free BPN plate element described in Section 5.8.3 is extended now to shell analysis. Figure 8.30 shows a typical *cell-vertex* control domain surrounding a node and the corresponding patch of BPN shell triangles. The following coordinate systems are defined:

Global system: x, y, z, defining the global displacements u, v, w.

Local element system: x', y', z', defining the element curvatures. This coincides with the local system for the BST element (Figure 8.25).

Nodal system: $\bar{x}_i, \bar{y}_i, \bar{z}_i$, defining the constant curvature field over the control domain. Here $\bar{z}_i$ is the average normal direction at node i, $\bar{x}_i$ is defined as orthogonal to $\bar{z}_i$ and lying on the global plane x, z (if $\bar{z}_i$ coincides with the global y axis, then $\bar{x}_i = z$) and the $\bar{y}_i$ direction is taken as the cross product of unit vectors in the $\bar{z}_i$ and $\bar{x}_i$ directions.

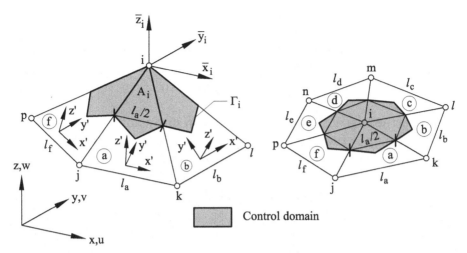

Fig. 8.30 BSN element. Control domain and coordinate systems

A constant curvature field is assumed over each control domain. For convenience the curvatures are defined in the nodal coordinate system. From simple transformation rules we can write

$$\bar{\varepsilon}_b = \mathbf{R}_1 \boldsymbol{\kappa} = \mathbf{R}_1 \mathbf{R}_2 \varepsilon'_b \tag{8.124}$$

In above

$$\bar{\varepsilon}_b = \left[\frac{\partial^2 \bar{w}}{\partial \bar{x}^2}, \frac{\partial^2 \bar{w}}{\partial \bar{y}^2}, 2 \frac{\partial^2 \bar{w}}{\partial \bar{x} \partial \bar{y}} \right]^T \tag{8.125}$$

is the nodal curvature vector

$$\varepsilon'_b = \left[\frac{\partial^2 w'}{\partial x'^2}, \frac{\partial^2 w'}{\partial y'^2}, 2 \frac{\partial^2 w'}{\partial x' \partial y'} \right]^T \tag{8.126}$$

is the element curvature vector and $\boldsymbol{\kappa}$ is an auxiliary "global" curvature vector used to simplify the transformation from element to nodal curvatures. The transformation matrices $\mathbf{R}_1$ and $\mathbf{R}_2$ are [OZ]

$$\mathbf{R}_1 = \begin{bmatrix} c_{\bar{x}x}^2 & c_{\bar{x}y}^2 & c_{\bar{x}z}^2 & c_{\bar{x}x}c_{\bar{x}y} & c_{\bar{x}x}c_{\bar{x}z} & c_{\bar{x}y}c_{\bar{x}z} \\ c_{\bar{y}x}^2 & c_{\bar{y}y}^2 & c_{\bar{y}z}^2 & c_{\bar{y}x}c_{\bar{y}y} & c_{\bar{y}x}c_{\bar{y}z} & c_{\bar{y}y}c_{\bar{y}z} \\ 2c_{\bar{x}x}2c_{\bar{y}x} & 2c_{\bar{x}y}c_{\bar{y}y} & 2c_{\bar{x}z}c_{\bar{y}z} & c_{\bar{x}y}c_{\bar{y}x}+c_{\bar{x}x}c_{\bar{y}y} & c_{\bar{x}z}c_{\bar{y}x}+c_{\bar{x}x}c_{\bar{y}z} & c_{\bar{x}z}c_{\bar{y}y}+c_{\bar{x}y}c_{\bar{y}z} \end{bmatrix}$$

$$\mathbf{R}_2 = \begin{bmatrix} c_{x'x}^2 & c_{y'x}^2 & c_{x'x}c_{y'x} \\ c_{x'y}^2 & c_{y'y}^2 & c_{x'y}c_{y'y} \\ c_{x'z}^2 & c_{y'z}^2 & c_{x'z}c_{y'z} \\ 2c_{x'x}c_{x'y} & 2c_{y'x}c_{y'y} & c_{x'y}c_{y'x}+c_{x'x}c_{y'y} \\ 2c_{x'x}c_{x'z} & 2c_{y'x}c_{y'z} & c_{x'z}c_{y'x}+c_{x'x}c_{y'z} \\ 2c_{x'y}c_{x'z} & 2c_{y'y}c_{y'z} & c_{x'z}c_{y'y}+c_{x'y}c_{y'z} \end{bmatrix} \tag{8.127}$$

where $c_{\bar{x}x}$ is the cosine of the angle between the $\bar{x}$ and x axes, etc.

Eq.(8.124) is written in average integral form for each control domain as

$$\int_{A_i} [\bar{\varepsilon}_b - \mathbf{R}_1 \mathbf{R}_2 \hat{\varepsilon}_b'] dA = 0 \tag{8.128}$$

where A_i is the area of the i-th control domain surrounding node i. Integration by parts gives (noting that the curvatures $\bar{\varepsilon}_b$ and the transformation matrix $\mathbf{R}_1$ are constant within the control domain)

$$\bar{\varepsilon}_b^i = \frac{1}{A_i} \mathbf{R}_1^i \int_{\Gamma_i} \mathbf{R}_2 \mathbf{T} \nabla' w' d\Gamma \tag{8.129}$$

where Γ_i is the control domain boundary (Figure 8.30) and $\mathbf{T}$ is given by Eq.(8.103) and $\mathbf{R}_1^i$ is the transformation matrix for the ith control domain. The changes of matrix $\mathbf{R}_2$ across the element sides have been neglected in Eq.(8.129). These changes tend to zero as the mesh is refined.

Eq.(8.129) is computed as

$$\bar{\varepsilon}_b^i = \frac{1}{A_i} \mathbf{R}_1^i \sum_{j=1}^{n_i} \frac{l_j}{2} \mathbf{R}_2^j \mathbf{T}_j \nabla' w' \tag{8.130}$$

where the sum extends over the n_i elements pertaining to the i-th control domain (for instance $n_i = 6$ for the patch of Figure 8.30), l_j is the external side of element j and A_i is computed as $A_i = \frac{1}{3} \sum_{j=1}^{n_i} A_j^{(i)}$, where $A_j^{(i)}$ is the area of the j-th triangle contributing to the control domain. In Eq.(8.130) summation numbers $1, 2, \cdots, n_i$ corresponde to actual element numbers $a, b, \cdots, f$ (Figure 8.30).

Substituting the standard linear interpolation for the normal deflection w' within each triangle into Eq.(8.130) gives

$$\bar{\varepsilon}_b^i = \mathbf{S}_i w_i' \quad \text{with} \quad \mathbf{S}_i = \overset{\displaystyle 1 \quad 2 \quad \dots \quad n_i}{[\mathbf{S}_i^a, \mathbf{S}_i^b, \dots, \mathbf{S}_i^r]} \tag{8.131}$$

where superindexes $a, b, \ldots, r$ refer to global element numbers. Matrix $\mathbf{S}_i^{(k)}$ is expressed as

$$\mathbf{S}_i^k = \mathbf{F}_i^k [\mathbf{G}_1^k, \mathbf{G}_2^k, \mathbf{G}_3^k] \quad \text{with} \quad \mathbf{F}_i^k = \frac{l_k}{2A_i} \mathbf{R}_1^i \mathbf{R}_2^k \mathbf{T}_k \tag{8.132}$$

and

$$\mathbf{G}_i^k = \nabla' N_i^k = \frac{1}{2A^k} \begin{Bmatrix} b_i^k \\ c_i^k \end{Bmatrix}, \quad b_i^k = x_j'^k - x_k'^k, \quad c_i^k = y_k'^k - y_j'^k \tag{8.133}$$

Vector $\boldsymbol{w}'_i$ in Eq.(8.131) is

$$\boldsymbol{w}'_i = \begin{Bmatrix} \boldsymbol{w}'^a \\ \boldsymbol{w}'^b \\ \vdots \\ \boldsymbol{w}'^r \end{Bmatrix} \begin{matrix} 1 \\ 2 \\ \\ p_n \end{matrix}, \quad \boldsymbol{w}'^k = [w_1'^k, w_2'^k, w_3'^k]^T \tag{8.134}$$

where upper index k denotes the element number.

The final step is transforming vector $\boldsymbol{w}'_i$ to global axes. The process is as explained for the BST element, i.e.

$$\boldsymbol{w}'_i = \mathbf{C}_i \mathbf{a}_i \tag{8.135}$$

with

$$\mathbf{a}_i^T = [\mathbf{u}_i^T, \mathbf{u}_j^T, \mathbf{u}_k^T, \dots, \mathbf{u}_{p_n}^T]^T, \quad \mathbf{u}_i = [u_i, v_i, w_i]^T \tag{8.136}$$

In Eqs.(8.134) and (8.136) p_n is the number of nodes in the patch associated to the i-th control domain (i.e. $p_n = 7$ in Figure 8.30).

The transformation matrix $\mathbf{C}_i$ depends on the numbering of nodes in the patch. A simple scheme is taking the central node as the first node in the patch and the edge nodes in anticlockwise order.

The curvature matrix for the ith control domain is obtained by substituting Eq.(8.135) into (8.131) giving

$$\bar{\boldsymbol{\varepsilon}}_b^i = \mathbf{B}_{b_i} \mathbf{a}_i \quad \text{with} \quad \mathbf{B}_{b_i} = \mathbf{S}_i \mathbf{C}_i \tag{8.137}$$

The bending stiffness matrix for the i-th control domain is obtained by

$$\mathbf{K}_{b_i} = A_i \mathbf{B}_{b_i}^T \hat{\mathbf{D}}_b'^i \mathbf{B}_{b_i} \tag{8.138}$$

where $\hat{\mathbf{D}}_b'^i$ is given by Eq.(8.116).

BSN element. Membrane stiffness matrix

The membrane stiffness contribution to a nodal control domain, $\mathbf{K}_{m_i}$, can be obtained from the stiffness matrix of the CST element. This is not so straightforward in the BSN element as cell-vertex control domains do not coincide with triangles as for the BST element [OZ].

An alternative is to obtain *directly* the membrane stiffness matrix for each control domain following a similar procedure as for the bending stiffness matrix. Details are given in [OZ].

BSN element. Stiffness matrix and nodal force vector

The stiffness matrix for the ith control domain characterizing a BSN element is obtained by adding the membrane and bending contributions as

$$\mathbf{K}_i = \mathbf{K}_{b_i} + \mathbf{K}_{m_i} \tag{8.139}$$

Recall that in the BSN formulation control domains do not coincide with individual elements as for the BST element. The stiffness matrix $\mathbf{K}_i$ of Eq.(8.139) assembles all the contributions to a single node. The assembly process can be implemented on a node to node basis as explained for the BPN rotation-free plate triangle (Section 5.8.3).

The equivalent nodal force vector can be obtained in identical form as for the BST element, i.e. a uniformly distributed load is split into three equal parts and assigned to each element node and nodal point loads are directly assigned to the node at global level.

Boundary conditions for the BSN element

The prescribed displacements are imposed at the equation solution level after the global assembly process.

The conditions on the prescribed rotations at the edges follow a similar process as for the BPN plate element (Section 5.8.3). Free boundary edges are treated by noting that the free edge is a part of the control domain boundary (Figure 5.28b). The condition of zero rotation along an edge is imposed when forming the curvature matrix by making zero the appropriate row in matrix $\mathbf{G}_j^k$ of Eq.(8.133) [OZ].

The nodal definition of the curvatures allows us to prescribe a zero bending moment at free and simply supported boundaries by making zero the appropriate rows of the constitutive matrix, as explained for the BPN plate element (Section 5.8.3 and [OZ]).

8.14 FLAT SHALLOW SHELL ELEMENTS

A shallow shell has surface slopes less than five degrees. Marguere theory [Ma3] is used to formulate shallow shell elements in global axes. This theory is useful for shallow roofs, curved bridges and for studying imperfections in steel plates.

Figure 8.31 shows a shallow shell discretized in 4-noded flat shell rectangles. The local x' axis is defined by the intersection of each element with the global xz axis. As the x' and x direction coincide then

$$\lambda_{x'z} = \frac{\partial z}{\partial x}, \quad \lambda_{y'z'} = \frac{\partial z}{\partial y} \quad , \quad \lambda_{x'y} = \lambda_{z'y} = \lambda_{y'x} = 0$$

$$\theta_{x'} = -\theta_y, \quad \theta_{y'} = \theta_x; \quad w'_0 = w_0 \tag{8.140}$$

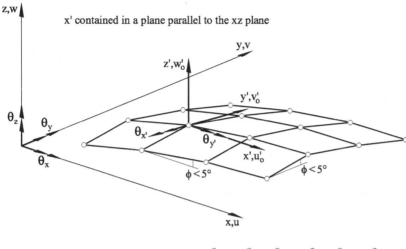

$$w_0' = w_0, \; \theta_{y'} = \theta_x, \; \theta_{x'} = -\theta_y \; ; \; \frac{\partial}{\partial x'} \simeq \frac{\partial}{\partial x}, \; \frac{\partial}{\partial y'} \simeq \frac{\partial}{\partial y}, \; \frac{\partial}{\partial z'} \simeq \frac{\partial}{\partial z}$$

Fig. 8.31 Shallow shell discretized in 4-noded flat shell rectangles

The kinematics of the normal vector follow Reissner-Mindlin assumptions. Thus, the local strain vector of Eq.(8.3) is written in terms of the global displacements using the transformation (8.39) and above assumptions as

$$\varepsilon' = \left\{ \begin{array}{c} \dfrac{\partial u_0}{\partial x} + \dfrac{\partial z}{\partial x}\dfrac{\partial w_0}{\partial x} + z'\dfrac{\partial \theta_y}{\partial x} \\[2mm] \dfrac{\partial v_0}{\partial x} + \dfrac{\partial z}{\partial y}\dfrac{\partial w_0}{\partial y} - z'\dfrac{\partial \theta_x}{\partial y} \\[2mm] \dfrac{\partial u_0}{\partial y} + \dfrac{\partial v_0}{\partial x} + \dfrac{\partial z}{\partial y}\dfrac{\partial w_0}{\partial x} + \dfrac{\partial z}{\partial x}\dfrac{\partial w_0}{\partial y} - z'(\dfrac{\partial \theta_x}{\partial x} - \dfrac{\partial \theta_y}{\partial x}) \\[2mm] \dfrac{\partial w_0}{\partial x} + \theta_y \\[2mm] \dfrac{\partial w_0}{\partial y} - \theta_x \end{array} \right\} \qquad (8.141)$$

The local generalized strain vectors are deduced from Eq.(8.141) as

$$\hat{\varepsilon}'_m = \left\{ \begin{array}{c} \dfrac{\partial u_0}{\partial x} + \dfrac{\partial z}{\partial x}\dfrac{\partial w_0}{\partial x} \\[2mm] \dfrac{\partial v_0}{\partial y} + \dfrac{\partial z}{\partial y}\dfrac{\partial w_0}{\partial y} \\[2mm] \dfrac{\partial u_0}{\partial y} + \dfrac{\partial v_0}{\partial x} + \dfrac{\partial z}{\partial y}\dfrac{\partial w_0}{\partial x} + \dfrac{\partial z}{\partial x}\dfrac{\partial w_0}{\partial y} \end{array} \right\}$$

$$\hat{\varepsilon}_b' = \left[-\frac{\partial \theta_y}{\partial x}, \frac{\partial \theta_x}{\partial y}, \left(\frac{\partial \theta_x}{\partial x} - \frac{\partial \theta_y}{\partial x} \right) \right]^T \quad ; \quad \hat{\varepsilon}_s' = \left[\frac{\partial w_0}{\partial x} + \theta_y, \frac{\partial w_0}{\partial y} - \theta_x \right]^T$$

$$(8.142)$$

The signs in the components of the generalized local strain vectors have been chosen so as to preserve the form of the transformation matrix $\mathbf{S}$ of Eq.(8.8).

A standard C^0 interpolation is chosen for the global displacement variables $(u_0, v_0, w_0, \theta_x, \theta_y)$. Substituting the interpolation into (8.142) gives the generalized strain matrix as

$$\mathbf{B}_i = \left\{ \begin{array}{c} \mathbf{B}_{m_i} \\ \cdots \\ \mathbf{B}_{b_i} \\ \cdots \\ \mathbf{B}_{s_i} \end{array} \right\} = \begin{bmatrix} \frac{\partial N_i}{\partial x} & 0 & \frac{\partial z}{\partial x}\frac{\partial N_i}{\partial x} & 0 & 0 \\ 0 & \frac{\partial N_i}{\partial y} & \frac{\partial z}{\partial y}\frac{\partial N_i}{\partial y} & 0 & 0 \\ \frac{\partial N_i}{\partial y} & \frac{\partial N_i}{\partial x} & \left(\frac{\partial z}{\partial y}\frac{\partial N_i}{\partial x} + \frac{\partial z}{\partial x}\frac{\partial N_i}{\partial y}\right) & 0 & 0 \\ \cdots & & \cdots & & \\ 0 & 0 & 0 & 0 & -\frac{\partial N_i}{\partial x} \\ 0 & 0 & 0 & \frac{\partial N_i}{\partial y} & 0 \\ 0 & 0 & 0 & \frac{\partial N_i}{\partial x} & -\frac{\partial N_i}{\partial y} \\ \cdots & & \cdots & & \cdots \\ 0 & 0 & \frac{\partial N_i}{\partial x} & 0 & N_i \\ 0 & 0 & \frac{\partial N_i}{\partial y} & -N_i & 0 \end{bmatrix}$$

$$(8.143)$$

The cartesian derivatives of the shape functions are obtained by the standard transformation

$$\left[\frac{\partial(\cdot)}{\partial x}, \frac{\partial(\cdot)}{\partial y} \right]^T = \left[\mathbf{J}^{(e)} \right]^{-1} \left[\frac{\partial(\cdot)}{\xi}, \frac{\partial(\cdot)}{\partial \eta} \right]^T \qquad (8.144)$$

The resulting elements have five DOFs per node defined in the global coordinate system $(u_{0_i}, v_{0_i}, w_{0_i}, \theta_{x_i}, \theta_{y_i})$. No singularity arises in the global stiffness matrix as all nodes are coplanar and the contribution of the local rotations to the in-plane rotation θ_z has been neglected.

The choice of shell element follows the criteria explained in Section 8.10. Thin (Kirchhoff) shallow shell elements can be directly derived from above expressions simply by making $\theta_y = -\frac{\partial w_0}{\partial x}$ and $\theta_x = \frac{\partial w_0}{\partial y}$ and neglecting the contributions from the transverse shear strains $\hat{\varepsilon}'_s$. C^1 continuity is now required for the bending approximation and the same recommendations of Section 8.12 apply now [CB,CLO].

Above expressions will be used in Chapter 8 to derive an axisymmetric shallow shell element, also applicable to shallow arches.

8.15 FLAT MEMBRANE ELEMENTS

A membrane is a very thin shell structure with a negligible flexure resistance. Typical examples of membranes are balloons, parachutes, textile covers and inflatable structures [BA,OFM,OK,OK2].

Many shell structures such as spherical domes, cylindrical tanks, etc. behave as quasi-membranes under specific loads (i.e. self weight, pressure loading, etc.). Flexural effects are localized in regions such as supports and membrane theory provides a good estimate of the overall structural behaviour in these cases.

Membrane theory can be readily derived from flat shell theory by neglecting flexural terms (bending and transverse shear) in the formulation. The resulting kinematic, constitutive and equilibrium equations are expressed in terms of the three nodal displacements only. This explains why membrane theory is attractive for obtaining analytical solutions approximating the behaviour of shell structures.

The formulation of flat membrane elements is straightforward. The element stiffness matrix in local axes is given by matrix $\mathbf{K}_m'^{(e)}$ of Eq.(8.37). The transformation to global axes follows the procedure of Section 8.5 noting simply that only translational DOFs are now involved. The resulting membrane element has three global displacement DOF's per node.

Numerical problems can however arise in the direct application of the membrane finite element formulation. The lack of flexural stiffness makes membrane elements not applicable in the presence of loads inducing bending behaviour. These loads can also occur, for instance, due to discretization errors in pure membrane loading situations, such as an internal pressure acting on cylindrical or spherical shells.

These problems can be overcome by adding some flexural stiffness while preserving the translational character of the membrane formulation. A better alternative is to introduce drilling DOFs as explained in Section

8.9.3 [YY]. The drilling rotation can be expressed in terms of nodal displacements leading to a displacement formulation only.

Membrane structures can be analyzed using shell elements incorporating membrane and flexural stiffness. This however has two problems. First, the number of DOFs increases as the nodal rotations are now involved. This problem can be overcome by prescribing all the nodal rotations to zero.

The second problem is the ill-conditioning of the equilibrium equations due to the large difference between the membrane and flexural stiffness terms for small thickness in presence of loads inducing bending behaviour. The problem is identical to that explained above for the pure membrane formulation. Once again the use of drilling rotations overcomes the difficulty. A simpler alternative is to use an artificially large thickness to compute the flexural stiffness terms. As the solution is mainly driven by the membrane stiffness, the artificial flexural stiffness does not affect the numerical results. This is, in fact, equivalent to introducing some stabilizing flexural stiffness in the original membrane formulation. Note that the rotation DOFs are now involved leading to a larger number of variables.

An elegant solution for the analysis of both shell and membrane-type structures is to use the rotation-free shell triangles of Section 8.13. These elements have translational DOFs only and, therefore, their cost is similar to that of "pure" membrane elements. The eventual ill-conditioning of the equations for small thickness is overcome as the elements introduce "naturally" a bending stiffness that has a stabilization effect. Examples of the good behaviour of the BST and EBST elements for analysis of membrane structures can be found in [OF,FO,FO2,5].

8.16 HIGHER ORDER COMPOSITE LAMINATED FLAT SHELL ELEMENTS

Modelling of composite laminated shells can be enhanced by using higher order approximations across the shell thickness. Here the layer-wise and zigzag theories explained for composite laminated beams and plates (Sections 3.14 and 7.6) can be readily extended by using higher order approximations for the *local in-plane displacements* across the thickness.

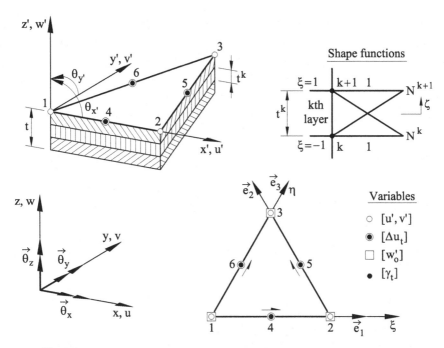

Fig. 8.32 TLQL flat shell element with layer-wise approximation

8.16.1 Layer-wise TLQL element

The TLQL plate triangle of Section 6.8.2 can be extended to laminated shells. The kinematics of the element are defined with respect to a local cartesian system x', y', z' (Figure 8.32). The local axes x' and y' define the directions of the inplane displacements u', v', whereas z' defines the normal displacement w'. Once the stiffness matrix is obtained in this local system, then it is transformed to global axes in the usual way.

As for plates it is possible to define n_l analysis layers and $n+1$ interfaces (Figure 8.32). The displacements u', v' in the element plane corresponding to the k-th layer are interpolated by

$$\begin{Bmatrix} u' \\ v' \end{Bmatrix} = \sum_{i=1}^{3} N_i(\xi, \eta) \left[\begin{Bmatrix} u'_{0_i} \\ v'_{0_i} \end{Bmatrix} + N^k(\zeta) \begin{Bmatrix} u'^k_i \\ v'^k_i \end{Bmatrix} + N^{k+1}(\zeta) \begin{Bmatrix} u'^{k+1}_i \\ v'^{k+1}_i \end{Bmatrix} \right]$$

$$+ \sum_{i=4}^{6} N_i(\xi, \eta) \mathbf{e}_{i-3} \left[N^k(\zeta) \Delta u^k_{t_i} + N^{k+1}(\zeta) \Delta u^{k+1}_{t_i} \right] \qquad (8.145)$$

where $\left(u'_{0_i}, v'_{0_i} \right)$ are constant (rigid-body) in-plane displacements across the laminate thickness, $\left(u'^k_i, v'^k_i \right)$ are the in-plane displacements which vary over the thickness and $\Delta u_{t_i}^k$ are the in-plane displacement increments at the midside nodes of the triangle in the directions defined by the tangent side vectors $\mathbf{e}_1$, $\mathbf{e}_2$, $\mathbf{e}_3$ (Figure 8.32).

The normal displacement w' is assumed to be constant across the thickness. Following this hypothesis we can write

$$w' = \sum_{i=1}^{3} N_i(\xi, \eta) w_i' \tag{8.146}$$

In Eq.(8.146) N_i are the standard linear shape functions for the 3-noded triangle (Eq.(6.162)) and $N^k(\xi) = \frac{1-\zeta}{2}$ and $N^{k+1}(\xi) = \frac{1+\zeta}{2}$.

Eqs.(8.145) and (8.146) define a quadratic interpolation over every interface for the in-plane displacements u' and v' and a linear interpolation for the displacement w'.

The local strains for the k-th layer are written

$$\varepsilon_b' = \left[\frac{\partial u'}{\partial x'}, \frac{\partial v'}{\partial y'}, \left(\frac{\partial u'}{\partial y'} + \frac{\partial v'}{\partial x'}\right)\right]^T = \mathbf{B}_b a' \tag{8.147}$$

$$\varepsilon_s' = \left[\left(\frac{\partial w'}{\partial x'} + \frac{\partial u'}{\partial z'}\right), \left(\frac{\partial w'}{\partial y'} + \frac{\partial v'}{\partial z'}\right)\right]^T = \mathbf{B}_s a' \tag{8.148}$$

where ε_b' are the local strains accounting for membrane and bending effects and ε_s' are the transverse shear strains with

$$\mathbf{B}_b = [\mathbf{B}_b^k, \mathbf{B}_b^{k+1}, \mathbf{B}_b^0] \quad ; \quad \mathbf{B}_b^k = [\mathbf{B}_{b_1}^k, \mathbf{B}_{b_2}^k, \mathbf{B}_{b_3}^k, \bar{\mathbf{B}}_{b_4}^k, \bar{\mathbf{B}}_{b_5}^k, \bar{\mathbf{B}}_{b_6}^k] \tag{8.149}$$

and

$$\mathbf{B}_{b_i}^0 = \mathbf{B}_{m_i}' \quad , \quad \mathbf{B}_{b_i}^k = N^k \mathbf{B}_{b_i}^0 \quad i = 1, 2, 3$$
$$\mathbf{B}_{b_i}^k = \mathbf{B}_{b_{i-3}}^k \mathbf{e}_{i-3} \quad\quad\quad\quad i = 4, 5, 6 \tag{8.150}$$

In Eq.(8.150) $\mathbf{B}_{m_i}'$ is given by Eq.(8.31). On the other hand,

$$\underset{2\times 33}{\mathbf{B}_s} = \mathbf{J}^{-1}\mathbf{P}[\underset{3\times 12}{\mathbf{B}_s^k}, \underset{3\times 12}{\mathbf{B}_s^{k+1}}, \underset{3\times 12}{\mathbf{B}_w}] \tag{8.151}$$

with

$$\mathbf{B}_s^k = \begin{bmatrix} a_{12} & b_{12} & 0 \vdots & a_{12} & b_{12} & 0 \vdots & 0 & 0 & 0 \vdots & c_{12} & 0 & 0 \\ 0 & 0 & 0 \vdots & a_{23} & b_{23} & 0 \vdots & a_{23} & b_{23} & 0 \vdots & 0 & c_{23} & 0 \\ a_{13} & b_{13} & 0 \vdots & 0 & 0 & 0 \vdots & a_{32} & b_{32} & 0 \vdots & 0 & 0 & c_{32} \end{bmatrix} \tag{8.152}$$

$$\mathbf{B}_s^{k+1} = -\mathbf{B}_s^k \quad , \quad \mathbf{B}_w = \begin{bmatrix} 0\ 0 -1 \vdots 0\ 0\ \ 1 \ \vdots 0\ 0\ 0 \\ 0\ 0\ \ 0\ \vdots 0\ 0 -1 \vdots 0\ 0\ 1 \\ 0\ 0 -1 \vdots 0\ 0\ \ 0\ \vdots 0\ 0\ 1 \end{bmatrix}$$

$$\mathbf{P} = \begin{bmatrix} 1 - \eta & -\eta & \eta \\ \xi & \xi & 1 - \xi \end{bmatrix} \quad , \quad a_{ij} = -\frac{C_i l^{ij}}{2t^k}, \quad b_{ij} = -\frac{S_i l^{ij}}{2t^k}, \quad c_{ij} = -\frac{2l^{ij}}{3t^k}$$

where $l^{ij} =$ is the length of the side ij, t^k is the thickness of the kth layer and $C_i, S_i =$ are the components of the unit side vector $\mathbf{e}_i = [C_i, S_i]^T$. In Eq.(8.151) $\mathbf{J}$ is the Jacobian matrix.

The local displacement vector $\mathbf{a}'$ is written for the k-th layer as

$$\mathbf{a}' = \begin{Bmatrix} \mathbf{a}'^k \\ \mathbf{a}'^{k+1} \\ \mathbf{a}'_0 \end{Bmatrix} \tag{8.153}$$

where

$$\mathbf{a}'^k = \left[u'^k_1, v'^k_1, w'^k_1, u'^k_2, v'^k_2, w'^k_2, u'^k_3, v'^k_3, w'^k_3, \Delta u'^k_{t4}, \Delta u'^k_{t5}, \Delta u'^k_{t6} \right]^T$$

$$\mathbf{a}'_0 = \left[u'_{01}, v'_{01}, w'_{01}, u'_{02}, v'_{02}, w'_{02}, u'_{03}, v'_{03}, w'_{03} \right]^T$$

$$\tag{8.154}$$

The normal displacements w'_{oi} have been introduced $\mathbf{a}'_0$ to simplify the transformation process although its contribution to the local strain matrices is zero.

The local displacements $\mathbf{a}'$ are transformed to global axes by

$$\mathbf{a}' = \bar{\mathbf{T}} \mathbf{a} \tag{8.155a}$$

where

$$\mathbf{a} = \begin{Bmatrix} \mathbf{a}^k \\ \mathbf{a}^{k+1} \\ \mathbf{a}^0 \end{Bmatrix} \tag{8.155b}$$

and

$$\mathbf{a}^k = [u^k_1, v^k_1, w^k_1, u^k_2, v^k_2, w^k_2, u^k_3, v^k_3, w^k_3, \Delta u^k_{t4}, \Delta u^k_{t5}, \Delta u^k_{t6}]^T$$

$$\mathbf{a}_0 = [u_{01}, v_{01}, w_{01}, u_{02}, v_{02}, w_{02}, u_{03}, v_{03}, w_{03}]^T \tag{8.156}$$

The transformation matrix is

$$\bar{\mathbf{T}} = \begin{bmatrix} \hat{\mathbf{T}} & 0 & 0 \\ 0 & \hat{\mathbf{T}} & 0 \\ 0 & 0 & \hat{\mathbf{T}}' \end{bmatrix} \quad \text{with} \quad \hat{\mathbf{T}} = \begin{bmatrix} \hat{\mathbf{T}}' & 0 \\ 0 & \mathbf{I}_3 \end{bmatrix} \tag{8.157}$$

where $\hat{\mathbf{T}}$ is the $\mathbf{I}_3$ identity 3×3 matrix and

$$\hat{\mathbf{T}}' = \begin{bmatrix} \mathbf{T} & 0 \\ & \mathbf{T} \\ 0 & \mathbf{T} \end{bmatrix}, \qquad \mathbf{T} = \begin{bmatrix} \lambda_{x'x} & \lambda_{x'y} & \lambda_{x'z} \\ \lambda_{y'x} & \lambda_{y'y} & \lambda_{y'z} \\ \lambda_{z'x} & \lambda_{z'y} & \lambda_{z'z} \end{bmatrix} \tag{8.158}$$

and $\lambda_{x'x}$ is the cosine of the angle between axes x' and x etc.

The global stiffness matrix is obtained by the standard transformation

$$\mathbf{K}^{(e)} = \bar{\mathbf{T}}^T \mathbf{K}'^{(e)} \bar{\mathbf{T}} \tag{8.159}$$

where the local stiffness matrix is given by

$$\mathbf{K}^{(e)} = \iint_{A^{(e)}} \left(\sum_{k=1}^{n_l} \int_{t^k} \mathbf{B}^T \hat{\mathbf{D}}'^k \mathbf{B} dz \right) dA \tag{8.160}$$

where $\mathbf{B} = \begin{Bmatrix} \mathbf{B}_b \\ \mathbf{B}_s \end{Bmatrix}$ and $\hat{\mathbf{D}}'^k$ is the local constitutive matrix for the kth (deduced from Eq.(8.18a)).

The integration across the thickness is performed explicitly, whereas the integration over the surface of every interface is made by a three point Gauss quadrature. For details see [BOM].

A condensation technique across the thickness can be used for reducing the amount of calculations. The procedure consists in eliminating the global displacements at the lower interlaminar surface $\mathbf{a}^k$ for every layer k in terms of $\mathbf{a}^{k+1}$ and $\mathbf{a}^o$ by an expression similar to (7.68) [BOM].

8.16.2 Composite laminated flat shell elements based on the refined zigzag theory

We present the derivation of composite laminated fat shell element based on an extension of the refined zigzag theory (RZT) described in Section 7.8.

Following the arguments explained in Section 7.8 for laminated plates, the displacements in RZ flat shell theory are expressed in the local coordinate system as

$$\begin{aligned} u'^k(x', y', z') &= u'_0(x', y') - z'\theta_{x'}(x', y') + u'^k(x', y', z') \\ v'^k(x', y', z') &= v'_0(x', y') - z'\theta_{y'}(x', y') + v'^k(x', y', z') \\ w'(x, y, z) &= w'_0(x', y') \end{aligned} \tag{8.161a}$$

with

$$\bar{u}'^{k}(x', y', z') = \phi_{x'}^{k}(z)\psi_{x'}(x', y')$$
$$\bar{v}'^{k}(x', y', z') = \phi_{y'}^{k}(z)\psi_{y'}(x', y')$$

(8.161b)

The terms in Eq.(8.161) have the same meaning as in Eq.(7.70) for composite laminated plates. The $\phi_{x'}^{k}, \phi_{y'}^{k}$ functions represent the through-the-thickness piecewise-linear zigzag functions and $\psi_{x'}, \psi_{y'}$ are the spatial amplitudes of the zigzag displacements. These two amplitude together with the five standard kinematic variables $(u'_0, v'_0, w'_0, \theta_{x'}, \theta_{y'})$ are the problem unknowns.

The zigzag displacements u'^{k} and v'^{k} may be regarded as a correction to the in-plane displacement of standard Reissner-Mindlin theory due to the laminate heterogeneity.

The in-plane and transverse shear strains consistent with Eqs.(8.161) are

$$\varepsilon_{x'}^{k} = \frac{\partial u_0}{\partial x'} - z'\frac{\partial \theta_{x'}}{\partial x'} + \phi_{x'}^{k}\frac{\partial \psi_{x'}}{\partial x'}$$

$$\varepsilon_{y'}^{k} = \frac{\partial v_0}{\partial x'} - z'\frac{\partial \theta_{y'}}{\partial y'} + \phi_{y'}^{k}\frac{\partial \psi_{y'}}{\partial y'}$$

$$\gamma_{x'y'}^{k} = \frac{\partial u_0}{\partial y'} + \frac{\partial v_0}{\partial x'} - z'\left(\frac{\partial \theta_{x'}}{\partial y'} + \frac{\partial \theta_{y'}}{\partial x'}\right) + \phi_{x'}^{k}\frac{\partial \psi_{x'}}{\partial y'} + \phi_{y'}^{k}\frac{\partial \psi_{y'}}{\partial x'}$$

$$\gamma_{\alpha z'}^{k} = \gamma_\alpha + \beta_\alpha^k \psi_\alpha \ , \quad \alpha = x', y'$$

(8.162a)

where

$$\gamma_\alpha = \frac{\partial w'_0}{\partial \alpha} - \theta_\alpha \ , \quad \beta_\alpha^k = \frac{\partial \phi_\alpha^k}{\partial z'} \ , \quad \alpha = x', y'$$

(8.162b)

In Eq.(8.162b) β_α^k are piecewise constant functions that are uniform through the thickness of each individual lamina and the γ_α represent the average transverse shear strains through the laminate thickness and coincide with the expression of standard Reissner-Mindlin flat shell theory.

The generalized Hooke law for the kth orthotropic lamina, whose principal material directions are arbitrary with respect to the middle place reference coordinates x', y' is written as (disregarding initial stresses)

$$\sigma_p'^{k} = \begin{Bmatrix} \sigma_{x'} \\ \sigma_{y'} \\ \tau_{x'y'} \end{Bmatrix}^k = \mathbf{D}_p'^{k}\begin{Bmatrix} \varepsilon_{x'} \\ \varepsilon_{y'} \\ \gamma_{x'y'} \end{Bmatrix}^k = \mathbf{D}_p'^{k}\varepsilon_p'^{k}$$

$$\sigma_s'^{k} = \begin{Bmatrix} \tau_{x'z'} \\ \tau_{y'z'} \end{Bmatrix}^k = \mathbf{D}_s'^{k}\begin{Bmatrix} \gamma_{x'z'} \\ \gamma_{y'z'} \end{Bmatrix}^k = \mathbf{D}_s'^{k}\varepsilon_s'^{k}$$

(8.163)

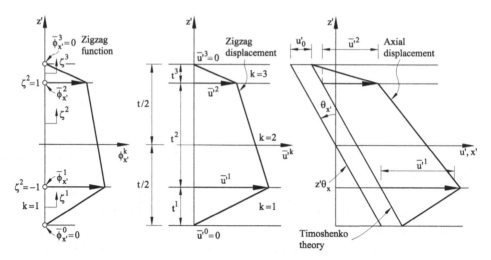

Fig. 8.33 Notation for a three-layered in a flat shell laminate and $\phi_{x'}^k$ zigzag function. Same applies for $\phi_{y'}^k$ by interchanging x' for y' and u_i' by v_i'

where the constitutive matrices $\mathbf{D}_p'^k$ and $\mathbf{D}_s'^k$ for the kth lamina are obtained as explained in Section 7.2.3.

The following linear and C^o continuous zigzag function within each layer is assumed

$$\phi_{x'}^k = \frac{1}{2}(1 - \zeta^k)\bar{\phi}_{x'}'^{k-1} + \frac{1}{2}(1 + \zeta^k)\bar{\phi}_{x'}'^{k}$$
$$\phi_{y'}^k = \frac{1}{2}(1 - \zeta^k)\bar{\phi}_{y'}'^{k-1} + \frac{1}{2}(1 + \zeta^k)\bar{\phi}_{y'}'^{k} \tag{8.164a}$$

where $(\bar{\phi}_{x'}^k, \bar{\phi}_{y'}^k)$ and $(\bar{\phi}_{x'}^{k-1}, \bar{\phi}_{y'}^{k-1})$ are the zigzag function values at the k and $k-1$ interfaces, respectively, and

$$\zeta^k = \frac{2(z' - z_{k-1}')}{t^k} - 1 \quad , \quad k = 1, \cdots, n_l \tag{8.164b}$$

with the first layer begining at $z_0' = -t/2$, the last n_lth layer ending at $z_{n_l}' = t/2$ and the kth layer ending at $z_k' = z_{k-1}' + t^k$ where t^k is the layer thickness. Figure 8.33 shows the graphic representation of the zigzag function across the thickness of a three-layered laminate.

The zigzag function values (and hence the interfacial displacements) at the bottom and top surface of the laminate are set herein to vanish identically, i.e.,

$$\phi_{x'}'^0 = \phi_{x'}'^{n_l} = \phi_{y'}'^0 = \phi_{y'}'^{n_l} = 0 \tag{8.165a}$$

and hence

$$\bar{u}'^0 = \bar{u}'^{n_l} = \bar{v}'^0 = \bar{v}'^{n_l} \tag{8.165b}$$

The procedure for computing the piecewise constant functions β_α^k follows precisely the arguments of Section 7.8.1.3 for composite laminate plates giving

$$\left\{\begin{matrix} \beta_{x'}^k \\ \beta_{y'}^k \end{matrix}\right\} = \left\{\begin{matrix} \dfrac{G_{x'}}{D_{s_{11}}^{\prime k}} - 1 \\ \dfrac{G_{y'}}{D_{s_{22}}^{\prime k}} - 1 \end{matrix}\right\} \quad \text{with} \quad \left\{\begin{matrix} G_{x'} \\ G_{y'} \end{matrix}\right\} = \left\{\begin{matrix} \left(\dfrac{1}{t}\displaystyle\sum_{k=1}^{n_e} \dfrac{t^k}{D_{s_{11}}^{\prime k}}\right)^{-1} \\ \left(\dfrac{1}{t}\displaystyle\sum_{k=1}^{n_e} \dfrac{t^k}{D_{s_{22}}^{\prime k}}\right)^{-1} \end{matrix}\right\} \tag{8.166}$$

In Eq.(8.166) $G_{x'}$ and $G_{y'}$ are weighted-average transverse shear stiffness coefficients of their respective layer level coefficients $D_{s_{11}}^{\prime k}$ and $D_{s_{22}}^{\prime k}$ (obtained from Eq.(8.12)). For homogeneous material $G_{x'} = D_{s_{11}}^{\prime k}$ and $G_{y'} = D_{s_{22}}^{\prime k}$, $\beta_\alpha^k = 0$ and the kinematic and constitutive equations coincide with those of standard Reissner-Mindlin theory for flat shells.

The explicit form of the zigzag function in terms of β_α^k is given by Eq.(7.88), simply interchanging the global axes x, y by x', y'.

The PVW for a distributed load f_z (expressed in global axes) is written as

$$\iint_A \delta\hat{\varepsilon}^T \hat{\sigma}\, dA = \iint_A \delta w_0 f_z\, dA \tag{8.167}$$

where

$$\hat{\varepsilon}' = \left\{\begin{matrix} \hat{\varepsilon}'_m \\ \hat{\varepsilon}'_b \\ \hat{\varepsilon}'_t \end{matrix}\right\} \quad , \quad \hat{\sigma}' = \left\{\begin{matrix} \hat{\sigma}'_m \\ \hat{\sigma}'_b \\ \hat{\sigma}'_t \end{matrix}\right\} \tag{8.168}$$

are the local generalized strain vector and the local resultant stress vector, respectively.

In Eqs.(8.168) the membrane strain vector $\hat{\varepsilon}'_m$ and the membrane stress vector $\hat{\sigma}'_m$ coincide with the expressions given in Eqs.(8.69) and (8.16), respectively and

$$\hat{\varepsilon}'_b = \left[\frac{\partial\theta_{x'}}{\partial x'}, \frac{\partial\theta_{y'}}{\partial y'}, \left(\frac{\partial\theta_{x'}}{\partial y'} + \frac{\partial\theta_{y'}}{\partial x'}\right), \frac{\partial\psi_{x'}}{\partial x'}, \frac{\partial\psi_{x'}}{\partial y'}, \frac{\partial\psi_{y'}}{\partial y'}, \frac{\partial\psi_{y'}}{\partial x'}\right]^T$$

$$\hat{\varepsilon}'_t = \left[\frac{\partial w}{\partial x'} - \theta_{x'}, \frac{\partial w}{\partial y'} - \theta_{y'}, \psi_{x'}, \psi_{y'}\right]^T \tag{8.169}$$

$$\hat{\sigma}'_b = [M_{x'}, M_{y'}, M_{x'y'}, M_{x'}^\phi, M_{y'}^\phi, M_{y'}^\phi, M_{y'x'}^\phi]^T$$

$$\hat{\sigma}'_t = [Q_{x'}, Q_{y'}, Q_{x'}^\phi, Q_{y'}^\phi]^T$$

The resultant constitutive equations are expressed in matrix form as

$$\left\{ \begin{array}{c} \hat{\sigma}'_m \\ \hat{\sigma}'_b \\ \hat{\sigma}'_t \end{array} \right\} = \left[\begin{array}{ccc} \hat{\mathbf{D}}'_m & \hat{\mathbf{D}}'_{mb} & 0 \\ \hat{\mathbf{D}}'^T_{mb} & \hat{\mathbf{D}}'_b & 0 \\ 0 & 0 & \hat{\mathbf{D}}'_t \end{array} \right] \left\{ \begin{array}{c} \hat{\varepsilon}'_m \\ \hat{\varepsilon}'_b \\ \hat{\varepsilon}'_t \end{array} \right\} \tag{8.170}$$

Matrices $\hat{\mathbf{D}}'_m$, $\hat{\mathbf{D}}'_{mb}$, $\hat{\mathbf{D}}'_b$ and $\hat{\mathbf{D}}'_t$ are deduced from Eqs.(7.93) simply substituting the global coordinates x, y, z by x', y', z'.

Composite laminated flat shell elements based on the RZT can be derived by adding functions $\psi_{x'}$ and $\psi_{y'}$ to the standard five local kinematic variables $(u'_0, v'_0, w', \theta_{x'}, \theta_{y'})$. The expressions for the generalized strain matrices for a general C^0 continuous flat shell element are deduced from Eqs.(7.99) changing the shape functions derivatives $\frac{\partial N_i}{\partial x}$ and $\frac{\partial N_i}{\partial y}$ by $\frac{\partial N_i}{\partial x'}$ and $\frac{\partial N_i}{\partial y'}$, respectively.

The nodal displacement vector is

$$\mathbf{a}_i^{(e)} = [u'_{0_i}, v'_{0_i}, w'_{0_i}, \theta_{x'_i}, \theta_{y'_i}, \psi_{x'_i}, \psi_{y'_i}]^T \tag{8.171a}$$

The stiffness matrix is obtained as explained in Section 7.8.16. The different matrices are integrated with the same quadratures as for Reissner-Mindlin flat shell elements. Transformation of the local stiffness matrix to global axes follows the rules explained in Section 8.5. The nodal zigzag displacement $\psi_{x'_i}, \psi_{y'_i}$ are expressed in terms of global components $(\psi_{x_i}, \psi_{y_i}, \psi_{z_i})$ as

$$\left\{ \begin{array}{c} \psi_{x'_i} \\ \psi_{y'_i} \end{array} \right\} = \bar{\boldsymbol{\lambda}}^{(e)} \left\{ \begin{array}{c} \psi_{x_i} \\ \psi_{y_i} \\ \psi_{z_i} \end{array} \right\} \quad \text{with} \quad \bar{\boldsymbol{\lambda}}^{(e)} = \left[\begin{array}{ccc} \lambda_{x'x} & \lambda_{x'y} & \lambda_{x'z} \\ \lambda_{y'x} & \lambda_{y'y} & \lambda_{y'z} \end{array} \right] \tag{8.171b}$$

Shear and membrane locking can be avoided using any of the techniques explained in Section 8.11.

QLRZ flat shell element for composite laminated shell analysis based on the RZT can be derived as explained in Section 7.8.2.

The expression of the equivalent nodal force vector for a distributed load f_z is

$$\mathbf{f}^{(e)} = \iint_{A^e} N_i f_z [0, 0, 1, 0, 0, 0, 0]^T dA \tag{8.172}$$

The global stiffness equations are assembled in the standard manner.

The boundary conditions for the new global nodal rotation variables $\psi_{x_i}, \psi_{y_i}, \psi_{z_i}$ are

$$\begin{aligned}
\psi_{a_i} &= 0 & &\text{at a clamped edge } (\alpha = x, y, z) \\
\psi_{\alpha_i} &= 0 & &\text{if } \alpha \text{ is a symmetry axis} \\
\psi_{y_i} &= 0 & &\text{if } x \text{ is a SS edge (likewise for } \psi_{x_i})
\end{aligned} \tag{8.173}$$

8.17 EXAMPLES

8.17.1 Comparison of different flat shell elements

The performance of a selected number of flat shell elements is tested in the analysis of three problems of the so-called "shell obstacle course" proposed by Belytschko *et al.* [BSL+]. These problems include a cylindrical shell (Figure 8.34), an open spherical dome (Figure 8.35)) and a pinched cylinder (Figure 8.36). Details of the geometrical, mechanical and loading conditions are given in the figures. The elements studied are the TLQL, TLLL, QLQL and QLLL from the Reissner-Mindlin family (Table 8.1) and the rotation-free BST and EBST1 triangles from the Kirchhoff thin shell family (Sections 8.13.1 and 8.13.2). Results show the convergence of a characteristic displacement value with the mesh size. Some resultant stress diagrams along specific lines are also plotted for a fixed mesh. Note the accuracy of all the elements considered. In particular, the simplest triangular and quadrilateral elements of the Reissner-Mindlin family (TLLL and QLLL) give excellent results, even for relatively coarse meshes. It is also remarkable the good behaviour of the rotation-free BST and EBST1 triangles which have a considerable less number of DOFs. This makes these elements very attractive for practical purposes.

More examples of the performance of different triangular and quadrilateral shell elements can be found in [BBH,BBt,BD2,BD6,Bel,BSL+,Hu2, MH2, SBCK,SFR,ZT2].

8.17.2 Adaptive mesh refinement of analysis cylindrical shells

We present two examples of the analysis of cylindrical shells using adaptive mesh refinement (AMR). The problems were solved using two AMR criteria. The first one is based on the equidistribution of the global energy error in the mesh (criterion A) and the second one is based on the equidistribution of the error density (criterion B). The description of these AMR criteria can be found in [OB] and in Section 9.9.4 of [On4]. The resultant

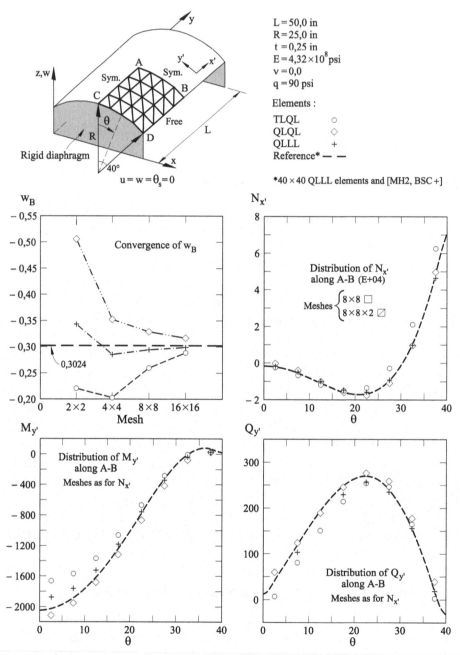

Fig. 8.34 Cylindrical shell under uniform load. Convergence of vertical displacement w_B (in) at the center and diagrams of $N_{x'}$ (lb/in), $M_{y'}$ (lb×in/in) and $Q_{y'}$ (lb/in) along line $A - B$. Results for TLQL, QLQL and QLLL elements

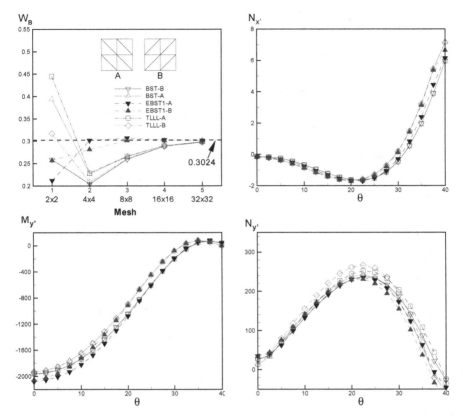

Fig. 8.34 (Continued). Cylindrical shell under uniform load. Results for TLLL, BST and EBST1 elements

stresses have been smoothed in order to obtain nodal values using a local coordinate system for each node common to all the elements sharing the node [OCK]. A permissible error $\eta = 10\%$ was taken for both problems which are solved using TLQL flat shell triangles.

The first example is the analysis of a classical cylindrical dome supported at the two ends under self weight loading. Figure 8.37 shows the initial mesh of 48 TLQL elements. The global error parameter for the initial mesh is $\xi_g = 6.2295$.

Figure 8.37 shows the sequence of the meshes generated using the two AMR criteria A and B considered. Criterion B based on the error density concentrates more elements in the zones where higher gradients of the axial force $N_{x'}$ exist, while larger elements are generated in the upper part of the cylinder where the gradients are smaller.

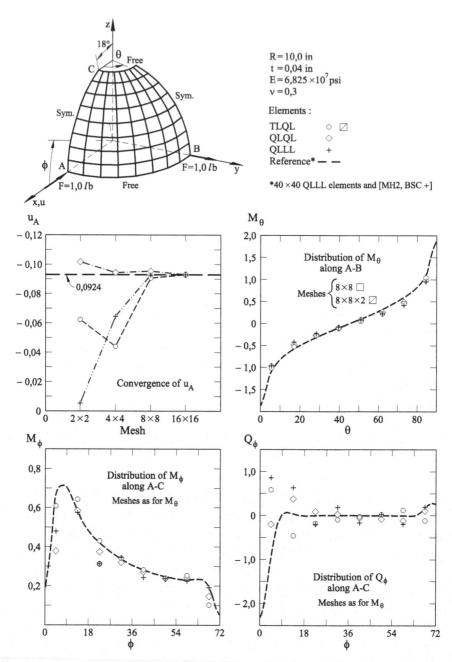

Fig. 8.35 Open spherical dome under radial point loads. Convergence of radial displacement at point A, u_A (in), and diagrams of M_θ and M_ϕ (lb×in/in) and Q_ϕ (lb/in) along different lines. Results for TLQL, QLQL and QLLL elements

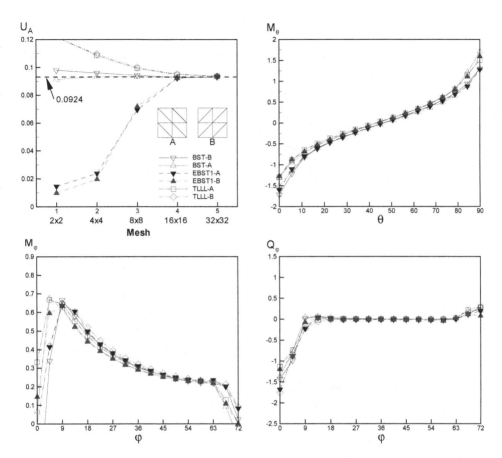

Fig. 8.35 (Continued). Open spherical dome under radial point loads. Results for TLLL, BST and EBST1 elements

The second example shown in Figure 8.38 is the analysis of a cylindrical shell with a central hole. The shell is supported along the two vertical sides and it is loaded by equal and opposite traction forces acting on the upper and lower edges. A quarter of the geometry has been analyzed due to symmetry. The global error parameter for the initial mesh of 96 TLQL elements is 1.776. Both mesh adaption criteria A and B lead to a higher density of elements in the vicinity of the central hole where the stress gradients are higher. Once again, the number of elements generated using the AMR criterion B based on the error density is larger than for criterion A based on the equal distribution of the global error. The same conclusion was found in other problems solved with the two AMR criteria [OB,OCK].

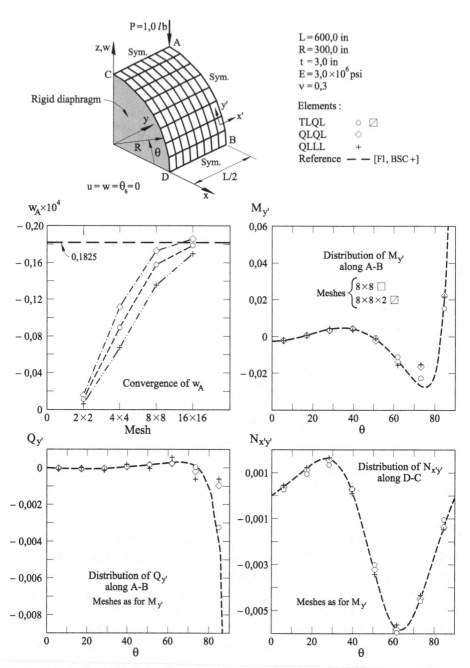

Fig. 8.36 Cylinder under central point load. Convergence of vertical displacement w_A (in) at the center and diagrams of $M_{y'}$ (lb×in/in), $Q_{y'}$ (lb/in) and $N_{x'y'}$ (lb/in) along different lines. Results for TLQL, QLQL and QLLL elements

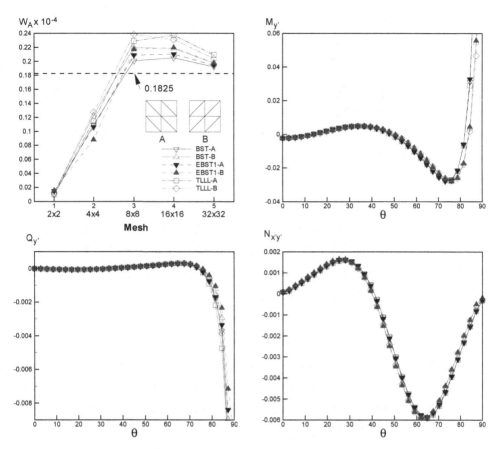

Fig. 8.36 (Continued). Cylinder under central point load. Results for TLLL, BST and EBST1 elements

8.17.3 Examples of application

Figures 8.39–8.43 show examples of application of the QLLL flat shell element to cellular structures including a bogie of train car, a ship hull and a car body. Figures 8.44–8.47 show examples of application of the rotation-free EBST element to the analysis of a sheet stamping problem, an inflatable pavilion, an aircraft wing structure and the hull and appendages of a racing sailboat. Figures 8.48 and 8.49 show finally results of the analysis of a parachute and the sails of a sailboat with 3-noded membrane triangles.

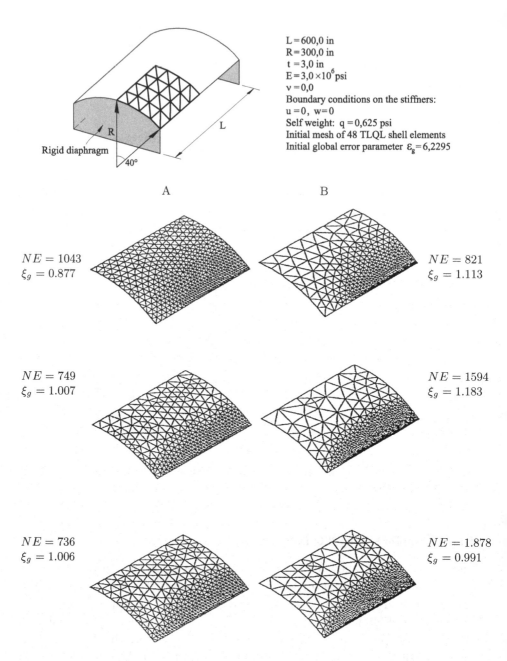

L = 600,0 in
R = 300,0 in
t = 3,0 in
E = 3,0 × 10⁶ psi
ν = 0,0
Boundary conditions on the stiffners:
u = 0, w = 0
Self weight: q = 0,625 psi
Initial mesh of 48 TLQL shell elements
Initial global error parameter ε_g = 6,2295

Fig. 8.37 Cylindrical shell under self weight. Sequence of meshes obtained with mesh adaption strategies based on: a) Uniform distribution of global error; b) Uniform distribution of error density; Target error: $\eta = 10\%$,; NE = number of elements. Dimensions in inches

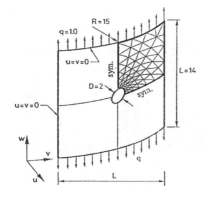

$E = 3.0 \times 10^6, \nu = 0.0$
Initial mesh of 96 TLQL flat shell elements
Initial global error parameter $\xi_g = 1.776$

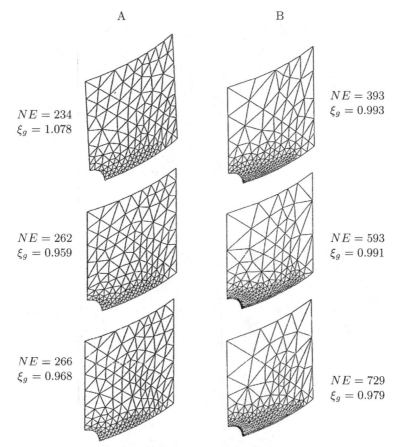

A   B

$NE = 234$
$\xi_g = 1.078$

$NE = 393$
$\xi_g = 0.993$

$NE = 262$
$\xi_g = 0.959$

$NE = 593$
$\xi_g = 0.991$

$NE = 266$
$\xi_g = 0.968$

$NE = 729$
$\xi_g = 0.979$

Fig. 8.38 Cylindrical shell with central hole under traction load. Sequence of meshes obtained with mesh adaption strategies based on: a) Uniform distribution of the global error; b) Uniform distribution of the error density; Target error: $\eta = 5\%$; $NE =$ number of elements. Dimensions in inches, E in psi and line load in lb/in

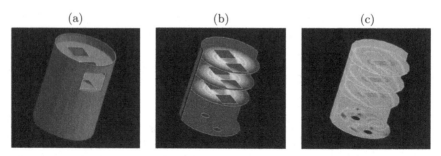

Fig. 8.39 Cellular structure under distributed load acting on interior plates. (a) Mesh of QLLL flat shell elements. Contours of (b) $M_{x'}$ and (c) $M_{y'}$

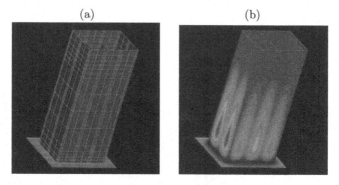

Fig. 8.40 Cellular structure under pressure acting on the interior. (a) Mesh of QLLL flat shell elements. (b) Contours of displacement vector modulus

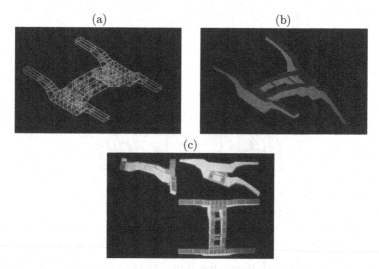

Fig. 8.41 Bogie of train car. (a) Mesh of QLLL flat shell elements. (b) and (c) Amplified deformation under point load acting at one end keeping the other three ends fixed

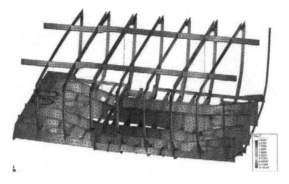

Fig. 8.42 Discretization of ship hull structure using QLLL flat shell elements. Courtesy of Compass Ingeniería y Sistemas SA (www.compassis.com)

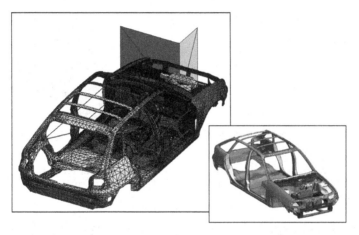

Fig. 8.43 Car body discretized with QLLL flat shell triangles and 2-noded beam elements

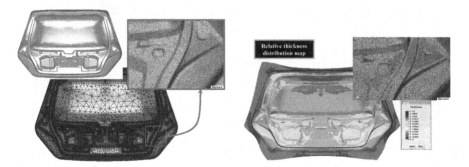

Fig. 8.44 Discretization of the backdoor of a car using rotation-free EBST shell triangles. Results of sheet stamping analysis [OFN]. Courtesy of Quantech ATZ SA (www.quantech.es)

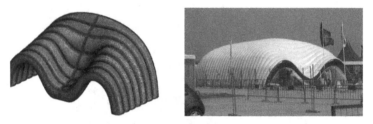

Fig. 8.45 Inflatable pavilion discretized using rotation-free EBST shell triangles. Courtesy of BuildAir SA (www.buildair.com) [OFM]

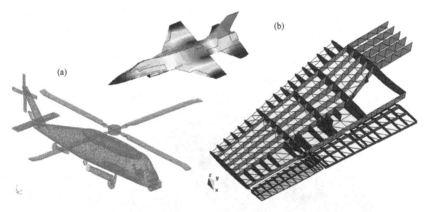

Fig. 8.46 Discretization of helicopter body and blades (a) and aircraft wing (b) using rotation-free EBST shell triangles (Courtesy of GiD team, www.gidhome.com)

Fig. 8.47 Underwater view of racing sail boat hull and appendages discretized with rotation-free EBST shell triangles. Courtesy of Compass Ingeniería y Sistemas SA (www.compassis.com) [GO,OG,OGI]

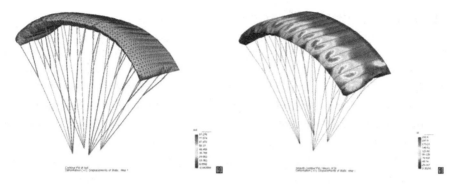

Fig. 8.48 Parachute discretized with 3-noded membrane triangles. Displacement modulus contours [FOO]

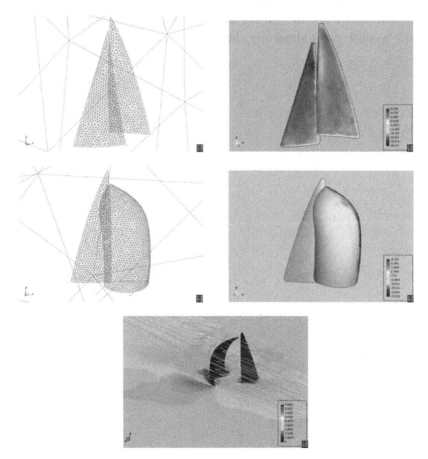

Fig. 8.49 Discretization of sails of sailboat with 3-noded membrane triangles. Stresses on sails from aerodynamic analysis. Wind particles around sails [POM]

8.18 CONCLUDING REMARKS

Flat shell elements can be understood as a blending of plane stress and plate elements. A variety of thick and thin flat shell elements can be derived by combining plane stress triangles and quadrilaterals with the adequate plate elements using Reissner-Mindlin or Kirchhoff theories. The behaviour of flat shell elements depends in most cases on the accuracy of the individual plane stress and plate elements selected.

The practical use of flat shell elements requires good knowledge of concepts such as membrane-bending coupling for composite laminated material, local-global stiffness transformation, the treatment of coplanar nodes, shear and membrane locking, and the incompatibility between membrane and bending fields for thin situations.

Flat shell elements are simpler than curved shell elements (Chapter 10) and can be used to analyze any shell structure.

9

AXISYMMETRIC SHELLS

9.1 INTRODUCTION

Many shell structures of practical interest have axisymmetric forms. Examples are water and oil tanks, grain silos, cooling towers, nuclear containment shells, spherical and conical roofs and other structures outside the civil construction industry such as pressure vessels, missiles, airplane and spacecraft fuselages, etc. (Figure 9.1).

An axisymmetric shell is a particular case of an axisymmetric solid and, therefore, it can be analyzed using the procedures described in Chapter 6 of [On4]. It is also possible to perform a full three-dimensional (3D) analysis using 3D solid elements (Chapter 7 of [On4]), or flat or curved shell elements (Chapters 8 and 10). However, the coincidence of axial symmetry and thin thickness allows *axisymmetric shell elements* to be employed. These elements are one-dimensional (1D) and this simplifies the discretization process and reduces the computational cost.

This chapter deals with axisymmetric shells under *axisymmetric loading*. The computations can be simplified by analyzing the deformation of the "middle" line of a meridional section only (hereafter termed *generating line* or *generatrix*). The 1D nature of the problem can still be preserved for arbitrary loading by expanding the loads and the displacements in Fourier series in the circumferential direction. The non-symmetric response is obtained by superposing a number of axisymmetric solutions. This problem will be studied in Chapter 11.

The simplest way for analysing axisymmetric shells with the FEM is to discretize the generatrix using conical fustrum shell elements (also called troncoconical shell elements). The process is analogous to the analysis of curved shells by flat elements (Chapter 8). Troncoconical elements were the first choice for analysis of axisymmetric shells under symmetric and arbitrary loading in the early 1960's [GS,PPL] and have become very popular since then. Curved elements are also available and can be useful

E. Oñate, *Structural Analysis with the Finite Element Method. Linear Statics:*
Volume 2: Beams, Plates and Shells, Lecture Notes on Numerical Methods
in Engineering and Sciences, DOI 10.1007/978-1-4020-8743-1_9,
© International Center for Numerical Methods in Engineering (CIMNE), 2013

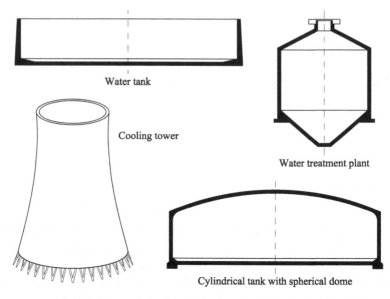

Water tank

Cooling tower

Water treatment plant

Cylindrical tank with spherical dome

Fig. 9.1 Examples of axisymmetric shell structures

in some cases [JS]. A comprehensive list of references of axisymmetric shell analysis by the FEM can be found in [AG,Ga,Ga2,Go,JS2,ZT].

In this chapter both troncoconical shell elements and curved axisymmetric shell elements will be studied. Elements which do not satisfy the normal orthogonality condition will be considered first. These elements basically follow the assumptions of Reissner-Mindlin theory for plates and flat shells and, hence, account for shear deformation effects. We will find that troncoconical shell elements have many similarities with Timoshenko beam elements (Chapter 2). Here again the use of reduced integration and assumed shear fields are essential to ensure their good performance for both thick and thin situations. Axisymmetric thin shell elements based in Kirchhoff thin shell theory will also be presented and two families of rotation-free troncoconical thin shell elements will be described.

The axisymmetric shell formulation will be simplified for the analysis of *axisymmetric plates, shallow axisymmetric shells* and *arches*. An axisymmetric plate can be viewed as a particular axisymmetric shell with a horizontal generatrix and the finite element formulation is simply derived by neglecting membrane effects in the general theory. The formulation for arches also emerges from the axisymmetric case by ignoring meridional effects. The subsequent study of shallow axisymmetric shells will allow us to reinterprete some concepts of membrane locking.

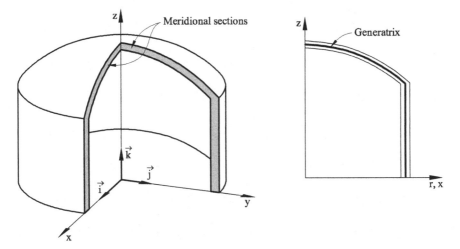

Fig. 9.2 Geometrical description of an axisymmetric shell

The chapter concludes with the formulation of axisymmetric shell elements by degeneration of axisymmetric solid elements and a description of some higher order theories for axisymmetric shells with composite laminated material. Here we will detail the formulation of a simple 2-noded axisymmetric shell element based on the refined zigzag theory.

9.2 GEOMETRICAL DESCRIPTION

Let us consider an axisymmetric shell defined in a global cartesian system x, y, z with associated unit vectors $\vec{i}$, $\vec{j}$, $\vec{k}$, respectively (Figure 9.2). A section in a plane rz, where r is an arbitrary radial direction, containing the axis of symmetry is termed a "meridional section". A generatrix is therefore the middle line of the meridional section.

A meridional cartesian coordinate system r, z is defined so that the plane rz contains the meridional section. The unit vector $\vec{i_r}$ is associated to the radial direction r (Figure 9.3). *In the following, coordinates x and r will be indistinguishably used for the global horizontal axis in the meridional section.*

A local coordinate system x', y', z' is defined at each point of the generatrix. The x' axis defines the direction of the unit tangent vector $\vec{t}$, z' is the direction across the thickness defining the unit normal vector $\vec{n}$ and y' is a direction orthogonal to the meridional plane. The unit vector $\vec{a}$ along y' is obtained by the cross product of $\vec{n}$ and $\vec{t}$ (Figure 9.3a). The

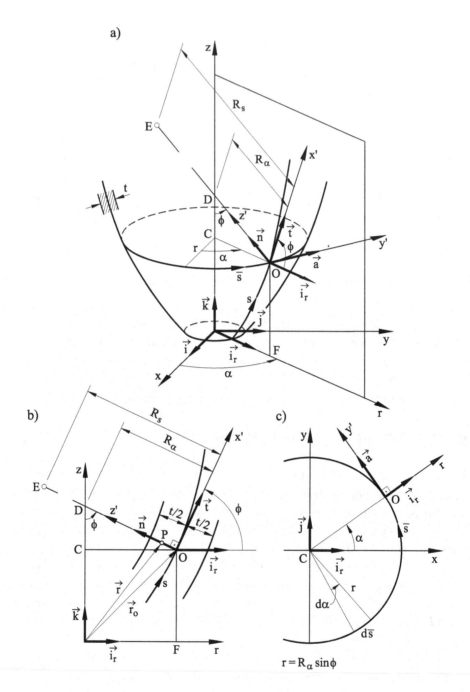

Fig. 9.3 Axisymmetric shell. Geometrical parameters

global components of the triad $\vec{t}, \vec{a}, \vec{n}$ are expressed as

$$\vec{t} = \frac{\partial x'}{\partial x}\vec{i} + \frac{\partial x'}{\partial y}\vec{j} + \frac{\partial x'}{\partial z}\vec{k} = \cos\phi\cos\alpha\,\vec{i} + \cos\phi\sin\alpha\,\vec{j} + \sin\phi\,\vec{k}$$

$$\vec{a} = \frac{\partial y'}{\partial x}\vec{i} + \frac{\partial y'}{\partial y}\vec{j} + \frac{\partial y'}{\partial z}\vec{k} = -\sin\alpha\,\vec{i} + \cos\alpha\,\vec{j}$$

$$\vec{n} = \frac{\partial z'}{\partial x}\vec{i} + \frac{\partial z'}{\partial y}\vec{j} + \frac{\partial z'}{\partial z}\vec{k} = -\sin\phi\cos\alpha\,\vec{i} - \sin\phi\sin\alpha\,\vec{j} + \cos\phi\,\vec{k}$$

$$(9.1)$$

where ϕ is the angle formed by the tangent direction x' with the radial direction $\vec{i_r}$ and α is the angle between vector $\vec{i_r}$ and the global x axis (Figure 9.3a).

The positive direction of arches is defined so that

$$ds = R_s d\phi \qquad\qquad d\bar{s} = r d\alpha = (R_\alpha \sin\phi) d\alpha \qquad (9.2)$$

where R_s and R_α are the curvature radii of the curves defined by the generatrix and the circumferential line respectively and s and $\bar{s}$ are arch length parameters along these lines (Figure 9.3a).

Let us consider a point P on the meridional section. Its position can be expressed as

$$\vec{r} = \vec{r_0} + \overline{OP}\,\vec{n} \qquad (9.3)$$

where vector $\vec{r_0}$ defines the position of point O on the generatrix and $\overline{OP}$ is the distance between points O and P (Figure 9.3b). As $\overline{OP}$ is an arbitrary distance along the normal direction, in the following we will take $\overline{OP} \equiv z'$.

Let us compute the derivative $\partial\vec{r}/\partial s$ at point P. Making use of Eqs.(9.1)–(9.3) gives (with $\overline{OP} \equiv z'$)

$$\frac{\partial\vec{r}}{\partial s} = \frac{\partial\vec{r_0}}{\partial s} + z'\frac{\partial\vec{n}}{R_s\partial\phi} = \left(1 - \frac{z'}{R_s}\right)\vec{t} \qquad (9.4)$$

where $\vec{t} = \partial\vec{r_0}/\partial s$ is the tangent vector at point O. Eq.(9.4) allows the following derivative to be computed

$$\frac{\partial x'}{\partial s} = \frac{\partial x'}{\partial x}\frac{\partial x}{\partial s} + \frac{\partial x'}{\partial y}\frac{\partial y}{\partial s} + \frac{\partial x'}{\partial z}\frac{\partial z}{\partial s} = \vec{t}\cdot\frac{\partial\vec{r}}{\partial s} = (1 - \frac{z'}{R_s}) \equiv C_s \qquad (9.5)$$

i.e.

$$dx' = C_s ds \quad \text{with} \quad C_s = 1 - \frac{z'}{R_s} \qquad (9.6)$$

The derivative $\partial \vec{r}/\partial \bar{s}$ is obtained using Eqs.(9.1)–(9.3) as

$$\frac{\partial \vec{r}}{\partial \bar{s}} = \frac{\partial \vec{r_0}}{\partial \bar{s}} + z' \frac{\partial \vec{n}}{R_\alpha \sin \phi \partial \alpha} = (1 - \frac{z'}{R_\alpha}) \vec{a} \qquad (9.7)$$

where the definition $\vec{a} = \partial \vec{r_0}/\partial \bar{s}$ has been used. Also

$$\frac{\partial y'}{\partial \bar{s}} = \frac{\partial y'}{\partial x} \frac{\partial x}{\partial \bar{s}} + \frac{\partial y'}{\partial y} \frac{\partial y}{\partial \bar{s}} + \frac{\partial y'}{\partial z} \frac{\partial z}{\partial \bar{s}} = \vec{a} \cdot \frac{\partial \vec{r}}{\partial \bar{s}} = (1 - \frac{z'}{R_\alpha}) \equiv C_\alpha \qquad (9.8)$$

i.e.

$$dy' = C_\alpha d\bar{s} \quad \text{with} \quad C_\alpha = 1 - \frac{z'}{R_\alpha} = 1 - \frac{z' \sin \alpha}{r} \qquad (9.9)$$

where the relationship $r = R_\alpha \sin \phi$ has been used (Figure 9.3c).

Eqs.(9.5) and (9.8) will be useful for deriving the expressions for the strains in the next section.

For shallow or thin shells $t \ll R_m$, with $m = s$, α and $C_s = C_\alpha = 1$, which implies $dx' = ds$ and $dy' = d\bar{s}$. These equations are accepted for many practical cases.

9.3 AXISYMMETRIC SHELL THEORY BASED ON REISSNER–MINDLIN ASSUMPTIONS

The axisymmetric shell theory derived next is based in the main following assumptions:

1. Loads are axisymmetric.
2. The thickness does not change with deformation.
3. The normal stress $\sigma_{z'}$ is zero.
4. Lines normal to the generatrix before deformation remain straight but not necessarily orthogonal to the generatrix after deformation.

These assumptions are identical to those used in Reissner-Mindlin theory for plates and flat shells (Chapters 6 and 8). In a later section the Kirchhoff orthogonality condition for the rotation of the normal will be assumed.

9.3.1 Displacement field

Due to the axial-symmetry only the deformation of a meridional section needs to be considered. The movement of a point in the meridional plane

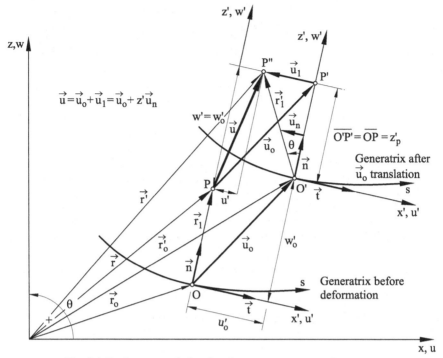

Fig. 9.4 Definition of the displacement vector of a point

is perfectly defined by the displacements u and w in the radial and vertical directions v and z, respectively.

Let us consider a point O over the generatrix and a point P on the normal direction at O and at a distance $z'_p = \overline{OP}$ from O. Points O and P move to the position O' and P'', respectively (Figure 9.4).

The displacement vector $\vec{u}$ joining points P and P'' can be split as the sum of the (rigid body) translation vector $\vec{u}_0$ (joining points O and P to O' and P' respectively) and a rotation vector $\vec{u}_1$ (joining point P' and P'') induced by the angle θ rotated by the normal $\vec{n}$ (Figure 9.4).

Using assumption 4 for the rotation of the normal we can write

$$\vec{u}(s, z') = \vec{u}_0(s) + \vec{u}_1(s, z') = \vec{u}_0(s) + z'_p \vec{u}_n(s) \tag{9.10}$$

where $\vec{u}_n$ is the vector defining the displacement of the end of the normal vector $\vec{n}$ and z'_p is the distance $\overline{OP}$.

The components of the displacement vectors of Eq.(9.10) are written in the local axes x', z' as

$$\vec{u} = u'\vec{t} + w'\vec{n}, \quad \vec{u}_0 = u'_0\vec{t} + w'_0\vec{n}, \quad \vec{u}_n = -\theta\vec{t} \tag{9.11}$$

where θ is the rotation of the normal vector (defined as positive in the anticlockwise direction) and $(\cdot)'$ denotes local displacement components. Combining Eqs.(9.10) and (9.11) and noting that z'_p is an arbitrary distance (i.e. $z'_p \equiv z'$) gives

$$u'(s, z') = u'_0(s) - z'\theta(s)$$
$$w'(s, z') = w'_0(s)$$

(9.12)

Eqs.(9.12) are the one-dimensional version of Eqs.(8.1) for flat shells. Note that the tangent displacement u' is the sum of the in-plane contribution u'_0 plus the bending term $z'\theta$ due to the rotation of the normal. The normal displacement w' is constant across the thickness.

The *local displacement vector* of a point on the generatrix is

$$\mathbf{u}' = [u'_0, \, w'_0, \, \theta]^T$$

(9.13)

The *global displacement vector* $\mathbf{u} = [u_0, \, w_0, \, \theta]^T$ is related to the local displacements $\mathbf{u}'$ by the following transformation

$$\mathbf{u} = \mathbf{L}^T \mathbf{u}' \qquad \text{with} \qquad \mathbf{L} = \begin{bmatrix} \cos\phi & \sin\phi & 0 \\ -\sin\phi & \cos\phi & 0 \\ 0 & 0 & 1 \end{bmatrix}$$

(9.14)

Recall that u_0 and w_0 are displacement components along the global axes r and z, respectively and ϕ is the angle formed by the tangent and radial vectors (Figure 9.3).

Eq.(9.14) allows us to relate the local and global displacements of an arbitrary point P giving (using Eq.(9.12))

$$u = u'\cos\phi - w'\sin\phi = u'_0\cos\phi - w'_0\sin\phi - z'\theta\cos\phi \qquad (9.15a)$$

$$w = u'\sin\phi - w'\cos\phi = u'_0\sin\phi + w'_0\cos\phi - z'\theta\sin\phi \qquad (9.15b)$$

The above expression of u is useful for deriving the circumferential strain.

9.3.2 Strain vector

Let us obtain the strains referred to the local axes. The tangential strains $\gamma_{x'y'}$ and $\gamma_{y'z'}$ are zero due to axial symmetry. Additionally $\sigma_{z'} = 0$ due to assumption 3 of Section 9.3 and, therefore, $\varepsilon_{z'}$ does not contribute to the internal work, similarly as it happens for plates and flat shells. The significant local strains are

$$\varepsilon_{x'} = \frac{\partial u'}{\partial x'}, \qquad \varepsilon_{y'} = \frac{\partial v'}{\partial y'}, \qquad \gamma_{x'z'} = \frac{\partial u'}{\partial z'} + \frac{\partial w'}{\partial x'} \qquad (9.16)$$

The local strain vector is defined as

$$\varepsilon = \begin{Bmatrix} \varepsilon_{x'} \\ \varepsilon_{y'} \\ \cdots \\ \gamma_{x'z'} \end{Bmatrix} = \begin{Bmatrix} \varepsilon'_p \\ \cdots \\ \varepsilon'_s \end{Bmatrix} \qquad (9.17)$$

where $\varepsilon'_p = [\varepsilon_{x'}, \varepsilon_{y'}]^T$ is the in-plane strain vector containing the axial strain $\varepsilon_{x'}$ and the circumferential strain $\varepsilon_{y'}$, and $\varepsilon'_s = \gamma_{x'z'}$ is the transverse shear strain.

From Eqs.(9.5) and (9.8) we deduce

$$
\begin{aligned}
\varepsilon_{x'} &= \frac{\partial u'}{\partial x'} = \frac{\partial s}{\partial x'}\frac{\partial u'}{\partial s} = \frac{1}{C_s}\frac{\partial u'}{\partial s} \\
\varepsilon_{y'} &= \frac{\partial v'}{\partial y'} = \frac{\partial \bar{s}}{\partial y'}\frac{\partial v'}{\partial \bar{s}} = \frac{1}{C_\alpha}\frac{\partial v'}{\partial \bar{s}} \\
\gamma_{x'z'} &= \frac{\partial u'}{\partial z'} + \frac{\partial s}{\partial x'}\frac{\partial w'}{\partial s} = \frac{\partial u'}{\partial z'} + \frac{1}{C_s}\frac{\partial w'}{\partial s}
\end{aligned}
\qquad (9.18)
$$

The derivatives $\partial u'/\partial s$ and $\partial w'/\partial s$ are defined as the tangential and normal components of vector $\partial \vec{u}/\partial s$. Using Eqs.(9.1), (9.2), (9.10) and (9.11) yields (noting that $\dfrac{\partial \vec{t}}{\partial s} = \dfrac{1}{R_s}\vec{n}$ and $\dfrac{\partial \vec{n}}{\partial t} = -\dfrac{1}{R_s}\vec{t}$)

$$
\begin{aligned}
\frac{\partial \vec{u}}{\partial s} &= \frac{\partial}{\partial s}(\vec{u_0} + z'\vec{u_n}) = \frac{\partial}{\partial s}(u'_0 \vec{t} + w'_0 \vec{n}) - \frac{\partial}{\partial s}(z'\theta)\vec{t} = \\
&= \left(\frac{\partial u'_0}{\partial s} - \frac{w'_0}{R_s} - z'\frac{\partial \theta}{\partial s}\right)\vec{t} + \left(\frac{\partial w'_0}{\partial s} + \frac{u'_0}{R_s} - z'\frac{\theta}{R_s}\right)\vec{n} \quad (9.19)
\end{aligned}
$$

which gives

$$\frac{\partial u'}{\partial s} = \frac{\partial u'_0}{\partial s} - \frac{w'_0}{R_s} - z'\frac{\partial \theta}{\partial s} \qquad \text{and} \qquad \frac{\partial w'}{\partial s} = \frac{\partial w'_0}{\partial s} + \frac{u'_0}{R_s} - z'\frac{\theta}{R_s} \qquad (9.20)$$

Similarly, the derivative $\partial u'/\partial z'$ is defined as the tangential component of vector $\partial \vec{u}/\partial z'$. From Eqs.(9.10) and (9.11)

$$\frac{\partial \vec{u}}{\partial z'} = -\theta \vec{t} \qquad \text{and therefore} \qquad \frac{\partial u'}{\partial z'} = -\theta \qquad (9.21)$$

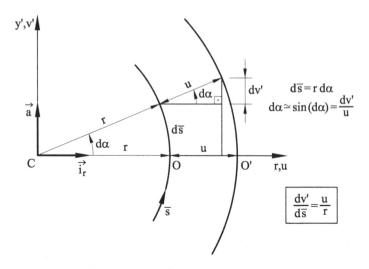

Fig. 9.5 Circumferential changes after deformation

Substituting Eqs.(9.20) and (9.21) into the expression of $\varepsilon_{x'}$ and $\gamma_{x'z'}$ of (9.18) and noting that $\partial v'/\partial \bar{s} = u/r$ (Figure 9.5), gives

$$\varepsilon_{x'} = \frac{1}{C_s}\left(\frac{\partial u_0'}{\partial s} - \frac{w_0'}{R_s} - z'\frac{\partial \theta}{\partial s}\right) \tag{9.22}$$

$$\varepsilon_{y'} = \frac{1}{C_\alpha}\left(\frac{u_0' \cos\phi - w_0' \sin\phi - z'\theta \cos\phi}{r}\right) \tag{9.23}$$

$$\gamma_{x'z'} = \frac{1}{C_s}\left(\frac{\partial w_0'}{\partial s} + \frac{u_0'}{R_s} - \theta\right) \tag{9.24}$$

The expression of u in Eq.(9.15a) has been used in the derivation of Eq.(9.23). The *local strain vector* is split as

$$\boldsymbol{\varepsilon}' = \left\{\begin{array}{c} \varepsilon_{x'} \\ \varepsilon_{y'} \\ \cdots \\ \gamma_{x'z'} \end{array}\right\} = \left\{\begin{array}{c} \dfrac{1}{C_s}\left(\dfrac{\partial u_0'}{\partial s} - \dfrac{w_0'}{R_s}\right) \\ \dfrac{u_0' \cos\phi - w_0' \sin\phi}{rC_\alpha} \\ \cdots\cdots\cdots \\ 0 \end{array}\right\} + \left\{\begin{array}{c} -\dfrac{z'}{C_s}\dfrac{\partial \theta}{\partial s} \\ -\dfrac{z'\theta \cos\phi}{rC_\alpha} \\ \cdots\cdots\cdots \\ \dfrac{1}{C_s}\left(\dfrac{\partial w_0'}{\partial s} + \dfrac{u_0'}{R_s} - \theta\right) \end{array}\right\} =$$

$$= \left\{\begin{array}{c} \mathbf{S}\hat{\boldsymbol{\varepsilon}}_m' \\ \cdots \\ 0 \end{array}\right\} + \left\{\begin{array}{c} -z'\mathbf{S}\hat{\boldsymbol{\varepsilon}}_b' \\ \cdots\cdots \\ \dfrac{1}{C_s}\hat{\varepsilon}_s' \end{array}\right\} = \mathbf{S}_1\left\{\begin{array}{c} \hat{\boldsymbol{\varepsilon}}_m' \\ \hat{\varepsilon}_b' \\ \hat{\varepsilon}_s' \end{array}\right\} = \mathbf{S}_1\hat{\boldsymbol{\varepsilon}}' \tag{9.25}$$

where

$$S_1 = \begin{bmatrix} S & -z'S & 0 \\ 0 & 0 & \dfrac{1}{C_s} \end{bmatrix} \quad \text{and} \quad S = \begin{bmatrix} \dfrac{1}{C_s} & 0 \\ 0 & \dfrac{1}{C_\alpha} \end{bmatrix} \qquad (9.26)$$

In Eq.(9.25)

$$\hat{\varepsilon}'_m = \left\{ \begin{array}{c} \dfrac{\partial u'_0}{\partial s} - \dfrac{w'_0}{R_s} \\[2mm] \dfrac{u'_0 \cos\phi - w'_0 \sin\phi}{r} \end{array} \right\} \qquad (9.27a)$$

is the *membrane strain* vector and

$$\hat{\varepsilon}'_b = \left\{ \begin{array}{c} \dfrac{\partial \theta}{\partial s} \\[2mm] \dfrac{\theta \cos\phi}{r} \end{array} \right\} , \qquad \hat{\varepsilon}'_s = \left\{ \dfrac{\partial w'_0}{\partial s} + \dfrac{u'_0}{R_s} - \theta \right\} \qquad (9.27b)$$

are the *bending* and *transverse shear strain* vectors respectively. Vector

$$\hat{\varepsilon}' = \left\{ \begin{array}{c} \hat{\varepsilon}'_m \\ \hat{\varepsilon}'_b \\ \hat{\varepsilon}'_s \end{array} \right\} \qquad (9.28)$$

is the generalized local strain vector containing the membrane, bending and transverse shear contributions.

The two components of $\hat{\varepsilon}'_m$ are the meridional (axial) and circumferential elongations of the generatrix, respectively. The components of $\hat{\varepsilon}'_b$ are the curvatures of this line along the meridional and circumferential directions.

The relationship between the local in-plane and transverse shear strains and the generalized strains is deduced from Eq.(9.25) as

$$\varepsilon'_p = [\varepsilon'_x, \varepsilon'_y]^T = S(\hat{\varepsilon}'_m - z'\hat{\varepsilon}'_b) \quad ; \quad \varepsilon'_s = \gamma_{x'z'} = \dfrac{1}{C_s}\hat{\varepsilon}'_s \qquad (9.29)$$

Eqs.(9.25) and (9.29) simplify for $\frac{t}{R_\alpha}$ and $\frac{t}{R_s} \ll 1$ as then $C_s = C_\alpha = 1$ and S is the 2×2 unit matrix.

9.3.3 Stresses and resultant stresses

The local stress vector is

$$\sigma' = \left\{ \begin{array}{c} \sigma_{x'} \\ \sigma_{y'} \\ \cdots \\ \tau_{x'z'} \end{array} \right\} = \left\{ \begin{array}{c} \sigma'_p \\ \cdots \\ \sigma'_s \end{array} \right\} \qquad (9.30)$$

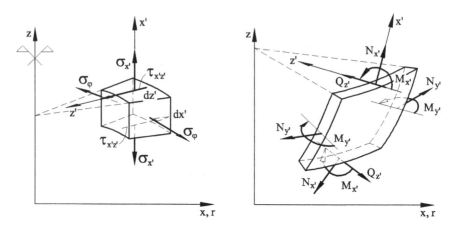

Fig. 9.6 Sign convention for local stresses and local resultant stresses in an axisymmetric shell

where $\boldsymbol{\sigma}'_p = [\sigma_{x'}, \sigma_{y'}]^T$ is the local in-plane stress vector containing the radial stress $\sigma_{x'}$ and the circumferential stress $\sigma_{y'}$ and $\sigma'_s = \tau_{x'z'}$ is the transverse shear stress. For sign convention see Figure 9.6.

9.3.3.1 Constitutive equations

The stress-strain constitutive relationship is obtained from that of 3D elasticity expressed in local axes x', z', by imposing that the normal stress $\sigma_{z'}$ and the shear strains $\gamma_{x'y'}$ and $\gamma_{x'z'}$ are zero. The result is

$$\boldsymbol{\sigma}'_p = \mathbf{D}'_p \boldsymbol{\varepsilon}'_p \quad , \quad \sigma'_s = G_{sz'} \varepsilon'_s \tag{9.31a}$$

In compact form

$$\boldsymbol{\sigma}' = \mathbf{D}' \boldsymbol{\varepsilon}' \quad \text{with} \quad \mathbf{D}' = \begin{bmatrix} \mathbf{D}'_p & 0 \\ 0 & G_{sz'} \end{bmatrix} \tag{9.31b}$$

If s, z' and $\bar{s}$ are directions of material orthotropy (with $1 \equiv s$ and $2 \equiv \alpha$), then

$$\mathbf{D}'_p = \begin{bmatrix} d'_1 & \nu_{21}d'_1 \\ \nu_{12}d'_2 & d'_2 \end{bmatrix} \quad ; \quad \text{with} \quad d'_1 = \frac{E_1}{1 - \nu_{12}\nu_{21}} \quad , \quad d'_2 = \frac{E_2}{1 - \nu_{12}\nu_{21}} \tag{9.32}$$

In the above E_1 and E_2 are the Young moduli in the principal directions $1 \equiv s$, $2 \equiv \theta$, ν_{12} and ν_{21} are the corresponding Poisson ratios (with $\nu_{12}E_2 =$

$\nu_{21}E_1$) and $G_{sz'}$ is the shear modulus in the transverse direction. For isotropic material

$$\mathbf{D}'_p = \frac{E}{1-\nu^2}\begin{bmatrix}1 & \nu \\ \nu & 1\end{bmatrix}, \qquad G_{sz'} = \frac{E}{2(1+\nu)} \tag{9.33}$$

The *local resultant stresses* are defined by

$$\hat{\boldsymbol{\sigma}}' = \left\{\begin{matrix}\hat{\boldsymbol{\sigma}}'_m \\ \cdots \\ \hat{\boldsymbol{\sigma}}'_b \\ \cdots \\ \hat{\boldsymbol{\sigma}}'_s\end{matrix}\right\} = \left\{\begin{matrix}N_{x'} \\ N_{y'} \\ \cdots \\ M_{x'} \\ M_{y'} \\ \cdots \\ Q_{z'}\end{matrix}\right\} = \int_{-\frac{t}{2}}^{\frac{t}{2}}\left\{\begin{matrix}C_\alpha\sigma_{x'} \\ C_s\sigma_{y'} \\ \cdots \\ -z'C_\alpha\sigma_{x'} \\ -z'C_s\sigma_{y'} \\ \cdots \\ C_\alpha\tau_{x'z'}\end{matrix}\right\}dz' = \int_{-\frac{t}{2}}^{\frac{t}{2}}\mathbf{S}_1^T\boldsymbol{\sigma}'C_sC_\alpha dz' \tag{9.34}$$

Subscripts m, b and s in Eq.(9.34) denote the axial (membrane) forces $N_{x'}, N_{y'}$, the bending moments $M_{x'}, M_{y'}$ and the transverse shear force $Q_{z'}$, respectively. For sign convention see Figure 9.6.

The curvature terms C_s and C_α in Eq.(9.34) are a consequence of $N_{x'}$, $M_{x'}$ and $Q_{z'}$ being defined per *unit circumferential length*, whereas $N_{y'}$ and $M_{y'}$ are defined per *unit meridional length*. For instance (Figure 9.7)

$$N_{x'}d\bar{s} = N_{x'}rd\alpha = \int_{-\frac{t}{2}}^{\frac{t}{2}}\sigma_{x'}d\bar{s}_p dz' = \int_{-\frac{t}{2}}^{\frac{t}{2}}\sigma_{x'}r_p d\alpha dz' = \left(\int_{-\frac{t}{2}}^{\frac{t}{2}}\sigma_{x'}C_\alpha dz'\right)rd\alpha \tag{9.35a}$$

and, as $d\bar{s} = rd\alpha$

$$N_{x'} = \int_{-\frac{t}{2}}^{\frac{t}{2}}\sigma_{x'}C_\alpha dz' \tag{9.35b}$$

Similarly for the meridional force (Figure 9.7)

$$N_{y'}ds = \int_{-\frac{t}{2}}^{\frac{t}{2}}\sigma_{y'}ds_p dz' = \left(\int_{-\frac{t}{2}}^{\frac{t}{2}}\sigma_{y'}C_s dz'\right)ds \tag{9.36a}$$

and, thus

$$N_{y'} = \int_{-\frac{t}{2}}^{\frac{t}{2}}\sigma_{y'}C_s dz' \tag{9.36b}$$

Substituting the expression for $\boldsymbol{\sigma}'$ of Eq.(9.31b) into (9.34) and using Eq.(9.25) yields the relationship between the local resultant stresses and

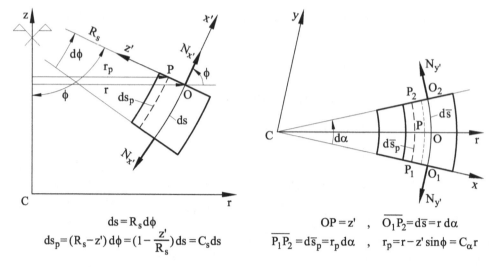

$$ds = R_s d\phi$$
$$ds_p = (R_s - z') d\phi = (1 - \frac{z'}{R_s}) ds = C_s ds$$

$$OP = z' \quad , \quad \overline{O_1 P_2} = d\overline{s} = r \, d\alpha$$
$$\overline{P_1 P_2} = d\overline{s}_p = r_p \, d\alpha \quad , \quad r_p = r - z' \sin\phi = C_\alpha r$$

Fig. 9.7 Axial resultant stresses $N_{x'}$ (left) and $N_{y'}$ (right)

the generalized local strains as

$$\hat{\sigma}' = \hat{\mathbf{D}}' \hat{\varepsilon}' \qquad \text{with} \qquad \hat{\mathbf{D}}' = \int_{-\frac{t}{2}}^{\frac{t}{2}} \mathbf{S}_1^T \mathbf{D}' \mathbf{S}_1 C_s C_\alpha dz' \tag{9.37}$$

In expanded form

$$\hat{\mathbf{D}}' = \int_{-\frac{t}{2}}^{\frac{t}{2}} \begin{bmatrix} \bar{\mathbf{D}}'_p & -z'\bar{\mathbf{D}}'_p & 0 \\ -z'\bar{\mathbf{D}}'_p & z'^2\bar{\mathbf{D}}'_p & 0 \\ 0 & 0 & \frac{C_\alpha}{C_s} G_{sz'} \end{bmatrix} dz' = \begin{bmatrix} \hat{\mathbf{D}}'_m & \hat{\mathbf{D}}'_{mb} & 0 \\ \hat{\mathbf{D}}'_{mb} & \hat{\mathbf{D}}'_b & 0 \\ 0 & 0 & \hat{D}'_s \end{bmatrix} \tag{9.38a}$$

with

$$\bar{\mathbf{D}}'_p = \begin{bmatrix} \frac{C_\alpha}{C_s} d'_1 & \nu_{21} d'_1 \\ \nu_{12} d'_2 & \frac{C_\alpha}{C_\alpha} d'_2 \end{bmatrix} \tag{9.38b}$$

From simple observation of Eq.(9.38a) we deduce

$$(\hat{\mathbf{D}}'_m, \hat{\mathbf{D}}'_{mb}, \hat{\mathbf{D}}'_b) = \int_{-\frac{t}{2}}^{\frac{t}{2}} \left(\bar{\mathbf{D}}'_p, -z'\bar{\mathbf{D}}'_p, z'^2 \bar{\mathbf{D}}'_p \right) dz' \tag{9.38c}$$

$$\hat{D}'_s = \int_{-\frac{t}{2}}^{\frac{t}{2}} \frac{C_\alpha}{C_s} G_{sz'} dz' = \hat{G}_{sz'} \qquad (9.38\text{d})$$

where d'_1 and d'_2 are given in Eq.(9.32). For $C_s = C_\alpha = 1$, then $\bar{\mathbf{D}}'_p = \mathbf{D}'_p$ (see Eqs.(9.32) and (9.38b) like for plates and flat shells. If, in addition, the material properties are homogeneous or they are symmetrical with respect to the generatrix, then $\hat{\mathbf{D}}'_{mb} = 0$ and

$$\hat{\mathbf{D}}'_m = t\mathbf{D}'_p, \qquad \hat{\mathbf{D}}'_b = \frac{t^3}{12}\mathbf{D}'_p, \qquad \hat{D}'_s = tG_{sz'} \qquad (9.39)$$

The resultant stress $N_{x'}$ of Eq.(9.35b) is undefined for $r \to 0$ (if $\alpha \neq 0$) as $C_\alpha \to \infty$ in this case (Eq.(9.9)). This problem can be overcome by defining $N_{x'}$ *by unit radian* and introducing r within the integral of (9.35a) [BD6]. In practice $N_{x'} = N_{y'}$ is taken at $r = 0$, due to symmetry.

9.3.3.2 Shear correction factor

The constant distribution for the transverse shear stress across the thickness assumed in the previous theory does not satisfy the condition $\tau_{x'z'} = 0$ on the shell surface. A shear correction factor is introduced to take into account the "exact" thickness distribution for the transverse shear stresses $\tau_{x'z'}$ in a similar way as for beams and plates. This correction is introduced by modifying the shear rigidity $\hat{D}'_s$ as

$$\hat{D}'_s = k\hat{G}_{sz'} \qquad (9.40)$$

with $\hat{G}_{sz'}$ defined by Eq.(9.38d).

The shear correction factor k is usually taken equal to 5/6 for homogeneous material and $t/R_{\min} \ll 1$. A more accurate expression for k involving the curvature radii can be found in [BD6].

9.3.3.3 Layered composite material

For an axisymmetric shell formed by a layered composite material (Figure 9.8) matrix $\hat{\mathbf{D}}'$ can be obtained from Eq.(9.37) as

$$\hat{\mathbf{D}}' = \sum_{i=1}^{n_l} \int_{z'_i}^{z'_{i+1}} \mathbf{S}_{1_i}^T \mathbf{D}'_i \mathbf{S}_{1_i} C_s C_\alpha dz' \qquad (9.41\text{a})$$

where n_l es the number of layers and subscript i refers to the properties of the ith layer defined by $z'_i \leq z' \leq z'_{i+1}$. The following explicit form for

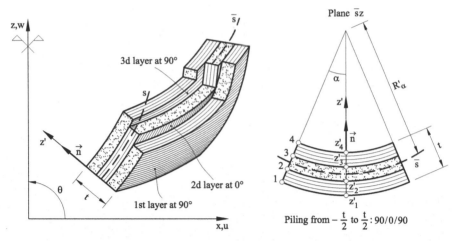

Fig. 9.8 Layered composite axisymmetric shell

the submatrices in Eq.(9.41a) is obtained for constant material properties within each layer and $C_s = C_\alpha = 1$

$$\hat{\mathbf{D}}'_m = \sum_{i=1}^{n_l} \mathbf{D}_{p'_i} \Delta z'_i, \quad \hat{\mathbf{D}}'_b = \frac{1}{3} \sum_{i=1}^{n_l} \mathbf{D}'_{p_i} \Delta z'^3_i,$$

$$\hat{\mathbf{D}}'_{mb} = -\frac{1}{2} \sum_{i=1}^{n_l} \mathbf{D}'_{p_i} \Delta z'^2_i, \quad \hat{D}'_s = \sum_{i=1}^{n_l} \hat{D}'_{s_i} \Delta z'_i \quad , \quad \Delta z'_i = z'_{i+1} - z'_i$$

$$(9.41b)$$

Details of the FEM analysis of laminated axisymmetric shells are given in [BD6,Go,NP,PN].

9.3.3.4 Initial stresses

The effect of initial stresses can be taken into account by modifying the stress-strain relationship (9.31b) as

$$\sigma' = \mathbf{D}'\varepsilon' + \sigma'^0 \tag{9.42a}$$

where $\sigma'^0 = [\sigma^0_{x'}, \sigma^0_{y'}, \sigma^0_{x'y'}]^T$ is the initial stress vector. If these stresses are due to thermal effects, then (for orthotropic material)

$$\sigma'^0 = -\Delta t \left[d'_1(\alpha^t_1 + \alpha^t_2 \nu_{21}), d'_2(\alpha^t_1 \nu_{12} + \alpha^t_2), 0 \right]^T$$

where Δt is the temperature increment, α^t_1 and α^t_2 are the coefficients of thermal expansion in the orthotropy directions $1(\equiv s)$ and $2(\equiv \alpha)$, respectively and d'_1 and d'_2 are defined in Eq.(9.32).

For isotropic material $\alpha_1^t = \alpha_2^t = \alpha^t$, and

$$\boldsymbol{\sigma}'^0 = -\frac{E\alpha^t \Delta t}{(1 - \nu)}[1, 1, 0]^T \qquad (9.42b)$$

The relationship between resultan stresses and generalized strains (Eq.(9.37)) is modified for the case of initial stresses as

$$\hat{\boldsymbol{\sigma}}' = \hat{\mathbf{D}}'\hat{\boldsymbol{\varepsilon}}' + \hat{\boldsymbol{\sigma}}'^0 \qquad (9.43a)$$

where

$$\hat{\boldsymbol{\sigma}}'^0 = \left[N_{x'}^0, N_{y'}^0, M_{x'}^0, M_{y'}^0, M_{z'}^0\right]^T =$$

$$= \int_{-\frac{t}{2}}^{\frac{t}{2}} \left[C_\alpha \sigma_{x'}^0, C_s \sigma_{y'}^0, -z'C_\alpha \sigma_{x'}^0, -z'C_s \sigma_{y'}^0, C_\alpha \tau_{x'z'}^0\right]^T dz' \quad (9.43b)$$

For initial thermal stresses $\tau_{x'z'}^0 = 0$ in the above expression.

9.3.4 Principle of virtual work

The PVW is written as

$$\iiint_V \delta\boldsymbol{\varepsilon}'^T \boldsymbol{\sigma}' dV = \iiint_V \delta\mathbf{u}^T \mathbf{b} dV + \iint_A \delta\mathbf{u}^T \mathbf{t} dA + \sum_i \oint_l \delta\mathbf{u}_i^T \mathbf{p}_i ds \quad (9.44a)$$

where V and A are the shell volume and the area of the meridional section, respectively and l is the length of the generatrix.

The virtual internal work is expressed in terms of the local stresses and strains, while all other vectors are written in global axes for convenience. In the above, $\mathbf{u} = [u_0, w_0, \theta]^T$ is the global displacement vector and

$$\mathbf{b} = [b_x, b_z, m]^T, \quad \mathbf{t} = [f_x, f_z, m_s]^T, \quad \mathbf{p}_i = [P_{x_i}, P_{z_i}, M_i]^T \qquad (9.44b)$$

are axisymmetric body force, distributed load and point load vectors, respectively, defined in *global axes* and per *unit circumferential length* (Figure 9.9). In Eq. (9.44b), m and m_s are distributed couples per unit volume and per unit surface, respectively. Initial stresses and strains have been neglected here for simplicity.

Eq.(9.44a) is simplified by expressing

$$dV = dx'dy'dz' = C_sC_\alpha \, d\bar{s} \, ds \, dz' = rC_sC_\alpha \, d\alpha \, ds \, dz' \qquad (9.45a)$$

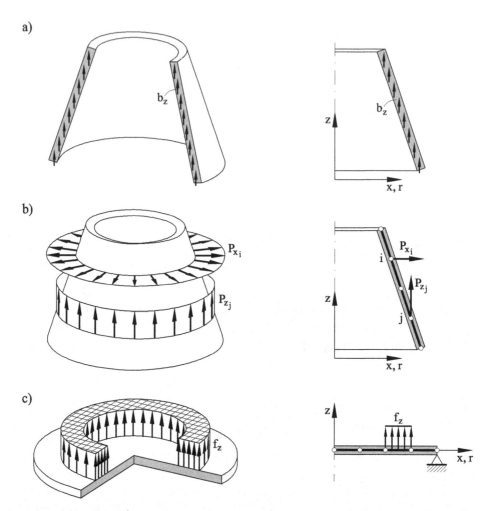

Fig. 9.9 External forces in axisymmetric shells: a) body forces (gravity), b) point loads, c) distributed loads

$$dA = dx'dy' = C_s C_\alpha d\bar{s}\, d\alpha = r C_s C_\alpha\, d\alpha\, ds$$

$$d\Gamma = dy' = C_\alpha\, d\bar{s} = r C_\alpha\, d\alpha$$

Integrating over a circumference gives

$$2\pi \iint_A \delta\boldsymbol{\varepsilon}'^T \boldsymbol{\sigma}' C_s C_\alpha r\, ds\, dz' = 2\pi \iint_A \delta\mathbf{u}^T \mathbf{b} C_s C_\alpha r\, ds\, dz' +$$
$$+ 2\pi \int_l \delta\mathbf{u}^T \mathbf{t} C_s C_\alpha r\, ds + 2\pi \sum_i r_i C_{\alpha_i} \delta\mathbf{u}_i^T \mathbf{q}_i \tag{9.45b}$$

Like we do for axisymmetric solids (Chapter 6 of [On4]) we retain the 2π factor on both sides of Eq.(9.45b). This will remind us that the forces are defined per circumferential length.

The internal work can be expressed in terms of the resultant stresses and the generalized strains using Eqs.(9.25) and (9.34) as

$$2\pi \iint_A \delta\varepsilon'^T \boldsymbol{\sigma}' C_s C_\alpha r \, ds \, dz' = 2\pi \int_l \delta\hat{\boldsymbol{\varepsilon}}'^T \underbrace{\left[\int_{-\frac{t}{2}}^{\frac{t}{2}} \mathbf{S}_1^T \boldsymbol{\sigma}' C_s C_\alpha dz' \right]}_{\hat{\boldsymbol{\sigma}}'} r \, ds =$$

$$= 2\pi \int_l \delta\hat{\boldsymbol{\varepsilon}}'^T \hat{\boldsymbol{\sigma}}' r \, ds$$

$$(9.45c)$$

The PVW is finally written in terms of curvilinear integrals as

$$2\pi \int_l \delta\hat{\boldsymbol{\varepsilon}}'^T \hat{\boldsymbol{\sigma}}' r \, ds = 2\pi \left[\int_l \delta\mathbf{u}^T \bar{\mathbf{b}} r \, ds + \int_l \delta\mathbf{u}^T \mathbf{t} C_s C_\alpha r \, ds + \sum_i r_i C_{\alpha_i} \delta\mathbf{u}_i^T \mathbf{q}_i \right]$$

$$(9.46)$$

where $\bar{\mathbf{b}}$ contains generalized body forces acting on the generatrix and is given by

$$\bar{\mathbf{b}} = [\bar{b}_r, \bar{b}_z, \bar{m}]^T = \int_{-\frac{t}{2}}^{\frac{t}{2}} \mathbf{b} C_s C_\alpha dz' \qquad (9.47)$$

Eq.(9.46) simplifies for $C_s = C_\alpha = 1$. If the body forces are constant then $\bar{\mathbf{b}} = t\mathbf{b}$.

Surface loads $\mathbf{t}^{(e)}$ due to an *internal pressure* typically require a transformation to global axes (Figure 9.12). The contribution of a point load acting *on the axis of symmetry* is simply given by the value of the force (i.e. the $2\pi r_i$ factor is not required in this case).

The effect of initial stresses can be simply taken into account by substituting the expression of $\hat{\boldsymbol{\sigma}}'$ of Eq.(9.43b) into the l.h.s. of Eq.(9.46).

The integrals in Eq.(9.46) contain first derivatives of the displacements only. This allows C^0 continuous axisymmetric shell elements to be used. A simple choice are the troncoconical elements studied in the next section.

9.4 TRONCOCONICAL REISSNER-MINDLIN ELEMENTS

Figures 9.10 and 9.11 show the discretization of an axisymmetric shell in troncoconical elements. The discretization process is extremely simple and it merely involves dividing the generatrix into straight segments, as is done for a plane frame or an arch.

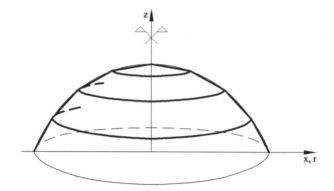

Fig. 9.10 Discretization of an axisymmetric shell in troncoconical shell elements

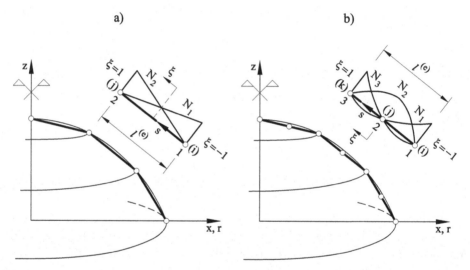

Fig. 9.11 Troncoconical shell elements: a) linear (2-noded) and b) quadratic (3-noded). Letters in brackets denote global node numbers

9.4.1 Displacement and strain interpolation

The local displacement field within a troncoconical element with n nodes is written as

$$\mathbf{u}' = \begin{Bmatrix} u'_0 \\ w'_0 \\ \theta \end{Bmatrix} = \sum_{i=1}^{n} \mathbf{N}_i \, \mathbf{a}_i'^{(e)} = \mathbf{N} \, \mathbf{a}'^{(e)} \tag{9.48}$$

with

$$\mathbf{N} = [\mathbf{N}_1, \mathbf{N}_2, \ldots, \mathbf{N}_n]; \qquad \mathbf{a}'^{(e)} = \begin{Bmatrix} \mathbf{a}_1'^{(e)} \\ \mathbf{a}_2'^{(e)} \\ \vdots \\ \mathbf{a}_n'^{(e)} \end{Bmatrix} \tag{9.49}$$

where

$$\mathbf{N}_i = \begin{bmatrix} N_i(\xi) & 0 & 0 \\ 0 & N_i(\xi) & 0 \\ 0 & 0 & N_i(\xi) \end{bmatrix}, \qquad \mathbf{a}_i'^{(e)} = \begin{Bmatrix} u_{0_i}' \\ w_{0_i}' \\ \theta_i \end{Bmatrix} \tag{9.50}$$

and $N_i(\xi)$ are the shape functions of 1D Lagrange elements (Chapter 2 of [On4]).

Since the element is straight, $R_s = \infty$ and $C_s = 1$ which implies $\partial/\partial s = \partial/\partial x'$. For simplicity we will also assume in the following $C_\alpha = 1$. The generalized local strain vector is found as (Eqs.(9.27-9.28))

$$\hat{\boldsymbol{\varepsilon}}' = \begin{Bmatrix} \hat{\boldsymbol{\varepsilon}}_m' \\ \cdots \\ \hat{\varepsilon}_b' \\ \cdots \\ \hat{\varepsilon}_s' \end{Bmatrix} = \begin{Bmatrix} \dfrac{\partial u_o'}{\partial s} \\ \dfrac{u_o' \cos\phi - w_o' \mathrm{sen}\phi}{r} \\ \cdots \\ \dfrac{\partial \theta}{\partial s} \\ \dfrac{\theta \cos\phi}{r} \\ \cdots \\ \dfrac{\partial w_0'}{\partial s} - \theta \end{Bmatrix} \tag{9.51}$$

The generalized local strain vector includes the axial stretching $\frac{\partial u_o'}{\partial s}$, the (pseudo) curvature $\frac{\partial \theta}{\partial s}$ and the shear angle $\left(\frac{\partial w_o'}{\partial s} - \theta\right)$, as for a plane frame or an arch. It also incorporates the circumferential stretching $\frac{u_0}{r}$ and the circumferential curvature $\frac{\theta \cos\phi}{r}$.

Substituting Eq.(9.48) into (9.51) gives

$$\hat{\boldsymbol{\varepsilon}}' = \sum_{i=1}^{n} \mathbf{B}_i' \, \mathbf{a}_i'^{(e)} = \mathbf{B}' \, \mathbf{a}'^{(e)} \tag{9.52}$$

with

$$\mathbf{B}' = [\mathbf{B}_1', \mathbf{B}_2', \ldots, \mathbf{B}_n'] \tag{9.53}$$

and

$$
\mathbf{B}'_i = \left\{ \begin{array}{c} \mathbf{B}'_{m_i} \\ \text{---} \\ \mathbf{B}'_{b_i} \\ \text{---} \\ \mathbf{B}'_{s_i} \end{array} \right\} =
\left[\begin{array}{ccc}
\dfrac{\partial N_i}{\partial s} & 0 & 0 \\[2mm]
\dfrac{N_i\cos\phi}{r} & -\dfrac{N_i \operatorname{sen}\phi}{r} & 0 \\[2mm]
\hline
0 & 0 & \dfrac{\partial N_i}{\partial s} \\[2mm]
0 & 0 & \dfrac{N_i\cos\phi}{r} \\[2mm]
\hline
0 & \dfrac{\partial N_i}{\partial s} & -N_i
\end{array} \right]
\tag{9.54}
$$

where subscripts m, b and s denote the membrane, bending and transverse shear strain matrices, respectively.

9.4.2 Local stiffness matrix

Substituting Eqs.(9.43a), (9.48) and (9.52) into the PVW (Eq.(9.46)) gives the equilibrium equation for the element as

$$
\mathbf{K}'^{(e)}\mathbf{a}'^{(e)} - \mathbf{f}'^{(e)} = \mathbf{q}'^{(e)}
\tag{9.55}
$$

where

$$
\mathbf{K}'^{(e)}_{ij} = 2\pi \int_{l^{(e)}} \mathbf{B}'^{T}_i \; \hat{\mathbf{D}} \; \mathbf{B}'_j r \, ds \quad , \quad i,j = 1, n
\tag{9.56}
$$

is a typical contribution to the *local stiffness matrix* of a troncoconical shell element of length $l^{(e)}$, $\mathbf{f}^{(e)}$ is the equivalent nodal force vector and $\mathbf{q}'^{(e)}$ is the equilibrating nodal forces vector. For convenience, the components of both force vectors are expressed in local axes.

Making use of Eqs.(9.38a) and (9.54) the local stiffness matrix for the troncoconical element is written as

$$
\mathbf{K}'^{(e)}_{ij} = 2\pi \int_{l^{(e)}} \Big[\underbrace{\mathbf{B}'^{T}_{m_i} \; \hat{\mathbf{D}}'_m \mathbf{B}'_{m_j}}_{(1)} + \underbrace{\mathbf{B}'^{T}_{b_i} \hat{\mathbf{D}}'_b \mathbf{B}'_{b_j}}_{(2)} + \underbrace{\mathbf{B}'^{T}_{s_i} \hat{\mathbf{D}}'_s \mathbf{B}'_{s_j}}_{(3)} +
$$
$$
+ \underbrace{\Big(\mathbf{B}'^{T}_{m_i} \; \hat{\mathbf{D}}'_{mb} \; \mathbf{B}'_{b_j} + \mathbf{B}'^{T}_{b_i} \; \hat{\mathbf{D}}'_{mb} \; \mathbf{B}'_{m_j} \Big)}_{(4)} \Big] r \, ds =
$$
$$
= \underbrace{\mathbf{K}'^{(e)}_{m_{ij}}}_{(1)} + \underbrace{\mathbf{K}'^{(e)}_{b_{ij}}}_{(2)} + \underbrace{\mathbf{K}'^{(e)}_{s_{ij}}}_{(3)} + \underbrace{\mathbf{K}'^{(e)}_{mb_{ij}} + [\mathbf{K}'^{(e)}_{mb_{ij}}]^T}_{(4)}
\tag{9.57}
$$

where subscripts m, b, s and mb denote the stiffness contributions due to membrane, bending, shear and coupled membrane-bending effects.

Eq.(9.57) is analogous to (8.36) for flat shell elements. The coupling of the local membrane and bending stiffness at element level via matrix $\mathbf{K}_{mb}^{\prime(e)}$ is a distinct feature of composite laminated axisymmetric shells. For $\hat{\mathbf{D}}_{mb}^{\prime}=0$, which occurs for particular cases, such as symmetric laminates or homogeneous material, then $\mathbf{K}_{mb}^{\prime(e)}=0$ and the membrane and bending stiffness are *uncoupled at element level*. Membrane-bending coupling invariably occurs at structural level when the local stiffness equations of non coplanar elements are assembled in global axes, as for flat shell elements.

9.4.3 Transformation to global axes

The stiffness transformation process is very similar to that explained for flat shells in Section 8.5 and the details will not be repeated here. The global stiffness matrix for the element is

$$
\mathbf{K}^{(e)} = [\mathbf{T}^{(e)}]^T \mathbf{K}^{(e)} \mathbf{T}^{(e)} \quad \text{with} \quad \mathbf{T}^{(e)} = \begin{bmatrix} \mathbf{L}_1^{(e)} & & & 0 \\ & \mathbf{L}_2^{(e)} & & \\ & & \ddots & \\ 0 & & & \mathbf{L}_n^{(e)} \end{bmatrix} \tag{9.58}
$$

A typical submatrix is given by

$$
\mathbf{K}_{ij}^{(e)} = \left[\mathbf{L}_i^{(e)}\right]^T \mathbf{K}_{ij}^{\prime(e)} \ \mathbf{L}_j^{(e)} \tag{9.59}
$$

where $\mathbf{L}_i^{(e)}$ coincides with matrix $\mathbf{L}$ of Eq.(9.14). The nodal transformation matrices are identical for all the element nodes as the element is straight.

As explained in Section 8.5, it is generally more convenient to transform first the local strain matrix $\mathbf{B}_i'$ as

$$
\mathbf{B}_i = \begin{Bmatrix} \mathbf{B}_{m_i} \\ \mathbf{B}_{b_i} \\ \mathbf{B}_{s_i} \end{Bmatrix} = \mathbf{B}_i' \left[\mathbf{L}_i^{(e)}\right]^T = \begin{bmatrix} \dfrac{\partial N_i}{\partial s}\cos\phi & \dfrac{\partial N_i}{\partial s}\operatorname{sen}\phi & 0 \\[2mm] \dfrac{N_i}{r} & 0 & 0 \\[2mm] 0 & 0 & \dfrac{\partial N_i}{\partial s} \\[2mm] 0 & 0 & \dfrac{N_i\cos\phi}{r} \\[2mm] -\dfrac{\partial N_i}{\partial s}\operatorname{sen}\phi & \dfrac{\partial N_i}{\partial s}\cos\phi & -N_i \end{bmatrix} \begin{matrix} \mathbf{B}_{m_i} \\[6mm] \mathbf{B}_{b_i} \\[6mm] \mathbf{B}_{s_i} \end{matrix} \tag{9.60}
$$

The global stiffness matrix is now directly obtained as

$$\mathbf{K}_{ij}^{(e)} = 2\pi \int_{l^{(e)}} \mathbf{B}_i^T \hat{\mathbf{D}}' \mathbf{B}_j r ds = \mathbf{K}_{m_{ij}}^{(e)} + \mathbf{K}_{b_{ij}}^{(e)} + \mathbf{K}_{s_{ij}}^{(e)} + \mathbf{K}_{mb_{ij}}^{(e)} + [\mathbf{K}_{mb_{ij}}'^{(e)}]^T$$

(9.61)

where the different stiffness matrices are obtained by substituting $\mathbf{B}_{m_i}$, $\mathbf{B}_{b_i}$ and $\mathbf{B}_{s_i}$ of Eq.(9.60) into (9.57).

The element equilibrium equation in global axes is

$$\mathbf{K}^{(e)} \mathbf{a}^{(e)} - \mathbf{f}^{(e)} = \mathbf{q}^{(e)}$$

(9.62)

The equivalent nodal force vector is given in global axes by

$$\mathbf{f}_i^{(e)} = 2\pi \int_{l^{(e)}} \mathbf{N}_i^T \mathbf{b} t r ds + 2\pi \int_{l^{(e)}} \mathbf{N}_i^T \mathbf{t} r ds + 2\pi r_i \mathbf{p}_i^{(e)} - 2\pi \int_{l^{(e)}} \mathbf{B}_i'^T \hat{\boldsymbol{\sigma}}'^0 r ds$$

(9.63)

where the last integral accounts for the effect of initial stresses.

Figure 9.12 shows the transformation required to obtain the global components of the distributed load vector $\mathbf{t}$ for an internal pressure acting on a mesh of 2-noded troncoconical elements.

Matrix $\mathbf{K}_{ij}^{(e)}$ is computed numerically with a 1D Gauss quadrature as

$$\mathbf{K}_{ij}^{(e)} = 2\pi \int_{-1}^{+1} \mathbf{B}_i^T \hat{\mathbf{D}}' \mathbf{B}_j r J^{(e)} d\xi = \sum_{q=1}^{n_q} \left(\mathbf{B}_i^T \hat{\mathbf{D}}' \mathbf{B}_j r J^{(e)} \right)_q W_q =$$

$$= \sum_{q_m=1}^{n_m} \left(\mathbf{I}_m^{(e)} \right)_{q_m} W_{q_m} + \sum_{q_b=1}^{n_b} \left(\mathbf{I}_b^{(e)} \right)_{q_b} W_{q_b} +$$

$$+ \sum_{q_s=1}^{n_s} \left(\mathbf{I}_s^{(e)} \right)_{q_s} W_{q_s} + \sum_{q_{mb}=1}^{n_{mb}} \left(\mathbf{I}_{mb}^{(e)} \right)_{q_{mb}} W_{q_{mb}}$$

(9.64)

where $J^{(e)} = \frac{ds}{d\xi}$ and

$$\mathbf{I}_a^{(e)} = 2\pi \mathbf{B}_{a_i}^T \hat{\mathbf{D}}'_{a_i} \mathbf{B}_{a_j} r J^{(e)}, \qquad a = m, b, s$$

(9.65)

$$\mathbf{I}_{mb}^{(e)} = 2\pi \left(\mathbf{B}_{m_i}^T \hat{\mathbf{D}}'_{mb} \mathbf{B}_{b_j} + \mathbf{B}_{b_i}^T \hat{\mathbf{D}}'_{mb} \mathbf{B}_{m_j} \right) r J^{(e)}$$

Typically $J^{(e)} = \frac{l^{(e)}}{2}$ for straight elements (see Section 2.3.2 and Chapter 3 of [On4]).

In Eq.(9.64) n_i, W_{q_i}, $i = m, b, s, mb$ are the number of integration points and the corresponding weights for computing the membrane, bending, shear and membrane-bending stiffness matrices. The selection of the quadrature order is discussed later.

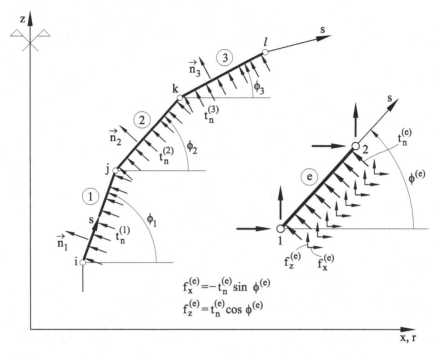

Fig. 9.12 Internal pressure acting on an axisymmetric shell discretized with 2-noded troncoconical element. Transformation to global axes

A similar quadrature is used for integrating the equivalent nodal forces. Grouping terms in Eq.(9.63) gives

$$\mathbf{f}_i^{(e)} = 2\pi \sum_{q=1}^{n_q} \left[\mathbf{N}_i^T \left(\mathbf{b}t + \mathbf{t} \right) - \mathbf{B}_i'^T \hat{\boldsymbol{\sigma}}'^0 \right] r J^{(e)} W_q + 2\pi r_i \mathbf{p}_i^{(e)} \qquad (9.66)$$

9.5 SHEAR AND MEMBRANE LOCKING

9.5.1 Transverse shear locking

Let us write the global stiffness equation for an axisymmetric shell with constant thickness as

$$[t(\bar{\mathbf{K}}_m + \bar{\mathbf{K}}_s) + t^3 \bar{\mathbf{K}}_b \mathbf{a}] = \mathbf{f} \qquad (9.67)$$

where $\overline{(\cdot)}$ denotes the membrane, transverse shear and bending stiffness matrices once the thickness has been taken out as shown. For simplicity the coupling membrane-bending stiffness matrix $\mathbf{K}_{mb}$ has been ignored.

Let us consider a circular plate under lateral loading only. Now the membrane and bending effects are decoupled and Eq.(9.67) simplifies to

$$[t\bar{\mathbf{K}}_s + t^3\bar{\mathbf{K}}_b]\mathbf{a} = \mathbf{f} \tag{9.68}$$

The exact analytical solution in this case is proportional to $1/t^3$ [BD6,Go]. Dividing Eq.(9.68) by t^3 gives

$$[\alpha\bar{\mathbf{K}}_s + \bar{\mathbf{K}}_b]\mathbf{a} = \frac{1}{t^3}\mathbf{f} = \bar{\mathbf{f}} \tag{9.69}$$

where $\alpha = 1/t^2$ and $\bar{\mathbf{f}}$ is a vector *of the order* of magnitude of the exact solution. For $t \to \infty$ then $\alpha \to 0$ and the bending stiffness plays no role in Eq.(9.69). Thus, the solution tends to the following limit value

$$\bar{\mathbf{K}}_s\mathbf{a} = \frac{1}{\alpha}\mathbf{f} \to 0 \tag{9.70}$$

Clearly as the thickness reduces, the solution stiffness (locks) at a rate proportional to t^2 with respect to the exact value, giving zero displacements in the limit thin case $(t = 0)$. Eq.(9.70) shows that the existance of a non-trivial solution requires $\mathbf{K}_s$ to be singular. Here the singularity rule of Eq.(2.50) applies again. This singularity can be achieved by using reduced integration for $\mathbf{K}_s$.

9.5.2 Membrane locking

Membrane terms can contribute to increase locking behaviour in axisymmetric shells. High membrane stiffness values relative to the bending ones can introduce a "parasitic" membrane stiffness leading to membrane locking. This effect, is generally of less importance than shear locking and can be understood by observing Eq.(9.67). Dividing by t^3 gives

$$[\alpha(\bar{\mathbf{K}}_m + \bar{\mathbf{K}}_s) + \bar{\mathbf{K}}_b]\mathbf{a} = \bar{\mathbf{f}} \tag{9.71a}$$

For the limit case of $t \to 0$ and $\alpha \to \infty$ we have

$$(\bar{\mathbf{K}}_m + \bar{\mathbf{K}}_s)\mathbf{a} = \frac{1}{\alpha}\bar{\mathbf{f}} \to 0 \tag{9.71b}$$

Clearly the existence of a non-zero solution requires the singularity of the sum of the transverse shear and membrane matrices. In practice this implies using reduced integration for $\mathbf{K}_s^{(e)}$ and $\mathbf{K}_m^{(e)}$. However, this condition is less strict than that required for $\mathbf{K}_s^{(e)}$ to avoid transverse shear locking. This is so because in shells the coupling between membrane and flexural effects is generally weaker than that between shear and bending effects.

Membrane locking can be fully avoided if the displacement approximation can reproduce a null membrane strain field without modifying the bending approximation. The so called "inextensional mode" (zero membrane strains under a pure bending state) does not exist in axisymmetric shells, as the circumferential strains are always non-zero. Membrane locking is less severe when the displacement approximation allows a zero radial membrane strain $\frac{\partial u'_0}{\partial s} - \frac{w'_0}{R_s}$ to be represented without constraining the flexural approximation [Cr].

In troncoconical Reissner-Mindlin shell elements $1/R_s = 0$ and, hence, the radial membrane strain is fully decoupled from the bending and shear strains at element level. This practically eliminates membrane locking in troncoconical elements, as the membrane-bending coupling is induced by the transformation of the stiffness equations to global axes only. This is not so for curved axisymmetric shell elements for which membrane and bending effects are coupled at element level and hence they require a compatible approximation for u'_o and w'_o, or reduced integration for the membrane stiffness (Section 9.15).

This situation worsens if coupling between transverse shear, membrane and bending behaviour exists. Eq.(9.71) reads in this case

$$[\alpha(\bar{\mathbf{K}}_m + \bar{\mathbf{K}}_s) + \beta\bar{\mathbf{K}}_{mb} + \bar{\mathbf{K}}_b]\mathbf{a} = \bar{\mathbf{f}} \tag{9.72}$$

where $\beta = 1/t$.

Eq.(9.72) shows that matrix $\mathbf{K}_{mb}^{(e)}$ introduces a coupling between membrane and bending effects at element level than can also induce locking as $t \rightarrow 0$. The effect of $\mathbf{K}_{mb}^{(e)}$ is less relevant in terms of locking than that of $\mathbf{K}_s^{(e)}$ and $\mathbf{K}_m^{(e)}$. It is however recommended in practice to use reduced integration for $\mathbf{K}_{mb}^{(e)}$ in order to prevent membrane locking in composite shells.

9.5.3 Other techniques to avoid locking in Reissner-Mindlin troncoconical shell elements

Shear locking can also be avoided by combining an assumed transverse shear strain field with adequate (compatible) approximations for displacements and rotations. The application of this technique follows the lines detailed for Timoshenko beam elements (Chapter 2).

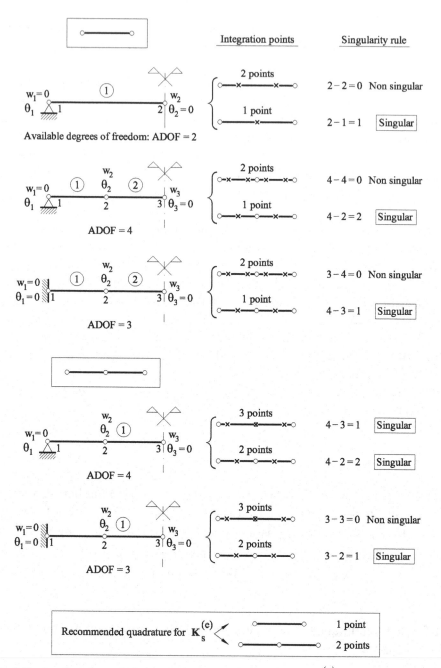

Fig. 9.13 Application of the singularity rule (2.50) for $\mathbf{K}_s^{(e)}$ in 2 and 3-noded Reissner-Mindlin troncoconical shell elements

For the 2-noded troncoconical element, combining an assumed constant field for $\gamma_{x',z'}$ with a linear approximation for u'_0, w'_0 and θ is equivalent to using one-point reduced quadrature for $\mathbf{K}_s^{(e)}$ in the original element.

The same equivalence is found between the two-point reduced integration for the quadratic 3-noded troncoconical element and using an assumed linear shear strain field. This is similar to what happens for the 2-noded Timoshenko beam element (Example 2.9).

Similar techniques can be devised to avoid membrane locking. However, their interpretation is less obvious. See Section 9.11 for more details.

9.6 INTEGRATION RULES FOR THE LINEAR AND QUADRATIC REISSNER-MINDLIN TRONCOCONICAL ELEMENTS

9.6.1 Quadrature for the 2-noded Reissner-Mindlin troncoconical element

Shear locking in the linear (2-noded) Reissner-Mindlin troncoconical element can be avoided by using a *one point reduced quadrature* for $\mathbf{K}_s^{(e)}$. Examples are shown in Figure 9.13.

The one-point quadrature for the all the stiffness terms preserves the correct rank in the overall element stiffness matrix and leads to a simple expression as

$$\mathbf{K}_{ij}^{(e)} = 2\pi\bar{\mathbf{B}}_i\hat{\bar{\mathbf{D}}}'\bar{\mathbf{B}}_j\bar{r}l^{(e)} \tag{9.73}$$

where $(\bar{\cdot})$ denotes element midpoint values. Figure 9.14 shows the explicit form for the stiffness matrix neglecting coupled membrane-bending effects. The reduced one-point quadrature for all the stiffness matrix terms eliminates membrane and shear locking (Figure 9.15).

The equivalent nodal force vector is integrated with a two-point quadrature for the general case of arbitrary loading (Eq.(9.66)). A simple analytical expression for constant body forces, uniformly distributed loading and zero initial stresses is found as

$$\mathbf{f}_i^{(e)} = \frac{\pi l^{(e)}c_i}{3}[\mathbf{tb} + \mathbf{t}] + 2\pi x_i \mathbf{p}_i^{(e)}, \quad i = 1, 2 \tag{9.74}$$

with $c_1 = 2r_1^{(e)} + r_2^{(e)}$, $c_2 = 2r_2^{(e)} + r_1^{(e)}$ where $r_1^{(e)}$, $r_2^{(e)}$ are the radial coordinates of the two element nodes. Due to the axial-symmetry, nodes at a greater distance from the axis have larger nodal force values.

The 2-noded Reissner-Mindlin troncoconical element with a single integrating point was originally developed by Zienkiewicz *et al.* [ZBMO] and

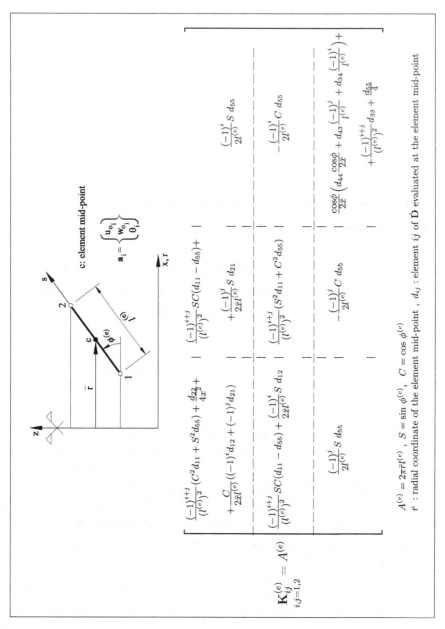

Fig. 9.14 Stiffness matrix for the 2-noded Reissner-Mindlin troncoconical shell element with uniform one-point reduced integration. Coupled membrane-bending effects are neglected

it is the simplest and most popular axisymmetric shell element. Examples of its good performance are given in the next section.

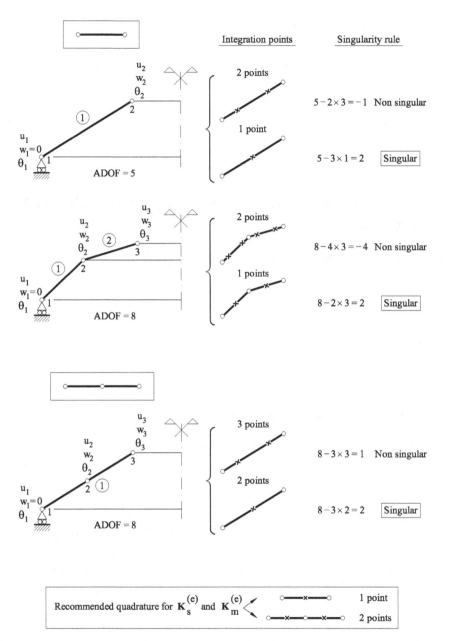

Fig. 9.15 Applications of the singularity rule (2.50) for 2- and 3-noded troncoconical elements using reduced integration for $\mathbf{K}_m^{(e)}$ and $\mathbf{K}_s^{(e)}$

9.6.2 Quadrature for the 3-noded Reissner-Mindlin troncoconical element

The 3-noded (quadratic) Reissner-Mindlin troncoconical element requires a reduced two-point quadrature for $\mathbf{K}_s^{(e)}$ to avoid shear locking. The singularity rule (2.50) is satisfied using a three-point quadrature for $\mathbf{K}_s^{(e)}$ in some cases (Figure 9.13). The rest of the stiffness terms can be also integrated with the reduced two-point rule without perturbing the correct rank of the element stiffness matrix. The two-point quadrature is also typically used for integrating vector $\mathbf{f}^{(e)}$ (Eq.(9.66)).

The reduced quadrature for both $\mathbf{K}_m^{(e)}$ and $\mathbf{K}_s^{(e)}$ satisfies the singularity rule (2.50) for the sum of the two matrices. This ensures that membrane and transverse shear locking is avoided (Figure 9.15).

9.7 APPLICATIONS OF THE TWO-NODED REISSNER-MINDLIN TRONCOCONICAL ELEMENT

9.7.1 Clamped spherical dome under uniform pressure

Figure 9.16 shows the geometry of the dome, the material properties and the loading. A uniform mesh of ten 2-noded troncoconical elements is used.

Figure 9.16 displays the diagrams of radial bending moment $M_{x'}$ and circumferential force $N_{y'}$. Very good agreement is obtained with the theoretical values [TW] as well as with numerical results using curved axisymmetric elements based on Kirchhoff thin shell theory [Del].

9.7.2 Toroidal shell under internal pressure

The geometry of the toroidal shell is shown in Figure 9.17, where details of the material properties and the loading are given. One half of the meridional section has been analyzed due to symmetry. A mesh of eighteen 2-noded troncoconical shell elements has been used.

Results for the distribution of the radial displacement and the diagrams of axial forces $N_{x'}$ and $N_{y'}$ are compared in Figure 9.17 with alternative axisymmetric solutions obtained using Kirchhoff theory [ChF,Del,GM]. Good agreement is found in all cases despite the relative coarseness of the mesh. Other FE solutions to this problem can be found in [JO,SL].

9.7.3 Cylindrical tank with spherical dome under internal pressure

This example coincides with the cylindrical concrete tank studied in Section 7.7.2 of [On4] using axisymmetric solid elements. The geometry and the material properties can be seen in Figure 7.9 of [On4]. A constant internal pressure of 1 T/m^2 acts on the cylindrical wall and the dome.

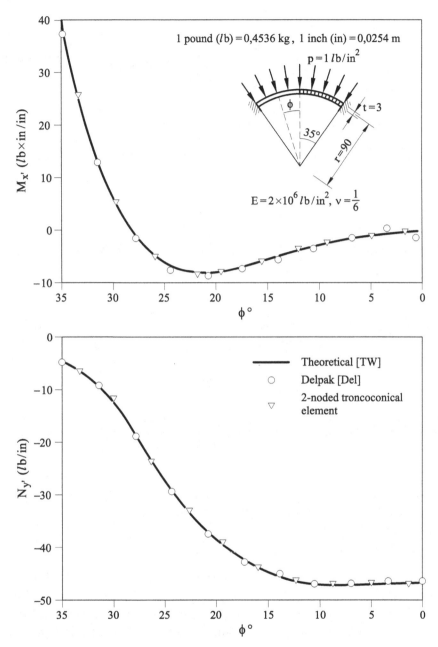

Fig. 9.16 Spherical dome under uniform pressure. Diagrams of $M_{x'}$ and $N_{y'}$ obtained with a mesh of ten 2-noded Reissner-Mindlin troncoconical elements

A mesh of thirty nine 2-noded troncoconical elements has been used as shown in Figure 9.18. The axial forces and bending moments diagrams

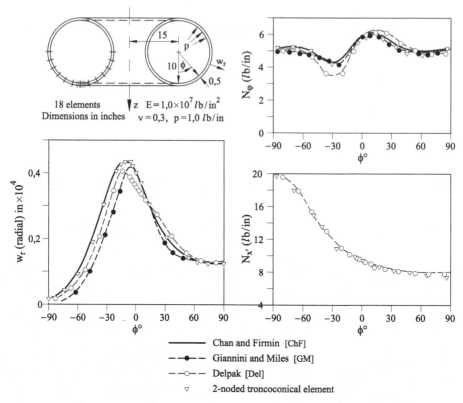

Fig. 9.17 Toroidal shell under internal pressure. Diagrams of w_r, $N_{x'}$ and $N_{y'}$ using eighteen 2-noded Reissner-Mindlin troncoconical shell elements

are displayed in the figure. The reader can verify that the distribution of circumferential stresses $\sigma_{y'}$ shown in Figure 7.9 of [On4] is obtained from the diagramas of $N_{y'}$ and $M_{y'}$ computed in this example.

9.7.4 Elevated water tank

The last example is the analysis of the elevated water tank shown in Figure 9.19. The tank is supported by a central cylindrical thin wall and lateral columns. The effect of the discrete columns has been modeled by an equivalent cylindrical wall of 2 mm thickness. The tank is loaded by the weight of internal water as shown in the figure. Two different analysis with meshes of forty and eighty 2-noded troncoconical elements for the discretization of the tank and the central cylinder were performed. The side wall and the lateral cylindrical wall were discretized in both cases with one and ten elements, respectively.

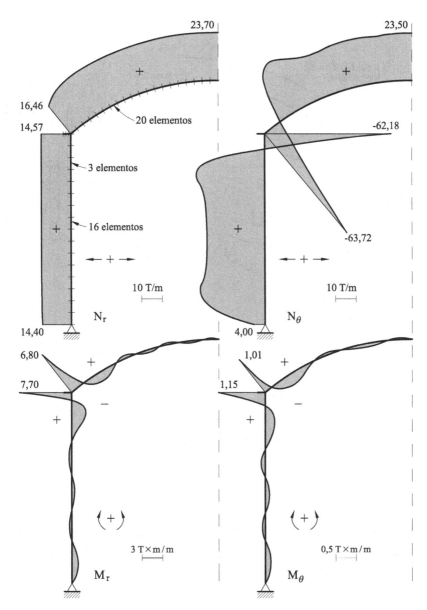

Fig. 9.18 Cylindrical tank with spherical dome analyzed with 39 2-noded Reissner-Mindlin troncoconical elements. Diagrams of axial force $N_{y'}$ and bending moment $M_{y'}$. Details of the geometry are shown in Figure 7.9 of [On4]

Figures 9.19a and b show the deformed shape of the tank and the diagrams of axial forces $N_{x'}$ and $N_{y'}$ for the two meshes. Note the coincidence of the results which indicates the accuracy of the solution. The

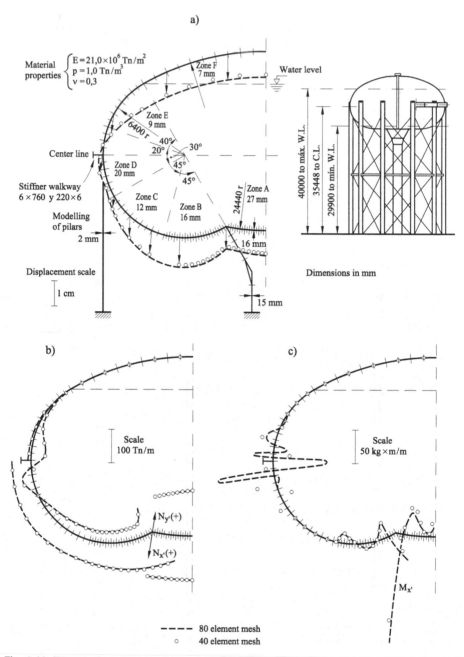

Fig. 9.19 Elevated water tank analyzed with two meshes of 40 and 80 2-noded Reissner-Mindlin troncoconical elements. (a) Geometry and deformed shape. (b) Diagrams of axial forces $N_{x'}$ and $N_{y'}$. (c) Diagram of bending moment $M_{x'}$

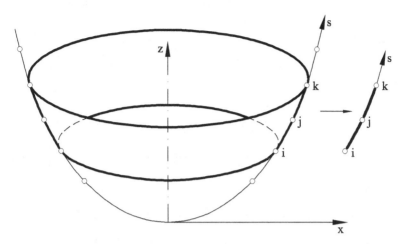

Fig. 9.20 Discretization of axisymmetric shell using 3-noded curved elements

radial bending moment $M_{x'}$ diagram is plotted in Figure 9.19c. This moment takes large values in the vicinity of the lower folded part and at the joints with the cylindrical supports. The difference in the local bending moments obtained with the two meshes indicates that finer refirement is required in these zones. Details of this example can be found in [ZBMO].

9.8 CURVED AXISYMMETRIC SHELL ELEMENTS OF THE REISSNER-MINDLIN FAMILY

Curved axisymmetric elements following Reissner-Mindlin theory can be derived starting from the generalized strains and resultant stresses of Eqs.(9.27), (9.28) and (9.34) including the curvature terms. These terms have also an influence on the virtual work of the nodal forces (Eq.(9.46)).

Figure 9.20 shows the discretization of the generatrix using curved elements. The lower member of the C° continuous family is the 3-noded quadratic axisymmetric shell element.

9.8.1 Displacement and load generalized strain fields

The local displacement field is expressed by Eq.(9.48) as for troncoconical elements. The element geometry is expressed in isoparametric form as

$$x = \sum_{i=1}^{n} N_i(\xi)x_i, \quad z = \sum_{i=1}^{n} N_i(\xi)z_i \tag{9.75}$$

where n is the number of element nodes.

The standard relationship between local generalized strains and local displacements (Eq.(9.52)) is obtained by substituting Eq.(9.48) into (9.27) and (9.28), giving the generalized strain matrix as

$$
\mathbf{B}'_i = \left\{ \begin{array}{c} \mathbf{B}'_m \\ -\,-\,- \\ \mathbf{B}'_b \\ -\,-\,- \\ \mathbf{B}'_s \end{array} \right\} = \begin{bmatrix} \dfrac{\partial N_i}{\partial s} & \boxed{\dfrac{-N_i}{R_s}} & 0 \\[2ex] \dfrac{N_i\cos\phi}{r} & \dfrac{-N_i\sin\phi}{r} & 0 \\[2ex] \hline 0 & 0 & \dfrac{-\partial N_i}{\partial s} \\[1ex] 0 & 0 & \dfrac{-N_i\cos\phi}{r} \\[2ex] \hline \boxed{\dfrac{N_i}{R_s}} & \dfrac{\partial N_i}{\partial s} & -N_i \end{bmatrix} \tag{9.76}
$$

where the terms contributed by the shell curvature have been framed.

9.8.2 Computation of curvilinear derivatives and curvature radius

The isoparametric expression (9.75) yields

$$
\frac{\partial x}{\partial \xi} = \sum_{i=1}^{n} \frac{dN_i}{d\xi} x_i; \qquad \frac{d^2 x}{d\xi^2} = \sum_{i=1}^{n} \frac{d^2 N_i}{d\xi^2} x_i
$$

$$
\frac{dz}{d\xi} = \sum_{i=1}^{n} \frac{dN_i}{d\xi} z_i; \qquad \frac{d^2 z}{d\xi^2} = \sum_{i=1}^{n} \frac{d^2 N_i}{d\xi^2} z_i \tag{9.77}
$$

Angle ϕ defining the tangent direction at each point of the generatrix (Figure 9.21) is obtained from the expression

$$
\tan \phi = \frac{dz}{dx} = \frac{dz/d\xi}{dx/d\xi} = \sum_{i=1}^{n} \frac{\dfrac{dN_i}{d\xi} z_i}{\dfrac{dN_i}{d\xi} x_i} \tag{9.78}
$$

The curvature radius R_α is computed as (Figure 9.21 and Eq.(9.2))

$$
R_\alpha = \frac{r}{\sin \phi} \tag{9.79}
$$

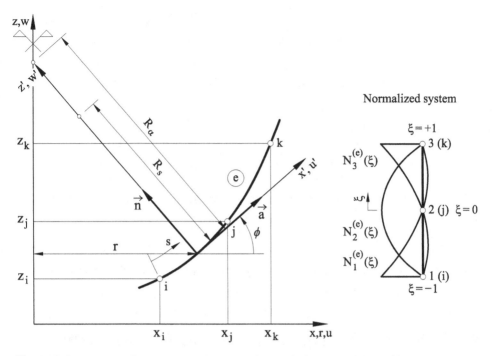

Fig. 9.21 Geometric description of a curved 3-noded Reissner-Mindlin axisymmetric shell element

The curvature radius R_s is computed as (Eq.(9.2))

$$\frac{1}{R_s} = \frac{d\phi}{ds} = \frac{d\phi/d\xi}{ds/d\xi} \tag{9.80}$$

From Eq.(9.77) we have

$$\frac{ds}{d\xi} = \frac{\sqrt{dx^2 + dz^2}}{d\xi} = \left[\left(\frac{dx}{d\xi}\right)^2 + \left(\frac{dz}{d\xi}\right)^2\right]^{1/2} =$$
$$= \sum_{i=1}^{n} \left[\left(\frac{dN_i}{d\xi}x_i\right)^2 + \left(\frac{dN_i}{d\xi}y_i\right)^2\right]^{1/2} = J^{(e)} \tag{9.81}$$

Substituting Eq.(9.81) and the derivative of ϕ from Eq.(9.80) into (9.79) gives finally

$$\frac{1}{R_s} = \frac{\frac{d^2z}{d\xi^2}\frac{dx}{d\xi} - \frac{dz}{d\xi}\frac{d^2x}{d\xi^2}}{\left[\left(\frac{dx}{d\xi}\right)^2 + \left(\frac{dz}{d\xi}\right)^2\right]^{3/2}} = \sum_{i=1}^{n} \frac{\left(\frac{d^2N_i}{d\xi^2}z_i\right)\left(\frac{dN_i}{d\xi}x_i\right) - \left(\frac{dN_i}{d\xi}z_i\right)\left(\frac{d^2N_i}{d\xi^2}x_i\right)}{\left[\left(\frac{dN_i}{d\xi}x_i\right)^2 + \left(\frac{dN_i}{d\xi}z_i\right)^2\right]^{3/2}} \tag{9.82}$$

The curvilinear derivative of the shape function N_i is computed by

$$\frac{dN_i}{ds} = \frac{\dfrac{dN_i}{d\xi}}{\dfrac{ds}{d\xi}} = \frac{1}{J^{(e)}} \frac{dN_i}{d\xi} \tag{9.83}$$

The global stiffness matrix and the load vector are computed by expressions identical to Eqs.(9.64) and (9.66) with $J^{(e)}$ given by Eq.(9.81). The transformation matrix $\mathbf{L}_i^{(e)}$ of Eq.(9.59) is now different for each node and is computed by Eq.(9.14) with ϕ deduced from Eq.(9.78). The effect of the curvature makes it difficult to derive simple explicit forms for the element matrices and numerical integration is needed.

Locking in curved axisymmetric Reissner–Mindlin elements can be avoided by similar techniques to those explained for troncoconical elements. The simplest remedy for the quadratic 3-noded element is the uniform two-point reduced quadrature.

9.9 AXISYMMETRIC THIN SHELL ELEMENTS BASED ON KIRCHHOFF ASSUMPTIONS

9.9.1 Introduction

Most analytical solutions for axisymmetric shells are based on Kirchhoff assumption for the orthogonality of the normal rotation [Kr,TW,WOK]. This hypothesis, though only acceptable for thin shell situations, can be applied to many problems of practical interest and analytical solutions are available for cylindrical reservoirs, spherical and conical domes, circular plates etc. [TW].

The early applications of the FEM to axisymmetric shells were also based on Kirchhoff theory [AG,Ga,Go,GS,Jor,JS,JS2,ZT2] and many elements of this kind are found in commercial FE codes. This alone justifies the study of Kirchhoff axisymmetric thin shell elements which have historic, didactic and practical interest. In addition, they are an excellent introduction to the analysis of thin circular plates and slender arches.

Kirchhoff assumptions are also the starting point for deriving a family of rotation-free thin troncoconical element following a similar approach as for rotation-free beam elements in Chapter 1, as shown in a next section.

9.9.2 Basic formulation

The key difference between Kirchhoff and Reissner-Mindlin theories is the assumption made for the rotation of the normal. Kirchhoff theory establishes that, as the thickness is small, the normals to the generatrix remain straight and orthogonal to the generatrix after deformation. Hence, the normal rotation coincides with the slope of the generatrix at each point.

In mathematical form we can write

$$\theta = \frac{\partial w'}{\partial s}\Big|_{z'=0} \tag{9.84}$$

Substituting the expression for $\frac{\partial w'}{\partial s}$ of Eq.(9.20) into (9.84) gives

$$\theta = \frac{\partial w'_0}{\partial s} + \frac{u'_0}{R_s} \tag{9.85}$$

Introducing this equation into (9.24) yields

$$\gamma_{x'z'} = \frac{1}{C_s}\left(\frac{\partial w'_0}{\partial s} + \frac{u'_0}{R_s} - \left(\frac{\partial w'_0}{\partial s} + \frac{u'_0}{R_s}\right)\right) = 0 \tag{9.86}$$

i.e. the Kirchhoff orthogonality condition is equivalent to neglecting the effect of transverse shear deformation, as expected.

The local displacement vector is now defined as

$$\mathbf{u}' = \left[u'_0, w'_0, \frac{\partial w'_0}{\partial s}\right]^T \tag{9.87}$$

The expressions for the axial and circumferential strains are deduced by substituting Eq.(9.85) into Eqs.(9.22) and (9.23) to give

$$\begin{aligned}
\varepsilon_{x'} &= \frac{1}{C_s}\left[\frac{\partial u'_0}{\partial s} - \frac{w'_0}{R_s} - z'\left(\frac{\partial^2 w'_0}{\partial s^2} + \frac{\partial}{\partial s}\left(\frac{u'_0}{R_s}\right)\right)\right] \\
\varepsilon_{y'} &= \frac{1}{C_\alpha}\left[\frac{u'_0\cos\phi - w'_0\sin\phi}{r} - \frac{z'\cos\phi}{r}\left(\frac{\partial w'_0}{\partial s} + \frac{u'_0}{R_s}\right)\right]
\end{aligned} \tag{9.88}$$

The generalized strain vector is

$$\hat{\boldsymbol{\varepsilon}}' = \mathbf{S}_2\left\{\begin{matrix} \hat{\varepsilon}'_m \\ \hat{\varepsilon}'_b \end{matrix}\right\} \quad ; \quad \mathbf{S}_2 = [\mathbf{S}, -z'\mathbf{S}] \tag{9.89}$$

where $\mathbf{S}$ is given by Eq.(9.26) and the membrane and bending generalized strains are

$$
\hat{\boldsymbol{\varepsilon}}'_m = \left\{ \begin{array}{c} \dfrac{\partial u'_0}{\partial s} - \dfrac{w'_0}{R_s} \\[2mm] \dfrac{u'_0\cos\phi - w'_0\sin\phi}{r} \end{array} \right\} ; \quad \hat{\boldsymbol{\varepsilon}}'_b = \left\{ \begin{array}{c} \dfrac{\partial^2 w'_0}{\partial s^2} + \dfrac{\partial}{\partial s}\left(\dfrac{u'_0}{R_s}\right) \\[2mm] \dfrac{\cos\phi}{r}\left(\dfrac{\partial w'_0}{\partial s} + \dfrac{u'_0}{R_s}\right) \end{array} \right\} \tag{9.90}
$$

The local resultant stresses $\boldsymbol{\sigma}'_m$ and $\boldsymbol{\sigma}'_b$ are given by Eq.(9.34) after substitution of $\mathbf{S}_1$ by $\mathbf{S}_2$. The transverse shear forces do not contribute to the internal work and must be computed "a posteriori" from the equilibrium equation, as for thin plates (Section 4.2.6)). The PVW is obtained by neglecting the transverse shear terms in Eq.(9.46). The internal virtual work now contains second derivatives of the normal displacement w'_0. Hence, C^1 continuity is needed for the approximation of w_0.

9.9.3 Troncoconical shell elements based on Kirchhoff theory

Troncoconical elements have $R_s = \infty$ and the generalized strain vectors of Eq.(9.90) simplify to

$$
\hat{\boldsymbol{\varepsilon}}'_m = \left\{ \begin{array}{c} \dfrac{\partial u'_0}{\partial s} \\[2mm] \dfrac{u'_0\cos\phi - w'_0\sin\phi}{r} \end{array} \right\} ; \quad \hat{\boldsymbol{\varepsilon}}'_b = \left\{ \begin{array}{c} \dfrac{\partial^2 w'_0}{\partial s^2} \\[2mm] \dfrac{\cos\phi}{r}\dfrac{\partial w'_0}{\partial s} \end{array} \right\} \tag{9.91}
$$

As mentioned above, a C^1 continuous interpolation must be used for the normal displacement w'_0 to satisfy element conformity. A simpler C° continuous Lagrange approximation can however be employed for the tangential displacement u'_0.

Also as the element is straight $\partial w'_0/\partial s = \partial w'_0/\partial x'$ and $\partial^2 w'_0/\partial s^2 = \partial^2 w'_0/\partial x'^2$.

9.9.3.1 Two-noded Kirchhoff troncoconical element

The simplest troncoconical element based on Kirchhoff theory has two nodes (Figure 9.22). The tangential displacement is linearly interpolated as

$$
u'_0 = \sum_{i=1}^{2} N_i^u\, u'_{0_i} \quad \text{with} \quad N_i^u = \frac{1 + \xi\xi_i}{2} \tag{9.92}
$$

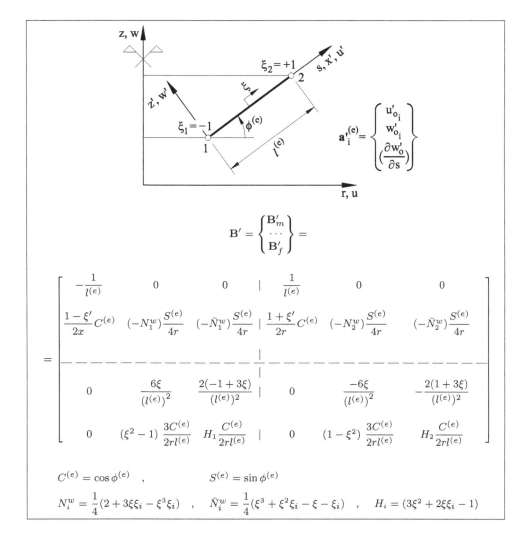

$C^{(e)} = \cos \phi^{(e)}$, $S^{(e)} = \sin \phi^{(e)}$

$N_i^w = \frac{1}{4}(2 + 3\xi\xi_i - \xi^3\xi_i)$, $\bar{N}_i^w = \frac{1}{4}(\xi^3 + \xi^2\xi_i - \xi - \xi_i)$, $H_i = (3\xi^2 + 2\xi\xi_i - 1)$

Fig. 9.22 Two-noded Kirchhoff troncoconical element. Local strain matrix

The following C^1 continuous approximation is chosen for w_0'

$$w_0' = \sum_{i=1}^{2}\left[N_i^w \, w_{o_i}' + \bar{N}_i^w \left(\frac{\partial w_0'}{\partial s}\right)_i\right] \tag{9.93}$$

where N_i^w and $\bar{N}_i^w$ are the cubic 1D Hermite shape functions (Eq.(1.11)).
The local generalized strain matrix is written as

$$\hat{\varepsilon}' = [\mathbf{B}_1', \mathbf{B}_2']\mathbf{a}'^{(e)} = \mathbf{B}' \, \mathbf{a}'^{(e)} \tag{9.94}$$

with

$$
\mathbf{B}'_i = \left\{ \frac{\mathbf{B}'_{m_i}}{\mathbf{B}'_{b_i}} \right\} =
\begin{bmatrix}
\dfrac{\partial N_i^u}{\partial s} & 0 & 0 \\[2ex]
\dfrac{N_i^u \cos\phi}{r} & \dfrac{-N_i^w \sin\phi}{r} & \dfrac{-\bar{N}_i^w \sin\phi}{r} \\[2ex]
\hline
0 & \dfrac{\partial^2 N_i^w}{\partial s^2} & \dfrac{\partial^2 \bar{N}_i^w}{\partial s^2} \\[2ex]
0 & \dfrac{\cos\phi}{r}\dfrac{\partial N_i^w}{\partial s} & \dfrac{\cos\phi}{r}\dfrac{\partial \bar{N}_i^w}{\partial s}
\end{bmatrix}
\tag{9.95}
$$

The explicit form for $\mathbf{B}'$ is shown in Figure 9.22.

The local stiffness matrix is obtained by eliminating the contribution of the transverse shear stiffness terms in Eq.(9.57) giving

$$
\mathbf{K}_{ij}^{\prime(e)} = \mathbf{K}_{m_{ij}}^{\prime(e)} + \mathbf{K}_{b_{ij}}^{\prime(e)} + \mathbf{K}_{mb_{ij}}^{\prime(e)} + [\mathbf{K}_{mb_{ij}}^{\prime(e)}]^T
\tag{9.96}
$$

where all matrices have identical expressions to those given in Eq.(9.57). A two-point quadrature is recommended for computing the integrals containing rational terms. A more accurate expression for the stiffness matrix using a seven-point quadrature can be found in [Kl]. Good results are obtained however with the simplest reduced one-point quadrature. This is equivalent to making $\xi = 0$ and $r = r_m$ in $\mathbf{B}'$ of Figure 9.22. A finer mesh is then needed in zones where the stress gradients are high. Details and examples are given in [GS,ZT2].

The stiffness transformation to global axes follows the rules of Section 9.4.3. The Hermite approximation for w_0' introduces bending moments in the equivalent nodal force vector, like for Euler-Bernoulli beam elements (Section 1.2.2).

9.9.3.2 Curved Kirchhoff axisymmetric shell elements

Much work has been done since the early 1960's on the derivation of curved axisymmetric shell elements based on Kirchhoff theory [AG,As,ASR,Bat2, BD6,CC,Da3,Del,DG,Ga,Ga2,JS,JS2,SNP,ZT2]. The main difficulty is finding compatible approximations for the geometry and the displacement field so that C^1 continuity is preserved. Delpak [Del] proposed a Hermite approximation for the coordinates and the displacements in global axes for 2-noded elements as

$$
\mathbf{x} = \left\{ \begin{array}{c} x \\ z \end{array} \right\} = \sum_{i=1}^{2} \left[N_i \, \mathbf{x}_i + \bar{N}_i \left(\frac{\partial \mathbf{x}}{\partial s} \right)_i \right]
$$

$$\mathbf{u} = \left\{ \begin{matrix} u \\ w \end{matrix} \right\} = \sum_{i=1}^{2} \left[N_i \, \mathbf{u}_i + \bar{N}_i \Big(\frac{\partial \mathbf{u}}{\partial s} \Big)_i \right] \qquad (9.97)$$

where N_i and $\bar{N}_i$ are the cubic Hermite polynomials of Eq.(1.11). Both displacement components now vary as a cubic. This introduces an additional nodal variable $\big(\frac{\partial u}{\partial s} \big)_i$ which contributes to the continuity of the slope of the generatrix. A difficulty arises for obtaining the local strain matrix where the derivatives of the local displacements are needed and the curvature radius varies from point to point. This element is described in [Del,ZT2].

Curved Kirchhoff axisymmetric shell elements typically suffer from membrane locking. The reason is that the displacement approximation can not usually represent a zero membrane strain state without polluting the bending approximation. Elimination of membrane locking requires compatible (higher order) approximations for the tangential and normal displacements, or reduced integration as explained in Section 9.5.2.

Different locking–free curved axisymmetric thin shell elements have been reported in [BD6,Cr,Cr4,Del,SNP,ZT2].

9.10 AXISYMMETRIC MEMBRANE ELEMENTS

A membrane analysis can be applied in cases when bending and transverse shear effects are negligible, or if these effects are concentrated in zones near the support and/or point loads. Examples of pure axisymmetric membrane situations are found in water bags, thin tubes under internal pressure and inflatable structures such as balloons, inflatable pavilions, air-supported domes and air-beams (see examples in www.buildair.com). The differential equations for an axisymmetric membrane are simple and many analytical solutions are available. Typical examples are the analysis of hydrostatic tanks under water load or spherical domes under self-weight where the amount of reinforcing steel and the prestressing can be estimated from the membrane solution. A full bending analysis is typically performed at the end of the design process for verification purposes and for defining the lay-out of steel reinforcement near the supports.

The formulation of axisymmetric membrane elements is straightforward as they are a particular case of the general axisymmetric shell formulation. The displacement field is expressed in terms of the two local displacements as

$$\mathbf{u}_0' = [u_0', \ w_0'] \qquad (9.98)$$

The local strain vector is given by the elongation vector $\hat{\boldsymbol{\varepsilon}}'_m$ of Eq.(9.27a) while the axial force–elongation relationship is deduced from Eq.(9.34). The PVW coincides with Eq.(9.46), neglecting the bending and shear contributions.

The displacement field is discretized using a standard C^0 continuous approximation as

$$\mathbf{u}' = \sum_{i=1}^{n} N_i \mathbf{u}'_i \qquad (9.99)$$

The local membrane strain matrix is deduced from Eq.(9.76) as

$$\mathbf{B}'_{m_i} = \begin{bmatrix} \dfrac{\partial N_i}{\partial s} & -\dfrac{N_i}{R_s} \\[2mm] \dfrac{N_i \cos\phi}{r} & \dfrac{-N_i \sin\phi}{r} \end{bmatrix} \qquad (9.100)$$

The curvature term $-\dfrac{N_i}{R_s}$ is zero for troncoconical membrane elements. The local stiffness matrix contains membrane contributions only, i.e.

$$\mathbf{K}'^{(e)}_{ij} \equiv \mathbf{K}'^{(e)}_{m_{ij}} = 2\pi \int_{l^{(e)}} \mathbf{B}'^{T}_{m_i} \hat{\mathbf{D}}'_m \mathbf{B}'_{m_j} \, r \, d s \qquad (9.101)$$

The transformation to global axes follows the procedure explained in Section 9.4.3. The global stiffness matrix is given by

$$\mathbf{K}^{(e)}_{m_{ij}} = 2\pi \int_{l^{(e)}} \mathbf{B}^{T}_{m_i} \hat{\mathbf{D}}'_m \mathbf{B}_{m_j} \, r \, d s \qquad (9.102a)$$

where

$$\mathbf{B}_{m_i} = \mathbf{B}'_{m_i} \hat{\mathbf{L}}_i ; \quad \hat{\mathbf{L}}_i = \begin{bmatrix} \cos\phi & \sin\phi \\ -\sin\phi & \cos\phi \end{bmatrix} \qquad (9.102b)$$

The stiffness matrix is computed using numerical integration. Three- and two-point quadratures are typically used for the quadratic (3-noded) and linear (2-noded) axisymmetric membrane elements, respectively.

Note that an axisymmetric membrane can only analyzed with the present formulation if subject to tension axial forces only.

Troncoconical membrane elements may suffer from numerical instabilities due to undesirable flexural loads arising from discretization errors. The solution to this problem, also discussed in Section 8.14 for flat membrane elements, is to introduce a fictitious bending stiffness. This, however, increases in one rotation the nodal DOFs.

General purpose axisymmetric shell elements preserving two displacement DOFs per node and applicable to thin axisymmetric membrane and shell structures can be derived via the rotation-free formulation described in the following section.

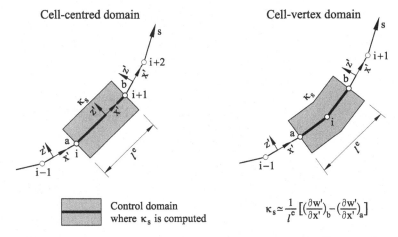

Fig. 9.23 Computation of average meridional curvature in cell-centred and cell-vertex troncoconical domains

9.11 ROTATION-FREE AXISYMMETRIC SHELL ELEMENTS

Rotation-free axisymmetric thin (Kirchhoff) shell elements can be derived as explained for rotation-free beam elements in Section 1.4. The radial (constant) curvature over a control domain is expressed in terms of the displacements at selected points in the vicinity of the control domain. For simplicity we will consider troncoconical elements only in the following.

The average meridional curvature over a control domain in a thin troncoconical shell can be estimated as (in the following we will take $s \equiv x'$ and skip index 0 in u'_0 and w'_0, for simplicity)

$$\kappa_s = \left(\frac{\partial^2 w'}{\partial x'^2} \right)^e \simeq \frac{1}{l^e} \left[\left(\frac{\partial w'}{\partial x'} \right)_b - \left(\frac{\partial w'}{\partial x'} \right)_a \right] \qquad (9.103)$$

where l^e is the length of the ith control domain and $(\cdot)_j$, $j = a, b$ denotes values computed at the end points of the domain (Figure 9.23).

Eq.(9.103) is analogous to Eq.(1.28) defining the curvature in a beam segment from the difference of the slopes at the segment ends.

Eq.(9.103) yields the curvature $\frac{\partial^2 w'}{\partial x'^2}$ in terms of the derivative $\frac{\partial w'}{\partial x'}$ at the end points of the control domain. If a C^0 interpolation is chosen for w', then $\frac{\partial w'}{\partial x'}$ is not continuous at the element ends and this poses a problem if the control domain coincides with an element. This difficulty is overcome by computing $\frac{\partial w'}{\partial x'}$ at a node using the average value of the derivatives contributed by the two elements meeting at the node.

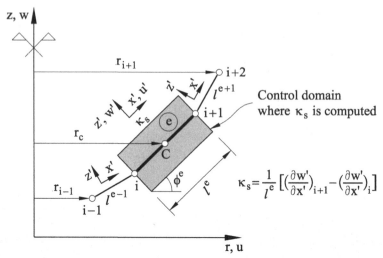

Fig. 9.24 Cell-centred rotation-free troncoconical element (ACC). Computation of average meridional curvature

Two options for choosing the control domain leading to two different rotation-free troncoconical elements are described next.

9.11.1 Cell-centred rotation-free troncoconical element (ACC)

9.11.1.1 ACC element matrices

The control domain coincides with a typical 2-noded C° troncoconical element (Figure 9.24). A linear interpolation for the two local displacements is used. Application of Eq.(9.103) gives

$$
\kappa_s = \left(\frac{\partial^2 w'}{\partial x'^2}\right)^e = \frac{1}{l^e}\left[\left(\frac{\partial w'}{\partial x'}\right)_{i+1} - \left(\frac{\partial w'}{\partial x'}\right)_i\right] =
$$

$$
= \frac{1}{l^e}\left[\frac{1}{2}\left(\frac{w'^{e+1}_{i+2} - w'^{e+1}_{i+1}}{l^{e+1}} + \frac{w'^e_{i+1} - w'^e_i}{l^e}\right) - \frac{1}{2}\left(\frac{w'^e_{i+1} - w'^e_i}{l^e} + \frac{w'^{e-1}_i - w'^{e-1}_{i-1}}{l^{e-1}}\right)\right]
$$

$$
= \frac{1}{2l^e l^{e-1} l^{e+1}}[l^{e+1}, -l^{e+1}, -l^{e-1}, l^{e-1}]\begin{Bmatrix} w'^{e-1}_{i-1} \\ w'^{e-1}_i \\ w'^{e+1}_{i+1} \\ w'^{e+1}_{i+2} \end{Bmatrix} = \mathbf{B}_{b_1}\mathbf{w}'^e \qquad (9.104a)
$$

where

$$
\mathbf{B}'_{b_1} = \frac{1}{2l^e l^{e-1} l^{e+1}}\left[l^{e+1}, -l^{e+1}, -l^{e-1}, l^{e-1}\right] \ , \ \mathbf{w}'^e = \left[w'^{e-1}_{i-1}, w'^{e-1}_i, w'^{e+1}_{i+1}, w'^{e+1}_{i+2}\right]^T
$$

$$
(9.104b)
$$

The transformation of the deflection $w_i'^e$ to global axes is written as

$$w_i'^e = -u_i \sin \phi^e + w_i \cos \phi^e \tag{9.105}$$

where u_i, w_i are the global displacements of node i and ϕ^e is the angle between the x' and x axis for element e (Figure 9.24).

Transformation of the r.h.s. of Eq.(9.104a) to global axes gives

$$\kappa_s = \mathbf{B}_{b_1} \mathbf{a}^e \quad \text{with} \quad \mathbf{B}_{b_1} = \mathbf{B}'_{b_1} \mathbf{T} \tag{9.106}$$

where

$$\mathbf{T} = \begin{bmatrix} \mathbf{n}^{e-1} & \mathbf{0} & \mathbf{0} & \mathbf{0} \\ \mathbf{0} & \mathbf{n}^{e-1} & \mathbf{0} & \mathbf{0} \\ \mathbf{0} & \mathbf{0} & \mathbf{n}^{e+1} & \mathbf{0} \\ \mathbf{0} & \mathbf{0} & \mathbf{0} & \mathbf{n}^{e+1} \end{bmatrix}, \quad \mathbf{n}^e = [-\sin \phi^e, \cos \phi^e], \quad \mathbf{0} = [0, 0]$$

$$\tag{9.107a}$$

and

$$\mathbf{a}^e = \begin{Bmatrix} \mathbf{a}_{i-1} \\ \mathbf{a}_i \\ \mathbf{a}_{i+1} \\ \mathbf{a}_{i+2} \end{Bmatrix} \quad \text{with} \quad \mathbf{a}_i = \begin{Bmatrix} u_i \\ w_i \end{Bmatrix} \tag{9.107b}$$

The (constant) circumferential curvature for element e is computed as

$$\kappa_\theta = \frac{\cos \phi^e}{r} \frac{\partial w'}{\partial x'} = \frac{\cos \phi^e}{rl^e}[-1, 1] \begin{Bmatrix} w_i'^e \\ w_{i+1}'^e \end{Bmatrix} =$$

$$= \underbrace{\frac{\cos \phi^e}{rl^e}[\mathbf{0}, -\mathbf{n}^e, \mathbf{n}^e, \mathbf{0}]}_{\mathbf{B}_{b_2}} \mathbf{a}^e = \mathbf{B}_{b_2} \mathbf{a}^e \tag{9.108}$$

with $\mathbf{n}^e$ and $\mathbf{0}$ hereonwards defined as in Eq.(9.107a).

The meridional membrane strain is expressed as

$$\lambda_s = \frac{\partial u'}{\partial x'} = \frac{1}{l^e}[-1, 1] \begin{Bmatrix} u_i'^e \\ u_{i+1}'^e \end{Bmatrix} = \underbrace{\frac{1}{l^e}[\mathbf{0}, -\mathbf{t}^e, \mathbf{t}^e, \mathbf{0}]}_{\mathbf{B}_{m_1}} \mathbf{a}^e = \mathbf{B}_{m_1} \mathbf{a}^e \tag{9.109}$$

with $\mathbf{t}^e = [\cos \phi^e, \sin \phi^e]$.

Finally, the circumferential membrane strain is computed as

$$\lambda_\theta = \frac{u}{r} = \frac{u_i + u_{i+1}}{2r_c} = \underbrace{\frac{1}{2r_c}[0, 0, 1, 0, 1, 0, 0, 0]}_{\mathbf{B}_{m_2}} \mathbf{a}^e = \mathbf{B}_{m_2} \mathbf{a}^e \tag{9.110}$$

where r_c is the radius of the element mid-point, $r_c = \frac{r_i + r_{i+1}}{2}$ (Figure 9.24).
The generalized strain matrix is expressed as

$$
\mathbf{B} = \left\{ \begin{matrix} \mathbf{B}_m \\ \cdots \\ \mathbf{B}_b \end{matrix} \right\} = \left\{ \begin{matrix} \mathbf{B}_{m_1} \\ \mathbf{B}_{m_2} \\ \cdots \\ \mathbf{B}_{b_1} \\ \mathbf{B}_{b_2} \end{matrix} \right\} \tag{9.111}
$$

The global element stiffness matrix is computed as

$$
\mathbf{K}^{(e)} = \mathbf{K}_m^{(e)} + \mathbf{K}_b^{(e)} + \mathbf{K}_{mb}^{(e)} + [\mathbf{K}_{mb_{ij}}^{(e)}]^T \tag{9.112a}
$$

where

$$
\mathbf{K}_\alpha^{(e)} = 2\pi \mathbf{B}_\alpha^T \hat{\mathbf{D}}_\alpha' \mathbf{B}_\alpha r_c l^{(e)} \quad , \quad \alpha = m, b, mb \tag{9.112b}
$$

The size of $\mathbf{K}^{(e)}$ is 8×8, as it involves the eight displacements of nodes $i-1, i, i+1$ and $i+2$ (two displacements per node).

The expression for the equivalent nodal force vector for a uniformly distributed loading is simply

$$
\mathbf{f}_i^{(e)} = \frac{l^e r_c}{2} \mathbf{t} \quad \text{with } \mathbf{t} = \left\{ \begin{matrix} t_x \\ t_y \end{matrix} \right\} \tag{9.113}
$$

Nodal point loads are directly assembled into the global expression of $\mathbf{f}$ as usual.

9.11.1.2 Boundary conditions for the ACC element

The conditions on prescribed nodal displacements are imposed in the standard manner when solving the global equations system $\mathbf{Ka} = \mathbf{f}$. The conditions on simply supported (SS), clamped or symmetry nodes involving the nodal rotation are implemented when building up the generalized strain matrix $\mathbf{B}_{b_1}$, as it is usual in rotation-free elements. The form of matrix $\mathbf{B}_{m_1}'$ for each boundary condition is given next.

Free or SS node

Let us consider a SS or free node i placed at the left-hand end of a mesh (Figure 9.25a). The curvature at the prescribed node is zero. Hence the

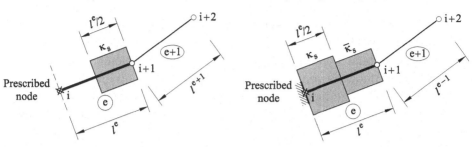

Fig. 9.25 Cell-centred rotation-free troncoconical element (ACC). Control domains (in shadow) in elements with boundary node. (a) SS or free end node. (b) Clamped or symmetry node

average curvature at the element adjacent to the prescribed node is computed as

$$\kappa_s = \frac{2}{l^e}\left[\left(\frac{\partial w'}{\partial x'}\right)_{i+1} - \left(\frac{\partial w'}{\partial x'}\right)_i\right] \tag{9.114}$$

with $\left(\frac{\partial w'}{\partial x'}\right)_{i+1}$ computed as in Eq.(9.104a) and $\left(\frac{\partial w'}{\partial x'}\right)_i = \frac{w'^e_{i+1}-w'^e_i}{l^e}$.

Substituting these expressions for $\frac{\partial w'}{\partial x'}$ into Eq.(9.114) gives

$$\kappa_s = \underbrace{\frac{1}{(l^e)^2 l^{e+1}}[l^{e+1}, -l^{e+1}, -l^e, l^e]}_{\mathbf{B}'_{b_1}}\begin{Bmatrix} w'^e_i \\ w'^e_{i+1} \\ w'^e_{i+1} \\ w'^e_{i+2} \end{Bmatrix} = \mathbf{B}'_{b_1}\mathbf{w}'^e = \mathbf{B}_{b_1}\mathbf{a}^e \tag{9.115a}$$

where vector $\mathbf{a}^e$ has the form of Eq.(9.107b) and $\mathbf{B}_{b_1}$ is obtained as

$$\mathbf{B}_{b_1} = \mathbf{B}'_{b_1}\mathbf{T} \quad \text{with} \quad \mathbf{T} = \begin{bmatrix} 0 & \mathbf{n}^e & 0 & 0 \\ 0 & 0 & \mathbf{n}^e & 0 \\ 0 & 0 & \mathbf{n}^{e+1} & 0 \\ 0 & 0 & 0 & \mathbf{n}^{e+1} \end{bmatrix} \tag{9.115b}$$

The terms of $\mathbf{K}_b^{(e)}$ involving $\mathbf{K}_{b_1}^{(e)}$ are computed over half the element length $\left(\frac{l^e}{2}\right)$ in Eq.(8.112b).

For a SS node the global vertical displacement w_i is prescribed to zero. Note finally that $\mathbf{a}_{i-1}$ are "ghost" displacements in vector $\mathbf{a}^e$ as node $i-1$ has been introduced simply for preserving the dimensions of matrices $\mathbf{B}_{b_1}$ and $\mathbf{K}^{(e)}$. Consequently $\mathbf{a}_{i-1}$ is prescribed to zero when solving the system of equations.

A right-hand SS or free node is treated in a similar manner [JO].

Clamped or symmetry node

Let us consider a clamped or symmetry node i placed at the left-hand end of a mesh where the rotation is prescribed to a zero value (Figure 9.25b)).

The meridional bending stiffness matrix for the element adjacent to the prescribed node is computed as the sum of the contributions from the meridional curvature fields κ_s and $\bar{\kappa}_s$ shown in Figure 9.25b.

The meridional curvature at the element adjacent to the prescribed node is computed as

$$\kappa_s = \frac{2}{l^e}\left[\frac{w'^e_{i+1} - w'^e_i}{l^e}\right] = \underbrace{\frac{2}{(l^e)^2}[-1, 1]}_{\mathbf{B}'_{b_1}}\left\{\begin{array}{c} w'^e_i \\ w'^e_{i+1} \end{array}\right\} = \mathbf{B}'_{b_1}\mathbf{w}'^{(e)} = \mathbf{B}_{b_1}\mathbf{a}^e$$

(9.116)

with $\mathbf{a}^e$ is given by Eq.(9.107b) and

$$\mathbf{B}_{b_1} = \mathbf{B}'_{b_1}\mathbf{T} \quad \text{with} \quad \mathbf{T} = \begin{bmatrix} 0 & n^e & 0 & 0 \\ 0 & 0 & n^e & 0 \end{bmatrix}$$

(9.117)

The meridional curvature $\bar{\kappa}_s$ is computed as

$$\bar{\kappa}_s = \frac{2}{l^e}\left[\left(\frac{\partial w'}{\partial x'}\right)_{i+1} - \frac{w'^e_{i+1} - w'^e_i}{l^e}\right]$$

(9.118)

with $\left(\frac{\partial w'}{\partial x'}\right)_{i+1}$ computed as in Eq.(9.104). After small algebra we obtain

$$\bar{\kappa}_s = \underbrace{\frac{2}{(l^e)^2 l^{e+1}}[l^{e+1}, -l^{e+1}, -l^e, l^e]}_{\bar{\mathbf{B}}'_{b_1}}\left\{\begin{array}{c} w'^e_i \\ w'^e_{i+1} \\ w'^{e+1}_{i+1} \\ w'^{e+1}_{i+2} \end{array}\right\} = \bar{\mathbf{B}}'_{b_1}\mathbf{w}' = \bar{\mathbf{B}}_{b_1}\mathbf{a}^e \quad (9.119)$$

with $\bar{\mathbf{B}}_{b_1} = \bar{\mathbf{B}}'_{b_1}\mathbf{T}$, and $\mathbf{T}$ as in Eq.(9.115b).

The rest of the generalized strain matrices have the same expressions as in Eq.(9.111). The displacements at a clamped node are prescribed to zero, as usual.

The element stiffness matrix is obtained as

$$\mathbf{K}^{(e)} = \mathbf{K}_1^{(e)} + \bar{\mathbf{K}}_1^{(e)}$$

(9.120)

where $\mathbf{K}_1^{(e)}$ and $\bar{\mathbf{K}}_1^{(e)}$ are given by Eq.(9.112a) using $\mathbf{B}_{b_1}$ and $\bar{\mathbf{B}}_{b_1}$ in $\mathbf{K}_b^{(e)}$, respectively. The terms involving $\mathbf{B}_{b_1}$ and $\bar{\mathbf{B}}_{b_1}$ are computed over $\frac{l^e}{2}$.

In addition, the displacements of the "ghost node" $(\mathbf{a}_{i-1})$ are prescribed to zero for the same reasons explained for the free or SS node.

A clamped or symmetry node placed at the right-hand end of the mesh is treated in the same manner [JO].

9.11.2 Cell-vertex rotation-free troncoconical element

9.11.2.1 ACV element matrices

The stiffness matrix for the so-called ACV element is expressed as the sum of the stiffness contributions from the two subdomains 1 and 2 which form the element (Figures 9.23 and 9.26), i.e.

$$\mathbf{K}^{(e)} = \pi l^e \left[\mathbf{B}_i^T \hat{\mathbf{D}}' \mathbf{B}_i r_1 + \mathbf{B}_{i+1}^T \hat{\mathbf{D}}' \mathbf{B}_{i+1} r_2 \right] \tag{9.121}$$

The meridional curvature at node i is

$$
\begin{aligned}
\kappa_{s_i} &= \frac{2}{l^e + l^{e-1}} \left[\left(\frac{\partial w'}{\partial x'} \right)_B - \left(\frac{\partial w'}{\partial x'} \right)_A \right] = \\
&= \frac{2}{l^e + l^{e-1}} \left[\frac{w_{i+1}'^e - w_i'^e}{l^e} - \frac{w_i'^{e-1} - w_{i-1}'^{e-1}}{l^{e-1}} \right] = \\
&= \underbrace{\frac{1}{l^e l^{e-1} (l^e + l^{e-1})} [l^e, -l^e, -l^{e-1}, l^{e-1}]}_{\bar{\mathbf{B}}_{b_i}'} \begin{Bmatrix} w_{i-1}'^{e-1} \\ w_i'^{e-1} \\ w_{i+1}'^e \\ w_{i+1}'^e \end{Bmatrix} = \mathbf{B}_{b_i}' \mathbf{w}' = \mathbf{B}_{b_i} \mathbf{a}^e
\end{aligned} \tag{9.122a}
$$

where $\mathbf{a}^e$ is given by Eq.(9.107b) and

$$\mathbf{B}_{b_i} = \mathbf{B}_{b_i}' \mathbf{T} \quad \text{with} \quad \mathbf{T} = \begin{bmatrix} \mathbf{n}^{e-1} & 0 & 0 & 0 \\ 0 & \mathbf{n}^{e-1} & 0 & 0 \\ 0 & 0 & \mathbf{n}^e & 0 \\ 0 & 0 & 0 & \mathbf{n}^e \end{bmatrix} \tag{9.122b}$$

Similarly for node $i + 1$

$$
\begin{aligned}
\kappa_{s_{i+1}} &= \frac{2}{l^e + l^{e+1}} \left[\left(\frac{\partial w'}{\partial x'} \right)_C - \left(\frac{\partial w'}{\partial x'} \right)_B \right] = \\
&= \frac{2}{l^e + l^{e+1}} \left[\frac{w_{i+2}'^{e+1} - w_{i+1}'^e}{l^{e+1}} - \frac{w_{i+1}'^e - w_i'^e}{l^e} \right] =
\end{aligned} \tag{9.123a}
$$

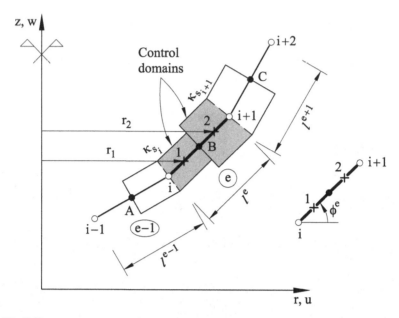

Fig. 9.26 Cell-vertex rotation-free troncoconical element (ACV). Control domains for computing the average radial curvature. A, B and C are element mid-points

$$
= \underbrace{\frac{2}{l^e l^{e+1}(l^e + l^{e+1})}[l^{e+1}, -l^{e+1}, -l^e, l^e]}_{\bar{\mathbf{B}}_{b_{i+1}}}\begin{Bmatrix} w_i'^e \\ w_{i+1}'^e \\ w_{i+1}'^{e+1} \\ w_{i+1} \\ w_{i+2}'^{e+1} \end{Bmatrix} = \mathbf{B}_{b_{i+1}}' \mathbf{w}' = \mathbf{B}_{b_{i+1}} \mathbf{a}^e
$$

with

$$
\mathbf{B}_{b_{i+1}} = \mathbf{B}_{b_{i+1}}' \mathbf{T} \tag{9.123b}
$$

and $\mathbf{T}$ as in Eq.(9.115b).

Matrices $\mathbf{B}_i$ and $\mathbf{B}_{i+1}$ are

$$
\mathbf{B}_i = \begin{Bmatrix} \mathbf{B}_{m_1} \\ \mathbf{B}_{m_2} \\ \mathbf{B}_{b_i} \\ \mathbf{B}_{b_2} \end{Bmatrix} \quad , \quad \mathbf{B}_{i+1} = \begin{Bmatrix} \mathbf{B}_{m_1} \\ \bar{\mathbf{B}}_{m_2} \\ \mathbf{B}_{b_{i+1}} \\ \mathbf{B}_{b_2} \end{Bmatrix} \tag{9.124}
$$

with $\mathbf{B}_{m_1}$ and $\mathbf{B}_{b_2}$ given by Eqs.(9.109) and (9.108). The circumferential membrane strain matrices at points 1 and 2 (Figure 9.26) are obtained by

$$
\lambda_{\theta_1} = \frac{u_1}{r_1} = \frac{1}{r_1}\left[\frac{3}{4}u_i + \frac{1}{4}u_{i+1}\right] = \underbrace{\left[0, 0, \frac{3}{4r_1}, 0, \frac{1}{4r_1}, 0, 0,\right]}_{\mathbf{B}_{m_2}} \mathbf{a}^e \tag{9.125}
$$

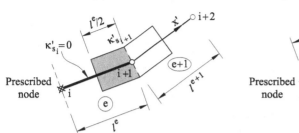
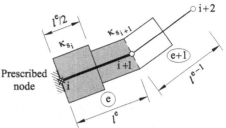

Fig. 9.27 Cell-vertex rotation-free troncoconical element (ACV). Control domains for elements with boundary node. (a) SS or free end node. (b) Clamped or symmetry node

$$\lambda_{\theta_2} = \frac{u_2}{r_2} = \frac{1}{r_2}\left[\frac{u_i}{4} + \frac{3}{4}u_{i+1}\right] = \underbrace{\left[0, 0, \frac{1}{4r_2}, 0, \frac{3}{4r_2}, 0, 0,\right]}_{\bar{\mathbf{B}}_{m_2}} \mathbf{a}^e. \qquad (9.126)$$

The size of $\mathbf{K}^{(e)}$ for the ACV element is 8×8 as it involves the DOFs of nodes $i-1, i, i+1$ and $i+2$, similarly as for the ACC element.

The assembly process follows the general rule. We note that the assembled stiffness matrix for the whole mesh can be directly obtained from the nodal expression of the generalized strain matrices, as explained for the CVB beam element in Section 1.4.2 and Example 1.6.

9.11.2.2 Boundary conditions for the ACV element

Free or SS end node

The radial curvature κ_s is zero at a free or simply supported (SS) end node i. This is simply enforced by making matrix $\mathbf{B}_{b_i}$ equal to zero in Eq.(9.124). Similarly $\mathbf{B}_{b_{i+1}}$ is made zero if node $i+1$ is prescribed.

Clamped or symmetry edge

Let us consider a clamped or symmetry node at the left-end of a mesh (Figure 9.27b) where the rotation is prescribed to a zero value.

The radial curvature at node i is obtained as

$$\kappa_{s_i} = \frac{2}{l^e}\left[\frac{w'^e_{i+1} - w'^e_i}{l^e}\right] = \frac{2}{(l^e)^2}[0, -1, 1, 0]\begin{Bmatrix} w'^{e-1}_{i-1} \\ w'^{e-1}_i \\ w'^e_i \\ w'^e_{i+1} \end{Bmatrix} = \mathbf{B}'_{b_i}\mathbf{w}' = \mathbf{B}_{b_i}\mathbf{a}^e$$

$$(9.127a)$$

where

$$\mathbf{B}_{b_i} = \mathbf{B}'_{b_i} \mathbf{T} \tag{9.127b}$$

and $\mathbf{T}$ as in Eq.(9.122b).

Matrix $\mathbf{B}_{b_{i+1}}$ has the same expression as in Eq.(9.123b).

A similar process is followed for a clamped or symmetry node at the right-hand end, leading to a modification of $\mathbf{B}_{b_{i+1}}$.

As usual, the global displacements $\mathbf{a}_i$ at clamped node (or the vertical displacement w_i at a SS node) are prescribed to zero when solving the global system of equations.

9.11.3 Example. Analysis of a dome under internal pressure

Figures 9.28 shows the performance of the rotation-free ACC and ACV elements in the analysis of a dome under uniform pressure. The same problem was solved with the 2-noded troncoconical Reissner-Mindlin element in Figure 9.16.

The figures show the better performance of the ACV element for the same number of elements. This is consistent with the higher accuracy of the cell-vertex rotation-free beam element (Section 1.4.3).

Both ACC and ACV elements are useful for solving practical problems were axisymmetric shell elements need to be coupled to 2D axisymmetric solid elements. The lack of rotational DOFs simplifies the coupling of the stiffness equations involving displacement DOFs only.

9.12 AXISYMMETRIC PLATES

An axisymmetric plate is a particular case of an axisymmetric shell with an horizontal generatrix. The formulation of axisymmetric plate elements is deduced from that of axisymmetric shells by making the angle $\phi = 0$ and ignoring the membrane contributions (we will assume that the in-plane forces are zero). Let us consider, for instance, an axisymmetric plate element with n nodes based on Reissner-Mindlin theory. The displacement field is expressed in terms of the deflection and the rotation as

$$\mathbf{u} = \begin{Bmatrix} w \\ \theta \end{Bmatrix} = \sum_{i=1}^{n} N_i \underset{2\times2}{\mathbf{I}} \mathbf{a}_i^{(e)}; \qquad \mathbf{a}_i^{(e)} = \begin{Bmatrix} w_i \\ \theta_i \end{Bmatrix} \tag{9.128}$$

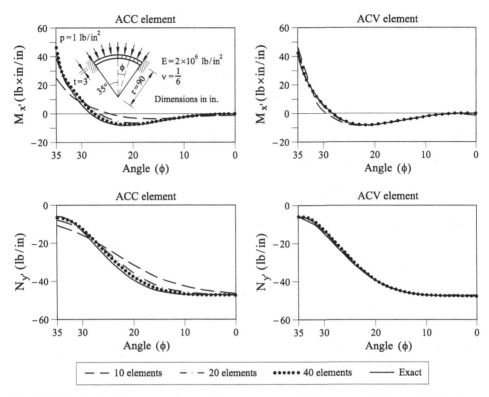

Fig. 9.28 Spherical dome under uniform pressure. Diagrams of $N_{x'}$ and $N_{y'}$ for different meshes of rotation-free troncoconical elements ACC and ACV

The generalized strain matrix is deduced from Eqs.(9.51) and (9.54) as

$$
\hat{\varepsilon} = \left\{ \begin{array}{c} \hat{\varepsilon}_b \\ \cdots \\ \hat{\varepsilon}_s \end{array} \right\} = \left\{ \begin{array}{c} \dfrac{\partial \theta}{\partial r} \\[2mm] \dfrac{\theta}{r} \\[2mm] \cdots\cdots \\[1mm] \dfrac{\partial w_0}{\partial r} - \theta \end{array} \right\} = \sum_{i=1}^{n} \mathbf{B}_i \, \mathbf{a}_i^{(e)} \tag{9.129}
$$

with

$$
\mathbf{B}_i = \left\{ \begin{array}{c} \mathbf{B}_{b_i} \\ --- \\ \mathbf{B}_{s_i} \end{array} \right\} = \left[\begin{array}{cc} 0 & \dfrac{\partial N_i}{\partial r} \\[2mm] 0 & \dfrac{N_i}{r} \\[1mm] - & -- \\[1mm] \dfrac{\partial N_i}{\partial r} & -N_i \end{array} \right] \tag{9.130}
$$

Since the element is horizontal, the displacement and strain fields can be directly expressed in global axes. Note that the local coordinate s has

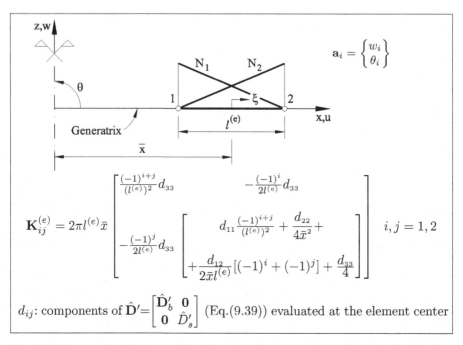

Fig. 9.29 Two–noded Reissner-Mindlin axisymmetric plate element. Stiffness matrix using uniform reduced one-point integration

been replaced by r. The global element stiffness matrix is obtained from Eq.(9.57) as

$$\mathbf{K}_{ij}^{(e)} = 2\pi \int_{l^{(e)}} \left[\mathbf{B}_{b_i}^T \, \hat{\mathbf{D}}_b \, \mathbf{B}_{b_j} + \mathbf{B}_{s_i}^T \, \hat{\mathbf{D}}_s \, \mathbf{B}_{s_j} \right] r \, d \, r \; = \mathbf{K}_b^{(e)} + \mathbf{K}_s^{(e)} \quad (9.131)$$

Shear locking is avoided by any of the techniques mentioned in Section 9.5. Figure 9.29 shows the stiffness-matrix for the 2-noded axisymmetric plate element using one-point uniform reduced integration.

The equivalent nodal force vector for a distributed loading f_z and point loads $\mathbf{p}_i = [P_{zi}, \; M_i]^T$ is

$$\mathbf{f}_i^{(e)} = 2\pi \int_{l^{(e)}} \mathbf{N}_i \begin{Bmatrix} f_z \\ 0 \end{Bmatrix} r \, d \, r + 2\pi r_i \mathbf{p}_i \quad (9.132)$$

The above integral can be computed exactly for a uniform loading $f_z = q$ to give

$$\mathbf{f}_i^{(e)} = \frac{q\pi l^{(e)}}{3} \left[(2r_1^{(e)} + r_2^{(e)}), 0, (2r_2^{(e)} + r_1^{(e)}), 0 \right]^T \quad (9.133)$$

Example 9.1: Obtain the deflection and radial bending moment in a clamped circular plate under uniform distributed loading $f_z = q$ (Figure 9.30), using meshes of 1, 2, 4, 8 and 16 2-noded Reissner-Mindlin axisymmetric plate elements with one-point reduced integration.

- *Solution*

One-element mesh

The only unknown in this case is the central deflection w_1 (as $\theta_1 = w_2 = \theta_2 = 0$). The equilibrium equation is deduced from Figure 9.29 and Eq.(9.133) as

$$2\pi l \bar{r} \frac{d_{33}}{l^2} w_1 = \frac{\pi l q}{3}(2r_1 + r_2)$$

and

$$w_1 = \frac{q l^2}{6 d_{33} \bar{r}}(2r_1 + r_2)$$

substituting $r_1 = 0$, $r_2 = 10.0$ m, $l = 10$ m, $\bar{r} = 5$ m, $q = -1.0$ T $/\text{m}^2$ and $d_{33} = \frac{\alpha E T}{2(1+\nu)}$, with $\alpha = 5/6$, $E = 10^7$ T$/\text{m}^2$ and $\nu = 0.3$ gives

$$w_1 = -0.104 \times 10^{-3} \text{ m}$$

The solution is a far from the analytical value of -0.171 m [TW] due to the simplicity of the mesh leading to an over–stiff result. A zero bending moment field is obtained, as the rotations are zero.

Two–element mesh

Assembling the stiffness matrices and eliminating the constrained DOFs at the clamping end ($w_3 = \theta_3 = 0$) gives

$$\begin{bmatrix} \mathbf{K}_{11}^{(1)} & | & \mathbf{K}_{12}^{(1)} \\ & | & \\ \mathbf{K}_{21}^{(1)} & | & \mathbf{K}_{22}^{(1)} + \mathbf{K}_{11}^{(2)} \end{bmatrix} \begin{Bmatrix} w_1 \\ \theta_1 \\ - - - \\ w_2 \\ \theta_2 \end{Bmatrix} = \begin{Bmatrix} \mathbf{f}_1^{(1)} \\ - - - \\ \mathbf{f}_2^{(1)} + \mathbf{f}_1^{(2)} \end{Bmatrix}$$

Substituting the dimensions, material properties and load values gives

$$10^5 \begin{bmatrix} 10.1 & 2.52 & | & -10.1 & 25.2 \\ 25.2 & 63.0 & | & -25.2 & 62.9 \\ - & - & - & - & - \\ -10.1 & -25.2 & | & 40.0 & 50.3 \\ 25.2 & 62.9 & | & 50.3 & 252.0 \end{bmatrix} \begin{Bmatrix} w_1 \\ \theta_1 \\ w_2 \\ \theta_2 \end{Bmatrix} = \begin{Bmatrix} -26.17 \\ 0.0 \\ -157.16 \\ 0.0 \end{Bmatrix}$$

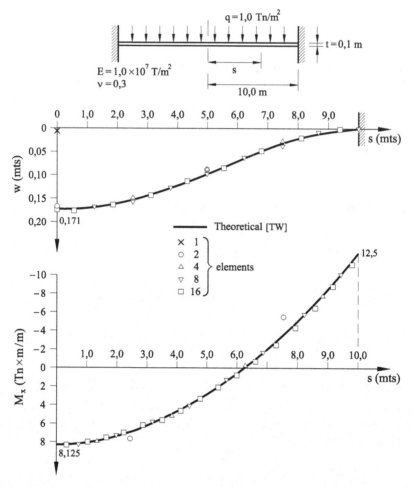

Fig. 9.30 Clamped circular plate under uniform loading. Deflection and radial bending moment distribution using 1, 2, 4, 8 and 16 2-noded Reissner-Mindlin axisymmetric plate elements

Making $\theta_1 = 0$ and solving for w_1, w_2 and θ_2 yields

$$w_1 = -0.170 \text{ m}, \quad w_2 = -0.0853 \text{ m} \quad \text{and} \quad \theta_2 = 0.03 \text{ rad}$$

The error in the central deflection with respect to to the exact solution [TW] is only 0.6 %. The bending moment M_r at the element midpoints are

Element 1 : $M_x = 8.124 \text{ T} \times \text{m/m}$

Element 2 : $M_x = -5.62 \text{ T} \times \text{ m/m}$

The moments are in good agreement with the theoretical values as shown in Figure 9.30. Results for meshes of 4, 8 and 16 elements are also plotted.

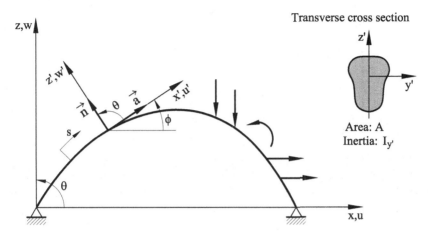

Fig. 9.31 Description of a plane arch

Axisymmetric plate elements based on Kirchhoff theory can be formulated by direct simplification of the corresponding axisymmetric shell element. This includes the ACC and ACV rotation-free troncoconical elements of Section 9.11. The reader is invited to develop one of these axisymmetric plate elements as an exercise.

9.13 PLANE ARCHES

The finite element formulation for plane arches is developed here as a particular case of axisymmetric shells ignoring circumferential effects.

Figure 9.31 shows a plane arch defined by the middle line of the transverse sections. The displacement field of a point is defined by the tangential and normal displacements and the rotation of the normal vector. The general case of non-orthogonal rotation of the normal will be considered first. The local generalized strain vector contains the elongation of the middle line λ, the curvature κ and the transverse shear strain γ defined as

$$\hat{\varepsilon}' = \left\{ \begin{array}{c} \lambda \\ \kappa \\ \gamma \end{array} \right\} = \left\{ \begin{array}{c} \dfrac{\partial u'_0}{\partial s} - \dfrac{w'_0}{R_s} \\[2mm] -\dfrac{\partial \theta}{\partial s} \\[2mm] \dfrac{\partial w'_0}{\partial s} + \dfrac{u'_0}{R_s} - \theta \end{array} \right\} \tag{9.134}$$

Eq.(9.183) is obtained by ignoring the circumferential effects in Eqs.(9.27).

For straight elements

$$\hat{\varepsilon}' = \left\{ \begin{array}{c} \dfrac{\partial u_0'}{\partial s} \\[2mm] -\dfrac{\partial \theta}{\partial s} \\[2mm] \dfrac{\partial w_0'}{\partial s} - \theta \end{array} \right\} \begin{array}{l} \text{elongation: axial strain} \\[2mm] \left. \begin{array}{l} \text{curvature} \\[2mm] \text{transverse shear} \end{array} \right\} \begin{array}{l} \text{bending} \\ \text{strains} \end{array} \end{array} \qquad (9.135)$$

Vector $\hat{\varepsilon}'$ of Eq.(9.135) contains the axial strain of a bar element [On4] and the two flexural strains of the Timoshenko beam element of Chapter 2. For the homogeneous material case considered here, axial and flexural effects are uncoupled *at element level*. Coupling occurs after the global assembly of the local element stiffness matrices.

The discretization follows standard procedures. For a n-noded element

$$\mathbf{u} = [u_0', w_0', \theta]^T = \sum_{i=1}^{n} \mathbf{N}_i \mathbf{a}_i^{(e)} \qquad (9.136a)$$

$$\hat{\varepsilon}' = \sum_{i=1}^{n} \mathbf{B}_i' \, \mathbf{a}_i'^{(e)} = \mathbf{B}' \mathbf{a}'^{(e)} \qquad (9.136b)$$

with

$$\mathbf{B}_i' = \begin{bmatrix} \dfrac{\partial N_i}{\partial s} & 0 & 0 \\[2mm] 0 & 0 & \dfrac{-\partial N_i}{\partial s} \\[2mm] 0 & \dfrac{\partial N_i}{\partial s} & -N_i \end{bmatrix} ; \qquad \mathbf{a}_i'^{(e)} = \left\{ \begin{array}{c} u_{0i}' \\ w_{0i}' \\ \theta_i \end{array} \right\} \qquad (9.137)$$

The constitutive equation for homogeneous isotropic material is

$$\hat{\sigma}' = \left\{ \begin{array}{c} N \\ M \\ Q \end{array} \right\} = \begin{bmatrix} EA & 0 & 0 \\ 0 & EI_{y'} & 0 \\ 0 & 0 & kGA \end{bmatrix} \hat{\varepsilon}' = \hat{\mathbf{D}}' \, \hat{\varepsilon}' \qquad (9.138)$$

where A is the area of the transverse cross section and $I_{y'}$ the moment of inertia with respect to the transverse coordinate axis y' (Figure 9.31). The sign convenion for the resultant stresses coincides with that of $N_{x'}$, $M_{x'}$ and $Q_{z'}$ of Figure 9.6.

The local stiffness matrix is

$$\mathbf{K}_{ij}'^{(e)} = \int_{l^{(e)}} \mathbf{B}_i'^T \, \hat{\mathbf{D}}' \, \mathbf{B}_j' \, dx' \qquad (9.139)$$

An alternative expression is

$$\mathbf{K}'^{(e)}_{ij} = \begin{bmatrix} \mathbf{K}^{(e)}_{AB_{ij}} & \mathbf{0} \\ {\scriptstyle 1\times 1} & \\ \mathbf{0} & \mathbf{K}^{(e)}_{TB_{ij}} \\ & {\scriptstyle 2\times 2} \end{bmatrix} \tag{9.140}$$

where $\mathbf{K}^{(e)}_{AB_{ij}}$ and $\mathbf{K}^{(e)}_{TB_{ij}}$ are the stiffness matrices of the *axial bar element* and the *Timoshenko beam* element of Chapters 2 of [On4] and Chapter 2 of this book, respectively.

The stiffness transformation to global axes follows the rules of Section 9.4.3 giving

$$\mathbf{K}^{(e)}_{ij} = \int_{l^{(e)}} \mathbf{B}^T_i\, \hat{\mathbf{D}}'\, \mathbf{B}_j\, dx \tag{9.141}$$

with $\mathbf{B}_i = \mathbf{B}'_i \left[\mathbf{L}^{(e)}_i\right]^T$ and $\mathbf{L}^{(e)}_i = \mathbf{L}$ is given by Eq.(9.14).

Shear locking is avoided by the same techniques as for Reissner-Mindlin axisymmetric shell elements or Timoshenko beam elements. The simplest plane arch element is the *2-noded arch element* with one-point uniform reduced integration. The global stiffness matrix is given by

$$\mathbf{K}^{(e)}_{ij} = \bar{\mathbf{B}}^T_i \bar{\hat{\mathbf{D}}}' \bar{\mathbf{B}}_j l^{(e)} \tag{9.142}$$

where $\overline{(\cdot)}$ denotes values at the element mid-point. The explicit form of the stiffness matrix is shown in Figure 9.32.

The equivalent nodal force vector is

$$\mathbf{f}^{(e)}_i = \int_{l^{(e)}} \mathbf{N}_i \mathbf{t}^{(e)} dx' + \mathbf{p}^{(e)}_i \tag{9.143}$$

where $\mathbf{t}^{(e)}$ and $\mathbf{p}^{(e)}_i$ are distributed load and point load vectors given by $\mathbf{t}^{(e)}_i = [f_x, f_z, m]^T$ and $\mathbf{p}^{(e)}_i = [P_{x_i}, P_{z_i}, M_i]^T$. If the load $\mathbf{t}^{(e)}$ is uniformly distributed, Eq.(9.143) simplifies to

$$\mathbf{f}^{(e)}_i = \frac{l^{(e)}}{2}\mathbf{t}^{(e)} + \mathbf{p}^{(e)}_i \tag{9.144}$$

Curved arch elements can be derived following the arguments given in Section 9.8. The generalized strain matrix now includes the effect of the

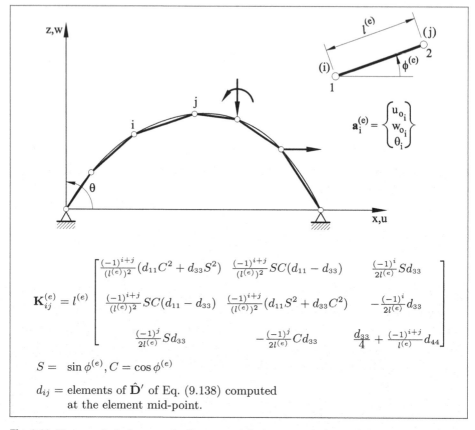

$$\mathbf{K}_{ij}^{(e)} = l^{(e)} \begin{bmatrix} \frac{(-1)^{i+j}}{(l^{(e)})^2}(d_{11}C^2 + d_{33}S^2) & \frac{(-1)^{i+j}}{(l^{(e)})^2}SC(d_{11} - d_{33}) & \frac{(-1)^i}{2l^{(e)}}Sd_{33} \\ \frac{(-1)^{i+j}}{(l^{(e)})^2}SC(d_{11} - d_{33}) & \frac{(-1)^{i+j}}{(l^{(e)})^2}(d_{11}S^2 + d_{33}C^2) & -\frac{(-1)^i}{2l^{(e)}}d_{33} \\ \frac{(-1)^j}{2l^{(e)}}Sd_{33} & -\frac{(-1)^j}{2l^{(e)}}Cd_{33} & \frac{d_{33}}{4} + \frac{(-1)^{i+j}}{l^{(e)}}d_{44} \end{bmatrix}$$

$S = \sin\phi^{(e)}, C = \cos\phi^{(e)}$

$d_{ij} =$ elements of $\hat{\mathbf{D}}'$ of Eq. (9.138) computed at the element mid-point.

Fig. 9.32 Two-noded plane arch element with transverse shear deformation. Global stiffness matrix using uniform one-point reduced integration

arch curvature, i.e.

$$\mathbf{B}_i' = \begin{bmatrix} \dfrac{\partial N_i}{\partial s} - \dfrac{N_i}{R_s} & 0 \\ 0 & 0 & \dfrac{-\partial N_i}{\partial s} \\ \dfrac{N_i}{R_s} & \dfrac{\partial N_i}{\partial s} & -N_i \end{bmatrix} \tag{9.145}$$

The computation of R_s and the curvilinear derivatives follow the steps of Section 9.8.2.

Imposing the *orthogonality* of the normals leads to the more classical formulation for thin arches. The transverse shear strain is then zero and the generalized strain vector (for the curved case) is

$$
\hat{\varepsilon}' = \left\{
\begin{array}{c}
\dfrac{\partial u_0'}{\partial s} - \dfrac{w_0'}{R_s} \\[2mm]
-\dfrac{\partial^2 w_0'}{\partial s^2} - \dfrac{\partial}{\partial s}\left(\dfrac{u_0'}{R_s}\right)
\end{array}
\right\}
\tag{9.146}
$$

Many authors have proposed thin curved arch elements using specific approximations for u_0' and w_0'. Some of these elements are described in [AG,BFS2,CB,Da2,Da3,DG,Ga2,Go,Ya].

If 2-noded straight elements are used, $R_s = \infty$ and the local stiffness matrix can be simply written as a contribution of the axial and bending stiffness in a form identical to Eq.(9.140), where $\mathbf{K}_{TB}$ is substituted by $\mathbf{K}_{EB}$, this now being the stiffness matrix for the Euler-Bernoulli beam element of Chapter 1.

Arch elements can be applied directly to the analysis of plane frame structures.

9.14 SHALLOW AXISYMMETRIC SHELLS AND ARCHES

The simpler shallow axisymmetric shell theory [Ma3] is applicable when the curvature of the generatrix is small (i.e. R_s is large and $t/R_s \simeq 0$). The advantage is that all variables are expressed in the global coordinate system. The main assumption is that the tangent angle $\phi \leq 5°$ (Figure 9.33) and, thus, $\cos\phi \simeq 1$, $\sin\phi \simeq \frac{\partial z}{\partial x}$, $w_0' \simeq w_0$ and $\frac{\partial}{\partial x'} \simeq \frac{\partial}{\partial x}$, where x denotes the global horizontal axis (i.e. the radial axis). This formulation is a particular case of that presented in Section 8.14 for shallow flat shell elements.

Using these simplifications and Reissner-Mindlin assumptions, the local strain vector is deduced from Eq.(9.25) as (taking $C_s = C_\alpha = 1$)

$$
\varepsilon' = \left\{
\begin{array}{c}
\varepsilon_{x'} \\[1mm]
\varepsilon_{y'} \\[1mm]
\gamma_{x'z'}
\end{array}
\right\} = \left\{
\begin{array}{c}
\dfrac{\partial u_0}{\partial x} + \dfrac{\partial z}{\partial x}\dfrac{\partial w_0}{\partial x} + z'\dfrac{\partial \theta}{\partial x} \\[3mm]
\dfrac{u_0}{x} + z'\dfrac{\theta}{x} \\[3mm]
\dfrac{\partial w_0}{\partial x} - \theta
\end{array}
\right\}
\tag{9.147}
$$

The membrane, bending and transverse shear strain vectors are

$$
\hat{\varepsilon}_m' = \left\{
\begin{array}{c}
\dfrac{\partial u_0}{\partial x} + \dfrac{\partial z}{\partial x}\dfrac{\partial w_0}{\partial x} \\[3mm]
\dfrac{u_0}{x}
\end{array}
\right\}, \quad
\hat{\varepsilon}_b' = \left\{
\begin{array}{c}
\dfrac{\partial \theta}{\partial x} \\[3mm]
\dfrac{\theta}{x}
\end{array}
\right\}, \quad
\hat{\varepsilon}_s' = \left\{ \dfrac{\partial w_0}{\partial x} - \theta \right\}
\tag{9.148}
$$

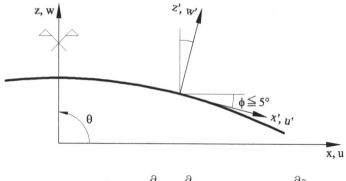

$$\cos \phi \simeq 1, \ w' \simeq w, \ \frac{\partial}{\partial x'} \simeq \frac{\partial}{\partial x}, \ \sin \phi \simeq \tan \phi \simeq \frac{\partial z}{\partial x}$$

Fig. 9.33 Description of an axisymmetric shallow shell (or arch)

The generatrix can be discretized into straight or curved elements. The generalized strain matrix is expressed in terms of the global displacements in both cases by

$$\hat{\varepsilon}' = \left\{ \begin{matrix} \hat{\varepsilon}'_m \\ \hat{\varepsilon}'_b \\ \hat{\varepsilon}'_s \end{matrix} \right\} = \sum_{i=1}^{n} \mathbf{B}_i \mathbf{a}_i^{(e)} \quad , \quad \mathbf{a}_i^{(e)} = \left\{ \begin{matrix} u_0 \\ w_0 \\ \theta \end{matrix} \right\} \tag{9.149a}$$

$$\mathbf{B}_i = \left\{ \begin{matrix} \mathbf{B}_{m_i} \\ ---- \\ \mathbf{B}_{b_i} \\ ---- \\ \mathbf{B}_{s_i} \end{matrix} \right\} = \begin{bmatrix} \dfrac{\partial N_i}{\partial x} & \left(\dfrac{\partial z}{\partial x} \dfrac{\partial N_i}{\partial x} \right) & 0 \\ \dfrac{N_i}{x} & 0 & 0 \\ \hline 0 & 0 & \dfrac{\partial N_i}{\partial x} \\ 0 & 0 & \dfrac{N_i}{x} \\ \hline 0 & \dfrac{\partial N_i}{\partial x} & -N_i \end{bmatrix} \tag{9.149b}$$

The derivatives $\frac{\partial z}{\partial x}$ and $\frac{\partial N_i}{\partial x}$ are computed from the isoparametric description as

$$\frac{\partial N_i}{\partial x} = \frac{\partial N_i}{\partial x} \frac{\partial \xi}{\partial x} = \left[\sum_{i=1}^{n} \frac{\partial N_i}{\partial x} x_i \right]^{-1} \frac{\partial N_i}{\partial \xi} \tag{9.150}$$

The computations are simplified if 2-noded troncoconical elements are chosen. For this case

$$\frac{\partial z}{\partial x} = \frac{z_2^{(e)} - z_1^{(e)}}{l^{(e)}} \quad ; \quad \frac{\partial N_i}{\partial x} = \frac{2}{l^{(e)}}(-1)^i \qquad (9.151)$$

The global stiffness matrix is given by Eq.(9.61) for the general case and by Eq.(9.73) for 2-noded troncoconical elements.

The equivalent formulation for *shallow arches* is readily derived by ignoring circumferential terms in the previous expressions. The local generalized strain vector and the corresponding $\mathbf{B}_i$ matrix are

$$\hat{\varepsilon}' = \left\{ \begin{array}{c} \dfrac{\partial u_0}{\partial x} + \dfrac{\partial z}{\partial x}\dfrac{\partial w_0}{\partial x} \\[2mm] \dfrac{\partial \theta}{\partial x} \\[2mm] \dfrac{\partial w_0}{\partial x} - \theta \end{array} \right\} ; \quad \mathbf{B}_i = \begin{bmatrix} \dfrac{\partial N_i}{\partial x} & \left(\dfrac{\partial z}{\partial x}\dfrac{\partial N_i}{\partial x}\right) & 0 \\[2mm] 0 & 0 & \dfrac{\partial N_i}{\partial x} \\[2mm] 0 & \dfrac{\partial N_i}{\partial x} & -N_i \end{bmatrix} \qquad (9.152)$$

The global stiffness matrix is given by Eq.(9.141) for the general case and by Eq.(9.142) for 2-noded arch elements.

This formulation can be particularized for Kirchhoff theory by making $\theta = \frac{\partial w_0}{\partial x}$ and eliminating the transverse shear deformation term. This introduces the second derivative term $\frac{\partial^2 w_0}{\partial x^2}$ which requires C^1 a continuous interpolation, for w_0 as explained in Section 9.9.

9.15 MORE ABOUT MEMBRANE LOCKING

The arguments given in Section 9.5.2 to explain membrane locking can be clarified by considering the case of shallow axisymmetric shells and arches of previous section.

Let us consider an arch subjected to loads giving a pure bending state. Obviously the axial strain should be zero in this case. The analogous axisymmetric shell situation will require zero meridional membrane strains under a bending field (note that the circumferential strains are always non-zero). Let us take for the sake at simplicity a shallow arch or a shallow axisymmetrix shell based on Reissner-Mindlin theory. The reproduction of the situation as described previously requires in both cases that the elements satisfy the following condition

$$\hat{\varepsilon}'_m = \frac{\partial u_0}{\partial x} + \frac{\partial w_0}{\partial x}\sin\phi = 0 \quad \text{, with } \sin\phi = \frac{dz}{dx} \qquad (9.153)$$

Naturally, elements giving $\hat{\varepsilon}'_m \neq 0$ in a bending situation will introduce spureous parasitic membrane strains. These strains can have a big influence on the overall element stiffness (recall that membrane terms are proportional to the thickness t whereas the bending terms vary as t^3) and this leads to an overstiff solution (membrane locking).

The condition $\hat{\varepsilon}'_m = 0$ is analogous to that of the zero transverse shear strain in the Reissner-Mindlin formulation. Similar procedures can therefore be used to satisfy both conditions.

Satisfaction of Eq.(9.153) in displacement–based elements requires that the polynomial terms contained in $\frac{\partial u_0}{\partial x}$ equal those in $\frac{\partial w_0}{\partial x} \sin \phi$. This is automatically satisfied for straight elements using equal interpolations for u_0 and w_0, as $\sin \phi = \frac{\partial z}{\partial x}$ is constant. However, angle ϕ varies from point to point for curved elements and different approximations for u_0 and w_0 are then required in order to match the polynomial terms in $\frac{\partial u_0}{\partial x}$ and $\frac{\partial w_0}{\partial x} \sin \phi$. The approximations for u_0 and w_0 will naturally depend on that used for the geometry. Some examples are given next.

Consider, for instance, a quadratic isoparametric element where both the displacements and the geometry are approximated quadratically. Eq.(9.153) can now be written as

$$\hat{\varepsilon}'_m = A + B\xi + C\xi^2 \tag{9.154}$$

where the coefficients A and B are functions of u_i, w_i, z_i and C of the products $w_i z_j, \quad i, j = 1, 2, 3$.

It can be checked that $\hat{\varepsilon}'_m$ is zero for a rigid body displacement [Cr]. Also, coefficients A and B vanish for a "pure" bending state. Unfortunately, the coefficient C is not zero in this case and this introduces undesirable membrane strains and leads to membrane locking.

The solution to this problem follows the lines explained previously to treat transverse shear locking. A two-point reduced quadrature for the axial strain leads to a linear variation of $\hat{\varepsilon}'_m$ within the element which is able to satisfy the condition $\hat{\varepsilon}'_m = 0$. The axial forces are also sampled at these quadrature points for higher accuracy. An alternative procedure is assuming a compatible polynomial axial field for $\hat{\varepsilon}'_m$ with coefficients defined as linear functions of u_0 and w_0. These coefficients can vanish to satisfy $\hat{\varepsilon}'_m = 0$ by adequate contributions of both displacement variables. This technique will be used in the next chapter to derive membrane locking-free curved shell elements of arbitrary shape.

Membrane locking can be even more severe for Kirchhoff elements as the compatibility between the polynomials defining $\frac{\partial u_0}{\partial x}$ and $\frac{\partial w_0}{\partial x} \sin \phi$ is more difficult in this case. Take, for instance, a straight element ($\phi =$ constant) with a linear C^0 approximation for u_0 and a cubic (Hermite) one for w_0. The expression for $\hat{\varepsilon}'_m$ now reads [Cr]

$$\hat{\varepsilon}'_m = A(u_{0_i}, w_{0_i}) + B\left(w_{0_i}, \frac{\partial w_{0_i}}{\partial x}\right)\xi + C\left(w_{0_i}, \frac{\partial w_{0_i}}{\partial x}\right)\xi^2 = 0 \qquad (9.155)$$

Coefficient A vanishes for a pure bending state; however coefficients B and C do not. The spurious terms $B\xi$ and $C\xi^2$ can be eliminated by using a one-point reduced quadrature to compute the axial membrane stiffness. This is equivalent to assuming a constant membrane strain field $\hat{\varepsilon}'_m = A$ within the element.

Let us finally consider a 2-noded curved element with an Hermite approximation for w_0 and z. The term $\frac{\partial w_0}{\partial x}\sin\phi$ is now a quartic polynomial and hence the satisfaction of condition (9.153) requires that u_o varies as a quintic polynomial [Cr,Da2,Da3,Del]. Fortunately, the two-point reduced integration alleviates this problem, allowing a lower order quadratic interpolation for u_0 [Cr].

Further information on membrane locking in shells can be found in [CBS,CP,Cr,Cr4,Mo3,Mo5,MS,SB,SBCK,ZT2].

9.16 AXISYMMETRIC SHELL ELEMENTS OBTAINED FROM DEGENERATED AXISYMMETRIC SOLID ELEMENTS

A family of axisymmetric shell elements can be derived by a consistent modification (degeneration) of axisymmetric solid elements using the assumptions of shell theory. This class of shell elements are typically called "degenerated shell elements". The derivation of 3D degenerated shell elements is presented in Chapter 10. Figure 9.34 shows an axisymmetric solid element and the corresponding "degenerated" axisymmetric shell element.

The degeneration steps are the following: a) a *reference line* is defined; b) the normals to this line are assumed to remain straight during deformation, and c) the normal stress $\sigma_{z'}$ is taken as zero (plane stress). The element geometry is defined by the coordinates of the n nodes on the reference line and the normal distance ζ as

$$\mathbf{x} = \begin{Bmatrix} x \\ z \end{Bmatrix} = \sum_{i=1}^{n} N_i(\xi)\left(\mathbf{x}_i + \zeta\frac{t_i}{2}\mathbf{v}_{3_i}\right) \qquad (9.156)$$

where $\mathbf{x}_i = [x_i, z_i]^T$ and i; t_i are the coordinates vector and the thickness of node i; $N_i(\xi)$ is the 1D (Lagrange) shape function and $\mathbf{v}_{3_i} =$

$[-\sin\phi_i, \cos\phi_i]^T$ is the unit nodal vector at a node (Figure 9.34). Vectors $\overrightarrow{v_{1_i}}$ and $\overrightarrow{v_{3_i}}$ coincide with vectors $\overrightarrow{t}$ and $\overrightarrow{n}$ as defined in Figure 9.3.

Let us consider now the case of axisymmetric loading only. The axisymmetric displacement field is written as

$$\mathbf{u} = \begin{Bmatrix} u \\ w \end{Bmatrix} = \sum_{i=1}^{n} N_i(\xi)(\mathbf{u}_{0_i} - \frac{t_i}{2}\zeta\mathbf{v}_{1_i}\theta_i) \tag{9.157}$$

where $\mathbf{u}_{0_i} = [u_{0_i}, w_{0_i}]^T$ contains the global cartesian components of the ith node and θ_i is the nodal rotation. Eq.(9.157) is rewritten as

$$\mathbf{u} = \sum_{i=1}^{n} \mathbf{N}_i \, \mathbf{a}_i^{(e)} \tag{9.158}$$

with

$$\mathbf{a}_i^{(e)} = \begin{Bmatrix} u_{0_i} \\ w_{0_i} \\ \theta_i \end{Bmatrix} ; \qquad \mathbf{N}_i = N_i[\mathbf{I}_2, -\frac{\zeta t_i}{2}\mathbf{v}_{1_i}] \tag{9.159}$$

where $\mathbf{I}_2$ is the 2×2 unit matrix.

The strains in local axes x', z' are related to their global components by

$$\boldsymbol{\varepsilon}' = \begin{Bmatrix} \dfrac{\partial u'}{\partial x'} \\[2mm] \dfrac{u}{r} \\[2mm] \dfrac{\partial u'}{\partial z'} + \dfrac{\partial v'}{\partial x'} \end{Bmatrix} = \begin{bmatrix} (v_1^x)^2 & (v_1^z)^2 & 0 & v_1^x v_1^z \\ 0 & 0 & 1 & 0 \\ 2v_1^x v_3^x & 2v_1^z v_3^z & 0 & v_1^x v_3^z + v_1^x v_3^z \end{bmatrix} \begin{Bmatrix} \dfrac{\partial u}{\partial x} \\[2mm] \dfrac{\partial w}{\partial z} \\[2mm] \dfrac{u}{r} \\[2mm] \dfrac{\partial u}{\partial z} + \dfrac{\partial w}{\partial x} \end{Bmatrix} = \mathbf{Q}\,\boldsymbol{\varepsilon} \tag{9.160}$$

where upper indexes x, z denote the components of vectors $\overrightarrow{v_1}$ and $\overrightarrow{v_2}$ in global axes.

The derivation of the strain matrix follows the steps given below. The curvilinear derivatives of the displacements are first computed as

$$\frac{\partial \mathbf{u}}{\partial \xi} = \sum_{i=1}^{n} \frac{\partial N_i}{\partial \xi}[\mathbf{I}_2, -\frac{\zeta t_i}{2}\mathbf{v}_{1_i}]\mathbf{a}_i^{(e)}$$

$$\frac{\partial \mathbf{u}}{\partial \zeta} = \sum_{i=1}^{n} N_i[0, -\frac{t_i}{2}\mathbf{v}_{1_i}]\mathbf{a}_i^{(e)} \tag{9.161}$$

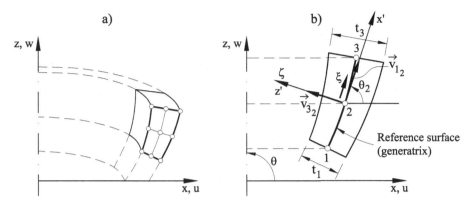

Fig. 9.34 Axisymmetric shell element (b) obtained by consistent modification (degeneration) of an axisymmetric solid element (a)

The global cartesian and curvilinear derivatives are related by the standard expression

$$\left\{\begin{array}{c} \dfrac{\partial}{\partial x} \\ \dfrac{\partial}{\partial z} \end{array}\right\} = [\mathbf{J}^{(e)}]^{-1} \left\{\begin{array}{c} \dfrac{\partial}{\partial \xi} \\ \dfrac{\partial}{\partial \zeta} \end{array}\right\}, \qquad \mathbf{J}^{(e)} = \begin{bmatrix} \dfrac{\partial x}{\partial \xi} & \dfrac{\partial y}{\partial \xi} \\ \dfrac{\partial x}{\partial \zeta} & \dfrac{\partial y}{\partial \zeta} \end{bmatrix} \qquad (9.162)$$

The terms of the jacobian matrix $\mathbf{J}^{(e)}$ are computed using Eq.(9.156) as

$$\frac{\partial \mathbf{x}}{\partial \xi} = \sum_{i=1}^{n} \frac{\partial N_i}{\partial \xi} \left[\left(\mathbf{x}_i + \zeta \frac{t_i}{2} \mathbf{v}_{3_i} \right) \right]$$

$$\frac{\partial \mathbf{x}}{\partial \zeta} = \sum_{i=1}^{n} N_i \frac{t_i}{2} \mathbf{v}_{3_i} \qquad (9.163)$$

Combining now Eqs.(9.161)–(9.163) gives

$$\frac{\partial \mathbf{u}}{\partial x_j} = \sum_{i=1}^{n} [N_i^j \mathbf{I}_2, \mathbf{g}_i^j] \mathbf{a}_i^{(e)} \quad \text{with} \quad j = 1, 2 \quad (x_1 = x, x_2 = z) \qquad (9.164a)$$

with

$$N_i^j = J_{j1}^{-1} \frac{\partial N_i}{\partial \xi} \qquad \text{and} \qquad \mathbf{g}_i^j = -\frac{t_i}{2} \left(\zeta N_i^j + J_{j2}^{-1} N_i \right) \mathbf{v}_{1_i} \qquad (9.164b)$$

where J_{ij}^{-1} is the inverse element ij of the jacobian matrix.

$$\varepsilon = \sum_{i=1}^{n} \mathbf{B}_i \mathbf{a}_i = [\mathbf{B}_1, \mathbf{B}_2, \cdots, \mathbf{B}_n]\mathbf{a}^{(e)} = \mathbf{B}\mathbf{a}^{(e)}$$

$$\mathbf{B}_i = \left\{ \begin{array}{c} \mathbf{B}_{p_i} \\ \cdots \\ \mathbf{B}_{s_i} \end{array} \right\} = \begin{bmatrix} N_i^1 & 0 & (\mathbf{g}_i^1)_{11} \\ 0 & N_i^2 & (\mathbf{g}_i^2)_{21} \\ \dfrac{N_i^2}{x} & 0 & \bar{g}_i \\ \hline N_i^1 & N_i^2 & (\mathbf{g}_i^1)_{21} + (\mathbf{g}_i^2)_{11} \end{bmatrix}$$

$$N_i^j = J_{j_1}^{-1}\frac{\partial N_i}{\partial \xi}, \quad \bar{g}_i = -\frac{\zeta t_i N_i}{2x}\cos\phi_i$$

$$\mathbf{g}_i^j = -\frac{t_i}{2}\left(\zeta J_{j_1}^{-1}\cdot\frac{\partial N_i}{\partial \xi} + J_{j_2}^{-1}N_i\right)\mathbf{v}_{1_i}$$

Box 9.1 Global strain matrix for a degenerated axisymmetric shell element

Eqs.(9.164) can be used to obtain the terms of the global strain matrix $\mathbf{B}$ relating global strains and global displacements which expression is shown in Box 9.1. Substituting this expression into Eq.(9.160) gives finally

$$\boldsymbol{\varepsilon}' = \mathbf{Q}\,\boldsymbol{\varepsilon} = \mathbf{Q}\,\mathbf{B}\mathbf{a}^{(e)} = \mathbf{B}'\mathbf{a}^{(e)} \tag{9.165}$$

The local strain matrix is

$$\mathbf{B}' = \mathbf{Q}\,\mathbf{B} = \left\{ \begin{array}{c} \mathbf{B}'_p \\ \mathbf{B}'_s \end{array} \right\} \quad \text{with } \mathbf{B}'_p = \mathbf{Q}_1\mathbf{B}, \quad \mathbf{B}'_s = \mathbf{Q}_2\,\mathbf{B} \tag{9.166}$$

In Eq.(9.166) $\mathbf{B}'_p$ and $\mathbf{S}'_s$ are the in-plane and transverse shear strain matrices and $\mathbf{Q}_1$ and $\mathbf{Q}_2$ contain the first two rows and the last row of the strain transformation $\mathbf{Q}$ of Eq.(9.160), respectively.

The constitutive equation relating the three non-zero stresses $\sigma_{x'}$, $\sigma_{z'}$, $\tau_{x'z'}$ and their conjugate local strains is defined by Eq.(9.31a). The local constitutive matrix coincides with that given in Eq.(9.31b). The resultant stresses $N_{x'}, N_{y'}, M_{x'}, M_{y'}, Q_{x'}$ can be computed "a posteriori" from the local stresses by Eq.(9.34).

The expressions for the global element stiffness matrix and the global equivalent nodal force vector (in absence of initial stresses) are

$$\mathbf{K}_{ij}^{(e)} = 2\pi \iint_{A^{(e)}} \mathbf{B}_i'^T \mathbf{D}' \mathbf{B}_j' x\,dx\,dz =$$

$$= 2\pi \iint_{A^{(e)}} (\mathbf{B}_{p_i}'^T \mathbf{D}_p' \mathbf{B}_{p_j}' + \mathbf{B}_{s_i}'^T G_{sz'} \mathbf{B}_{s_j}')x\,dx\,dz =$$

$$= \mathbf{K}_{p_{ij}}^{(e)} + \mathbf{K}_{s_{ij}}^{(e)} \tag{9.167a}$$

$$\mathbf{f}_i^{(e)} = 2\pi \iint_{A^{(e)}} \mathbf{N}_i^T \mathbf{b} r \, dx' \, dz' + 2\pi \int_{l^{(e)}} \mathbf{N}_i^T \mathbf{t} r \, dz' + 2\pi r_i \mathbf{p}_i \qquad (9.167\text{b})$$

where $\mathbf{K}_p^{(e)}$ and $\mathbf{K}_s^{(e)}$ are the in-plane and transverse shear stiffness matrices, respectively. Vectors $\mathbf{b}$, $\mathbf{t}$ and $\mathbf{p}_i$ denote the standard body force, distributed load and point load vectors. Recall that all loads are assumed to be axisymmetric.

The stiffness matrix is computed via an integral over the 2D domain of the original parent solid element. The explicit thickness integration is difficult except for straight element sides and constant thickness. In practice all integrals are computed using a 2D Gauss quadrature as

$$\mathbf{K}_{ij}^{(e)} = 2\pi \sum_{p=1}^{n_\xi} \sum_{q=1}^{n_\zeta} (\mathbf{B}_i'^T \, \mathbf{D}' \, \mathbf{B}_j' \, x |\mathbf{J}^{(e)}|)_{p,q} W_p W_q$$

$$\mathbf{f}_i^{(e)} = 2\pi \sum_{p=1}^{n_\xi} \sum_{q=1}^{n_\zeta} (\mathbf{N}_i^T \, \mathbf{b} \, x |\mathbf{J}^{(e)}|)_{p,q} W_p W_q \qquad (9.168)$$

where n_ξ and n_ζ are the integration points in directions ξ and ζ, respectively. Typically, $n_\zeta = 2$ is taken, except for laminate shells. These elements suffer from transverse shear locking, as well as from membrane locking (in their curve form). Shear locking can be avoided by using a reduced quadrature for $\mathbf{K}_s^{(e)}$. Membrane locking can also be alleviated by underintegrating $\mathbf{K}_p'^{(e)}$ at the expense of underevaluating the bending stiffness too. A more precise selective integration technique would involve splitting the membrane and bending contribution following the lines explained in Section 10.14 for 3D degenerated shell elements. Both shear and membrane locking can also be avoided by assumed strain procedures.

The simplest degenerated axisymmetric shell elements are the 2-noded troncoconical and the curved 3-noded elements with linear and quadratic shape functions, respectively (Figure 9.35). For homogeneous material $n_\xi = 1$ and $n_\zeta = 2$ for the 2-noded element, and $n_\xi = n_\zeta = 2$ for the 3-noded element are recommended.

These elements are an alternative to standard Reissner–Mindlin axisymmetric shell elements. In the later, the internal virtual work is expressed in terms of magnitudes resulting from the a priori integration in the thickness direction, such as resultant stresses and generalized strains. In degenerated axisymmetric shell elements the PVW is expressed in terms of stresses and strains and the thickness integration is performed numerically when computing the stiffness matrix. Which formulation to choose

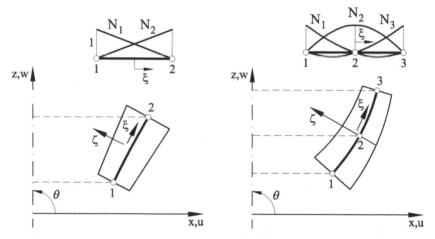

Fig. 9.35 Linear (2-noded) and quadratic (3-noded) axisymmetric shell elements obtanied by degeneration of the corresponding axisymmetric solid element

is a question of taste and basically depends on the experience of the user. The two approaches share the same basic assumptions and even the element expressions are the same in some cases, such as for the 2-noded linear element with constant thickness.

9.17 HIGHER ORDER COMPOSITE LAMINATED AXISYMMETRIC SHELL ELEMENTS

9.17.1 Layer-wise theory

An enhancement in the prediction of the correct shear and axial stresses for composite laminated axisymmetric shells can be achieved by using *layer-wise* theory. As explained in Section 3.15, in layer-wise theory the thickness coordinate is split into a number of *analysis layers* that may or not coincide with the number of laminate plies. The kinematics are independently described within each layer and certain physical continuity requirements are enforced.

For laminated axisymmetric shells the local displacement field in layer-wise theory is described as

$$u'(s, z') = \sum_{j=0}^{N} N_j(z') u'^j(s)$$

$$(9.169)$$

$$w'(s, z') = w'_0(s)$$

where N is the number of analysis layers, u' and w' are the axial and lateral displacements respectively and N_j are linear shape functions for each layer.

A well known drawback of layer-wise theory is that the number of kinematic variables depends on the number of analysis layers. The layer displacements u'^j can be however condensed at each section in terms of the axial displacement for the top layer during the equation solution process, as described in Section 7.7.3 for laminated plates.

9.17.2 Zigzag theory

Zigzag theories assume a zigzag pattern for the axial displacements across the laminate depth. Importantly, the number of kinematic variables in zigzag theories is *independent of the number of layers*.

In this section we will derive a two-noded axisymmetric shell element based on an extension of the refined zigzag theory (RZT) presented in Section 3.17. This theory was used in Sections 3.18, 7.8 and 8.16.2 for deriving zigzag beam, plate and flat shell elements, respectively.

9.17.2.1 Displacement, strain and stress fields

The kinematic field in the RZT for axisymmetric shells is written as

$$u'^k(s, z') = u'_0(s) - z'\theta(s) + \bar{u}'^k(s, z')$$

$$w'(s, z') = w'_0(s)$$

$$(9.170a)$$

with

$$\bar{u}'^k(s, z') = \phi^k(z')\Psi(s) \qquad (9.170b)$$

is the zigzag displacement. In above equations superscript k indicates quantities within the kth layer with $z'_k \leq z' \leq z'_{k+1}$ and z'_k is the thickness coordinate of the kth interface. In Eq.(9.170) the uniform axial displacement $u'_0(x)$, the rotation $\theta(x)$ and the transverse deflection $w'_0(x)$ are the primary kinematic variables of the underlying equivalent single-layer Reissner-Mindlin axisymmetric theory studied in the previous sections. Function $\phi^k(z')$ denotes a piecewise linear zigzag function and $\Psi(s)$ is a primary kinematic variable that defines the amplitude of the zigzag function along the generatrix. The interfacial axial displacement field has the zigzag distribution as shown in Figure 9.36.

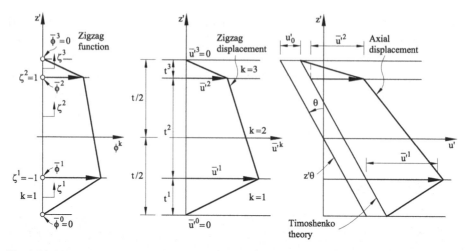

Fig. 9.36 Notation for a three-layered laminate and zigzag function in an axisymmetric shell

Assuming that elements are troncoconical we write the local strain-local displacement relationship using Eqs.(9.18), (9.25) and (9.170a) as

$$
\varepsilon_{x'} = \frac{\partial u'}{\partial x'} = \frac{\partial u'_0}{\partial s} - z'\frac{\partial \theta}{\partial s} + \phi^k \frac{\partial \Psi}{\partial s}
$$

$$
\varepsilon_{y'} = \frac{\partial v'}{\partial y'} = \frac{u'_0 \cos\phi - w'_0 \sin\phi}{r} - \frac{z'\theta\cos\phi}{r} + \frac{\phi^k \cos\phi}{r}\Psi \qquad (9.171)
$$

$$
\gamma_{x'z'} = \frac{\partial u'}{\partial z'} + \frac{\partial w'}{\partial x'} = \frac{\partial w'_0}{\partial s} - \theta + \beta^k \Psi
$$

where $\beta^k = \frac{\partial \phi^k}{\partial z'}$.

The local strains and the local generalized strains are related as

$$
\varepsilon' = \left\{ \begin{array}{c} \varepsilon_{x'} \\ \varepsilon_{y'} \\ \varepsilon_{x'z'} \end{array} \right\} = \begin{bmatrix} 1 & 0 & -z' & 0 & \phi^k & 0 & 0 & 0 \\ 0 & 1 & 0 & -z' & 0 & \phi^k & 0 & 0 \\ 0 & 0 & 0 & 0 & 0 & 0 & 1 & \beta^k \end{bmatrix} \left\{ \begin{array}{c} \hat{\varepsilon}'_m \\ \hat{\varepsilon}'_b \\ \hat{\varepsilon}'_s \end{array} \right\} = \mathbf{S}\hat{\varepsilon}' \qquad (9.172)
$$

when $\hat{\varepsilon}'_m$ is deduced from Eq.(9.51) and

$$
\hat{\varepsilon}'_b = \left[\frac{\partial\theta}{\partial s}, \frac{\theta\cos\phi}{r}, \frac{\partial\Psi}{\partial s}, \frac{\Psi\cos\phi}{r} \right]^T \quad , \quad \hat{\varepsilon}'_s = \left[\frac{\partial w'_0}{\partial s} - \theta, \Psi \right]^T \qquad (9.173)
$$

The local resultant stresses are defined as

$$
\hat{\sigma}' = \left\{ \begin{array}{c} \hat{\sigma}'_m \\ \hat{\sigma}'_b \\ \hat{\sigma}'_s \end{array} \right\} = \int_{-t/2}^{t/2} \mathbf{S}^T \sigma' dz' \qquad (9.174)
$$

The axial force vector $\hat{\sigma}'_m$ has the standard form of Eq.(9.34) while

$$\hat{\sigma}'_b = [M_{x'}, M_{y'}, M^{\phi}_{x'}, M^{\phi}_{y'}]^T \quad , \quad \hat{\sigma}'_s = [Q_{z'}, Q^{\phi}]^T \tag{9.175}$$

where $Q_{z'}$ is the standard shear force and $M^{\phi}_{x'}, M^{\phi}_{y'}, Q^{\phi}$ are additional local resultant stresses emanating from the zigzag theory.

A convenient form of matrix $\mathbf{S}$ is

$$\mathbf{S} = \begin{bmatrix} \mathbf{I}_2 \, \mathbf{S}_{\phi} & \mathbf{0} \\ \mathbf{0} & \mathbf{0} \, \mathbf{S}_{\beta} \end{bmatrix} \quad \text{with } \mathbf{S}_{\phi} = \begin{bmatrix} -z' & 0 & \phi^k & 0 \\ 0 & -z' & 0 & \phi^k \end{bmatrix} \quad , \quad \mathbf{S}_{\beta} = [1, \beta^k] \tag{9.176}$$

and $\mathbf{I}_2$ is the 2×2 unit matrix. The form of the null matrices in $\mathbf{S}$ is deduced from the observation of $\mathbf{S}_{\phi}$ and $\mathbf{S}_{\beta}$.

Substituting Eqs.(9.31b) and (9.172) into (9.174) gives the generalized constitutive expression as

$$\hat{\sigma}' = \hat{\mathbf{D}}'\hat{\varepsilon}' \quad \text{with } \hat{\mathbf{D}}' = \int_{-t/2}^{t/2} \mathbf{S}^T \mathbf{D}'^k \mathbf{S} dz' \tag{9.177}$$

A simpler form of the generalized constitutive matrix $\hat{\mathbf{D}}'$ can be obtained using the expression for $\mathbf{D}'$ of Eq.(9.31b) and Eq.(9.176) as

$$\hat{\mathbf{D}}' = \begin{bmatrix} \hat{\mathbf{D}}'_m & \hat{\mathbf{D}}'_{mb} & \mathbf{0} \\ [\hat{\mathbf{D}}'_{mb}]^T & \hat{\mathbf{D}}'_b & \mathbf{0} \\ \mathbf{0} & \mathbf{0} & \hat{D}'_s \end{bmatrix} \tag{9.178}$$

where $\hat{\mathbf{D}}'_m$ coincides with the expression of Eq.(9.38a) and

$$\hat{\mathbf{D}}'_{mb} = \int_{-t/2}^{t/2} \mathbf{D}'_p \mathbf{S}_{\phi} dz' \,, \; \hat{\mathbf{D}}'_b = \int_{-t/2}^{t/2} \mathbf{S}^T_{\phi} \mathbf{D}'_p \mathbf{S}_{\phi} dz' \,, \; \hat{D}'_s = \int_{-t/2}^{t/2} G_{sz'} \mathbf{S}^T_{\beta} \mathbf{S}_{\beta} dz' \tag{9.179}$$

where $\mathbf{D}'_p$ is given in Eq.(9.32a). Note that a shear correction factor is not needed, as it usual in zigzag theory [OEO2,TDG,TDG2].

If zigzag effects are neglected, then $\phi^k = 0$ and $\beta^k = 0$ and the constitutive expressions of classical Reissner-Mindlin laminated axisymmetric shell theory are recovered (see Eqs.(9.37) and (9.38)).

9.17.2.2 Computation of the zigzag function

Within each layer the zigzag function is expressed as

$$\phi^k = \frac{1}{2}(1 - \zeta^k)\bar{\phi}^{k-1} + \frac{1}{2}(1 + \zeta^k)\bar{\phi}^k = \frac{\bar{\phi}^k + \bar{\phi}^{k-1}}{2} + \frac{\bar{\phi}^k - \bar{\phi}^{k-1}}{2}\zeta^k \tag{9.180}$$

where $\bar{\phi}^k$ and $\bar{\phi}^{k-1}$ are the zigzag function values of the k and $k-1$ interface, respectively with $\bar{\phi}^0 = \bar{\phi}^N = 0$ and $\zeta^k = \frac{2(z'-z'^{k-1})}{t^k} - 1$ where t^k is the thickness of the kth layer.

The above form of ϕ^k is identical to that used in Section 3.17.1 for composite laminated beams.

From Eq.(9.180) and the conditions $\bar{\phi}^0 = \bar{\phi}^N = 0$ we deduce

$$\beta^k = \frac{\partial \phi^k}{\partial z'} = \frac{\bar{\phi}^k - \bar{\phi}^{k-1}}{t^k} \quad \text{and} \quad \int_{-t/2}^{t/2} \beta^k dz' = 0 \qquad (9.181)$$

The computation of function β^k follows precisely the arguments given for composite laminated beams in Section (3.17). The result is

$$\beta^k = \frac{G}{G_{s_{z'}}^k} - 1 \quad \text{with } G = \left[\frac{1}{t}\int_{-t/2}^{t/2}\frac{dz'}{G_{s_{z'}}^k}\right]^{-1} = \left[t\sum_{k=1}^{n_l}\frac{t^k}{G_{s_{z'}}^k}\right]^{-1} \qquad (9.182)$$

From Eqs.(9.181) the following recursion relation for the zigzag function value at the layer interface is obtained

$$\bar{\phi}_k = \sum_{i=1}^{k} h^i \beta^i \quad \text{with} \quad \bar{\phi}^0 = \bar{\phi}^N = 0 \qquad (9.183)$$

Introducing Eq.(9.183) into (9.180) gives the expression for the zigzag function as

$$\phi^k = \frac{t^k \beta^k}{2}(\xi^k - 1) + \sum_{i=1}^{k} t^i \beta^i \qquad (9.184)$$

Function Ψ can be interpreted as a weighted-average shear strain quantity [TDG] and acts as an additional rotation. Ψ should therefore be prescribed to zero at a clamped edge and left unprescribed at a free edge.

For homogeneous material $G_{s_{z'}}^k = G$, $\beta^k = 0$, the zigzag function ϕ^k vanishes and we recover the kinematic and constitutive expression of the Reissner-Mindlin axisymmetric shell theory previously studied.

9.17.2.3 Two-noded zigzag axisymmetric shell element

The kinematic variables are u_0', w_0', θ and Ψ. They can be discretized using C° linear 2-noded axisymmetric shell elements in the standard form as

$$\mathbf{u} = \left[u_0', w_0', \theta, \Psi\right]^T = \sum_{i=1}^{2} N_i \mathbf{a}_i \quad \text{with} \quad \mathbf{a}_i = \left[u_{0_i}', w_{0_i}', \theta_i, \Psi_i\right]^T \qquad (9.185)$$

The local generalized strain matrix is

$$\mathbf{B}' = [\mathbf{B}'_1, \mathbf{B}'_2] \tag{9.186a}$$

$$
\mathbf{B}'_i =
\begin{bmatrix}
\dfrac{\partial N_i}{\partial s} & 0 & 0 & 0 \\[2mm]
\dfrac{N_i \cos\phi}{r} & -\dfrac{N_i \sin\phi}{r} & 0 & 0 \\
\cdots & \cdots & \cdots & \cdots \\[2mm]
0 & 0 & \dfrac{\partial N_i}{\partial s} & 0 \\[2mm]
0 & 0 & \dfrac{N_i \cos\phi}{r} & 0 \\[2mm]
0 & 0 & 0 & \dfrac{\partial N_i}{\partial s} \\[2mm]
0 & 0 & 0 & \dfrac{N_i \cos\phi}{r} \\
\cdots & \cdots & \cdots & \cdots \\[2mm]
0 & \dfrac{\partial N_i}{\partial z'} & -N_i & 0 \\[2mm]
0 & 0 & 0 & N_i
\end{bmatrix}
=
\begin{Bmatrix}
\mathbf{B}_{m_i} \\
\mathbf{B}_{b_i} \\
\mathbf{B}_{s_i}
\end{Bmatrix}
\tag{9.186b}
$$

where indexes m, b and s denote as usual the contribution form the local generalized strains vectors $\hat{\boldsymbol{\varepsilon}}'_m, \hat{\boldsymbol{\varepsilon}}'_b$ and $\hat{\boldsymbol{\varepsilon}}'_s$, respectively.

The local element stiffness matrix is obtained by Eq.(9.57) with $\mathbf{B}_i$ and $\hat{\mathbf{D}}'$ given by Eqs.(9.186b) and (9.178), respectively. One single point quadrature is used to compute all the terms of $\mathbf{K}_{ij}$ and this eliminates shear locking. Transformation to global axes follows precisely the steps of Section 9.4.3 with the transformation matrix $\mathbf{L}_i^{(e)}$ given by

$$\mathbf{L}_i^{(e)} = \begin{bmatrix} \mathbf{L} & 0 \\ 0 & 1 \end{bmatrix} \tag{9.187}$$

where $\mathbf{L}$ is given in Eq.(9.14) and the additional row and column take into account the "rotational" DOF ψ_i emanating from the zigzag theory.

Applications of the two-noded zigzag axisymmetric shell element to composite laminated shells can be found in [OEO4].

9.18 FINAL REMARKS

The formulation of axisymmetric shell elements based on Reissner-Mindlin and Kirchhoff assumptions has been studied in some detail. The 2-noded troncoconical element based on Reissner-Mindlin theory with just one integration point is the simplest choice for practical purposes.

The simplification of the general axisymmetric shell theory for membranes, circular plates, plane arches and shallow shells has shown the possibilities of the general methodology. Here again the 2-noded Reissner-Mindlin troncoconical element is very competitive for the analysis of the different structures considered.

Particular emphasis has been put on explaining the reasons for transverse shear and membrane locking and the different alternatives to eliminate these spurious effects. The ideas presented will be useful when studying 3D shell elements of arbitrary shape in the next chapter.

The ACC and ACV rotation-free troncoconical elements are an interesting option for analysis of thin axisymmetric shells and membranes using the displacements as the only nodal variables.

Axisymmetric shell elements obtained from the degeneration of 2D solid elements are an alternative to the standard shell formulation. The approach will be generalized for developing a family of 3D degenerated shell elements in the next chapter.

The refined zigzag theory opens new possibilities for a more accurate analysis of axisymmetric shells with composite laminated material.

10

CURVED 3D SHELL ELEMENTS AND SHELL STIFFNERS

10.1 INTRODUCTION

This chapter studies the derivation of curved shell elements for analysis of shells of arbitrary shape. Curved shell elements are an alternative to the "flat" shell elements studied in Chapter 8 and their formulation is interesting both from the theoretical and practical points of view.

Considerable literature exists on the topic of general shell analysis [Cal,Cas,Go2,Mo6,Ni,No,Ug,Vi,We,We2,We3,WOK,ZT2]. As early as in 1989, Noor [No2] compiled 411 books, 98 conference proceedings and 23 synthesis papers on different topics related to the analytical and numerical analysis of shells including a number of relevant publications dedicated to finite element shell analysis only. A review of shell theories and finite element models can be found in [BWBR,JS2,SBL].

An "ideal" curved shell element should be capable to model complex curved shell geometries and also account for both membrane and flexural effects. It should be able to satisfy the rigid body and constant strain patch tests, be free of membrane and shear locking as well as of spureous internal mechanisms, be easily combined with other element types and also give accurate results for coarse meshes. Finally, the element formulation should be simple, avoiding whenever possible displacement derivatives as nodal variables. The satisfaction of all these conditions is still a major and, to some extent, unsolved challenge. Nevertheless, this chapter presents a finite element formulation for analysis of shells of arbitrary geometry fulfilling a large number of these ideal requirements.

There are basically three options for analysis of arbitrary shape shells with curved elements:

1. To develop curved shell elements based on classical shell theory.

2. To use 3D solid elements of small thickness.

3. To develop curved shell elements by degeneration of solid elements.

Option 1 faces the intrinsic complexity and diversity of the shell theories proposed by different authors [BS2,CC,Fl,Ko,Kr,Na,Ni,No,SF,SFR, TW,Vl2]. A possibility is to use the simpler shallow shell theory [Ma3], the drawback being now that this is only applicable to small curvature shells.

Option 2 eliminates most problems associated to classical shell theories. It has only three drawbacks in practice. First, the thickness discretization introduces additional nodal variables which increase the computational cost. However, for the simplest linear interpolation through the thickness the cost is basically the same as for shell elements (6 nodal DOFs for 3D solid elements versus 5 DOFs per node for shell elements). Second, the thickness stiffness can be substantially smaller than the rest of stiffness terms for very thin shells and this can lead to ill-conditioning of the global equation system. This problem can be overcome by using double precision in the computation and very thin shells ($a/t = 10^2, 10^3$) have been analyzed with this simple remedy. Smaller thicknesses can be treated by choosing the relative displacements between the upper and lower element faces as nodal variables [WTDG,WZ]. The third drawback of solid elements is the difficulty in generating 3D meshes versus the simpler option of generating surface meshes only.

General 3D elasticity theory does not benefit from the simplifications intrinsic to shell theory, such as the inextensible straight normals and zero normal stress. These assumptions can be adequately introduced in the formulation of solid elements to yield the so-called "degenerated shell elements" [AIZ2]. Elements of this type where introduced in the previous chapter for deriving a particular class of axisymmetric shell elements (Section 8.16). 3D degenerated shell elements have enjoyed much popularity in past decades [Ah,AIZ,AIZ2,BD6,BR,ChR,FV,Ga,Ga2,HL,Ka,Pa2,Paw, ZT2]. These elements can be considered as a generalization of Reissner–Mindlin shell theory regarding the non–orthogonality of the surface normals and, therefore, they suffer from transverse shear locking. In their curved shapes they can also suffer from membrane locking. Shear and membrane locking in degenerated shell elements can be eliminated by using selective/reduced integration or assumed strain fields.

The name "degenerated shell element" is somehow misleading as, in fact, the degeneration process effectively occurs in the parent 3D solid element. We will however retain the denomination to distinguish degenerated shell elements from curved shell elements derived from shell theory.

Buechter and Ramm [BR] have shown that there is a practical coincidence between degenerated shell elements and those based on standard Reissner-Mindlin/Naghdi type shell theory. This favours using the former for practical purposes as their formulation is generally simpler.

The organization of the chapter is as follows. The first part focuses on the derivation of degenerated shell elements from 3D solid elements. We will find that the PVW in these shell elements is written in terms of stresses and strains, as for standard 3D solid elements, and the element stiffness is computed by integrals over the 3D solid element volume.

In the second part of the chapter we present first the development of "truly" curved shell elements based on the so-called continuum-based resultant (CBR) shell theory. CBR theory introduces simplifications in degenerated shell elements, so that the PVW can be expressed in terms of resultant stresses and generalized strains and the element stiffness matrix can be computed by integrals over the shell surface only.

After that we present an overview of curved shell elements based on the Discrete-Kirchhoff assumptions. Then we briefly describe the pros and cons of shell elements based on a 6 and 7 parameters formulation. This is followed an introduction to the so-called *isogeometric shell* analysis [CHB,HCB].

In the last part of the chapter we present a finite element formulation for analysis of arbitrary shaped shells stiffened with 3D beams.

10.2 DEGENERATED SHELL ELEMENTS. BASIC CONCEPTS

Figure 10.1 shows a 20-noded quadratic hexahedron and the corresponding degenerated shell element. The upper, middle and lower surfaces are curved whereas the transverse sections are limited by straight lines (fibers). The degeneration process implies the definition of a *reference surface* (generally coinciding with the shell middle surface) with respect to which all displacements of the shell points are defined.

The displacement field is specified assuming that the fibers remain straight and inextensible after deformation. In addition, the stress normal to the reference surface is ignored (plane stress condition). The first assumption introduces transverse shear deformation and it also allows using a C^0 continuous interpolation for all the kinematic variables (displacement and rotations). It is interesting that this formulation is identical to Reissner-Mindlin plate theory when particularized for flat surfaces.

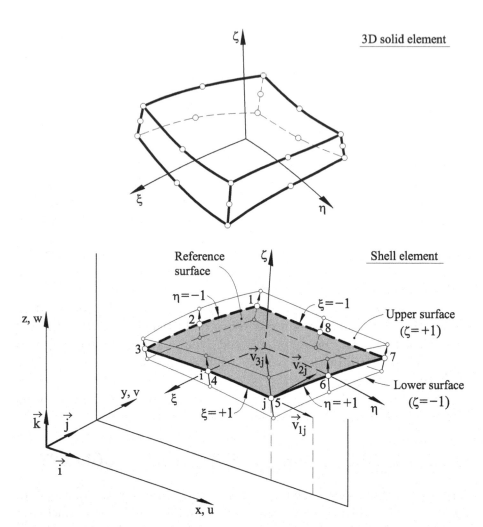

Fig. 10.1 Modification of a 20-noded hexahedron into a 8-noded degenerated shell element. Global, nodal and curvilinear coordinate systems

The reference surface is discretized into C^0 continuous curved shell elements with n nodes. Each node typically has five DOFs: three cartesian displacements and two rotations defining the motion of the normal to the shell surface. A third rotation may be necessary in folded shells, similarly as explained for flat shell elements.

Before proceeding any further we will define the different coordinate systems necessary for the analysis.

10.3 COORDINATE SYSTEMS

10.3.1 Global cartesian coordinate system x, y, z

The geometry of the shell is defined with respect to a global cartesian coordinate system x, y, z with associated unit vectors $\vec{i}, \vec{j}, \vec{k}$ (Figure 10.1). This system defines the directions for the global displacements u, v, w associated with the axes x, y, z, respectively; the global stiffness matrix $\mathbf{K}^{(e)}$ and the global equivalent nodal force vector $\mathbf{f}^{(e)}$ for each element.

10.3.2 Nodal coordinate system $\mathbf{v}_{1_i}, \mathbf{v}_{2_i}, \mathbf{v}_{3_i}$

A cartesian coordinate system formed by unit vectors $\vec{v}_{1_i}, \vec{v}_{2_i}, \vec{v}_{3_i}$ is defined at each node on the reference surface (Figure 10.1). This system is used to define the nodal rotations (Section 10.5). The fiber (or pseudo-normal) vector $\vec{v}_{3_i}$ can be chosen in one of the following three ways:

a) The components of vector $\vec{V}_{3_i}$ (with associated unit vector $\vec{v}_{3_i}$) in the global coordinate system are obtained from the coordinates of the two nodes placed on the upper and lower surfaces of the fiber associated with node i lying on the reference surface (Figure 10.2), i.e.

$$\vec{V}_{3_i} = \vec{r}_i^{\,\mathrm{up}} - \vec{r}_i^{\,\mathrm{lo}} \tag{10.1}$$

where

$$\vec{r}_i = x_i\,\vec{i} + y_i\,\vec{j} + z_i\,\vec{k} \tag{10.2}$$

In cartesian component form

$$\mathbf{r}_i = [x_i, y_i, z_i]^T, \quad \mathbf{V}_{3_i} = \left[V_{3_i}^x, V_{3_i}^y, V_{3_i}^z\right]^T \tag{10.3}$$

Vector $\mathbf{V}_{3_i}$ defines the fiber or thickness direction of node i while its modulus defines the nodal thickness. *The fiber direction is not necessarily normal to the reference surface* at i. This definition guarantees geometrical compatibility between folded elements (Figure 10.3). The drawback is that the coordinates of two points are necessary. Alternatively, the coordinates of node i, the fiber direction and the nodal thickness are required.

b) For shells with a smooth surface and uniform thickness it is simpler to define $\mathbf{V}_{3_i}$ as a vector *in the normal direction* at node i with a modulus equal to the shell thickness. The direction of $\mathbf{V}_{3_i}$ can be obtained by the cross product of two vectors which are tangent to the reference surface at i (see Eq.(10.16)).

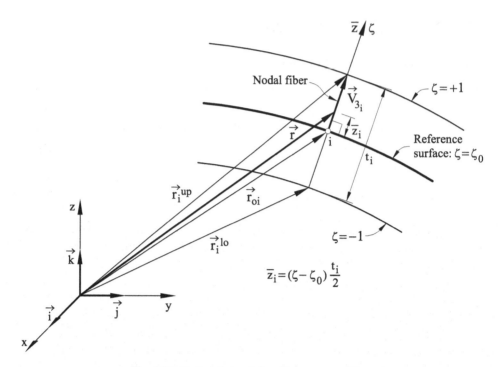

Fig. 10.2 Definition of the fiber vector $\mathbf{V}_{3_i}$

c) A third alternative is to define the fiber vector $\mathbf{V}_{3_i}$ as the average of all the normal vectors at node i corresponding to the elements sharing node i. This ensures geometric compatibility for coarse meshes and folded shells (Figure 10.3).

Vector $\mathbf{V}_{1_i}$ is defined as perpendicular to $\mathbf{V}_{3_i}$ and *contained within a plane parallel to the global plane xz*. Thus

$$\mathbf{V}_{1_i} = \mathbf{j} \wedge \mathbf{V}_{3_i} = V_{3_i}^z \mathbf{i} - V_{3_i}^x \mathbf{k} \tag{10.4}$$

If $\mathbf{V}_{3_i}$ coincides with the global y axis then $V_{3_i}^x = V_{3_i}^z = 0$ and $\mathbf{V}_{1_i}$ is taken in the direction of the x axis, i.e.

$$\mathbf{V}_{1_i} = -V_{3_i}^y \mathbf{i} \tag{10.5}$$

Finally, vector $\mathbf{V}_{2_i}$ is obtained by the cross product of $\mathbf{V}_{3_i}$ and $\mathbf{V}_{1_i}$ as

$$\mathbf{V}_{2_i} = \mathbf{V}_{3_i} \wedge \mathbf{V}_{1_i} \tag{10.6}$$

The above definition of $\mathbf{V}_{2_i}$ and $\mathbf{V}_{3_i}$ ensures the coincidence of these vectors for adjacent elements sharing node i in smooth shells. This defines

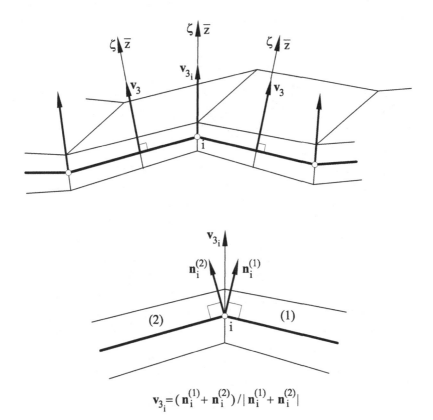

$$\mathbf{v}_{3_i} = (\mathbf{n}_i^{(1)} + \mathbf{n}_i^{(2)}) / |\mathbf{n}_i^{(1)} + \mathbf{n}_i^{(2)}|$$

Fig. 10.3 Geometric compatibility in a folded shell. Computation of unit pseudonormal (fiber) vector at a node

the nodal rotations uniquely and avoids the global transformation of the rotations for assembly purposes (Section 10.10).

The unit vectors associated to $\mathbf{V}_{1_i}$, $\mathbf{V}_{2_i}$ and $\mathbf{V}_{3_i}$ are $\mathbf{v}_{1_i}$, $\mathbf{v}_{2_i}$ and $\mathbf{v}_{3_i}$, respectively.

10.3.3 Curvilinear parametric coordinate system ξ, η, ζ

A normalized curvilinear system ξ, η, ζ is defined such that ζ is a linear coordinate in the thickness direction at each point on the reference surface (Figures 10.1 and 10.2). ζ takes the values $+1$ and -1 at the upper and lower surfaces, respectively and zero at the middle surface. The thickness direction at a point is obtained by standard interpolation of the nodal thickness (fiber) directions. Clearly, the direction of ζ coincides with that of $\mathbf{V}_{3_i}$ at each node (Figures 10.4 and 10.5). ξ and η are curvilinear para-

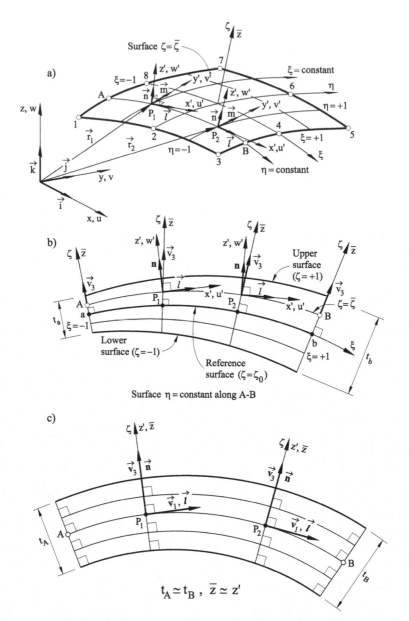

Fig. 10.4 (a) Lamina and curvilinear coordinate systems across the shell thickness. (b) Tapered shell with skew edges. (c) Quasi-uniform thickness (mild taper) shell with normal edges

metric coordinates describing the element surface as

$$\mathbf{x} = x(\xi, \eta)\mathbf{i} + y(\xi, \eta)\mathbf{j} + z(\xi, \eta)\mathbf{k} \tag{10.7}$$

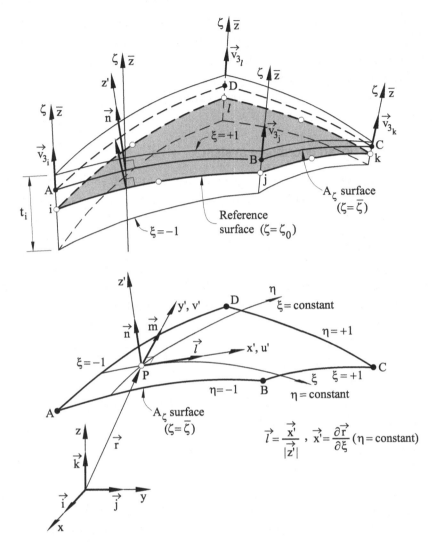

Fig. 10.5 Local coordinate system at a point P across the shell thickness

The direction for ξ is defined by the numbering of the first three nodes of the element (Figures 10.1 and 10.4). The relationship between the curvilinear and global coordinates is governed by the isoparametric description of the element geometry (Section 10.4).

A new *thickness coordinate* $\bar{z}$ in the direction of ζ is defined as

$$\bar{z} = (\zeta - \zeta_0)\frac{t}{2} \tag{10.8}$$

where t is the shell thickness and ζ_0 is the value of ζ at the reference surface ($\zeta_0 = 0$ if the reference surface coincides with the shell middle surface, as usually happens). Coordinate $\bar{z}$ measures the distance of a point to the reference surface along the thickness direction ζ. If the thickness is constant across the shell (mild taper, Figure 10.4c), then ζ coincides with the local axis z' and then $\bar{z} = z'$.

10.3.4 Lamina (local) coordinate system $\mathbf{x'}, \mathbf{y'}, \mathbf{z'}$

A cartesian coordinate system x', y', z' is chosen at each shell point defining the direction for the local displacements u', v', w'. The so called *lamina (or local) coordinate system* changes from point to point in the shell. The local system is used for defining and computing the local strain and stress fields. The unit vectors associated to directions x', y', z' are $\mathbf{l}$, $\mathbf{m}$, $\mathbf{n}$, respectively. Direction z' is perpendicular to the surface $\xi = $ constant (Figure 10.4). The normal vector $\mathbf{z'}$ is obtained as the cross product of two tangent vectors to the curves $\xi = $ constant and $\eta = $ constant at each point, i.e.

$$\mathbf{z'}_{\zeta=\bar{\zeta}} = \left(\frac{\partial \mathbf{r}}{\partial \xi}\right)_{\bar{\zeta}, \eta=\bar{\eta}} \wedge \left(\frac{\partial \mathbf{r}}{\partial \eta}\right)_{\bar{\zeta}, \xi=\bar{\xi}} \tag{10.9}$$

where $\mathbf{r} = [x, y, z]^T$ is the position vector of a shell point (Figure 10.2). The unit normal vector $\mathbf{n}$ is obtained as

$$\mathbf{n} = \frac{1}{|\mathbf{z'}|}\mathbf{z'} \tag{10.10}$$

The direction of x' is taken as tangent to the curvilinear coordinate $\xi = $ constant at each point (Figures 10.4 and 10.5). Thus

$$\mathbf{x'} = \left(\frac{\partial \mathbf{r}}{\partial \xi}\right)_{\eta=\bar{\eta}} = \left[\frac{\partial x}{\partial \xi}, \frac{\partial y}{\partial \xi}, \frac{\partial z}{\partial \xi}\right]^T_{\eta=\bar{\eta}} \tag{10.11}$$

and the associated unit vector is $\mathbf{l} = \frac{1}{|\mathbf{x'}|}\mathbf{x'}$.

The unit vector $\mathbf{m'}$ associated to y' is obtained by the cross product of vectors $\mathbf{n}$ and $\mathbf{l}$, i.e. $\mathbf{m} = \mathbf{n} \wedge \mathbf{l}$. Note that the y' direction is not tangent to the curvilinear coordinate η defined by $\xi = $ constant (Figure 10.5).

The local coordinate system changes at each point. If the shell thickness is constant, then the x', y', z' system is also constant across the thickness.

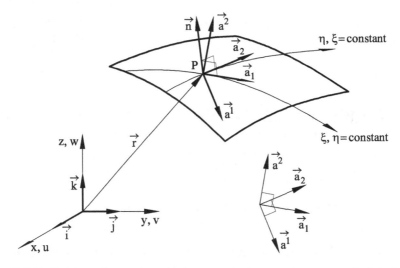

Fig. 10.6 Definition of covariant and contravariant systems at a shell point P

10.3.5 Covariant and contravariant coordinate systems

Two additional coordinate systems are usually defined for shell analysis. The *covariant* system $\mathbf{a}_1$, $\mathbf{a}_2$, $\mathbf{n}$ where $\mathbf{a}_1 = \frac{d\mathbf{r}}{d\xi}$ and $\mathbf{a}_2 = \frac{d\mathbf{r}}{d\eta}$ are vectors tangent to the lines $\eta = $ constant and $\xi = $ constant, respectively and $\mathbf{n}$ is the normal vector. Note that $\mathbf{a}_1$ coincides with the local vector x' (Figure 10.6). In general $\mathbf{a}_1$ and $\mathbf{a}_2$ are not orthogonal.

The *contravariant* system (or dual system) $\mathbf{a}^1$, $\mathbf{a}^2$, $\mathbf{n}$ with

$$\mathbf{a}^1 = \frac{1}{a}(a_{22}\mathbf{a}_1 - a_{12}\mathbf{a}_2) \quad ; \quad \mathbf{a}^2 = \frac{1}{a}(-a_{21}\mathbf{a}_1 - a_{11}\mathbf{a}_2) \tag{10.12}$$

where $a = a_{11}a_{22} - a_{12}a_{21}$ and a_{ij} are the components of vectors $\mathbf{a}_1$ and $\mathbf{a}_2$. Vector $\mathbf{a}^1$ and $\mathbf{a}^2$ satisfy

$$\begin{aligned} \mathbf{a}^i\mathbf{a}_j &= 0 \quad i \neq j \\ &= 1 \quad i = j \quad i,j = 1,2 \end{aligned} \tag{10.13a}$$

i.e. $\mathbf{a}^1$ and $\mathbf{a}^2$ are orthogonal to $\mathbf{a}_1$ and $\mathbf{a}_2$, respectively (Figure 10.6). Also [BD6,Cal,No]

$$d\xi = [\mathbf{a}^1]^T d\mathbf{r}, \quad d\eta = [\mathbf{a}^2]^T d\mathbf{r} \tag{10.13b}$$

The covariant and contravariant systems are useful for developing consistent expressions for the shell curvatures and the equilibrium equations in classical shell theory. Some authors have used these coordinate systems to formulate curved shell elements [BD6,BD,DB,JP,JP2]. Their use in this text will be limited to a few particular cases, such as the definition of assumed transverse shear strain fields for some shell elements.

10.4 GEOMETRIC DESCRIPTION

The position of a point within an element can be obtained by the standard isoparametric interpolation of the nodal coordinates [On4] as

$$\mathbf{r} = \sum_{i=1}^{n} N_i(\xi, \eta) \frac{1+\zeta}{2} \mathbf{r}_i^{\text{up}} + \sum_{i=1}^{n} N_i(\xi, \eta) \frac{1-\zeta}{2} \mathbf{r}_i^{\text{lo}} \tag{10.14}$$

where $\mathbf{r}_i^{\text{up}}$ and $\mathbf{r}_i^{\text{lo}}$ are the position vectors for the upper and lower surfaces at each node i; $N_i(\xi, \eta)$ is the 2D shape function of node i (Figure 10.2; ζ defines the position of the point in the thickness direction and n is the number of element nodes. If the geometry is defined by the reference surface (as it is the usual case) Eq.(10.14) is changed to (Figure 10.7)

$$\vec{r} = \vec{r}_0 + \bar{z}_p \vec{v}_3 \tag{10.15a}$$

where $\vec{r}_0$ defines the position vector of a point on the reference surface, $\vec{v}_3$ is the pseudo-normal vector at O and $\bar{z}_p$ is the coordinate of point P across the shell thickness. The first term in the r.h.s. in Eq.(10.15a) defines the position of a point on the reference surface, whereas the second defines the position of the point in the thickness direction. Recall that $\mathbf{v}_{3_i}$ is not necessarily normal to the reference surface (Section 10.3.2).

Vectors $\vec{r}_0$ and the product $\bar{z}_p \vec{v}_3$ can be interpolated from the nodal values to give (in component form)

$$\mathbf{r} = \sum_{i=1}^{n} N_i(\xi, \eta) \mathbf{r}_i = \sum_{i=1}^{n} N_i(\xi, \eta) \left[\mathbf{r}_{0_i} + \bar{z}_i \mathbf{v}_{3_i} \right] \tag{10.15b}$$

where $\mathbf{r}_{0_i}$ is the vector defining the position of node i, $\mathbf{v}_{3_i}$ is the unit normal nodal vector at the node and

$$\bar{z}_i = (\zeta - \zeta_0) \frac{t_i}{2} \tag{10.15c}$$

with t_i being the nodal thickness (Figures 10.5 and 10.7).

The computation of the normal direction $\mathbf{z}'$ at each point is easily performed combining Eqs.(10.9) and (10.15b). In particular, the normal

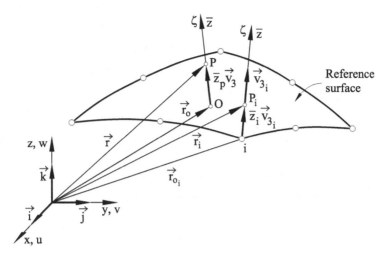

Fig. 10.7 Position vector of an arbitrary point $P(\mathbf{r})$ on a normal direction $\mathbf{V}_3$ and of a point P_i on the thickness direction at node i ($\mathbf{r}_i$)

vector to the reference surface at a node ($\bar{z}_i = 0$) is obtained by

$$\mathbf{z}'_{0_i} = \sum_{j=1}^{n} \left(\frac{\partial N_j(\xi_i, \eta_i)}{\partial \xi} \mathbf{r}_{0_j} \wedge \frac{\partial N_j(\xi_i, \eta_i)}{\partial \eta} \mathbf{r}_{0_j} \right) \quad (10.16)$$

This expression is particularly useful if the fiber vector $\mathbf{V}_{3_i}$ is taken to coincide with $\mathbf{z}'_{0_i}$ (Section 10.3.2).

10.5 DISPLACEMENT FIELD

The three global displacements of a 3D shell point can be expressed in terms of the displacements of the reference surface and the rotation of the normal vector. Let us take, for instance, point A of Figure 10.8 located on the normal direction $\mathbf{v}_{3_i}$ at a distance z_i from node i ($\bar{z}_i \equiv \overline{iA}$). The displacement vector of point A can be written as

$$\mathbf{u}_i = \mathbf{u}_{0_i} + \bar{z}_i \mathbf{u}_{n_i} \quad (10.17)$$

where

$$\mathbf{u}_i = [u_i, v_i, w_i]^T \quad , \quad \mathbf{u}_{0_i} = [u_{0_i}, v_{0_i}, w_{0_i}]^T \quad (10.18)$$

are the displacement vectors of point A and node i, respectively, with

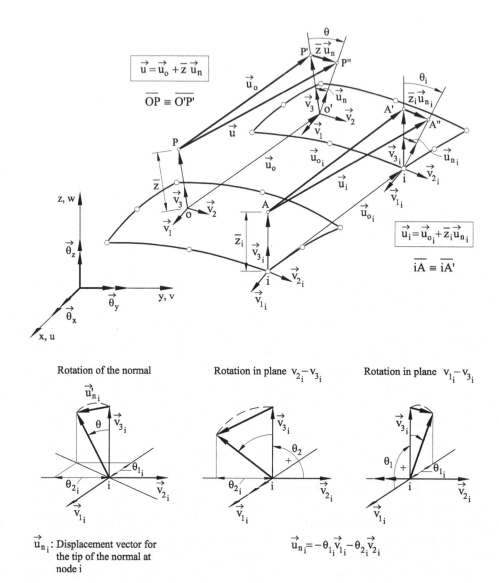

$\vec{u}_{n_i}$: Displacement vector for the tip of the normal at node i

$\vec{u}_{n_i} = -\theta_{1_i}\vec{v}_{1_i} - \theta_{2_i}\vec{v}_{2_i}$

Fig. 10.8 Displacement vector of a point and sign convention for nodal rotations

components in *the global coordinate system*, i.e.

$$\vec{u}_i = u_i\,\vec{i} + v_i\,\vec{j} + z_i\,\vec{k} \quad , \quad \vec{u}_{0_i} = u_{0_i}\,\vec{i} + v_{0_i}\,\vec{j} + w_{0_i}\,\vec{k} \qquad (10.19)$$

Index 0 in the above expression denotes points on the reference surface. Vector $\mathbf{u}_{n_i}$ in Eq.(10.17) defines the displacement of the end point of the unit normal vector $\mathbf{v}_{3_i}$. Thus $\bar{z}_i\mathbf{u}_{n_i}$ is the vector defining the relative displacement of point A with respect to node i. For simplicity we express the components of $\mathbf{u}_{n_i}$ in the nodal coordinate system. From Figure 10.8 it is deduced

$$\boxed{\mathbf{u}_{n_i} = -\theta_{1_i}\mathbf{v}_{1_i} - \theta_{2_i}\mathbf{v}_{2_i}} \tag{10.20}$$

where θ_{1_i} and θ_{2_i} are the two *local rotations* of the normal vector at node i. Rotations θ_{1_i} and θ_{2_i} are taken as *positive if they rotate in an anticlockwise direction* in the planes $\mathbf{v}_{1_i} - \mathbf{v}_{3_i}$ and $\mathbf{v}_{2_i} - \mathbf{v}_{3_i}$, respectively, i.e. when the normal end point moves in the opposite direction to the nodal vectors $\mathbf{v}_{1_i}$ and $\mathbf{v}_{3_i}$. This criterion is the same one as for the rotations in beams, plates and shells in previous chapters.

Vector $\mathbf{u}_{0_i}$ in Eq.(10.17) can be also interpreted as a rigid body translation of the element, so that the original fiber iA at a node i moves to the position iA' (Figure 10.8). Vector $\bar{z}_i\mathbf{u}_{n_i}$ provides the additional displacement so that the point A' reaches the final position A''.

The displacements of an arbitrary point P within an element (Figure 10.8) are expressed in terms of the nodal displacements by a standard interpolation. From Eqs.(10.17) and (10.20) we write

$$\mathbf{u} = \begin{Bmatrix} u \\ v \\ w \end{Bmatrix} = \sum_{i=1}^{n} N_i\mathbf{u}_i = \sum_{i=1}^{n} N_i \left[\mathbf{u}_{0_i} - \bar{z}_i[\mathbf{v}_{1_i}, \mathbf{v}_{2_i}] \begin{Bmatrix} \theta_{1_i} \\ \theta_{2_i} \end{Bmatrix} \right] = \sum_{i=1}^{n} \mathbf{N}_i\mathbf{a}_i^{(e)} \tag{10.21a}$$

where

$$\mathbf{a}_i^{(e)} = [u_{0_i}, v_{0_i}, w_{0_i}, \theta_{1_i}, \theta_{2_i}]^T \tag{10.21b}$$

is the *displacement vector* of node i containing the *three cartesian displacements* u_{0_i}, v_{0_i} and w_{0_i} and the *two local rotations* θ_{1_i} and θ_{2_i} defined in the nodal coordinate system,

$$\mathbf{N}_i = [\mathbf{I}_3, -\bar{z}_i\mathbf{C}_i] \tag{10.22a}$$

where $\mathbf{I}_3$ is the 3×3 unit matrix, and

$$\mathbf{C}_i = [\mathbf{v}_{1_i}, \mathbf{v}_{2_i}] \tag{10.22b}$$

contains the *cartesian components* of vectors $\mathbf{v}_{1_i}$ and $\mathbf{v}_{2_i}$.

10.6 LOCAL STRAIN MATRIX

The plane stress assumption ($\sigma_{z'} = 0$) allows us to simplify the constitutive equation for 3D elasticity written in the local axes x', y', z' by eliminating the normal strain $\varepsilon_{z'}$ (similarly as for flat shell elements). The local strain vector has the following five significant components

$$\varepsilon' = \left\{ \begin{array}{c} \varepsilon_{x'} \\ \varepsilon_{y'} \\ \gamma_{x'y'} \\ \cdots \\ \gamma_{x'z'} \\ \gamma_{y'z'} \end{array} \right\} = \left\{ \begin{array}{c} \dfrac{\partial u'}{\partial x'} \\ \dfrac{\partial v'}{\partial y'} \\ \dfrac{\partial u'}{\partial y'} + \dfrac{\partial v'}{\partial x'} \\ \cdots\cdots\cdots \\ \dfrac{\partial u'}{\partial z'} + \dfrac{\partial w'}{\partial x'} \\ \dfrac{\partial v'}{\partial z'} + \dfrac{\partial w'}{\partial y'} \end{array} \right\} = \left\{ \begin{array}{c} \varepsilon'_p \\ \cdots\cdots \\ \varepsilon'_s \end{array} \right\} \qquad (10.23)$$

where u', v', w' are the displacements in the local directions x', y', z', respectively and ε'_p and ε'_s are the *in-plane* strain vector and the *transverse shear* strain vector, respectively. Vector ε'_p includes both membrane and bending strains which are difficult to decouple for curved shells.

The components of ε' are expressed in terms of the nodal displacements as follows. First, the local derivatives of Eq.(10.23) are written in terms of the derivatives of the global displacements u, v, w with respect to the global coordinates x, y, z using the chain derivative rule as

$$\frac{\partial u'}{\partial x'} = \frac{\partial u'}{\partial x}\frac{\partial x}{\partial x'} + \frac{\partial u'}{\partial y}\frac{\partial y}{\partial x'} + \frac{\partial u'}{\partial z}\frac{\partial z}{\partial x'} = \frac{\partial u'}{\partial x}l^x + \frac{\partial u'}{\partial y}l^y + \frac{\partial u'}{\partial z}l^z \quad (10.24)$$

where l^x, l^y, l^z are the global components of the local vector $\mathbf{l}$ (Section 10.3.4). As usual l^x is the cosine of the angle formed by axes x' and x etc.
Also

$$\frac{\partial u'}{\partial x} = \frac{\partial u}{\partial x}\frac{\partial u'}{\partial u} + \frac{\partial v}{\partial x}\frac{\partial u'}{\partial v} + \frac{\partial w}{\partial x}\frac{\partial u'}{\partial w} = \frac{\partial u}{\partial x}l^x + \frac{\partial v}{\partial x}l^y + \frac{\partial w}{\partial x}l^z \quad (10.25)$$

Performing similar transformations for $\frac{\partial u'}{\partial y}$ and $\frac{\partial v'}{\partial z}$ gives, after substituting in (10.24)

$$\frac{\partial u'}{\partial x'} = \frac{\partial u}{\partial x}(l^x)^2 + \frac{\partial u}{\partial y}(l^y)^2 + \frac{\partial u}{\partial z}(l^z)^2 + \left(\frac{\partial u}{\partial y} + \frac{\partial v}{\partial x}\right)l^x l^y +$$
$$+ \left(\frac{\partial u}{\partial z} + \frac{\partial w}{\partial x}\right)l^x l^z + \left(\frac{\partial v}{\partial z} + \frac{\partial w}{\partial y}\right)l^y l^z \qquad (10.26)$$

Following the same procedure for all the terms in ε' gives

$$\varepsilon' = \mathbf{Q}\varepsilon \tag{10.27}$$

where

$$\varepsilon = \left[\varepsilon_x, \varepsilon_y, \varepsilon_z, \gamma_{xy}, \gamma_{xz}, \gamma_{yz}\right]^T = \left[\frac{\partial u}{\partial x}, \frac{\partial v}{\partial y}, \frac{\partial w}{\partial z}, \frac{\partial u}{\partial y} + \frac{\partial v}{\partial x}, \frac{\partial u}{\partial z} + \frac{\partial w}{\partial x}, \frac{\partial v}{\partial z} + \frac{\partial w}{\partial y}\right]^T \tag{10.28}$$

is the *global* strain vector and

$$\mathbf{Q} = \begin{bmatrix} (l^x)^2 & (l^y)^2 & (l^z)^2 & l^x l^y & l^x l^z & l^y l^z \\ (m^x)^2 & (m^y)^2 & (m^z)^2 & m^x m^y & m^y m^z & n^y n^z \\ 2l^x m^x & 2l^y m^y & 2l^z m^z & (l^x m^y + l^y m^x) & (l^z m^x + l^x m^z) & (l^y m^z + l^z m^y) \\ 2l^x n^x & 2l^y n^y & 2l^z n^z & (l^x n^y + l^y n^x) & (l^x n^z + l^z n^x) & (l^y n^z + l^z n^y) \\ 2m^x n^x & 2m^y n^y & 2m^z n^z & (m^x n^y + m^y n^x) & (m^x n^z + m^z n^x) & (m^y n^z + m^z n^y) \end{bmatrix} \tag{10.29}$$

Note that the global strain ε_z is non zero in Eq.(10.28).

The global cartesian derivatives in vector ε are next expressed in terms of the curvilinear derivatives of the displacement by

$$\begin{bmatrix} \dfrac{\partial u}{\partial x} & \dfrac{\partial v}{\partial x} & \dfrac{\partial w}{\partial x} \\ \dfrac{\partial u}{\partial y} & \dfrac{\partial v}{\partial y} & \dfrac{\partial w}{\partial y} \\ \dfrac{\partial u}{\partial z} & \dfrac{\partial v}{\partial z} & \dfrac{\partial w}{\partial z} \end{bmatrix} = \left[\mathbf{J}^{(e)}\right]^{-1} \begin{bmatrix} \dfrac{\partial u}{\partial \xi} & \dfrac{\partial v}{\partial \xi} & \dfrac{\partial w}{\partial \xi} \\ \dfrac{\partial u}{\partial \eta} & \dfrac{\partial v}{\partial \eta} & \dfrac{\partial w}{\partial \eta} \\ \dfrac{\partial u}{\partial \zeta} & \dfrac{\partial v}{\partial \zeta} & \dfrac{\partial w}{\partial \zeta} \end{bmatrix} \tag{10.30}$$

where $\mathbf{J}^{(e)}$ is the jacobian matrix (sometimes simply called "the Jacobian")

$$\mathbf{J}^{(e)} = \begin{bmatrix} \dfrac{\partial x}{\partial \xi} & \dfrac{\partial y}{\partial \xi} & \dfrac{\partial z}{\partial \xi} \\ \dfrac{\partial x}{\partial \eta} & \dfrac{\partial y}{\partial \eta} & \dfrac{\partial z}{\partial \eta} \\ \dfrac{\partial x}{\partial \zeta} & \dfrac{\partial y}{\partial \zeta} & \dfrac{\partial z}{\partial \zeta} \end{bmatrix} \tag{10.31}$$

The derivation of the local strain matrix therefore includes the following steps:

1. Compute the derivatives of the global displacements with respect to

the curvilinear coordinates ξ, η, ζ. From Eq.(10.21a) we obtain

$$\frac{\partial \mathbf{u}}{\partial \xi} = \left[\frac{\partial u}{\partial \xi}, \frac{\partial v}{\partial \xi}, \frac{\partial w}{\partial \xi} \right]^T = \sum_{i=1}^{n} \frac{\partial N_i}{\partial \xi} \left[\mathbf{I}_3, -\bar{z}_i \mathbf{C}_i \right] \mathbf{a}_i^{(e)}$$

$$\frac{\partial \mathbf{u}}{\partial \eta} = \left[\frac{\partial u}{\partial \eta}, \frac{\partial v}{\partial \eta}, \frac{\partial w}{\partial \eta} \right]^T = \sum_{i=1}^{n} \frac{\partial N_i}{\partial \eta} \left[\mathbf{I}_3, -\bar{z}_i \mathbf{C}_i \right] \mathbf{a}_i^{(e)}$$

$$\frac{\partial \mathbf{u}}{\partial \zeta} = \left[\frac{\partial u}{\partial \zeta}, \frac{\partial v}{\partial \zeta}, \frac{\partial w}{\partial \zeta} \right]^T = \sum_{i=1}^{n} N_i \left[\mathbf{0}, -\frac{t_i}{2} \mathbf{C}_i \right] \mathbf{a}_i^{(e)} \qquad (10.32)$$

2. Compute the terms in $\mathbf{J}^{(e)}$. From Eq.(10.15b)

$$\frac{\partial \mathbf{r}}{\partial \xi} = \left[\frac{\partial x}{\partial \xi}, \frac{\partial y}{\partial \xi}, \frac{\partial z}{\partial \xi} \right]^T = \sum_{i=1}^{n} \frac{\partial N_i}{\partial \xi} \left[\mathbf{r}_{0_i} + \bar{z}_i \mathbf{v}_{3_i} \right]$$

$$\frac{\partial \mathbf{r}}{\partial \eta} = \left[\frac{\partial x}{\partial \eta}, \frac{\partial y}{\partial \eta}, \frac{\partial z}{\partial \eta} \right]^T = \sum_{i=1}^{n} \frac{\partial N_i}{\partial \eta} \left[\mathbf{r}_{0_i} + \bar{z}_i \mathbf{v}_{3_i} \right]$$

$$\frac{\partial \mathbf{r}}{\partial \zeta} = \left[\frac{\partial x}{\partial \zeta}, \frac{\partial y}{\partial \zeta}, \frac{\partial z}{\partial \zeta} \right]^T = \sum_{i=1}^{n} N_i \frac{t_i}{2} \mathbf{v}_{3_i} \qquad (10.33)$$

3. Compute the global cartesian derivatives. From Eqs.(10.32) and (10.33)

$$\frac{\partial \mathbf{u}}{\partial x_j} = \sum_{i=1}^{n} \left[N_i^j \mathbf{I}_3, \mathbf{G}_i^j \right] \mathbf{a}_i^{(e)} \qquad (10.34)$$

with $x_j = x, y, z$ for $j = 1, 2, 3$, respectively and

$$N_i^j = J_{j1}^{-1} \frac{\partial N_i}{\partial \xi} + J_{j2}^{-1} \frac{\partial N_i}{\partial \eta} \quad ; \quad \mathbf{G}_i^j = -\left(\frac{t_i}{2} J_{j3}^{-1} N_i + \bar{z}_i N_i^j \right) \mathbf{C}_i$$

$$(10.35)$$

where J_{ij}^{-1} is the element ij of the inverse of the Jacobian $\mathbf{J}^{(e)}$. The global strain vector is expressed in terms of the global displacements as

$$\varepsilon = \sum_{i=1}^{n} \mathbf{B}_i \mathbf{a}_i^{(e)} = \mathbf{B} \mathbf{a}^{(e)} \qquad (10.36)$$

where $\mathbf{B}$ is the global strain matrix. Box 10.1 shows the explicit form for $\mathbf{B}_i$.

4. Compute the local strain vector. Substituting Eq.(10.36) into the transformation (10.27) gives finally

$$\varepsilon' = \mathbf{Q} \varepsilon = \sum_{i=1}^{n} \mathbf{Q} \mathbf{B}_i \mathbf{a}_i^{(e)} = \sum_{i=1}^{n} \mathbf{B}_i' \mathbf{a}_i^{(e)} = \mathbf{B}' \mathbf{a}^{(e)} \qquad (10.37)$$

$$\mathbf{B} = [\mathbf{B}_1, \mathbf{B}_2, \cdots\cdots, \mathbf{B}_n]$$

$$\mathbf{B}_i = \begin{bmatrix} N_i^1 & 0 & 0 & (\mathbf{G}_i^1)_{11} & (\mathbf{G}_i^1)_{12} \\ 0 & N_i^2 & 0 & (\mathbf{G}_i^2)_{21} & (\mathbf{G}_i^2)_{22} \\ 0 & 0 & N_i^3 & (\mathbf{G}_i^3)_{31} & (\mathbf{G}_i^3)_{32} \\ N_i^2 & N_i^1 & 0 & [(\mathbf{G}_i^2)_{11} + (\mathbf{G}_i^1)_{21}] & [(\mathbf{G}_i^2)_{12} + (\mathbf{G}_i^1)_{22}] \\ N_i^3 & 0 & N_i^1 & [(\mathbf{G}_i^3)_{11} + (\mathbf{G}_i^1)_{31}] & [(\mathbf{G}_i^3)_{12} + (\mathbf{G}_i^1)_{32}] \\ 0 & N_i^3 & N_i^2 & [(\mathbf{G}_i^3)_{21} + (\mathbf{G}_i^2)_{31}] & [(\mathbf{G}_i^3)_{22} + (\mathbf{G}_i^2)_{32}] \end{bmatrix}$$

$$N_i^j = J_{j_1}^{-1}\frac{\partial N_i}{\partial \xi} + J_{j_2}^{-1}\frac{\partial N_i}{\partial \eta} ; \qquad \mathbf{G}_i^j = -(\frac{t_i}{2}J_{j_3}^{-1}N_i + \bar{z}_i N_i^j)\mathbf{C}_i$$

Box 10.1 Global strain matrix for 3D degenerated shell elements

$\mathbf{B}'$ is the *local strain matrix* with

$$\mathbf{B}'_i = \mathbf{Q}\mathbf{B}_i \tag{10.38a}$$

$$\underset{5\times5}{\mathbf{B}'_i} = \begin{Bmatrix} \mathbf{B}'_{p_i} \\ \mathbf{B}'_{s_i} \end{Bmatrix} ; \qquad \underset{3\times5}{\mathbf{B}'_{p_i}} = \underset{3\times6}{\mathbf{Q}_1}\underset{6\times5}{\mathbf{B}_i} \quad ; \quad \underset{2\times5}{\mathbf{B}'_{s_i}} = \underset{2\times6}{\mathbf{Q}_2}\underset{6\times5}{\mathbf{B}_i} \tag{10.38b}$$

where $\mathbf{B}'_{p_i}$ and $\mathbf{B}'_{s_i}$ are the in-plane and transverse shear contributions to the local strain matrix, respectively. In Eq.(10.38b) $\mathbf{Q}_1$ and $\mathbf{Q}_2$ contain the first three files and the last two files of the transformation matrix $\mathbf{Q}$ of Eq.(10.29), respectively. Matrix $\mathbf{Q}$ must be constructed at each point where the strains are evaluated using the global components of the local vectors $\mathbf{l}$, $\mathbf{m}$ and $\mathbf{n}$ at the point.

10.7 STRESS VECTOR AND CONSTITUTIVE EQUATION

The stress vector at a shell point is written in the local coordinate system x', y', z', taking into account the plane stress assumption ($\sigma_{z'} = 0$) as

$$\boldsymbol{\sigma}' = \begin{Bmatrix} \sigma_{x'} \\ \sigma_{y'} \\ \tau_{x'y'} \\ \cdots\cdots \\ \tau_{x'z'} \\ \tau_{y'z'} \end{Bmatrix} = \begin{Bmatrix} \boldsymbol{\sigma}'_p \\ \cdots\cdots \\ \boldsymbol{\sigma}'_s \end{Bmatrix} \tag{10.39}$$

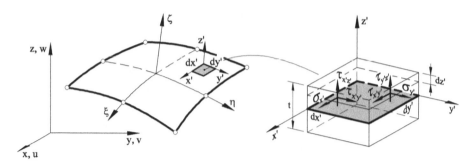

Fig. 10.9 Definition of local stresses at a shell point

where vectors σ'_p and σ'_s contain the in-plane and transverse shear stresses, respectively (Figure 10.9).

The relationship between the local stresses and strains can be written in the usual form

$$\sigma' = D'\varepsilon' + \sigma'_0 \tag{10.40}$$

where ε' was defined in Eq.(10.23) and σ'_0 is an initial stress vector. For initial stresses due to thermal effects and isotropic material [On4]

$$\sigma'_0 = -\frac{E\alpha\Delta T}{1 - \nu^2}[1, 1, 0, 0, 0]^T \tag{10.41}$$

where α is the thermal expansion coefficient and ΔT is the temperature increment at each point. The local constitutive matrix D' for an isotropic material coincides with that given in Section 8.3.3.

Let us consider a composite laminated shell formed by a number of orthotropic layers of variable thickness (Figure 10.10). It is assumed for each layer that the material is orthotropic with orthotropy axes 1,2,3 with $z' \equiv 3$ and satisfies the plane anisotropy conditions [BD6]. Consequently the membrane/bending effects are decoupled from the transverse shear effects on the plane x', y'. Under these assumptions the constitutive equation at a point of each layer is written in the orthotropy axes 1,2,3 as

$$\sigma_I = D_I \varepsilon_I \tag{10.42}$$

where

$$\sigma_I = \begin{Bmatrix} \sigma_1 \\ \sigma_2 \\ \tau_{12} \\ \cdots \\ \tau_{13} \\ \tau_{23} \end{Bmatrix} = \begin{Bmatrix} \sigma_1 \\ \cdots \\ \sigma_2 \end{Bmatrix} ; \qquad \varepsilon_I = \begin{Bmatrix} \varepsilon_1 \\ \varepsilon_2 \\ \gamma_{12} \\ \cdots \\ \gamma_{13} \\ \gamma_{23} \end{Bmatrix} = \begin{Bmatrix} \varepsilon_1 \\ \cdots \\ \varepsilon_2 \end{Bmatrix} \tag{10.43}$$

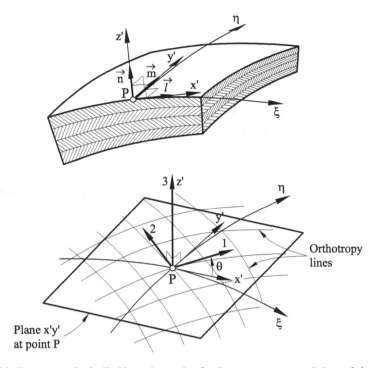

Fig. 10.10 Laminated shell (four layers). Orthotropy axes 1,2,3 and local axes x', y', z' with $z' \equiv 3$ at a point P

and
$$\mathbf{D}_I = \begin{bmatrix} \mathbf{D}_1 & \mathbf{0} \\ \mathbf{0} & \mathbf{D}_2 \end{bmatrix} \tag{10.44}$$

The orthotropy constitutive matrix $\mathbf{D}_I$ coincides with that of Eq.(7.8b) for composite plates. Transformation of $\mathbf{D}_I$ to the local coordinate system x', y', z' follows the rules explained in Section 7.3.3 leading to

$$\mathbf{D}' = \begin{bmatrix} \mathbf{D}'_p & \mathbf{0} \\ \mathbf{0} & \mathbf{D}'_s \end{bmatrix} \tag{10.45}$$

where
$$\mathbf{D}'_p = \mathbf{T}_1^T \mathbf{D}_1 \mathbf{T}_1 \quad \text{and} \quad \mathbf{D}'_s = \mathbf{T}_2^T \mathbf{D}_2 \mathbf{T}_2 \tag{10.46}$$

with $\mathbf{T}_1$ and $\mathbf{T}_2$ given by Eq.(6.15b).

The initial stress vectors $\boldsymbol{\sigma}'^0_p$ and $\boldsymbol{\sigma}'^0_s$ for an orthotropic material can be directly deduced from Eqs.(7.14a).

The transverse shear stresses obtained using Eq.(10.40) do not allow to satisfy the condition $\boldsymbol{\sigma}'_s = 0$ at the upper and lower faces of the shell or the continuity conditions at the interfaces of a composite laminated shell.

Transverse shear correction factors are introduced in $\mathbf{D}'_s$, similarly as in thick plates and flat shells for improving the representation of the transverse shear stresses. A simple approach is to neglect the influence of the curvature and to use the shear correction factors k_{ij} derived for composite laminated plates in Section 7.3. A further simplification is to assume cylindrical bending along x' and y' to define the two coefficients k_{11} and k_{22} that weight the diagonal terms of $\mathbf{D}'_s$. For an homogeneous shell we will use the classical coefficient $k_{11} = k_{22} = k = 5/6$ [BD6,Co1HO,NB,OF2].

10.8 STIFFNESS MATRIX AND EQUIVALENT NODAL FORCE VECTOR

The PVW for a 3D shell element is written as [On4,ZT2]

$$\iiint_{V^{(e)}} \delta\boldsymbol{\varepsilon}'^T \boldsymbol{\sigma}' \, dV = \iiint_{V^{(e)}} \delta\mathbf{u}^T \mathbf{b} \, dV + \iint_{A^{(e)}} \delta\mathbf{u}^T \mathbf{t} \, dA + \sum_{i=1}^{n} \left[\delta\mathbf{a}_i^{(e)}\right]^T \mathbf{q}^{(e)}$$

(10.47)

where the internal virtual work is expressed in terms of the local stresses and strains and

$$\mathbf{b} = [b_x, b_y, b_z]^T, \quad \mathbf{t} = [t_x, t_y, t_z]^T \tag{10.48a}$$

$$\mathbf{q}_i^{(e)} = [X_i, Y_i, Z_i, M_{1_i}, M_{2_i}]^T \tag{10.48b}$$

are the standard body forces, surface loads and equilibrating nodal force vectors, respectively. The load components are expressed in the global axes except the nodal bending moments M_{1_i} and M_{2_i} which are expressed in the local axes as they are conjugate to the local nodal rotations θ_{1_i} and θ_{2_i}. $V^{(e)}$ and $A^{(e)}$ in Eq.(10.47) denote the element volume and the area of the surface where body forces and distributed loads $\mathbf{t}$ act, respectively.

Substituting Eqs.(10.21a), (10.37) and (10.40) into (10.47) and following the usual process of expressing the virtual strains in terms of the virtual displacements yields, after standard algebra, the element stiffness matrix and the equivalent nodal force vectors as

$$\mathbf{K}_{ij}^{(e)} = \iiint_{V^{(e)}} \mathbf{B}_i'^T \mathbf{D}' \mathbf{B}_j' \, dV \tag{10.49a}$$

$$\mathbf{f}_i^{(e)} = \iiint_{V^{(e)}} \mathbf{N}_i^T \mathbf{b} \, dV + \iint_{A^{(e)}} \mathbf{N}_i^T \mathbf{t} \, dA - \iiint_{V^{(e)}} \mathbf{B}_i'^T \boldsymbol{\sigma}_0' \, dV \tag{10.49b}$$

The element stiffness matrix can be expressed in terms of the global strain matrix $\mathbf{B}$ of Eq.(10.36). Substituting Eq.(10.38a) into (10.49a) gives

$$\mathbf{K}_{ij}^{(e)} = \iiint_{V^{(e)}} \mathbf{B}_i^T \mathbf{D} \mathbf{B}_j \, dV \quad \text{wid } \mathbf{D} = \mathbf{Q}^T \mathbf{D}' \mathbf{Q} \tag{10.50}$$

Eq.(10.49a) is quite convenient as it avoids the double matrix product for computing the "global" constitutive matrix $\mathbf{D}$.

Matrix $\mathbf{B}'$ is also useful for computing "a posteriori" the local stresses in terms of the global displacement vector $\mathbf{a}_i^{(e)}$. Substituting Eq.(10.37) into (10.40) gives

$$\boldsymbol{\sigma}' = \mathbf{D}' \sum_{i=1}^{n} \mathbf{B}_i' \mathbf{a}^{(e)} + \boldsymbol{\sigma}_0' \tag{10.51}$$

The element integrals contain rational polynomials in ξ, η, ζ. The dependence on ζ is inherent to degenerated shell theory, whereas the dependence on ξ, η depends on the element type.

The element stiffness matrix can be split into membrane-bending and transverse-shear contributions as

$$\mathbf{K}^{(e)} = \iiint_{V^{(e)}} [\mathbf{B}_p'^T \mathbf{D}_p' \mathbf{B}_p' + \mathbf{B}_s'^T \mathbf{D}_s' \mathbf{B}_s'] dV = \mathbf{K}_p^{(e)} + \mathbf{K}_s^{(e)} \tag{10.52}$$

This allows us using a reduced quadrature for $\mathbf{K}_s^{(e)}$ to avoid transverse-shear locking.

The computation of the element matrices is performed using Gauss quadratures for 3D solid elements [Hu2,On4,ZT2,ZTZ] as follows.

Quadrilateral degenerated shell elements

$$\mathbf{K}_{ij} = \sum_{p=1}^{n_\xi} \sum_{q=1}^{n_\eta} \sum_{r=1}^{n_\zeta} \mathbf{I}_{ij}(\xi_p, \, \eta_q, \, \zeta_r) W_p W_q W_r \tag{10.53}$$

Triangular degenerated shell elements

$$\mathbf{K}_{ij} = \sum_{p=1}^{n_p} \sum_{r=1}^{n_\zeta} \mathbf{I}_{ij}(\xi_p, \, \eta_p, \, \zeta_r) W_p W_r \tag{10.54a}$$

where

$$\mathbf{I}_{ij} = \mathbf{B}_i'^T \mathbf{D}' \mathbf{B}_j' |\mathbf{J}^{(e)}| \tag{10.54b}$$

In above $|\mathbf{J}^{(e)}|$ is the jacobian determinant n_ξ, n_η (n_p for triangles) are the number of integration points on the element surface, n_ζ are the integration points along the thickness direction and W_i are the corresponding weights. As a general rule, the surface integration requires the same quadrature as for the analogous Reissner–Mindlin plate elements (Section 10.11.1 and Table 10.1). For homogeneous isotropic material a two–point quadrature is typically chosen for the thickness integration.

Composite laminated shells

Composite laminated shells require an independent quadrature for each layer. If ζ_l and ζ_{l+1} define the thickness position of the lth layer (Figure 10.11) then

$$\zeta = \zeta_l + \frac{t_l}{t}(1+\zeta'); \quad d\zeta = \frac{t_l}{t}\,d\zeta' \tag{10.55}$$

where t_l is the layer thickness and $-1 \le \zeta' \le 1$ for $\zeta_l \le \zeta \le \zeta_{l+1}$.

The element stiffness matrix is written as

$$\mathbf{K}^{(e)} = \sum_{l=1}^{n_l} \int_{A_r^{(e)}} \int_{-1}^{+1} \left(\mathbf{B}'^T \mathbf{D}' \mathbf{B}' \frac{t_l}{t}\, d\zeta'\right) d\xi\, d\eta \tag{10.56}$$

where $A_r^{(e)}$ is the area of the reference surface of the element.

Expression (10.56) is integrated over each layer using a 2D Gauss quadrature. One integration point for each layer suffices if the number of layers is large and the material properties are constant within the layer. The form of Eq.(10.56) for this case with $\zeta' = 0$ and $W_{\zeta'} = 2$ is

$$\mathbf{K}^{(e)} = \sum_{l=1}^{n_l} \int_{A_r^{(e)}} (\mathbf{B}'^T \mathbf{D}' \mathbf{B}')_{\zeta'=0}\, 2\frac{t_l}{t}\, d\xi\, d\zeta \tag{10.57}$$

where n_l is the number of layers.

The numerical integration over the reference surface gives finally

Quadrilaterals

$$\mathbf{K}_{ij}^{(e)} = \sum_{l=1}^{n_l} \sum_{p=1}^{n_q} \sum_{q=1}^{n_q} \mathbf{I}_{ij}\left(\xi_p,\, \eta_q,\, \zeta_l + \frac{t_l}{t}\right) 2\frac{t_l}{t} W_p W_q \tag{10.58a}$$

Triangles

$$\mathbf{K}_{ij}^{(e)} = \sum_{l=1}^{n_l} \sum_{p=1}^{n_p} \mathbf{I}_{ij}\left(\xi_p,\, \eta_p,\, \zeta_l + \frac{t_l}{t}\right) 2\frac{t_l}{t} W_p \tag{10.58b}$$

where n_p and n_q are the integration points over the reference surface and $\mathbf{I}_{ij}$ is defined in Eq.(10.54b).

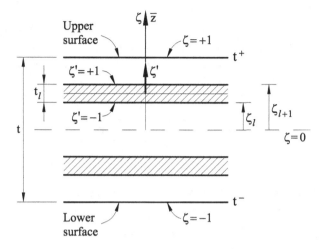

Fig. 10.11 Coordinate axes for layer integration

Surface quadratures for several degenerated shell quadrilateral elements are given in Section 10.11.1.

The numerical integration of the volume integrals in the equivalent nodal force vector is performed by an identical expression to Eqs.(10.54b) with $\mathbf{I}_{ij}$ given now by $\mathbf{N}_i^T \mathbf{b}|\mathbf{J}^{(e)}|$ and $-\mathbf{B}'^T \boldsymbol{\sigma}_0'|\mathbf{J}^{(e)}|$ for body forces and initial stresses, respectively.

Distributed loads acting on element faces are treated as for 3D solid elements (Section 7.10 of [On4]).

10.9 COMPUTATION OF STRESS RESULTANTS

The (local) resultant stress vector is obtained from the local stresses as

$$
\hat{\boldsymbol{\sigma}}' = \left\{ \begin{array}{c} \hat{\boldsymbol{\sigma}}'_m \\ \hat{\boldsymbol{\sigma}}'_b \\ \hat{\boldsymbol{\sigma}}'_s \end{array} \right\} = \int_{t^-}^{t^{\,|}} \left\{ \begin{array}{c} \boldsymbol{\sigma}'_p \\ \cdots \\ \bar{z}\boldsymbol{\sigma}'_p \\ \cdots \\ \boldsymbol{\sigma}'_s \end{array} \right\} d\bar{z} \tag{10.59}
$$

where t^+ and t^- are the thickness coordinates of the upper and lower surfaces of the shell (Figure 10.11) and

$$
\hat{\boldsymbol{\sigma}}'_m = \left\{ \begin{array}{c} N_{x'} \\ N_{y'} \\ N_{x'y'} \end{array} \right\} \quad , \quad \hat{\boldsymbol{\sigma}}'_b = \left\{ \begin{array}{c} M_{x'} \\ M_{y'} \\ M_{x'y'} \end{array} \right\} \quad , \quad \hat{\boldsymbol{\sigma}}'_s = \left\{ \begin{array}{c} Q_{x'} \\ Q_{y'} \end{array} \right\} \tag{10.60}
$$

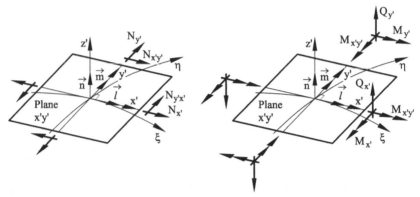

Fig. 10.12 Sign convention for stress resultants in a curved shell

The sign convention for the resultant stresses coincides with that given in Figure 8.9 for flat shells and it is again shown in Figure 10.12. Typically the resultant stresses are first computed at the Gauss points which are optimal for evaluation of the stresses [On4,ZTZ]. The nodal resultant stresses can be simply obtained by extrapolation of the Gauss point values using any of the smoothing techniques explained in Chapter 9 of [On4].

Expression (10.57) is only *approximate* since it neglects the effects of the shell curvature. A more precise (and complex) expression for the stress resultants can be found in [BD6].

The integral in Eq.(10.59) is usually computed numerically at each Gauss point over the surface at $\xi = \bar{\xi}$, $\eta = \bar{\eta}$ as

$$[\hat{\boldsymbol{\sigma}}']_{\bar{\xi}\bar{\eta}} = \int_{-1}^{+1} \left\{ \begin{array}{c} \boldsymbol{\sigma}'_p \\ \cdots \\ \bar{z}\boldsymbol{\sigma}'_p \\ \cdots \\ \boldsymbol{\sigma}'_s \end{array} \right\}_{\bar{\xi}\bar{\eta}} \frac{t}{2} d\zeta = \sum_{r=1}^{n_\zeta} \frac{t}{2} W_r \left\{ \begin{array}{c} \mathbf{D}'_p \mathbf{B}'_p \\ \cdots \\ \bar{z}\mathbf{D}'_p \mathbf{B}'_p \\ \cdots \\ \mathbf{D}'_s \mathbf{B}'_s \end{array} \right\}_{\bar{\xi}\bar{\eta}} \mathbf{a}^{(e)} \qquad (10.61)$$

In practice, $n_\zeta = 2$ is taken. For composite laminated shells an integration over the number of the thickness layers should be performed.

10.10 FOLDED CURVED SHELLS

Folded curved shells with kinks and/or branching surfaces containing non-coplanar elements can be also treated with the five nodal DOFs formulation (three global displacements and two rotations per node) presented. This requires that the nodal fiber vector $\mathbf{v}_{3i}$ is defined so as to ensure geometric compatibility as described in options b) and c) of Section 10.3.2.

Otherwise, if the nodal coordinate system varies between adjacent elements, *then the nodal rotations must be transformed to a common coordinate system.* This introduces a third in-plane rotational variable as described in Section 8.9 for flat shell elements. The transformation of the displacements and forces to the global cartesian system x, y, z is written as

$$\underset{5\times1}{\mathbf{a}_i^{(e)}} = \underset{5\times6}{\mathbf{L}_i} \underset{6\times1}{\bar{\mathbf{a}}_i^{(e)}}; \qquad \underset{5\times1}{\mathbf{f}_i^{(e)}} = \underset{5\times6}{\mathbf{L}_i} \underset{6\times1}{\bar{\mathbf{f}}_i^{(e)}} \tag{10.62}$$

where $(\bar{\cdot})$ denotes displacements and forces with rotations and moments expressed now in global axes and

$$\mathbf{L}_i^{(e)} = \begin{bmatrix} \mathbf{I}_3 & \mathbf{0} \\ \mathbf{0} & \hat{\boldsymbol{\lambda}}_i^{(e)} \end{bmatrix} \tag{10.63a}$$

where $\mathbf{I}_3$ is the 3×3 unit matrix and

$$\hat{\boldsymbol{\lambda}}_i^{(e)} = \begin{bmatrix} -\mathbf{v}_{2_i}^T \\ \mathbf{v}_{1_i}^T \end{bmatrix} \tag{10.63b}$$

The global stiffness matrix is written after transformation as

$$\bar{\mathbf{K}}_{ij}^{(e)} = \left[\mathbf{L}_i^{(e)}\right]^T \mathbf{K}_{ij}^{(e)} \mathbf{L}_j^{(e)} \tag{10.64}$$

The nodal transformation matrix $\mathbf{L}_i^{(e)}$ is different for each node due to the curvature of the element.

The computation of $\bar{\mathbf{K}}_{ij}^{(e)}$ can be simplified by combining Eqs.(10.49a) and (10.64) to give

$$\bar{\mathbf{K}}_{ij} = \iiint_{V^{(e)}} \bar{\mathbf{B}}_i \mathbf{D}' \bar{\mathbf{B}}_j dV \quad \text{with} \quad \bar{\mathbf{B}}_i = \mathbf{B}_i' \mathbf{L}_i \tag{10.65}$$

A similar procedure can be used if $\mathbf{K}_{ij}^{(e)}$ is computed using Eq.(10.50).

The above stiffness transformation is recommended for *non-coplanar* nodes only to avoid stiffness singularities of the kind described in Section 8.9 for flat shells. Otherwise these singularities can be overcome using the techniques there explained.

10.11 TECHNIQUES FOR ENHANCING THE PERFORMANCE OF DEGENERATED SHELL ELEMENTS

Degenerated shell elements suffer from the same defects as thick Reissner–Mindlin type plate and curved shell elements, i.e.

a) It is difficult to reproduce the thin shell conditions (zero transverse shear strains). This leads to over-stiff shear dominated solutions (shear locking). This defect can be overcome using selective/reduced integration [OHG,ZTT] and assumed transverse shear strain fields [DB,HH3].

b) It is difficult to reproduce a pure bending solution without introducing spureous membrane strains (membrane locking). This defect leads to additional stiffening of the solution. Elimination of membrane locking is possible via reduced integration of the membrane terms in $\mathbf{K}^{(e)}$ and also by using assumed membrane strain fields [BWS,HH4,HH5].

10.11.1 Selective/reduced integration techniques

Transverse shear locking defects can be alleviated by using a reduced quadrature for the transverse shear stiffness terms. This requires splitting the element stiffness matrix

$$\mathbf{K}^{(e)} = \mathbf{K}_p^{(e)} + \mathbf{K}_s^{(e)} \tag{10.66}$$

where $\mathbf{K}_p^{(e)}$ and $\mathbf{K}_s^{(e)}$ contain the in-plane (membrane–bending) and transverse shear contributions, respectively. The number of reduced integration points for $\mathbf{K}_s^{(e)}$ is chosen so that condition (2.50) is satisfied. The performance of the Serendipity and Lagrange degenerated shell elements with respect to transverse shear locking is very similar to the analogous Reissner–Mindlin plate elements of Chapter 6.

It is difficult to avoid membrane locking in degenerated shell elements via selective integration due to the complexity of splitting the membrane and bending contributions in $\mathbf{K}_p^{(e)}$. A possibility is using a reduced integration rule for $\mathbf{K}_p^{(e)}$. Unfortunately, a uniform reduced integration for $\mathbf{K}_p^{(e)}$ and $\mathbf{K}_s^{(e)}$ introduces spurious mechanisms in most elements (with the exception of the QL12, see Table 10.1) and it should be handled with extreme care. An alternative is to use a formulation where $\mathbf{K}_p^{(e)}$ is split into the sum of membrane ($\mathbf{K}_m^{(e)}$), bending ($\mathbf{K}_m^{(e)}$) and coupled membrane-bending ($\mathbf{K}_{mb}^{(e)}$) matrices (Section 10.14). A reduced quadrature for $\mathbf{K}_m^{(e)}$, $\mathbf{K}_{mb}^{(e)}$ and $\mathbf{K}_s^{(e)}$ can be chosen in these cases while preserving the full integration for $\mathbf{K}_b^{(e)}$, as for flat shell elements (Section 8.11).

Table 10.1 shows the typical ($n_\xi \times n_\eta$) quadratures over the surface for various degenerated shell quadrilaterals. The spurious mechanisms introduced by each integration rule elements are shown in the last three columns of Table 10.1.

	Full integration (FI)		Reduced integration (RI)		Selective integration (SI)		Spurious Mechanics (FI) (RI) (SI)		
	$\mathbf{K}_p^{(e)}$	$\mathbf{K}_s^{(e)}$	$\mathbf{K}_p^{(e)}$	$\mathbf{K}_s^{(e)}$	$\mathbf{K}_p^{(e)}$	$\mathbf{K}_s^{(e)}$			
Q4	2×2	2×2	1×1	1×1	2×2	1×1	0	6	2
QS8	3×3	3×3	2×2	2×2	3×3	2×2	0	2	0
QL9	3×3	3×3	2×2	2×2	3×3	2×2	0	7	1
QS12	4×4	4×4	3×3	3×3	4×4	3×3	0	0	0
QL16	4×4	4×4	3×3	3×3	4×4	3×3	0	4	1

Table 10.1 Full, reduced and selective integration quadratures over the surface $(n_p \times n_q)$ for degenerated shell quadrilaterals. The last three columns show the spurious mechanics induced by each integration rule

As a general rule:

The *4-noded quadrilateral* (Q4) with 2×2 full integration (FI) presents severe shear locking. The uniform one-point reduced integration (RI) introduces six spureous mechanisms. The uniform selective integration (SI), 2×2 for $\mathbf{K}_p^{(e)}$ and 1×1 for $\mathbf{K}_s^{(e)}$, brings the number of mechanisms down to two. These mechanisms can be eliminated by using stabilization techniques [BT,BTL].

The *8-noded Serendipity quadrilateral* (QS8) presents shear and membrane locking with 3×3 FI. The 2×2 RI eliminates locking in most situations although it introduces two spureous mechanisms. Fortunately, these mechanisms disappear after assembly of the global stiffness matrix and the element can be considered safe for practical purposes. SI is free of mechanisms although shear and membrane locking occur in some situations as the singularity rule (2.50) is not always satisfied.

The *9-noded lagrangian quadrilateral* (QL9) satisfies the singularity rule with RI and SI. However, RI introduces seven spureous mechanisms and it should be avoided. SI still induces one internal mechanism which can pollute the solution in some situations. Unless this mechanism is stabilized the QL9 element can not be considered safe for practical purposes [Pa].

Oñate *et al.* [OHG] developed a stable and locking–free version of the QL9 element by adding hierarchical nodes to the QS8 element as we did as for the QHG plate element (Section 6.5.5). Similar elements can be obtained using the Heterosis approximation (Section 6.5.6) [HL,Hu] although the element still has an internal spureous mode with RI in this case [Cr,Cr2].

The *12-noded Serendipity quadrilateral* (QS12) is free of spureous mechanisms with both RI and SI rules, although it does not always satisfy the singularity rule (2.50) and the absence of locking can not be fully guaranteed [ZT2].

The *16-noded lagrangian quadrilateral* (QL16) has four spurious and propagable mechanisms with RI and one with SI. The singularity rule is satisfied for both quadratures, but the element is not robust for practical purposes [ZT2].

10.11.2 Assumed fields for the transverse shear and membrane strains

Assumed strain fields have been successfully used for developing robust locking–free curved shell elements. The first applications were aimed to deriving Discrete-Kirchhoff degenerated shell elements by imposing the condition of zero transverse shear strains at a number of element points [BD6,Dh,Dh2,DMM,Mo3,We,WOK]. An example is the *semi-Loof* element developed by Irons [He,IA,Ir2,Mo3] which has enjoyed much popularity despite its relative complexity (Section 10.12.5). Assumed transverse shear and membrane strain fields were first introduced in the late 1980's to develop shear-locking free versions of the Q4, QS8 and QL9 degenerated shell elements [BD,DB,HH4,HH5,HL2,JP,JP2]. These techniques can be viewed as an extension of the assumed strain procedures for beams and plates explained in previous chapters (Sections 2.8.4 and 6.6).

In the next section we present a extension of the Q4 degenerated shell element using an assumed linear transverse shear strain field. The derivation of an enhanced QL9 degenerated shell element with assumed transverse shear and membrane fields will be presented in Section 10.15.

10.11.3 4-noded degenerated shell quadrilateral (Q4) with assumed linear transverse shear strain field

Due to the element curvature it is convenient to express the assumed transverse shear strain field in the covariant components $\gamma_{\xi\zeta}$ and $\gamma_{\eta\zeta}$ and then transform the resulting expressions to the lamina coordinate system.

The definition of the assumed transverse shear strain field follows the same arguments as for Reissner-Mindlin plate elements in Section 6.6. The aim is to ensure the compatibility between the derivatives of the normal displacement with respect to the curvilinear coordinates ξ and η and the local rotations so that the thin shell conditions of vanishing transverse shear strains are satisfied. In practice, this means that the assumed transverse shear strain field must contain just the polynomial terms emanating from the approximation of the curvilinear derivatives of the deflection.

For the bi-linear 4-noded element this implies that the transverse shear strains $\gamma_{\xi\zeta}$ and $\gamma_{\eta\zeta}$ should vary linearly with η and ξ, respectively.

In consequence, the following interpolation for the transverse shear strains expressed in the *covariant* system $\mathbf{a}_1, \mathbf{a}_2, \mathbf{n}$ (Figure 10.6) is used

$$\gamma_{\xi\zeta} = \frac{1-\eta}{2}\gamma_{\xi\zeta}^1 + \frac{1-\eta}{2}\gamma_{\xi\zeta}^3$$

$$\gamma_{\eta\zeta} = \frac{1+\xi}{2}\gamma_{\eta\zeta}^2 + \frac{1-\xi}{2}\gamma_{\eta\zeta}^4 \tag{10.67}$$

The above interpolation coincides with that used for the QLLL Reissner-Mindlin plate element (Section 6.7.1 and Eq.(6.93)). Points $1, 2, 3, 4$ on the element sides are shown in Figure 6.16. Note that due to the mild taper assumption (Figure 10.4c) the normal direction and the ζ direction coincide. The values $\gamma_{\xi\zeta}^1, \gamma_{\xi\zeta}^3, \gamma_{\eta\zeta}^2$ and $\gamma_{\eta\zeta}^4$ are transformed to the lamina system by

$$\gamma^k = \left\{ \begin{matrix} \gamma_{\xi\zeta}^k \\ \gamma_{\eta\zeta}^k \end{matrix} \right\} = \mathbf{L}^k \left\{ \begin{matrix} \gamma_{x'z}^k \\ \gamma_{y'z}^k \end{matrix} \right\} = \mathbf{L}^k [\mathbf{B}_s']^k \mathbf{a}^{(e)} \tag{10.68a}$$

with

$$\mathbf{L}^k = [\mathbf{a}_1^k, \mathbf{a}_2^k] \quad \text{and} \quad k = 1, 2, 3, 4 \tag{10.68b}$$

Eq.(10.67) can therefore be written as

$$\gamma = \left\{ \begin{matrix} \gamma_{\xi\zeta} \\ \gamma_{\eta\zeta} \end{matrix} \right\} = \mathbf{B}^\gamma \mathbf{a}^{(e)} \tag{10.69}$$

where

$$\mathbf{B}^\gamma = \frac{1}{2} \begin{bmatrix} (1 - \eta)[\mathbf{B}'_s]^1 + (1 + \eta)[\mathbf{B}'_s]^3 \\ (1 - \eta)[\mathbf{B}'_s]^2 + (1 + \eta)[\mathbf{B}'_s]^4 \end{bmatrix} \tag{10.70}$$

The final transformation of the transverse shear strains from the co-variant to the lamina system yields

$$\varepsilon'_s = \hat{\mathbf{L}}^T \gamma = \hat{\mathbf{L}}^T \mathbf{B}^\gamma \mathbf{a}^{(e)} = \bar{\mathbf{B}}'_s \mathbf{a}^{(e)} \quad , \quad \bar{\mathbf{B}}'_s = \hat{\mathbf{L}}^T \mathbf{B}^\gamma \tag{10.71a}$$

where

$$\hat{\mathbf{L}} = [\mathbf{a}_1, \mathbf{a}_2] \tag{10.71b}$$

Eq.(10.71a) accounts for the fact that the transverse shear strains transform as a vector, since it is assumed that the covariant and lamina (local) bases are coplanar.

Matrix $\bar{\mathbf{B}}'_s$ substitutes matrix $\mathbf{B}'_s$ in the computation of the transverse shear strain stiffness matrix $\mathbf{K}_s^{(e)}$ in Eq.(10.52). This matrix is computed using a 2×2 Gauss quadrature.

The membrane and bending strains and the stiffness matrix $\mathbf{K}_p^{(e)}$ are computed as explained in Sections 10.6 and 10.8, respectively.

10.12 EXPLICIT THICKNESS INTEGRATION OF THE STIFFNESS MATRIX

Degenerated shell elements are typically formulated in terms of strains and stresses as it is the case 3D solid elements. This is a difference with flat shell elements where the PVW is expressed in terms of generalized strains and resultant stresses resulting from a pre-integration across the thickness direction. This allows to identify the membrane and flexural contributions to the element stiffness matrix. However, the stiffness matrix terms in degenerated shell elements depend on the thickness coordinate and, hence an integration over the element volume is required. This can be quite expensive, specially for composite laminated shells, which often require many integration points through the thickness. It is therefore desirable to convert the volume integrals to surface integrals via pre-integration through the thickness. This is however not straightforward and different attempts have been reported [BD6,CMPW,Cr,Cr4,MSch,PC,Sta,Vl3].

To better understand the difficulties for the explicit integration let us list the type of thickness dependencies of the stiffness terms.

a) *Dependence with z' of the local coordinate system.*

The orientation of the local coordinate system changes across the thickness (except for constant thickness) (Figures 10.4 and 10.5). Hence, matrix $\mathbf{Q}$ of Eq.(10.29) is a function of the thickness coordinate ζ

b) *Strain z'-dependence* The dependence of the strains with ζ comes from the derivatives of the global displacements with respect to the natural coordinate system (Eq.(10.32)) and from the terms of the jacobian (Eq.(10.33)). The jacobian matrix depends linearly on ζ. Hence, the jacobian determinant is quadratic on ζ and its inverse is a rational algebraic function.

c) *Surface area z'-dependence* The expression for the differential of volume in curvilinear coordinates is

$$dV = dx' \, dy' \, dz' = |\mathbf{J}^{(e)}| \, d\xi \, d\eta \, d\zeta \qquad (10.72)$$

Clearly dV depends quadratically on ζ through the jacobian determinant.

d) *z'-dependence of the material properties* The constitutive matrix for heterogeneous material or a composite laminated shell varies across the thickness.

In the following section we present a continuum-based resultant shell formulation based on a pre-integration across the thickness of the degenerated shell equations, under certain assumptions which are acceptable for most thin and moderately thick shells. The process basically follows the ideas proposed by Stanley *et al.* [Sta,SPH].

10.13 CONTINUUM-BASED RESULTANT (CBR) SHELL THEORY

To facilitate thickness pre-integration of the degenerated shell equations the following simplifying assumptions are made:

Assumption 1: Mild Taper. Variations in the shell thickness, t, with respect to the surface coordinates (ξ, η) (Figure 10.4c) may be neglected for purposes of spatial *differentiation* and in the expression of the *zero normal stress* hypothesis. Thickness changes need not be neglected for the purposes of integration.

The *mild taper* assumption is mathematically equivalent to the following three statements

$$\mathbf{n}(z') \equiv \mathbf{n}(0) \equiv \mathbf{v}_3 \tag{10.73a}$$

$$\frac{\partial z'}{\partial \xi} = \frac{\partial z'}{\partial \eta} = 0 \tag{10.73b}$$

$$\frac{\partial \bar{z}}{\partial \xi} = \frac{\partial \bar{z}}{\partial \eta} = 0 \tag{10.73c}$$

The first statement (10.73a) arbitrary fixes the definition of the lamina basis at the *reference surface*. The second statement (10.73b) implies that the covariant basis vectors $\mathbf{a}_1$ and $\mathbf{a}_2$ which are tangent to ξ and η, respectively (Figure 10.6), are also tangent to the reference surface, that is

$$\mathbf{a}_1^T \mathbf{n}\Big|_{\bar{z}=0} = 0 \quad , \quad \mathbf{a}_2^T \mathbf{n}\Big|_{\bar{z}=0} = 0 \tag{10.74}$$

It is obvious from Figure 10.4c that this is true only if taper is neglected. Finally, (10.73c) follows from (10.73a), since derivatives of $\bar{z}$ with respect to ξ, η are proportional to the derivatives of the thickness t, which are neglected by the assumption made.

A justification for the *mild taper* assumption is that taper is easily circumvented computationally by the use of a *piecewise-constant* (element-by-element) thickness variation. Thus, even areas of severe taper may be accommodated through local grid refinement.

The mild taper assumption also implies

$$\frac{\partial(\cdot)}{\partial z'} = \frac{\partial(\cdot)}{\partial \bar{z}} \quad , \quad dz' = d\bar{z} \quad \text{and} \quad z' \equiv \bar{z} \tag{10.75}$$

These expressions are useful for computing the strain and stress resultants.

Assumption 2: Normality. The shell pseudo-normal vectors $\mathbf{v}_3$ may be approximated as perfectly *normal* to the reference surface when evaluating the Jacobian matrix.

The normality assumption, in conjunction with Eqs.(10.73), is mathematically equivalent to the following statements

$$\frac{\partial x'}{\partial \xi} = \frac{\partial y'}{\partial \eta} = 0 \quad , \quad \frac{\partial z'}{\partial \eta} = \frac{t}{2} \tag{10.76}$$

which follows by noting that $\left(\frac{\partial x'}{\partial \xi}, \frac{\partial y'}{\partial \eta}, \frac{\partial z'}{\partial \eta}\right)$ are just the lamina-based components of the pseudo-normal vector, $\vec{v}_3$, which can be viewed as normal to the reference surface. Clearly, $\mathbf{v}_3 \equiv \mathbf{n}$ in this case.

10.13.1 Geometric and kinematic description

The position vector of a point is expressed in the lamina (local) coordinate system x', y', z' as

$$\mathbf{r}' = \mathbf{r}_0' + \bar{z}\mathbf{v}_3' \tag{10.77}$$

where

$$\mathbf{r}' = [x', y', z']^T \quad , \quad \mathbf{r}_0' = [x_0', y_0', z_0']^T \quad , \quad \mathbf{v}_3' = [\bar{x}_0', \bar{y}_0', \bar{z}_0']^T \tag{10.78}$$

define the components of vectors $\vec{r}, \vec{r}_0$ and $\vec{v}_3$ in the lamina coordinate system.

For simplicity of the notation the components of $\mathbf{r}_0'$ and $\mathbf{v}_3'$ are written with similar symbols.

The displacement of a point is expressed as

$$\mathbf{u}' = \mathbf{u}_0' + \bar{z}\mathbf{u}_n' \tag{10.79}$$

where

$$\mathbf{u}' = [u', v', w']^T \quad , \quad \mathbf{u}_0' = [u_0', v_0', w_0']^T \quad , \quad \mathbf{u}_n' = [\bar{u}_n', \bar{v}_n', \bar{w}_n']^T \tag{10.80}$$

define the components of vectors $\vec{u}, \vec{u}_0$ and $\vec{u}_n$ in the lamina system.

10.13.2 Computation of the CBR shell Jacobian

The Jacobian is simplified using Eqs.(10.76) as

$$\mathbf{J} = \begin{bmatrix} \dfrac{\partial x'}{\partial \xi} & \dfrac{\partial y'}{\partial \xi} & \dfrac{\partial z'}{\partial \xi} \\[6pt] \dfrac{\partial x'}{\partial \eta} & \dfrac{\partial y'}{\partial \eta} & \dfrac{\partial z'}{\partial \eta} \\[6pt] \dfrac{\partial x'}{\partial \zeta} & \dfrac{\partial y'}{\partial \zeta} & \dfrac{\partial z'}{\partial \zeta} \end{bmatrix} = \begin{bmatrix} \dfrac{\partial x'}{\partial \xi} & \dfrac{\partial y'}{\partial \xi} & 0 \\[6pt] \dfrac{\partial x'}{\partial \eta} & \dfrac{\partial y'}{\partial \eta} & 0 \\[6pt] 0 & 0 & \dfrac{t}{2} \end{bmatrix} = \begin{bmatrix} \mathbf{J}_s & 0 \\[4pt] 0 & \dfrac{t}{2} \end{bmatrix} \tag{10.81}$$

where the CBR hypothesis has eliminated all coupling between in-plane and out-of-plane differentials. The 2×2 surface Jacobian $\mathbf{J}_s$ can be decomposed using Eqs.(10.79) as

$$\mathbf{J}_s = \mathbf{J}_0 + \bar{z}\mathbf{J}_1 \tag{10.82}$$

where

$$\mathbf{J}_0 = \begin{bmatrix} \dfrac{\partial x_0'}{\partial \xi} & \dfrac{\partial y_0'}{\partial \xi} \\[6pt] \dfrac{\partial x_0'}{\partial \eta} & \dfrac{\partial y_0'}{\partial \eta} \end{bmatrix} \quad , \quad \mathbf{J}_1 = \begin{bmatrix} \dfrac{\partial \bar{x}_0'}{\partial \xi} & \dfrac{\partial \bar{y}_0'}{\partial \xi} \\[6pt] \dfrac{\partial \bar{x}_0'}{\partial \eta} & \dfrac{\partial \bar{y}_0'}{\partial \eta} \end{bmatrix} \tag{10.83}$$

The inverse of $\mathbf{J}$ in (10.81) is

$$\mathbf{J}^{-1} = \begin{bmatrix} \mathbf{J}_s^{-1} & 0 \\ 0 & \dfrac{t}{2} \end{bmatrix} \tag{10.84}$$

From Eqs.(10.82) and (10.83)

$$\mathbf{J}_s^{-1} = \frac{1}{J(\bar{z})} \left[\mathbf{J}_0^{-1} + \frac{\bar{z}}{J_0} \mathbf{J}_1^{-1} \right] \tag{10.85}$$

with

$$\mathbf{J}_0^{-1} = \frac{1}{J_0} \begin{bmatrix} \dfrac{\partial y_0'}{\partial \eta} & -\dfrac{\partial y_0'}{\partial \xi} \\[2mm] -\dfrac{\partial x_0'}{\partial \eta} & \dfrac{\partial x_0'}{\partial \xi} \end{bmatrix} \quad , \quad \mathbf{J}_1^{-1} = \begin{bmatrix} \dfrac{\partial \bar{y}_0'}{\partial \eta} & -\dfrac{\partial \bar{y}_0'}{\partial \xi} \\[2mm] -\dfrac{\partial \bar{x}_0'}{\partial \eta} & \dfrac{\partial \bar{x}_0'}{\partial \xi} \end{bmatrix} \tag{10.86}$$

The quantity $J(\bar{z})$ in Eq.(10.85) is the normalized determinant of the surface Jacobian, i.e.

$$J(\bar{z}) = \frac{|\mathbf{J}_s|}{J_0} = 1 + \left(\frac{J_1}{J_0} \right) \bar{z} + \left(\frac{J_2}{J_0} \right) \bar{z}^2 \tag{10.87}$$

where

$$J_0 = |\mathbf{J}_0| \quad , \quad J_2 = |\mathbf{J}_1| \tag{10.88}$$

and

$$J_1 = \left(\frac{\partial \bar{x}_0'}{\partial \xi} \frac{\partial y_0'}{\partial \eta} - \frac{\partial \bar{x}_0'}{\partial \eta} \frac{\partial y_0'}{\partial \xi} \right) + \left(\frac{\partial x_0'}{\partial \xi} \frac{\partial \bar{y}_0'}{\partial \eta} - \frac{\partial x_0'}{\partial \eta} \frac{\partial \bar{y}_0'}{\partial \xi} \right) \tag{10.89}$$

Note that the coefficients J_0, J_1 and J_2 are all independent of $\bar{z}$.

10.13.3 CBR Strains

10.13.3.1 Displacement derivatives

Using the new Jacobian-inverse (Eq.(10.84)), the lamina based displacement derivatives may be re-rewritten such that the $\bar{z}$-dependence is explicit. For instance

$$\frac{\partial u'}{\partial \mathbf{x'}} = \left\{ \begin{array}{c} \dfrac{\partial u'}{\partial \mathbf{x}_p'} \\[2mm] \dfrac{\partial u'}{\partial z'} \end{array} \right\} = \frac{1}{J(\bar{z})} \left\{ \overline{\dfrac{\partial u_0'}{\partial \mathbf{x}_p'}} + \bar{z} \left(\overline{\dfrac{\partial u_k'}{\partial \mathbf{x}_p'}} + \widehat{\dfrac{\partial u_0'}{\partial \mathbf{x}_p'}} \right) + \bar{z}^2 \widehat{\dfrac{\partial u_n'}{\partial \mathbf{x}_p'}} \right\} \tag{10.90a}$$

with

$$\mathbf{x}'_p = [x', y']^T \quad \text{and} \quad \frac{\partial(\cdot)}{\partial \mathbf{x}'_p} = \left\{ \begin{array}{c} \dfrac{\partial(\cdot)}{\partial x'} \\[2mm] \dfrac{\partial(\cdot)}{\partial y'} \end{array} \right\} \tag{10.90b}$$

Similar expressions are found for $\frac{\partial v'}{\partial \mathbf{x}'}$ and $\frac{\partial w'}{\partial \mathbf{x}'}$.

The wide bars and hats in Eq.(10.90a) indicate association with $\mathbf{J}_0$ and $\mathbf{J}_1$, respectively, i.e.

$$\frac{\overline{\partial(\cdot)}}{\partial \mathbf{x}'_p} := \mathbf{J}_0^{-1} \frac{\partial(\cdot)}{\partial \boldsymbol{\xi}_p} \quad , \quad \frac{\widehat{\partial(\cdot)}}{\partial \mathbf{x}'_p} = \frac{1}{J_0} \mathbf{J}_1^{-1} \frac{\partial(\cdot)}{\partial \boldsymbol{\xi}_p} \tag{10.91}$$

with $\boldsymbol{\xi}_p = [\xi, \eta]^T$.

10.13.3.2 Generalized strains

The local strain vector can be written in the lamina basis (using Eqs.(10.23) and (10.90)) as

$$\boldsymbol{\varepsilon}' = \left\{ \begin{array}{c} \boldsymbol{\varepsilon}'_p \\ \boldsymbol{\varepsilon}'_s \end{array} \right\} = \frac{1}{J(\bar{z})} \left\{ \begin{array}{c} \boldsymbol{\varepsilon}_0^p + \bar{z}\boldsymbol{\varepsilon}_1^p + \bar{z}^2\boldsymbol{\varepsilon}_2^p \\ \boldsymbol{\varepsilon}_0^s + \bar{z}\boldsymbol{\varepsilon}_1^s + \bar{z}^2\boldsymbol{\varepsilon}_2^s + J(\bar{z})\boldsymbol{\varepsilon}_3^s \end{array} \right\} \tag{10.92}$$

where $\boldsymbol{\varepsilon}'_p = [\varepsilon_{x'}, \varepsilon_{y'}, \gamma_{x'y'}]^T$ and $\boldsymbol{\varepsilon}'_s = [\gamma_{x'z'}, \gamma_{y'z'}]^T$ are in-plane and transverse-shear strain components, respectively and the $\bar{z}$-independent quantities $\boldsymbol{\varepsilon}_i^p$ and $\boldsymbol{\varepsilon}_i^s$, $i = 0, 1, 2$ are defined as

$$\boldsymbol{\varepsilon}_0^p = \left\{ \begin{array}{c} \dfrac{\overline{\partial u'_0}}{\partial x'} \\[3mm] \dfrac{\overline{\partial v'_0}}{\partial y'} \\[3mm] \dfrac{\overline{\partial v'_0}}{\partial y'} + \dfrac{\overline{\partial u'_0}}{\partial x'} \end{array} \right\} \quad , \quad \boldsymbol{\varepsilon}_0^s = \left\{ \begin{array}{c} \dfrac{\overline{\partial w'_0}}{\partial x'} \\[3mm] \dfrac{\overline{\partial w'_0}}{\partial y'} \end{array} \right\}$$

$$\boldsymbol{\varepsilon}_1^p = \left\{ \begin{array}{c} \dfrac{\overline{\partial u'_n}}{\partial x'} + \dfrac{\widehat{\partial u'_0}}{\partial x'} \\[3mm] \dfrac{\overline{\partial v'_n}}{\partial y'} + \dfrac{\widehat{\partial v'_0}}{\partial y'} \\[3mm] \left(\dfrac{\overline{\partial v'_n}}{\partial x'} + \dfrac{\overline{\partial u'_n}}{\partial y'} \right) + \left(\dfrac{\widehat{\partial v'_0}}{\partial x'} + \dfrac{\widehat{\partial u'_n}}{\partial y'} \right) \end{array} \right\} \quad , \quad \boldsymbol{\varepsilon}_1^s = \left\{ \begin{array}{c} \dfrac{\overline{\partial w'_n}}{\partial x'} + \dfrac{\widehat{\partial w'_0}}{\partial x'} \\[3mm] \dfrac{\overline{\partial w'_n}}{\partial y'} + \dfrac{\widehat{\partial w'_0}}{\partial y'} \end{array} \right\}$$

$$\varepsilon_2^p = \begin{Bmatrix} \widehat{\dfrac{\partial u_n'}{\partial x'}} \\[2mm] \widehat{\dfrac{\partial v_n'}{\partial y'}} \\[2mm] \widehat{\dfrac{\partial u_n'}{\partial y'}} + \widehat{\dfrac{\partial v_n'}{\partial x'}} \end{Bmatrix} \quad , \quad \varepsilon_2^s = \begin{Bmatrix} \widehat{\dfrac{\partial w_n'}{\partial x'}} \\[2mm] \widehat{\dfrac{\partial w_n'}{\partial y'}} \end{Bmatrix} \quad , \quad \varepsilon_3^s = \begin{Bmatrix} u_n' \\ v_n' \end{Bmatrix} \tag{10.93}$$

It is useful to factor out the $\bar{z}$ dependence in Eq.(10.92) and re-write it as

$$\varepsilon' = \mathbf{S}\hat{\varepsilon}' \tag{10.94}$$

where $\hat{\varepsilon}'$ can be viewed as a resultant strain vector with partitions

$$\hat{\varepsilon}' = \begin{Bmatrix} \hat{\varepsilon}_p' \\ \hat{\varepsilon}_s' \end{Bmatrix} \quad \text{where} \quad \hat{\varepsilon}_p' = \begin{Bmatrix} \hat{\varepsilon}_0^p \\ \hat{\varepsilon}_1^p \\ \hat{\varepsilon}_2^p \end{Bmatrix} \quad , \quad \hat{\varepsilon}_s' = \begin{Bmatrix} \hat{\varepsilon}_0^s \\ \hat{\varepsilon}_1^s \\ \hat{\varepsilon}_2^s \\ \hat{\varepsilon}_3^s \end{Bmatrix} \tag{10.95}$$

and $\mathbf{S}$ is a sparse matrix containing all the $\bar{z}$ dependence terms, i.e.

$$\mathbf{S} = \begin{bmatrix} \mathbf{S}_p & \mathbf{0} \\ \mathbf{0} & \mathbf{S}_s \end{bmatrix} \tag{10.96a}$$

$$\mathbf{S}_p = \frac{1}{J(\bar{z})}[\mathbf{I}_3, \bar{z}\mathbf{I}_3, \bar{z}^2\mathbf{I}_3] \quad , \quad \mathbf{S}_s = \frac{1}{J(\bar{z})}[\mathbf{I}_2, \bar{z}\mathbf{I}_2, \bar{z}^2\mathbf{I}_2, J(\bar{z})\mathbf{I}_2] \tag{10.96b}$$

10.13.4 PVW, stress resultants and generalized constitutive matrix

By substituting the resultant strains into the PVW expression (10.47) and noting that $dV = J(\bar{z})dAd\bar{z}$, where dA is the differential surface area at the shell reference surface, we obtain the following expression for the virtual internal work

$$\iiint_V \delta\varepsilon'\sigma' dV = \iint_A [\delta\hat{\varepsilon}']^T \left[\int_{-t/2}^{t/2} \mathbf{S}^T \sigma' J(\bar{z}) d\bar{z} \right] dA = \iint_A [\delta\hat{\varepsilon}']^T \hat{\sigma}' dA \tag{10.97}$$

where

$$\hat{\sigma}' = \begin{Bmatrix} \hat{\sigma}_p' \\ \hat{\sigma}_s' \end{Bmatrix} = \int_{-t/2}^{t/2} \mathbf{S}^T \sigma' J(\bar{z}) d\bar{z} \tag{10.98}$$

is the local *stress resultant* vector with

$$
\hat{\sigma}'_p = \begin{Bmatrix} \hat{\sigma}'_m \\ \hat{\sigma}'_b \\ \hat{\sigma}'_{b_1} \end{Bmatrix} = \int_{-t/2}^{t/2} \begin{Bmatrix} \sigma'_p \\ z'\sigma'_p \\ z'^2\sigma'_p \end{Bmatrix} dz'
$$

$$
\hat{\sigma}'_s = \begin{Bmatrix} \hat{\sigma}'_s \\ \hat{\sigma}'_{s1} \\ \hat{\sigma}'_{s2} \\ \hat{\sigma}'_{s3} \end{Bmatrix} = \int_{-t/2}^{t/2} \begin{Bmatrix} \sigma'_s \\ z'\sigma'_s \\ z'^2\sigma'_s \\ J(\bar{z})\sigma'^s \end{Bmatrix} dz'
$$

(10.99)

In Eq.(10.99) $\hat{\sigma}'_m, \hat{\sigma}'_b$ and $\hat{\sigma}'_s$ can be interpreted as the standard vectors of axial forces, bending moments and shear forces (in local axes) studied in flat shell theory (Chapter 8). The other vectors in Eq.(10.99) contain additional resultant stresses that have no direct physical meaning.

Note that the initial volume integral in Eq.(10.97) has been transformed into a *surface* integral (over the reference surface area) and all $\bar{z}$-dependence is embodied in the definition of the stress resultants.

A constitutive relationship between the stress and strain resultants is obtained as

$$
\hat{\sigma} = \int_{-t/2}^{t/2} \mathbf{S}^T \sigma' J(\bar{z})d\bar{z} = \int_{-t/2}^{t/2} \mathbf{S}^T \mathbf{D}' \varepsilon' J(\bar{z})d\bar{z} = \left[\int_{-t/2}^{t/2} \mathbf{S}^T \mathbf{D}' \mathbf{S} J(\bar{z})d\bar{z} \right] \hat{\varepsilon}' = \hat{\mathbf{D}}' \hat{\varepsilon}'
$$

(10.100)

where the *generalized constitutive matrix* $\hat{\mathbf{D}}'$ is

$$
\hat{\mathbf{D}}' = \begin{bmatrix} \hat{\mathbf{D}}'_p & \mathbf{0} \\ \mathbf{0} & \hat{\mathbf{D}}'_s \end{bmatrix} = \int_{-t/2}^{t/2} \mathbf{S}^T \mathbf{D}' \mathbf{S} J(\bar{z})d\bar{z}
$$

(10.101)

with

$$
\hat{\mathbf{D}}'_p = \int_{-t/2}^{t/2} \mathbf{S}_p^T \mathbf{D}'_p \mathbf{S}_p J(\bar{z})d\bar{z} = \int_{-t/2}^{t/2} \frac{1}{J(\bar{z})} \begin{bmatrix} \mathbf{D}'_p & z\mathbf{D}'_p & \bar{z}^2\mathbf{D}'_p \\ & \bar{z}^2\mathbf{D}'_p & \bar{z}^3\mathbf{D}'_p \\ \text{Sym.} & & \bar{z}^4\mathbf{D}'_p \end{bmatrix} d\bar{z}
$$

$$
\hat{\mathbf{D}}'_s = \int_{-t/2}^{t/2} \mathbf{S}_s^T \mathbf{D}'_s \mathbf{S}_s J(\bar{z})d\bar{z} = \int_{-t/2}^{t/2} \frac{1}{J(\bar{z})} \begin{bmatrix} \mathbf{D}'_s & z\mathbf{D}'_s & \bar{z}^2\mathbf{D}'_s & J\mathbf{D}'_s \\ & \bar{z}^2\mathbf{D}'_s & \bar{z}^3\mathbf{D}'_s & \bar{z}J\mathbf{D}'_s \\ \text{Sym.} & & \bar{z}^4\mathbf{D}'_s & \bar{z}^2 J\mathbf{D}'_s \\ & & & J^2\mathbf{D}'_s \end{bmatrix} d\bar{z}
$$

(10.102)

10.13.5 Enhanced transverse shear deformation matrix

Shear correction factors may be introduced in order to improve the representation of the transverse shear deformations, specially for thick shells. This can be simply accomplished by replacing $\mathbf{S}_s$ in Eq.(10.96a) by

$$\mathbf{S}_s = [k]\mathbf{S}_s \tag{10.103}$$

with

$$[k] = \begin{bmatrix} k_{x'}(\bar{z}) & 0 \\ 0 & k_{y'}(\bar{z}) \end{bmatrix} \tag{10.104}$$

where $k_{x'}$ and $k_{y'}$ are transverse shear correction parameters for $\gamma_{x'z'}$ and $\gamma_{y'z'}$, respectively. For thick monocoque shells the following parabolic profile can be chosen [DZ,SPH,St]

$$k_{x'} = k_{y'} = 1 - \frac{4(\bar{z} - \bar{z}_0)^2}{t^2} \tag{10.105}$$

A simpler alternative is to choose the conventional constant parameters $k_{x'} = k_{y'} = k = \sqrt{5/6}$. This simply implies substituting $\mathbf{D}'_s$ in Eq.(10.102) by $k^2 \mathbf{D}'_s$. This option is exact for analysis of thick homogeneous plates with uniform thickness.

10.13.6 CBR shell elements

The shell surface is discretized into standard isoparametric elements similarly as described for degenerated shell elements.

The lamina displacements are interpolated from the nodal values as

$$\mathbf{u}'_0 = \sum_{i=1}^{n} N_i(\xi, \eta)\mathbf{u}'_{0_i} \quad , \quad \mathbf{u}'_n = \sum_{i=1}^{n} N_i(\xi, \eta)\mathbf{u}'_{n_i} \tag{10.106}$$

where

$$\mathbf{u}'_{0_i} = [u'_{0_i}, v'_{0_i}, w'_{0_i}]^T \quad , \quad \mathbf{u}'_{n_i} = [u'_{n_i}, v'_{n_i}, w'_{n_i}]^T \tag{10.107}$$

Grouping terms we have

$$\mathbf{d}' = \left\{ \begin{matrix} \mathbf{u}'_0 \\ \mathbf{u}'_n \end{matrix} \right\} = \sum_{i=1}^{n} N_i \mathbf{d}'_i \quad \text{with} \quad \mathbf{d}'_i = \left\{ \begin{matrix} \mathbf{u}'_{0_i} \\ \mathbf{u}'_{n_i} \end{matrix} \right\} \tag{10.108a}$$

The lamina displacement components are related to their global Cartesian components by

$$\mathbf{d}'_i = \mathbf{L} \left\{ \begin{matrix} \mathbf{u}'_{0_i} \\ \mathbf{u}'_{n_i} \end{matrix} \right\} = \mathbf{L}\mathbf{d}_i \quad \text{where} \quad \mathbf{L} = [\mathbf{l}, \mathbf{m}, \mathbf{n}]^T \tag{10.108b}$$

Substituting Eq.(10.108a) into (10.95) gives the relationship between the local generalized strains and the nodal lamina displacement components in the global axes as

$$\hat{\varepsilon}' = \left\{ \begin{array}{c} \hat{\varepsilon}'_p \\ \hat{\varepsilon}'_s \end{array} \right\} = \mathbf{B}' \mathbf{d}^{(e)} \tag{10.109}$$

where

$$\mathbf{d}^{(e)} = \left\{ \begin{array}{c} \mathbf{d}_1 \\ \mathbf{d}_2 \\ \vdots \\ \mathbf{d}_n \end{array} \right\} \quad ; \quad \mathbf{B}' = [\mathbf{B}'_1, \mathbf{B}'_2 \cdots, \mathbf{B}'_n] \quad \text{with} \quad \mathbf{B}'_i = \left\{ \begin{array}{c} \mathbf{B}'^p_i \\ \mathbf{B}'^s_i \end{array} \right\} \tag{10.110}$$

and

$$\mathbf{B}'^p_i \atop 9 \times 6 = \begin{bmatrix} \overline{N_{i,x'}}\mathbf{l}^T & \mathbf{0} \\ \overline{N_{i,y'}}\mathbf{m}^T & \mathbf{0} \\ \overline{N_{i,y'}}\mathbf{l}^T + \overline{N_{i,x'}}\mathbf{m}^T & \mathbf{0} \\ \cdots\cdots\cdots & \cdots\cdots\cdots \\ \widehat{N_{i,x'}}\mathbf{l}^T & \overline{N_{i,x'}}\mathbf{l}^T \\ \widehat{N_{i,y'}}\mathbf{m}^T & \overline{N_{i,y'}}\mathbf{m}^T \\ (\widehat{N_{i,y'}}\mathbf{l}^T + \widehat{N_{i,x'}}\mathbf{m}^T) & (\overline{N_{i,y'}}\mathbf{l}^T + \overline{N_{i,x'}}\mathbf{m}^T) \\ \cdots\cdots\cdots & \cdots\cdots\cdots \\ \mathbf{0} & \overline{N_{i,x'}}\mathbf{l}^T \\ \mathbf{0} & \overline{N_{i,y'}}\mathbf{m}^T \\ \mathbf{0} & (\widehat{N_{i,y'}}\mathbf{l}^T + \widehat{N_{i,x'}}\mathbf{m}^T) \end{bmatrix} = \begin{bmatrix} \mathbf{B}'^p_{0_i} \\ \cdots \\ \mathbf{B}'^p_{1_i} \\ \cdots \\ \mathbf{B}'^p_{2_i} \end{bmatrix} \tag{10.111}$$

$$\mathbf{B}'^s_i \atop 8 \times 6 = \begin{bmatrix} \overline{\nabla N_i}\mathbf{n}^T & \mathbf{0} \\ \widehat{\nabla N_i}\mathbf{n}^T & \overline{\nabla N_i}\mathbf{n}^T \\ \mathbf{0} & \widehat{\nabla N_i}\mathbf{n}^T \\ \mathbf{0} & N_i \left\{ \begin{array}{c} \mathbf{l}^T \\ \mathbf{m}^T \end{array} \right\} \end{bmatrix} \tag{10.112}$$

where

$$\overline{\nabla N_i} = \left\{ \begin{array}{c} \overline{N_{i,x'}} \\ \overline{N_{i,y'}} \end{array} \right\} = \mathbf{J}_0^{-1} \left\{ \begin{array}{c} N_{i,\xi} \\ N_{i,\eta} \end{array} \right\} \quad , \quad \widehat{\nabla N_i} = \left\{ \begin{array}{c} \widehat{N_{i,x'}} \\ \widehat{N_{i,y'}} \end{array} \right\} = \frac{1}{J_0} \mathbf{J}_1^{-1} \left\{ \begin{array}{c} N_{i,\xi} \\ N_{i,\eta} \end{array} \right\} \tag{10.113}$$

and

$$(\cdot)_{x'} = \frac{\partial(\cdot)}{\partial x'} \quad , \quad (\cdot)_{y'} = \frac{\partial(\cdot)}{\partial y'} \tag{10.114}$$

Recall that in above derivations we have assumed that vectors $\mathbf{v}_1, \mathbf{v}_2, \mathbf{v}_3$ are coincident with $\mathbf{l}, \mathbf{m}, \mathbf{n}$ (mild taper assumption; Figure 10.4c).

The local strains are related to the global nodal displacements $\mathbf{u}'_{0_i}$ and the local nodal rotations $\boldsymbol{\theta}_i = [\theta_{1_i}, \theta_{2_i}]^T$ by the following transformation

$$\mathbf{d}_i = \left\{ \begin{matrix} \mathbf{u}_{0_i} \\ \mathbf{u}_{n_i} \end{matrix} \right\} = \begin{bmatrix} \mathbf{I}_3 & 0 \\ 0 & -\mathbf{C}_i \end{bmatrix} \left\{ \begin{matrix} \mathbf{u}_{0_i} \\ \boldsymbol{\theta}_i \end{matrix} \right\} = \mathbf{T}_i \mathbf{a}_i \qquad (10.115)$$

with

$$\mathbf{T}_i = \begin{bmatrix} \mathbf{I}_3 & 0 \\ 0 & -\mathbf{C}_i \end{bmatrix} \quad \text{and} \quad \mathbf{a}_i = \left\{ \begin{matrix} \mathbf{u}_{0_i} \\ \boldsymbol{\theta}_i \end{matrix} \right\} \qquad (10.116)$$

where $\mathbf{C}_i$ is given by Eq.(10.22b) and $\mathbf{I}_3$ is the 3×3 unit matrix.

In the previous derivation we have used Eq.(10.20) relating the displacements of the end of the normal vector and the local rotations (Figure 10.8).

It is important to note that matrix $\mathbf{L}$ relating the displacement components in the lamina and global coordinate systems *is defined at each integration point*, whereas matrix $\mathbf{C}_i$ *is defined at each node*.

Substituting Eq.(10.115) into (10.109) gives

$$\hat{\boldsymbol{\varepsilon}}' = \mathbf{B}\mathbf{a}^{(e)} \qquad (10.117)$$

where

$$\mathbf{a}^{(e)} = \left\{ \begin{matrix} \mathbf{a}_1 \\ \vdots \\ \mathbf{a}_n \end{matrix} \right\} \quad , \quad \mathbf{B} = [\mathbf{B}_1, \overset{.}{.}, \mathbf{B}_n] \quad \text{and} \quad \mathbf{B}_i = \mathbf{B}'_i \mathbf{T}_i \qquad (10.118)$$

Matrix $\mathbf{B}_i$ can be split as

$$\mathbf{B}_i = \left\{ \begin{matrix} \mathbf{B}_i^p \\ \mathbf{B}_i^s \end{matrix} \right\} \quad \text{with} \quad \mathbf{B}_i^p = \mathbf{B}'^p_i \mathbf{T}_i \quad , \quad \mathbf{B}_i^s = \mathbf{B}'^s_i \mathbf{T}_i \qquad (10.119)$$

with $\mathbf{B}'^p_i$ and $\mathbf{B}'^s_i$ given by Eqs.(10.111) and (10.112), respectively.

10.13.6.1 CBR element stiffness matrix and equivalent nodal force vectors

The element stiffness matrix for the CBR element is obtained *in the global coordinate system* as

$$\mathbf{K}_{ij}^{(e)} = \iint_{A^{(e)}} \mathbf{B}_i^T \hat{\mathbf{D}}' \mathbf{B}_j dA = \iint_{A^{(e)}} \left[[\mathbf{B}_i^p]^T \hat{\mathbf{D}}'_p \mathbf{B}_j^p + [\mathbf{B}_i^s]^T \hat{\mathbf{D}}'_s \mathbf{B}_j^s \right] dA = \mathbf{K}_{p_{ij}}^{(e)} + \mathbf{K}_{s_{ij}}^{(e)} \qquad (10.120)$$

where $A^{(e)}$ is the element reference surface area and $i, j = 1, n$.

The equivalent nodal force vector is expressed by surface integrals as

$$\mathbf{f}_i^{(e)} = \iint_{A^{(e)}} N_i \bar{\mathbf{b}} dA + \iint_{A^{(e)}} N_i \bar{\mathbf{t}} dA \qquad (10.121)$$

where the resultant body forces $\bar{\mathbf{b}}$ and surface loads $\bar{\mathbf{t}}$ are

$$\bar{\mathbf{b}} = \int_{-t/2}^{t/2} \left\{ \begin{array}{c} \mathbf{b} \\ \bar{z}\mathbf{b} \end{array} \right\} J(\bar{z}) dz \quad , \quad \bar{\mathbf{t}} = \left\{ \begin{array}{c} \mathbf{t} \\ \bar{z}^*\mathbf{t} \end{array} \right\} \qquad (10.122)$$

with $\mathbf{b} = [b_x, b_y, b_z]^T$, $\mathbf{t} = [t_x, t_y, t_z]^T$, $J(\bar{z})$ is given by Eq.(10.87) and $\bar{z}^*$ is the thickness coordinate of the loaded (top or bottom) surface.

Computation of the element stiffness matrix terms requires selective integration to avoid transverse shear/membrane locking. As a general rule the terms of $\mathbf{K}_{p_{ij}}^{(e)}$ generated by rows 1–3 of the in-plane submatrix $\bar{\mathbf{B}}_i'^p$ (Eq.(10.111)) and the entire transverse shear stiffness matrix $\mathbf{K}_{s_{ij}}^{(e)}$ are integrated by a reduced quadrature.

Full integration can be used for the rest of the terms in $\mathbf{K}_{p_{ij}}^{(e)}$. The full and reduced quadratures for CBR shell elements coincide with those defined in Table 10.1 for degenerated shell elements.

An alternative to selective integration is the assumed strain technique described in Section 10.15.

10.14 CBR-S SHELL ELEMENTS

The CBR shell formulation can be simplified by assuming that the shell is thin enough with respect to the current radii of curvature so that the through-thickness variation of the Jacobian may be neglected. The simplified CBR shell elements will be termed CBR-S.

The CBR-S assumption eliminates the $\bar{z}$-dependence in the surface Jacobian $\mathbf{J}_s$, via the approximation

$$\mathbf{J}_s(\xi, \eta, \bar{z}) \cong \mathbf{J}_0(\xi, \eta) \qquad (10.123)$$

Hence the inverse Jacobian is now both uncoupled and constant, i.e.

$$\mathbf{J}^{-1} = \begin{bmatrix} \mathbf{J}_0^{-1} & \mathbf{0} \\ \mathbf{0} & 2/t \end{bmatrix} \qquad (10.124)$$

The formulation is now very similar to that presented for flat shell elements in Chapter 8. The generalized strains are defined as

$$\hat{\varepsilon}' = \begin{Bmatrix} \hat{\varepsilon}'_m \\ \hat{\varepsilon}'_b \\ \hat{\varepsilon}'_s \end{Bmatrix} = \mathbf{S}\varepsilon' \tag{10.125}$$

with

$$\hat{\varepsilon}'_m = \left[\frac{\partial u'_0}{\partial x'}, \frac{\partial v'_0}{\partial y'}, \frac{\partial u'_0}{\partial y'} + \frac{\partial v'_0}{\partial x'}\right]^T \quad ; \quad \hat{\varepsilon}'_b = \left[\frac{\partial u'_n}{\partial x'}, \frac{\partial v'_n}{\partial y'}, \frac{\partial u'_n}{\partial y'} + \frac{\partial v'_n}{\partial x'}\right]^T$$

$$\hat{\varepsilon}'_s = \left[\frac{\partial w'_0}{\partial x'} + u'_n, \frac{\partial w'_0}{\partial y'} + v'_n\right]^T \quad ; \quad \mathbf{S} = \begin{bmatrix} \mathbf{I}_3 & \bar{z}\mathbf{I}_3 & \mathbf{0} \\ \mathbf{0} & \mathbf{0} & [k] \end{bmatrix}$$

$$\tag{10.126}$$

where $\hat{\varepsilon}'_m$, $\hat{\varepsilon}'_b$ and $\hat{\varepsilon}'_s$ are local membrane, bending and transverse shear strain vectors and $[k]$ is a diagonal matrix of transverse correction parameters (Eq.10.104)).

The resultant stress vector is

$$\hat{\sigma}' = \begin{Bmatrix} \hat{\sigma}'_m \\ \hat{\sigma}'_b \\ \hat{\sigma}'_s \end{Bmatrix} = \int_{-t/2}^{t/2} \begin{Bmatrix} \sigma'_p \\ z\sigma'_p \\ \sigma'_s \end{Bmatrix} dz \tag{10.127}$$

where σ'_p and σ'_s are given in Eq.(10.39).

The generalized constitutive matrix is

$$\hat{\mathbf{D}}' = \begin{bmatrix} \hat{\mathbf{D}}_m & \hat{\mathbf{D}}'_{mb} & \mathbf{0} \\ \hat{\mathbf{D}}'_{mb} & \hat{\mathbf{D}}'_b & \mathbf{0} \\ \mathbf{0} & \mathbf{0} & \hat{\mathbf{D}}'_s \end{bmatrix} = \int_{-t/2}^{t/2} \mathbf{S}^T \mathbf{D}' \mathbf{S} dz \tag{10.128}$$

and

$$(\hat{\mathbf{D}}'_m, \hat{\mathbf{D}}'_{mb}, \hat{\mathbf{D}}'_b, \hat{\mathbf{D}}'_s) = \int_{-t/2}^{t/2} (\mathbf{D}'_p, \bar{z}\mathbf{D}'_p, \bar{z}^2\mathbf{D}'_p, \mathbf{D}'_s) d\bar{z} \tag{10.129}$$

The discretization follows the procedure explained for the CBR element in the previous section. The lamina-based strain matrix is now written as

$$\mathbf{B}'_i = \begin{Bmatrix} \mathbf{B}'_{m_i} \\ \mathbf{B}'_{b_i} \\ \mathbf{B}'_{s_i} \end{Bmatrix} \quad \text{with} \quad \mathbf{B}'_{m_i} = \begin{bmatrix} \dfrac{\partial N_i}{\partial x'}\mathbf{l}^T & \mathbf{0} \\[2mm] \dfrac{\partial N_i}{\partial y'}\mathbf{m}^T & \mathbf{0} \\[2mm] \left(\dfrac{\partial N_i}{\partial y'}\mathbf{l}^T + \dfrac{\partial N_i}{\partial x'}\mathbf{m}^T\right) & \mathbf{0} \end{bmatrix} \tag{10.130}$$

$$\mathbf{B}'_{b_i} = \begin{bmatrix} 0 & \dfrac{\partial N_i}{\partial x'}\mathbf{1}^T \\ 0 & \dfrac{\partial N_i}{\partial y'}\mathbf{m}^T \\ 0 & \left(\dfrac{\partial N_i}{\partial y'}\mathbf{1}^T + \dfrac{\partial N_i}{\partial x'}\mathbf{m}^T\right) \end{bmatrix} \quad , \quad \mathbf{B}'_{s_i} = \begin{bmatrix} \dfrac{\partial N_i}{\partial x'}\mathbf{n}^T & N_i\mathbf{1}^T \\ \dfrac{\partial N_i}{\partial y'}\mathbf{n}^T & N_i\mathbf{m}^T \end{bmatrix}$$

The "global" generalized strain matrix is

$$\mathbf{B}_i = \mathbf{B}'_i \mathbf{T}_i \tag{10.131}$$

with

$$\mathbf{B}_i = \begin{Bmatrix} \mathbf{B}_{m_i} \\ \mathbf{B}_{b_i} \\ \mathbf{B}_{s_i} \end{Bmatrix} \quad , \quad \mathbf{B}_{m_i} = \mathbf{B}'_{m_i}\mathbf{T}_i \quad , \quad \mathbf{B}_{b_i} = \mathbf{B}'_{b_i}\mathbf{T}_i \quad , \quad \mathbf{B}_{s_i} = \mathbf{B}'_{s_i}\mathbf{T}_i$$

$$\tag{10.132}$$

The element stiffness matrix is finally computed as

$$\mathbf{K}_{ij}^{(e)} = \iint_{A^e} \mathbf{B}_i^T \hat{\mathbf{D}}' \mathbf{B}_j \, dA = \mathbf{K}_{m_{ij}}^{(e)} + \mathbf{K}_{b_{ij}}^{(e)} + \mathbf{K}_{s_{ij}}^{(e)} + \mathbf{K}_{mb_{ij}}^{(e)} + [\mathbf{K}_{mb_{ij}}^{(e)}]^T$$

$$\tag{10.133}$$

with

$$\begin{aligned} \mathbf{K}_{a_{ij}}^{(e)} &= \mathbf{B}_{a_i}^T \hat{\mathbf{D}}'_a \mathbf{B}_{a_j} \quad , \quad a = m, b, s \\ \mathbf{K}_{mb_{ij}}^{(e)} &= \mathbf{B}_{m_i}^T \hat{\mathbf{D}}'_{m_b} \mathbf{B}_{b_i} \end{aligned} \tag{10.134}$$

For homogeneous or symmetric material properties in the thickness direction, then the coupling membrane-bending matrix $\mathbf{K}_{mb}$ is zero as usual. The expression for the equivalent nodal force vector coincides with Eq.(10.121).

For very curved thick shells the CBR-S formulation is approximate. It is however accurate for moderately curved and shallow thick/thin shells and thin curved shells [Sta,SPH]. We note finally that for flat shell surfaces the CBR-S formulation coincides with that for flat shell elements presented in Chapter 8.

10.15 QL9 CBR-S SHELL ELEMENT WITH ASSUMED MEMBRANE AND TRANSVERSE SHEAR STRAINS

Membrane locking in curved shell elements is characterized by the incapacity of the element to reproduce pure bending modes without introducing spureous membrane strains. This problem is similar to the inability of some elements to satisfy the condition of zero transverse shear strains in the thin limit which leads to shear locking. We present here a locking-free 9-noded quadrilateral shell element based on the assumed strain technique and the CBR-S formulation of the previous section.

10.15.1 Assumed quadratic membrane strain field

A way to ensure the satisfaction of the condition of zero membrane strains is to assume "a priori" a polynomial membrane strain field with coefficients expressed as a linear combination of the nodal displacements. The difficulty in curved shell elements is splitting the membrane strains from the bending strains. A way to do this is to adopt the CBR-S shell formulation described in Section 10.14 where a local membrane strain vector, $\hat{\varepsilon}'_m$, is defined (see Eq.(10.126)).

From the shape functions of the 9-noded Lagrange quadrilateral elements [On4] we can find the polynomial terms included in the derivatives of the displacement with respect to the parametric coordinates ξ, η as

$$\frac{\partial u'_0}{\partial \xi} : (1, \xi, \eta, \xi\eta, \xi\eta^2, \eta^2)$$

$$\frac{\partial v'_0}{\partial \eta'} : (1, \xi, \eta, \xi\eta, \eta\xi^2, \xi^2) \tag{10.135}$$

$$\frac{\partial u'_0}{\partial \eta} + \frac{\partial v'_0}{\partial \xi} : (1, \xi, \eta, \xi\eta, \xi\eta^2, \eta\xi^2, \xi^2, \eta^2)$$

The assumed membrane strain field should include these polynomial terms. Consequently, we choose the following interpolation for the components of $\hat{\varepsilon}'_m$

$$\frac{\partial u'_0}{\partial x'} = \sum_{i=1}^{3}\sum_{j=1}^{2} L_j(\xi)\, H_i(\eta) \left(\frac{\partial u'_0}{\partial x'}\right)_{ij}$$

$$\frac{\partial v'_0}{\partial y'} = \sum_{i=1}^{3}\sum_{j=1}^{2} H_i(\xi)\, L_j(\eta) \left(\frac{\partial v'_0}{\partial y'}\right)_{ij}$$

$$\frac{\partial u'_0}{\partial y'} + \frac{\partial v'_0}{\partial x'} = \frac{1}{2}\sum_{i=1}^{3}\sum_{j=1}^{2} L_j(\xi)\, H_i(\eta) \left(\frac{\partial u'_0}{\partial y'}\right)_{ij} + \tag{10.136}$$

$$+ \frac{1}{2}\sum_{i=1}^{3}\sum_{j=1}^{2} H_i(\xi)\, L_j(\eta) \left(\frac{\partial v'_0}{\partial x'}\right)_{ij}$$

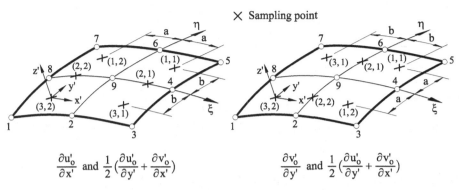

Fig. 10.13 Sampling points for interpolating the membrane strains in the QL9 CBR-S shell element. The same points are used for interpolating the transverse shear strains for $b = 1$

where $(\cdot)_{ij}$ and $(\cdot)^{ij}$ denote values at the i, j sampling point, and

$$H_1(z) = z\frac{\left(\frac{z}{b}+1\right)}{2b}, \quad H_2(z) = 1 - \left(\frac{z}{b}\right)^2, \quad H_3(z) = z\frac{\left(\frac{z}{b}-1\right)}{2b}$$

$$L_1(z) = \frac{\left(1+\frac{z}{a}\right)}{2}, \quad L_2(z) = \frac{\left(1-\frac{z}{a}\right)}{2} \quad z = \xi, \eta$$

$$(10.137)$$

Note that $\frac{\partial u'_o}{\partial x'}$ varies linearly in x' and quadratically in y'; $\frac{\partial v'_o}{\partial y'}$ is linear in y' and quadratic in x' and $\left(\frac{\partial u'_o}{\partial y'} + \frac{\partial v'_o}{\partial x'}\right)$ is quadratic in both directions. Figure 10.13 shows the sampling points for each membrane strain.

The relation between the derivatives $\left(\frac{\partial u'_o}{\partial x'}\right)_{ij}$, etc and the nodal displacements is deduced from Section 10.14 (Eqs.(10.130) and (10.131)) as

$$\left\{ \begin{array}{c} \left(\frac{\partial u'_o}{\partial x'}\right)_{ij} \\ \left(\frac{\partial v'_o}{\partial y'}\right)_{ij} \\ \left(\frac{\partial u'_o}{\partial y'}\right)_{ij} \\ \left(\frac{\partial v'_o}{\partial x'}\right)_{ij} \end{array} \right\} = \sum_{k=1}^{9} \begin{bmatrix} N_{k,x'}\mathbf{l}^T & \mathbf{0} \\ N_{k,y'}\mathbf{m}^T & \mathbf{0} \\ N_{k,y'}\mathbf{l}^T & \mathbf{0} \\ N_{k,x'}\mathbf{m}^T & \mathbf{0} \end{bmatrix}_{ij} \mathbf{d}_k = \mathbf{G}_{ij}\mathbf{d}^{(e)} \qquad (10.138)$$

Combining Eq.(10.138) and (10.136) yields

$$\hat{\boldsymbol{\varepsilon}}'_m = \bar{\mathbf{B}}'_{m_i}\mathbf{d}^{(e)} \qquad (10.139)$$

$$\bar{\mathbf{B}}'_{m_i} = \sum_{i=1}^{3} \sum_{j=1}^{2} \begin{bmatrix} [L_j(\xi)H_i(\eta)] & 0 & 0 & 0 \\ 0 & [H_i(\xi)L_j(\eta)] & 0 & 0 \\ 0 & 0 & \frac{1}{2}[L_j(\xi)H_i(\eta)] & \frac{1}{2}[H_i(\xi)L_j(\eta)] \end{bmatrix} \mathbf{G}_{ij}$$

$$(10.140)$$

Matrix $\bar{\mathbf{B}}'_{m_i}$ substitutes $\mathbf{B}'_{m_i}$ in Eq.(10.130).

A similar procedure can be followed for deriving an enhanced membrane strain matrix $\mathbf{B}'^p_{0_i}$ for the CBR element.

10.15.2 Assumed quadratic transverse shear strain field

The transverse shear strains in the covariant system for the 9-noded Lagrange quadrilateral element contain the following terms

$$\gamma_{\xi\zeta} : f_1(1, \xi, \eta, \xi\eta, \eta^2, \xi\eta^2, \mathbf{u}'_0, \mathbf{u}'_n) + f_2(\xi^2\eta, \xi^2\eta^2, \mathbf{u}'_n)$$
$$\gamma_{\eta\zeta} : g_1(1, \xi, \eta, \xi\eta, \xi^2, \xi^2\eta, \mathbf{u}'_0, \mathbf{u}'_n) + g_2(\xi\eta^2, \eta^2, \xi^2\eta^2, \mathbf{u}'_n)$$

$$(10.141)$$

The spurious terms in f_2 and g_2 involving the rotational displacements $\mathbf{u}'_n$ only are eliminated by assuming the following quadratic interpolations for the transverse shear strains

$$\gamma_{\xi\zeta} = \sum_{i=1}^{3} \sum_{j=1}^{2} L_j(\xi)H_i(\eta)\gamma_{\xi\zeta}^{ij} \quad ; \quad \gamma_{\eta\zeta} = \sum_{i=1}^{3} \sum_{j=1}^{2} H_i(\xi)L_j(\eta)\gamma_{\eta\zeta}^{ij} \quad (10.142)$$

The sampling points i, j for the transverse shear strains are the same as for the QQQQ-L element (Section 6.7.3). The points coincide with those shown in Figure 10.13 for $b = 1$. Functions $L_i(\xi), H_i(\eta)$ are deduced from Eqs.(10.137) making $b = 1$.

The transverse shear strains $\gamma_{\xi\zeta}^{ij}, \gamma_{\eta\zeta}^{ij}$ at the sampling point i, j are expressed in terms of the nodal displacement as

$$\boldsymbol{\gamma}^{ij} = \begin{Bmatrix} \gamma_{\xi\zeta}^{ij} \\ \gamma_{\eta\zeta}^{ij} \end{Bmatrix} = \hat{\mathbf{L}}^{ij} \begin{Bmatrix} \gamma_{x'z'}^{ij} \\ \gamma_{y'z'}^{ij} \end{Bmatrix} = \hat{\mathbf{L}}^{ij}[\mathbf{B}'_s]^{ij}\mathbf{d}^{(e)} \qquad (10.143)$$

with $\hat{\mathbf{L}}$ and $\mathbf{B}'_s$ given in Eqs.(10.71b) and (10.130), respectively.

The lamina-Cartesian components of the transverse shear strains at the integration points are obtained from the covariant components as

$$\hat{\boldsymbol{\varepsilon}}'_s = \begin{Bmatrix} \gamma_{x'z'} \\ \gamma_{y'z'} \end{Bmatrix} = \hat{\mathbf{L}}^T \begin{Bmatrix} \gamma_{\xi\zeta} \\ \gamma_{\eta\zeta} \end{Bmatrix} = \bar{\mathbf{B}}'_s\mathbf{d}^{(e)} \qquad (10.144)$$

where the substitute shear strain matrix is

$$\bar{\mathbf{B}}'_s = \sum_{i=1}^{3} \sum_{j=1}^{2} \begin{bmatrix} L_j(\xi)H_i(\eta) \\ H_i(\xi)L_j(\eta) \end{bmatrix} \hat{\mathbf{L}}^{ij} [\mathbf{B}'_s]^{ij} \tag{10.145}$$

$\bar{\mathbf{B}}'_s$ replaces $\mathbf{B}'_s$ in Eq.(10.130) for computing the shear stiffness matrix for the CBR-S element.

A similar procedure can be followed for deriving the substitute transverse shear strain matrix for the CBR element in Section 10.13.

10.16 DK CURVED SHELL ELEMENTS

Several authors have derived Discrete Kirchhoff (DK) thin curved shell elements [Dh,Dh2,DMM,DMMT,DV,IA,Ir2,MG,We,WOK]. The procedure is very similar to that explained in Section 5.10 for DK plate elements, i.e. the transverse shear strains are constrained to a zero value at a number of points within the element. In this manner the element can effectively approximate the thin shell conditions of zero transverse shear strain. A popular DK curved shell element was derived by Irons [IA,Ir4] and named *semi-Loof*. The element is briefly presented next.

10.16.1 Semi-Loof curved shell element

The starting point is the 43 DOFs degenerated shell element of Figure 10.14. The eight tangent rotations at the side nodes and the three central DOFs are eliminated using the following conditions:

a) The tangential shear stress is constrained to zero at the eight mid-side nodes marked with a circle in Figure 10.14a (eight conditions)

b)
$$\iint_{A^{(e)}} (\mathbf{v}_{1_9})^T \boldsymbol{\gamma} \, dA = \iint_{A^{(e)}} (\mathbf{v}_{2_9})^T \boldsymbol{\gamma} \, dA$$

with
$$\boldsymbol{\gamma} = [\gamma_{xz}, \gamma_{yz}]^T \quad \text{(2 conditions)} \tag{10.146}$$

c)
$$\oint_{l^{(e)}} \gamma_n ds = 0 \quad \text{(1 condition)} \tag{10.147}$$

In above $\mathbf{v}_{1_9}$ and $\mathbf{v}_{2_9}$ are the nodal vectors of the central node; γ_n is the transverse shear strain normal to a side and $l^{(e)}$ is the side length.

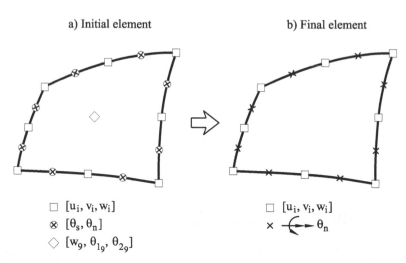

Fig. 10.14 Semi-Loof shell element. Initial (a) and final (b) nodal variables

Figure 10.14a shows the final 32 DOFs semi-loof element (3-nodal displacements at the eight square nodes and the normal rotation at the side nodes). The semi-Loof element has an excellent performance for thin shells. Further details of the element formulation and its application can be found in [He,IA,Ir4].

10.17 PERFORMANCE OF DEGENERATED SHELL ELEMENTS WITH ASSUMED STRAIN FIELDS

The Q4 degenerated shell element based on a linear assumed transverse shear strain and 2×2 full integration (Section 10.11.3) is an accurate and robust element for shell analysis. Its only drawback is its incapacity to model exactly curved geometries. The flat version of this element coincides with the QLLL plate/flat shell element presented in previous chapters. A selection of examples showing the good performance of this element can be found in [DB].

The QL9 CBR-S shell element with assumed quadratic membrane and transverse strain fields (Section 10.15) performs well for curved shell analysis. The assumed membrane strain field enhances the accuracy of the element for coarse meshes as shown for two standard shell test problems: a cylindrical roof under self weight and a pinched cylinder (Figure 10.15).

a) Cylindrical shell under uniform load

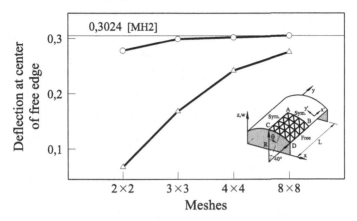

b) Cylinder under central point load

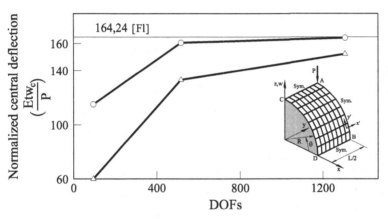

△ QL9 CBR-S element with assumed transverse shear strains

○ QL9 CBR-S element with assumed transverse shear and membrane strains

Fig. 10.15 Performance of QL9 CBR-S shell element. a) Cylindrical shell under uniform loading (Figure 8.34). b) Pinched cylinder (Figure 8.22)

Figure 10.16 shows the good performance of the Q4 degenerated shell element and the QL9 CBR-S shell element for analysis of an hyperbolic shell under uniform pressure.

Huang and Hinton [HH5] studied the membrane behaviour of an element very similar to the QL9 CBR-S shell element for two different positions of the sampling points for the assumed membrane field: $a = \frac{1}{\sqrt{3}}$ and $b = 1$, and $a = \frac{1}{\sqrt{3}}$ and $b = \left(\frac{3}{5}\right)^{\frac{1}{2}}$ (Figure 10.13). Good performance

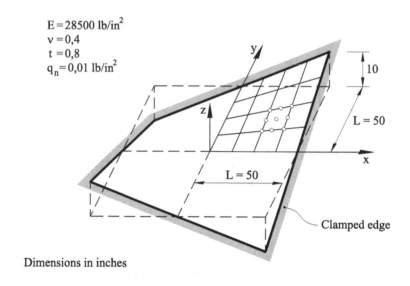

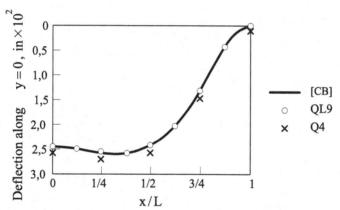

○ QL9 element with assumed transverse shear and membrane strains

Fig. 10.16 Clamped hyperbolic shell under uniform pressure. Vertical deflection along central line $(y = 0)$ obtained with the Q4 degenerated shell element and the QL9 CBR-S shell element using an assumed strain approach

was found for both cases. It is interesting that the effect of the assumed membrane strain field is equivalent to using a 2×3 reduced quadrature in the original CBR-S formulation (Section 10.14) for the stiffness matrix terms emanating from $\frac{\partial u'_0}{\partial x'}$ and $\frac{1}{2}\left(\frac{\partial u'_0}{\partial y'} + \frac{\partial v'_0}{\partial x'}\right)$, and a 3×2 quadrature for those stiffness terms emanating from $\frac{\partial v'_0}{\partial y'}$ and $\frac{1}{2}\left(\frac{\partial u'_0}{\partial y'} + \frac{\partial v'_0}{\partial x'}\right)$.

A similar technique was introduced in [HH4] for enhancing the membrane and transverse shear behaviour of the QS8 CBR-S shell element using a five point quadratic membrane strain interpolation with the same sampling points chosen for the assumed transverse shear strains in the QQQQ-S plate element of Section 6.7.2.

The performance of the QS8 element with assumed membrane and transverse shear strain fields has been found to be poorer than the QL9 element [HH4,HH5].

More information on the behaviour of degenerated and "truly" shell elements can be found in [BR,DB,FV,HH4,HH5,HL2,Hu2,JP2,OHG,ZT2].

10.18 DEGENERATED FLAT SHELL AND PLATE ELEMENTS

Degenerated shell elements can be easily particularized for folded plate structures discretized with flat shell elements. Thickness integration is straightforward and the resulting elements are identical to the Reissner-Mindlin flat shell elements of Chapter 8. Also, when particularized for plates the formulation coincides with that described for Reissner-Mindlin plate elements in Chapter 6.

10.19 SHELL ELEMENTS BASED ON SIX AND SEVEN PARAMETER MODELS

In the shell theory studied in this and the previous chapters the 3D-constitutive equations have been considered by employing the assumption of vanishing stress in the thickness direction (the so-called plane stress condition, $\sigma_{z'} = 0$). The thickness stretch (i.e. the deformation $\varepsilon_{z'}$) is not involved in the PVW and it can be computed a posteriori. However, this procedure is not satisfactory in certain cases that require to resort to a complete unmodified 3D constitutive relationship. An example is when local effects are relevant in shell structures, as the accurate prediction of stresses and strains in the thickness direction is important in those cases. This is typical for reinforced concrete shells under concentrated loading, or in the prediction of delamination in composite shells. Another reason might be that for certain constitutive models (i.e. for non linear material models) the condensation of $\varepsilon_{z'}$ cannot be done explicitly or renders elaborate strain equations.

A way to circumvent this problem, thereby allowing the use of full 3D constitutive laws, is to account for the thickness stretch and stress in the shell equations.

This is the basis of the 6-parameter shell formulation that leads to shell elements with 6 DOF per node: the three displacements and the three components of vector $\Delta\mathbf{V}_{3i} = \mathbf{V}_{3i}^r - \mathbf{V}_{3i}^c$ where $\mathbf{V}_{3i}$ is the normal vector defined in Section 10.3.2 and upper indices r and c denote the reference and current configuration of the shell geometry, respectively.

A problem of the 6-parameter shell theory is that the displacement field is linear in the thickness direction and, hence, the $\varepsilon_{z'}$ strain is constant across the thickness. This introduces errors in the solution of the order of ν^2 for bending dominated cases [BRR,SR2].

The deficiency of the 6-parameter shell formulation can be overcome by introducing an additional nodal displacement DOF which originates a quadratic displacement field in the thickness direction and, consequently, yields a linear distribution of the thickness strain (and stress). This is the basis of the 7-parameter shell formulation that has been successfully used by different authors [BR3,San2].

An interesting alternative for removing the deficiency of the 6-parameter formulation while still preserving 6 DOF in the final shell element, is to enrich the strain field of the 6-parameter theory by an additional linear component of the thickness strain. This extra strain can be introduced via the so-called enhanced assumed strain approach and then it can be eliminated at the element level after discretization [SR2]. This procedure has been used by different authors for deriving 3D shell elements compatible with 3D constitutive equations for a variety of non linear problems [BBR,BR2,BRR,SR2].

10.20 ISOGEOMETRIC SHELL ANALYSIS

10.20.1 Isogeometric analysis

The name *isogeometric* refers to using the same basis functions for representing the geometry in computer-aided design (CAD), computer graphics (CG) and animation as for the shape functions in finite element computations. Thus, the FE analysis works on a geometrically exact model and no meshing is necessary [CHB,HCB]. For the first application of the isoparametric methodology, non-uniform rational B-Splines (NURBS) were chosen as the basis function, due to their relative simplicity and ubiquity in the world of CAD, CG and animation.

NURBS and isogeometric analysis fundamentals

Non-uniform rational B-splines (NURBS) are a standard tool for describing and modeling curves and surfaces in computer aided design and computer graphics [CRE,PT2,Ro]. The aim of this section is to introduce NURBS briefly and to present an overview of isogeometric analysis, for which an extensive account has been given in [CHB,HCB].

A B-spline is a non-interpolating, piecewise polynomial curve. It is defined by a set of control points $\mathbf{P}_i$, $i = 1, 2 \cdots, n$, the degree p of the B-spline polynomial (also called the *order* of the B-spline) and a so-called knot vector. A knot vector is a set of non-decreasing real numbers representing coordinates in the parametric [0,1] space, written as

$$\Xi := \{\xi_1, \xi_2, \cdots, \xi_{n+p+1}\} \tag{10.148}$$

where ξ_i is the ith *knot* and i is the knot index ranging from one to the number of knots $n_k = n + p + 1$. The interval $[\xi_1, \xi_{n+p+1}]$ is called a patch.

A B-spline basis function is C^∞ continuous inside a knot span region (i.e. between two distinct knots) and C^{p-1} continuous at a single knot. A knot value can appear more than one time and is called a *multiple knot*. At a knot of multiplicity k the continuity is C^{p-k}.

If the first and the last knot have the multiplicity $p + 1$, the knot vector is called open [CHB,CRE,DJS,HCB]. In a B-spline with an open know vector the first and the last control points are interpolated and the curve is tangential to the control polygon at the start and the end of the curve. Open knot vectors are standard in CAD applications and are assumed for the remainder of this section.

Basis functions

Given a knot vector, B-spline basis functions are defined recursively by the Cox-de Boor formula [Co,DeB] starting with $p = 0$ (piecewise constants):

$$N_{j,0}(\xi) = \begin{cases} 1 & \text{if } \xi_j \leq \xi < \xi_{j+1} \\ 0 & \text{otherwise} \end{cases} \tag{10.149}$$

For $p = 1, 2, 3, \cdots$, the basis is defined by the following recursion formula

$$N_{j,p}(\xi) = \frac{\xi - \xi_j}{\xi_{j+p} - \xi_j} N_{j,p}(\xi) + \frac{\xi_{j+p+1} - \xi}{\xi_{j+p+1} - \xi_{j+1}} N_{j+1,p-1}(\xi) \tag{10.150}$$

Figure 10.17 shows an example of cubic B-spline basis functions with an open knot vector.

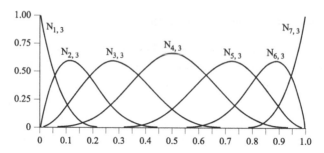

Fig. 10.17 Cubic B-spline basis functions with open knot vector $\Xi =$ $[0, 0, 0, 0, 0.25, 0.5, 0.75, 1, 1, 1, 1]$

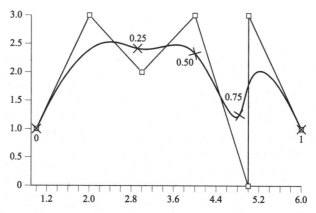

Fig. 10.18 Cubic B-spline with the open knot vector of Figure 10.17. The dashed lines represent the control polygon; the small crosses are the knots on the curve. The first and the last control point are interpolated and the curve is tangential to the control polygon at its start and end [KBLW]

B-spline curves

A B-spline curve of degree p is computed by the linear combination of control points and the respective basis function

$$C(\xi) = \sum_{i=1}^{n} N_{i,p}(\xi)\mathbf{P}_i \qquad (10.151)$$

Figure 10.18 shows an example of a cubic B-spline with the open knot vector of Figure 10.17. Due to the open knot vector the first and last control point (P_1 and P_7) are interpolated and it can be seen that the curve is tangential to the control polygon at its start and end.

B-spline surfaces

Using tensor products, B-spline surfaces can be constructed starting from knot vectors $\Xi = \{\xi_1, \xi_2, \cdots, \xi_{n+p+1}\}$ and a $\mathcal{H} = \{\eta_1, \eta_2, \cdots, \eta_{m+q+1}\}$ and a $n \times m$ net of control points $\mathbf{P}_{jk}$, also called the *control mesh*. One-dimensional basis functions $N_{j,p}$ and $M_{k,q}$ (with $j = 1, \cdots, n$ and $k = 1, \cdots, m$) of order p and q, respectively, are defined from the corresponding knot vectors, and the B-spline surface is constructed as:

$$S(\xi, \eta) = \sum_{j=1}^{n} \sum_{k=1}^{m} N_{j,p} M_{k,q}(\eta) \mathbf{P}_{jk} \qquad (10.152)$$

The patch for the surface is now the domain $[\xi_1, \xi_{n+p+1}] \times [\eta_1, \eta_{m+q+1}]$. Identifying the logical coordinates (j, k) of the B-spline surface with the traditional notation of a node i and the Cartesian product of the associated basis functions with the shape function $N_i(\xi, \eta) = N_{j,p}(\xi) M_{k,q}(\eta)$ the standard finite element notation is recovered, namely,

$$S(\xi, \eta) = \sum_{i=1}^{nm} N_i(\xi, \eta) \mathbf{P}_i \quad \text{with} \quad nm = n \times m \qquad (10.153)$$

Non-uniform rational B-splines (NURBS) are obtained by augmenting every point in the control mesh $\mathbf{P}_i$ with the homogenous coordinate w_i, then dividing the Eq.(10.153) by the weighting function $w(\xi, \eta) = \sum_{i=1}^{nm} N_i(\xi, \eta) w_i$, giving the final spatial surface definition,

$$S(\xi, \eta) = \frac{\sum_{i=1}^{nm} N_i(\xi, \eta) w_i \mathbf{P}_i}{w(\xi, \eta)} = \sum_{i=1}^{nm} \bar{N}_i(\xi, \eta) \mathbf{P}_i \qquad (10.154)$$

In Eq.(10.154), $\bar{N}_i(\xi, \eta) = N_i(\xi, \eta) w_i / w(\xi, \eta)$ are the rational basis functions. These functions are pushed forward by the surface mapping $S(\xi, \eta)$ to form the approximation space for NURBS-based shell analysis. The w_i's are not treated as solution variables; they are data coming from the description of the NURBS surface.

Figures 10.19 and 10.20 show examples of NURBS surfaces.

The detailed description of NURBS falls outside the objective of this text and it can be found in many specialized books [CRE,HCB,PT2,Ro].

A challenge in the definition of surfaces with NURBS is the treatment of isolated surface patches (the so-called trimmed surfaces) obtained by intersection of the original surface with polyhedra (Figures 10.19 and 10.21). Details on this topic can be found in CAD publications [CRE,PT2,Ro].

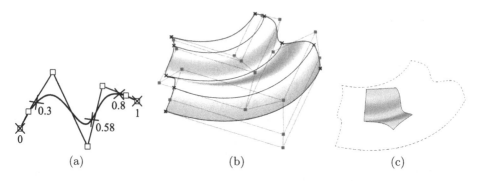

(a)	(b)	(c)

Fig. 10.19 (a) NURBS line with 7 control points (squares), order $= 3$ (cubic) and knot vector (KV) : $\{0, 0, 0, 0, 0.3, 0.58, 0.8, 1, 1, 1, 1\}$. (b) Cubic-quadratic NURBS surface created by revolving the curve (a) around an axis. (c) Trimmed surface obtained by cutting the NURBS surface (b) with a pentagonal prism. Crosses show the position of knots on the (parametrized) lines and define "elements" (4 elements in (a) and 4×1 elements in (b)). Courtesy of GiD team [GiD]

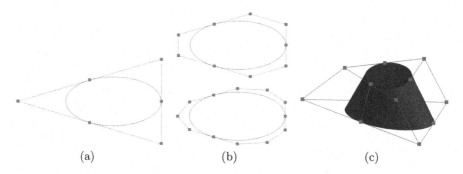

(a)	(b)	(c)

Fig. 10.20 Ellipse defined by NURBS (a) using 6 control points (squares) and (b) using 9 and 12 control points (knot insertion). (c) Cylinder with elliptical (lower) and circular (upper) faces defined by a NURBS surface created by extruding the ellipse of (b) to a circle. Courtesy of GiD team [GiD]

10.20.2 NURBS as a basis for isogeometric FE shell analysis

The shell is represented by a collection of NURBS patches. Each NURBS patch defines a subdomain on the shell surface and the elements are defined by the *knot spans* of this patch (i.e. the regions between two different knot values). This means that each basis function has support on a small number of elements (depending on the polynomial degree). Figures 10.19, 10.22 and 10.23 show examples of control points, knot vectors and elements for several lines and surface modelled with NURBS of different order. The isoparametric concept [Hu,On4,ZTZ] is adopted, i.e. the NURBS basis functions chosen for representing the shell geometry are

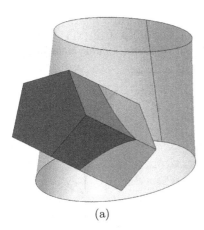
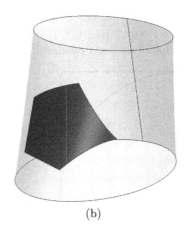

(a)   (b)

Fig. 10.21 Trimmed surface obtained by interesting the cylinder of Figure 10.24c with a pentagon prism. (a) View of cylinder and intersecting prism. (b) Trimmed surface (in shadow). Courtesy of GiD team [GiD]

used as shape functions for describing the displacement field in the shell. A Gauss quadrature is used for the numerical integration of the element stiffness matrix and the equivalent nodal force vector [HRS].

The formulation of a degenerated isogeometric shell element simply implies replacing the standard shape functions N_i (i.e. in Eq.(10.15b)) by the NURBS functions $\bar{N}_i$ of Eq.(10.154). The rest of the element formulation follows as explained in Sections 10.6–10.8. The same technique would apply for the formulation of CBR and CBR-S isogeometric shell elements.

It is important to note that the "nodes" in the FEM are replaced now by the control points and these do not necessarily lay on the shell surface. The displacement of the control points are the degrees of freedom of the structure. The computation of the actual displacements of the shell surface requires the interpolation of the control point values via Eq.(10.21a) with $\bar{N}_i$ replacing the standard displacement shape functions N_i.

Mesh refinement

There are two ways of mesh refinement analogous to standard FE, namely knot insertion and order elevation.

The knot spans can be divided into smaller ones by inserting new knots [CHB,DJS,HCB]. This introduces new and smaller elements and it corresponds to h-refinement in classical FE analysis [ZTZ]. Inserting knots does neither change the geometry nor the parametrization (i.e. the polynomial order is maintained with $C^{order\,-1}$ continuity between "elements").

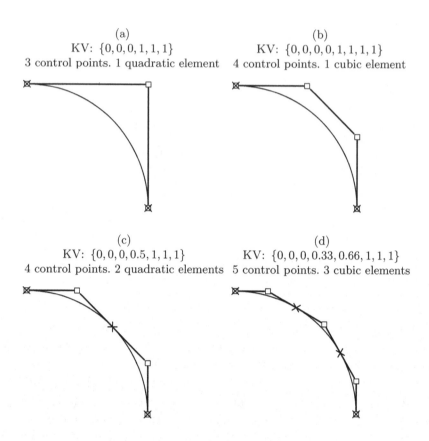

Fig. 10.22 Quadratic line modelled with 1D NURBS. (a) One quadratic element, 3 control points. (b) One cubic element, 4 control points. (c) Two quadratic elements, 4 control points. (d) Three cubic elements, 5 control points. The knot vector (KV) is shown for each case. Crosses show element divisions

Analogous to p-refinement [ZTZ] the polynomial degree of the basis functions can be increased. While increasing the order, existing knots have to be repeated so that the continuity between them remains unchanged. Increasing the order does not necessarily increase the member of elements (knot spans). Also, similarly to knot insertion, order elevation does neither change the geometry nor the parametrization.

For both knot insertion and order elevation new control points have to be introduced in accordance to the relationship $n_k = n + p + 1$. In both cases the computation of the refined NURBS curve has to be done in the projective space with homogeneous coordinates. For details see [PT2].

Figures 10.22 and 10.23 show examples of h and p-refinement. For more details see [CHB].

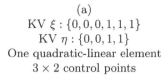

(a)

KV ξ : $\{0, 0, 0, 1, 1, 1\}$

KV η : $\{0, 0, 1, 1\}$

One quadratic-linear element

3×2 control points

(b)

KV ξ : $\{0, 0, 0, 0, 0, 1, 1, 1, 1, 1\}$

KV η : $\{0, 0, 0, 0, 1, 1, 1, 1\}$

One quartic-quartic element

5×5 control points

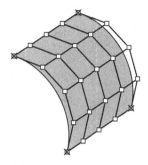

(c)

KV ξ : $\{0, 0, 0, 0.5, 1, 1, 1\}$

KV η : $\{0, 0, 1, 1\}$

2×1 quadratic-linear elements

4×2 control points

(d)

KV ξ : $\{0, 0, 0, 0.5, 1, 1, 1\}$

KV η : $\{0, 0, 0.5, 1, 1\}$

2×2 quadratic-quadratic elements

4×3 control points

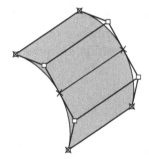

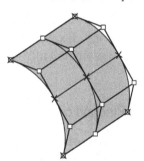

Fig. 10.23 Cylindrical surfaces modelled with 2D NURBS obtained by extruding a quadratic NURBS curve along a line. (a) One quadratic-linear element, 3×2 control points. (b) One quartic-quartic element, 5×5 control points. (c) Two × one quadratic-linear elements, 4×2 control points. (d) Two × two quadratic-quadratic elements, 4×4 control points

Benson *et al.* [BBHH] have presented a degenerated isogeometric shell element based on the degenerated shell formulation proposed by Hughes and Liu [HL]. The good performance of the elements is demonstrated for a set of linear elastic and nonlinear elastoplastic shell problems.

Kiendl *et al.* [KBLW] have derived isogeometric thin shell elements based on Kirchhoff thin shell theory. The C^1 continuity requirement typ-

a)   b)   c)

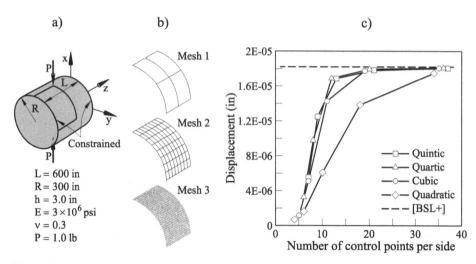

Fig. 10.24 Pinched cylinder analyzed with isogeometric degenerated shell elements [BBHH]. (a) Geometry and load. (b) Mesh 1, 2 and 3. (c) Convergence of vertical displacement under load

ical of thin shell formulations is overcome by the smooth, higher order NURBS basis functions. Examples of the good performance of this element for linear and geometrically non-linear shell analysis are presented in [KBLW].

10.20.3 Example. Pinched cylinder

We present an example of the accuracy and convergence of the isogeometric shell approach in the linear elastic analysis of the so-called pinched cylinder problem [BSL+]. The problem set up is illustrated in Figure 10.24 and also in Figure 8.36 where the same problem was solved using different flat shell elements. The displacement under the point loads acting at the diametrically opposite location on the cylinder surface is the quantity of interest in this case.

This problem has been solved by Benson *et al.* [BBHH] using an isogeometric degenerated shell formulation and by Kiendl *et al.* [KBLW] with isogeometric shell elements based on Kirchhoff-Love thin shell theory. Other isogeometric-type solutions to this problem were reported by Hughes *et al.* [HCB] using volumetric NURBS and Bazilevs *et al.* [BCC+] using T-splines.

Figure 10.24 shows the results for the vertical displacement under the load reported in [BBHH]. A sequence of five meshes obtained by global *h*-

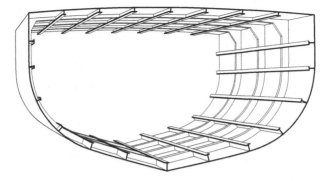

Fig. 10.25 Ship hull stiffened with straight beams

refinement were used. The first, third and fifth mesh form the sequence are
shown. Quadratic through quintic NURBS are employed in each case. One
eight of the geometry is modelled with symmetry boundary conditions.

Figure 10.24 shows that quadratic NURBS exhibit locking which is
gradually alleviated with the increasing order and continuity of NURBS
[BBHH]. These results are essentially identical to those reported in [BCC+,
HCB,KBLW].

10.21 SHELL STIFFENERS

Shell stiffeners are used to increase the strength of shell structures. Examples of stiffened shells are common in the fusselage of airplanes and
aircrafts, in ship hulls and in structural elements of land transport vehicles. Figure 10.25 shows an example of a ship hull stiffened with 3D beams.
Earlier formulation of stiffened plate elements can be found in [MS2,PF].

In this section we will derive a stiffened shell element by coupling a
3D beam element and a degenerated shell element The same procedure
applies for flat shell elements (Chapter 8).

Let us consider for generality a curve degenerated shell element and a
3D beam element connected to the former at a nodal line. The beam can
be placed at the upper or lower shell surfaces as shown in Figure 10.26.
The two following simplifying assumptions are made:

a) The stiffener is rigidly connected to the shell. This implies that the
   transverse section of the beam A_i rotates precisely as the shell node i
   to which it is linked, and

b) The thickness nodal vector $\mathbf{V}_{3i}$ intersects point O of the beam defining
   the position of the neutral axis. The local axes x', y' of the beam

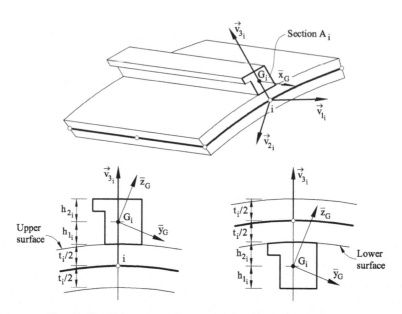

Fig. 10.26 3D beam element acting as a shell stiffener

section need not to coincide with the nodal vectors $\mathbf{V}_{2i}$ and $\mathbf{V}_{3i}$ of the shell nodes.

The global displacements of point O_i in the beam can now be expressed in terms of the global DOFs of node i in the shell element as

$$
\begin{aligned}
\mathbf{u}_{G_i} &= \mathbf{u}_i + \boldsymbol{\theta}_i \times \alpha_i \mathbf{v}_{3_i} = \mathbf{u}_i + \mathbf{A}_i \boldsymbol{\theta}_i \\
\boldsymbol{\theta}_{G_i} &= \boldsymbol{\theta}_i
\end{aligned}
\tag{10.155}
$$

with

$$
\mathbf{u}_{G_i} = [u_{G_i}, v_{G_i}, w_{G_i}]^T \quad ; \quad \mathbf{u}_i = [u_i, v_i, w_i]^T \quad ; \quad \mathbf{A}_i = \alpha_i \begin{bmatrix} 0 & v_{3_i}^z & -v_{3_i}^y \\ -v_{3_i}^z & 0 & v_{3_i}^x \\ v_{3_i}^y & -v_{3_i}^x & 0 \end{bmatrix}
$$

$$
\boldsymbol{\theta}_{G_i} = \boldsymbol{\theta}_i = [\theta_{x_i}, \theta_{y_i}, \theta_{z_i}]^T
\tag{10.156}
$$

and

$$
\alpha_i = \begin{cases} \left(\frac{t_i}{2} + h_{1_i}\right) & \text{- if the stiffener is at the} \\ & \quad \text{upper shell surface} \\ -\left(\frac{t_i}{2} + h_{2_i}\right) & \text{- if the stiffener is at the} \\ & \quad \text{lower shell surface} \end{cases}
\tag{10.157}
$$

In above t_i is the shell thickness at node i and h_{1_i} and h_{2_i} are the distances of the beam axis position G_i to the upper and lower shell surfaces, respectively (Figure 10.26).

Eq.(10.148) allows us to define the movement of a beam section in terms of the corresponding shell node as

$$\mathbf{a}_i^B = \begin{Bmatrix} \mathbf{u}_{0_i} \\ \boldsymbol{\theta}_{G_i} \end{Bmatrix} = \begin{bmatrix} \mathbf{I} & \mathbf{A}_i \\ \mathbf{0} & \mathbf{I} \end{bmatrix} \begin{Bmatrix} \mathbf{u}_i \\ \boldsymbol{\theta}_i \end{Bmatrix} = \mathbf{R}_i \mathbf{a}_i^S \qquad (10.158)$$

where upper indexes B and S denote displacement vectors of the beam and shell nodes, respectively.

The local generalized strains in the beam stiffener can be expressed in terms of the shell DOFs. From Eqs.(4.77), (4.87a) and (10.158) we find

$$\hat{\boldsymbol{\varepsilon}}' = \sum_{i=1}^n \mathbf{B}_i \mathbf{a}_i^B = \sum_{i=1}^n \mathbf{B}_i \mathbf{R}_i \mathbf{a}_i^S = \sum_{i=1}^n \mathbf{B}_i^B \mathbf{a}_i^S \qquad (10.159)$$

The global stiffness matrix for the stiffener element is obtained by substituting matrix $\mathbf{B}_i$ by $\mathbf{B}_i^B = \mathbf{B}_i \mathbf{R}_i$ in Eqs.(4.87b). This can also be interpreted as computing the stiffness matrix of the stiffener element as

$$\mathbf{K}_{ij}^R = \mathbf{R}_i^T \mathbf{K}_{ij}^{(e)} \mathbf{R}_j \qquad (10.160)$$

where $\mathbf{K}_{ij}^{(e)}$ is given by Eq.(4.87b). This allows us to assemble the stiffness equation for the beam stiffener and shell elements without increasing the number of nodal variables.

This formulation requires working with the global rotations at the shell nodes which are linked to the beam stiffener nodes. This does not pose any problem, *even if the shell nodes are originally co-planar*, since the stiffener contributes the necessary rotational stiffness to avoid singularity of the global stiffness matrix. In fact, shell nodes linked to the beam stiffener nodes behave as non-coplanar nodes.

Arbitrarily oriented beam stiffeners with respect to the shell surface are treated similarly. The only difficulty is the definition of the relative position of point G_i in each beam section with respect to the corresponding ith shell node, the distance h_{G_i} between points i and G_i and the unit vector $\mathbf{e}_{G_i}$ linking points i and G_i (Figure 10.27). h_{Gi} and $\mathbf{e}_{Gi}$ substitute the distance α_i and the normal vector $\mathbf{v}_{3i}$ in Eq.(10.155), respectively. The rest of the assembly process follows the lines explained above.

The procedure holds for Euler-Bernouilli beam elements linked to Kirchhoff flat shell elements. A displacement incompatibility leading to errors in the numerical solution may arise if the axial displacement in the beam varies linearly and a cubic Hermite approximation (giving a quadratic rotation field) is chosen for the flat shell element [Cr,ZT2]. This

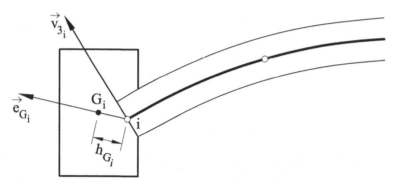

Fig. 10.27 Arbitrarily oriented stiffener with respect to the shell surface

problem is similar to that discussed for folded plate elements in Section 8.12.4. This error can be eliminated by introducing additional DOFs in the beam so as to ensure a quadratic variation of the axial displacement [Cr].

10.22 SLAB-BEAM BRIDGES

The formulation described previously can be particularized for bridges formed by assembly of a flat slab and a number of beams. Figure 10.28 shows the discretization of a rectangular slab into 4-noded flat shell elements. Simple 2-noded straight beam elements are chosen to model the excentric rectangular beam stiffeners which are assumed to be placed at the lower surface of the slab.

The movement of each beam section is expressed in terms of the displacements of the associated slab node as

$$
u_{G_i} = u_i - \frac{1}{2}(t_i + h_i)\theta_{y_i} \;\; ; \;\; v_{G_i} = v_i + \frac{1}{2}(t_i + h_i)\theta_{x_i}
$$
$$
w_{G_i} = w_i \;\; ; \;\; \theta_{xG_i} = \theta_{x_i} \;\; ; \;\; \theta_{yG_i} = \theta_{y_i} \;\; ; \;\; \theta_{zG_i} = \theta_{z_i}
$$

(10.161)

Matrix $\mathbf{A}_i$ of Eq.(10.156) is now

$$
\mathbf{A}_i = \begin{bmatrix} 0 & -\frac{1}{2}(t_i + h_i) & 0 \\ \frac{1}{2}(t_i + h_i) & 0 & 0 \\ 0 & 0 & 0 \end{bmatrix}
$$

(10.162)

The signs in $\mathbf{A}_i$ should be reversed if the beam stiffeners are located at the upper slab surface.

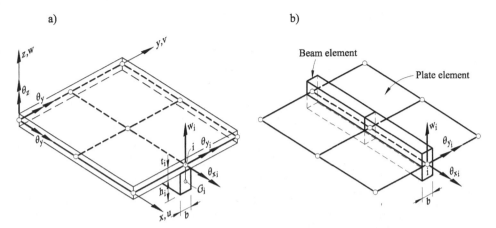

a)   b)

Fig. 10.28 Assembly of plate and beam elements. (a) Excentric beam. (b) The beam eccentricity is neglected

The computation of the beam stiffness matrix follows the steps explained in the previous section.

The process requires the definition of the nodal rotations at the slab nodes in *global axes*. The local definition of the rotation can still be kept at the "free" co-planar slab nodes to avoid singularity.

A further simplification is possible by neglecting the effect of beam excentricity. Standard *plate* elements can be used to model the slab behaviour, whereas simple *beam* elements involving the deflection w, the bending rotation $\theta_{y'}$ and the torsional rotation $\theta_{x'}$ are only needed (Figure 10.28b).

Above procedure is applicable irrespectively of the formulation chosen for the slab and beam elements.

Figure 10.29 shows an example of a slab-beam bridge simply supported at four points of two opposite edges. A uniform vertical distributed loading $q = 10$ kN/m^2 is considered. $E = 10^7$ kN/m^2 and $\nu = 0.3$ are assumed for the beam and the slab. A quarter of the structure is analyzed only due to the double symmetry using the following three element choices:

1. Four-node QLLL flat shell elements (Section 8.10) discretizing *both* the slab and the beams. The 32 element mesh is shown in Figure 10.29a. The thickness and width have been taken equal to 0.2 m and 0.3 m for the slab and the beam elements, respectively.
2. A mesh of 24 QLLL flat shell elements for the slab stiffened with off-centered 2-noded Timoshenko beam elements (Figure 10.29b).

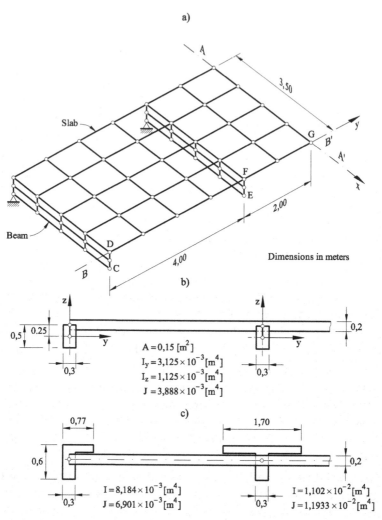

Fig. 10.29 Slab-beam bridge simply supported at eight points under uniformly distributed loading: a) Discretization of one fourth of the structure into 32 QLLL flat shell elements. b) Detail of central section for flat shell analysis using QLLL elements and off-centered 2-noded Timoshenko beam stiffeners. c) Idem using QLLL plate elements stiffened with 2-noded Timoshenko beam elements

3. 24 QLLL plate elements (Section 6.7.1) stiffened with 2-noded Timo-shenko beam elements including torsional effects (Figure 10.29c).

Figure 10.30 shows the distribution of the axial stress σ_x in the central section obtained with each of above three procedures. The stresses in the stiffeners and the compression stresses in the upper surface of the slab

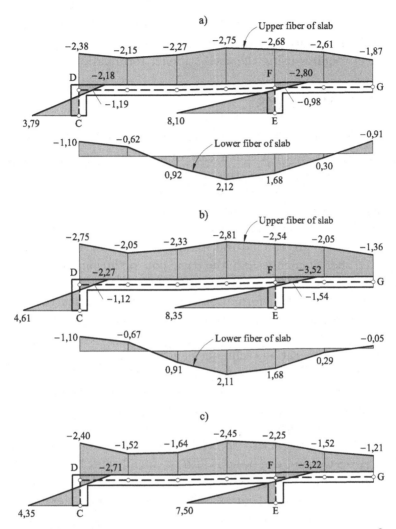

Fig. 10.30 SS slab beam bridge. Distribution of axial stress σ_x (N/mm^2) in the central section using: a) QLLL flat shell elements. b) QLLL flat shell elements stiffened with off-centered 2-noded Timoshenko rod stiffeners. c) QLLL plate elements stiffened with 2-noded Timoshenko beam elements

obtained with the simplest third procedure do not exceed in 20 % the values given by the more precise first two options. The distribution of stresses at the lower surface of the slab is less accurate with the simpler plate model (option 3) giving considerably higher tension stresses in some zones. Further information on this example can be obtained in [PF].

In conclusion, the stiffened plate model is adequate for pre-dimensioning purposes. More accurate stress values and the precise design of the reinforcing steel requires using the shell-stiffener formulation.

10.23 FINAL REMARKS

The 3D degenerated shell elements studied in the first part of this chapter can be considered as a particular case of solid elements under kinematic (straight normals) and plane stress constraints. The internal virtual work is expressed in terms of the stresses and strains and the element matrices involve volume integrals like solid elements do. Magnitudes resulting from a thickness integration such as resultant stresses are computed "a posteriori", whereas they are the natural unknowns in classical shell theory. 3D degenerated shell elements share the basic non–orthogonality condition for the normal rotation of Reissner–Mindlin theory. As a consequence, they include transverse shear deformation effects and, consequently, they suffer from shear locking. They are also affected by membrane locking in their curve forms. Both shear and membrane locking can be alleviated by selective/reduced integration techniques and by using assumed strain fields. The more popular 3D degenerated shell element is probably the 4-noded quadrilateral with a linear assumed transverse shear strain field.

The derivation of 3D shell elements based on a continuum-based resultant (CBR) shell formulation allows one the explicit integration of the element stiffness terms across the thickness. The stiffness matrix for the simpler CBR-S element can be split into the membrane, bending and shear contributions, which facilitates using selective integration techniques.

A procedure for deriving curved shell elements based on assumed membrane and transverse shear strains fields has also been detailed.

A finite element formulation for analysis of stiffened shells can be implemented by coupling 3D beam elements and degenerated or flat shell elements, as explained in the last part of the chapter.

The isogeometric shell theory opens new possibilities for integrating curved shell elements with CAD data.

11

PRISMATIC STRUCTURES. FINITE STRIP AND FINITE PRISM METHODS

11.1 INTRODUCTION

Many structures have uniform geometry in a particular "prismatic" direction. Examples are plates and bridges with rectangular or curved plant. Also the meridional section of an axisymmetric shell does not change along the circumference (Figure 11.1). The name *prismatic structure* will refer hereonwards to a structure with *uniform geometrical and material properties* in the prismatic direction, while the loads can act at any point of the structure. The analysis of prismatic structures can be greatly simplified using finite elements and Fourier expansions to model the transverse and longitudinal behavior, respectively.

The combination of FEM and Fourier series goes back to the 1960's, following the semi-analytical method proposed by Kantorovitch and Krylov [KK] and used for studying of arbitrarily loaded axisymmetric shells and solids by Grafton and Strone [GS], Ahmad *et al.* [AIZ] and Wilson [Wi].

The extension to the analysis of prismatic thin (Kirchhoff) plates and shells was first developed by Cheung [Ch–Ch6] and termed the *finite strip method*. The finite strip method based on Reissner-Mindlin plate/shell theory was developed by Oñate [On] and Suárez [Su], Many refinements of the finite strip method for structural analysis have been reported and the main contributions are listed in [Ch6,Ch7,ChT,Fra,LC,OS,OS2,OS3]. A bibliography of finite strip methods up to the year 2000 can be found in [Fri].

Zienkiewicz and Too [ZT] applied the same ideas to the analysis of prismatic solids in structural and geotechnical engineering problems which they termed the *finite prism method*. Here 2D solid elements are used to discretize the transverse cross section, whereas the longitudinal behavior is modeled by Fourier expansions.

Finite strip and finite prism methods just require to discretize the transverse cross section of a structure using simple 1D and 2D elements

E. Oñate, *Structural Analysis with the Finite Element Method. Linear Statics: Volume 2: Beams, Plates and Shells*, Lecture Notes on Numerical Methods in Engineering and Sciences, DOI 10.1007/978-1-4020-8743-1_11,
© International Center for Numerical Methods in Engineering (CIMNE), 2013

a)   a)   c)

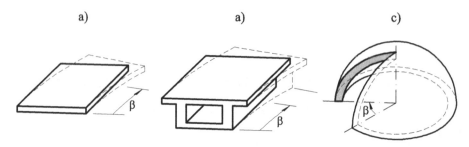

Fig. 11.1 Examples of straight and curved prismatic structures

only, which simplifies the preparation of analysis data enormously. The problem is solved by adding up a number of standard finite element solutions for each harmonic term of the Fourier expansions chosen, each one involving the DOFs of the transverse section only. This allows engineers solving complex 3D structures with very small computational effort.

Finite strip and finite prism methods fail in the category of the so-called *reduction methods*. These methods have become very popular for solving complex problems in mechanics using a reduced set of variables over the original geometry or simplified forms of it [ChLC,HTF,OA].

In this chapter, the finite strip formulation for rectangular plates will be considered first as an introduction to the general formulation for folded shell structures. It will be shown that the finite strip formulation for right folded plates and axisymmetric shells can be derived as a special case of the formulation for folded plates with curved platforms.

The last part of the chapter presents the finite prism formulation for the analysis of prismatic 3D solids and axisymmetric solids under arbitrary loading. Extensions of the finite strip and finite prism methods to a wider class of prismatic structures are also briefly described. Finally the possibilities of finite strip and finite prism methods are shown in the analysis of different structures.

Before discussing the basis of the finite strip method, we will introduce the key concepts of Fourier series for structural analysis. This is done in the next section for the simple case of a beam.

11.2 ANALYSIS OF A SIMPLY SUPPORTED BEAM BY FOURIER SERIES

Consider the simply supported beam shown in Figure 11.2 under arbitrary loading $f_z(y)$. The Total Potential Energy (TPE) for a beam in bending

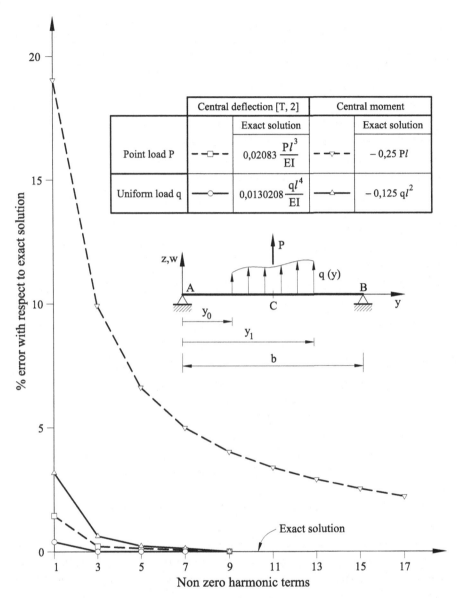

Fig. 11.2 Simple supported beam analyzed by Fourier series. Convergence of central deflection and central moment with the number of non zero harmonic terms for central point load and uniform load

is written using Euler-Bernoulli theory (Chapter 1) as [ZT2]

$$\Pi(w) = \frac{EI}{2} \int_0^b \left(\frac{d^2w}{dy^2}\right)^2 dy - \int_{y_o}^{y_1} f_z w \, dy \qquad (11.1)$$

where E and I are the Young modulus and the modulus of inertia of the beam cross-section with respect to the x axis, respectively, and w is the lateral deflection which must satisfy the following boundary conditions

$$w = \frac{d^2w}{dy^2} = 0 \quad \text{at} \quad y = 0 \quad \text{and} \quad y = b \tag{11.2}$$

The above conditions are satisfied by the following Fourier series

$$w = \sum_{l=1}^{\infty} w^l \sin l \frac{\pi y}{b} \tag{11.3}$$

where b is the beam length, l refers to a particular harmonic term, i.e. $l = 1, 2, 3$, etc. And w^l is the unknown deflection amplitude for the l-th harmonic (also called the modal deflection amplitude).

The loading $f_z(y)$ is also expanded in Fourier series as

$$f_z(y) = \sum_{l=1}^{\infty} f_z^l \sin \frac{l\pi}{b} y \tag{11.4}$$

where f_z^l is the loading amplitude for the l-th harmonic term. This can be obtained using Euler formula for Fourier series by

$$f_z^l = \frac{\displaystyle\int_{y_0}^{y_1} f_z(y) \sin \frac{l\pi}{b} y \, dy}{\displaystyle\int_0^b \sin^2 \frac{l\pi}{b} y \, dy} \tag{11.5}$$

where the load is applied in the zone from $y = b_0$ to $y = b_1$, as shown in Figure 11.2.

The value of f_z^l is easily obtained if the product $f_z(y) \sin \frac{l\pi y}{b}$ is integrable. Thus, for a given load harmonic the problem is one of finding the unknown amplitude w^l, which uniquely describes the deflected beam profile for that harmonic. Substituting Eqs.(11.3) and (11.4) into (11.1) we can write

$$\pi(w) = \sum_{l=1}^{\infty} \left(\frac{EI}{4} (w^l)^2 \frac{l^4 \pi^4}{b^3} - \frac{b}{2} f_z^l w^l \right) \tag{11.6}$$

The value of w^l is obtained by minimizing the TPE with respect to w^l, i.e.

$$\frac{\partial \Pi}{\partial w^l} = 0 \quad \text{which leads to} \quad w^l = \frac{f_z^l b^4}{EI l^4 \pi^4} \tag{11.7}$$

The deflection profile is obtained by performing the summation of Eq.(11.3). The beam curvature, and hence the bending moment distribution, may be calculated from the deformed shape of the beam as

$$M = EI \frac{d^2 w}{dy^2} = -EI \left(\frac{\pi}{b}\right)^2 \sum_{l=1}^{\infty} w^l l^2 \sin\frac{l\pi}{b} \tag{11.8}$$

An alternative to the TPE approach is the Principle of Virtual Work used throughout this book. In this chapter, however, the TPE method [ZT,ZTZ] will be chosen for deriving the finite strip and finite prism equations.

As an example of application we consider the beam shown in Figure 11.2 for two loading cases: a uniformly distributed loading of intensity $f_z = q$ acting along the whole beam length (i.e. $y_0 = 0$ and $y_1 = b$ in Eq.(11.5)) and a vertical point load P acting at the mid-span. The Fourier coefficient f_z^l, the vertical deflection and the bending moment distribution for each loading case are:

Uniform loading   **Central point load**

$$f_z^l = \frac{2q}{l\pi}(1 - \cos l\pi)$$

$$w = \frac{2qb^4}{EI\pi^5} \sum_{l=1}^{\infty} \frac{1 - \cos l\pi}{l^5} \sin\frac{l\pi}{b} y$$

$$M = -\frac{2qb^2}{\pi^3} \sum \frac{1}{l^3}(1 - \cos l\pi)\sin\frac{l\pi}{b} y$$

$$f_z^l = \frac{2P}{b} \sin\frac{l\pi}{2}$$

$$w = \frac{2Pb^3}{EI\pi^4} \sum_{l=1}^{\infty} \frac{1}{l^4} \sin\frac{l\pi}{2} \sin\frac{l\pi}{b} y$$

$$M = -\frac{2\pi b}{\pi^2} \sum \frac{1}{l^2} \sin\frac{l\pi}{2} \sin\frac{l\pi}{b} y \tag{11.9}$$

Note that in Eq.(11.9) the even harmonic terms are zero. This is due to the symmetry of the loading about the center of the beam.

Figure 11.2 shows the percentage of error versus the exact solution for the vertical deflection and the bending moment at the mid-span taking different non-zero harmonic terms are shown in Figure 11.2. The following conclusions can be drawn:

a) Convergence to the theoretical value [Ti2] for the deflection and the bending moment is much faster for uniform loading than for the point load.
b) Convergence of the bending moment is slower than the vertical deflection for both loading cases.

The convergence of the solution is therefore loading dependent. As a rule, solutions for uniform loads converge more rapidly than those for point loads. Also, the number of harmonic terms needed for achieving a degree of accuracy for the bending moments is larger than for the displacements.

These practical rules, deduced for a simple case, apply to the general finite strip and finite prism formulations described in the following sections.

11.3 BASIC CONCEPTS OF FINITE STRIP AND FINITE PRISM METHODS

Finite strip and finite prism methods share many basic concepts, which can be summarized as follows.

The first step in both procedures is to expand the displacements in Fourier series along the prismatic direction. For a generic displacement $w(x, y)$ for 2D problems, or $w(x, y, z)$ for 3D problems, we write

$$
\begin{aligned}
w(x, y) &= \sum_{l=1}^{m} \left(w^l(x)\sin\frac{l\pi}{b}y + \bar{w}^l(x)\cos\frac{l\pi}{b}y \right) && \text{for 2D} \\
w(x, y, z) &= \sum_{l=1}^{m} \left(w^l(x, z)\sin\frac{l\pi}{b}y + \bar{w}^l(x, z)\cos\frac{l\pi}{b}y \right) && \text{for 3D}
\end{aligned}
\tag{11.10}
$$

where y is the prismatic direction, w^l and $\bar{w}^l$ are the l-th modal displacement amplitudes and m is the number of harmonic terms chosen for the analysis. The sine and cosine functions are chosen such that the displacement field satisfies "a priori" the boundary conditions at the end sections $y = 0$ and $y = b$, where b is the length of the structure in the prismatic direction. Eqs.(11.10) can be generalized as

$$
\mathbf{u}(x, y, z) = \sum_{l=1}^{m} \mathbf{S}^l(y)\mathbf{u}^l(x, z)
\tag{11.11}
$$

where $\mathbf{u}$ is the displacement vector, $\mathbf{u}^l$ is the corresponding l-th modal displacement amplitude and $\mathbf{S}^l$ contains trigonometric functions.

The second step is the interpolation of the modal displacement amplitude field $\mathbf{u}^l$ across the transverse cross section. This is performed by discretizing the transverse section into finite elements. This allows us to write the standard interpolations within each n-noded element as

$$
\mathbf{u}^l(x, z) = \sum_{i=1}^{n} \mathbf{N}_i(x, z)\mathbf{a}_i^l
\tag{11.12}
$$

where $\mathbf{a}_i^l$ is the modal displacement amplitude vector of node i for the l-th harmonic term. Substituting Eq.(11.12) into (11.11) gives

$$\mathbf{u}(x, y, z) = \sum_{l=1}^{m} \sum_{i=1}^{n} \mathbf{S}^l(y) \mathbf{N}_i(x, z) \mathbf{a}_i^l \qquad \text{for 3D} \qquad (11.13)$$

Eq.(11.13) can be interpreted as splitting the displacement function $\mathbf{u}(x, y, z)$ into the product of the shape function $\mathbf{N}_i(x, z)$ defining the interpolation for the displacements in the transverse cross section, the analytical function $\mathbf{S}^l(y)$ describing the behavior in the prismatic direction and the modal parameters $\mathbf{a}_i^{(e)}$. This explains why finite strip and finite prism procedures are termed *semi-analytical* methods.

In the finite strip method the transverse section is discretized in simple 1D finite elements, while 2D elements are used for discretizing the transverse section in the finite prism method.

Figure 11.3 shows an example of above discretization steps for the analysis of a rectangular plate using 2-noded finite strips.

Substituting Eq.(11.13) into the expression for the strain vector gives

$$\boldsymbol{\varepsilon}(x, y, z) = \mathbf{L}\mathbf{u} = \sum_{l=i}^{m} \sum_{i=1}^{n} \hat{\mathbf{S}}^l(y) \mathbf{B}_i^l(x, z) \mathbf{a}_i^l \qquad (11.14)$$

where $\mathbf{L}$ is the strain operator, $\mathbf{B}_i^l$ is the nodal strain matrix for the l-th harmonic term and $\hat{\mathbf{S}}^l$ contains trigonometric functions.

The constitutive equation is written as

$$\boldsymbol{\sigma}(x, y, z) = \mathbf{D}\boldsymbol{\varepsilon} = \mathbf{D} \sum_{l=1}^{m} \sum_{i=1}^{n} \hat{\mathbf{S}}^l(y) \mathbf{B}_i^l(x, z) \mathbf{a}_i^l \qquad (11.15)$$

The loads are also expanded in Fourier series using the same expansions as for the displacement field, i.e.

$$\mathbf{b}(x, y, z) = \sum_{l=1}^{m} \mathbf{S}^l(y) \mathbf{b}^l(x, z) \qquad (11.16)$$

where the modal force amplitudes $\mathbf{b}^l$ are obtained from the external force values as

$$\mathbf{b}^l(x, z) = \frac{\displaystyle\int_{y_1}^{y_2} \mathbf{S}^l(y) \mathbf{b}(x, z) \, dy}{\displaystyle\int_0^b [\mathbf{S}^l(y)]^2 \, dy} \qquad (11.17)$$

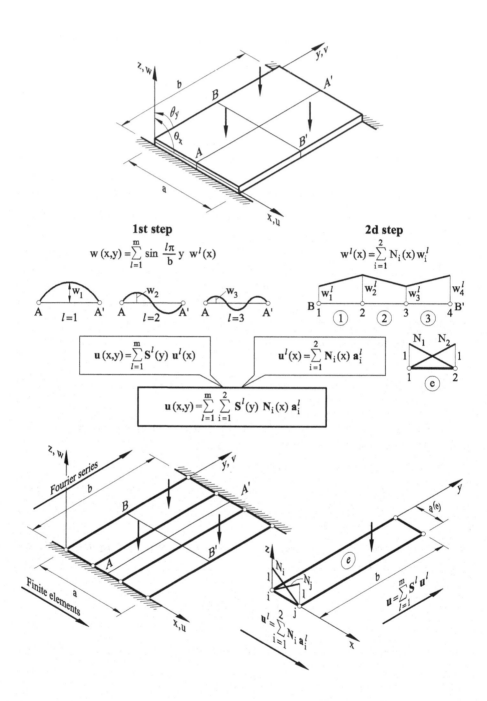

Fig. 11.3 Discretization steps for the finite strip analysis of a rectangular plate using 2-noded strips

Substituting Eqs.(11.13)-(11.16) into the TPE (or the PVW) and accounting for the orthogonal properties of the trigonometric functions, i.e.

$$
\left.
\begin{array}{l}
\displaystyle\int_0^b \sin\frac{l\pi}{b}y \, \sin\frac{m\pi}{b}y \, dy \\[4mm]
\displaystyle\int_0^b \cos\frac{l\pi}{b}y \, \cos\frac{m\pi}{b}y \, dy
\end{array}
\right\}
=
\begin{array}{ll}
\dfrac{b}{2} & \text{for} \quad l = m \\[3mm]
0 & \text{for} \quad l \neq m
\end{array}
\tag{11.18}
$$

the following system of equations is obtained

$$
\begin{bmatrix}
\mathbf{K}^{11} & & & & & \\
& \mathbf{K}^{22} & & \mathbf{0} & & \\
& & \ddots & & & \\
& & & \mathbf{K}^{ll} & & \\
& & & & \ddots & \\
& \mathbf{0} & & & & \mathbf{K}^{mm}
\end{bmatrix}
\begin{Bmatrix}
\mathbf{a}^1 \\ \mathbf{a}^2 \\ \vdots \\ \mathbf{a}^l \\ \vdots \\ \mathbf{a}^m
\end{Bmatrix}
=
\begin{Bmatrix}
\mathbf{f}^1 \\ \mathbf{f}^2 \\ \vdots \\ \mathbf{f}^l \\ \vdots \\ \mathbf{f}^m
\end{Bmatrix}
\tag{11.19}
$$

where $\mathbf{a}^l$ and $\mathbf{f}^l$ respectively contain the modal displacement amplitudes and the equivalent modal force amplitudes for all the nodes discretizing the transverse sections and the lth harmonic term.

Note that the solution for each harmonic term is decoupled from the rest. Consequently, the nodal modal displacement amplitude vectors for the different harmonics can be independently computed as

$$
\mathbf{K}^{ll}\mathbf{a}^l = \mathbf{f}^l \quad ; \quad l = 1, 2, \ldots, m
\tag{11.20}
$$

where $\mathbf{K}^{ll}$ is the stiffness matrix for the l-th harmonic term. $\mathbf{K}^{ll}$ and $\mathbf{f}^l$ and obtained by the standard assembly of the contributions from each of the strip (or prism) elements as

$$
[\mathbf{K}_{ij}^{ll}]^{(e)} = \frac{b}{2}\int_{\Omega^{(e)}} [\mathbf{B}_i^l]^T \mathbf{D}\mathbf{B}_j^l \, d\Omega
\tag{11.21}
$$

$$
[\mathbf{f}_i^l]^{(e)} = \frac{b}{2}\int_{\Omega^{(e)}} \mathbf{N}_i^T \mathbf{b}^l \, d\Omega
\tag{11.22}
$$

where $\Omega^{(e)}$ is the area or length of the prism or strip element, respectively.

Once the nodal modal displacement amplitude vectors $\mathbf{a}^l$ have been obtained, the displacements, strains and stress at any point of the structure can be computed using Eqs.(11.13), (11.14) and (11.15), respectively.

These concepts will be detailed in the chapter for analysis of different prismatic structures with the finite strip and the finite prism methods.

11.4 FINITE STRIP METHOD FOR RECTANGULAR REISSNER-MINDLIN PLATES

The finite strip formulation for Reissner-Mindlin plates follows similar steps to those explained for the simple beam problem of previous section. The displacements are expanded in truncated Fourier series along the prismatic direction y, along which both the material and geometrical properties of the plate are uniform, i.e.

$$
\begin{aligned}
w(x,y) &= \sum_{l=1}^{m} w^l(x) \sin\gamma y \\
\theta_x(x,y) &= \sum_{l=1}^{m} \theta_x^l(x) \sin\gamma y \qquad \text{with} \qquad \gamma = \frac{l\pi}{b} \\
\theta_y(x,y) &= \sum_{l=1}^{m} \theta_y^l(x) \cos\gamma y
\end{aligned}
\qquad (11.23)
$$

where b is the plate length, w^l, θ_x^l and θ_y^l are the modal amplitudes for the vertical deflection and the rotations for the l-th harmonic term and m is the number of harmonic terms chosen for the analysis.

The harmonic expansions satisfy the following boundary conditions

$$
\left.\begin{aligned}
w &= \theta_x = 0 \\
\frac{\partial w}{\partial x} &= \frac{\partial \theta_x}{\partial x} = \frac{\partial \theta_y}{\partial y} = 0
\end{aligned}\right\} \quad \text{for} \quad y = 0 \quad \text{and} \quad y = b
\qquad (11.24)
$$

Eqs.(11.24) imply that the plate is simply supported at the two ends along the prismatic direction. Other boundary conditions can be reproduced by appropriate selection of the expansions in Eq.(11.23) [CH6,ChT,Fra,LC].

The next step is to discretize the modal displacement amplitudes (which are a function of the x coordinate only) using a standard 1D finite element interpolation along the transverse direction of the plate. The modal displacement amplitudes are expressed within an element as

$$
[w^l(x), \theta_x^l(x), \theta_y^l(x)] = \sum_{i=1}^{n} N_i(x)[w_i^l, \theta_{x_i}^l, \theta_{y_i}^l]
\qquad (11.25)
$$

where w_i^l, $\theta_{x_i}^l$ and $\theta_{y_i}^l$ are the modal displacement amplitudes for the i-th node, $N_i(x)$ is the 1D shape function of node i and n is the number of nodes in the element.

The above process is equivalent to dividing the plate into longitudinal elements (or strips) so that each strip has a certain number of nodes (or more precisely, nodal lines). The displacement field is defined longitudinally by the Fourier expansion of Eq.(11.23) and transversely by the finite element discretization of Eq.(11.25) (Figure 11.3).

Substituting Eq.(11.25) into (11.23) gives

$$\mathbf{u}(x,y) = \sum_{l=1}^{m}\sum_{i=1}^{n} \mathbf{S}^l(y)\mathbf{N}_i(x)\mathbf{a}_i^l \tag{11.26}$$

where

$$\mathbf{u}(x,y) = [w(x,y), \theta_x(x,y), \theta_y(x,y)]^T \quad ; \quad \mathbf{a}_i^l = [w_i^l, \theta_{x_i}^l, \theta_{y_i}^l]^T$$

$$\mathbf{N}_i = N_i(x)\mathbf{I}_3 \quad ; \quad \mathbf{S}^l = \begin{bmatrix} S^l & & \mathbf{0} \\ & S^l & \\ \mathbf{0} & & C^l \end{bmatrix} \quad ; \quad \begin{array}{l} S^l(y) = \sin\gamma y \\ C^l = \cos\gamma y \end{array} \tag{11.27}$$

The generalized strains and the resultant stresses in a strip are obtained substituting Eq.(11.27) into the expressions of $\hat{\boldsymbol{\varepsilon}}$ and $\hat{\boldsymbol{\sigma}}$ of Reissner-Mindlin plate theory (Section 6.2.2). This gives

$$\hat{\boldsymbol{\varepsilon}}(x,y) = \sum_{l=1}^{m}\sum_{i=1}^{n} \hat{\mathbf{S}}^l(y)\mathbf{B}_i^l(x)\mathbf{a}_i^l \quad , \quad \hat{\boldsymbol{\sigma}}(x,y) = \hat{\mathbf{D}}\sum_{l=1}^{m}\sum_{i=1}^{n} \hat{\mathbf{S}}^l(y)\mathbf{B}_i^l(x)\mathbf{a}_i^l \tag{11.28}$$

where $\hat{\mathbf{D}}$ is the generalized constitutive matrix of Eq.(6.23), $\mathbf{B}_i^l$ is the nodal strain-displacement matrix for the l-th harmonic term given by

$$\mathbf{B}_i^l(x) = \left\{ \begin{array}{c} \mathbf{B}_{b_i}^l \\ \cdots \\ \mathbf{B}_{s_i}^l \end{array} \right\} = \begin{bmatrix} 0 & -\dfrac{\partial N_i}{\partial x} & 0 \\ 0 & 0 & N_i\gamma \\ 0 & -N_i\gamma & -\dfrac{\partial N_i}{\partial x} \\ \cdots & \cdots & \cdots \\ \dfrac{\partial N_i}{\partial x} & -N_i & 0 \\ N_i\gamma & 0 & -N_i \end{bmatrix} \tag{11.29}$$

and

$$\hat{\mathbf{S}}^l(y) = \begin{bmatrix} S^l & & & \\ & S^l & & \mathbf{0} \\ & & C^l & \\ & & & S^l \\ \mathbf{0} & & & C^l \end{bmatrix} \tag{11.30}$$

$\mathbf{B}_{b_i}^l$ and $\mathbf{B}_{s_i}^l$ in Eq.(11.29) are the contributions to the nodal strain matrix due to bending and shear, respectively for the l-th harmonic term.

Matrix $\hat{\mathbf{D}}$ can account for a layered composite material as described in Section 7.2.4. Recall that the finite strip formulation requires a uniform distribution of the material properties along the prismatic (longitudinal) direction.

The loads are expanded along the longitudinal direction similarly as for the displacements. A uniformly distributed load $f_z = q$ is represented by

$$q(x, y) = \sum_{l=1}^{m} S^l(y) q^l \tag{11.31}$$

The (constant) modal amplitudes for the load are obtained by

$$q^l = \frac{\displaystyle\int_{y_o}^{y_1} q \sin\gamma y \, dy}{\displaystyle\int_o^b \sin^2\gamma y \, dy} = \frac{2q}{l\pi} \left[\cos\gamma y_o - \cos\gamma y_1\right] \tag{11.32}$$

In Eq.(11.32) y_0 and y_1 are the limits of application of the load along the longitudinal direction (Figure 11.4). If the load acts over the whole plate then $y_0 = 0$, $y_1 = b$ and

$$q^l = \frac{2q}{l\pi}(1 - \cos l\pi) \tag{11.33}$$

Note that q^l is equal to zero for the even harmonic terms in Eq.(11.33), which simplifies the computations.

The uniform loading q in Eq.(11.31) can represent the value of the self-weight per unit area wit $q = \rho g t$, where ρ is the density, g the gravity constant and t the plate thickness.

The discretized equations can be obtained from the PVW in the usual manner. The alternative procedure chosen here is starting from the TPE as for the beam example of Section 11.2. The TPE is written for a distributed loading f_z as [ZT2]

$$\Pi = \frac{1}{2} \iint_A \hat{\boldsymbol{\varepsilon}}^T \hat{\boldsymbol{\sigma}} \, dA - \iint_A w f_z \, dA \tag{11.34}$$

Substituting Eqs.(11.27), (11.29) and (11.32) into (11.34) gives

$$\Pi = \sum_{e=1}^{n_e} \Pi^e \tag{11.35a}$$

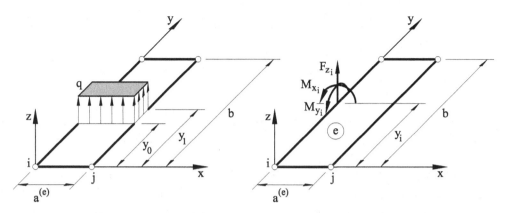

Fig. 11.4 Distributed load and point load for a 2-noded strip

where Π^e is the potential energy for a strip element given by

$$\Pi^e = \frac{1}{2}\sum_{l=1}^{m}\sum_{q=1}^{m}\sum_{i=1}^{n}\sum_{j=1}^{n}[\mathbf{a}_i^l]^T[\mathbf{K}_{ij}^{ll}]^{(e)}\mathbf{a}_j^m - \sum_{l=1}^{m}\sum_{i=1}^{n}[\mathbf{a}_i^l]^T[\mathbf{f}_i^l]^{(e)} \qquad (11.35\text{b})$$

Minimizing Eq.(11.35a) with respect to the modal displacement amplitudes of all the nodes leads to the standard stiffness equation system

$$\frac{\partial\Pi}{\partial\mathbf{a}_i^l} = \mathbf{0} \quad \Rightarrow \quad \mathbf{Ka} = \mathbf{f} \qquad (11.36)$$

The system of equations (11.36) is *uncoupled* for each harmonic term due to the orthogonality properties of the trigonometric functions chosen (Eq.(11.18)). The decoupled system is identical to that shown in Eq.(11.19). The stiffness matrix and the equivalent nodal modal force vector can be assembled from the individual strip contributions for each harmonic term given by

$$[\mathbf{K}_{ij}^{ll}]^{(e)} = \frac{b}{2}\int_{a^{(e)}}[\mathbf{B}_i^l]^T\hat{\mathbf{D}}\mathbf{B}_j^l\,dx \qquad (11.37)$$

$$[\mathbf{f}_i^l]^{(e)} = \frac{b}{2}\int_{a^{(e)}}\mathbf{N}_i^T[f_z^l,0,0]^T\,dx \qquad (11.38)$$

where $a^{(e)}$ is the strip width (Figure 11.4). The evaluation of the integrals requires a simple 1D Gauss quadrature. The exact analytical integration is possible for many cases.

Point loads are treated in the same manner and the expression of vector $\mathbf{f}_i^l$ is now simply

$$\mathbf{f}_i^l = \left\{ \begin{array}{c} F_{z_i}\sin\gamma y_i \\ M_{x_i}\sin\gamma y_i \\ M_{y_i}\cos\gamma y_i \end{array} \right\} \tag{11.39}$$

where F_{z_i}, M_{x_i} and M_{y_i} are respectively the intensities of the point load and the two bending moments acting at node i at a distance y_i from the simple supported end (Figure 11.4).

Solution of Eq.(11.20) yields the nodal modal displacement amplitudes $\mathbf{a}^l$, $l = 1, 2, 3, \ldots, m$. The displacements at any point in the plate are computed by Eq.(11.26), while the generalized strains and the resultant stresses are obtained by Eq.(11.28).

The decoupling of the stiffness equation depends on the Fourier expansions chosen in Eq.(11.23). The reproduction of other boundary conditions different from the simple supported case typically leads to terms as $S^l C^m$ in the stiffness matrix which do not satisfy the orthogonality conditions (11.18) [Ch6,LC]. The global stiffness matrix is then a *full matrix* and the solution of the stiffness equations (11.36) requires iterative techniques for preserving the competitiveness of the finite strip method versus the FEM [SMO,Su].

11.4.1 Reduced integration for Reissner-Mindlin plate strips

Plate strip elements derived in the previous section suffer from the shear locking defect inherent to Reissner-Mindlin plate elements (Chapter 6). Hence, erroneous results are obtained for thin plates unless some precautions are taken.

Similarly as for plates, the simplest remedy for avoiding shear locking in plate strip elements is the reduced integration of the shear stiffness terms leading to the singularity of the shear stiffness matrix. The strip stiffness matrix can be split into the bending and shear contributions using Eqs.(6.23), (11.29) and (11.37) as

$$[\mathbf{K}_{ij}^{ll}]^{(e)} = [\mathbf{K}_{b_{ij}}^{ll}]^{(e)} + [\mathbf{K}_{s_{ij}}^{ll}]^{(e)} \tag{11.40}$$

where

$$[\mathbf{K}_{b_{ij}}^{ll}]^{(e)} = \frac{b}{2} \int_{a^{(e)}} \mathbf{B}_{b_i}^{l\,T} \hat{\mathbf{D}}_b \, \mathbf{B}_{b_j}^{l} \, dx$$

$$[\mathbf{K}_{s_{ij}}^{ll}]^{(e)} = \frac{b}{2} \int_{a^{(e)}} \mathbf{B}_{s_i}^{l\,T} \hat{\mathbf{D}}_s \mathbf{B}_{s_j}^{l} \, dx \tag{11.41}$$

	Integration	Full		Selective		Reduced	
Strip		$\mathbf{K}_b^{ll}$	$\mathbf{K}_s^{ll}$	$\mathbf{K}_b^{ll}$	$\mathbf{K}_s^{ll}$	$\mathbf{K}_b^{ll}$	$\mathbf{K}_s^{ll}$
○━━━━━━━○		2	2	2	1	1	1
○━━━●━━━○		3	3	3	2	2	2
○━━●━━●━━○		4	4	4	3	3	3

Fig. 11.5 Linear, quadratic and cubic Reissner-Mindlin plate strip elements. Full, selective and reduced integration quadratures

Figure 11.5 shows the full, selective and reduced integration quadratures for computing the bending and the shear stiffness matrices for linear, quadratic and cubic plate strip elements.

Oñate and Suárez [OS] have verified that the required singularity of $\mathbf{K}_s^{ll}$ for the plate strip elements of Figure 11.5 is guaranteed if reduced or selective integration is used. Full integration is also applicable for the cubic strip element, although the selective and reduced integration rules are recommended in practice.

The simplest Reissner-Mindlin plate strip element is the linear 2-noded strip with *a uniform one point reduced integration*. The stiffness matrix of Eq.(11.37) is then simply computed as

$$[\mathbf{K}_{ij}^{ll}]^{(e)} = \frac{ba^{(e)}}{2} [\bar{\mathbf{B}}_i^l]^T \bar{\hat{\mathbf{D}}} \bar{\mathbf{B}}_j^l \tag{11.42}$$

where $(\bar{\cdot})$ denotes values at the strip mid-point. Matrix $\bar{\mathbf{B}}_i^l$ is obtained from Eq.(11.29) substituting $\frac{dN_i}{dx}$ by $\frac{(-1)^i}{a^e}$ and N_i by $\frac{1}{2}$. The strip stiffness matrix with uniform one point integration is shown in Figure 11.6.

This formulation can be readily extended for composite laminated plates if the material properties have a uniform distribution along the prismatic direction. This requires taking into account the in-plane displacements in the element kinematics and introducing the effect of membrane and membrane-bending effects in the element stiffness matrix as described in Chapter 7. The composite-laminated plate strip formulation can be also derived as a particular case of the folded plate strip element presented in the following section.

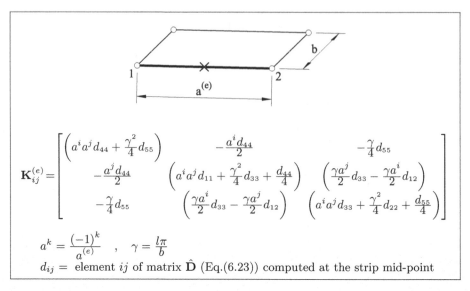

$$\mathbf{K}_{ij}^{(e)} = \begin{bmatrix} \left(a^i a^j d_{44} + \frac{\gamma^2}{4} d_{55}\right) & -\frac{a^i d_{44}}{2} & -\frac{\gamma}{4} d_{55} \\ -\frac{a^j d_{44}}{2} & \left(a^i a^j d_{11} + \frac{\gamma^2}{4} d_{33} + \frac{d_{44}}{4}\right) & \left(\frac{\gamma a^j}{2} d_{33} - \frac{\gamma a^i}{2} d_{12}\right) \\ -\frac{\gamma}{4} d_{55} & \left(\frac{\gamma a^i}{2} d_{33} - \frac{\gamma a^j}{2} d_{12}\right) & \left(a^i a^j d_{33} + \frac{\gamma^2}{4} d_{22} + \frac{d_{55}}{4}\right) \end{bmatrix}$$

$$a^k = \frac{(-1)^k}{a^{(e)}} \quad , \quad \gamma = \frac{l\pi}{b}$$

$d_{ij} =$ element ij of matrix $\hat{\mathbf{D}}$ (Eq.(6.23)) computed at the strip mid-point

Fig. 11.6 Two-noded linear Reissner-Mindlin plate strip element. Stiffness matrix with uniform one point reduced integration

11.5 FINITE STRIP METHOD FOR STRAIGHT PRISMATIC FOLDED PLATE STRUCTURES

11.5.1 General formulation

We will extend the plate strip formulation of the previous section to the analysis of straight prismatic folded plate structures with simply supported ends. Examples of application include bridges with cellular cross section and prismatic shells discretized by flat strip elements (Figure 11.7).

The Reissner-Mindlin theory for flat shell analysis used here was presented in Chapter 8. The finite strip formulation follows the same steps as explained for plates in the previous sections.

The first step is to divide the structure in longitudinal strips as shown in Figure 11.7. The local displacement vector (in axes x', y', z') within a strip with n nodes is expressed as

$$\mathbf{u}'(x', y) = \sum_{l=1}^{m} \sum_{i=1}^{n} \mathbf{S}^l(y) \mathbf{N}_i(x') \mathbf{a}_i^{\prime l} \tag{11.43}$$

with

$$\mathbf{u}'(x', y) = [u'_{0_i}, v'_{0_i}, w'_{0_i}, \theta_{x'}, \theta_{y'}]^T \quad , \quad \mathbf{a}_i^{\prime l} = [u'^l_{0_i}, v'^l_{0_i}, w'^l_{0_i}, \theta^l_{x'_i}, \theta^l_{y'_i}]^T$$

Fig. 11.7 Discretization of straight prismatic shells using 2-noded strips

$$\mathbf{N}_i(x') = N_i(x')\mathbf{I}_5 \quad \text{and} \quad \mathbf{S}^l(y) = \begin{bmatrix} S^l & & & & \\ & C^l & & \mathbf{0} & \\ & & S^l & & \\ & & & S^l & \\ \mathbf{0} & & & & C^l \end{bmatrix} \tag{11.44}$$

where functions S^l, C^l were defined in Eq.(11.27) and $N_i(x')$ are the standard 1D Lagrange shape functions which depend on the local coordinate x' (Figure 11.8). Subscript "0" in Eqs.(11.44) denotes the displacements of the strip mid-plane as usual.

The harmonic functions chosen satisfy the condition of zero movement on the planes $y = 0$ and $y = b$. This formulation therefore holds for simply supported folded plates with rigid diaphragms at the two ends. This is a typical boundary condition in cellular bridges [Ch6,CT,LC].

The local generalized strains are expressed in terms of the modal displacement amplitudes of the nodes. Substituting Eq.(11.43) into $\hat{\boldsymbol{\varepsilon}}'$ in Eq.(8.8) gives

$$\hat{\boldsymbol{\varepsilon}}'(x', y) = \sum_{l=1}^{m}\sum_{i=1}^{n} \hat{\mathbf{S}}^l(y)\mathbf{B}_i''(x')\mathbf{a}_i'^l \tag{11.45}$$

Figure 11.9 shows the local generalized strain matrix $\mathbf{B}_i''$ and the harmonic matrix $\hat{\mathbf{S}}^l$.

The local resultant stresses are obtained from Eq.(8.17) (neglecting initial stresses) as

$$\hat{\boldsymbol{\sigma}}'(x', y) = \hat{\mathbf{D}}'\hat{\boldsymbol{\varepsilon}}' = \hat{\mathbf{D}}' \sum_{l=1}^{m}\sum_{i=1}^{n} \hat{\mathbf{S}}^l(y)\mathbf{B}'_i{}''(x')\mathbf{a}_i'^l \tag{11.46}$$

The constitutive matrix $\hat{\mathbf{D}}'$ is given in (8.18). The sign convention for the local resultant stresses is shown in Figure 11.13.

Composite laminated materials can also be analyzed with the present finite strip method using the correct expression for $\hat{\mathbf{D}}'$ (Section 8.3.4), as *long as the material properties are uniform in the prismatic direction.*

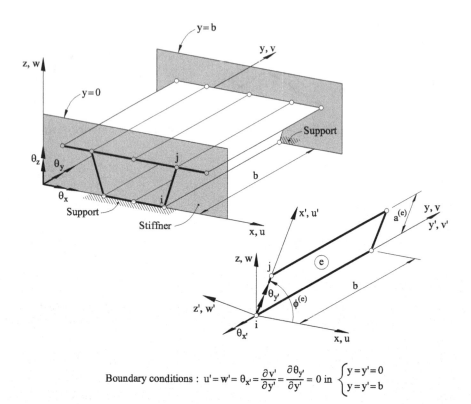

Boundary conditions : $u'=w'=\theta_{x'}=\dfrac{\partial v'}{\partial y'}=\dfrac{\partial \theta_{y'}}{\partial y'}=0$ in $\begin{cases} y=y'=0 \\ y=y'=b \end{cases}$

Fig. 11.8 Discretization of a box girder bridge in 2-noded strips. Local and global axes and boundary conditions at the end sections

The derivation of the stiffness equations follows the same process as for plates. The harmonic expansions chosen guarantee the decoupling of the stiffness equations for the different harmonic terms. Consequently, the global stiffness matrix for the l-th harmonic term is obtained by assembling the individual contributions from the strips. The stiffness matrix for the strip in local axes is

$$[\mathbf{K}_{ij}''^l]^{(e)} = \frac{b}{2} \int_{a^{(e)}} [\mathbf{B}_i''^l]^T \hat{\mathbf{D}}' \mathbf{B}_j''^l \, dx' \tag{11.47}$$

The membrane, bending and shear contribution to the local stiffness matrix can be identified in the usual manner as explained in Section 8.4.3. For homogeneous or symmetric material properties we obtain

$$[\mathbf{K}_{ij}''^l]^{(e)} = \frac{b}{2} \int_{a^{(e)}} \left([\mathbf{B}_{m_i}''^l]^T \hat{\mathbf{D}}_m' \mathbf{B}_{m_j j}''^l + [\mathbf{B}_{b_i}''^l]^T \hat{\mathbf{D}}_b' \mathbf{B}_{b_j}''^l + [\mathbf{B}_{s_i}''^l]^T \hat{\mathbf{D}}_s' \mathbf{B}_{s_j}''^l \right) dx' =$$

$$= [\mathbf{K}_{m_{ij}}''^l]^{(e)} + [\mathbf{K}_{b_{ij}}''^l]^{(e)} + [\mathbf{K}_{s_{ij}}''^l]^{(e)} \tag{11.48}$$

$$\hat{\varepsilon}' = \sum_{l=1}^{m}\sum_{i=1}^{n} \hat{\mathbf{S}}^l \, \mathbf{B}_i'^l \, \mathbf{a}_i^l$$

$$\hat{\varepsilon}' = \left\{ \begin{matrix} \hat{\varepsilon}_m' \\ \hat{\varepsilon}_b' \\ \hat{\varepsilon}_s' \end{matrix} \right\} \; ;$$

$$\hat{\varepsilon}_m' = \left\{ \begin{matrix} \dfrac{\partial u_0'}{\partial x'} \\[2mm] \dfrac{\partial v_0'}{\partial y'} \\[2mm] \dfrac{\partial u_0'}{\partial y'} + \dfrac{\partial v_0'}{\partial x'} \end{matrix} \right\}$$

$$\hat{\varepsilon}_f' = \left\{ \begin{matrix} \dfrac{\partial \theta x'}{\partial x'} \\[2mm] \dfrac{\partial \theta y'}{\partial y'} \\[2mm] \left(\dfrac{\partial \theta x'}{\partial y'} + \dfrac{\partial \theta y'}{\partial x'} \right) \end{matrix} \right\}$$

$$\hat{\varepsilon}_s' = \left\{ \begin{matrix} \dfrac{\partial w_0'}{\partial x'} - \theta x' \\[2mm] \dfrac{\partial w_0'}{\partial y'} - \theta y' \end{matrix} \right\}$$

$$\mathbf{B}_i'^l = \left\{ \begin{matrix} \mathbf{B}_{m_i}'^l \\ \mathbf{B}_{b_i}'^l \\ \mathbf{B}_{s_i}'^l \end{matrix} \right\} \; ;$$

$$\mathbf{B}_{m_i}'^l = \begin{bmatrix} \dfrac{\partial N_i}{\partial x'} & 0 & 0 & 0 & 0 \\[2mm] 0 & -N_i\gamma & 0 & 0 & 0 \\[2mm] N_i\gamma & \dfrac{\partial N_i}{\partial x'} & 0 & 0 & 0 \end{bmatrix}$$

$$\mathbf{B}_{b_i}'^l = \begin{bmatrix} 0 & 0 & 0 & \dfrac{\partial N_i}{\partial x'} & 0 \\[2mm] 0 & 0 & 0 & 0 & N_i\gamma \\[2mm] 0 & 0 & 0 & N_i\gamma & \dfrac{\partial N_i}{\partial x'} \end{bmatrix}$$

$$\mathbf{B}_{s_i}'^l = \begin{bmatrix} 0 & 0 & \dfrac{\partial N_i}{\partial x'} & -N_i & 0 \\[2mm] 0 & 0 & N_i\gamma & 0 & -N_i \end{bmatrix}$$

$$\hat{\mathbf{S}}^l = \begin{bmatrix} \mathbf{S}_m^l & & 0 \\ & \mathbf{S}_b^l & \\ 0 & & \mathbf{S}_s^l \end{bmatrix} \; ; \; \mathbf{S}_m^l = \mathbf{S}_b^l = \begin{bmatrix} S^l & 0 \\ & S^l \\ 0 & C^l \end{bmatrix} \; ; \; \mathbf{S}_s^l = \begin{bmatrix} S^l & 0 \\ 0 & C^l \end{bmatrix}$$

$$S^l = \sin\gamma y \; ; \quad C^l = \cos\gamma y, \quad \text{with } \gamma = \frac{l\pi}{b}$$

Fig. 11.9 Local generalized strain matrix for a flat strip element for straight prismatic folded plate structures

For *composite laminated* material the membrane-bending coupling stiffness matrix $[\mathbf{K}_{mb}^{ll}]^{(e)}$ and its transpose must be added to Eq.(11.48) (Section 8.4.3). Recall again that the finite strip analysis requires uniform material properties along the prismatic direction.

Similarly as for flat shell elements the local stiffness matrix (11.48) (for homogeneous material) can be formed by assembling directly the membrane and flexural stiffness terms as

$$
[\mathbf{K}_{ij}^{ll}]^{(e)}_{5\times5} = \begin{bmatrix} \mathbf{K}_{PS_{ij}}^{ll} & 0 \\ {\scriptstyle 2\times2} & \\ 0 & \mathbf{K}_{RM_{ij}}^{ll} \\ & {\scriptstyle 3\times3} \end{bmatrix}^{(e)}
\tag{11.49}
$$

where $\mathbf{K}_{RM_{ij}}^{ll}$ coincides with the stiffness matrix of the Reissner-Mindlin plate strip element (Eq.(11.37)) and $\mathbf{K}_{PS_{ij}}^{ll}$ is the stiffness matrix of the plane stress strip element given by

$$
[\mathbf{K}_{PS_{ij}}^{ll}]^{(e)} = \frac{b}{2} \int_{a^{(e)}} \bar{\mathbf{B}}_{m_i}^{llT} \hat{\mathbf{D}}_m' \bar{\mathbf{B}}_{m_j}' \, dx'
\tag{11.50}
$$

where

$$
\bar{\mathbf{B}}_{m_i}' = \begin{bmatrix} \dfrac{\partial N_i}{\partial x'} & 0 \\ 0 & -N_i\gamma \\ N_i\gamma & \dfrac{\partial N_i}{\partial x'} \end{bmatrix}
\tag{11.51}
$$

and $\hat{\mathbf{D}}_m'$ is the membrane constitutive matrix of Eq.(8.18a).

11.5.2 Assembly of strip equations. Transformation to global axes

The assembly of the stiffness equations in folded plate structures requires transforming the local vectors and matrices to a global coordinate system. The process is as explained for flat shell elements in Section 8.5. The transformation of the local modal displacement amplitudes is written as

$$
\mathbf{a}_i^{ll} = \mathbf{L}_i^{(e)} \mathbf{a}_i^l
\tag{11.52}
$$

where

$$
\mathbf{a}_i^l = [u_{0_i}^l, \ v_{0_i}^l, \ w_{0_i}^l, \ \theta_{x_i}^l, \ \theta_{y_i}^l, \ \theta_{z_i}^l]
\tag{11.53}
$$

is the *global modal displacement amplitude* vector for the l-th harmonic term where the rotations are defined in vector form along the global axes

x, y, z and the transformation matrix $\mathbf{L}_i^{(e)}$ is

$$
\mathbf{L}_i^{(e)} = \begin{bmatrix} C & 0 & S & \vdots & & \\ 0 & 1 & 0 & \vdots & \mathbf{0} & \\ -S & 0 & C & \vdots & & \\ \cdots & \cdots & \cdots & \vdots & \cdots & \cdots \\ & & & \vdots & 0 & -1 & 0 \\ & \mathbf{0} & & \vdots & C & 0 & S \end{bmatrix} \tag{11.54}
$$

with $S = \sin\phi^{(e)}$ and $C = \cos\phi^{(e)}$ and $\phi^{(e)}$ is the angle between the strip and the global x axis (Figure 11.8). Note that the global rotation $\theta_{z_i}^l$ has been included in Eq.(11.53). This is necessary for assembly of the stiffness equation in non-coplanar nodes (Section 8.5).

The global stiffness matrix for the strip is obtained by

$$
[\mathbf{K}_{ij}^{ll}]^{(e)} = \underset{6\times 5}{[\mathbf{L}_i^{(e)}]^T} \ \underset{5\times 5}{\mathbf{K}_{ij}^{'ll}} \ \underset{5\times 6}{\mathbf{L}_j^{(e)}} \tag{11.55}
$$
$$
\underset{6\times 6}{}
$$

As the element is flat, then $\mathbf{L}_i^{(e)} = \mathbf{L}_j^{(e)}$.

A simpler expression for the global stiffness matrix is found by observation of Eqs.(11.46) and (11.55) as

$$
[\mathbf{K}_{ij}^{ll}]^{(e)} = \frac{b}{2} \int_{a^{(e)}} [\mathbf{B}_i^l]^T \hat{\mathbf{D}}' \mathbf{B}_j^l \ dx' \tag{11.56}
$$

where

$$
\mathbf{B}_i^l = \mathbf{B}_i^{'l} \mathbf{L}_i^{(e)} \tag{11.57}
$$

Matrix $\mathbf{B}_i^l$ allows us to compute the local resultant stresses from the global displacement as

$$
\hat{\boldsymbol{\sigma}}' = \hat{\mathbf{D}}' \sum_{l=1}^m \sum_{i=1}^n \hat{\mathbf{S}}^l \mathbf{B}_i^{'l} \mathbf{a}_i^{'l} = \hat{\mathbf{D}}' \sum_{l=1}^m \sum_{i=1}^n \hat{\mathbf{S}}^l \mathbf{B}_i^l \mathbf{a}_i^l \tag{11.58}
$$

Some problems might arise in the assembly process for *coplanar* nodes leading to the singularity of the global stiffness equations. This problem can be overcome using any of the techniques explained in Section 8.9 for flat shell elements.

11.5.3 Equivalent nodal force vector

The equivalent nodal modal force amplitude vector for the strip for a distributed loading is given (in global axes) by

$$[\mathbf{f}_i^l]^{(e)} = \frac{b}{2} \int_{a^{(e)}} N_i \, \mathbf{t}^l \, dx' \qquad i = 1, 2 \tag{11.59a}$$

where $\mathbf{t}^l$ is the amplitude vector for a distributed load for the l-th harmonic term given by

$$\mathbf{t}^l = \frac{2}{l\pi} \left[f_x(C_0^l - C_1^l), f_y(S_1^l - S_0^l), f_z(C_0^l - C_1^l), m_x(C_0^l - C_1^l), m_y(S_1^l - S_0^l), 0 \right]^T \tag{11.59b}$$

For a point load acting at a node with global number i we have

$$\mathbf{f}_i^l = \mathbf{p}_i^l \quad \text{with} \quad \mathbf{p}_i^l = \frac{2}{b} \left\{ F_{x_i} \, S_i^l, F_{y_i} \, C_i^l, F_{z_i} \, S_i^l, M_{x_i} \, S_i^l, M_{y_i} \, C_i^l, 0 \right\}^T \tag{11.60}$$

In Eqs.(11.60)

$$\begin{array}{lll} C_0^l = \cos\gamma y_0 \;\; ; & C_1^l = \cos\gamma y_1 \;\; ; & C_i^l = \cos\gamma y_i \\ S_0^l = \sin\gamma y_0 \;\; ; & S_1^l = \sin\gamma y_1 \;\; ; & S_i^l = \sin\gamma y_i \end{array} \tag{11.61}$$

Coordinates y_0, y_1 and y_i coincide with those of Figure 11.4.

11.5.4 Flat shell strips. Two-noded strip with reduced integration

Flat shell strips elements based on Reissner-Mindlin theory suffer from shear locking as the shell finite elements of Chapter 8 do. The simpler procedure to avoid shear locking is the reduced/selective quadrature. Here again the rules of Figure 11.5 apply for the linear, quadratic and cubic strips. The reduced integration of the membrane stiffness terms improves the in-plane behaviour and it also eliminates membrane locking that may appear in same special cases (i.e. for composite laminated shells). This requires integrating $\mathbf{K}_m^{ll}$ with one, two and three point quadratures for linear, quadratic and cubic strips, respectively.

The 2-noded strip with a single point quadrature is the simplest one and the stiffness matrix is directly obtained by Eq.(11.42). Figure 11.10 shows the generalized strain matrix computed at the element mid-point.

The equivalent nodal modal force vector for the 2-noded strip for uniformly distributed loading and the lth harmonic term is

$$[\mathbf{f}_i^l]^{(e)} = \frac{ba^{(e)}}{4} \, \mathbf{t}^l \tag{11.62}$$

where $\mathbf{t}^l$ is obtained by Eq.(11.59b). Nodal point loads contribute directly to the global equivalent nodal force vector via Eq.(11.60).

$$[\mathbf{K}_{ij}^{\prime ll}]^{(e)} = \frac{ba^{(e)}}{2} [\bar{\mathbf{B}}_i^l]^T \ \bar{\mathbf{D}} \bar{\mathbf{B}}_j^l$$

$(\bar{\cdot}) \equiv$ values at the strip mid-point.

$$\bar{\mathbf{B}}_i^l = \left\{ \begin{array}{c} \bar{\mathbf{B}}_{m_i}^l \\ ----- \\ \bar{\mathbf{B}}_{b_i}^l \\ ----- \\ \bar{\mathbf{B}}_{s_i}^l \end{array} \right\} = \begin{bmatrix} A_iC & 0 & A_iS & 0 & 0 & 0 \\ 0 & -\frac{\gamma}{2} & 0 & 0 & 0 & 0 \\ \frac{\gamma}{2}C & A_i & \frac{\gamma}{2}S & 0 & 0 & 0 \\ \cdots & \cdots & \cdots & \cdots & \cdots & \cdots \\ 0 & 0 & 0 & 0 & -A_i & 0 \\ 0 & 0 & 0 & -\frac{\gamma}{2}C & 0 & -\frac{\gamma}{2}S \\ 0 & 0 & 0 & A_iC & -\frac{\gamma}{2} & -A_iS \\ \cdots & \cdots & \cdots & \cdots & \cdots & \cdots \\ A_iS & 0 & A_iC & 0 & \frac{1}{2} & 0 \\ -\frac{\gamma}{2}S & 0 & -\frac{\gamma}{2}C & -\frac{1}{2}C & 0 & -\frac{1}{2}S \end{bmatrix}$$

$$A_i = \frac{(-1)^i}{a^{(e)}} \quad ; \quad S = \sin\phi^{(e)} \quad ; \quad C = \cos\phi^{(e)} \quad ; \quad \gamma = \frac{l\pi}{b}$$

Fig. 11.10 Two-noded flat shell strip element. Stiffness matrix computed using one single point reduced integration

11.5.5 Simplification for composite laminated plate strip element

The formulation of strip elements for composite laminated plates can be simply derived from the flat shell strip element formulation previously presented for the particular case of $\phi^{(e)} = 0$, i.e. all elements lay on the same horizontal plane.

11.6 ANALYSIS OF CURVED PRISMATIC SHELLS BY THE FINITE STRIP METHOD

11.6.1 General formulation

The flat shell strip formulation of previous section can be extended for prismatic shells with *circular plant* defined by the angle α. The Fourier expansions are written now in terms of the angle β defining the position of the points over the circular arch along the circular prismatic direction (Figure 11.11). Examples of these structures are found in plates and bridges of circular plant and in troncoconical shells.

The finite strip formulation follows the same steps as for rectangular plates and flat shells. The basic expressions are derived starting from Reissner-Mindlin *troncoconical shell theory*. This is an extension of the

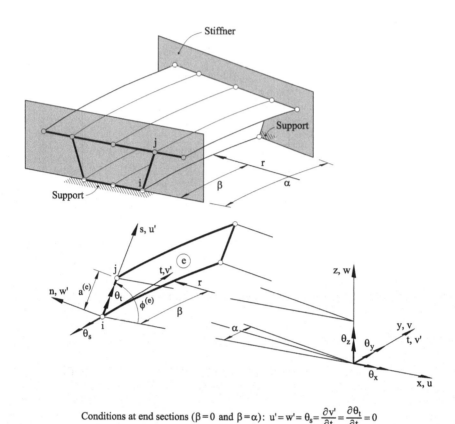

Conditions at end sections ($\beta = 0$ and $\beta = \alpha$): $u' = w' = \theta_s = \dfrac{\partial v'}{\partial t} = \dfrac{\partial \theta_t}{\partial t} = 0$

Fig. 11.11 Discretization of a circular box girder bridge using 2-noded troncoconical strips. Local and global axes and boundary conditions at the end sections

axisymmetric shell theory studied in Chapter 9 for arbitrary loading [BD6,No,Su]. For simplicity the assumption $t/R_\alpha \ll 1$ will be made; (i.e. $C_\alpha = 1$ in Eq.(9.9) (see R_α in Figure 9.3).

Figure 11.11 shows a circular box girder bridge discretized in 2-noded troncoconical strips. A typical strip is shown together with the local (s, t, n) and global (x, y, z) axes. Note that the prismatic direction t and the global direction y coincide.

The local displacements are expanded in Fouries series along the circular (circumferential) direction. 1D finite elements are used to discretize the transversal (meridional) direction. The local displacement vector is written as

$$\mathbf{u}'(s, \beta) = \sum_{l=1}^{m} \sum_{i=1}^{n} \mathbf{S}^l(\beta) \mathbf{N}_i(s) \mathbf{a}_i^{\prime l} \tag{11.63}$$

where

$$\mathbf{u}'(s, \beta) = [u_0', v_0', w_0', \theta_s, \theta_t]^T \quad , \quad \mathbf{a}_i'^l = [u_{0_i}'^l, v_{0_i}'^l, w_{0_i}'^l, \theta_{s_i}^l, \theta_{t_i}^l]^T \quad (11.64)$$

In (11.63) $\mathbf{S}^l(\beta)$ and $\mathbf{N}^l(\beta)$ coincide with expressions (11.44) for straight folded plate structures simply interchanging the local coordinate x' by s and the prismatic coordinate y and the length b by the angles β and α, respectively (Figures 11.8 and 11.11).

θ_s and θ_t in Eq.(11.64) are the rotations of the normal vector contained in the planes sn and st, respectively (see sign convention in Figure 11.11).

The discretization process leads to the following relationship between the local generalized strains and the nodal modal displacement amplitudes

$$\hat{\varepsilon}'(s, \beta) = \sum_{l=1}^{m} \sum_{i=1}^{n} \hat{\mathbf{S}}^l(\beta) \mathbf{B}_i'^l(\beta) \mathbf{a}_i'^l \quad (11.65)$$

The components of $\hat{\varepsilon}'$ and of matrix $\mathbf{B}_i'^l$ are shown in Figure 11.12. The harmonic transformation matrix $\hat{\mathbf{S}}^l$ coincides with that given in Figure 11.9 for $b = \alpha$. By making $\gamma = r\gamma$, $s = x'$ and the coordinate r equal to a large number (so that $\frac{1}{r} \cong 0$) matrix $\mathbf{B}_i'^l$ coincides with that for the straight case (Figure 11.9). The resultant stresses and the constitutive equation are written as in Eq.(11.46) for the straight case. The only difference is that the resultant stresses are referred now to the local axes s, t, n (Figure 11.13).

Following the standard discretization procedure, the uncoupled system of stiffness equations is obtained. The local stiffness matrix and the equivalent nodal modal force amplitude vector for distributed loading for a troncoconical strip are

$$[\mathbf{K}_{ij}'^{ll}]^{(e)} = \frac{\alpha}{2} \int_{a^{(e)}} [\mathbf{B}_i'^l]^T \hat{\mathbf{D}}' \mathbf{B}_j'^l r \, ds \quad (11.66)$$

$$[\mathbf{f}_i^l]^{(e)} = \frac{\alpha}{2} \int_{a^{(e)}} N_i \mathbf{t}^l r \, ds \quad (11.67)$$

The only difference with the expressions for the straight case is that the radius r appears within the integrals, similarly as for axisymmetric solids and shells. Nodal point loads are directly assembled in the global equivalent nodal modal force vector in the standard manner.

The nodal modal amplitude vectors for distributed forces $\mathbf{t}^l$ and point loads $\mathbf{p}^l$ are computed by Eqs.(11.59) and (11.60) simply substituting the prismatic coordinate y and the length b by angles β and α, respectively.

$$\hat{\varepsilon}' = \sum_{l=1}^{m} \sum_{i=1}^{n} \hat{\mathbf{S}}^l \mathbf{B}_i''^l \mathbf{a}_i^l$$

$$\hat{\varepsilon}' = \left\{ \begin{array}{c} \hat{\varepsilon}_m' \\ \hat{\varepsilon}_b' \\ \hat{\varepsilon}_s' \end{array} \right\} ; \quad \hat{\varepsilon}_m' = \left\{ \begin{array}{c} \dfrac{\partial u_0'}{\partial s} \\[2mm] \dfrac{1}{r}\dfrac{\partial v_0}{\partial \beta} + \dfrac{u_0'}{r}C - \dfrac{w_0'}{r}S \\[2mm] \dfrac{\partial v_0'}{\partial s} - \dfrac{1}{r}\dfrac{\partial u_0'}{\partial \beta} - \dfrac{v_0'}{r}C \end{array} \right\}$$

$$\hat{\varepsilon}_b' = \left\{ \begin{array}{c} \dfrac{\partial \theta_s}{\partial s} \\[2mm] \dfrac{1}{r}\dfrac{\partial \theta_t}{\partial \beta} + \dfrac{\theta_s}{r}C \\[2mm] \dfrac{\partial \theta_t}{\partial s} + \dfrac{1}{r}\dfrac{\partial \theta_s}{\partial \beta} - \dfrac{\theta_t}{r}C \end{array} \right\}$$

$$\hat{\varepsilon}_s' = \left\{ \begin{array}{c} \dfrac{\partial w_0'}{\partial s} - \theta_s \\[2mm] \dfrac{1}{r}\dfrac{\partial w_0'}{\partial \beta} + \dfrac{v_0'}{r}S - \theta_t \end{array} \right\}$$

$$\mathbf{B}_i''^l = \left\{ \begin{array}{c} \mathbf{B}_{m_i}''^l \\ \mathbf{B}_{b_i}''^l \\ \mathbf{B}_{s_i}''^l \end{array} \right\} ; \quad \mathbf{B}_{m_i}''^l = \left[\begin{array}{ccccc} \dfrac{\partial N_i}{\partial s} & 0 & 0 & 0 & 0 \\[2mm] \dfrac{N_i}{r}C & -\dfrac{N_i}{r}\gamma & -\dfrac{N_i}{r}S & 0 & 0 \\[2mm] \dfrac{N_i}{r}\gamma & \left(\dfrac{\partial N_i}{\partial s} - \dfrac{N_i}{r}C\right) & 0 & 0 & 0 \end{array} \right]$$

$$\mathbf{B}_{b_i}''^l = \left[\begin{array}{ccccc} 0 & 0 & 0 & \dfrac{\partial N_i}{\partial s} & 0 \\[2mm] 0 & 0 & 0 & \dfrac{N_i}{r}C & -\dfrac{N_i}{r}\gamma \\[2mm] 0 & 0 & 0 & \dfrac{N_i}{r}\gamma & \left[\dfrac{\partial N_i}{\partial s} - \dfrac{N_i}{r}C\right] \end{array} \right]$$

$$\mathbf{B}_{s_i}''^l = \left[\begin{array}{ccccc} 0 & 0 & \dfrac{\partial N_i}{\partial s} & -N_i & 0 \\[2mm] 0 & \dfrac{N_i}{r}S & \dfrac{N_i}{r}\gamma & 0 & -N_i \end{array} \right]$$

$S = \sin\phi^{(e)}, C = \cos\phi^{(e)};$

$\hat{\mathbf{S}}^l$ as in Figure 11.9 for straight prismatic shells with $\gamma = \dfrac{l\pi}{\alpha}$ and $y = \beta$

Fig. 11.12 Local generalized strain vectors and matrices for a troncoconical strip

The transformation of the strip equations to global axes follows the steps of Section 11.5.2 for straight structures. The global stiffness matrix

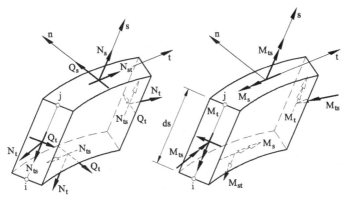

Fig. 11.13 Sign convention for resultant stresses in a troncoconical shell. For straight prismatic shells change s, t, n by x', y', z'

is computed as

$$[\mathbf{K}_{ij}^{ll}]^{(e)} = \frac{\alpha}{2} \int_{a^{(e)}} [\mathbf{B}_i^l]^T \hat{\mathbf{D}}' \mathbf{B}_j^l r ds \qquad (11.68)$$

with

$$\mathbf{B}_i^l = \mathbf{B}_i' \mathbf{L}_i^{(e)} \qquad (11.69)$$

where $\mathbf{L}_i^{(e)}$ is the transformation matrix of Eq.(11.54). Matrix $\mathbf{B}_i^l$ is useful for computing the local resultant stresses in terms of the global displacement by Eq.(11.58). The problems associated to coplanar nodes are treated as explained for flat shell elements in Section 8.8.

The 2-noded troncoconical strip with a uniform reduced one-point quadrature has an excellent performance for thin and moderately thick curved folded plate structures [On,OS2,OS3,Su]. The stiffness matrix and the equivalent nodal modal force amplitude vector for uniformly distributed loading for the 2-noded strip and the lth harmonic term are given by

$$[\mathbf{K}_{ij}^{ll}]^{(e)} = \frac{\alpha a^{(e)}}{2} [\bar{\mathbf{B}}_i^l]^T \tilde{\bar{\mathbf{D}}}' \bar{\mathbf{B}}_j^l \bar{r} \qquad (11.70)$$

$$[\mathbf{f}_i^l]^{(e)} = \frac{\alpha \bar{r} a^{(e)}}{4} \mathbf{t}^l \qquad (11.71)$$

where again $(\bar{\cdot})$ denotes values at the strip mid-point.

Figure 11.14 shows matrix $\bar{\mathbf{B}}_i^l$. Note that by making $\gamma = \bar{r}\gamma$ and $\frac{1}{r} = 0$ the expression of $\bar{\mathbf{B}}_i^l$ for straight prismatic shell structures is obtained (see Figure 11.10).

Above coincidences make it easy to organize the finite strip computations for straight and curved prismatic shell structures in a single code.

$$\mathbf{\bar{B}}_i^l = \left\{ \begin{array}{c} \mathbf{\bar{B}}_{m_i}^l \\ ---- \\ \mathbf{\bar{B}}_{b_i}^l \\ ---- \\ \mathbf{\bar{B}}_{s_i}^l \end{array} \right\} = \begin{bmatrix} A_i C & 0 & A_i S & 0 & 0 & 0 \\ \dfrac{1}{2\bar{r}} & -\dfrac{\gamma}{2\bar{r}} & 0 & 0 & 0 & 0 \\ \dfrac{\gamma C}{2\bar{r}} & (A_i - \dfrac{C}{2\bar{r}}) & \dfrac{\gamma S}{2\bar{r}} & 0 & 0 & 0 \\ \cdots & \cdots & \cdots & \cdots & \cdots & \cdots \\ 0 & 0 & 0 & 0 & -A_i & 0 \\ 0 & 0 & 0 & \dfrac{\gamma C}{2\bar{r}} & \dfrac{C}{2\bar{r}} & -\dfrac{\gamma S}{2\bar{r}} \\ 0 & 0 & 0 & \left(A_i - \dfrac{C}{2\bar{r}}\right) & C - \dfrac{\gamma}{2\bar{r}} & \left(A_i - \dfrac{C}{2\bar{r}}\right) S \\ \cdots & \cdots & \cdots & \cdots & \cdots & \cdots \\ -A_i S & 0 & A_i C & 0 & 1/2 & 0 \\ \dfrac{\gamma S}{2\bar{r}} & \dfrac{S}{2\bar{r}} & \dfrac{\gamma C}{2\bar{r}} & -\dfrac{C}{2} & 0 & -\dfrac{S}{2} \end{bmatrix}$$

$$A_i = \frac{(-1)^i}{a^{(e)}}; S = \sin\phi^{(e)}; C = \cos\phi^{(e)}; \gamma = \frac{l\pi}{\alpha}$$

Fig. 11.14 Matrix $\mathbf{\bar{B}}_i^l$ for explicit computation of the stiffness matrix for the 2-noded troncoconical strip element

11.6.2 Circular plate strips

Circular plate strips can be derived as a simplification of the troncoconical strips (Figure 11.15a) by making $\phi = 0°$ and neglecting the membrane terms. The circular strip stiffness matrix is given by Eq.(11.68) with

$$\mathbf{B}_i^l = \left\{ \begin{array}{c} \mathbf{B}_{b_i}^l \\ \cdots \\ \mathbf{B}_{s_i}^l \end{array} \right\} = \begin{bmatrix} 0 & -\dfrac{\partial N_i}{\partial x} & 0 \\ 0 & -\dfrac{N_i}{r} & \dfrac{N_i}{r}\gamma \\ 0 & -\dfrac{N_i}{r}\gamma & \left(\dfrac{N_i}{r} - \dfrac{\partial N_i}{\partial x}\right) \\ \cdots & \cdots & \cdots \\ \dfrac{\partial N_i}{\partial s} & -N_i & 0 \\ \dfrac{N_i}{r}\gamma & 0 & -N_i \end{bmatrix} \tag{11.72}$$

and $\mathbf{\hat{D}}'$ given by Eq.(8.18a) for a general composite laminated material.

The equivalent nodal modal force amplitude vector is given by Eq.(11.67). Note that only vertical forces and moments M_x and M_y are involved.

Making $\gamma = \bar{r}\gamma$ and $\frac{1}{r} = 0$ in Eq.(11.72) yields the generalized strain matrix for rectangular plates. The explicit form of $\mathbf{\bar{B}}_i^l$ in Eq.(11.70) for the

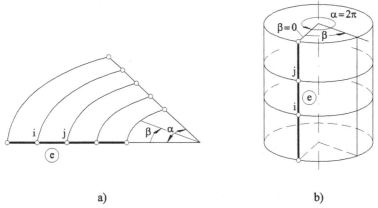

Fig. 11.15 Discretization of circular plate and axisymmetric shell in 2-noded strips

2-noded circular plate strip with one point reduced quadrature is obtained by making $\partial N_i/\partial x = (-1)^i/a^{(e)}$, $N_i = 1/2$ and $r = \bar{r}$ in Eq.(11.72).

11.7 AXISYMMETRIC SHELLS UNDER ARBITRARY LOADING

The troncoconical shell formulation can be extended for analysis of axisymmetric shells under arbitrary loading. The expressions for the displacements, generalized strains and resultant stresses are identical to those given in Section 11.6. The only difference is that the *"length" of the structure is a whole circumference* and, therefore, angle α is replaced by 2π.

Figure 11.15b shows an axisymmetric shell discretized in circular strips. The displacements are expanded in Fourier series along the circumferential direction. It is convenient for this purpose to split the displacement field in symmetric and anti-symmetric components with respect to the meridional plane at $\beta = 0$. For a n-noded strip with n nodes we have

$$\mathbf{u}' = \sum_{l=0}^{m} \sum_{i=1}^{n} \mathbf{N}_i (\bar{\mathbf{S}}^l \bar{\mathbf{a}}_i'^l + \bar{\bar{\mathbf{S}}}^l \bar{\bar{\mathbf{a}}}_i'^l) \tag{11.73}$$

where $\mathbf{u}'$ is the displacement vector of Eq.(11.64) (see also Figure 11.11) and $(\bar{\cdot})$ and $(\bar{\bar{\cdot}})$ denote the symmetric and anti-symmetric components of the displacements, respectively. It is interesting that $\bar{\bar{\mathbf{S}}}^l$ coincides with matrix $\mathbf{S}^l$ of Eq.(11.44), while $\bar{\mathbf{S}}^l$ is obtained from $\bar{\bar{\mathbf{S}}}^l$ simply interchanging S^l by C^l and viceversa. Also $\gamma = l$ is taken in all equations. Finally, the components of $\bar{\mathbf{a}}_i'^l$ and $\bar{\bar{\mathbf{a}}}_i'^l$ coincide with those of $\mathbf{a}_i'^l$ in Eq.(11.64).

Eq.(11.73) indicates that the out of plane displacement v' and the rotation θ_t are zero for a symmetric displacement field. Conversely, the displacement components u', w' and the rotation θ_s contained in the symmetry plane are zero for an anti-symmetric mode.

The zero harmonic term in Eq.(11.73) corresponds to an *axisymmetric* deformation where all the meridional sections ($\beta = $ constant) have the same displacement components u', w' and θ_s.

The loads are expanded in Fourier series using the same harmonic functions as for the displacements, i.e.

$$t = \sum_{l=0}^{m}(\bar{\mathbf{S}}^l\bar{\mathbf{t}}^l + \bar{\bar{\mathbf{S}}}^l\bar{\bar{\mathbf{t}}}^l) \tag{11.74}$$

where $\bar{\mathbf{S}}^l$ and $\bar{\bar{\mathbf{S}}}^l$ correspond to symmetric and anti-symmetric loads, respectively. The load amplitudes $\bar{\mathbf{t}}^l$ and $\bar{\bar{\mathbf{t}}}^l$ are obtained from expressions as shown in Section 11.5.3 [On,OS2,OS3,Su].

The analysis can be simplified by computing *independently* the symmetric and anti-symmetric solutions. The symmetric results are found first by retaining the terms $(\bar{\cdot})$ from Eq.(11.73) and (11.74) only. The anti-symmetric field can be obtained next by considering the terms $(\bar{\bar{\cdot}})$ in those equations.

The finite strip formulation for the symmetric and anti-symmetric cases can be treated in a unified manner. The local generalized strain matrix is identical for both cases and given by the expression of Figure 11.12 for the troncoconical strip simply interchanging

$$\gamma \text{ by } -l \quad \text{for the symmetric case}$$

$$\gamma \text{ by } l \quad \text{for the anti-symmetric case}$$

The constitutive matrix $\hat{\mathbf{D}}'$ relating the local resultant stresses and the generalized strains is the same as for the troncoconical case. Once again a decoupled system of equations for the different harmonic terms is found. The local stiffness matrix for an axisymmetric strip element is given by

$$[\mathbf{K}_{ij}^{ll}]^{(e)} = C \int_{a^{(e)}} [\mathbf{B}_i^{ll}]^T \hat{\mathbf{D}}' \mathbf{B}_j^{ll} r \, ds \tag{11.75}$$

where

$$C = \begin{cases} 2\pi & \text{for } l = 0 \\ \pi & \text{for } l \neq 0 \end{cases} \tag{11.76}$$

Transformation to global axes follows the steps of Section 11.5.2. The

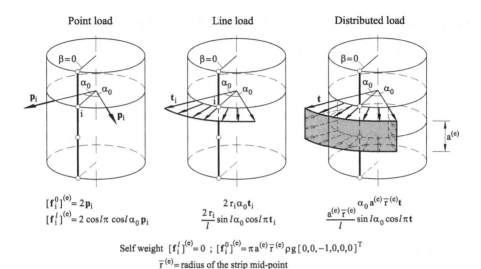

$$[\mathbf{f}_i^0]^{(e)} = 2\mathbf{p}_i$$
$$[\mathbf{f}_i^l]^{(e)} = 2\cos l\pi \, \cos l\,\alpha_0 \, \mathbf{p}_i$$

$$2\,\mathbf{r}_i\,\alpha_0\,\mathbf{t}_i$$
$$\frac{2\,\mathbf{r}_i}{l}\sin l\,\alpha_0 \cos l\,\pi\,\mathbf{t}_i$$

$$\alpha_0\,\mathbf{a}^{(e)}\,\overline{\mathbf{r}}^{(e)}\mathbf{t}$$
$$\frac{\mathbf{a}^{(e)}\,\overline{\mathbf{r}}^{(e)}}{l}\sin l\,\alpha_0 \cos l\,\pi\,\mathbf{t}$$

Self weight $[\mathbf{f}_i^l]^{(e)} = 0$; $[\mathbf{f}_i^0]^{(e)} = \pi\,\mathbf{a}^{(e)}\,\overline{\mathbf{r}}^{(e)}\rho\mathbf{g}\,[0,0,-1,0,0,0]^{\mathrm{T}}$
$\overline{\mathbf{r}}^{(e)} =$ radius of the strip mid-point

Fig. 11.16 Equivalent nodal modal force amplitude vectors for 2-noded troncoconical strip

global stiffness matrix is obtained as

$$[\mathbf{K}_{ij}^{ll}]^{(e)} = C \int_{a^{(e)}} [\mathbf{B}_i^l]^T \hat{\mathbf{D}}' \mathbf{B}_j^l \, r ds \qquad (11.77)$$

where $\mathbf{B}_i = \mathbf{B}_i'' \mathbf{L}_i^{(e)}$ and $\mathbf{L}_i^{(e)}$ is the transformation matrix of Eq.(11.54).

Axisymmetric strip elements based on Reissner-Mindlin theory require reduced/selective quadratures to prevent locking. The linear, quadratic and cubic strips use the quadratures shown in Figure 11.5. We note the simplicity of the 2-noded axisymmetric strip with one integration point. Its stiffness matrix is obtained by computing the integrant of Eq.(11.77) at the strip mid-point. The final expression is identical to Eq.(11.70) if $\frac{\alpha}{2}$ is substituted by C. Matrix $\bar{\mathbf{B}}_i^l$ is obtained by Figure 11.14 substituting γ by $-l$ and l for the symmetric and anti-axisymmetric cases, respectively.

Figure 11.16 shows the expression for the equivalent nodal force vector for the 2-noded strip for different symmetric loads. Other loading types can be found in [On,OS2,OS3,Su].

In conclusion, the troncoconical formulation of Section 11.6 yields all the strip matrices and vectors for curved and straight prismatic shells and plates and axisymmetric shells in a unified form [OS3]. This is useful for the organization of a computer code for analysis of all these structures.

Note finally that the stiffness and loading expressions for $l = 0$ (full axisymmetric solution) coincide with those obtained for Reissner-Mindlin troncoconical shell elements in Chapter 9.

11.8 TRONCOCONICAL STRIP ELEMENTS BASED ON KIRCHHOFF SHELL THEORY

Kirchhoff troncoconical strip elements can be derived by introducing the normal orthogonality condition in the kinematic field. This leads, as usual, to neglecting the effect of the transverse shear strains in the analysis. The formulation is therefore applicable to thin shell problems only. Making $\hat{\varepsilon}'_s = \mathbf{0}$ in Figure 11.12 we obtain

$$\theta_s = \frac{\partial w'_0}{\partial s} \quad , \quad \theta_t = \frac{1}{r}\frac{\partial w'_0}{\partial \beta} + \frac{v'_0}{r}S \tag{11.78}$$

Substituting these rotations into vector $\hat{\varepsilon}'_b$ of Figure 11.12 gives the expression for the local curvatures (Figure 11.17). Note that the second derivative of the w'_0 with respect to the local coordinate s is now involved. This introduces the need for C^1 continuity for the interpolation of w'_0 along s.

The Fourier expansions for u'_0, v'_0 and w'_0 are the same as for the Reissner-Mindlin formulation (Eq.(11.63)). Figure 11.17 shows the generalized strain matrix where N_i are the C^0 continuous Lagrange shape functions and H_i and $\bar{H}_i$ are C^1 continuous Hermite shape functions for 1D elements. The local displacement vector has four components: $\mathbf{a}^l_i = [u'^l_{0_i}, v'^l_{0_i}, w'^l_{0_i}, \frac{\partial w'^l_0}{\partial s}]^T$.

The derivation of the stiffness matrix and the equivalent nodal force vector follows the same process as for Reissner-Mindlin troncoconical elements. The 2-noded Kirchhoff strip is again the simplest option. The shape functions N_i are standard linear functions, while H_i and $\bar{H}_i$ are cubic Hermite polynomials (Eq.(8.11a)), respectively. The integration of the membrane and bending stiffness terms requires a two-point quadrature.

Full details on the formulation of troncoconical strip elements using Kirchhoff thin shell theory can be found in [Ch6,LC].

11.9 THE FINITE PRISM METHOD

Zienkiewicz and Too [ZT] extended the finite strip method to the analysis of prismatic solids. The so-called *finite prism* method combines Fourier expansions along the prismatic direction and 2D solid elements for discretizing the transverse cross section (Figure 11.18). The finite prism method is useful for analysis of prismatic solid structures, such as thick

$$\hat{\varepsilon}' = \sum_{l=1}^{m}\sum_{i=1}^{n}\hat{\mathbf{S}}^l\mathbf{B}_i'^l\mathbf{a}_i'^l \quad , \quad \mathbf{a}_i'^l = \left[u_{o_i}'^l, \ v_{o_i}'^l, \ w_{o_i}'^l, \ \frac{\partial w_o'^l}{\partial s}\right]^T$$

$$\hat{\varepsilon}' = \left\{\begin{array}{c}\hat{\varepsilon}_m'\\ \hat{\varepsilon}_b'\end{array}\right\} \quad ; \quad
\hat{\varepsilon}_m' = \left\{\begin{array}{c}\dfrac{\partial u_o'}{\partial s}\\[2mm] \dfrac{1}{r}\dfrac{\partial v_o'}{\partial \beta} + \dfrac{u_o'}{r}C - \dfrac{w_o'}{r}S\\[2mm] \dfrac{\partial v_o'}{\partial s} + \dfrac{1}{r}\dfrac{\partial u_o'}{\partial \beta} - \dfrac{v_o'}{r}C\end{array}\right\}$$

$$\hat{\varepsilon}_b' = \left\{\begin{array}{c}\dfrac{\partial^2 w_o'}{\partial s^2}\\[2mm] \dfrac{1}{r^2}\dfrac{\partial^2 w_o'}{\partial \beta^2} + \dfrac{S}{r^2}\dfrac{\partial v_o'}{\partial \beta} + \dfrac{C}{r}\dfrac{\partial w_o'}{\partial s}\\[2mm] \dfrac{2}{r}\dfrac{\partial^2 w_o'}{\partial s\partial \beta} - \dfrac{2CS}{r^2}v_o' - \dfrac{2C}{r^2}\dfrac{\partial w_o'}{\partial \beta} + \dfrac{S}{r}\dfrac{\partial v_o'}{\partial s}\end{array}\right\}$$

$$\mathbf{B}_i'^l = \left\{\begin{array}{c}\mathbf{B}_{m_i}'^l\\ \mathbf{B}_{b_i}'^l\end{array}\right\} ;$$

$$\mathbf{B}_{m_i}'^l = \begin{bmatrix}\dfrac{\partial N_i}{\partial s} & 0 & 0 & 0\\[2mm] \dfrac{N_i}{r}C & -\dfrac{N_i}{r}\gamma & -\dfrac{H_i'}{r}S & -\dfrac{\bar{H}_i}{r}S\\[2mm] \dfrac{N_i}{r}\gamma & \left(\dfrac{\partial N_i}{\partial s} - \dfrac{N_i}{r}C\right) & 0 & 0\end{bmatrix}$$

$$\mathbf{B}_{b_i}'^l = \begin{bmatrix}0 & 0 & \dfrac{\partial^2 H_i}{\partial s^2} & \dfrac{\partial^2 \bar{H}_i}{\partial s^2}\\[3mm] 0 & \dfrac{N_i}{r^2}S\gamma & \left[\dfrac{C}{r}\dfrac{\partial H_i}{\partial s} - \left(\dfrac{\gamma}{r}\right)^2 H_i\right] & \left[\dfrac{C}{r}\dfrac{\partial \bar{H}_i}{\partial s} - \left(\dfrac{\gamma}{r}\right)^2 \bar{H}_i\right]\\[3mm] 0 & \left(\dfrac{S}{r}\dfrac{\partial N_i}{\partial s} - \dfrac{2N_i}{r^2}CS\right) & \left(\dfrac{2\gamma}{r}\dfrac{\partial H_i}{\partial s} - \dfrac{2H_i}{r^2}C\gamma\right) & \left(\dfrac{2\gamma}{r}\dfrac{\partial \bar{H}_i}{\partial s} - \dfrac{2\bar{H}_i}{r^2}C\gamma\right)\end{bmatrix}$$

N_i: Lagrange shape functions; $H_i, \bar{H}_i$: Hermite shape functions

Fig. 11.17 Local generalized strain vectors and matrices for the Kirchhoff tronco-conical strip

box girder bridges, ground foundations and other soil deformation problems in geotechnics [ChT]. Wong and Vardy [WV] have applied the finite prism method to thick plates and shells with beam stiffners.

Let us consider a prismatic solid as shown in Figure 11.18. The solid is discretized in a number of prisms. The 3D displacement field is expressed within each prism as

$$\mathbf{u}(x,y,z) = \left\{\begin{array}{c}u\\ v\\ w\end{array}\right\} = \sum_{l=1}^{m}\sum_{i=1}^{n}\mathbf{S}^l(y)\mathbf{N}_i(x,z)\mathbf{a}_i^l \qquad (11.79)$$

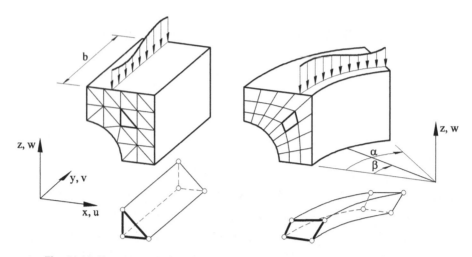

Fig. 11.18 Prismatic solids discretized in straight and circular prisms

with

$$\mathbf{a}_i^l = [u_i^l, v_i^l, w_i^l]^T \tag{11.80}$$

$$\mathbf{N}_i(x, z) = N_i(x, z)\mathbf{I}_3 \quad , \quad \mathbf{S}^l(y) = \begin{bmatrix} S^l & & \mathbf{0} \\ & C^l & \\ \mathbf{0} & & S^l \end{bmatrix}, \quad \begin{matrix} S^l = \sin\gamma y \\ C^l = \cos\gamma y \end{matrix}$$

where $\gamma = \frac{l\pi}{b}$ for straight prisms; $\gamma = \frac{l\pi}{\alpha}$, $y = \beta$ for circular prisms.

In Eq.(11.80) $N_i(x, z)$ are the shape functions of the 2D solid elements with n nodes discretizing the transverse cross section.

Eq.(11.79) satisfies the following boundary conditions

$$u = w = \frac{dv}{dy} = 0 \quad \text{at} \quad y = 0 \quad \text{and} \quad y = b \tag{11.81}$$

Above conditions correspond to simply supported end sections with a rigid diaphragm in the plane x, z. For curved prisms $y = \beta$ and $b = \alpha$ (Figure 11.18).

An alternative definition of matrix $\mathbf{S}^l$ is

$$\mathbf{S}^l(y) = \begin{bmatrix} C^l & & \mathbf{0} \\ & S^l & \\ \mathbf{0} & & C^l \end{bmatrix} \tag{11.82}$$

This reproduces the "quasi" plane strain conditions at the two-end sections of the solid (i.e. $v = \left(\frac{\partial u}{\partial y} + \frac{\partial v}{\partial x}\right) = \left(\frac{\partial v}{\partial x} + \frac{\partial w}{\partial y}\right) = 0$ at $y = 0$ and

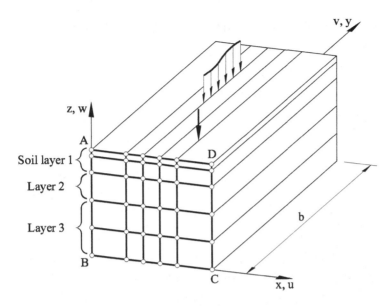

Natural boundary conditions : $v = \gamma_{xy} = \gamma_{yz} = 0$ at $y = 0$ and $y = b$

Constrained boundary conditions : $u = 0$ on lines AB y DC
$w = 0$ on line C

Fig. 11.19 Application of the finite prism method to a soil mechanics problem

$y = b$). This is of interest for many soil mechanics problems (Figure 11.19). The strain vector is deduced from 3D elasticity theory [On4,ZTZ] as

$$
\boldsymbol{\varepsilon}(x, y, z) =
\begin{Bmatrix}
\varepsilon_x \\
\varepsilon_y \\
\varepsilon_z \\
\gamma_{xy} \\
\gamma_{xz} \\
\gamma_{yz}
\end{Bmatrix}
=
\begin{Bmatrix}
\dfrac{\partial u}{\partial x} \\[4pt]
\dfrac{\partial v}{\partial y} \\[4pt]
\dfrac{\partial w}{\partial z} \\[4pt]
\dfrac{\partial u}{\partial y} + \dfrac{\partial v}{\partial x} \\[4pt]
\dfrac{\partial u}{\partial z} + \dfrac{\partial w}{\partial x} \\[4pt]
\dfrac{\partial v}{\partial z} + \dfrac{\partial w}{\partial y}
\end{Bmatrix}
= \sum_{l=1}^{m} \sum_{i=1}^{n} \hat{\mathbf{S}}^{l}(y) \mathbf{B}_{i}^{l}(x, z) \mathbf{a}_{i}^{l} \quad (11.83)
$$

Matrices $\mathbf{B}_i^l$ and $\mathbf{S}^l$ are deduced from the displacement field in Eq.(11.79) as

$$
\mathbf{B}_i^l(x,z) = \begin{bmatrix} \dfrac{\partial N_i}{\partial x} & 0 & 0 \\[2mm] 0 & -\gamma N_i & 0 \\[2mm] 0 & & \dfrac{\partial N_i}{\partial z} \\[2mm] \gamma N_i & \dfrac{\partial N_i}{\partial x} & 0 \\[2mm] \dfrac{\partial N_i}{\partial z} & 0 & \dfrac{\partial N_i}{\partial x} \\[2mm] 0 & \dfrac{\partial N_i}{\partial z} & \gamma N_i \end{bmatrix} \quad ; \quad \hat{\mathbf{S}}^l(y) = \begin{bmatrix} S^l & & & & \\ & S^l & & & \mathbf{0} \\ & & S^l & & \\ & & & C^l & \\ \mathbf{0} & & & & S^l \\ & & & & & C^l \end{bmatrix} \quad (11.84)
$$

Using matrix $\mathbf{S}^l$ given by Eq.(11.82) for the "quasi" plane strain case simply implies interchanging γ for $-\gamma$ and S^l by C^l in Eq.(11.84).

Figure 11.20 shows the 3D strain vector and matrix $\mathbf{B}_i^l$ for circular prisms. By making $\gamma = r\gamma$ and $\frac{N_i}{r} = 0$ in Figure 11.20, the strain matrix of Eq.(11.84) for straight prisms is recovered (with $r = z$).

The stress-strain relationship deduced from elasticity theory is

$$
\boldsymbol{\sigma} = [\sigma_x, \sigma_y, \sigma_z, \tau_{xy}, \tau_{xz}, \tau_{yz}]^T = \mathbf{D} \sum_{l=1}^{m} \sum_{i=1}^{n} \hat{\mathbf{S}}^l \mathbf{B}_i^l \mathbf{a}_i^l \tag{11.85}
$$

where $\mathbf{D}$ is the constitutive matrix of 3D elasticity theory [On4,ZTZ].

Indexes x, y, z are substituted by r, β, z, respectively for circular prisms.

Substituting Eqs.(11.75)-(11.85) into the 3D expression of the TPE (or the PVW) [On4,ZTZ] and using the orthogonal properties of the harmonic functions yields the decoupled system of equations (11.19) for the different harmonic terms with

$$
[\mathbf{K}_{ij}^{ll}]^{(e)} = C \iint_{A^{(e)}} [\mathbf{B}_i^l]^T \mathbf{D} \mathbf{B}_j^l \, dA \tag{11.86}
$$

$$
[\mathbf{f}_i^l]^{(e)} = C \iint_{A^{(e)}} \mathbf{N}_i^T \mathbf{b}^l \, dA + C \oint_{S^{(e)}} \mathbf{N}_i^T \mathbf{t}^l \, dA + \mathbf{p}_i^l \tag{11.87}
$$

where

$$
\text{straight prisms} \qquad \begin{cases} C = \dfrac{b}{2} \\[2mm] dA = dx\,dz \end{cases} \tag{11.88}
$$

$$
\text{circular prisms} \qquad \begin{cases} C = \dfrac{\alpha}{2} \\[2mm] dA = r\,dr\,dz \end{cases} \tag{11.89}
$$

$$\varepsilon = \begin{Bmatrix} \varepsilon_r \\ \varepsilon_\beta \\ \varepsilon_z \\ \varepsilon_{r\beta} \\ \varepsilon_{rz} \\ \varepsilon_{\beta z} \end{Bmatrix} = \begin{Bmatrix} \dfrac{\partial u}{\partial r} \\[2mm] \dfrac{1}{r}\dfrac{\partial v}{\partial \beta} + \dfrac{u}{r} \\[2mm] \dfrac{\partial w}{\partial z} \\[2mm] \dfrac{1}{r}\dfrac{\partial u}{\partial \beta} + \dfrac{\partial v}{\partial r} - \dfrac{v}{r} \\[2mm] \dfrac{\partial u}{\partial z} + \dfrac{\partial w}{\partial r} \\[2mm] \dfrac{1}{r}\dfrac{\partial w}{\partial \beta} + \dfrac{\partial v}{\partial z} \end{Bmatrix} = \sum_{l=1}^{m}\sum_{i=1}^{n} \hat{\mathbf{S}}^l \mathbf{B}_i^l \mathbf{a}_i^l$$

$$\mathbf{B}_i^l = \begin{bmatrix} \dfrac{\partial N_i}{\partial x} & 0 & 0 \\[2mm] \dfrac{N_i}{r} & -\dfrac{\gamma N_i}{r} & 0 \\[2mm] 0 & 0 & \dfrac{\partial N_i}{\partial z} \\[2mm] \dfrac{\gamma N_i}{r} & \left(\dfrac{\partial N_i}{\partial r} - \dfrac{N_i}{r}\right) & 0 \\[2mm] \dfrac{\partial N_i}{\partial z} & 0 & \dfrac{\partial N_i}{\partial r} \\[2mm] 0 & \dfrac{\partial N_i}{\partial z} & \dfrac{\gamma N_i}{r} \end{bmatrix}$$

$\hat{\mathbf{S}}^l$ as in Eq.(11.84) for rectangular prisms with $\gamma = \dfrac{l\pi}{\alpha}$

Fig. 11.20 Strain vector and matrix for a circular prism

and $\mathbf{b}^l$, $\mathbf{t}^l$ and $\mathbf{p}_i^l$ are the nodal modal amplitudes for body forces, distributed forces and point loads, respectively. Typical expressions are:

Self-weight

$$\mathbf{b}^l = \frac{2\rho}{l\pi}(1 - (-1)^l)[g_x, g_y, g_z]^T \tag{11.90}$$

where ρ is the density and g_i the intensity of the gravity in the direction of the ith axis.

Uniformly distributed loading

$$\mathbf{t}^l = \frac{2}{l\pi}[f_x(C_0 - C_1), f_y(S_1 - S_0), f_z(C_0 - C_1)]^T \tag{11.91}$$

$$C_0 = \cos\gamma y_0, \; C_1 = \cos\gamma y_1$$

$$S_0 = \sin\gamma y_0, \; S_1 = \sin\gamma y_1$$

where y_0 and y_1 are the limits of the distributed loading along the prismatic direction.

Point loads

$$\mathbf{p}_i^l = \frac{2}{b}[F_{x_i}\sin\gamma y_i, F_{y_i}\cos\gamma y_i, F_{z_i}\sin\gamma y_i]^T \qquad (11.92)$$

where y_i is the prismatic coordinate of the point of application of the load.

For circular prisms the coordinates x and y are substituted by r and β, respectively and distance b by angle α.

If the loads are symmetric with respect to the center of the prism, then the contribution from the even harmonic terms is zero in Eqs.(11.90)–(11.92), as for the finite strip method.

The computation of the element matrices and vectors in Eqs.(11.86) and (11.87) typically requires a Gauss quadrature. The exception is the simple 3-noded triangular prism for which analytic forms of the integrals can be easily found. This case is shown in the following example.

Example 11.1: Compute the stiffness matrix and the self-weight equivalent nodal modal force amplitude vector for a 3-noded triangular straight prism.

- Solution

The faces of the prism are linear triangles (Figure 11.21). The shape functions are

$$N_i = \frac{1}{2A^{(e)}}[a_i + b_i x + c_i z]$$

where a_i, b_i and c_i are given by Eq.(4.32b) of [On4] with x, z for x, y. The element strain matrix is

$$\mathbf{B}_i^l = \frac{1}{2A^{(e)}}\begin{bmatrix} b_i & 0 & 0 \\ 0 & -\gamma N_i & 0 \\ 0 & 0 & c_i \\ \gamma N_i & b_i & 0 \\ c_i & 0 & b_i \\ 0 & c_i & \gamma N_i \end{bmatrix} \qquad \text{con}\gamma = \frac{l\pi}{b}$$

The product $\mathbf{B}_i^{l^T}\mathbf{D}\mathbf{B}_j^l$ contains terms such as N_i and $N_i N_j$. Analytical integration leads to

$$[\mathbf{K}_{ij}^{ll}]^{(e)} = \frac{b}{8A^{(e)}} \times$$

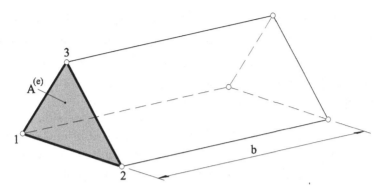

Fig. 11.21 Three-noded triangular prism

$$
\begin{bmatrix}
(d_{11}b_ib_j + d_{55}c_ic_j+ & \frac{\gamma A^{(e)}}{3}(-d_{12}b_i + d_{44}b_j) & (d_{13}b_ic_j + d_{55}c_ib_j) \\
+d_{44}\alpha_{ij}) & & \\
\frac{\gamma A^{(e)}}{3}(-d_{21}b_j + c_{44}b_i) & (d_{44}b_ib_j + d_{66}c_ic_j+ & \frac{\gamma A^{(e)}}{3}(d_{66}c_i - d_{23}c_j) \\
& +d_{22}\alpha_{ij}) & \\
(d_{31}c_ib_j + d_{55}b_ic_j) & \frac{\gamma A^{(e)}}{3}(d_{66}c_j - d_{32}c_i) & (d_{33}c_ic_j + d_{55}b_ib_j+ \\
& & +d_{66}\alpha_{ij})
\end{bmatrix}
$$

with

$$
\alpha_{ij} = \frac{\gamma^2 A^{(e)}}{6} \quad \text{if} \quad i = j \quad \text{or} \quad \frac{\gamma^2 A^{(e)}}{12} \quad \text{if} \quad i \neq j
$$

The equivalent nodal modal force amplitude vector for self-weight is obtained from Eqs.(11.87)–(11.90). For $\rho = $ constant, $g_x = g_y = 0$ and $g_z = -g$,

$$
\mathbf{f}_i^l = \frac{b}{2}\mathbf{b}^l \iint_{A^{(e)}} \mathbf{N}_i^T dA = \frac{bA^{(e)}}{6}\mathbf{b}^l = \frac{b\rho g}{3l\pi}[1 - (-1)^l]\begin{Bmatrix} 0 \\ 0 \\ -1 \end{Bmatrix}
$$

11.10 AXISYMMETRIC SOLIDS UNDER ARBITRARY LOADING

An axisymmetric solid can be viewed as a circular prism describing a circle. Axisymmetric solids under arbitrary loading can be therefore analyzed via a simple extension of the finite prism formulation of the previous section. The problem is analogous to the analysis of axisymmetric shells under arbitrary loading using troncoconical strips studied in Section 11.7.

The displacement field in an axisymmetric solid can be written in general form as

$$
\mathbf{u} = \begin{Bmatrix} u \\ v \\ w \end{Bmatrix} = \sum_{l=0}^{m}\sum_{i=1}^{n} \mathbf{N}_i(\bar{\mathbf{S}}^l\bar{\mathbf{a}}_i^l + \bar{\bar{\mathbf{S}}}^l\bar{\bar{\mathbf{a}}}_i^l) \tag{11.93}
$$

where $(\tilde{\cdot})$ and $(\bar{\cdot})$ denote the symmetric and anti-symmetric components of the displacement field with respect to a meridional plane. Matrix $\bar{\bar{\mathbf{S}}}^l$ coincides with $\mathbf{S}^l$ of Eq.(11.82) for $\gamma = l$, and $\bar{\mathbf{S}}^l$ is obtained from $\bar{\bar{\mathbf{S}}}^l$ simply by interchanging S^l for C^l. The term $l = 0$ represents an axisymmetric displacement field.

Loads are expanded in Fourier series by an expression similar to Eq.(11.93). The symmetric and anti-symmetric contributions are analyzed separately as explained in Section 11.7. The strain matrix for both cases is deduced from the expression in Figure 11.20 for circular prisms making

$$\gamma = -l \quad \text{for symmetric loads}$$
$$\gamma = l \quad \text{for anti-symmetric loads} \tag{11.94}$$

The stiffness matrix for the axisymmetric prism is given by an expression identical to Eq.(11.86) with

$$C = \begin{cases} 2\pi \text{ for } l = 0 \\ \pi \text{ for } l \neq 0 \end{cases} \tag{11.95}$$

The equivalent nodal modal force vector is given by Eq.(11.87) with C given by Eq.(11.95). For $l = 0$ the loading terms involving S^l are zero. Hence, for symmetric loads and $l = 0$ the problem reduces to the fully axisymmetric case and the unknowns are the meridional displacements u and w [On4,ZTZ]. For anti-symmetric loads and $l = 0$ the only unknown is the tangential displacement v. This problem is equivalent to torsion in circular bars, traditionally solved in elasticity theory using stress functions [ZTZ].

11.11 INTERMEDIATE SUPPORTS WITH RIGID DIAPHRAGMS

Intermediate supports in prismatic structures can be analyzed using a flexibility approach. Practical examples are multi-span bridges where rigid diaphragms are typically placed at the support cross-section to prevent in-plane displacements.

The flexibility method is schematically described in Figure 11.22. First the displacement constraints are released at the support points and the diaphragms points. The resulting isostatic structure is then analyzed under the external loads by the finite strip or the finite prism methods. The displacements at the released points are grouped in vector $\mathbf{a}^*$.

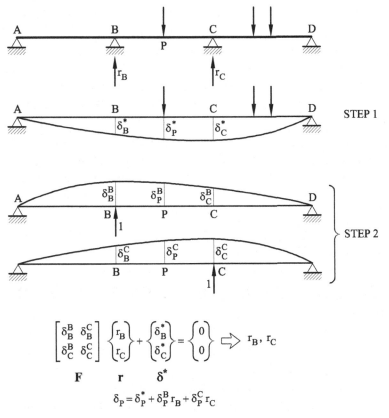

Fig. 11.22 Analysis of a three-span beam by the flexibility method

In the second step, the isostatic structure is analyzed under unit point loads (or unit moments) acting at each of the released points in the direction of the constrained displacements or rotations. A single load is considered at a time as shown in Figure 11.22. The computed displacements in the direction of the constraint movement are grouped in the flexibility matrix $\mathbf{F}$ [Li]. The geometry compatibility conditions are written as

$$\mathbf{Fr} + \mathbf{a}^* = \bar{\mathbf{a}} \tag{11.96}$$

where $\mathbf{r}$ is the vector or constraint forces and/or bending moments and $\bar{\mathbf{a}}$ are the prescribed displacements at the support points (usually $\bar{\mathbf{a}} = 0$).

Following the computation of $\mathbf{r}$ from Eq.(11.96) the displacements, strains and stresses at any point of the structure are computed by

$$\mathbf{a} = \mathbf{a}^* + \sum_{j=1}^{n_c} \mathbf{a}^j \mathbf{r}_j \quad ; \quad \hat{\boldsymbol{\varepsilon}} = \hat{\boldsymbol{\varepsilon}}^* + \sum_{j=1}^{n_c} \hat{\boldsymbol{\varepsilon}}^j \mathbf{r}_j \quad ; \quad \hat{\boldsymbol{\sigma}} = \hat{\boldsymbol{\sigma}}^* + \sum_{j=1}^{n_c} \hat{\boldsymbol{\sigma}}^j \mathbf{r}_j \tag{11.97}$$

where $(\cdot)^*$ denotes the values computed in the isostatic structure under the external loads, $(\cdot)^j$ are the values induced by a unit point load or a unit moment corresponding to the j-th constraint and n_c is the number of constraints.

Eq.(11.96) is built-up in finite strip and finite prism methods by adding-up the contributions from the different harmonics. Each element of $\mathbf{F}$ corresponds to a displacement (rotation) induced by a point load (or a bending moment). Consequently, a considerable number of harmonic terms has to be taken for computing $\mathbf{F}$ and $\mathbf{r}$ with enough accuracy [On]. The same number of harmonic terms should be taken to compute the displacements $\mathbf{a}^*$ for consistency.

The displacements, strains and stresses in (11.97) are obtained by adding up the contributions of the harmonic terms chosen as it is usual in finite strip and finite prism methods [On].

11.12 EXTENSION OF FINITE STRIP AND FINITE PRISM METHODS

A number of procedures have been proposed for overcoming the difficulties of the finite strip method in dealing with multi-span or column supported structures, or with structures with variables thickness along the prismatic direction. A popular option is to use spline functions to replace the trigonometric series. The so-called *spline finite strip* originally developed by Fan [Fa] uses B-3 spline functions to approximate the longitudinal behaviour and hyperbolic series in the interpolation function across the transverse direction of the structure. This allows us to achieving a higher order of continuity (C^2 continuity) which is useful in some cases. The spline finite strip method has been successfully applied to the analysis of plates, shells, bridges and tall buildings [ChT,Fa].

The greater versatility of the spline finite strip method is compensated by the loss of the decoupled form of the assembled system of equations. Despite of this drawback the method allows to solve complex shell and bridge structures with a smaller number of degrees of freedom than the standard FEM. A detailed description of the spline finite strip method and many applications are reported in [ChT].

The application of the finite strip method to structures with variable thickness along the longitudinal directions is possible by using the so-called *computed shape functions* [ChT]. These functions are defined along the longitudinal direction analyzing a unit width strip as a beam with the same variation of the longitudinal rigidity as the original strip. The approximation along the transverse direction is chosen as in the standard finite strip method. An alternative for introducing for a variable thickness in the longitudinal (prismatic) direction is to expand the thickness in Fourier series in terms of the longitudinal coordinate. This leads to a full stiffness matrix where the contribution of the different harmonic terms are coupled. The coupled terms can be introduced in the l.h.s. of the equilibrium equations and the problem can be still solved independently for each harmonic term in an iterative manner [SMO].

Applications of the finite strip method and the spline finite strip method to the non linear analysis of structures have been reported by a number of authors. Clearly, the uncoupled structure of the equation is now lost, although the approach is still competitive with the FEM in some cases. A comprehensive list of references can be found in [ChT].

Some authors have modified the finite prism method by introducing a second series approximation across the thickness of the structure. This is the basis of the *finite layer method* which reduces the computation of a 3D problem to a 1D one. Applications include the analysis of layered plates of uniform thickness [ChC,ChC2], layered pavements and foundations [ChF2,LS], soil consolidation problems [BS3,BS4] and pollutant irrigation in soil [RB]. More details can be found in [ChT].

The finite prism and finite strip methods have been successfully combined for the analysis of multi-layer sandwich plates and panels of non-uniform thickness. Applications of the so-called finite-prism-strip method are reported in [ChTC,TChC].

11.13 EXAMPLES

11.13.1 Simply supported square plate under uniformly distributed loading

The geometry of the plate and the material properties are shown in Figure 11.23. The solution is found using different meshes of 2-noded Reissner-Mindlin plate strips with uniform one point integration. Figure 11.23 shows the convergence of the vertical deflection and the bending moment M_x at the place center with the number of harmonic terms for a mesh of 15 strips. A non-zero harmonic term suffices to obtain an accurate solution (recall that the even harmonic terms are zero due to symmetry). Figure 11.23 also shows the convergence of the central displacement and the central bending moment M_x with the number of strips using 15 non-zero harmonic terms. 1 % error in both magnitudes is obtained using eight

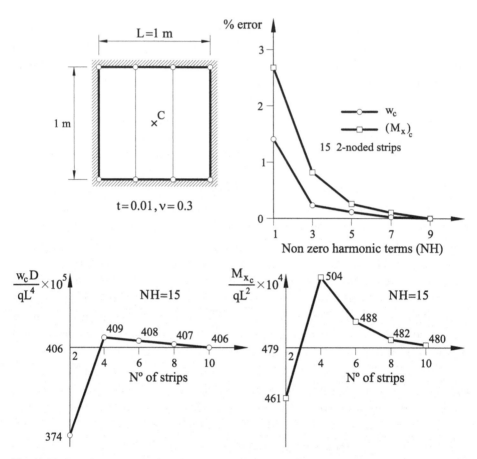

Fig. 11.23 Simple supported plate under uniformly distributed load q. Convergence of central deflection w_c and central bending moment M_{x_c} with the number of non zero harmonic terms (NH) and the number of 2-noded strips. $D = \dfrac{Et^3}{12(1 - \nu^2)}$

linear strips. The dimensions of the stiffness matrix for each harmonic solution is just 27×27 in this case [OS].

11.13.2 Curved simply supported plate

The curved thin plate analyzed is simply supported at the two ends and has a point load acting at three points in the mid-span section. Both experimental and numerical results are available for this problem.

The geometry of the plate, material properties, loading position and finite strip mesh used are given in Figure 11.24. Results for the mid-

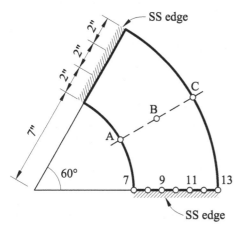

12 2-noded strips

6   non-zero harmonics

$E = 42,460 \ lb / in^2$

$v = 0,3$

$t = 0,172 \ in$

Loading: point load of 1 lb acting
in A, B and C

VERTICAL DEFLECTION (IN.)

Load position	Rad. inc.	Coull & Das Exp.	Coull & Das Theor.	Finite strip Thorpe	Finite strip Benson	Finite elements 2-noded strip	Finite elements Sawko	Finite elements Fam
A	13	0.876	0.752	0.882	0.995	0.874	0.851	0.880
	11	0.578	0.500	0.582	0.624	0.581	0.559	0.577
	9	0.353	0.300	0.356	0.388	0.357	0.344	0.353
	7	0.194	0.180	0.195	0.206	0.194	0.192	0.193
B	13	0.457	0.470	0.459	0.441	0.460	0.445	0.446
	11	0.342	0.370	0.343	0.329	0.348	0.333	0.342
	9	0.241	0.250	0.242	0.222	0.247	0.236	0.241
	7	0.155	0.170	0.156	0.147	0.158	0.154	0.154
C	13	0.195	0.150	0.195	0.180	0.195	0.192	0.193
	11	0.163	0.135	0.165	0.152	0.167	0.162	0.163
	9	0.157	0.125	0.153	0.149	0.155	0.151	0.151
	7	0.169	0.145	0.170	0.173	0.170	0.170	0.169
Reference		[CD]		[Tho]	[BH]	[OS2]	[SM]	[FT]

Fig. 11.24 Curved simply supported (SS) plate under vertical point load. Results for the deflection along the central line ABC for three positions of the load

span deflection obtained with the 2-noded Reissner-Mindlin plate strip with uniform one point integration using six non-zero harmonic terms [OS,OS2] are compared in Figure 11.24 with experimental and theoretical results obtained by Coull and Das [CD], Kirchhoff finite strip solutions [BH,Tho] and finite element solutions [FT,SM]. Results obtained with the 2-noded plate strip are very accurate.

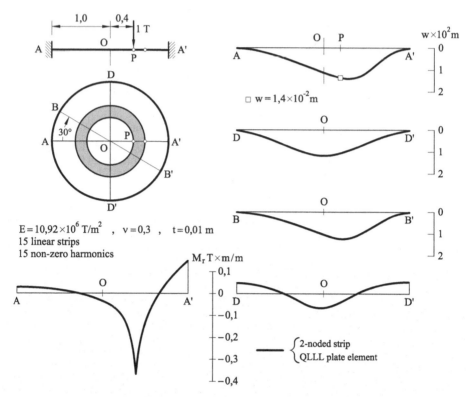

Fig. 11.25 Clamped circular plate under eccentric point load. Distribution of deflection and radial bending moment M_r along several diametral lines using ten 2-noded strips and 15 non-zero harmonic terms

11.13.3 Circular plate under eccentric point load

The next two examples show the possibilities of the 2-noded troncoconical Reissner-Mindlin strip with uniform one point integration for analysis of axisymmetric shells under arbitrary loading.

The first example presented in this section is the study of a clamped circular plate under a point load acting at a distance from the plate center. The geometry and the material properties are shown in Figure 11.25. The analysis is carried out with a mesh of ten 2-noded strips and 15 non-zero harmonic terms. A symmetric solution taking as symmetry axis the diameter containing the load suffices in this case.

Figure 11.25 shows the distribution of the deflection and the radial bending moments M_r along several diametral lines. Results compare very well with a solution obtained with a fine mesh of 4-noded QLLL plate

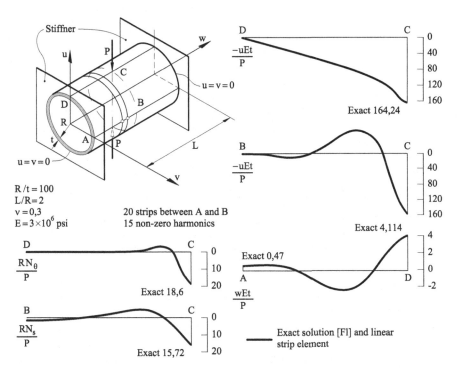

Fig. 11.26 Pinched cylinder. Distribution of displacements u and w and axial forces N_θ and N_s along several lines

elements (Section 6.7.1) [PHZ]. The deflection under the load also agrees with the theoretical solution [TW].

11.13.4 Pinched cylinder

This example was studied with flat and curved shell elements in Chapters 8 and 10 (Figures 8.36 and 10.15). The definition of the problem is shown in Figure 11.26. The analysis is performed with twenty 2-noded strips and 15 non-zero harmonic terms. Symmetry conditions with respect the horizontal meridional plane were imposed.

Figure 11.26 shows the distribution of the vertical and longitudinal displacements and the axial forces N_s and N_θ along different lines. The finite strip results are undistinguishable from the analytical solution [Fl].

11.13.5 Simply supported curved box girder-bridge

The geometry of the bridge and the material properties are shown in Figure 11.27. A point load acting at the center of the mid-span section is

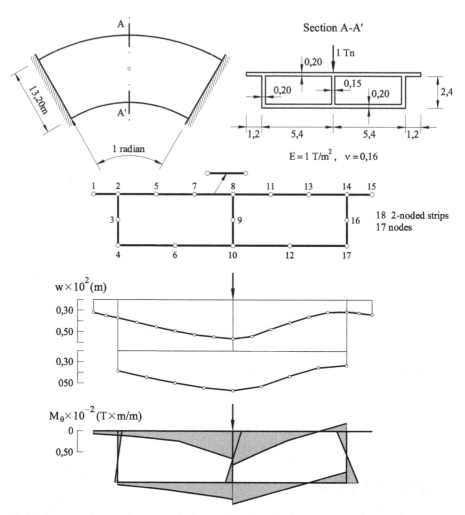

Fig. 11.27 SS circular box girder bridge under central point load. Distribution of vertical deflection and bending moment M_θ at mid-span section using eighteen 2-noded strips and fifteen non-zero harmonic terms [OS2,OS3]

considered. A mesh of eighteen 2-noded troncoconical Reissner-Mindlin strip with reduced integration has been used. Figure 11.27 displays the distribution of the vertical deflection and the circumferential bending moment M_θ at the mid-span section. All results have been obtained using 15 (odd) harmonic terms (for symmetry reasons). Numerical results agree with those obtained by Cheung [Ch3,Ch6,ChC3,ChT] using a Kirchhoff strip formulation. Further information can be obtained in [OS3,Su].

11.13.6 Two-span cellular bridge

The geometry of the bridge, the material properties and the loading is shown in Figure 11.28. A rigid diaphragm is placed at the mid-span section over the support. Figure 11.28 also displays the mesh of fifteen *3-noded quadratic Reissner-Mindlin flat shell strips* with selective integration used (Section 11.5.4) and the points of the upper and lower slabs of the mid-span section where the vertical displacement is constrained to zero in order to model a rigid cross section. Note that only two points are constrained at the upper slab next to the points of application of the loads where higher bending effects are expected. This suffices to model the rigid behaviour of the mid-span section.

The analysis was performed using 30 non-zero (odd) harmonic terms [On]. Figure 11.28 shows the distribution of the longitudinal axial force N_y for a cross section under the loads and for a section adjacent to the mid-span. Numerical results compare well with a Kirchhoff flat shell solution [KS].

11.13.7 Simple supported slab-beam bridge over a highway

This example shows the usefulness of the 2-noded strip for the analysis of a reinforced concrete bridge formed by a thick slab and cellular beams. The bridge was built in the early 1980s as part of the highway Nueve de Julio in the city of Buenos Aires (Argentina).

The geometry of the bridge, loading position, material properties and finite strip discretization are shown in Figure 11.29. Results for the vertical deflection, the transverse bending moment M_x and the longitudinal axial force N_y in the slab at the mid-section obtained with 15 non-zero harmonic terms are plotted in Figure 11.29. The diagrams shown are extrapolated from the finite strip results. The validity of the numerical solution was verified by an equilibrium check comparing the total longitudinal bending moment at the mid-span section with the value obtained using simple beam theory. The percentage of error was less than 5% which can be considered good for practical design purposes [OS,Su].

11.13.8 Circular bridge analyzed with finite strip and finite prism methods

This example compares the finite strip and finite prism methods for the analysis of a simple supported cellular bridge under two point loading cases. The definition of the bridge and the loading are shown in Figure 11.30. The finite strip analysis is performed using fourteen 3-noded

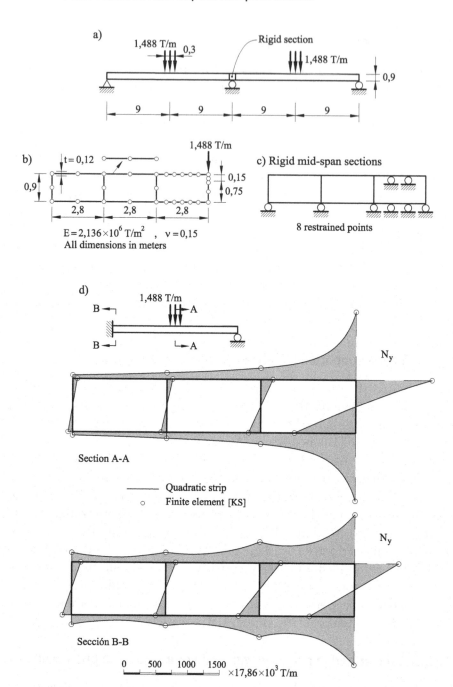

Fig. 11.28 Two-span cellular bridge. a) Geometry and loading; b) Discretization in fifteen 3-noded quadratic strips; c) Constrained points to model the rigid support cross section; d) Distribution of the axial force N_y in two sections [On]

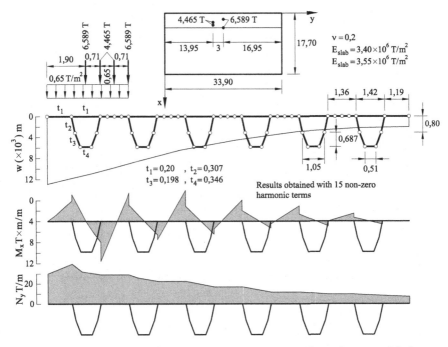

Fig. 11.29 Simply supported slab-beam highway bridge. Distribution of deflections, transversal bending moment M_x and longitudinal axial force N_y at the mid-span section using 58 2-noded strips and 15 non-zero harmonic terms

quadratic strips. Twenty 8-noded quadrilateral prisms have been chosen for the finite prism solution. 20 non-zero odd harmonic terms were used in both cases.

Figure 11.30 shows the deformation of the mid-span cross section for the two loading cases considered. Results for the finite strip and finite prism solutions compare well and also agree with those obtained with a finite element solution using Kirchhoff flat shell elements [On].

Figure 11.30 also shows the distribution of the longitudinal axial force N_y at the mid-span section. Results for N_y for the finite prism method have been obtained by integrating over the thickness the longitudinal stress σ_y. The agreement between of the finite strip and finite prism results is also very good.

The cost of the finite strip solution is clearly inferior to any other method. Assuming the cost to be proportional to the square of the number of DOFs gives that the finite strip solution is $\left(\frac{168}{300}\right)^2 \cong 31\ \%$ more economical than the finite prism solution.

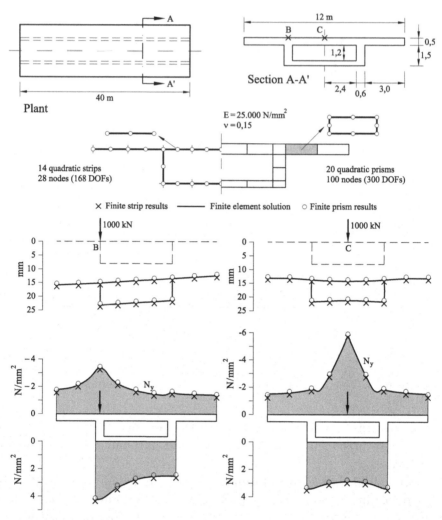

Fig. 11.30 SS box girder bridge analyzed with finite strip and finite prism methods. Both solutions agree with a finite element flat shell solution [On]

The finite strip solution for this problem can be simplified using 2-noded linear strips.

11.13.9 Simply supported thick box girder-bridge analyzed with the finite prism method

This example shows an application of the finite prism method to the analysis of a straight simply supported thick box girder bridge. The geometry of the cross section and the two meshes of 8-noded quadrilateral prisms

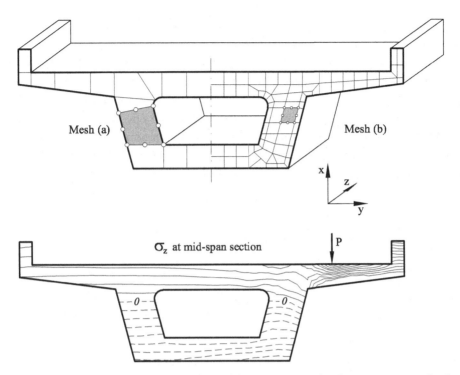

Fig. 11.31 SS thick bock girder bridge analyzed with the finite prism method

used are shown in Figure 11.31. A point local acting eccentrically at the mid-span section is considered. Figure 11.31 also shows the distribution of the longitudinal stresses for the finer mesh. Note that the finite prism method allows to capture the stress concentration in the vicinity of point loads and this is an adavantage versus the finite strip method. Further details on this example can be found in [ZT].

11.14 FINAL REMARKS

We have studied in this chapter the analysis of prismatic structures by combining the finite element method and Fourier series expansions. The finite strip method is adequate for thin-walled prismatic structures such as plates, shells and bridges with straight and circular plan. The finite strip method is also a competitive procedure for analysis of axisymmetric shells under arbitrary loading. The general finite strip equations for a wide range of prismatic thin-walled structures can be derived as a particular case of the general troncoconical strip formulation. The simple 2-noded strip element with uniform one point integration is the best candidate for analysis of prismatic *thin-walled* structures.

The finite prism method is applicable to prismatic solids which are typical in structural and geotechnical engineering. The finite prism formulation can be derived as a particular case of the circular prism theory. The finite prism method is adequate for studying local effects, such as stresses under point loads in slabs and bridges, foundations and other prismatic solids and structures. A competitive application of the finite prism method is the analysis of axisymmetric solids under arbitrary loading.

12

PROGRAMMING THE FEM FOR BEAM, PLATE AND SHELL ANALYSIS IN MAT-fem

written by Francisco Zárate[1]

12.1 INTRODUCTION

We present in this chapter the implementation of several of the elements for beam, plate and shell analysis studied in this book in the MAT-fem code environment written using the MATLAB® and GiD programming tools [On4]. MAT-fem includes several codes for FEM analysis of different structures. These codes can be freely downloaded from www.cimne.con/MAT-fem.

MATLAB® has been designed to work with matrices, facilitating the matrix algebra operations from the numerical and storage points of view, while providing a simple and easy way to handle complex routines.

Having an efficient analysis code is not the only requirement to work with the FEM. It is necessary to rely on a suitable interface to prepare the analysis data, to generate meshes in a simple and fast manner and to display the results so that their interpretation is clear. An ideal complement to MATLAB® for these purposes is the pre/postprocessor program GiD (www.gidhome.com and Appendix D of [On4]).

GiD allows users to treat any geometry via Computer Aided Design (CAD) tools and to easily assign to it the data needed for FE computations, i.e. material properties, boundary conditions, loads, etc. Different tasks, such as the mesh generation and data writing levels in a pre-defined format become a transparent task for the user with GiD.

The easy visualization of the analysis data and the numerical results with GiD allows users concentrating on their interpretation.

---

[1] Dr. F. Zárate can be contacted at zarate@cimne.upc.edu

E. Oñate, *Structural Analysis with the Finite Element Method. Linear Statics: Volume 2: Beams, Plates and Shells*, Lecture Notes on Numerical Methods in Engineering and Sciences, DOI 10.1007/978-1-4020-8743-1_12,
© International Center for Numerical Methods in Engineering (CIMNE), 2013

Structural type	Code name	GiD interface
Slender beams	Beam_EulerBernoulli	MAT-fem_Beams
Thick/slender beams	Beam_Timoshenko	MAT-fem_Beams
Thin plates	Plate_MZC	MAT-fem_Plates
Thick/thin plates	Plate_QLLL	MAT-fem_Plates
Thick/thin plates	Plate_Q4_Rect	MAT-fem_Plates
Thick/thin plates	Plate_Q4_Iso	MAT-fem_Plates
Thick/thin plates	Plate_TQQL	MAT-fem_Plates
3D shells with flat elements	Shell_QLLL	MAT-fem_Shells
Axisymmetric shells	Troncoconical_RM_Shell	MAT-fem_AxiShells

Table 12.1 Definition of the problem type, name of code and GiD interface

MAT-fem has been written thinking on the close interaction of GiD with MATLAB® for FEM analysis. GiD allows manipulating geometries and discretizations for writing the input data files required by MATLAB®. The calculation program is executed in MATLAB® without losing any of the MATLAB® advantages. Finally GiD gathers the output data files for graphical visualization and interpretation of results.

This scheme allows us understanding the development of a FEM program in detail, following each one of the code lines if desired, and making possible for users to solve examples that due to their dimensions would fall outside the capabilities of any program with educational aims.

We *have chosen to write a different MATLAB® code for the analysis of the different structures considered*, as shown in Table 12.1. The same programming strategy is followed and the variables have the same meaning in all cases. This simplifies the learning process and facilitates the modification of a specific code without introducing errors in the rest.

In the following sections the different MAT-fem codes for beam, plate and shell analysis are described in some detail. The description starts with the general input data file instructions, automatically generated by GiD, and follows with the information to understand the particular features of each code within MAT-fem.

Finally, the user interface implemented in GiD for each code is briefly described by means of examples of application.

12.2 MAT-fem

As already mentioned, MAT-fem contains several codes for analysis of beams, plates and shells (Table 12.1). All codes share the same programming philosophy and the differences are in the DOFs the constitutive

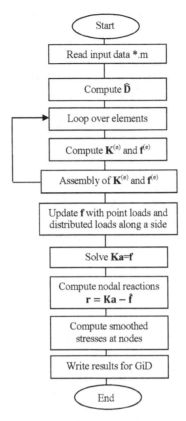

Fig. 12.1 MAT-fem flow chart for a typical problem type

matrix $\hat{\mathbf{D}}$, the generalized strain matrix $\mathbf{B}$ and the expressions for the element stiffness matrix and the equivalent nodal force vector.

MAT-fem is a top-down execution program. The program flow chart is shown in Figure 12.1. The input data module is implemented in the same file were the data is defined, as described in the next sections.

We consider that all elements have the same material properties. Hence, the constitutive matrix is evaluated outside the loop over the elements within which the element stiffness matrix and the equivalent nodal force vector are computed.

To save memory the stiffness matrix and the equivalent nodal force vector are assembled immediately after they are evaluated for each element.

Outside the element loop the equivalent nodal force vector is updated with the nodal point forces and the distributed loads acting along a side.

(a)
```
%
% Material Properties
% Beams
%
    young   =   4.761904762e+04 ;
    poiss   =   5.820105800e-02 ;
    denss   =   0.000000000e+00 ;
    area    =   1.600000000e-01 ;
    inertia=   2.133330000e-03 ;
```

(b)
```
%
% Material Properties
% Plates and Shells
%
    young   =   2.100000000e+11 ;
    poiss   =   3.000000000e-01 ;
    denss   =   0.000000000e+00 ;
    thick   =   1.000000000e-01 ;
```

Fig. 12.2 Input data file. Example of definition of material properties. a) Beams. b) Plates and shells

Once the unknown DOFs are found, the program evaluates the nodal reactions at the prescribed nodes and the smoothed stresses at the nodes. The final step is the writing of the numerical results to visualize them in GiD (Figure 12.1).

12.3 DATA FILES

Before executing MAT-fem it is necessary to feed it with information on the nodal coordinates, the mesh topology, the boundary conditions, the material properties and the loading.

The input data file uses MATLAB® syntax. The program variables are defined directly in that file. The name of the file will take the MATLAB® extension .m.

Inside the data file we distinguish three groups of variables: those associated to the material properties, those defining the topology of the mesh and those defining the boundary conditions.

12.3.1 Material data

With the intention of simplifying the code, an isotropic linear elastic material is used for all problems. Hence the material data appears only once in the data file.

Figure 12.2 shows the variables associated to the material data for each one of the structures considered. The definition of each variable is shown in Table 12.2.

We note that the program is free of data validation mechanisms. Hence it does not check up aspects such the Poisson's ratio rank ($0 <=$ poiss $<$ 0.5) and others. The reason is that these details, although they are important in practice, would hide the core of the FEM algorithm.

Variable	Description	Beams	Plates	3D Shells	Axisym. Shell
young	Young modulus	√	√	√	√
poiss	Poisson's ratio	√	√	√	√
dens	Density	√	√	√	√
area	Cross section area	√			
inertia	Cross section inertia moment	√			
thick	Thickness		√	√	√

Table 12.2 Material parameters for each structure

(a)

```
%
% Coordinates
%
global coordinates
coordinates = [
   0.00 ;
   0.50 ;

   2.00 ;
   2.50 ];
%
% Elements
%
elements = [
    1,    2 ;
    2,    3 ;

   17,   16 ;
   18,   17 ];
```

(b)

```
%
% Coordinates
%
global coordinates
coordinates = [
   0.00 , 0.00 ;
   0.50 , 0.00 ;

   2.00 , 2.50 ;
   2.50 , 2.50 ];
%
% Elements
%
elements = [
    1,    2,    5,    6 ;
    2,    3,    7,    5 ;

   17,   16,   20,   21 ;
   18,   17,   22,   20 ];
```

Fig. 12.3 Input data file. Topology definition. a) Beam. b) Plate

12.3.2 Mesh topology

The variable group that describes the mesh topology is defined with the attribute of a global variable that is accessible within the code by any subroutine. Figure 12.3 shows the definition of the coordinates and the nodal connectivities for a beam and a plate by means of the variables coordinates and elements.

coordinates is a matrix with as many rows as nodes in the mesh (npnod) and columns as the number of dimensions of the problem (1 for beams, 2 for plates and 3 for shells). For a beam the dimensions of the coordinates matrix are npnod×1. For a plate the coordinates x and y of each node are needed and coordinates has the dimensions npnod×3. For a shell, coordinates is a npnod×3 matrix. The number of a node corresponds to the position that its coordinates have in the coordinates matrix, i.e. node number 25 has the position 25 in coordinates.

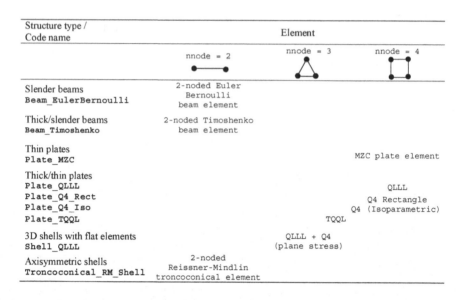

Structure type / Code name	Element		
	nnode = 2	nnode = 3	nnode = 4
Slender beams **Beam_EulerBernoulli**	2-noded Euler Bernoulli beam element		
Thick/slender beams **Beam_Timoshenko**	2-noded Timoshenko beam element		
Thin plates **Plate_MZC**			MZC plate element
Thick/thin plates **Plate_QLLL** **Plate_Q4_Rect** **Plate_Q4_Iso** **Plate_TQQL**		TQQL	QLLL Q4 Rectangle Q4 (Isoparametric)
3D shells with flat elements **Shell_QLLL**		QLLL + Q4 (plane stress)	
Axisymmetric shells **Troncoconical_RM_Shell**	2-noded Reissner-Mindlin troncoconical element		

Table 12.3 Element type for each structure

The **elements** matrix defines the number of elements and their nodal connectivities. **elements** has as many rows as the number of elements in the mesh and columns as the number of nodes on each element (**nelem** × **nnode**). Table 12.3 shows the number of nodes for the elements considered in MAT-fem. The number of an element corresponds with the row number where its nodes are stored in **elements**.

12.3.3 Boundary conditions

The last group of variables defines the boundary conditions of the problem, as shown in Figure 12.4.

The **fixnodes** matrix defines the DOFs prescribed for the particular problem to be solved. The number of rows in **fixnodes** corresponds to the number of prescribed DOFs and the number of columns describes in the following order: the prescribed node number, the fixed DOF parameter (1 if the node is fixed in the x direction and 2 if it is fixed in the y direction, etc.) and the prescribed DOF value. Hence, if a node is prescribed in both the x and y directions two lines are necessary to define this condition.

Table 12.4 shows the parameters associated to each prescribed DOF for the different problems considered.

```
%
% Prescribed nodes
%
fixnodes = [
      1, 1, 0.0 ;
      1, 2, 0.0 ;

     13, 1, 0.0 ;
     13, 2, 0.0 ];
%
% Point loads
%
pointload = [
              6, 2, -1.0 ;

             18, 2, -1.0 ];
%
% Side loads
%
uniload = sparse ( 24,1 );
uniload ( 1 ) =  -1.0000e+00  ;
uniload ( 2 ) =  -1.0000e+00  ;
uniload ( 3 ) =  -1.0000e+00  ;
uniload ( 4 ) =  -1.0000e+00  ;
```

Fig. 12.4 Input data file. Boundary conditions definition

DOF	u	v	w	dw/dx	θ	θ_x	θ_y	$\theta_{x'}$	$\theta_{y'}$
Associated point force	F_x	F_y	F_z	M	M	M_x	M_y	$M_{x'}$	$M_{y'}$
Slender beams	1		2						
Thick/slender beams	1				2				
Thin plates			1				2	3	
Thick/thin plates			1				2	3	
3D shells with flat elements	1	2	3					4	5
Axisymmetric shells	1	2				3			

Table 12.4 Parameters for the prescribed DOF and the associated point forces

12.3.4 Point and surface loads

The pointload matrix defines the point loads acting at a node. This is a matrix where the number of rows is the number of point loads acting on the structure and each of the three columns defines the number of the loaded node, the associated DOF and the magnitude of the load (Figure 12.4). Displacement DOFs are associated to point loads, while rotations are associated to bending moments. Point loads are defined in the global coordinate system. If there are no point loads, pointload is defined as an empty matrix by means of the command pointload = [];

```
%% MAT-fem
%
% Clear memory and variables.
  clear

  file_name = input('Enter the file name :','s');

  tic;                  % Start clock
  ttim = 0;             % Initialize time counter
  eval (file_name);     % Read input file

% Finds basic dimentions
  npnod  = size(coordinates,1);      % Number of nodes
  nelem  = size(elements,1);         % Number of elements
  nnode  = size(elements,2);         % Number of nodes per element
  dofpn  = 3;                        % Number of DOF per node
  dofpe  = nnode*dofpn;              % Number of DOF per element
  nndof  = npnod*dofpn;              % Number of total DOF

  ttim = timing('Time needed to read the input file',ttim);
```

Fig. 12.5 Program initialization and data reading

Finally, uniload is a sparse matrix that contains the information for uniformly distributed loads acting on the *normal direction* to each element. The distributed load is assumed to be constant for each element. Hence only the value of the load is needed for each element and the dimension of uniload is equal to nelem. If no uniform load acts, the uniload matrix remains empty with no memory consumption.

Figure 12.4 shows an example of the definition of uniform loads.

The name of the data file is up to the user. Nevertheless, the extension must be .m so that MATLAB® can recognize it.

12.4 START

MAT-fem begins by making all variables equal to zero with the clear command. Next it asks the user the name of the input data file that he/she will use (the .m extension in not included in the filename). Figure 12.5 shows the first lines of the code corresponding to the variables boot as well as the clock set up, which stores the total execution time in ttim.

Data reading, as previously said, is a direct variable allocation task in the program. From the data matrices it is possible to extract the basic dimensions of the problem, such as the number of nodal points, npnod, which corresponds to the number of lines in the **coordinates** matrix and the number of elements **nelem** which is equal to the number of lines in

```
% Dimension the global matrices.
  StifMat  = sparse ( nndof , nndof ); % Create the global stiffness matrix
  force    = sparse ( nndof , 1 );     % Create the global equivalent nodal force vector
  reaction = sparse ( nndof , 1 );     % Create the global reaction vector
  Str      = zeros  ( nelem , 2 );     % Create array for stresses
  u        = sparse (nndof, 1);        % Nodal variables
```

Fig. 12.6 Initialization of global stiffness matrix and equivalent nodal force vector

the `elements` matrix. The number of nodes for each element (Table 12.3) (`nnode`) is the number of columns in `elements`.

The total number of DOFs per element (`dofpe`) is equal to `nnode` multiplied by the number of DOFs for each node (`dofpn`).

Finally, the number of equations in the problem (`nndof`) is computed by multiplying the total number of nodes (`npnod`) by `dofpn`.

Note that these variables are defined in the data structure, which simplifies the code interpretation.

Throughout the program the `timing` routine is used to calculate the run time between two statements in the code. In this way the user can check the program modules that require higher computational effort. Inside `timing` the `tic` and `toc` MATLAB® commands are used.

12.5 STIFFNESS MATRIX AND EQUIVALENT NODAL FORCE VECTOR FOR SELF-WEIGHT AND DISTRIBUTED LOAD

12.5.1 Generalities

The code lines shown in Figure 12.6 define the global stiffness matrix (`stifMat`) and the equivalent nodal force (`force`) vector as a `sparse` matrix and vector, respectively. The reactions at the prescribed nodes are stored in `reaction`. Matrix `Str` and vector `u` are respectively used for storing the resultant stresses (at element level) and the nodal displacements. Table 12.5 shows the resultant stresses for each problem.

MAT-fem uses sparse matrices to optimize the memory using MATLAB® tools. In this manner and without additional effort, MAT-fem uses very powerful algorithms without losing its simplicity.

As the program's main purpose is to demonstrating the implementation of the FEM, some simplifications are made like using a single material for the whole structure. Consequently, the constitutive matrix does not vary between adjacent elements and it is evaluated before initiating the computation of the element stiffness matrix (Figure 12.1).

Structural type/ program name	N° of resultant stresses	Resultant stresses
Slender beams Beam_EulerBernoulli	1	M
Thick/slender beams Beam_Timoshenko	2	$M \quad Q$
Thin plates Plate_MZC	3	$M_x \quad M_y \quad M_{xy}$
Thick/thin plates Plate_QLLL Plate_Q4_Rect Plate_Q4_Iso Plate_TQQL	5	$M_x \quad M_y \quad M_{xy} \quad Q_x \quad Q_y$
3D shells with flat elements Shell_QLLL	6	$N_{x'} \quad M_{y'} \quad N_{x'y'} \quad M_{x'} \quad M_{y'} \quad M_{x'y'}$
Axisymmetric shells Troncoconical_RM_Shell	5	$N_{x'} \quad N_{y'} \quad M_{x'} \quad M_{y'} \quad Q_{z'}$

Table 12.5 Resultant stresses for each problem

The generalized constitutive matrix is typically split in the membrane (D_matm), bending (D_matb) and transverse shear (D_mats) components.

MAT-fem recalculates the values for each variable instead of storing them. The recalculation is performed in a fast manner and does not reduce significantly the code's efficiency. This leaves more memory for solving larger problems.

The definition of the Gauss point coordinates and weights is performed before entering the element loop for computing the element stiffness matrix $\mathbf{K}^{(e)}$ and the equivalent nodal force vector $\mathbf{f}^{(e)}$.

Figure 12.7 shows the element loop within which $\mathbf{K}^{(e)}$ and $\mathbf{f}^{(e)}$ are computed and assembled for uniformly distributed load and self-weight. The loop begins recovering the geometrical properties for each element. Vector lnods stores the nodal connectivities for the element. For 3D shells, the variables coor_x, coor_y and coor_z store the x, y, z coordinates for the nodes. For plates only the variables coor_x and coor_y are needed, while for beams just coor_x is used.

The next step is the computation of the element stiffness matrix. The same subroutine evaluates the equivalent nodal force vector for self-weight and uniformly distributed load for the element. The use of the same integration quadrature for integrating $\mathbf{K}^{(e)}$ and $\mathbf{f}^{(e)}$ allows us the organization

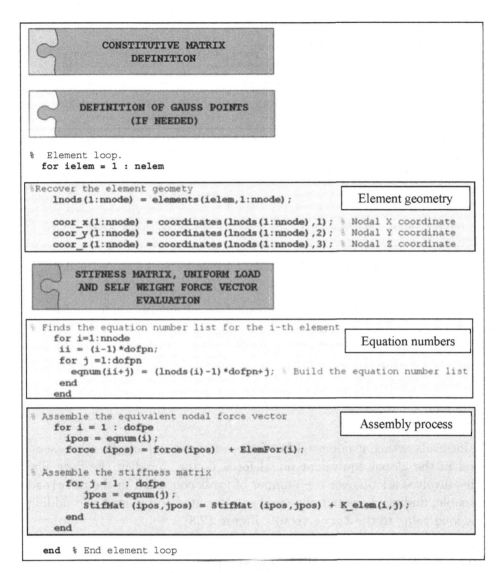

Fig. 12.7 Loop for computation and assembly of the stiffness matrix and the equivalent nodal force vector (uniform load and self-weight) for the element

of the code in this manner. The computation of $\mathbf{K}^{(e)}$ and $\mathbf{f}^{(e)}$ for each of the elements considered is detailed in the following sections.

Vector **eqnum** is defined before the assembly of the global equations. It contains the global number for each of the equations in the element stiffness matrix. This involves a loop over the number of nodes (**nnode**) and a second loop over the DOFs of each node (**dofpn**) (see Figure 12.7).

```
%  Add point loads to the equivalent nodal force vector
  for i = 1 : size(pointload,1)
    ieqn = (pointload(i,1)-1)* dofpn + pointload(i,2);   % Finds eq. number
    force(ieqn) = force(ieqn) + pointload(i,3);          % adds point load
  end
```

Fig. 12.8 Computation of the equivalent nodal force vector for point loads

```
%  Applies the Dirichlet conditions and adjusts the right-hand side.

  for i = 1 : size(fixnodes,1)
    ieqn = (fixnodes(i,1)-1)*dofpn+fixnodes(i,2);       % Finds eq. number
    u(ieqn) = fixnodes(i,3);                            % store the solution for u
    fix(i) = ieqn;                                      % mark the eq as a fix value
  end

  force = force - StifMat * u;       % adjust the rhs with the known values
```

Fig. 12.9 Update the equivalent nodal force vector for prescribed nodes

The assembly process is implemented by means of two loops from 1 to `dofpe` (number of equations for each element). In the first loop the equivalent nodal force vector is assembled and in the second one the element stiffness matrix is assembled. This scheme avoids storing the element matrices temporarily.

12.5.2 Point loads

Point loads acting at nodes (either forces or moments) are directly assembled in the global equivalent nodal force vector stored in the data file. This involves a loop over the number of loads contained in the `poinload` variable, finding the equations number associated to the load and adding the load value to the `force` vector (Figure 12.8).

12.6 PRESCRIBED DISPLACEMENTS

Figure 12.9 shows the loop over the prescribed DOFs and how their values, defined by the `fixnodes` matrix, are assigned to the nodal displacement vector u. Also the `fix` vector is defined to store the equation numbers for the prescribed DOFs.

Finally the `force` vector is updated with the product of the `StifMat` matrix and the u vector following the standard procedure [Chapter 1 of [On4]]. Vector u at this moment contains the values of the prescribed DOFs only.

```
%  Compute the solution by solving StifMat * u = force for the
%  unknown values of u.
   FreeNodes = setdiff ( 1:nndof, fix ); % Finds the free node list and
                                          % solve for it.
   u(FreeNodes) = StifMat(FreeNodes,FreeNodes) \ force(FreeNodes);
```

Fig. 12.10 Solution of the equations system

```
%  Compute the reactions at fixed nodes as R = StifMat * u - F

   reaction(fix) = StifMat(fix,1:nndof) * u(1:nndof) - force(fix);
```

Fig. 12.11 Computation of nodal reactions

12.7 SOLUTION OF THE EQUATIONS SYSTEM

The strategy used in MAT-fem basically consists in solving the global equation system without considering those DOFs whose values are known (i.e. prescribed). The **FreeNodes** vector contains the list of the equations to be solved (Figure 12.10).

The **FreeNodes** vector is used as a DOF index and allows us to write the solution of the equations system in a simple way. MATLAB® takes care of choosing the most suitable algorithm to solve the system. The routines implemented in MATLAB® nowadays compete in speed and memory optimization with the best existing algorithms.

12.8 NODAL REACTIONS

The solution of the equations system is stored in the u vector containing the nodal displacements (Figure 12.11). Nodal reactions at the prescribed nodes are computed by means of the expression: **reaction = StifMat*u - force** [On4]. In order to avoid unnecessary calculations we use vector **fix** which contains the list of the equations associated to the prescribed DOFs as shown in Figure 12.11.

12.9 RESULTANT STRESSES

12.9.1 Generalities

Once the nodal displacements have been found it is possible to evaluate the resultant stresses in the elements by means of the $\hat{D}Bu$ expression. Since the generalized strain matrix B was previously computed at the integration

```
% Compute the stresses for QLLL plate element

Strnod = Stress_Plate_QLLL(D_matb,D_mats,gauss_x,gauss_y,u);

ttim = timing('Time to  solve the  nodal stresses',ttim); %Reporting time
```

Fig. 12.12 Call the subroutine for computing the resultant stresses for the 4-noded Reissner-Mindlin QLLL plate element

points, the resultant stresses are also computed at these points which are also optimal for evaluation of stresses (Section 6.7 of [On4]). The next step is to transfer the values of the stresses from the integration points to the element nodes. This step is treated in the next section. Figure 12.12 shows the call for the subroutine for computing the resultant stresses for the QLLL plate element which are stored in the **Strnod** matrix.

12.9.2 Computation of the stresses at the nodes

Every element has it own subroutine for computing the resultant stresses. Typically, this subroutine requires the generalized constitutive matrix, the coordinates of the integration points and the nodal displacements.

Figure 12.13 shows the general form of the subroutine for computing the resultant stresses. The resultant stresses are first computed at the Gauss integration points within each element and then they are extrapolated to the nodes following a particular stress projection and smoothing scheme. The resultant stress computation starts with the definition of the variables **nelem**, **nnode**, **npnod**, **dofpn** and **dofpe**, as it was done at the beginning of the program (Figure 12.5). This avoids the transfer of these variables as arguments when the subroutine is called and preserves the clarity in the code. Similarly, the **Strnod** and **eqnum** matrices are dimensioned. **Strnod** contains the smoothed resultant stresses at the nodes and an additional parameter that defines the number of elements that surround each node that is required to perform the smoothing.

Next a loop over all the elements is performed for recovering the nodal connectivities (**lnods**), the nodal coordinates and the equation number associated to each DOF of the element nodes (**eqnum**). These operations were already performed for computing the element stiffness matrices and they are repeated here in order to save storing the data. Repeating some computations is typically a more efficient and faster procedure than storing the previously computed values in memory.

```
function Strnod = Stress_Plate (D_matb,D_mats,gauss_x,gauss_y,u)%
% Compute resultant stresses

global coordinates;
global elements;
% Basic variables definition
nelem  = size(elements,1);          % Number of elements
nnode  = size(elements,2);          % Number of nodes per element
npnod  = size(coordinates,1);       % Number of nodes
Strnod = zeros(npnod , 6 );         % Create array for stresses
dofpn  = 3;                         % Number of DOF per node
dofpe  = dofpn*nnode;               % Number of DOF per element
eqnum  = zeros(dofpe);              % Equation number list
```
Recover basic variables

```
%  Element loop.
for ielem = 1 : nelem

    %Recover the element geomety
    lnods(1:nnode) = elements(ielem,1:nnode);
```
Element geometry

```
    coor_x(1:nnode) = coordinates(lnods(1:nnode),1); % Elem. X coordinate
    coor_y(1:nnode) = coordinates(lnods(1:nnode),2); % Elem. Y coordinate

    % Finds the equation number list for the i-th element
    for i=1:nnode
      ii = (i-1)*dofpn;
      for j =1:dofpn
        eqnum(ii+j) = (lnods(i)-1)*dofpn+j; % Build the equation number list
      end
    end
```
Equation numbers

```
    % Recover the nodal displacements for the i-th element
    u_elem =u(eqnum);
```
Nodal displacements

```
    COMPUTE NODAL RESULTANT
    STRESSES FOR EACH ELEMENT
```

```
    for i = 1 : npnod
      Strnod(i,1:5) = Strnod(i,1:5)/Strnod(i,6);
    end
end
```
Nodal averaging

Fig. 12.13 Computation of resultant stresses

An important step is to identify the nodal displacements for the element. This is a simple task as the number of the equations associated to the element DOFs stored in `eqnum` allows us to recover the nodal displacement values in the `u_elem` vector.

The computation of the resultant stresses requires building the generalized strain matrix $\mathbf{B}$ at each Gauss point of the element. The resultant stresses are computed at each Gauss point as $\hat{\sigma} = \hat{\mathbf{D}}\mathbf{B}\mathbf{a}^{(e)}$ and are stored in `Strnod`.

The next step is to extrapolate the resultant stresses from the Gauss points to the nodes.

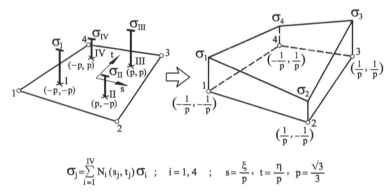

$$\sigma_j = \sum_{i=I}^{IV} N_i(s_j, t_j)\,\sigma_i \ ; \quad i = 1,4 \ ; \quad s = \frac{\xi}{p}, \ t = \frac{\eta}{p}, \ p = \frac{\sqrt{3}}{3}$$

Fig. 12.14 Extrapolation of the Gauss point stresses to the nodes for a 4-noded quadrilateral

For the 3-noded triangle the resultant stresses are constant over the element and nodal extrapolation is trivial. This is not the case for 4-noded quadrilateral where the resultant stresses typically have a bilinear variation over the element and the extrapolation to the nodes is performed using the shape functions as explained in Section 9.8.2 of [On4].

For instance, for the 4-noded quadrilateral the nodal value of each resultant stress component σ is obtained as

$$\sigma_j = \sum_{i=I}^{IV} N_i(s_j, t_j)\sigma_i \qquad j = 1,4 \tag{12.1}$$

where σ_j is the value of the stress at the jth node (j is the local number of the node), σ_i is the value of the stress component at each Gauss point and the coordinates s and t range from $1/p$ to $-1/p$ for the four element nodes as shown in Figure 12.14. For more details see [On4].

Once the resultant stresses have been extrapolated from the Gauss point to the nodes, an averaging of the nodal values contributed from each element sharing the node is performed to compute a smoothed resultant stress field at each node.

12.10 POSTPROCESSING STEP

Once the nodal displacements, the reactions and the resultant stresses have been calculated, their values are transferred to the postprocessing files from where GiD will be able to display them in graphical form. This is performed in the subroutine ToGiD (Figure 12.15).

```
% Graphic representation.
  ToGiD (file_name,u,reaction,Strnod);
```

Fig. 12.15 Call for the postprocessing step via GiD

Fig. 12.16 GUI for MAT-fem-Beams

12.11 GRAPHICAL USER INTERFACE

12.11.1 Preprocessing

In this section the Graphical User Interface (GUI) implemented in GiD is described. In order to access the GUI it is necessary to select from the GiD's Data menu the adequate module for each of the MAT-fem codes. When selected, an image similar to that shown in Figure 12.16 appears.

All the GiD capabilities are part of MAT-fem. These include geometry generation, import and handling, as well as a variety of meshing, input data and results visualization techniques. All this provides MAT-fem with capacities difficult to surpass for an educational code.

There is plenty of information on GiD available in Internet. We recommend visiting the GiD web site at www.gidhome.com.

Solving a problem with MAT-fem is very simple once the geometry has been defined. Just follow the icons of the MAT-fem graphical menu that appears when MAT-fem is activated (Figure 12.17).

The first button ⬆ in Figure 12.17 works to identify the geometrical entities (point or lines) that have nodes with prescribed displacements. When pressing on, an emergent window will appear to select the points or lines where the displacements are prescribed (Figure 12.18) The check

Fig. 12.17 MAT-fem graphical menu

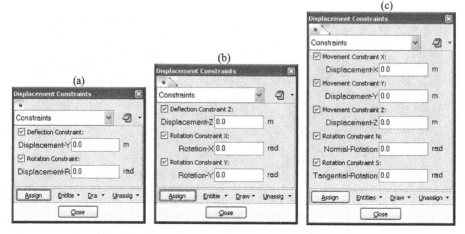

Fig. 12.18 Assign conditions at prescribed nodes. a) Beams. b) Plates. c) Shells

boxes identify the prescribed directions. Also it is possible to assign a non-zero value to the constraint.

The second button in Figure 12.17 is used for point loads allocation. When selected, an emergent window (Figure 12.19) allows introducing the point load values in the global coordinate system. Then it is necessary to select the nodes were the point load is applied.

Point loads act normal to the beam axis and the plate surface, or in an arbitrary direction for a 3D shell. Point bending moments are defined as positive if they act in an anti-clockwise sense.

The third button in Figure 12.17 is associated to uniformly distributed loads along the element sides and permits to assign this condition on geometry lines. The emergent window (Figure 12.20) allows introducing the value of the side load per unit length (or area). Uniform loads are assumed to act normal to the element surface (or the beam axis).

Material properties

The material properties are defined with the fourth button in Figure 12.17. This leads to the emergent window shown in Figure 12.21 which allows users defining the material parameters associated to each structure

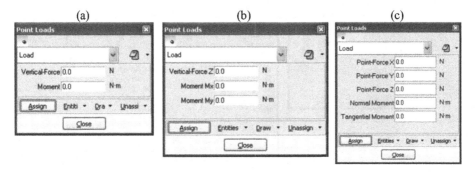

Fig. 12.19 Assign point load. a) Beams. b) Plates. c) Shells

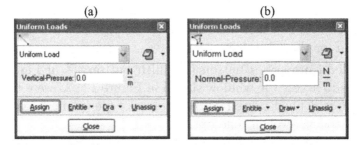

Fig. 12.20 Assign uniformly distributed load. a) Beams. b) Plates and shells

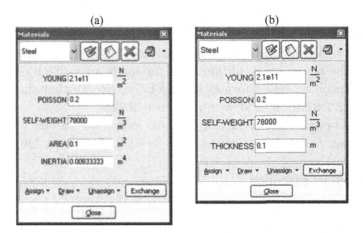

Fig. 12.21 Definition of material properties. a) Beams. b) Plates and shells

like the Young modulus, the Poisson's ratio, the density, the thickness, the transverse cross section, the inertia modulus, etc. It is necessary to assign these properties over the geometry entities that define the analysis domain (lines for beams and axisymmetric shells and surfaces for plates

(a) Problem title   (b) Data file

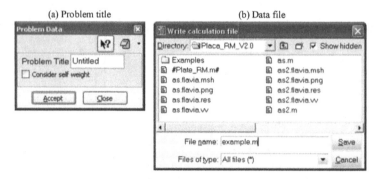

Fig. 12.22 (a) General title of the problem. (b) Writing of data file

and 3D shells). As mentioned earlier, only one type of material is allowed in MAT-fem for the sake of simplicity.

The general properties button (the fifth button of Figure 12.17) allows users to access the window shown in Figure 12.22a were the title of the problem is defined as well as the problem type (plane stress or plane strain) and the self-weight load option.

Once the boundary conditions and the material properties have been defined it is necessary to generate the mesh. The sixth button of Figure 12.17 is used to create the mesh with GiD.

The writing of the data file is made when pressing the last button of Figure 12.17. All the geometrical and material properties of the problem, as well as the boundary conditions and the loads are written on the data file in the specific reading format for MAT-fem. Recall that the file name needs the .m extension as shown in Figure 12.22b.

It is important to remark that the file extension .m can be set only when the box *Files of type* is set to *All files* (*) (Figure 12.22b).

12.11.2 Program execution

The problem calculation is performed with MATLAB® by using the appropriate code (i.e. Beam_EulerBernoulli, Plate_MZC, Shell_QLLL, etc.). The code execution does not have other complications than knowing the directory where the output file will be written. A good practice is to set this directory as the working directory were the postprocessing file will be also written.

During the code execution the total time used by the code will appear in the MATLAB® console as well as the time consumed in each subrou-

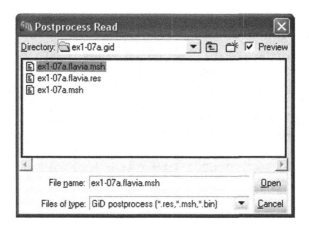

Fig. 12.23 Postprocessing file reading

tine. The largest time consumption in the academic problems solved with MAT-fem is invested in the calculation and assembly of the global stiffness matrix, while the solution of the equations system represents a small percentage of the consumed time. The opposite happens when solving larger structural problems.

Once the code execution is finished, the variables are still recorded inside MATLAB®, thus allowing users experiment with the collection of internal functions available.

12.11.3 Postprocessing

Once the problem execution in MATLAB® is concluded it is necessary to return to GiD for the postprocessing step in order to analyze the results. The next step is to open any of the generated files that contain the extension *.flavia.msh or *.flavia.res (Figure 12.23).

The results visualization step is performed using the GiD graphical possibilities which permit to visualize the results by means of iso-lines, cuts and graphs. This facilitates the interpretation of the MAT-fem results.

The spy(StifMat) command of MATLAB® displays the profile of the global stiffness matrix (see Figure 12.29b). Other MATLAB® commands allow users to find out the properties of this matrix, such as its rank, eigenvectors, determinant, etc.

In the following sections we describe the particular features of the different Mat-fem codes implemented.

```
% Material properties (Constant over the domain).
 D_matb = young*inertia;

 ttim = timing('Time needed to  set initial values',ttim); %Reporting time
```

Fig. 12.24 Constitutive matrix for 2-noded Euler Bernoulli beam element

```
    len = coor_x(2) - coor_x(1);
    const = D_matb/len^3;

    K_elem = [ 12    , 6*len , -12    , 6*len ;
              6*len, 4*len^2, -6*len,  2*len^2;
              -12   ,-6*len ,  12    , -6*len ;
              6*len, 2*len^2, -6*len,  4*len^2];

    K_elem = K_elem * const;

    f       = (-denss*area + uniload(ielem))*len/2;
    ElemFor = [ f, f*len/6, f,-f*len/6];
```

Fig. 12.25 Stiffness matrix and equivalent nodal force vector for 2-noded Euler-Bernoulli beam element

12.12 2-NODED EULER-BERNOUILLI BEAM ELEMENT

The formulation of this beam element can be found in Section 1.3.

We present next the parts of the Beam_EulerBernoulli code for computing the constitutive matrix, the stiffness matrix, the equivalent nodal force vector (for uniformly distributed load) and the bending moments. The rest of the code is identical to that explained in the previous section.

12.12.1 Stiffness matrix and equivalent nodal force vector

The constitutive matrix for the 2-noded Euler-Bernoulli beam element contains the bending stiffness (EI_y) only (Figure 12.24).

The element stiffness matrix is given explicitly in Eq.(1.20). The expression for the equivalent nodal force vector for a uniformly distributed loading is shown in Eq.(1.21b).

The computation of $\mathbf{K}^{(e)}$ (K_elem) and $\mathbf{f}^{(e)}$ (ElemFor) is shown in Figure 12.25.

12.12.2 Computation of bending moment

The bending moment within each element is first computed at the two Gauss points that integrate exactly the element stiffness matrix. The ben-

```
% Two gauss point for bending moment evaluation
  gaus1 =-1/sqrt(3);   % Gauss Point 1
  gaus2 = 1/sqrt(3);   % Gauss Point 2

  a = (1+sqrt(3))/2;
  b = (1-sqrt(3))/2;

  len2 = len^2;

% Bmat and bending moment at Gauss point 1
bmat_1=[6*gaus1/len2,(-1+3*gaus1)/len,-6*gaus1/len2,(1+3*gaus1)/len];
    Str_g1(ielem,1) = D_matb*(bmat_1*transpose(u_elem));

% Bmat and bending moment at Gauss point 2
bmat_2=[6*gaus2/len2,(-1+3*gaus2)/len,-6*gaus2/len2,(1+3*gaus2)/len];
    Str_g2(ielem,2) = D_matb*(bmat_2*transpose(u_elem));%

  Strnod(lnods(1),1) = Strnod(lnods(1),1 +a*Str_g1+b*Str_g2;
  Strnod(lnods(2),1) = Strnod(lnods(2),1)+b*Str_g1+a*Str_g2;
  Strnod(lnods(1),2) = Strnod(lnods(1),2)+1;
  Strnod(lnods(2),2) = Strnod(lnods(2),2)+1;
```

Fig. 12.26 Computation of the bending moment at the two Gauss points for the 2-noded Euler-Bernoulli beam element

ding moment at each Gauss point is computed as

$$M_i = EI_y \mathbf{B}_i \mathbf{a}^{(e)} \quad , \quad i = 1,2$$

where $\mathbf{B}_i^{(e)}$ is the curvature matrix of Eq.(1.16a) computed at the ith Gauss point. The bending moment at the Gauss points is stored in $\texttt{Str\_g1}$ and $\texttt{Str\_g2}$ for the subsequent nodal smoothing and visualization.

The computation of the bending moment is shown in Figure 12.26.

12.12.3 Example. Clamped slender cantilever beam under end point load

Figure 12.27 shows the beam geometry, the material properties and the load. The section is square and the beam slenderness ratio is $r = \frac{L}{h} = 100$.

The exact solution is a unit displacement at the free end and a bending moment of 50 N/m at the clamped end.

Figure 12.27 also shows the menus for defining the boundary conditions, the load and the material properties. The discretization and the writing of the input data file is carried out using the last two bottoms of the menu of Figure 12.17.

The problem has been solved with meshes of 2, 4, 8, 16, 32 and 64 2-noded Euler-Bernoulli beam elements. Figure 12.28 shows the numbering of elements and nodes for the 8 element mesh. This mesh is taken as the reference for the input data file also shown in Figure 12.28.

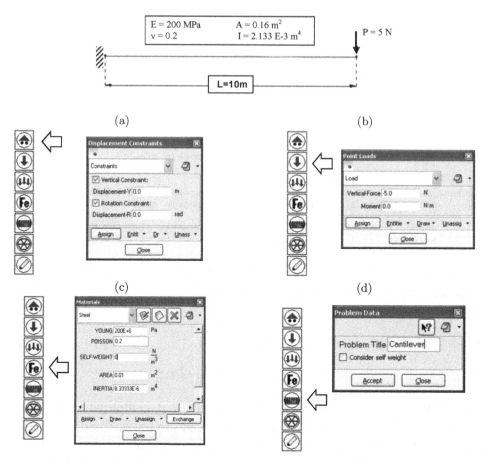

Fig. 12.27 Slender clamped cantilever beam ($r = 100$) under end point load. Definition of (a) boundary conditions, (b) loads, (c) materials and (d) problem characteristics

The code is executed within MATLAB® using the Beam_EulerBernoulli command, once the directory file where this code is located is selected. The total execution time for this problem (8 element mesh) is 0.010 sec.

Note that the assembly of the stiffness matrix takes most of the total execution time, as the internal indices of the sparse matrix need to be updated (Figure 12.29a).

Figure 12.29b shows the profile of the global stiffness matrix, as displayed by MATLAB®.

Figure 12.30 shows the deformed shape of the beam and the bending moment distribution. The *vertical displacement under the force is "exact"*

```
%=========================================================================
% MAT-fem_Beams 1.0  - MAT-fem is a learning tool for undestanding
%                      the Finite Element Method with MATLAB and GiD
%=========================================================================
% PROBLEM TITLE = Slender cantilever beam under end point load. Analysis
%                 with eight 2-noded Euler-Bernoulli beam elements
%  Material Properties
   young   =   2.000000000e+08 ;
   poiss   =   2.000000000e-01 ;
   denss   =   0.000000000e+00 ;
   area    =   1.000000000e-02 ;
   inertia=   8.333330000e-06 ;
% Coordinates
global coordinates
coordinates = [
   0.000000000e+00   ;
   1.250000000e+00   ;
   2.500000000e+00   ;
   3.750000000e+00   ;
   5.000000000e+00   ;
   6.250000000e+00   ;
   7.500000000e+00   ;
   8.750000000e+00   ;
   1.000000000e+01   ] ;
% Elements
global elements
elements = [
        1   ,      2   ;
        2   ,      3   ;
        3   ,      4   ;
        4   ,      5   ;
        5   ,      6   ;
        6   ,      7   ;
        7   ,      8   ;
        8   ,      9   ] ;
% Fixed Nodes
fixnodes = [
        1   , 1 ,   0.000000000e+00   ;
        1   , 2 ,   0.000000000e+00   ] ;
% Point loads
pointload = [
        9   , 1 ,  -5.000000000e+00   ;
        9   , 2 ,   0.000000000e+00   ] ;
% Distributed loads
uniload = sparse ( 8,1 );
```

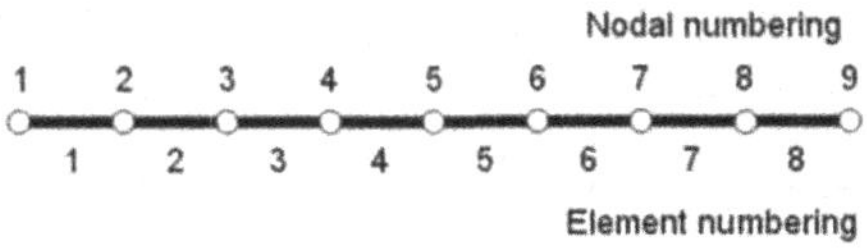

Nodal numbering

Element numbering

Fig. 12.28 Slender cantilever beam ($r = 100$) under end point load. Input data file for eight element mesh of 2-noded Euler-Bernoulli beam elements

for all the meshes considered. This is a particular feature of this problem, as explained in Section 1.3.4 (p. 19).

The bending moment at the clamped node is 50.0 Nxm. This value has been obtained from the reaction at that node which yields the exact solution for all meshes.

12.13 2-NODED TIMOSHENKO BEAM ELEMENT

The formulation of this beam element can be found in Section 2.3.

(a)   (b)

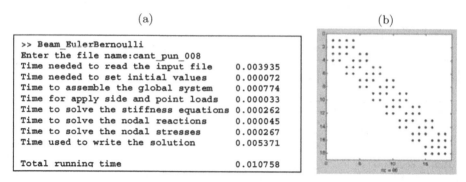

```
>> Beam_EulerBernoulli
Enter the file name:cant_pun_008
Time needed to read the input file       0.003935
Time needed to set initial values        0.000072
Time to assemble the global system       0.000774
Time for apply side and point loads      0.000033
Time to solve the stiffness equations    0.000262
Time to solve the nodal reactions        0.000045
Time to solve the nodal stresses         0.000267
Time used to write the solution          0.005371

Total running time                       0.010758
```

Fig. 12.29 Slender cantilever beam under end point load. (a) Execution time for eight element mesh. (b) Profile of stiffness matrix for the eight element mesh

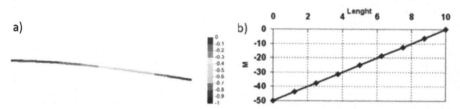

Fig. 12.30 Slender cantilever under end point load analized with eight 2-noded Euler-Bernoulli beam elements. a) Deformed shape. b) Moment distribution

```
%   Material properties (Constant over the domain).
  D_matb = young*inertia;
  D_mats = young/(2*(1+poiss))*area*5/6;

  ttim = timing('Time needed to  set initial values',ttim); %Reporting time
```

Fig. 12.31 2-noded Timoshenko beam element. Constitutive matrices

The structure of the code is very similar to that for the 2-noded Euler-Bernoulli beam element described in the previous section.

12.13.1 Stiffness matrix and equivalent nodal force vector

The constitutive matrix has been split in two parts: the bending term $\hat{D}_b$ (D_matb) and the transverse shear term $\hat{D}_s$ (D_mats). The shear correction factor has been taken equal to 5/6 (rectangular section). The computation of the constitutive matrices D_matb and D_mats are shown in Figure 12.31.

Figure 12.32 shows the explicit computation of the element stiffness matrix by sum of the bending and transverse shear contributions using a single integration point (see Eq.(2.25)).

```
len = coor_x(2) - coor_x(1);
const = D_matb/len;

K_b  = [ 0 ,  0 ,  0 ,  0 ;
         0 ,  1 ,  0 , -1 ;
         0 ,  0 ,  0 ,  0 ;
         0 , -1 ,  0 ,  1 ];          Bending stiffness matrix

K_b  = K_b * const;

const =   D_mats/len;

K_s = [   1   ,  len/2  ,   -1   ,  len/2 ;
        len/2 , len^2/3 , -len/2 , len^2/6 ;    Shear stiffness matrix
         -1   , -len/2  ,   1    , -len/2 ;
        len/2 , len^2/6 , -len/2 , len^2/3 ];

K_s = K_s * const;

K_elem = K_b + K_s;

% Equivalent nodal force vector

f       = (-denss*area + uniload(ielem))*len/2;
ElemFor = [ f, 0, f, 0];
```

Fig. 12.32 2-noded Timoshenko beam element. Computation of stiffness matrix and equivalent nodal force vector for self-weight and uniform load

The last lines of Figure 12.32 show the computation of the equivalent nodal force vector for a uniform load (uniload) and self-weight (denss*area) using a single integration point.

12.13.2 Computation of bending moment and shear force

Figure 12.33 shows the computation of the bending moment M and the shear force Q at the element center via Eq.(2.17).

The bending moment and the shear force are stored in Strnod for the subsequent representation.

12.13.3 Example. Thick cantilever beam under end-point load

Figure 12.34 shows the beam geometry of the cantilever, the material properties and the load. The section is square and the beam slenderness ratio is $r = \frac{L}{h} = 25$.

Table 12.6 shows the convergence of the deflection under the load and the bending moment and the shear force at the clamped end with the number of elements. The values for M_1 and Q_1 reported are the extrapolation of the values at the elements center. Note that M_1 and Q_1 can be

```
% One Gauss point for resultant stresses evaluation
   gaus0 =  0.0;
   bmat_b=[ 0, -1/len, 0, 1/len];
   bmat_s1=[-1/len,-(1-gaus0)/2, 1/len,-(1+gaus0)/2];

% Resultant stresses at Gauss points
   Str1_g0 = D_matb*(bmat_b *transpose(u_elem));
   Str2_g0 = D_mats*(bmat_s1*transpose(u_elem));

% Nodal extrapolation of resultant stresses
   Strnod(lnods(1),1) = Strnod(lnods(1),1)+Str1_g0;
   Strnod(lnods(2),1) = Strnod(lnods(2),1)+Str1_g0;
   Strnod(lnods(1),2) = Strnod(lnods(1),2)+Str2_g0;
   Strnod(lnods(2),2) = Strnod(lnods(2),2)+Str2_g0;
   Strnod(lnods(1),3) = Strnod(lnods(1),3)+1;
   Strnod(lnods(2),3) = Strnod(lnods(2),3)+1;
```

Fig. 12.33 2-noded Timoshenko beam element. Computation of bending moment M and shear force Q at the element center and the element nodes

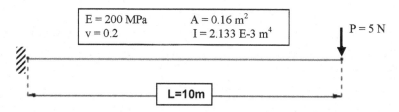

Fig. 12.34 Thick cantilever beam ($r = 25$) under end point load

Thick cantilever beam under end point load 2-noded Timoshenko beam element					
Elements	Nodes	DOF	w_N	M_1	Q_1
2	3	6	-0,00007	-0,68	5.00
4	5	10	-0,00027	-3,00	5.00
8	9	18	-0,00089	-10,68	5.00
16	17	34	-0,00212	-26,21	5.00
32	33	66	-0,00323	-40,61	5.00
64	65	130	-0,00371	-47,11	5.00
128	129	258	-0,00386	-49,15	5.00
Exact [Ti2]:			-0,00391	-50.00	5.00

Table 12.6 Thick cantilever beam ($r = 25$) under end point load analyzed with 2-noded Timoshenko beam elements. Convergence of free end deflection (w_N) and bending moment (M_1) and shear force (Q_1) at the clamped node

directly computed as the reactions at the clamped node *which would yield the exact solution* for all meshes.

```
%  Material properties (constant over the domain).

aux0 = thick^3 / 12 ;
aux1 = aux0*young/(1-poiss^2);
aux2 = poiss*aux1;
aux3 = aux0*young/2/(1+poiss);

D_matb  = [aux1,aux2,   0;
           aux2,aux1,   0;
              0,   0,aux3];
```

Fig. 12.35 Bending constitutive matrix for thin plate elements

```
gauss_x(1) =-1/sqrt(3);
gauss_y(1) =-1/sqrt(3);
gauss_x(2) = 1/sqrt(3);
gauss_y(2) =-1/sqrt(3);
gauss_x(3) = 1/sqrt(3);
gauss_y(3) = 1/sqrt(3);
gauss_x(4) =-1/sqrt(3);
gauss_y(4) = 1/sqrt(3);
```

Fig. 12.36 Local coordinates of Gauss points

12.14 4-NODED MZC THIN PLATE RECTANGLE

This plate element was studied in Section 5.4.1.

12.14.1 Element stiffness matrix and equivalent nodal force vector

Figure 12.35 shows the subroutine for computing the constitutive matrix for thin plate elements (Eq.(5.15b)) including bending terms only.

The Gauss point coordinates in the local axes s, t are shown in Figure 12.36. Recall that the weights for this quadrature are the unity.

Figure 12.37 shows the subroutine for computing $\mathbf{K}^{(e)}$ and $\mathbf{f}^{(e)}$ for the MZC plate rectangle. A 2×2 Gauss quadrature is used for the integration of $\mathbf{K}^{(e)}$. The computation of the bending strain matrix is shown in Figure 12.38.

Indeed, for this element the analytical expressions for the bending stiffness matrix shown in Box 5.1 could have been used directly. For didactic reasons, however, we have preferred to implement the numerical integration of $\mathbf{K}^{(e)}$ using a 2×2 Gauss quadrature.

The last lines of Figure 12.37 show the computation of the equivalent nodal force vector for self-weight (`denss*thick`) and a uniformly distributed load (`uniload`). The expression of $\mathbf{f}^{(e)}$ coincides with Eq.(5.47a).

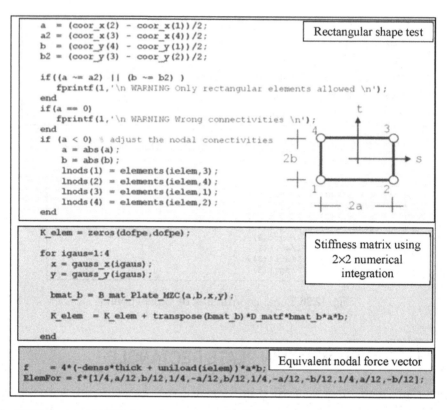

```
a   = (coor_x(2) - coor_x(1))/2;          Rectangular shape test
a2  = (coor_x(3) - coor_x(4))/2;
b   = (coor_y(4) - coor_y(1))/2;
b2  = (coor_y(3) - coor_y(2))/2;

if((a ~= a2) || (b ~= b2) )
    fprintf(1,'\n WARNING Only rectangular elements allowed \n');
end
if(a == 0)
    fprintf(1,'\n WARNING Wrong connectivities \n');
end
if (a < 0)  % adjust the nodal conectivities
    a = abs(a);
    b = abs(b);
    lnods(1) = elements(ielem,3);
    lnods(2) = elements(ielem,4);
    lnods(3) = elements(ielem,1);
    lnods(4) = elements(ielem,2);
end
```

```
K_elem = zeros(dofpe,dofpe);              Stiffness matrix using
                                          2×2 numerical
for igaus=1:4                             integration
    x = gauss_x(igaus);
    y = gauss_y(igaus);

    bmat_b = B_mat_Plate_MZC(a,b,x,y);

    K_elem = K_elem + transpose(bmat_b)*D_matf*bmat_b*a*b;

end
```

```
f      = 4*(-denss*thick + uniload(ielem))*a*b;   Equivalent nodal force vector
ElemFor = f*[1/4,a/12,b/12,1/4,-a/12,b/12,1/4,-a/12,-b/12,1/4,a/12,-b/12];
```

Fig. 12.37 4-noded MZC thin plate rectangle. Stiffness matrix and equivalent nodal force vector

12.14.2 Computation of bending moments

Figure 12.39 shows the subroutine for computing the bending moments for the 4-noded MZC plate rectangle. The moments are computed at the 2×2 Gauss point in Str1 and then are stored in Strnod for the subsequent nodal smoothing.

12.14.3 Example. Clamped thin square plate under uniform loading

Figure 12.40 shows the plate geometry, the material properties and the uniform load value. The analytical values for the deflection and the bending moments at the center are -0.12653 and -2.31, respectively. Units are in the International System (SI).

Figure 12.41 show the menus for introducing the boundary conditions, the material properties and the load.

```
function bmat = B_mat_Plate_MZC(a,b,x,y)

    d2N(1,1) = 3*( x - x*y )/(4*a^2);
    d2N(2,1) = 3*(-x + x*y )/(4*a^2);
    d2N(3,1) = 3*(-x - x*y )/(4*a^2);
    d2N(4,1) = 3*( x + x*y )/(4*a^2);
    d2N(1,2) = 3*( y - x*y )/(4*b^2);
    d2N(2,2) = 3*( y + x*y )/(4*b^2);
    d2N(3,2) = 3*(-y - x*y )/(4*b^2);
    d2N(4,2) = 3*(-y + x*y )/(4*b^2);
    d2N(1,3) = 2*(  1/2 - 3*x^2/8 - 3*y^2/8)/(a*b);
    d2N(2,3) = 2*( -1/2 + 3*x^2/8 + 3*y^2/8)/(a*b);
    d2N(3,3) = 2*(  1/2 - 3*x^2/8 - 3*y^2/8)/(a*b);
    d2N(4,3) = 2*( -1/2 + 3*x^2/8 + 3*y^2/8)/(a*b);
```

Evaluation of second derivatives of shape functions $\mathbf{N}$

```
    d2NN(1,1) = ( (3*a*x - 3*a*x*y - a + a*y)/4 )/a^2;
    d2NN(2,1) = ( (3*a*x - 3*a*x*y + a - a*y)/4 )/a^2;
    d2NN(3,1) = ( (3*a*x + 3*a*x*y + a + a*y)/4 )/a^2;
    d2NN(4,1) = ( (3*a*x + 3*a*x*y - a - a*y)/4 )/a^2;
    d2NN(1,2) = 0;
    d2NN(2,2) = 0;
    d2NN(3,2) = 0;
    d2NN(4,2) = 0;
    d2NN(1,3) = 2*( -3/8*a*x^2 + a*x/4 + a/8 )/(a*b);
    d2NN(2,3) = 2*( -3/8*a*x^2 - a*x/4 + a/8 )/(a*b);
    d2NN(3,3) = 2*(  3/8*a*x^2 + a*x/4 - a/8 )/(a*b);
    d2NN(4,3) = 2*(  3/8*a*x^2 - a*x/4 - a/8 )/(a*b);
```

Evaluation of second derivatives of shape functions $\overline{\mathbf{N}}$

```
    d2NNN(1,1) = 0;
    d2NNN(2,1) = 0;
    d2NNN(3,1) = 0;
    d2NNN(4,1) = 0;
    d2NNN(1,2) = ( (3*b*y - 3*b*x*y - b + b*x)/4 )/b^2;
    d2NNN(2,2) = ( (3*b*y + 3*b*x*y - b - b*x)/4 )/b^2;
    d2NNN(3,2) = ( (3*b*y + 3*b*x*y + b + b*x)/4 )/b^2;
    d2NNN(4,2) = ( (3*b*y - 3*b*x*y + b - b*x)/4 )/b^2;
    d2NNN(1,3) = 2*( -3/8*b*y^2 + b*y/4 + b/8 )/(a*b);
    d2NNN(2,3) = 2*(  3/8*b*y^2 - b*y/4 - b/8 )/(a*b);
    d2NNN(3,3) = 2*(  3/8*b*y^2 + b*y/4 - b/8 )/(a*b);
    d2NNN(4,3) = 2*( -3/8*b*y^2 - b*y/4 + b/8 )/(a*b);
```

Evaluation of second derivatives of shape functions $\overline{\overline{\mathbf{N}}}$

```
    bmat_1  = [ -d2N(1,1),-d2NN(1,1),-d2NNN(1,1)  ;
                -d2N(1,2),-d2NN(1,2),-d2NNN(1,2)  ;
                -d2N(1,3),-d2NN(1,3),-d2NNN(1,3)] ;
    bmat_2  = [ -d2N(2,1),-d2NN(2,1),-d2NNN(2,1)  ;
                -d2N(2,2),-d2NN(2,2),-d2NNN(2,2)  ;
                -d2N(2,3),-d2NN(2,3),-d2NNN(2,3)] ;
    bmat_3  = [ -d2N(3,1),-d2NN(3,1),-d2NNN(3,1)  ;
                -d2N(3,2),-d2NN(3,2),-d2NNN(3,2)  ;
                -d2N(3,3),-d2NN(3,3),-d2NNN(3,3)] ;
    bmat_4  = [ -d2N(4,1),-d2NN(4,1),-d2NNN(4,1)  ;
                -d2N(4,2),-d2NN(4,2),-d2NNN(4,2)  ;
                -d2N(4,3),-d2NN(4,3),-d2NNN(4,3)] ;

    bmat = [bmat_1,bmat_2,bmat_3,bmat_4];
```

Assembly of $\mathbf{B}_b$ matrix

Fig. 12.38 Bending strain matrix for the MZC plate rectangle

```
% Shape function matrix for extrapolation of bending moments to nodes
    aa = 1 + sqrt(3);
    bb = 1 - sqrt(3);
    mstres = [ aa*aa , aa*bb , bb*bb , aa*bb ;
               bb*aa , aa*aa , aa*bb , bb*bb ;
               bb*bb , aa*bb , aa*aa , bb*aa ;
               aa*bb , bb*bb , bb*aa , aa*aa ]/4;

% Bending moments at Gauss point
    for igaus=1:4
        x = gauss_x(igaus);
        y = gauss_y(igaus);

        bmat = B_mat_Plate_MZC(a,b,x,y);

        Str1=D_matb*bmat*transpose(u_elem);

        Strx(igaus)   = Str1(1);
        Stry(igaus)   = Str1(2);
        Strxy(igaus)  = Str1(3);

    End

% Nodal extrapolation of bending moments
    Str1 = mstres * transpose(Strx) ;
    Strnod(lnods(1:4),1) = Strnod(lnods(1:4),1)+ Str1(1:4);
    Str1 = mstres * transpose(Stry) ;
    Strnod(lnods(1:4),2) = Strnod(lnods(1:4),2)+ Str1(1:4);
    Str1 = mstres * transpose(Strxy) ;
    Strnod(lnods(1:4),3) = Strnod(lnods(1:4),3)+ Str1(1:4);
    Strnod(lnods(1:4),4) = Strnod(lnods(1:4),4)+ 1;
```

Fig. 12.39 Computation of bending moments for the MZC plate rectangle

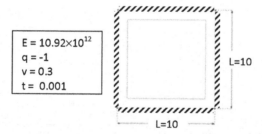

E = 10.92×10^{12}
q = -1
v = 0.3
t = 0.001

L=10

L=10

Fig. 12.40 Clamped thin square plate under uniformly distributed loading.Units are in the SI system

The discretization and the writing of the data file is performed with the last two bottoms of the MAT-fem-Plates menu. The problem has been solved with several meshes ranging from 2×2 to 20×20 MZC elements.

Figure 12.42 shows the input data file for the 2×2 mesh.

Table 12.7 shows the convergence of the central deflection and the bending moment M_x at the plate center with the number of elements. No advantage has been taken of the symmetry of the problem.

Figure 12.43 shows the contour plots for the vertical deflection and the bending moment M_x on the plate for a 10×10 element mesh.

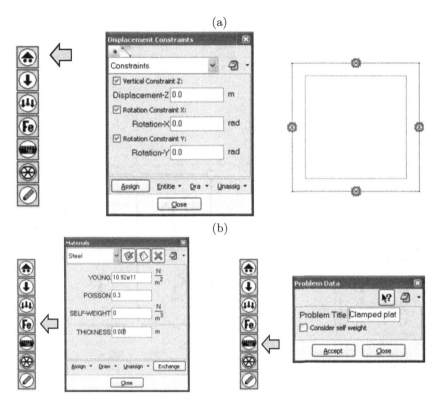

Fig. 12.41 MZC Definition of boundary conditions (a), material properties and distributed load (b) for the clamped thin square plate

Thin clamped square plate under uniform loading			
Mesh of MZC elements	Nodes	ω_c	M_{x_c}
2 × 2	9	-0.15625	-4.88
4 × 4	25	-0.14077	-2.80
6 × 6	49	-0.13333	-2.50
8 × 8	81	-0.13043	-2.41
10 × 10	121	-0.12904	-2.36
12 × 12	169	-0.12828	-2.34
14 × 14	225	-0.12782	-2.33
16 × 16	289	-0.12752	-2.32
18 × 18	361	-0.12731	-2.31
20 × 20	441	-0.12716	-2.31
Exact [TW]:		-0.12653	-2.31

Table 12.7 Thin clamped square plate under uniform load. Convergence of central deflection ω_c and central bending moment M_{x_c} using the MZC plate rectangle

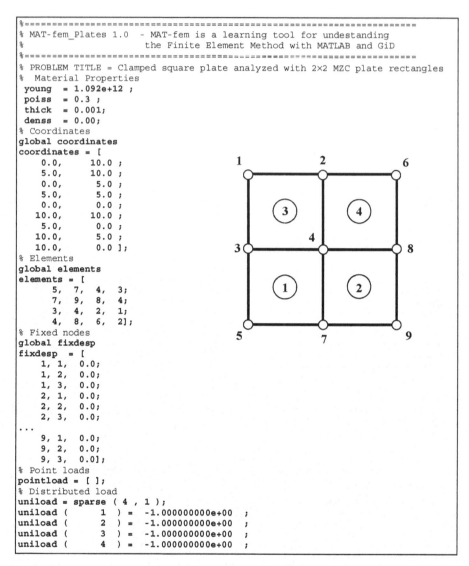

```
%=====================================================================
% MAT-fem_Plates 1.0   - MAT-fem is a learning tool for undestanding
%                        the Finite Element Method with MATLAB and GiD
%=====================================================================
% PROBLEM TITLE = Clamped square plate analyzed with 2×2 MZC plate rectangles
%  Material Properties
 young   = 1.092e+12 ;
 poiss   = 0.3 ;
 thick   = 0.001;
 denss   = 0.00;
% Coordinates
global coordinates
coordinates = [
     0.0,      10.0 ;
     5.0,      10.0 ;
     0.0,       5.0 ;
     5.0,       5.0 ;
     0.0,       0.0 ;
    10.0,      10.0 ;
     5.0,       0.0 ;
    10.0,       5.0 ;
    10.0,       0.0 ];
% Elements
global elements
elements = [
     5,   7,   4,   3;
     7,   9,   8,   4;
     3,   4,   2,   1;
     4,   8,   6,   2];
% Fixed nodes
global fixdesp
fixdesp  = [
     1,  1,   0.0;
     1,  2,   0.0;
     1,  3,   0.0;
     2,  1,   0.0;
     2,  2,   0.0;
     2,  3,   0.0;
 ...
     9,  1,   0.0;
     9,  2,   0.0;
     9,  3,   0.0];
% Point loads
pointload = [ ];
% Distributed load
uniload = sparse ( 4 , 1 );
uniload (      1  ) = -1.000000000e+00  ;
uniload (      2  ) = -1.000000000e+00  ;
uniload (      3  ) = -1.000000000e+00  ;
uniload (      4  ) = -1.000000000e+00  ;
```

Fig. 12.42 Input data for clamped square plate analyzed with 2×2 MZC plate rectangles

12.15 Q4 REISSNER-MINDLIN PLATE RECTANGLE

We present the key subroutines for the 4-noded Q4 Reissner-Mindlin (RM) plate element of Section 6.5.1 *in its rectangular form*. The stiffness matrix is computed with selective integration, i.e. a 2×2 quadrature for the bending stiffness terms and a reduced one point quadrature for the shear

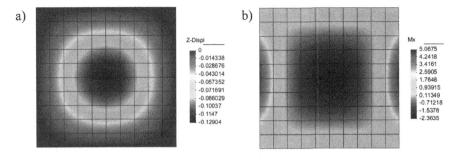

Fig. 12.43 Thin clamped square plate under uniform load. Contours of vertical deflection w (a) and M_x (b) for a mesh of 10×10 MZC plate rectangles

```
aux4 = (5/6)*thick*young/2/(1+poiss);

D_mats = [aux4,   0 ;
             0, aux4];
```

Fig. 12.44 Shear constitutive matrix for Reissner-Mindlin plate elements

stiffness terms. These subroutines can be easily extended for programming higher order RM plate rectangles based on selective integration techniques, as described in Section 6.5. The extension to non-rectangular shapes is straightforward using an isoparametric formulation. An isoparametric Q4 RM plate quadrilateral has been implemented in the `Plate_Q4_Iso` code.

12.15.1 Stiffness matrix and equivalent nodal force vector

The bending terms in the constitutive matrix coincide with those of Figure 12.35. The shear constitutive matrix $\hat{\mathbf{D}}_s$ of Eq.(6.24) is shown in Figure 12.44.

Figure 12.45 shows the subroutine for computing the stiffness matrix and the equivalent nodal force vector for the Q4 RM plate rectangle. Note the two loops for computing the bending and shear stiffness matrices using selective integration.

Figure 12.46 shows the computation of the generalized bending and shear strain matrices.

The last two rows of Figure 12.45 show the computation of the equivalent nodal force for self-weight (`dense*thick`) and a uniformly distributed load (`uniload`). As the element is a rectangle the nodal forces are simply computed as one fourth of the total force acting over the element.

```
% Local Gauss point coordinates

gauss_x(1) =-1/sqrt(3);
gauss_y(1) =-1/sqrt(3);
gauss_x(2) = 1/sqrt(3);
gauss_y(2) =-1/sqrt(3);
gauss_x(3) = 1/sqrt(3);
gauss_y(3) = 1/sqrt(3);
gauss_x(4) =-1/sqrt(3);
gauss_y(4) = 1/sqrt(3);
```

```
a  = (coor_x(2) - coor_x(1))/2;
a2 = (coor_x(3) - coor_x(4))/2;              Rectangular shape test
b  = (coor_y(4) - coor_y(1))/2;
b2 = (coor_y(3) - coor_y(2))/2;

if((a ~= a2) || (b ~= b2) )
    fprintf(1,'\n WARNING Only rectangular elements allowed \n');
end
if(a == 0)
    fprintf(1,'\n WARNING Wrong connectivities \n');
end
if (a < 0)  % adjust the nodal conectivities
    a = abs(a);
    b = abs(b);
    lnods(1) = elements(ielem,3);
    lnods(2) = elements(ielem,4);
    lnods(3) = elements(ielem,1);
    lnods(4) = elements(ielem,2);
end
```

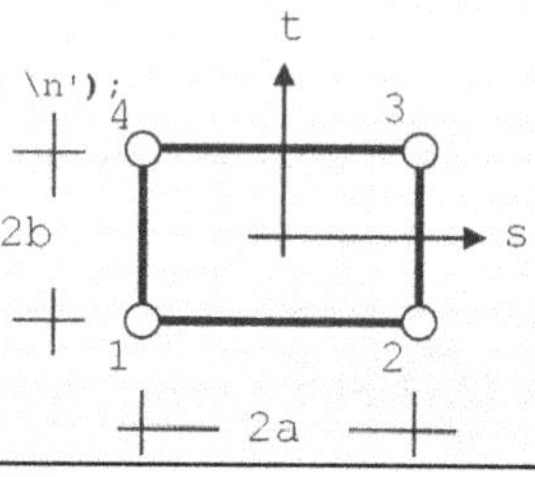

```
K_elem = zeros(dofpe,dofpe);
```

```
for igaus=1:4
    x = gauss_x(igaus);
    y = gauss_y(igaus);                       Numerical integration
                                              of stiffness matrix
    [bmat_b, ~] = B_mat_Plate_Q4_v2_3(a,b,x,y);
    K_b   = transpose(bmat_b)*D_matb*bmat_b*a*b;
    K_elem = K_elem + K_b;
end

% One gauss point for shear
x = 0;
y = 0;
[~,bmat_s] = B_mat_Plate_Q4_v2_3(a,b,x,y);
K_s   = transpose(bmat_s)*D_mats*bmat_s*4*a*b;
K_elem = K_elem + K_s;
```

```
f       = 4*(-denss*thick + uniload(ielem))*a*b;   Equivalent nodal
ElemFor = f*[1/4,0,0,1/4,0,0,1/4,0,0,1/4,0,0];       force vector
```

Fig. 12.45 Q4 Reissner-Mindlin plate rectangle. Stiffness matrix and equivalent nodal force vector

12.15.2 Computation of resultant stresses

Figure 12.47 shows the subroutine for computing the resultant stresses for the Q4 RM plate rectangle. The bending moment (Str1) are computed at the 2×2 Gauss points, while the shear forces (Str2) are computed at

```
function [bmat_b,bmat_s] = B_mat_Plate_Q4(a,b,x,y)

    x = x*a;
    y = y*b;

    N(1) = (1 - x/a)*(1 - y/b)/4;
    N(2) = (1 + x/a)*(1 - y/b)/4;
    N(3) = (1 + x/a)*(1 + y/b)/4;
    N(4) = (1 - x/a)*(1 + y/b)/4;

    dxN(1) = -( b - y )/(4*a*b);
    dxN(2) =  ( b - y )/(4*a*b);
    dxN(3) =  ( b + y )/(4*a*b);
    dxN(4) = -( b + y )/(4*a*b);

    dyN(1) = -( a - x )/(4*a*b);
    dyN(2) = -( a + x )/(4*a*b);
    dyN(3) =  ( a + x )/(4*a*b);
    dyN(4) =  ( a - x )/(4*a*b);
```

```
Bending strain matrix

    bmat_b1 = [ 0,-dxN(1),       0  ;
                0,       0,-dyN(1) ;
                0,-dyN(1),-dxN(1)];

    bmat_b2 = [ 0,-dxN(2),       0  ;
                0,       0,-dyN(2) ;
                0,-dyN(2),-dxN(2)];

    bmat_b3 = [ 0,-dxN(3),       0  ;
                0,       0,-dyN(3) ;
                0,-dyN(3),-dxN(3)];

    bmat_b4 = [ 0,-dxN(4),       0  ;
                0,       0,-dyN(4) ;
                0,-dyN(4),-dxN(4)];

    bmat_b = [bmat_b1,bmat_b2,bmat_b3,bmat_b4];
```

```
Transverse shear strain matrix

    bmat_s1 = [ dxN(1), -N(1),      0  ;
                dyN(1),     0, -N(1)];

    bmat_s2 = [ dxN(2), -N(2),      0  ;
                dyN(2),     0, -N(2)];

    bmat_s3 = [ dxN(3), -N(3),      0  ;
                dyN(3),     0, -N(3)];

    bmat_s4 = [ dxN(4), -N(4),      0  ;
                dyN(4),     0, -N(4)];
```

Fig. 12.46 Q4 Reissner-Mindlin plate rectangle. Generalized bending and transverse shear strain matrices

the element center. The resultant stresses are accumulated in `Strnod` for the subsequent nodal smoothing.

12.15.3 Example. Thick clamped square plate under uniformly load

The plate geometry, load and material properties are identical to those of Figure 12.40, with the exception of the thickness that now is $t = 1.0$.

```
% Shape Function matrix for extrapolation of resultant stresses to nodes
  aa = 1 + sqrt(3);
  bb = 1 - sqrt(3);
  mstres = [ aa*aa , aa*bb , bb*bb , aa*bb ;
             bb*aa , aa*aa , aa*bb , bb*bb ;
             bb*bb , aa*bb , aa*aa , bb*aa ;
             aa*bb , bb*bb , bb*aa , aa*aa ]/4;

    a  = (coor_x(2) - coor_x(1))/2;
    b  = (coor_y(4) - coor_y(1))/2;

% Resultant stresses at Gauss point
    for igaus=1:4
      x = gauss_x(igaus);
      y = gauss_y(igaus);
      [bmat_b, ~] = B_mat_Plate_Q4_v2_3(a,b,x,y);

      Str1=D_matb*bmat_b*transpose(u_elem);

      Strx(igaus)  = Str1(1);
      Stry(igaus)  = Str1(2);
      Strxy(igaus) = Str1(3);
    end

% Resultant stresses at nodes
    x = 0;
    y = 0;
    [~,bmat_s] = B_mat_Plate_Q4_v2_3(a,b,x,y);
    Str2=D_mats*bmat_s*transpose(u_elem);
    StrQx(1:4) = Str2(1);
    StrQy(1:4) = Str2(2);

    Str1 = mstres * transpose(Strx) ;
    Strnod(lnods(1:4),1) = Strnod(lnods(1:4),1)+ Str1(1:4);
    Str1 = mstres * transpose(Stry) ;
    Strnod(lnods(1:4),2) = Strnod(lnods(1:4),2)+ Str1(1:4);
    Str1 = mstres * transpose(Strxy) ;
    Strnod(lnods(1:4),3) = Strnod(lnods(1:4),3)+ Str1(1:4);
    Str2 = mstres * transpose(StrQx) ;
    Strnod(lnods(1:4),4) = Strnod(lnods(1:4),4)+ Str2(1:4);
    Str2 = mstres * transpose(StrQy) ;
    Strnod(lnods(1:4),5) = Strnod(lnods(1:4),5)+ Str2(1:4);
    Strnod(lnods(1:4),6) = Strnod(lnods(1:4),6)+ 1;
```

Fig. 12.47 Q4 Reissner-Mindlin plate rectangle. Computation of bending moments and shear forces

The reference solution for this problem is: central deflection, $w_c = -0.1505 \times 10^{-9}$, central bending moment $Mx_c = -2.31$ and maximum shear force at the clamped edge $Q_{y_m} = 4.12$ [TW].

The data input process is the same as for the MZC rectangle (Section 12.4). The input data file is similar as that shown in Figure 12.42.

Table 12.8 lists the results for w_c, M_{x_c} and Q_{y_m} for different meshes of Q4 RM plate rectangles.

12.16 QLLL REISSNER-MINDLIN PLATE QUADRILATERAL

We present the main parts of the Mat-fem code for the 4-noded QLLL Reissner-Mindlin plate quadrilateral studied in Section 6.7.1. The *element is derived using an isoparametric formulation* and, therefore, is not restricted to rectangular shapes.

Q4 Reissner Mindlin plate rectangle				
Mesh	Nodes	w_c	Mx_c	Q_{y_m}
2 × 2	9	-0.3571E-09	-0.00	1.250
4 × 4	25	-0.1458E-09	-2.262	2.524
6 × 6	49	-0.1486E-09	-2.408	3.159
8 × 8	81	-0.1494E-09	-2.339	3.374
10 × 10	121	-0.1498E-09	-2.337	3.526
12 × 12	169	-0.1500E-09	-2.331	3.626
14 × 14	225	-0.1501E-09	-2,329	3.700
16 × 16	289	-0.1502E-09	-2,327	3.753
18 × 18	361	-0.1502E-09	-2,325	3.796
20 × 20	441	-0.1503E-09	-2,324	3.830
Exact [TW]:		-0.1504E-09	-2.310	4.120

Table 12.8 Thick clamped square plate under uniform load. Central deflection w_c, central bending moment Mx_c and maximum shear force at the clamped edge Q_{y_m} for different meshes of Q4 Reissner-Mindlin plate rectangles

12.16.1 Stiffness matrix and equivalent nodal force vector

The *constitutive matrix* includes the bending and shear contributions as defined in Figures 12.35 and 12.44.

Figure 12.48 shows the subroutine for computing the bending and shear stiffness matrices (termed K_b and K_s, respectively). A 2 × 2 Gauss quadrature is used for the numerical integration.

The expression for the bending stiffness coincides precisely with $\mathbf{K}_b^{(e)}$ given in Eq.(6.39). The computation of the bending strain matrix is shown in Figure 12.49.

The shear stiffness matrix is obtained by substituting matrix $\mathbf{B}_s$ by $\bar{\mathbf{B}}_s$ in the expression of $\mathbf{K}_s^{(e)}$ of Eq.(6.39b). The derivation of the substitute shear strain matrix $\bar{\mathbf{B}}_s$ for the QLLL element is detailed in Section 6.7.1. The subroutine for computing $\bar{\mathbf{B}}_s$ is shown in Figure 12.50.

12.16.2 Computation of resultant stresses

Figure 12.51 shows the subroutine for computing the bending moments (Str1) and the shear forces (Str2) at the 2 × 2 Gauss point in the QLLL RM plate quadrilateral.

The resultant stresses are accumulated in Strnod for the subsequent smoothing.

```
K_elem = zeros(dofpe,dofpe);

for igaus=1:4
    x = gauss_x(igaus);
    y = gauss_y(igaus);

    [bmat_b,bmat_s,area] = B_mat_Plate_QLLL(x,y,coor_x,coor_y);

    K_b   = transpose(bmat_b)*D_matb*bmat_b*area;
    K_s   = transpose(bmat_s)*D_mats*bmat_s*area;

    K_elem = K_elem + K_f + K_s;
end
```
Numerical integration of stiffness matrix

```
f       = 4*(-denss*thick + uniload(ielem))*area;
ElemFor = f*[1/4,0,0,1/4,0,0,1/4,0,0,1/4,0,0];
```
Equivalent nodal force vector

Fig. 12.48 4-noded QLLL Reissner-Mindlin plate quadrilateral. Computation of the stiffness matrix and the equivalent nodal force vector

QLLL plate quadrilateral					TQQL plate triangle				
Mesh	Nodes	w_c	M_{x_c}	Q_{y_m}	Mesh	Nodes	w_c	M_{x_c}	Q_{y_m}
2 × 2	9	-0.026E-09	-0.000	1.875	2 × 2	25	-0.63E-09	-0.97	3.62
4 × 4	25	-0.143E-09	-2.364	3.253	4 × 4	81	-0.202E-09	-1.11	5.40
6 × 6	49	-0.147E-09	-2.367	3.331	6 × 6	169	-0.197E-09	-1.84	4.65
8 × 8	81	-0.148E-09	-2.343	3.494	8 × 8	289	-0.182E-09	-2.07	4.32
10 × 10	121	-0.149E-09	-2.334	3.602	10 × 10	441	-0.173E-09	-2.16	4.21
12 × 12	169	-0.149E-09	-2.330	3.680	12 × 12	625	-0.167E-09	-2.21	4.16
14 × 14	225	-0.149E-09	-2.327	3.740	14 × 14	841	-0.163E-09	-2.24	4.14
16 × 16	289	-0.150E-09	-2,325	3.786	16 × 16	1089	-0.160E-09	-2,26	4.13
18 × 18	361	-0.150E-09	-2,324	3.822	18 × 18	1369	-0.158E-09	-2,27	4.12
20 × 20	441	-0.150E-09	-2,323	3.852	20 × 20	1681	-0.157E-09	-2,28	4.11
Exact [TW]:		-0.150E-09	-2.31	4.55	Exact [TW]:		-0.150E-09	-2.31	4.55

Table 12.9 Thick clamped square plate under uniform loading. Results for deflection w_c and bending moment M_{x_c} at the plate center and maximum shear force Q_{y_m} at the clamped edge for different meshes of QLLL and TQQL plate elements

12.16.3 Example. Thick clamped plate under uniform distributed load

The example coincides with that used in Section 12.15.3 for testing the Q4 RM plate rectangle.

Table 12.9 shows the results for the deflection w_c and the bending moment Mx_c at the plate center and the maximum shear force Q_{y_m} at the clamped edge for different meshes of QLLL elements. Results for the same problem solved with TQQL elements are presented.

Note the excellent accuracy for the deflection and bending moment values for relatively coarse meshes.

```
function [bmat_b,bmat_s,area] = B_mat_Plate_QLLL(xgs,ygs,x,y)

dxN1(1) = (-1+ygs)/4;
dxN1(2) = ( 1-ygs)/4;
dxN1(3) = ( 1+ygs)/4;                    Cartesian
dxN1(4) = (-1-ygs)/4;                    derivatives of the
                                          shape functions
dyN1(1) = (-1+xgs)/4;
dyN1(2) = (-1-xgs)/4;
dyN1(3) = ( 1+xgs)/4;
dyN1(4) = ( 1-xgs)/4;

xjacm(1,1) = x(1)*dxN1(1) + x(2)*dxN1(2) + x(3)*dxN1(3) + x(4)*dxN1(4);
xjacm(1,2) = y(1)*dxN1(1) + y(2)*dxN1(2) + y(3)*dxN1(3) + y(4)*dxN1(4);
xjacm(2,1) = x(1)*dyN1(1) + x(2)*dyN1(2) + x(3)*dyN1(3) + x(4)*dyN1(4);
xjacm(2,2) = y(1)*dyN1(1) + y(2)*dyN1(2) + y(3)*dyN1(3) + y(4)*dyN1(4);

xjaci = inv(xjacm);

area = abs(xjacm(1,1)*xjacm(2,2) - xjacm(2,1)*xjacm(1,2));

dxN(1) = xjaci(1,1)*dxN1(1)+xjaci(1,2)*dyN1(1);
dxN(2) = xjaci(1,1)*dxN1(2)+xjaci(1,2)*dyN1(2);
dxN(3) = xjaci(1,1)*dxN1(3)+xjaci(1,2)*dyN1(3);
dxN(4) = xjaci(1,1)*dxN1(4)+xjaci(1,2)*dyN1(4);

dyN(1) = xjaci(2,1)*dxN1(1)+xjaci(2,2)*dyN1(1);
dyN(2) = xjaci(2,1)*dxN1(2)+xjaci(2,2)*dyN1(2);
dyN(3) = xjaci(2,1)*dxN1(3)+xjaci(2,2)*dyN1(3);
dyN(4) = xjaci(2,1)*dxN1(4)+xjaci(2,2)*dyN1(4);
%==================
bmat_b1 = [ 0,-dxN(1),      0   ;
            0,      0,-dyN(1)  ;           Bending strain
            0,-dyN(1),-dxN(1)];            matrix B_b
bmat_b2 = [ 0,-dxN(2),      0   ;
            0,      0,-dyN(2)  ;
            0,-dyN(2),-dxN(2)];
bmat_b3 = [ 0,-dxN(3),      0   ;
            0,      0,-dyN(3)  ;
            0,-dyN(3),-dxN(3)];
bmat_b4 = [ 0,-dxN(4),      0   ;
            0,      0,-dyN(4)  ;
            0,-dyN(4),-dxN(4)];

bmat_b = [bmat_b1,bmat_b2,bmat_b3,bmat_b4];
```

Fig. 12.49 Computation of the bending strain matrix for the 4-noded QLLL plate quadrilateral

12.17 TQQL REISSNER-MINDLIN PLATE TRIANGLE

The TQQL 6-noded plate element was studied in Section 6.8.1. We present next the main subroutines for programming the element using an isoparametric formulation.

12.17.1 Stiffness matrix and equivalent nodal force vector

Figure 12.52 shows the local coordinates of the 3-point Gauss quadrature used for the numerical integration of all the element matrices.

```
cx = [ 0 , 1 , 0 , -1 ];                    Coordinates for shear collocation points
cy = [-1 , 0 , 1 ,  0 ];

c     = zeros(8,8);
b_bar = [];
for i = 1 : 4
    N(1) = (1-cx(i))*(1-cy(i))/4 ;
    N(2) = (1+cx(i))*(1-cy(i))/4 ;
    N(3) = (1+cx(i))*(1+cy(i))/4 ;
    N(4) = (1-cx(i))*(1+cy(i))/4 ;

    dxN1(1) = (-1+cy(i))/4;
    dxN1(2) = ( 1-cy(i))/4;            Local shape functions and derivatives
    dxN1(3) = ( 1+cy(i))/4;                at shear collocation points
    dxN1(4) = (-1-cy(i))/4;

    dyN1(1) = (-1+cx(i))/4;
    dyN1(2) = (-1-cx(i))/4;
    dyN1(3) = ( 1+cx(i))/4;
    dyN1(4) = ( 1-cx(i))/4;

    xjacm(1,1) = x(1)*dxN1(1) + x(2)*dxN1(2) + x(3)*dxN1(3) + x(4)*dxN1(4);
    xjacm(1,2) = y(1)*dxN1(1) + y(2)*dxN1(2) + y(3)*dxN1(3) + y(4)*dxN1(4);
    xjacm(2,1) = x(1)*dyN1(1) + x(2)*dyN1(2) + x(3)*dyN1(3) + x(4)*dyN1(4);
    xjacm(2,2) = y(1)*dyN1(1) + y(2)*dyN1(2) + y(3)*dyN1(3) + y(4)*dyN1(4);

    jpos = [ i*2-1 , i*2 ];
    c(jpos,jpos) = xjacm;

    bmat_s1 = [ dxN(1), -N(1),    0  ;
                dyN(1),    0, -N(1)];
                                                    Shear strain matrix at
    bmat_s2 = [ dxN(2), -N(2),    0  ;              each collocation point
                dyN(2),    0, -N(2)];

    bmat_s3 = [ dxN(3), -N(3),    0  ;
                dyN(3),    0, -N(3)];

    bmat_s4 = [ dxN(4), -N(4),    0  ;
                dyN(4),    0, -N(4)];

    bmat_s = [bmat_s1,bmat_s2,bmat_s3,bmat_s4];

    b_bar = [ b_bar ;
              bmat_s];
end

T_mat = [ 1 , 0 , 0 , 0 , 0 , 0 , 0 , 0 ;
          0 , 0 , 0 , 1 , 0 , 0 , 0 , 0 ;                  T matrix
          0 , 0 , 0 , 0 , 1 , 0 , 0 , 0 ;
          0 , 0 , 0 , 0 , 0 , 0 , 0 , 1 ];

P_mat = [ 1 , -1 ,  0 ,  0 ;
          0 ,  0 ,  1 ,  1 ;                               P matrix
          1 ,  1 ,  0 ,  0 ;
          0 ,  0 ,  1 , -1 ];

A_mat = [ 1 , ygs , 0 ,  0 ;                               A matrix
          0 ,   0 , 1 , xgs ];

bmat_s = xjaci * A_mat * inv(P_mat) * T_mat * c * b_bar;
                                                Sustitutive transverse shear
                                                       strain matrix
```

Fig. 12.50 4-noded QLLL plate quadrilateral. Computation of the substitute transverse shear strain matrix $\bar{\mathbf{B}}_s$

```
% Shape Function matrix for extrapolation of resultant stresses to nodes
  aa = 1 + sqrt(3);
  bb = 1 - sqrt(3);
  mstres = [ aa*aa , aa*bb , bb*bb , aa*bb ;
             bb*aa , aa*aa , aa*bb , bb*bb ;
             bb*bb , aa*bb , aa*aa , bb*aa ;
             aa*bb , bb*bb , bb*aa , aa*aa ]/4;

% Resultant stresses at Gauss points
    for igaus=1:4
      x = gauss_x(igaus);
      y = gauss_y(igaus);

      [bmat_b,bmat_s,area] = B_mat_Plate_QLLL(x,y,coor_x,coor_y);

      Str1=D_matb*bmat_b*transpose(u_elem);
      Str2=D_mats*bmat_s*transpose(u_elem);

      Strx(igaus)  = Str1(1);
      Stry(igaus)  = Str1(2);
      Strxy(igaus) = Str1(3);
      StrQx(igaus) = Str2(1);
      StrQy(igaus) = Str2(2);
    end

% Resultant stresses at nodes
    Str1 = mstres * transpose(Strx) ;
    Strnod(lnods(1:4),1) = Strnod(lnods(1:4),1)+ Str1(1:4);
    Str1 = mstres * transpose(Stry) ;
    Strnod(lnods(1:4),2) = Strnod(lnods(1:4),2)+ Str1(1:4);
    Str1 = mstres * transpose(Strxy) ;
    Strnod(lnods(1:4),3) = Strnod(lnods(1:4),3)+ Str1(1:4);
    Str2 = mstres * transpose(StrQx) ;
    Strnod(lnods(1:4),4) = Strnod(lnods(1:4),4)+ Str2(1:4);
    Str2 = mstres * transpose(StrQy) ;
    Strnod(lnods(1:4),5) = Strnod(lnods(1:4),5)+ Str2(1:4);
    Strnod(lnods(1:4),6) = Strnod(lnods(1:4),6)+ 1;

    for i = 1 : npnod                          Nodal averaging of
      Strnod(i,1:5) = Strnod(i,1:5)/Strnod(i,6);   resultant stresses
    end
```

Fig. 12.51 Computation of resultant stresses for the QLLL plate quadrilateral

Figure 12.52 shows also the subroutine for computing the bending and shear stiffness matrices and the equivalent nodal force vector (ElemFor) for self-weight (denss*thick) and uniformly distributed load (uniload).

The bending stiffness matrix (K_b) is computed by Eq.(6.39). The computation of the bending strain matrix (bmat_b) is detailed in Figure 12.53.

The transverse shear stiffness matrix is obtained by using the substitute transverse shear strain matrix $\bar{\mathbf{B}}_s$ (Section 6.8.1) in the expression of $\mathbf{K}_s^{(e)}$ of Eq.(6.39b). The computation of $\bar{\mathbf{B}}_s$ is shown in Figure 12.54.

12.17.2 Computation of stress resultants

The bending moments (Str1) and the shear forces (Str2) are computed first at the 3 Gauss point used for the numerical integration of the stiffness

```
% Local coordinate of Gauss point
  xg(1) = 1.0/6.0;
  xg(2) = 2.0/3.0;
  xg(3) = 1.0/6.0;
  yg(1) = 1.0/6.0;
  yg(2) = 1.0/6.0;
  yg(3) = 2.0/3.0;

  K_elem = zeros(dofpe,dofpe);
  f      = zeros(1,6);
  f_e = -denss*thick + uniload(ielem);
```

```
  for igaus=1:3

    [bmat_b,bmat_s,N,area] =
    B_mat_Plate_TCCL_v1_0(x,y,xg(igaus),yg(igaus))

    K_b  = transpose(bmat_b)*D_matb*bmat_b*area*wg(igaus);
    K_s  = transpose(bmat_s)*D_mats*bmat_s*area*wg(igaus);

    K_elem  = K_elem + K_b + K_s;
```
| Numerical integration of stiffness matrix |

```
    f = f + f_e* area*wg(igaus)* N;
```
| Equivalent nodal force vector |

```
  end

  ElemFor = [f(1),0,0,f(2),0,0,f(3),0,0,f(4),0,0,f(5),0,0,f(6),0,0];
```

Fig. 12.52 TQQL plate triangle. Computation of stiffness matrix and equivalent nodal force vector

matrix. The Gauss point values for the resultant stresses are accumulated in **Strnod** for the subsequent nodal smoothing (Figure 12.55).

12.17.3 Example. Thick clamped square plate under uniform load

The example coincides with that used in Section 12.15.3 for testing the performance of the Q4 plate rectangle.

Table 12.9 in p. 768 shows the results for the deflection w_c and the bending moment M_{x_c} at the plate center and the maximum shear force Q_{y_m} at the clamped edge for different meshes of TQQL elements.

Note the excellent accuracy for the deflection and the bending moment values for relatively coarse meshes.

12.18 4-NODED QLLL FLAT SHELL ELEMENT

We present the key parts of the Mat-fem code for the 4-noded QLLL flat shell element. This element was studied in Chapter 8 (Sections 8.3–8.11). The element can be used for analysis of any 3D shell structure by discretizing the shell surface into 4-noded quadrilaterals.

The programming of other flat shell elements follows very similar steps as those described for the 4-noded QLLL flat shell element.

```
function [bmat_b,bmat_s,Ngp,area] = B_mat_Plate_TCCL_v1_0(x,y,xgs,ygs)
%==================
   L1=1.0-xgs-ygs;
   L2=xgs;
   L3=ygs;

   dxNloc(1)=1.0-4.0*L1;
   dxNloc(2)=4.0*(L1-L2);
   dxNloc(3)=4.0*L2-1.0;
   dxNloc(4)=4.0*L3;
   dxNloc(5)=0.0;
   dxNloc(6)=-4.0*L3;

   dyNloc(1)=1.0-4.0*L1;
   dyNloc(2)=-4.0*L2;
   dyNloc(3)=0.0;
   dyNloc(4)=4.0*L2;
   dyNloc(5)=4.0*L3-1.0;
   dyNloc(6)=4.0*(L1-L3);
```

Local derivatives of the quadratic shape functions

```
   xjacm(1,1) = x*dxNloc';
   xjacm(1,2) = y*dxNloc';
   xjacm(2,1) = x*dyNloc';
   xjacm(2,2) = y*dyNloc';

   xjaci = inv(xjacm);

   area2 = abs(xjacm(1,1)*xjacm(2,2) - xjacm(2,1)*xjacm(1,2));
   area  = area2/2;
```

Jacobian matrix definition

```
   dxN(1) = xjaci(1,1)*dxNloc(1)+xjaci(1,2)*dyNloc(1);
   dxN(2) = xjaci(1,1)*dxNloc(2)+xjaci(1,2)*dyNloc(2);
   dxN(3) = xjaci(1,1)*dxNloc(3)+xjaci(1,2)*dyNloc(3);
   dxN(4) = xjaci(1,1)*dxNloc(4)+xjaci(1,2)*dyNloc(4);
   dxN(5) = xjaci(1,1)*dxNloc(5)+xjaci(1,2)*dyNloc(5);
   dxN(6) = xjaci(1,1)*dxNloc(6)+xjaci(1,2)*dyNloc(6);

   dyN(1) = xjaci(2,1)*dxNloc(1)+xjaci(2,2)*dyNloc(1);
   dyN(2) = xjaci(2,1)*dxNloc(2)+xjaci(2,2)*dyNloc(2);
   dyN(3) = xjaci(2,1)*dxNloc(3)+xjaci(2,2)*dyNloc(3);
   dyN(4) = xjaci(2,1)*dxNloc(4)+xjaci(2,2)*dyNloc(4);
   dyN(5) = xjaci(2,1)*dxNloc(5)+xjaci(2,2)*dyNloc(5);
   dyN(6) = xjaci(2,1)*dxNloc(6)+xjaci(2,2)*dyNloc(6);
```

Cartesian derivatives of the shape functions.

```
   bmat_b1 = [ 0,-dxN(1),    0   ;
               0,     0,-dyN(1) ;
               0,-dyN(1),-dxN(1)];
   bmat_b2 = [ 0,-dxN(2),    0   ;
               0,     0,-dyN(2) ;
               0,-dyN(2),-dxN(2)];
   bmat_b3 = [ 0,-dxN(3),    0   ;
               0,     0,-dyN(3) ;
               0,-dyN(3),-dxN(3)];
   bmat_b4 = [ 0,-dxN(4),    0   ;
               0,     0,-dyN(4) ;
               0,-dyN(4),-dxN(4)];
   bmat_b5 = [ 0,-dxN(5),    0   ;
               0,     0,-dyN(5) ;
               0,-dyN(5),-dxN(5)];
   bmat_b6 = [ 0,-dxN(6),    0   ;
               0,     0,-dyN(6) ;
               0,-dyN(6),-dxN(6)];

   bmat_b = [bmat_b1,bmat_b2,bmat_b3,bmat_b4,bmat_b5,bmat_b6];
```

Bending strain matrix $\mathbf{B}_h$

Fig. 12.53 TQQL plate triangle. Computation of bending strain matrix

```
r3 = 1/sqrt(3);     ap = 0.5 - r3;
cx = [ 0.5-r3 , 0.5+r3,    ap,   1-ap ,    0 ,    0 ];
cy = [    0 ,    0 , 1.0-ap,    ap , 0.5+r3, 0.5-r3];

c     = zeros(12,12);
b_bar = [];                                     Coordinates for shear collocations points

for i = 1 : 6
  L1=1.0-cx(i)-cy(i);
  L2=cx(i);
  L3=cy(i);
  N(1) = (2*L1-1)*L1;
  N(2) = 4*L1*L2;                               Local shape functions and their derivates
  N(3) = (2*L2-1)*L2;
  N(4) = 4*L2*L3;
  N(5) = (2*L3-1)*L3;
  N(6) = 4*L1*L3;
  dxNloc(1)=1.0-4.0*L1;      dyNloc(1)=1.0-4.0*L1;
  dxNloc(2)=4.0*(L1-L2);     dyNloc(2)=-4.0*L2;
  dxNloc(3)=4.0*L2-1.0;      dyNloc(3)=0.0;
  dxNloc(4)=4.0*L3;          dyNloc(4)=4.0*L2;
  dxNloc(5)=0.0;             dyNloc(5)=4.0*L3-1.0;
  dxNloc(6)=-4.0*L3;         dyNloc(6)=4.0*(L1-L3);

  xjacm(1,1) = x*dxNloc';
  xjacm(1,2) = y*dxNloc';
  xjacm(2,1) = x*dyNloc';                        Jacobian definition and global
  xjacm(2,2) = y*dyNloc';                         shape function derivates

  xjacip = inv(xjacm);

  dxN(1) = xjacip(1,1)*dxNloc(1)+xjacip(1,2)*dyNloc(1);
  dxN(2) = xjacip(1,1)*dxNloc(2)+xjacip(1,2)*dyNloc(2);
  dxN(3) = xjacip(1,1)*dxNloc(3)+xjacip(1,2)*dyNloc(3);
  dxN(4) = xjacip(1,1)*dxNloc(4)+xjacip(1,2)*dyNloc(4);
  dxN(5) = xjacip(1,1)*dxNloc(5)+xjacip(1,2)*dyNloc(5);
  dxN(6) = xjacip(1,1)*dxNloc(6)+xjacip(1,2)*dyNloc(6);

  dyN(1) = xjacip(2,1)*dxNloc(1)+xjacip(2,2)*dyNloc(1);
  dyN(2) = xjacip(2,1)*dxNloc(2)+xjacip(2,2)*dyNloc(2);
  dyN(3) = xjacip(2,1)*dxNloc(3)+xjacip(2,2)*dyNloc(3);
  dyN(4) = xjacip(2,1)*dxNloc(4)+xjacip(2,2)*dyNloc(4);
  dyN(5) = xjacip(2,1)*dxNloc(5)+xjacip(2,2)*dyNloc(5);
  dyN(6) = xjacip(2,1)*dxNloc(6)+xjacip(2,2)*dyNloc(6);

  jpos = [ i*2-1 , i*2 ];
  c(jpos,jpos) = xjacm;
  bmat_s1 = [ dxN(1), -N(1),    0  ;
              dyN(1),    0, -N(1)];
  bmat_s2 = [ dxN(2), -N(2),    0  ;
              dyN(2),    0, -N(2)];
  bmat_s3 = [ dxN(3), -N(3),    0  ;
              dyN(3),    0, -N(3)];            Shear strain matrix at
  bmat_s4 = [ dxN(4), -N(4),    0  ;           each collocation point
              dyN(4),    0, -N(4)];
  bmat_s5 = [ dxN(5), -N(5),    0  ;
              dyN(5),    0, -N(5)];
  bmat_s6 = [ dxN(6), -N(6),    0  ;
              dyN(6),    0, -N(6)];
  bmat_s = [bmat_s1,bmat_s2,bmat_s3,bmat_s4,bmat_s5,bmat_s6];
  b_bar = [ b_bar ;
            bmat_s];
end
```

```
a = sqrt(2)/2;                                           T matrix
T_mat = [ 1,  0,  0,  0,  0,  0,  0,  0,  0,  0,  0,  0 ;
          0,  0,  1,  0,  0,  0,  0,  0,  0,  0,  0,  0 ;
          0,  0,  0,  0, -a,  a,  0,  0,  0,  0,  0,  0 ;
          0,  0,  0,  0,  0,  0, -a,  a,  0,  0,  0,  0 ;
          0,  0,  0,  0,  0,  0,  0,  0,  0,  1,  0,  0 ;
          0,  0,  0,  0,  0,  0,  0,  0,  0,  0,  0,  1 ];

P_mat = [ 1,    cx(1),   cy(1),  0,      0,      0  ;
          1,    cx(2),   cy(2),  0,      0,      0  ;            P matrix
         -a,-a*cx(3),-a*cy(3),   a, a*cx(3), a*cy(3) ;
         -a,-a*cx(4),-a*cy(4),   a, a*cx(4), a*cy(4) ;
          0,     0,      0,  1,  cx(5),  cy(5) ;
          0,     0,      0,  1,  cx(6),  cy(6) ];

A_mat = [ 1 , xgs, ygs, 0 ,  0 ,  0 ;
          0 ,  0,   0, 1 , xgs, ygs ];                          A matrix

bmat_s = xjaci * A_mat * inv(P_mat) * T_mat * c * b_bar;
                                        Substitutive shear strain matrix
```

Fig. 12.54 Computation of the substitute transverse shear strain matrix $\bar{\mathbf{B}}_s$ for the TQQL plate triangle

```
% Shape function matrix for extrapolation of resultant stresses to nodes
  mstres = [ 5 , -1 , -1 ;
            -1 ,  5 , -1 ;
            -1 , -1 ,  5 ]/3;
  g_or = [1,3,5]; % Vertex nodes

% Resultant stresses at Gauss points
    for igaus=1:3

        [bmat_b,bmat_s] = B_mat_Plate_TCCL_v1_0(x,y,xg(igaus),yg(igaus));

        Str1=D_matb*bmat_b*transpose(u_elem);
        Str2=D_mats*bmat_s*transpose(u_elem);

        Strx(igaus)  = Str1(1);
        Stry(igaus)  = Str1(2);
        Strxy(igaus) = Str1(3);
        StrQx(igaus) = Str2(1);
        StrQy(igaus) = Str2(2);
    end

% Resultant stresses at nodes
    Str1 = mstres * transpose(Strx) ;
    Strnod(lnods(g_or(1:3)),1) = Strnod(lnods(g_or(1:3)),1)+ Str1(1:3);
    Str1 = mstres * transpose(Stry) ;
    Strnod(lnods(g_or(1:3)),2) = Strnod(lnods(g_or(1:3)),2)+ Str1(1:3);
    Str1 = mstres * transpose(Strxy) ;
    Strnod(lnods(g_or(1:3)),3) = Strnod(lnods(g_or(1:3)),3)+ Str1(1:3);
    Str2 = mstres * transpose(StrQx) ;
    Strnod(lnods(g_or(1:3)),4) = Strnod(lnods(g_or(1:3)),4)+ Str2(1:3);
    Str2 = mstres * transpose(StrQy) ;
    Strnod(lnods(g_or(1:3)),5) = Strnod(lnods(g_or(1:3)),5)+ Str2(1:3);
    Strnod(lnods(g_or(1:3)),6) = Strnod(lnods(g_or(1:3)),6)+ 1;
```

```
    for i = 1 : npnod
        Strnod(i,1:5) = Strnod(i,1:5)/Strnod(i,6);     Nodal averaging of
    end                                                resultant stresses
```

Fig. 12.55 Computation of resultant stresses for the TQQL plate triangle

12.18.1 Generalized constitutive matrix

The local generalized constitutive matrix contains the membrane, bending and transverse shear contributions (Eqs.(8.18)). For simplicity, the coupling membrane-bending constitutive matrix $\hat{\mathbf{D}}'_{mb}$ will be neglected.

The computation of the generalized constitutive matrix is shown in Figure 12.56.

12.18.2 Stiffness matrix and equivalent nodal force vector

The first step in the computation of the element stiffness matrix is the definition of the local coordinate system x', y', z'.

The normal vector $\mathbf{v}_{z'}$ is computed as the cross product of vectors joining nodes 1,2 and 1,3. Vector $\mathbf{v}_{x'}$ is found by intersecting the element plane with the global xz plane as described in Section 8.8.2. Vector $\mathbf{v}_{y'}$ is finally found by cross product of $\mathbf{v}_{x'}$ and $\mathbf{v}_{z'}$.

```
aux1 = thick*young/(1-poiss^2);
aux2 = poiss*aux1;
aux3 = thick*young/2/(1+poiss);
```

`D_matm = [aux1,aux2,    0;`
 `        aux2,aux1,    0;`
 `           0,   0,aux3];`
	Membrane

`D_matb = D_matm*(thick^2/12);`	Bending

`aux4 = (5/6)*thick*young/2/(1+poiss);`
 `D_mats = [aux4,   0 ;`
 `            0,aux4];`
	Transverse shear

Fig. 12.56 4-noded QLLL flat shell element. Local generalized constitutive matrices

Figure 12.57 shows the computation of vectors $\mathbf{v}_{x'}$, $\mathbf{v}_{y'}$, $\mathbf{v}_{z'}$. These vectors are grouped in matrix Te.

All of the stiffness matrix terms are computed with by a 2×2 Gauss quadrature. The computation requires first transforming the global coordinates of the nodes to the local axes. The membrane and bending stiffness matrices are computed using matrices $\mathbf{B}'_m$ and $\mathbf{B}'_b$ of Eqs.(8.31). The transverse shear stiffness matrix is computed using the substitute transverse shear strain matrix $\bar{\mathbf{B}}'_s$ following the procedure explained for the QLLL plate element and shown in Figure 12.50.

The global stiffness matrix for the element is directly found by transforming the local generalized strain matrices to global axes as shown in Eqs.(8.47) and (8.48).

Figures 12.58 and 12.59 respectively show the steps for computing of the membrane and bending strain matrices and the substitute transverse shear strain matrix *in global axes*.

12.18.3 Computation of local resultant stresses

The computation of the local resultant stresses at the 2×2 Gauss points within an element follows the procedure explained in Eqs.(8.49).

The extrapolation of the Gauss point values to the nodes can be performed using Eq.(12.1). However, the nodal averaging of the local resultant stresses is not straightfogrward as the local axes of adjacent elements are not necessarily the same. An alternative is to transform the nodal resultant stresses contributed by each element to global axes, perform the nodal averaging and then transform back to a nodal coordinate system.

Figure 12.60 shows the nodal averaging of the local stresses assuming a negligible change of the local axes between adjacent elements.

```
function  Te = Rotation_system (cxyz)

    v12(1) = cxyz(2,1) - cxyz(1,1);
    v12(2) = cxyz(2,2) - cxyz(1,2);
    v12(3) = cxyz(2,3) - cxyz(1,3);

    v13(1) = cxyz(3,1) - cxyz(1,1);
    v13(2) = cxyz(3,2) - cxyz(1,2);
    v13(3) = cxyz(3,3) - cxyz(1,3);

    vze(1) = v12(2)*v13(3) -v12(3)*v13(2);
    vze(2) = v12(3)*v13(1) -v12(1)*v13(3);
    vze(3) = v12(1)*v13(2) -v12(2)*v13(1);

    dz = sqrt(vze(1)^2+vze(2)^2+vze(3)^2);

    % Unit vector normal to element surface
    vze(1) = vze(1) / dz;
    vze(2) = vze(2) / dz;
    vze(3) = vze(3) / dz;

    % XZ plane intesection with element surface
    vxe(1) =  1/sqrt(1+(vze(1)/vze(3))^2);
    vxe(2) =  0;
    vxe(3) = -1/sqrt(1+(vze(3)/vze(1))^2);

    dd = vxe(1)*vze(1) + vxe(3)*vze(3);
    if(abs(dd) > 1e-8)
        vxe(3) = -vxe(3);
    end

    if ((vze(3) == 0) && (vze(1) == 0))
      vxe(1) =  1;
      vxe(2) =  0;
      vxe(3) =  0;
    end

    % Vector product.
    vye(1) = vze(2)*vxe(3) - vxe(2)*vze(3);
    vye(2) = vze(3)*vxe(1) - vxe(3)*vze(1);
    vye(3) = vze(1)*vxe(2) - vxe(1)*vze(2);

    dy = sqrt(vye(1)^2+vye(2)^2+vye(3)^2);
    vye(1) = vye(1) / dy;
    vye(2) = vye(2) / dy;
    vye(3) = vye(3) / dy;

    Te = [ vxe(1), vxe(2), vxe(3) ;
           vye(1), vye(2), vye(3) ;
           vze(1), vze(2), vze(3) ];
```

Normal vector $\mathbf{v}_{z'}$ definition

Definition of vector $\mathbf{v}_{x'}$

Definition of vector $\mathbf{v}_{y'}$

Fig. 12.57 QLLL flat shell element. Definition of local axes

12.18.4 Examples

12.18.4.1 Clamped hyperbolic shell under uniform loading

We present results for the analysis of a thick hyperbolic shell clamped at the four edges under a vertical uniformly distributed load. The geometry, material properties and the load are shown in Figure 12.61a.

The problem has been solved with different meshes of QLLL flat shell elements. No advantage of the symmetry of the problem has been taken. Figures 12.61b show snapshots of the data input menus. The general struc-

```
function [bmat_b,bmat_m,bmat_s,area]=B_mat_Shell_CLLL(xgs,ygs,x,y,Te)

dxN1(1) = (-1+ygs)/4;   dyN1(1) = (-1+xgs)/4;      ┌─────────────────────────┐
dxN1(2) = ( 1-ygs)/4;   dyN1(2) = (-1-xgs)/4;      │ Shape function derivatives│
dxN1(3) = ( 1+ygs)/4;   dyN1(3) = ( 1+xgs)/4;      │ in local coordinate system│
dxN1(4) = (-1-ygs)/4;   dyN1(4) = ( 1-xgs)/4;      └─────────────────────────┘

xjacm(1,1) = x(1)*dxN1(1) + x(2)*dxN1(2) + x(3)*dxN1(3) + x(4)*dxN1(4);
xjacm(1,2) = y(1)*dxN1(1) + y(2)*dxN1(2) + y(3)*dxN1(3) + y(4)*dxN1(4);
xjacm(2,1) = x(1)*dyN1(1) + x(2)*dyN1(2) + x(3)*dyN1(3) + x(4)*dyN1(4);
xjacm(2,2) = y(1)*dyN1(1) + y(2)*dyN1(2) + y(3)*dyN1(3) + y(4)*dyN1(4);

xjaci = inv(xjacm);

area = abs(xjacm(1,1)*xjacm(2,2) - xjacm(2,1)*xjacm(1,2));

dxN(1) = xjaci(1,1)*dxN1(1)+xjaci(1,2)*dyN1(1);
dxN(2) = xjaci(1,1)*dxN1(2)+xjaci(1,2)*dyN1(2);    ┌──────────────────┐
dxN(3) = xjaci(1,1)*dxN1(3)+xjaci(1,2)*dyN1(3);    │   Isoparametric  │
dxN(4) = xjaci(1,1)*dxN1(4)+xjaci(1,2)*dyN1(4);    │  transformation  │
dyN(1) = xjaci(2,1)*dxN1(1)+xjaci(2,2)*dyN1(1);    └──────────────────┘
dyN(2) = xjaci(2,1)*dxN1(2)+xjaci(2,2)*dyN1(2);
dyN(3) = xjaci(2,1)*dxN1(3)+xjaci(2,2)*dyN1(3);
dyN(4) = xjaci(2,1)*dxN1(4)+xjaci(2,2)*dyN1(4);

bmat_b1  = [ 0, 0, 0,-dxN(1),      0  ;
             0, 0, 0,      0,-dyN(1) ;
             0, 0, 0,-dyN(1),-dxN(1)];
bmat_b2  = [ 0, 0, 0,-dxN(2),      0  ;
             0, 0, 0,      0,-dyN(2) ;      ┌─────────────────────┐
             0, 0, 0,-dyN(2),-dxN(2)];      │ Bending strain matrix│
bmat_b3  = [ 0, 0, 0,-dxN(3),      0  ;     └─────────────────────┘
             0, 0, 0,      0,-dyN(3) ;
             0, 0, 0,-dyN(3),-dxN(3)];
bmat_b4  = [ 0, 0, 0,-dxN(4),      0  ;
             0, 0, 0,      0,-dyN(4) ;
             0, 0, 0,-dyN(4),-dxN(4)];

bmat_b = [bmat_b1,bmat_b2,bmat_b3,bmat_b4];

bmat_m1d  = [ dxN(1),      0,  0 ;
                   0, dyN(1),  0 ;
              dyN(1), dxN(1),  0];
bmat_m2d  = [ dxN(2),      0,  0 ;
                   0, dyN(2),  0 ;          ┌──────────────────────┐
              dyN(2), dxN(2),  0];          │ Membrane strain matrix│
bmat_m3d  = [ dxN(3),      0,  0 ;          └──────────────────────┘
                   0, dyN(3),  0 ;
              dyN(3), dxN(3),  0];
bmat_m4d  = [ dxN(4),      0,  0 ;
                   0, dyN(4),  0 ;
              dyN(4), dxN(4),  0];

bmat_mir  = [ 0, 0 ;
              0, 0 ;
              0, 0];

bmat_m1 = [bmat_m1d*Te,bmat_mir];
bmat_m2 = [bmat_m2d*Te,bmat_mir];
bmat_m3 = [bmat_m3d*Te,bmat_mir];
bmat_m4 = [bmat_m4d*Te,bmat_mir];

bmat_m = [bmat_m1,bmat_m2,bmat_m3,bmat_m4];
```

Fig. 12.58 QLLL flat shell element. Computation of membrane and bending strain matrices in global axes

```
%== Colocation points:

cx = [ 0 , 1 , 0 , -1 ];          Coordinates for shear collocation points
cy = [-1 , 0 , 1 , 0 ];

c     = zeros(8,8);
b_bar = [];
for i = 1 : 4

    . . .

    bmat_s1  = [ dxN(1), -N(1),     0  ;
                 dyN(1),     0, -N(1)];
    bmat_s2  = [ dxN(2), -N(2),     0  ;
                 dyN(2),     0, -N(2)];          Shear strain matrix at
    bmat_s3  = [ dxN(3), -N(3),     0  ;         each collocation point
                 dyN(3),     0, -N(3)];
    bmat_s4  = [ dxN(4), -N(4),     0  ;
                 dyN(4),     0, -N(4)];

    bmat_s = [bmat_s1,bmat_s2,bmat_s3,bmat_s4];

    b_bar = [ b_bar ;
              bmat_s];

end

T_mat = [ 1 , 0 , 0 , 0 , 0 , 0 , 0 , 0 ;
          0 , 0 , 0 , 1 , 0 , 0 , 0 , 0 ;          T matrix
          0 , 0 , 0 , 0 , 1 , 0 , 0 , 0 ;
          0 , 0 , 0 , 0 , 0 , 0 , 0 , 1 ];
P_mat = [ 1 , -1 , 0 , 0 ;
          0 , 0 , 1 , 1 ;                          P matrix
          1 , 1 , 0 , 0 ;
          0 , 0 , 1 , -1 ];
A_mat = [ 1 , ygs , 0 , 0 ;
          0 , 0 , 1 , xgs ];                       A matrix

bmat_ss = xjaci * A_mat * inv(P_mat) * T_mat * c * b_bar;

bmat_s1 = [0 , 0 ,bmat_ss(1, 1);
           0 , 0 ,bmat_ss(2, 1)];
bmat_s2 = [0 , 0 ,bmat_ss(1, 4);
           0 , 0 ,bmat_ss(2, 4)];          Sustitutive shear strain matrix
bmat_s3 = [0 , 0 ,bmat_ss(1, 7);          for global displacements and
           0 , 0 ,bmat_ss(2, 7)];                   local rotations
bmat_s4 = [0 , 0 ,bmat_ss(1,10);
           0 , 0 ,bmat_ss(2,10)];

bmat_s1 = [bmat_s1*Te,bmat_ss(:, 2: 3)];
bmat_s2 = [bmat_s2*Te,bmat_ss(:, 5: 6)];
bmat_s3 = [bmat_s3*Te,bmat_ss(:, 8: 9)];
bmat_s4 = [bmat_s4*Te,bmat_ss(:,11:12)];

bmat_s = [bmat_s1,bmat_s2,bmat_s3,bmat_s4];
```

Fig. 12.59 QLLL flat shell element. Computation of substitute transverse shear strain matrix in global axes

ture of the input data file for the 4×4 element mesh is shown in Figure 12.62.

The reference solutions found with a mesh of 100×100 QLLL flat shell elements yield a central deflection of $w_c = -0.0245$, a bending moment at the center of $M_{x'} = 0.39$ and a shear force and the center of the clamped edge of $Q_{z'} = 0.15$ (units in International System).

```
% Shape Function matrix for extrapolation of resultant stresses to nodes
aa = 1 + sqrt(3);
bb = 1 - sqrt(3);
mstres = [ aa*aa , aa*bb , bb*bb , aa*bb ;
           bb*aa , aa*aa , aa*bb , bb*bb ;
           bb*bb , aa*bb , aa*aa , bb*aa ;
           aa*bb , bb*bb , bb*aa , aa*aa ]/4;
% Resultant stresses at Gauss points
for igs = 1 : 4
    x = gauss_x(igaus);
    y = gauss_y(igaus);
    [bmat_b,bmat_m,bmat_s,area]=B_mat_Shell_QLLL_v1_1(x,y,coor_x,coor_y,Te);

    Str1=D_matb*bmat_b*transpose(u_elem);
    Str2=D_matm*bmat_m*transpose(u_elem);
    Str3=D_mats*bmat_s*transpose(u_elem);

    StrMx(igaus)   = Str1(1);
    StrMy(igaus)   = Str1(2);
    StrMxy(igaus)  = Str1(3);
    StrNx(igaus)   = Str2(1);
    StrNy(igaus)   = Str2(2);
    StrNxy(igaus)  = Str2(3);
    StrQx(igaus)   = Str3(1);
    StrQy(igaus)   = Str3(2);

end
% Resultant stresses at nodes
Str1 = mstres * transpose(StrMx) ;
Strnod(lnods(1:4),1) = Strnod(lnods(1:4),1)+ Str1(1:4);
Str1 = mstres * transpose(StrMy) ;
Strnod(lnods(1:4),2) = Strnod(lnods(1:4),2)+ Str1(1:4);
Str1 = mstres * transpose(StrMxy) ;
Strnod(lnods(1:4),3) = Strnod(lnods(1:4),3)+ Str1(1:4);
Str2 = mstres * transpose(StrNx) ;
Strnod(lnods(1:4),4) = Strnod(lnods(1:4),4)+ Str2(1:4);
Str2 = mstres * transpose(StrNy) ;
Strnod(lnods(1:4),5) = Strnod(lnods(1:4),5)+ Str2(1:4);
Str2 = mstres * transpose(StrNxy) ;
Strnod(lnods(1:4),6) = Strnod(lnods(1:4),6)+ Str2(1:4);
Str3 = mstres * transpose(StrQx) ;
Strnod(lnods(1:4),7) = Strnod(lnods(1:4),7)+ Str3(1:4);
Str3 = mstres * transpose(StrQy) ;
Strnod(lnods(1:4),8) = Strnod(lnods(1:4),8)+ Str3(1:4);
Strnod(lnods(1:4),9) = Strnod(lnods(1:4),9)+ 1;
```

```
for i = 1 : npnod
    Strnod(i,1:5) = Strnod(i,1:5)/Strnod(i,6);
end
```
Nodal averaging of resultant stresses

Fig. 12.60 Computation of local resultant stresses at the Gauss points for the QLLL flat shell element

Table 12.10 shows the values for these three results for different meshes of QLLL flat shell elements.

Figure 12.63 shows contours of the vertical deflection and the bending moment $M_{x'}$ for the 16 × 16 mesh.

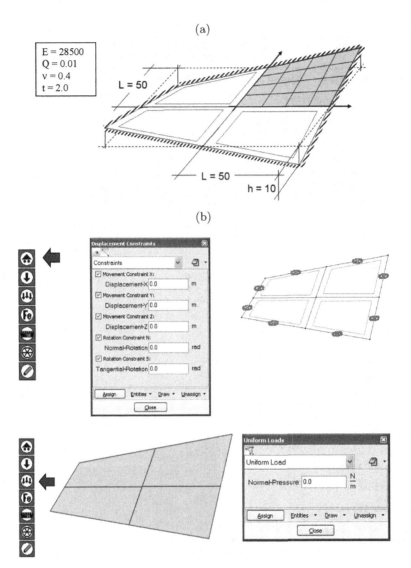

Fig. 12.61 Clamped hyperbolic shell under uniform distributed load. (a) Geometry, material properties and load. (b) Snapshots of data input menus

12.18.5 Scordelis roof

We present results for the analysis of a cylindrical shell with end diaphragms under uniform load (the so called *Scordelis roof*). The general description of the problem is shown in Figures 12.64 and 8.34.

Table 12.11 shows the results for the vertical displacement of point B using different meshes of QLLL flat shell elements.

```
%===================================================================
% MAT-fem_Shells 1.0 - MAT-fem is a learning tool for undestanding
%                       the Finite Element Method with MATLAB and GiD
%===================================================================
% PROBLEM TITLE = Clamped hyperbolic shell under uniform load

%  Material Properties
   young  =   2.850000000e+04 ;
   poiss  =   4.000000000e-01 ;
   denss  =   0.000000000e+00 ;
   thick  =   2.000000000e+00 ;
%
% Coordinates
global coordinates
coordinates = [
   0.000000000e+00   ,   1.000000000e+02   ,  -1.000000000e+01  ;
   5.000000000e+01   ,   1.000000000e+02   ,   0.000000000e+00  ;
....
   1.000000000e+02   ,   5.000000000e+01   ,   0.000000000e+00  ;
   1.000000000e+02   ,   0.000000000e+00   ,  -1.000000000e+01  ] ; %
% Elements
global elements
elements = [
      2   ,     6   ,     8   ,     4  ;
      3   ,     1   ,     2   ,     4  ;
      7   ,     5   ,     3   ,     4  ;
      8   ,     9   ,     7   ,     4  ] ; %
% Fixed Nodes
fixnodes = [
      1  ,  1 ,    0.000000000e+00  ;
      1  ,  2 ,    0.000000000e+00  ;
      1  ,  3 ,    0.000000000e+00  ;
      1  ,  4 ,    0.000000000e+00  ;
      1  ,  5 ,    0.000000000e+00  ;
...
      9  ,  1 ,    0.000000000e+00  ;
      9  ,  2 ,    0.000000000e+00  ;
      9  ,  3 ,    0.000000000e+00  ;
      9  ,  4 ,    0.000000000e+00  ;
      9  ,  5 ,    0.000000000e+00  ] ;
%
% Point loads
%
pointload = [ ] ;
%
% Side loadsss
%
uniload = sparse ( 4 , 1 );
uniload (      1  ) =   1.000000000e-02  ;
uniload (      2  ) =   1.000000000e-02  ;
uniload (      3  ) =   1.000000000e-02  ;
uniload (      4  ) =   1.000000000e-02  ;
```

Fig. 12.62 Clamped hyperbolic shell under uniform load. Example of data input for a 8-element mesh

Figure 12.67 shows results of $N_{y'}$, $M_{x'}$ and $Q_{x'}$ along the line CB obtained with a mesh of 32×32 QLLL flat shell elements. Figure 8.34 shows results for the same problem obtained with other flat shell elements.

Clamped hyperbolic shell				
QLLL Mesh	Nodes	w_c	$M_{x'}$	$Q_{y'}$
2 × 2	9	-1,068E-03	0,000	0.112
4 × 4	25	-1,751E-03	0,050	0.109
8 × 8	81	-1,266E-03	-0,093	-0.041
16 × 16	289	-1,655E-03	-0,108	-0.043
32 × 32	1089	-2,018E-03	-0,083	-0.047
64 × 64	4225	-2,212E-03	-0,065	-0.052
Reference [CB]		-2,451E-02		

Table 12.10 Clamped hyperbolic shell under uniform load. Results for deflection w_c and bending moment $M_{x'}$ at the center and shear force $Q_{y'}$ at the center of the clamped edge using different meshes of QLLL flat shell elements

(a)   (b)

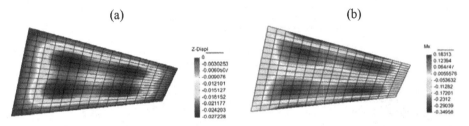

Fig. 12.63 Clamped hyperbolic shell. Contours of vertical deflection (a) and $M_{x'}$ (b)

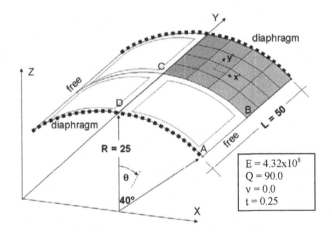

Fig. 12.64 Scordelis roof. Geometry, material properties and boundary conditions

12.19 2-NODED REISSNER-MINDLIN TRONCOCONICAL SHELL ELEMENT

This element was studied in Sections 9.4 and 9.6.1.

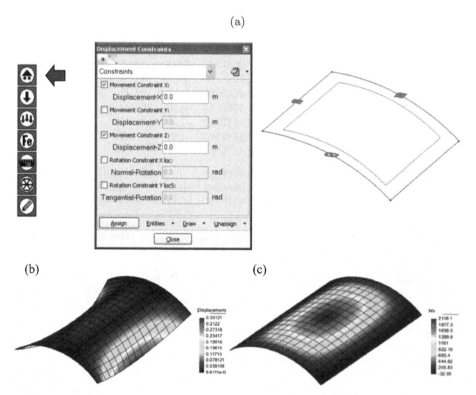

Fig. 12.65 Scordelis roof. (a) Boundary conditions. (b) Displacement contour. (c) Bending moment $M_{x'}$ contour for a 8×8 mesh of QLLL flat shell elements

	Scordelis roof			
QLLL elements	Mesh	Nodes	w_B	% Error
4	2×2	9	-0.3573	18.15
16	4×4	25	-0.2832	-6.34
64	8×8	81	-0.2943	-2.67
256	16×16	289	-0.3005	-0.63
1024	32×32	1089	-0.3026	-0.07
Reference [BSC+,MH2]			-0.3024	

Table 12.11 Scordelis roof shell under uniform load. Results for the vertical deflection of point B for different meshes of QLLL flat shell elements

12.19.1 Generalized constitutive matrix

Figure 12.67 shows the membrane (D_matm), bending (D_matb) and transverse shear (D_mats) generalized constitutive matrices of Eq.(9.39) and (9.40).

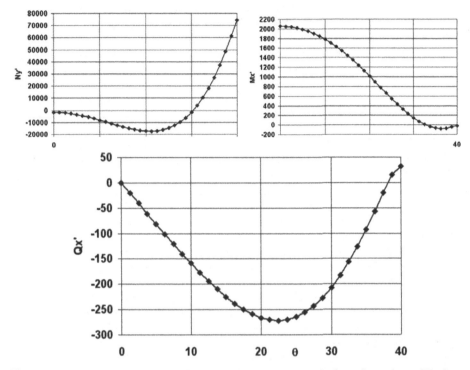

Fig. 12.66 Scordelis roof. Distribution of $N_{y'}$, $M_{x'}$ and $Q_{x'}$ along line CB for a mesh of 32×32 QLLL flat shell elements (see also Figure 8.34)

```
aux1 = thick*young/(1-poiss^2);
aux2 = poiss*aux1;
aux3 = thick*young/2/(1+poiss);
aux4 = (5/6)*thick*young/2/(1+poiss);

D_matm  = [aux1,aux2;
           aux2,aux1];

D_matb = D_matm*(thick^2/12);

D_mats = [aux4];

ttim = timing('Time needed to set initial values',ttim); %Reporting time
```

Fig. 12.67 2-noded RM troncoconical shell element. Membrane, bending and transverse shear generalized constitutive matrices

12.19.2 Stiffness matrix and equivalent nodal force vector

Figure 12.68 shows the computation of the transformation matrix **L** relating the global and the local nodal displacements (Eq.(9.14)).

The computation of the stiffness matrix and the equivalent nodal force vector for a uniform pressure is performed explicitly using a single inte-

```
function  Te = Rotation_system_RS (cxy)

    x = cxy(2,1) - cxy(1,1);
    y = cxy(2,2) - cxy(1,2);
    len = sqrt(x^2+y^2);

    co = x / len ;
    se = y / len ;

    Te = [  co, se, 0 ;
           -se, co, 0 ;
             0,  0, 1];
```

Fig. 12.68 2-noded troncoconical RM shell element. Displacement transformation matrix

```
cxy(1:nnode,:) = coordinates(lnods(1:nnode),:); % Element coordinates

Te = Rotation_system_RS(cxy);

r = (cxy(1,1)+cxy(2,1))/2; % Radius of the element
```

```
[bmat_b,bmat_s,bmat_m,len] = B_mat_Rev_Shell(cxy,Te);

K_b  = transpose(bmat_b)*D_matb*bmat_b*2*pi*r*len;       Stiffness matrix
K_s  = transpose(bmat_s)*D_mats*bmat_s*2*pi*r*len;     using 1 Gauss point
K_m  = transpose(bmat_m)*D_matm*bmat_m*2*pi*r*len;

K_elem  = K_elem + K_f + K_s + K_m;
```

```
c1 = 2*cxy(1,1)+  cxy(2,1);
c2 =   cxy(1,1)+2*cxy(2,1);                              Equivalent nodal
                                                          force vector
fx1 = uniload(ielem)* Te(2,1)*pi*len*c1/3;
fx2 = uniload(ielem)* Te(2,1)*pi*len*c2/3;

fy1 = (uniload(ielem)  * Te(1,1) - denss*thick)*pi*len*c1/3;
fy2 = (uniload(ielem)  * Te(1,1) - denss*thick)*pi*len*c2/3;

ElemFor =   [fx1,fy1,0,fx2,fy2,0];
```

Fig. 12.69 2-noded troncoconical RM shell element. Computation of stiffness matrix and equivalent nodal force vector

gration point, following Eqs.(9.73) and (9.74). The computation steps are shown in Figure 12.69. The membrane (K_m), bending (K_b) and transverse shear stiffness (K_s) matrices are computed separately and then added together in K_elem.

Figure 12.70 shows the subroutine for computing the membrane, bending and transverse shear generalized strain matrices.

12.19.3 Resultant stresses

The local resultant stresses are computed at the element center and then they are extrapolated to the nodes.

```
function [bmat_b,bmat_s,bmat_m,len] = B_mat_Troncoconical_RM_Shell(cxy,Te)

x = cxy(2,1) - cxy(1,1);
y = cxy(2,2) - cxy(1,2);                 Shape functions and their derivatives
len = sqrt(x^2+y^2);                        in local coordinate system

r = (cxy(1,1) + cxy(2,1)) / 2;
N(1) = 0.5 ;    % Shape functions and derivates at the gauss pt. = 0
N(2) = 0.5 ;

dxN(1) = -1/len;
dxN(2) =  1/len;

bmat_m1  = [              dxN(1),              0,  0 ;
              N(1)*Te(1,1)/r , -N(1)*Te(1,2)/r ,  0];   Membrane strain
bmat_m2  = [              dxN(2),              0,  0 ;       matrix
              N(2)*Te(1,1)/r , -N(2)*Te(1,2)/r ,  0];
bmat_m = [bmat_m1*Te,bmat_m2*Te];

bmat_b1  = [ 0, 0,             -dxN(1)    ;
             0, 0, -N(1)*Te(1,1)/r   ];
bmat_b2  = [ 0, 0,             -dxN(2)    ;       Bending strain matrix
             0, 0, -N(2)*Te(1,1)/r   ];
bmat_b = [bmat_b1*Te,bmat_b2*Te];

bmat_s1  = [ 0, dxN(1), -N(1)];
bmat_s2  = [ 0, dxN(2), -N(2)];       Shear strain matrix
bmat_s = [bmat_s1*Te,bmat_s2*Te];
```

Fig. 12.70 2-noded troncoconical RM shell element. Computation of $\mathbf{B}_m$, $\mathbf{B}_b$ and $\mathbf{B}_s$

```
[bmat_b,bmat_s,bmat_m,len] = B_mat_Troncoconical_RM_Shell(cxy,Te);

% Resultant stresses at element mid-point
Str1=D_matb*bmat_b*u_elem;
Str2=D_mats*bmat_s*u_elem;
Str3=D_matm*bmat_m*u_elem;
%
Mx(ielem) =Str1(1);
Mf(ielem) =Str1(2);
Qz(ielem) =Str2(1);
Nx(ielem) =Str3(1);
Nf(ielem) =Str3(2);
```

Fig. 12.71 2-noded troncoconical RM shell element. Computation of resultant stresses at the element mid-point

Figure 12.71 shows the steps for computing the bending moment $M_{x'}$ and $M_{y'}$ (stored in Str1), the transverse shear force Q (stored in Str2) and the axial forces $N_{x'}$ and $N_{y'}$ (stored in Str3).

12.19.4 Example. Thin spherical dome under uniform external pressure

Figure 12.72 shows the geometry of the dome, the material properties and the external pressure values. Further details are given in Section 9.7.1.

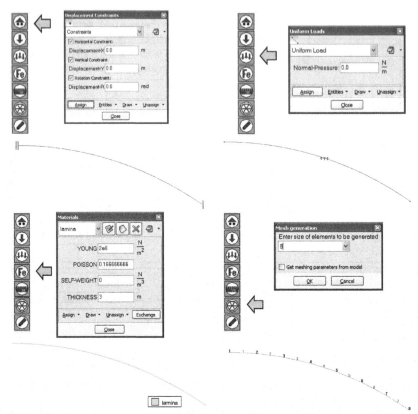

Fig. 12.72 2-noded troncoconical shell element. Data input menus for boundary conditions, pressure load, material properties and mesh

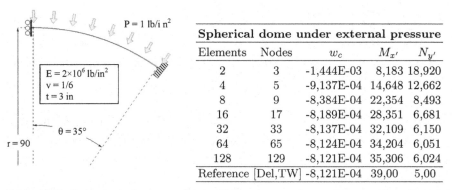

Spherical dome under external pressure				
Elements	Nodes	w_c	$M_{x'}$	$N_{y'}$
2	3	-1,444E-03	8,183	18,920
4	5	-9,137E-04	14,648	12,662
8	9	-8,384E-04	22,354	8,493
16	17	-8,189E-04	28,351	6,681
32	33	-8,137E-04	32,109	6,150
64	65	-8,124E-04	34,204	6,051
128	129	-8,121E-04	35,306	6,024
Reference [Del,TW]		-8,121E-04	39,00	5,00

Figure data: $P = 1$ lb/in^2; $E = 2 \times 10^6$ lb/in^2; $v = 1/6$; $t = 3$ in; $\theta = 35°$; $r = 90$

Table 12.12 Spherical dome under uniform external pressure analyzed with 2-noded troncoconical RM shell elements. Deflection at the symmetry node and radial bending moment and circumferential axial force at the center of the element adjacent to the clamped end for different meshes (see also Figure 8.16)

```
%========================================================================
% MAT-fem_RevShells 1.0   - MAT-fem is a learning tool for undestanding
%                            the Finite Element Method with MATLAB and GiD
%========================================================================
% PROBLEM TITLE = Thin spherical dome under uniform external pressure
%                 analyzed with 28 2-noded troncoconical RM shell elements
%
% Material Properties
  young   =   2.000000000e+06 ;
  poiss   =   1.666666660e-01 ;
  denss   =   0.000000000e+00 ;
  thikness=   3.000000000e+00 ;
%
% Coordinates
global coordinates
coordinates = [
   5.162187927e+01   ,    7.372368399e+01  ;
   4.994083848e+01   ,    7.487264288e+01  ;
  ...
   0.000000000e+00   ,    9.000000000e+01  ] ;
%
% Elements
global elements
elements = [
      28   ,       27  ;
      27   ,       26  ;
  ...
       2   ,        1   ] ;
%
% Fixed nodes
fixnodes = [
       1  , 1 ,   0.000000000e+00  ;
       1  , 2 ,   0.000000000e+00  ;
       1  , 3 ,   0.000000000e+00  ;
      28  , 1 ,   0.000000000e+00  ;
      28  , 3 ,   0.000000000e+00  ] ;
%
% Point loads
pointload = [ ] ;
%
% Side loads
uniload = sparse ( 27 );
uniload (      1  ) = -1.000000000e+00  ;
uniload (      2  ) = -1.000000000e+00  ;
..
uniload (     27  ) = -1.000000000e+00  ;
```

Fig. 12.73 Data input file for analysis of a thin spherical dome under uniform external pressure using 28 2-noded troncoconical RM elements

The reference solution has been obtained with a mesh of 60000 4-noded cubic troncoconical elements giving a deflection at the center of $w_c = 81211 \times 10^{-4}$ in, a radial bending moment at the clamped end $M_{x'} = 39$ lb×in/in and a circumferential axial force also at the clamped end $N_{y'} = 5.0$ lb/in. Other solution to this problem can be found in [Del,TW].

Figure 12.72 shows some of the data input menus corresponding to the boundary conditions, the pressure load, the material properties and the definition of the mesh.

The data input file for a 28 element mesh is shown in Figure 12.73.

Table 12.12 shows the convergence of the deflection at the symmetry node and the radial bending moment and the circumferential force at the center of the element adjacent to the clamped end for different meshes.

12.20 FINAL REMARKS

We have presented the basic concepts and the structure of the MAT-fem code environment for analysis of beams, plates and shells using some of the elements studied in the book.

Each element has been programmed as a separate code to facilitate the follow-up of the programming steps and the use of the code.

Each of the codes presented in this chapter, and other codes for several of the elements studied in the book, together with examples of applications, can be downloaded from www.cimne.com/MAT-fem.

For general questions about the use of MAT-fem please contact Dr. Francisco Zárate at zarate@cimne.upc.edu.

Appendix A

BASIC PROPERTIES OF MATERIALS

	E (GPa)	ν	ρ Kg/m^3	$\alpha \times 10^5$ °C^{-1}	Limit tensile stress MPa	ε failure %
Concrete	20-40	0.15	2400	2.0	4	
Carbon steel	207	0.30	7810	1.3	400-1600	1.8
Nickel stell	207	0.30	7750	1.3	400-1600	-
Stainless steel (18-8)	190	0.31	7750	1.6	400-1600	-
Alluminium (all alloys)	70	0.33	2710	2.2	140-600	-
Copper	110	0.33	8910	1.7	-	-
Cast iron gray	100	0.21	7200	1.1	-	-
Glass	46	0.25	2600	0.8	35-175	
Lead	37	0.43	11380	2.9	-	
Magnesium	45	0.35	1800	2.6	-	
Phosphor bronze	111	0.35	8170	1.8	-	
Wood (sense of fibers)	15	0.45	-	-	100	
Wood (transverse sense)	1	-	-	-	3.5	
Granit	60	0.27	-	-	4	
Diamant	1200	-	-	-		

Table A.1 CONVENTIONAL MATERIALS

	E MPa	ν
Unconsolidated sand	1034	0.3
Carbonates	2206	0.1
Shale	2413	0.1

Table A.2 SOILS

E. Oñate, *Structural Analysis with the Finite Element Method. Linear Statics: Volume 2: Beams, Plates and Shells*, Lecture Notes on Numerical Methods in Engineering and Sciences, DOI 10.1007/978-1-4020-8743-1,
© International Center for Numerical Methods in Engineering (CIMNE), 2013

	E (GPa)	ν	ρ Kg/m^3	$\alpha \times 10^{-5}$ °C^{-1}	Limit tensile stress MPa	ε failure %
E-Glass	72	0.25	2550	0.5	3400	4.5
S-Glass	86	0.20	2500	0.3	4600	1.5
Graphite	390	-	1900	-	2100	-
Boron (ϕ 0.1mm)	400	-	2600	0.4	3400	0.8
Aramid (Kplar 49)	130	-	1450	-	2700	-
Nylon	1.4	-	-	-	1000	-
Carbon	190	0.3 6	1410	0.05	1700	0.5
Carbon HR (high resistance)	230	0.3	1750	0.02	3200	1.3
Carbon HM (high modulus)	390	0.35	1800	0.08	2500	0.6

Table A.3 FIBERS

	E (GPa)	ν	ρ Kg/m^3	$\alpha \times 10^{-5}$ °C^{-1}	Limit tensile stress MPa	ε failure %
Epoxy resin	4-5	0.4	1200	9-13	130	3-6
Phelonic resin	3	0.4	1300	9-13	40	3-6
Polyester resin	4	0.4	1200	2	50-100	2.5
Polypropylene	1.1-1-4	0.4	900	-	25	-
Polycarbonate	2.4	0.1	1200	-	60	-
Polystyrene	0.020	0.4	280	-	-	-
Rubber	0.002-0.007	0.5	-	-	-	-

Table A.4 RESINS AND POLYMERS

	E (GPa)	ν	Limit tensile stress MPa
Phoetal cranial bone	$E_1 = 3.8$ $E_2 = 1.0$	0.22	-
Adult cranial bone	4.46	0.22	
Fresh bone	2.1	0.25	110
Human cartilage	0.024	-	3
Human tendon	0.6	-	82

Table A.5 BIOLOGICAL MATERIALS

The mechanical properties of other materials can be found in [Co2,PP4].

Appendix B

EQUILIBRIUM EQUATIONS FOR A SOLID

Let us consider the equilibrium of forces in a differential of area of a 2D solid under body forces (b_x, b_y) (Figure B.1a)

$$\sum F_x = 0 \quad : \quad (\sigma_x + d\sigma_x)dy + (\tau_{yx} + d\tau_{yx})dx + b_x dx dy - \sigma_x dy - \tau_{yx} dx = 0$$

$$\sum F_y = 0 \quad : \quad (\sigma_y + d\sigma_y)dx + (\tau_{xy} + d\tau_{xy})dy + b_y dx dy - \sigma_y dx - \tau_{xy} dy = 0$$

$$\text{(B.1)}$$

Noting that

$$d\sigma_x = \frac{\partial \sigma_x}{\partial x}dx \;, \;\; d\sigma_y = \frac{\partial \sigma_y}{\partial y}dy \;, \;\; d\tau_{xy} = \frac{\partial \tau_{xy}}{\partial x}dx \;, \;\; d\tau_{yx} = \frac{\partial \tau_{yx}}{\partial y}dy$$

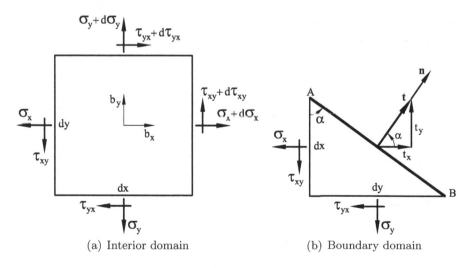

(a) Interior domain   (b) Boundary domain

Fig. B.1 Equilibrium of forces in an infinitesimal domain at the interior (a) and the boundary (b) of a 2D solid

E. Oñate, *Structural Analysis with the Finite Element Method. Linear Statics: Volume 2: Beams, Plates and Shells*, Lecture Notes on Numerical Methods in Engineering and Sciences, DOI 10.1007/978-1-4020-8743-1,
© International Center for Numerical Methods in Engineering (CIMNE), 2013

$$\tau_{xy} = \tau_{yx} \tag{B.2}$$

Substituting Eqs.(B.2) into (B.1) gives after small algebra

$$\frac{\partial \sigma_x}{\partial x} + \frac{\partial \tau_{xy}}{\partial y} + b_x = 0$$
$$\frac{\partial \sigma_y}{\partial y} + \frac{\partial \tau_{xy}}{\partial x} + b_y = 0 \tag{B.3}$$

B.1 EQUILIBRIUM AT A BOUNDARY SEGMENT

Let us consider the equilibrium of forces at an infinitesimal boundary segment AB of length L of a 2D solid under a traction force $\mathbf{t} = [t_x, t_y]^T$ acting in the normal direction $\mathbf{n} = [n_x, n_y]^T$ (Figure B.1b)

$$\sum F_x = 0 \quad : \quad t_x L - \sigma_x dy - \tau_{yx} dx = 0$$
$$\sum F_y = 0 \quad : \quad t_y L - \sigma_y dx - \tau_{xy} dy = 0 \tag{B.4}$$

Nothing that $\tau_{yx} = \tau_{xy}$ and $dx = L n_y$, $dy = L n_x$ with $n_x = \cos \alpha$ and $n_y = \sin \alpha$ and substituting these expressions into (B.4) yields

$$t_x = \sigma_x n_x + \tau_{xy} n_y$$
$$t_y = \sigma_y n_y + \tau_{xy} n_x \tag{B.5}$$

which are the sought equilibrium equations at the boundary.

Eqs.(B.3) and (B.5) can be readily extended to 3D solids as [ZT]

Equilibrium at the interior

$$\frac{\partial \sigma_x}{\partial x} + \frac{\partial \tau_{xy}}{\partial y} + \frac{\partial \tau_{xz}}{\partial z} + b_x = 0$$
$$\frac{\partial \tau_{xy}}{\partial x} + \frac{\partial \sigma_y}{\partial y} + \frac{\partial \tau_{yz}}{\partial z} + b_y = 0$$
$$\frac{\partial \tau_{xz}}{\partial x} + \frac{\partial \tau_{yz}}{\partial y} + \frac{\partial \sigma_z}{\partial z} + b_z = 0 \tag{B.6}$$

Equilibrium at the boundary

$$t_x = \sigma_x n_x + \tau_{xy} n_y + \tau_{xz} n_z$$
$$t_y = \tau_{xy} n_x + \sigma_y n_y + \tau_{yz} n_z$$
$$t_z = \tau_{xz} n_x + \tau_{yz} n_y + \sigma_z n_z \tag{B.7}$$

Appendix C

NUMERICAL INTEGRATION

C.1 1D NUMERICAL INTEGRATION

Let us assume that the integral of a function $f(x)$ in the interval [-1,1] is required, i.e.

$$I = \int_{-1}^{+1} f(\xi) \, d\xi \tag{C.1}$$

The Gauss integration rule, or Gauss *quadrature*, expresses the value of the above integral as a sum the function values at a number of known points multiplied by prescribed weights. For a quadrature of order q

$$I \simeq I_q = \sum_{i=1}^{q} f(\xi_i) W_i \tag{C.2}$$

where W_i is the weight corresponding to the ith sampling point located at $\xi = \xi_i$ and q the number of sampling points. A *Gauss quadrature of qth order integrates exactly a polynomial function of degree* $2q - 1$ [Ral]. The error in the computation of the integral is of the order $0(\triangle^{2q})$, where $\triangle$ is the spacing between the sampling points. Table C.1 shows the coordinates of the sampling points and their weights for the first eight 1D Gauss quadratures.

Note that the sampling points are all located within the normalized domain [-1,1]. This is useful for computing the element integrals expressed in terms of the natural coordinate ξ. The Gauss quadrature requires the minimum number of sampling points to achieve a prescribed error in the computation of an integral. Thus, it minimizes the number of times the integrand function is computed. The reader can find the details in [Dem,PFTV,Ral,WR].

E. Oñate, *Structural Analysis with the Finite Element Method. Linear Statics:*
Volume 2: Beams, Plates and Shells, Lecture Notes on Numerical Methods
in Engineering and Sciences, DOI 10.1007/978-1-4020-8743-1,
© International Center for Numerical Methods in Engineering (CIMNE), 2013

q	ξ_q	W_q
1	0.0	2.0
2	±0.5773502692	1.0
3	±0.774596697	0.5555555556
	0.0	0.8888888889
4	±0.8611363116	0.3478548451
	±0.3399810436	0.6521451549
5	±0.9061798459	0.2369268851
	±0.5384693101	0.4786286705
	0.0	0.5688888889
6	±0.9324695142	0.1713244924
	±0.6612093865	0.3607615730
	±0.2386191861	0.4679139346
7	±0.9491079123	0.1294849662
	±0.7415311856	0.2797053915
	±0.4058451514	0.3818300505
	0.0	0.4179591837
8	±0.9602898565	0.1012285363
	±0.7966664774	0.2223810345
	±0.5255324099	0.3137066459
	±0.1834346425	0.3626837834

Table C.1 Coordinates and weights for 1D Gauss quadratures

C.2 NUMERICAL INTEGRATION IN 2D

C.2.1 Numerical integration in quadrilateral domains

The integral of a term $g(\xi, \eta)$ over the normalized isoparametric quadrilateral domain can be evaluated using a 2D Gauss quadrature by

$$\int_{-1}^{+1} \int_{-1}^{+1} g(\xi, \eta) \, d\xi \, d\eta = \int_{-1}^{+1} d\xi \left[\sum_{q=1}^{n_q} g(\xi, \eta_q) W_q \right] = \sum_{p=1}^{n_p} \sum_{q=1}^{n_q} g(\xi_p, \eta_q) W_p W_q$$

(C.3)

where n_p and n_q are the number of integration points along each natural coordinate ξ and η respectively; ξ_p and η_q are the natural coordinates of the pth integration point and W_p, W_q are the corresponding weights.

The coordinates and weights for each natural direction are directly deduced from those given in Table C.1 for the 1D case. Let us recall that a 1D quadrature of qth order integrates *exactly* a polynomial of degree $q \leq 2n - 1$. Figure C.1 shows the more usual quadratures for quadrilateral elements.

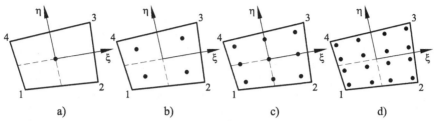

Fig. C.1 Gauss quadratures over quadrilateral elements, a) 1×1, b) 2×2, c) 3×3, d) 4×4 integration points

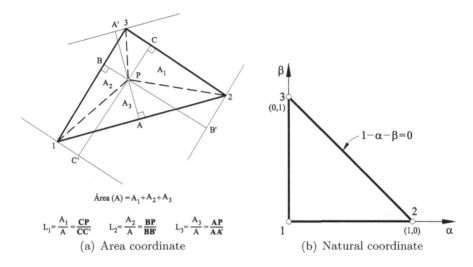

Área (A) = $A_1 + A_2 + A_3$

$$L_1 = \frac{A_1}{A} = \frac{CP}{CC'} \qquad L_2 = \frac{A_2}{A} = \frac{BP}{BB'} \qquad L_3 = \frac{A_3}{A} = \frac{AP}{AA'}$$

(a) Area coordinate   (b) Natural coordinate

Fig. C.2 Triangular element. (a) Area coordinates. (b) Natural coordinates

C.2.2 Numerical integration over triangles

The Gauss quadrature for triangles is written as

$$\int_0^1 \int_0^{1-L_3} f(L_1, L_2, L_3) \, dL_2 \, dL_3 = \sum_{p=1}^{n_p} f(L_{1_p}, L_{2_p}, L_{3_p}) \, W_p \qquad (C.4)$$

where n_p is the number of integration points: $L_{1_p}, L_{2_p}, L_{3_p}$ and W_p are the area coordinates (Figure C.2a) and the corresponding weights for the pth integration point [On4].

Figure C.3 shows the more usual coordinates and weights. The term "accuracy" in the figure refers to the highest degree polynomial which is exactly integrated by each quadrature. Figure C.3 is also of direct application for computing the integrals defined in terms of the natural coordinates for triangles α and β (Figure C.2b) defined as $L_2 = \alpha, L_3 = \beta$ and $L_1 = 1 - \alpha - \beta$ [On4].

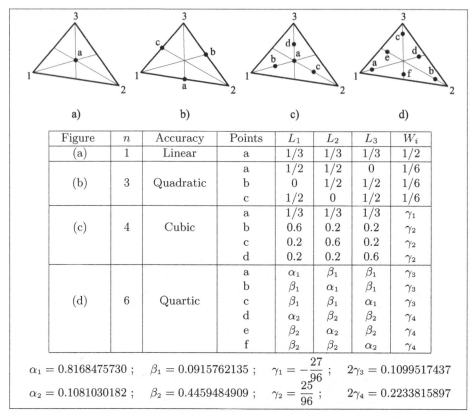

Figure	n	Accuracy	Points	L_1	L_2	L_3	W_i
(a)	1	Linear	a	$1/3$	$1/3$	$1/3$	$1/2$
(b)	3	Quadratic	a	$1/2$	$1/2$	0	$1/6$
			b	0	$1/2$	$1/2$	$1/6$
			c	$1/2$	0	$1/2$	$1/6$
(c)	4	Cubic	a	$1/3$	$1/3$	$1/3$	γ_1
			b	0.6	0.2	0.2	γ_2
			c	0.2	0.6	0.2	γ_2
			d	0.2	0.2	0.6	γ_2
(d)	6	Quartic	a	α_1	β_1	β_1	γ_3
			b	β_1	α_1	β_1	γ_3
			c	β_1	β_1	α_1	γ_3
			d	α_2	β_2	β_2	γ_4
			e	β_2	α_2	β_2	γ_4
			f	β_2	β_2	α_2	γ_4

$\alpha_1 = 0.8168475730$; $\beta_1 = 0.0915762135$; $\gamma_1 = -\dfrac{27}{96}$; $2\gamma_3 = 0.1099517437$

$\alpha_2 = 0.1081030182$; $\beta_2 = 0.4459484909$; $\gamma_2 = \dfrac{25}{96}$; $2\gamma_4 = 0.2233815897$

Fig. C.3 Coordinates and weights for the Gauss quadrature in triangular elements

The weights in Figure C.3 are normalized so that their sum is $1/2$. In many references this value is changed to the unity and this requires the sum of Eq.(D.4) to be multiplied by $1/2$ so that the element area is correctly computed in those cases [On4].

The quadrature of Figure C.2 can be extended for tetrahedral elements. For details see [On4].

C.3 NUMERICAL INTEGRATION OVER HEXAEDRA

Let us consider the integration of a function $f(x, y, z)$ over a hexahedral isoparametric element. The following transformations are required

$$\iiint_{V^{(e)}} f(x, y, z)\, dx\, dy\, dz = \int_{-1}^{1} \int_{-1}^{1} \int_{-1}^{1} f(\xi, \eta, \zeta)\, \left| \mathbf{J}^{(e)} \right|\, d\xi\, d\eta\, d\zeta =$$

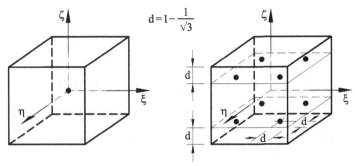

Fig. C.4 Gauss quadratures of $1 \times 1 \times 1$ and $2 \times 2 \times 2$ points in hexahedral elements

$$= \int_{-1}^{+1} \int_{-1}^{+1} \int_{-1}^{+1} g(\xi, \eta, \zeta) \, d\xi \, d\eta \, d\zeta \qquad (\text{C.5})$$

Gauss quadrature over the normalized cubic domain leads to

$$\int_{-1}^{+1} \int_{-1}^{+1} \int_{-1}^{+1} g(\xi, \eta, \zeta) \, d\xi \, d\eta \, d\zeta = \int_{-1}^{+1} \int_{-1}^{+1} \sum_{p=1}^{n_p} W_p \, g(\xi_p, \eta, \zeta) \, d\eta \, d\zeta =$$

$$= \int_{-1}^{+1} \sum_{q=1}^{n_q} \sum_{p=1}^{n_p} W_p W_q g(\xi_p, \eta_q, \zeta) \, d\zeta = \sum_{r=1}^{n_r} \sum_{q=1}^{n_q} \sum_{p=1}^{n_p} W_p W_q W_r g(\xi_p, \eta_q, \zeta_r)$$

$$(\text{C.6})$$

where n_p, n_q and n_r are the integration points via the ξ, η, ζ directions, respectively, ξ_p, η_q, ζ_r are the coordinates of the integration point (p, q, r) and W_p, W_q, W_r are the weights for each natural direction.

The local coordinates and weights for each quadrature are deduced from Table D.3 for the 1D case. We recall that a qth order quadrature integrates exactly a 1D polynomial of degree $2q - 1$. This rule helps us to identify the number of integration points in each natural direction. Figure C.4 shows the sampling points for the $1 \times 1 \times 1$ and $2 \times 2 \times 2$ quadratures.

Appendix D

COMPUTATION OF THE SHEAR CORRECTION PARAMETER FOR BEAMS

Let us consider the bending of a beam in the xz plane. The shear correction parameter k_z can be determined by comparing the transverse shear strain energy U_s associated to the theoretical (exact) distribution of the shear stresses τ_{xz} and τ_{xy} and the transverse shear strain U_s^T associated to the Timoshenko's model corrected by the coefficient k_z, i.e.

$$U_s = \frac{1}{2} \iint_A \left[\frac{\tau_{xy}^2}{G_{xy}} + \frac{\tau_{xz}^2}{G_{xz}} \right] dA \tag{D.1}$$

$$U_s^T = \frac{1}{2} \frac{Q_z^2}{k_z \bar{G}_{xz}} \quad , \quad \text{with} \quad Q_z = \iint_A \tau_{xz} dA \quad \text{and} \quad \bar{G}_{xz} = \iint_A G_{xz}(y, z) dA \tag{D.2}$$

If $U_s = U_s^T$, then

$$k_z = \frac{1}{2} \frac{Q_z^2}{\bar{G}_{xz} U_s} \tag{D.3}$$

The "exact" transverse strain energy U_s can be estimated as follows. The shear stresses satisfy the equilibrium equations (Eqs.(B.6) and (B.7) of Appendix B)

$$\frac{\partial \sigma_x}{\partial x} + \frac{\partial \tau_{xy}}{\partial y} + \frac{\partial \tau_{xz}}{\partial z} = 0 \quad \text{in} \quad A$$
$$\tau_{xn} = \tau_{xy} n_y + \tau_{xz} n_z = 0 \quad \text{in} \quad \Gamma_A \tag{D.4}$$

In Eq.(D.4) Γ_A represents the boundary of the section of area A where both σ_x and t_x are zero (Eq.(B.7)).

E. Oñate, *Structural Analysis with the Finite Element Method. Linear Statics: Volume 2: Beams, Plates and Shells*, Lecture Notes on Numerical Methods in Engineering and Sciences, DOI 10.1007/978-1-4020-8743-1,
© International Center for Numerical Methods in Engineering (CIMNE), 2013

The constitutive equations for a non homogeneous beam can be written for the general case as (see Eqs.(3.10) and (3.11))

$$N = \hat{D}_a \frac{\partial u_0}{\partial x} + \hat{D}_{ab} \frac{\partial \theta}{\partial x}$$

$$M = \hat{D}_{ab} \frac{\partial u_0}{\partial x} + \hat{D}_b \frac{\partial \theta}{\partial x}$$

(D.5)

with

$$\hat{D}_a = \iint_A E dA \quad , \quad \hat{D}_{ab} = - \iint_A E z dA \quad , \quad \hat{D}_b = \iint_A E z^2 dA \quad \text{(D.6)}$$

Inverting Eq.(D.5) gives

$$\frac{\partial u_0}{\partial x} = \frac{1}{\hat{D}}[\hat{D}_b N - \hat{D}_{ab} M] \quad ; \quad \frac{\partial \theta}{\partial x} = \frac{1}{\hat{D}}[-\hat{D}_{ab} N + \hat{D}_a M] \qquad \text{(D.7)}$$

with

$$\hat{D} = \hat{D}_a \hat{D}_b - \hat{D}_{ab}^2 \tag{D.8}$$

The axial stress σ_x is given by

$$\sigma_x = E \varepsilon_x = E \left[\frac{\partial u_0}{\partial x} - z \frac{\partial \theta}{\partial x} \right] \tag{D.9}$$

Substituting (D.7) into (D.9) gives

$$\frac{\partial \sigma_x}{\partial x} = \frac{E}{\hat{D}} \left[\hat{D}_b \frac{\partial N}{\partial x} - \hat{D}_{ab} \frac{\partial M}{\partial x} - z \left(-\hat{D}_{ab} \frac{\partial N}{\partial x} + \hat{D}_a \frac{\partial M}{\partial x} \right) \right] \tag{D.10}$$

Assuming that $\frac{\partial N}{\partial x} = 0$ using $\frac{\partial M_y}{\partial x} = -Q$ (Figure 1.7) then

$$\frac{\partial \sigma_x}{\partial x} = \frac{E}{\hat{D}}[\hat{D}_{ab} + z\hat{D}_a]Q = f(y, z) \tag{D.11}$$

Let us assume now the following displacement field

$$u(x, y, z) = u_0(x) - z\theta(x) + u_t(y, z)$$
$$v(x, y, z) = 0 \quad , \quad w(x, y, z) = w(x) \tag{D.12}$$

where $u_t(y, z)$ is the warping displacement due to torsion. Then

$$\tau_{xy} = G_{xy} \left(\frac{\partial u}{\partial y} + \frac{\partial v}{\partial x} \right) = G_{xy} \frac{\partial u_t}{\partial y}$$

$$\tau_{xz} = G_{xz} \left(\frac{\partial u}{\partial z} + \frac{\partial w}{\partial x} \right) = G_{xz} \left[\gamma_{xz} + \frac{\partial u_t}{\partial z} \right] \tag{D.13}$$

where $\gamma_{xz} = \frac{\partial w}{\partial x} - \theta$ is a function of the x coordinate only. From Eqs.(D.13) we deduce

$$\frac{\partial \tau_{xy}}{\partial t} - \frac{\partial \tau_{xz}}{\partial y} = 0 \tag{D.14}$$

Eq.(B.14) can be satisfied by introducing a function ϕ such that

$$\tau_{xy} = G_{xy}\frac{\partial\phi}{\partial y} \quad \text{and} \quad \tau_{xz} = G_{xz}\frac{\partial\phi}{\partial z} \tag{D.15}$$

The equilibrium equations (D.4) can be written using (D.11) and (D.14) as

$$\frac{\partial}{\partial y}\left(G_{xy}\frac{\partial\phi}{\partial y}\right) + \frac{\partial}{\partial z}\left(G_{xz}\frac{\partial\phi}{\partial z}\right) + f = 0 \quad \text{in } A$$
$$G_{xy}\frac{\partial\phi}{\partial y}n_y + G_{xz}\frac{\partial\phi}{\partial z}n_z = 0 \quad \text{in } \Gamma_A \tag{D.16}$$

Eqs.(D.15) define a Laplace problem of the same type that for the Saint-Venant torsion [OR]. These equations can be solved analytically (for simple sections) or numerically by the FEM for the values of $\phi(y, z)$ at each point of the section [ZTZ]. The expression of U_s to be used in Eq.(D.3) for the value of k_z can be computed in terms of ϕ as

$$U_s = \frac{1}{2}\iint_A\left[G_{xy}\left(\frac{\partial\phi}{\partial y}\right)^2 + G_{xz}\left(\frac{\partial\phi}{\partial z}\right)^2\right]dA = \frac{1}{2}\iint_A \phi f\, dA \tag{D.17}$$

Figure 2.3 shows the values of k_z for some cross sections using the procedure described above.

The same procedure can be followed for computing the shear parameter k_y for bending in the xy plane.

Appendix E

PROOF OF THE SINGULARITY RULE FOR THE STIFFNESS MATRIX

Let us consider the system of algebraic equations representing the equilibrium of a structure discretized with the FEM

$$\mathbf{f} = \mathbf{Ka} = \left[\int_A \mathbf{B}^T \, \mathbf{D} \, \mathbf{B} \, dA \right] \mathbf{a} \tag{E.1}$$

where $\mathbf{a} = \left\{ \begin{matrix} \mathbf{a}_1 \\ \vdots \\ \mathbf{a}_n \end{matrix} \right\}$ is a vector of $j = N \times d - r$ DOFs, N is the total number of nodes in the mesh, d is the number of DOFs per node and r the number of restrained DOFs.

The generalized strain matrix $\mathbf{B}$ in Eq.(E.1) can be written as $\mathbf{B} = [\mathbf{B}_1, \mathbf{B}_2, \cdots, \mathbf{B}_N]$ where $\mathbf{B}_i$ is the generalized strain matrix of node i that includes the global shape functions N_i^g. Functions N_i^g coincide with the standard shape functions $N_i^{(e)}$ within element e containing node i [ZT2,On4].

The equilibrium equations for the ith node reads

$$\mathbf{f}_i = \mathbf{K}_{i1}\mathbf{a}_1 + \mathbf{K}_{i2}\mathbf{a}_2 + \cdots + \mathbf{K}_{n}\mathbf{a}_n \tag{E.2}$$

Let us assume that the stiffness matrix is computed numerically using a Gauss quadrature with p points. Eq.(E.2) can be written in this case as

$$\begin{aligned} \mathbf{f}_i = {} & \left\{ \textstyle\sum_{m=1}^{p} [\mathbf{B}^T \, \mathbf{D} \, \mathbf{B} \, |\mathbf{J}|]_m W_m \right\} \mathbf{a} = \\ = {} & \bar{W}^{(1)} \left[\mathbf{B}_i^{(1)} \right]^T \left[\{ \mathbf{DB}_1^{(1)} \mathbf{a}_1 \} + \{ \mathbf{DB}_2^{(1)} \mathbf{a}_2 \} + \cdots + \{ \mathbf{DB}_n^{(1)} \mathbf{a}_n \} \right] + \\ + {} & \bar{W}^{(2)} \left[\mathbf{B}_i^{(2)} \right]^T \left[\{ \mathbf{DB}_1^{(2)} \mathbf{a}_1 \} + \{ \mathbf{DB}_2^{(2)} \mathbf{a}_2 \} + \cdots + \{ \mathbf{DB}_n^{(2)} \mathbf{a}_n \} \right] + \\ + {} & \cdots + \bar{W}^{(p)} \left[\mathbf{B}_i^{(p)} \right]^T \left[\{ \mathbf{DB}_1^{(p)} \mathbf{a}_1 \} + \{ \mathbf{DB}_2^{(p)} \mathbf{a}_2 \} + \cdots + \{ \mathbf{DB}_n^{(p)} \mathbf{a}_n \} \right] \end{aligned} \tag{E.3}$$

E. Oñate, *Structural Analysis with the Finite Element Method. Linear Statics: Volume 2: Beams, Plates and Shells*, Lecture Notes on Numerical Methods in Engineering and Sciences, DOI 10.1007/978-1-4020-8743-1, © International Center for Numerical Methods in Engineering (CIMNE), 2013

where $\bar{W}^{(j)}$ is the product of the integration weight and the Jacobian determinant at the jth integration point. From Eq.(E.3) we deduce that if s is the number of rows in $\mathbf{B}_i$ (i.e. the number of generalized strains), vector $\mathbf{f}_i$ is a combination of $k \times p$ linear relationships in $\mathbf{a}_1, \mathbf{a}_2, \cdots, \mathbf{a}_n$. It can be proven that these relationship are independent among themselves if p is less or equal to the minimum number of integration point that yields the exact expression for $\mathbf{K}$.

Eq.(E.3) can be rewritten in terms of the free DOFs, $a_1, a_2, \cdots, a_j$, after eliminating the prescribed DOFs, as

$$C_1^1(\alpha_1^1 a_1 + \cdots + \alpha_j^1 a_j) + C_2^1(\alpha_1^2 a_1 + \cdots + \alpha_j^2 a_j) + \cdots + C_{sp}^1(\alpha_1^j a_1 + \cdots + \alpha_j^j a_j) = f_1$$

$$C_1^2(\alpha_1^1 a_1 + \cdots + \alpha_j^1 a_j) + C_2^2(\alpha_1^2 a_1 + \cdots + \alpha_j^2 a_j) + \cdots + C_{sp}^q(\alpha_1^j a_1 + \cdots + \alpha_j^j a_j) = f_2$$

$$\vdots \qquad\qquad \vdots \qquad\qquad \vdots$$

$$C_1^q(\alpha_1^1 a_1 + \cdots + \alpha_j^1 a_j) + C_2^q(\alpha_1^2 a_1 + \cdots + \alpha_j^2 a_j) + \cdots + C_{sp}^q(\alpha_1^j a_1 + \cdots + \alpha_j^j a_j) = f_q$$

$$\vdots \qquad\qquad \vdots \qquad\qquad \vdots$$

$$C_1^j(\alpha_1^1 a_1 + \cdots + \alpha_j^1 a_j) + C_2^j(\alpha_1^2 a_1 + \cdots + \alpha_j^2 a_j) + \cdots + C_{sp}^j(\alpha_1^j a_1 + \cdots + \alpha_j^j a_j) = f_j$$

$$(E.4)$$

where $kp = k \times p$.

The matrix multiplying the variables $a_1, a_2, \cdots, a_j$ *will be singular* if the coefficients $C_1^i, C_2^i, \cdots, C_{k_p}^i$ of any of the above equations are a linear combination of the coefficients of any of the other rows. This is mathematically expressed as

$$
\begin{aligned}
C_1^i &= \beta_1 \, C_1^1 + \beta_2 \, C_1^2 + \cdots + \beta_r \, C_1^m \\
C_2^i &= \beta_1 \, C_2^1 + \beta_2 \, C_2^2 + \cdots + \beta_r \, C_2^m \\
&\vdots \qquad \vdots \qquad \vdots \qquad \vdots \qquad \vdots \\
C_{sp}^i &= \beta_1 \, C_{sp}^1 + \beta_2 \, C_{sp}^2 + \cdots + \beta_m \, C_{sp}^r
\end{aligned}
$$

$$(E.5)$$

with $r < j$. This system of equations can be solved for the unknowns $\beta_1, \beta_2, \cdots, \beta_r$, if the number of unknowns is greater or equal to the available equations. Hence, the sought singularity rule is

$$\boxed{r \geq s \times p} \qquad \text{or} \qquad \boxed{j - s \times p > 0} \qquad (E.6)$$

Eq.(E.6) coincides with Eq.(2.50).

Appendix F

COMPUTATION OF THE SHEAR CENTER AND THE WARPING FUNCTION IN THIN-WALLED OPEN COMPOSITE BEAM SECTIONS

We present a procedure for computing the shear center in beams with a thin-walled open section. The constrained torsion generates axial and shear stresses $\sigma_{x'}$ and $\tau_{x's}$ such that

$$
\begin{aligned}
\sigma_{x'}(x, s, \zeta) &= E\varepsilon_{x'}(x', s, \zeta) \\
\tau_{x's}(x, s', \zeta) &= G\gamma_{x's}(x', s, \zeta)
\end{aligned}
\tag{F.1}
$$

with $\varepsilon_{x'}$ and $\gamma_{x's}$ given by Eqs.(10.104).

The shear stresses due to torsion should only induce a torque $M_{\hat{x}'}$. Also, the axial stresses $\sigma_{x'}$ due to torsion must satisfy that the axial force N and the bending moments $M_{y'}$ and $M_{z'}$ are zero, i.e.

$$
N = \iint_A \sigma_{x'}\, dA = \frac{\partial^2 \theta_{\hat{x}'}}{\partial x'^2} \iint_A E\omega\, dA = 0
\tag{F.2a}
$$

$$
M_{y'} = \iint_A z'\sigma_{x'}\, dA = \frac{\partial^2 \theta_{\hat{x}'}}{\partial x'^2} \iint_A z'E\omega\, dA = 0
\tag{F.2b}
$$

$$
M_{z'} = -\iint_A y'\sigma_{x'}\, dA = -\frac{\partial^2 \theta_{\hat{x}'}}{\partial x'^2} \iint_A y'E\omega\, dA = 0
\tag{F.2c}
$$

where $dA = \left(1 - \frac{\zeta}{R}\right) ds\, d\zeta$ and ω is defined by Eq.(10.101).

Above conditions are satisfied if

$$
\iint_\Omega E\omega\, dA = \iint_A z'E\omega\, dA = \iint_\Omega y'E\omega\, dA = 0
\tag{F.3}
$$

E. Oñate, *Structural Analysis with the Finite Element Method. Linear Statics: Volume 2: Beams, Plates and Shells,* Lecture Notes on Numerical Methods in Engineering and Sciences, DOI 10.1007/978-1-4020-8743-1,
© International Center for Numerical Methods in Engineering (CIMNE), 2013

If $C(y'_c, z'_c)$ is not known, we can compute ω_D and the coordinates of C as follows. Taking C as a reference point, the coordinate vector of an arbitrary point p over the section middle line is

$$\mathbf{c} = \mathbf{r}_p - \mathbf{r}_c \tag{F.4}$$

with $\mathbf{r}_p = [y'_p, z'_p]^T$ and $\mathbf{r}_c = [y'_c, z'_c]^T$. From Eqs.(10.108), (10.90) and (10.91)

$$c_t = \mathbf{c}^T \mathbf{t} = c_t^0 - \left(y'_c \frac{\partial y'_p}{\partial s} + z'_c \frac{\partial z'_p}{\partial s} \right) \tag{F.5a}$$

$$c_n = \mathbf{c}^T \mathbf{n} = c_n^0 - \left(-y'_c \frac{\partial z'_p}{\partial s} + z'_c \frac{\partial y'_p}{\partial s} \right) \tag{F.5b}$$

with

$$c_t^0 = y'_p \frac{\partial y'_p}{\partial s} + z'_p \frac{\partial z'_p}{\partial s} \tag{F.6a}$$

$$r_n^0 = -y'_p \frac{\partial z'_p}{\partial s} + z'_p \frac{\partial y'_p}{\partial s} \tag{F.6b}$$

Substituting (10.113b) into (10.99) gives

$$\omega_s = \omega_s^0(s) + y'_c z'_p(s) - z'_c y'_p(s) \tag{F.7}$$

with

$$\omega_s^0(s) = \int_0^s r_n^0 ds = \int_D^p (-y'_p dz'_p + z'_p dy'_p) \tag{F.8}$$

where D is the end point of the section with $s = 0$.

Substituting Eq.(10.111) into (10.101a) and this into (10.106) gives

$$\iint_A E[(\omega_s^0 + y'_c z'_p + z'_c y'_p) + \omega_D - c_t \zeta] \, dA = 0 \tag{F.9a}$$

$$\iint_A z' E[(\omega_s^0 + y'_c z'_p + z'_c y'_p) + \omega_D - c_t \zeta] \, dA = 0 \tag{F.9b}$$

$$\iint_A y' E[(\omega_s^0 + y'_c z'_p + z'_c y'_p) + \omega_D - c_t \zeta] \, dA = 0 \tag{F.9c}$$

In above equations, z' and y' are given by Eq.(10.93) and c_t by Eq.(10.109a).

The system of Eqs.(10.125) allows us to compute ω_D and the coordinates y'_c, z'_c of the shear center C for thin-walled open composite beam.

For very thin walls $1 - \frac{\xi}{R} = 1$, $dA = ds d\xi$ and the terms in ζ can be neglected in above integrals.

If the material is homogeneous the constant ω_D can be computed from Eqs.(10.125a) as

$$\omega_D = -\frac{1}{A} \int_{l_s} \omega_s(s) t(s) ds \qquad (F.10)$$

where t is the wall thickness, l_s is the length of the wall member and ω_s is given by Eq.(10.111). ω_D can therefore be interpreted as *the average sectorial coordinate* (with opposite sign) over the wall member.

If in addition y', z' are the principal axes of inertia, we can compute y'_c, z'_c from Eqs.(10.125b,c) as

$$y'_c = -\frac{I_{\omega y'}}{I_{y'}} \quad , \quad z'_c = \frac{I_{\omega z'}}{I_{z'}} \qquad (F.11)$$

with

$$\begin{aligned} I_{\omega y'} &= \iint_A \omega_s^0 z' \, dA \simeq \int_{l_s} t \omega_s^0 z'_p \, ds \\ I_{\omega z'} &= \iint_A \omega_s^0 y' \, dA \simeq \int_{l_s} t \omega_s^0 y'_p \, ds \end{aligned} \qquad (F.12)$$

In the derivation of Eqs.(10.115) we have used the fact that ω_D is a constant and, hence,

$$\iint_A z' \omega_D dA = \omega_D \iint_A z' dA = 0 \quad \text{(same for } y'\text{)} \qquad (F.13)$$

In summary, the steps for computing the coordinates of the shear center $C(y'_c, z'_c)$ for an homogeneous section are:

- Define O and the principal inertia axes $I_{y'}, I_{z'}$ (Section 4.2.4).
- Define the coordinates $y_p(s)$ and $z_p(s)$ of a point on the middle line.
- Define ds (Section 4.10.1).
- Compute ω_s^0 (Eq.(C.8)).
- Compute $I_{\omega y'}$ and $I_{\omega z'}$ and then y'_z and z'_c (Eqs.(C.11)).

Once the coordinates of the shear center y'_c and z'_c are known, we can compute ω_s by (C.7) and then ω_D by Eq.(C.10) and g_s by Eq.(4.127).

Appendix G

STABILITY CONDITIONS FOR REISSNER-MINDLIN PLATE ELEMENTS BASED ON ASSUMED TRANSVERSE SHEAR STRAINS

Let us consider the governing equations for a Reissner-Mindlin (RM) plate written in the form [ZL,ZT2]

Definition of bending moments	:	$\mathbf{m} = \hat{\mathbf{D}}_b \, \mathbf{L} \, \boldsymbol{\theta}$	(G.1)
Equilibrium of bending moments	:	$\mathbf{L}^T \mathbf{m} + \mathbf{s} = 0$	(G.2)
Definition of shear forces	:	$\dfrac{1}{\beta}\mathbf{s} - \boldsymbol{\gamma} = 0$	(G.3)
Definition of shear strains	:	$\boldsymbol{\gamma} + \boldsymbol{\theta} - \boldsymbol{\nabla} w = 0$	(G.4)
Equilibrium of shear forces	:	$\boldsymbol{\nabla}^T \mathbf{s} + q = 0$	(G.5)

with

$$\beta = \alpha G t \quad , \quad \mathbf{m} = [M_x, M_y, M_{xy}]^T \quad , \quad \mathbf{s} = [Q_x, Q_y]^T$$

$$\boldsymbol{\theta} = [\theta_x, \theta_y]^T \quad , \quad \boldsymbol{\gamma} = [\gamma_x, \gamma_y]^T \quad , \quad \boldsymbol{\nabla} = [\frac{\partial}{\partial x}, \frac{\partial}{\partial y}]^T$$

and

$$\mathbf{L} = \begin{bmatrix} -\dfrac{\partial}{\partial x} & 0 & -\dfrac{\partial}{\partial y} \\[2mm] 0 & -\dfrac{\partial}{\partial y} & -\dfrac{\partial}{\partial x} \end{bmatrix}^T \qquad (G.6)$$

We retain the displacements $(w, \boldsymbol{\theta})$, the shear forces $\mathbf{s}$ and the transverse shear strains $\boldsymbol{\gamma}$ as the prime variables. Hence, substituting Eq.(G.1) into (G.2) gives

$$\mathbf{L}^T(\hat{\mathbf{D}}_f \, \mathbf{L} \, \boldsymbol{\theta}) + \mathbf{s} = 0 \qquad (G.7)$$

The resulting equations (G.1), (G.3), (G.4), (G.5) and (G.7) can be

E. Oñate, *Structural Analysis with the Finite Element Method. Linear Statics:*
Volume 2: Beams, Plates and Shells, Lecture Notes on Numerical Methods
in Engineering and Sciences, DOI 10.1007/978-1-4020-8743-1,
© International Center for Numerical Methods in Engineering (CIMNE), 2013

written in integral form using the weighted residual method [ZTZ,ZT2] as

$$\iint_A \mathbf{W}_1^T[\mathbf{L}^T(\hat{\mathbf{D}}_b\,\mathbf{L}\boldsymbol{\theta}) + \mathbf{s}]dA + \iint_A \mathbf{W}_2^T\left[\frac{1}{\beta}\mathbf{s} - \boldsymbol{\gamma}\right]dA +$$
$$+ \iint_A \mathbf{W}_3^T(\boldsymbol{\gamma} + \boldsymbol{\theta} - \nabla w)dA + \iint_A \mathbf{W}_{4^T}[\nabla^T\mathbf{s} + q]dA = 0 \ (\text{G.8})$$

The following FE approximation is used for the prime variables

$$\boldsymbol{\theta} = \mathbf{N}_\theta\bar{\boldsymbol{\theta}} \quad , \quad w = \mathbf{N}_w\,\bar{w} \quad , \quad \mathbf{s} = \mathbf{N}_s\,\bar{\mathbf{s}} \quad , \quad \boldsymbol{\gamma} = \mathbf{N}_\gamma\bar{\boldsymbol{\gamma}} \qquad (\text{G.9})$$

where $\mathbf{N}_\theta$, $\mathbf{N}_w$, $\mathbf{N}_s$ and $\mathbf{N}_\gamma$ are shape function matrices and $\bar{(.)}$ denotes nodal values.

The number of total "free" variables in the mesh for the rotations, the deflection, the transverse shear stresses and the transverse shear strains (after eliminating the prescribed DOFs) is n_θ, n_w, n_s and n_γ, respectively

Integrating by parts the first and the last integral in (G.7) and using the Galerkin method [ZTZ,ZT2] with $\mathbf{W}_1 = \mathbf{N}_\theta$, $\mathbf{W}_2 = \mathbf{N}_s$, $\mathbf{W}_3 = \mathbf{N}_\gamma$ and $\mathbf{W}_4 = \mathbf{N}_w$ yields the following system of equations

$$\begin{bmatrix} \mathbf{A} & \mathbf{B} & 0 & 0 \\ 0 & -\mathbf{E} & \frac{1}{\beta}\mathbf{S} & 0 \\ \mathbf{B}^T & & \mathbf{H} & \mathbf{C} \\ 0 & 0 & \mathbf{C}^T & 0 \end{bmatrix} \begin{Bmatrix} \bar{\boldsymbol{\theta}} \\ \bar{\boldsymbol{\gamma}} \\ \bar{\mathbf{s}} \\ \bar{w} \end{Bmatrix} = \begin{Bmatrix} \mathbf{f}_1 \\ 0 \\ 0 \\ \mathbf{f}_2 \end{Bmatrix} \qquad (\text{G.10})$$

where

$$\mathbf{A} = \iint_A [\mathbf{L}\mathbf{N}_\theta]^T\hat{\mathbf{D}}_b\mathbf{L}\mathbf{N}_\theta dA \ ; \ \mathbf{B} = \iint_A \mathbf{N}_\theta^T\mathbf{N}_\gamma dA \ ; \ \mathbf{S} = \iint_A \mathbf{N}_s^T\mathbf{N}_s dA$$
$$\mathbf{E} = \iint_A \mathbf{N}_s^T\mathbf{N}_\gamma dA \ ; \ \mathbf{H} = \iint_A \mathbf{N}_\gamma^T\mathbf{N}_\gamma dA \ ; \ \mathbf{C} = -\iint_A \mathbf{N}_\gamma^T\nabla\mathbf{N}_w dA$$
$$(\text{G.11})$$

and $\mathbf{f}_1$ and $\mathbf{f}_2$ depend on the external forces. In Eq.(G.10) we have eliminated the rows and columns associated to the prescribed DOFs.

The transverse shear strains $\bar{\boldsymbol{\gamma}}$ can be expressed in terms of the transverse shear stresses $\bar{\mathbf{s}}$ using the second row of Eq.(G.10) as

$$\bar{\boldsymbol{\gamma}} = \frac{1}{\beta}\mathbf{E}^{-1}\mathbf{S}\bar{\mathbf{s}} \qquad (\text{G.12})$$

Substituting this expression into the third row of Eq.(G.10) gives

$$\mathbf{B}^T\bar{\boldsymbol{\theta}} + \frac{1}{\beta}\hat{\mathbf{H}}\bar{\mathbf{s}} + \mathbf{C}\bar{w} = 0 \qquad (\text{G.13a})$$

with

$$\hat{\mathbf{H}} = \mathbf{H}\mathbf{E}^{-1}\mathbf{S} \qquad (G.13b)$$

Matrix $\mathbf{E}$ is invertible only if the number of transverse shear stress variables $\bar{\mathbf{s}}$ coincides with that of the transverse shear strains $\bar{\gamma}$ (i.e. $n_s = n_\gamma$). In this case $\mathbf{N}_s = \mathbf{N}_\gamma$ and, hence, $\mathbf{H} = \mathbf{E}$, $\mathbf{S} = \mathbf{H}$ and $\hat{\mathbf{H}} = \mathbf{H}$.

The system (G.10) can be therefore expressed in terms of the $\bar{\boldsymbol{\theta}}$, $\bar{\mathbf{s}}$ and $\bar{w}$ variables as

$$\begin{bmatrix} \mathbf{A} & \mathbf{B} & \mathbf{0} \\ \mathbf{B}^T & \frac{1}{\beta}\mathbf{H} & \mathbf{C} \\ \mathbf{0} & \mathbf{C}^T & \mathbf{0} \end{bmatrix} \begin{Bmatrix} \bar{\boldsymbol{\theta}} \\ \bar{\mathbf{s}} \\ \bar{w} \end{Bmatrix} = \begin{Bmatrix} \mathbf{f}_1 \\ \mathbf{0} \\ \mathbf{f}_2 \end{Bmatrix} \qquad (G.14)$$

The shear parameter $\beta = \alpha Gt$ is much larger than the bending parameter Et^3 for a thin plate (Section 0). In the thin limit the terms multiplying $1/\beta$ are irrelevant and Eq.(G.14) is equivalent to

$$\begin{bmatrix} \mathbf{A} & \mathbf{B} & \mathbf{0} \\ \mathbf{B}^T & \mathbf{0} & \mathbf{C} \\ \mathbf{0} & \mathbf{C}^T & \mathbf{0} \end{bmatrix} \begin{Bmatrix} \bar{\boldsymbol{\theta}} \\ \bar{\mathbf{s}} \\ \bar{w} \end{Bmatrix} = \begin{Bmatrix} \mathbf{f}_1 \\ \mathbf{0} \\ \mathbf{f}_2 \end{Bmatrix} \qquad (G.15)$$

The above system of equations can be regularized by adding $\beta\mathbf{C}$ times the third equation to the second one, where β is an arbitrary constant. This gives

$$\begin{bmatrix} \mathbf{A} & \mathbf{B} & \mathbf{0} \\ \mathbf{B}^T & \beta\mathbf{C}\mathbf{C}^T & \mathbf{C} \\ \mathbf{0} & \mathbf{C}^T & \mathbf{0} \end{bmatrix} \begin{Bmatrix} \bar{\boldsymbol{\theta}} \\ \bar{\mathbf{s}} \\ \bar{w} \end{Bmatrix} = \begin{Bmatrix} \mathbf{f}_1 \\ \beta\mathbf{C}\mathbf{f}_2 \\ \mathbf{f}_2 \end{Bmatrix} \qquad (G.16)$$

After elimination of $\bar{\boldsymbol{\theta}}$ we obtain

$$\begin{bmatrix} [\beta\mathbf{C}\mathbf{C}^T - \mathbf{B}^T\mathbf{A}^{-1}\mathbf{B}] & \mathbf{C} \\ \mathbf{C}^T & \mathbf{0} \end{bmatrix} \begin{Bmatrix} \bar{\mathbf{s}} \\ \bar{w} \end{Bmatrix} = \begin{Bmatrix} \beta\mathbf{C}\mathbf{f}_2 - \mathbf{B}^T\mathbf{A}^{-1}\mathbf{f}_1 \\ \mathbf{f}_2 \end{Bmatrix} \qquad (G.17)$$

If the number of rows in vector $\bar{w}$ (n_w) was greater than that in $\mathbf{s}$ (n_s), then the $\mathbf{s}$ variables could be directly obtained by the second row of (G.17) which would lead to infinity solutions for system. Hence, this system has a unique solution if

$$\boxed{n_s \geq n_w} \qquad (G.18)$$

We rewrite now (G.15) as

$$\begin{bmatrix} \mathbf{A} & \mathbf{0} & \mathbf{B}^T \\ \mathbf{0} & \mathbf{0} & \mathbf{C}^T \\ \mathbf{B}^T & \mathbf{C} & \mathbf{0} \end{bmatrix} \begin{Bmatrix} \bar{\boldsymbol{\theta}} \\ \bar{w} \\ \bar{\mathbf{s}} \end{Bmatrix} = \begin{Bmatrix} \mathbf{f}_1 \\ \mathbf{f}_2 \\ \mathbf{0} \end{Bmatrix} \qquad (G.19)$$

This matrix can be regularized similarly as explained above by multiplying the last row by $\beta \mathbf{B}$ and $\beta \mathbf{C}^T$ and adding the result to the first and second rows, respectively. This gives

$$
\left[\begin{array}{cc|c}
\mathbf{A} + \beta \mathbf{B}\mathbf{B}^T & \beta \mathbf{B}\mathbf{C} & \mathbf{B} \\
\beta \mathbf{C}^T \mathbf{B}^T & \beta \mathbf{C}^T \mathbf{C} & \mathbf{C}^T \\
\hline
\mathbf{B}^T & \mathbf{C} & 0
\end{array}\right]
\left\{\begin{array}{c} \bar{\theta} \\ \bar{w} \\ \bar{s} \end{array}\right\}
= \left\{\begin{array}{c} \mathbf{f}_1 \\ \mathbf{f}_2 0 \end{array}\right\}
\tag{G.20}
$$

Splitting the matrix in the form shown above, and using the same arguments used for obtaining Eq.(G.18) yields the condition for a unique solution as

$$
\boxed{n_\theta + n_w \geq n_s} \tag{G.21}
$$

which is the second condition sought.

Conditions (G.18) and (G.21) coincide with those of Eqs.(2.79) and (6.61) by making $n_s = n_\gamma$.

Demonstration of the singularity rule for $\mathbf{K}_s$

Let us consider the substitute transverse shear strain matrix for the whole mesh written as

$$
\hat{\mathbf{B}}_s = \underset{2\times j}{\mathbf{J}^{-1}} \; \underset{2\times 2}{\mathbf{A}} \; \underset{2\times n_\gamma n_\gamma}{\mathbf{P}^{-1}} \; \underset{\times n_\gamma n_\gamma \times 2 n_\gamma}{\mathbf{T}}
\left\{\begin{array}{c} \mathbf{B}_s^1 \\ \vdots \\ \mathbf{B}_s^{N_\gamma} \end{array}\right\}_{2 n_\gamma \times j}
= [JAP] \left\{\begin{array}{c} \overset{\circ}{\mathbf{B}}_{s_1} \\ \vdots \\ \overset{\circ}{\mathbf{B}}_{s_{N_\gamma}} \end{array}\right\}_{\substack{2\times n_\gamma \\ n_\gamma \times j}}
= [JAP]\overset{\circ}{\mathbf{B}}_s
\tag{G.22a}
$$

with

$$
\overset{\circ}{\mathbf{B}}_{s_i} = \mathbf{T}\mathbf{B}_s^i \tag{G.22b}
$$

where n_γ is the total number of assumed transverse shear strains and $j = n_\theta + n_w$ is the total number of displacement DOFs in the mesh (after eliminating the prescribed values). The transverse shear strain matrix is obtained as

$$
\underset{j\times j}{\mathbf{K}_s} = \iint_A \hat{\mathbf{B}}_s^T \hat{\mathbf{D}}_s \hat{\mathbf{B}}_s dA = \iint_A \overset{\circ}{\mathbf{B}}_s^T [JAP]^T \; \hat{\mathbf{D}}_s [JAP]\overset{\circ}{\mathbf{B}}_s dA =
$$

$$
= \overset{\circ}{\mathbf{B}}_s^T \underbrace{\left(\iint_A [JAP]^T \hat{\mathbf{D}}_s [JAP] dA \right)}_{\overset{\circ}{\mathbf{D}}_s} \overset{\circ}{\mathbf{B}}_s = \underset{j\times n_\gamma n_\gamma}{\overset{\circ}{\mathbf{B}}_s^T} \; \underset{\times n_\gamma n_\gamma \times j}{\overset{\circ}{\mathbf{D}}_s} \; \overset{\circ}{\mathbf{B}}_s \tag{G.23}
$$

where $\overset{\circ}{\mathbf{D}}_s$ represents the "exact" value of the integral shown.

The global equilibrium equations of a node (accounting only for the transverse shear stiffness contribution) is written as

$$\mathbf{f}_i = \sum_{k=1}^{N} \mathbf{K}_{s_{ik}}\,\mathbf{a}_k = \overset{\circ}{\mathbf{B}}{}^T_{s_i}\left[\overset{\circ}{\mathbf{D}}_s\,\overset{\circ}{\mathbf{B}}_{s_1}\mathbf{a}_1 + \overset{\circ}{\mathbf{D}}_s\overset{\circ}{\mathbf{B}}_{s_2}\mathbf{a}_2 + \cdots + \overset{\circ}{\mathbf{D}}_s\overset{\circ}{\mathbf{B}}_{s_N}\mathbf{a}_N\right]$$

$$(\text{G.24})$$

From Eq.(G.24) we deduce that vector $\mathbf{f}_i$ is a combination of n_γ linear relationships in $\mathbf{a}_1, \mathbf{a}_2, \cdots, \mathbf{a}_N$ where n_γ is the number of rows in $\overset{\circ}{\mathbf{B}}_{s_i}$ and N is the number of nodes in the mesh. Following the arguments of Appendix E we can conclude that the combination will yield a singular matrix if

$$\boxed{j = n_\theta + n_w > n_\gamma}\qquad\qquad(\text{G.25})$$

Recalling that $n_s = n_\gamma$, it is therefore concluded that the singularity condition (G.21) *incorporates* the singularity condition for the transverse shear stiffness matrix.

Appendix H

ANALYTICAL SOLUTIONS FOR ISOTROPIC THICK/THIN CIRCULAR PLATES

H.1 GOVERNING EQUATIONS

The governing equations for an isotropic thick/thin circular plate of radius R under uniform loading q (Reissner-Mindlin theory) can be written as [BD5,TW]

$$\frac{d}{dr}(Qr) + qr = 0 \quad ; \quad \frac{d}{dr}(rM_r) - M_\theta - rQ = 0 \qquad (H.1)$$

with

$$Q = D_s\left(\frac{d\omega_0}{dr} - \theta\right) \quad , \quad M_r = D\left(\frac{d\theta}{dr} + \nu\frac{\theta}{r}\right) \quad , \quad M_\theta = D\left(\nu\frac{d\theta}{dr} + \frac{\theta}{r}\right) \qquad (H.2)$$

and

$$D = \frac{Et^3}{12(1-\nu^2)} \quad , \quad D_s = k\frac{Et}{2(1+\nu)} \quad , \quad k = 5/6 \qquad (H.3)$$

These equations, with the appropriate conditions, can be integrated to give the following solutions.

H.2 CLAMPED PLATE UNDER UNIFORM LOADING q

Boundary conditions
$Q = 0, \quad \theta = 0 \quad$ at $\quad r = 0$
$w = \theta = 0, \quad$ at $\quad r = R$

Analytical solution

$$w = \frac{qR^4}{64D}(1 - \xi^2)[(1 - \xi^2) + \phi] \quad ; \quad \theta = \frac{1}{16}\frac{q}{D}r(R^2 - r^2)$$

E. Oñate, *Structural Analysis with the Finite Element Method. Linear Statics:*
Volume 2: Beams, Plates and Shells, Lecture Notes on Numerical Methods
in Engineering and Sciences, DOI 10.1007/978-1-4020-8743-1,
© International Center for Numerical Methods in Engineering (CIMNE), 2013

with

$$\phi = \frac{16D}{R^2 D_s} = \frac{16}{5}\left(\frac{t}{R}\right)^2 \frac{1}{1-\nu} \quad \text{and} \quad \xi = \frac{r}{R}$$

The deflection at the plate center is $w_c = \frac{qR^4}{64D}(1+\phi)$.
The Kirchhoff solution is found for $\phi = 0$.
The resultant stresses are

$$M_r = \frac{qR^2}{16}(1+\nu)\left(1 - \frac{3+\nu}{1+\nu}\xi^2\right)$$

$$M_\theta = \frac{qR^2}{16}(1+\nu)\left(1 - \frac{1+3\nu}{1+\nu}\xi^2\right) \quad , \quad Q = -\frac{rq}{2}$$

Note that M_r, M_θ and Q are independent of ϕ.

H.3 SIMPLY SUPPORTED PLATE UNDER UNIFORM LOAD q

Boundary conditions (soft support)

$Q = 0$ and $\theta = 0$ at $r = 0$
$w = 0$ and $M_r = 0$ at $r = R$

Analytical solution

$$w = \frac{qR^4}{64D}(1-\xi^2)\left[\frac{2(3+\nu)}{1+\nu} - (1+\xi^2) + \phi\right] \quad ; \quad \theta = \frac{qR^3}{16D}\left(\frac{3+\nu}{1+\nu}\xi - \xi^3\right)$$

The deflection at the center is $w_c = \frac{qR^4}{64D}\left(\frac{5+\nu}{1+\nu} + \phi\right)$.
For $\phi = 0$ (thin plate) then $w_c = \frac{qR^4}{64D}\left(\frac{5+\nu}{1+\nu}\right)$.
The maximum rotation is

$$\theta(r = R) = \frac{qR^3}{8D}\frac{1}{(1+\nu)}$$

The resultant stresses are

$$M_r = \frac{qR^2}{16}(3+\nu)(1-\xi^2)$$

$$M_\phi = \frac{qR^2}{16}(3+\nu)\left(1 - \frac{1+3\nu}{3+\nu}\xi\right) \quad , \quad Q = -\frac{q}{2}$$

H.4 CLAMPED PLATE UNDER POINT LOAD AT THE CENTER

The equilibrium equation for the vertical load P is [BD5,TW]

$$2n\frac{\partial(rQ)}{\partial r} + P\delta(r = 0) = 0$$

The boundary conditions are given in Section H.2.
The analytical solution is

$$w = \frac{PR^2}{16\pi D}\left(1 - \xi^2 + 2\xi^2\ln\xi - \frac{1}{2}\phi\ln\xi\right) \quad ; \quad \theta = \frac{P}{4\pi D}r\ln\frac{R}{r}$$

The resultant stresses are

$$M_r = -\frac{P}{4\pi}[1 + (1+\nu)\ln\xi] \quad , \quad M_\theta = -\frac{P}{4\pi}[\nu + (1+\nu)\ln\xi] \quad , \quad Q = -\frac{P}{2\pi r}$$

Note that M_r, M_θ and Q are theoretically infinite under the load. Also w tends toward the infinite as r tends to zero. This happens only due to the influence of the transverse shear term. The Kirchhoff solution at the center ($\phi = r = 0$) is finite $\left(w_c = \frac{PR^2}{46\pi D}\right)$. This shows that we should not consider plates under point loads for comparison with elements derived from Reissner-Mindlin plate theory.

H.5 SIMPLY SUPPORTED PLATE UNDER CENTRAL POINT LOAD

The boundary conditions are given in Section H.3.
The displacement solution is

$$w = \frac{PR^2}{16\pi D}\left[(1 - \xi^2)\frac{3+\nu}{1+\nu} + 2\xi^2\ln\xi - \frac{1}{2}\phi\ln\xi\right]$$

$$\theta = \frac{PR}{4\pi D}\xi\left(\frac{1}{1+\nu} - \ln\xi\right) \quad , \quad Q = -\frac{P}{2\pi r}$$

The resultant stresses are

$$M_r = -\frac{P}{4\pi}(1+\nu)\ln\xi \quad , \quad M_\theta = -\frac{P}{4\pi}[(1+\nu)\ln\xi - 1 + \nu]$$

For $\phi = 0$ and $\xi = 0$ (thin plate solution) we obtain $w_c = \frac{PR^2}{16\pi D}\left(\frac{3+\nu}{1+\nu}\right)$.

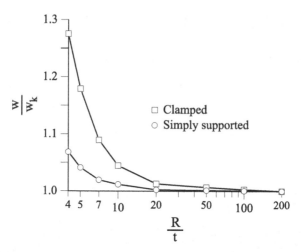

Fig. H.1 Circular plate under uniform loading. Influence of the ratio R/t on the ratio between the central deflection values obtained with Reissner-Mindlin (w) and Kirchhoff (w_k) theories

H.6 INFLUENCE OF THICKNESS IN THE SOLUTION FOR UNIFORM LOADED PLATE

Figure H.1 shows the influence of the ratio $\frac{R}{t}$ on the ratio $\frac{w}{w_k}$ at the plate center for isotropic circular plates under uniform loading q, $\nu = 0.3$ and $k = 5/6$. The analytical expressions are:

$$\text{Clamped:} \quad \left(\frac{w}{w_k}\right)_c = 1 + 4.57 \left(\frac{R}{t}\right)^{-2} \quad \text{with} \quad w_k = \frac{qR^4}{64D}$$

$$\text{Simply supported:} \quad \left(\frac{w}{w_k}\right)_c = 1 + 1.121 \left(\frac{R}{t}\right)^{-2} \quad \text{with} \quad w_k = \frac{qR^4}{64D}\left(\frac{5+\nu}{1+\nu}\right)$$

The central deflection for the thick (Reissner-Mindlin) solution is always greater that the thin (Kirchhoff) one. This is due to the higher deformation capacity of the thick plate formulation (as it accounts for transverse shear strain deformation), which leads to a more flexible solution than the traditional thin plate theory.

Appendix I

SHAPE FUNCTIONS FOR SOME C^0 CONTINUOUS TRIANGULAR AND QUADRILATERAL ELEMENTS

I.1 TRIANGULAR ELEMENTS

I.1.1 3-noded triangle

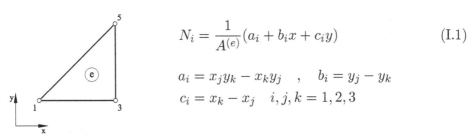

$$N_i = \frac{1}{A^{(e)}}(a_i + b_i x + c_i y) \tag{I.1}$$

$$a_i = x_j y_k - x_k y_j \quad , \quad b_i = y_j - y_k$$
$$c_i = x_k - x_j \quad i, j, k = 1, 2, 3$$

a_i, b_i and c_i are obtained by cyclic permutation of indexes i, j, k.

I.1.2 6-noded triangle (straight sides)

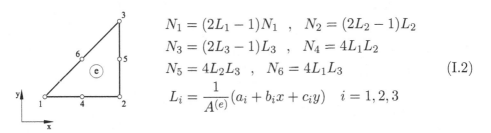

$$N_1 = (2L_1 - 1)N_1 \quad , \quad N_2 = (2L_2 - 1)L_2$$
$$N_3 = (2L_3 - 1)L_3 \quad , \quad N_4 = 4L_1 L_2$$
$$N_5 = 4L_2 L_3 \quad , \quad N_6 = 4L_1 L_3 \tag{I.2}$$
$$L_i = \frac{1}{A^{(e)}}(a_i + b_i x + c_i y) \quad i = 1, 2, 3$$

a_i, b_i and c_i as in Eq.(I.1).

E. Oñate, *Structural Analysis with the Finite Element Method. Linear Statics: Volume 2: Beams, Plates and Shells*, Lecture Notes on Numerical Methods in Engineering and Sciences, DOI 10.1007/978-1-4020-8743-1,
© International Center for Numerical Methods in Engineering (CIMNE), 2013

I.2 QUADRILATERAL ELEMENTS

I.2.1 4-noded rectangle

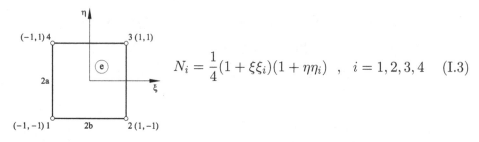

$$N_i = \frac{1}{4}(1 + \xi\xi_i)(1 + \eta\eta_i) \quad , \quad i = 1, 2, 3, 4 \quad (I.3)$$

I.2.2 8-noded Serendipity rectangle

$$N_i = \frac{1}{2}(1 + \xi\xi_i)(1 - \eta^2) \quad , \quad i = 4, 8$$

$$N_i = \frac{1}{2}(1 + \eta\eta_i)(1 - \xi^2) \quad , \quad i = 2, 6 \quad (I.4)$$

$$N_i = \frac{1}{4}(1 + \xi\xi_i)(1 + \eta\eta_i)(\xi\xi_i + \eta\eta_i - 1)$$
$$i = 1, 3, 5, 7$$

I.2.3 9-noded Lagrangian rectangle

$$N_i = \frac{1}{4}(\xi^2 + \xi\xi_i)(\eta^2 + \eta\eta_i) \quad , \quad i = 1, 3, 5, 7$$

$$N_i = \frac{1}{2}\eta_i^2(\eta^2 - \eta\eta_i)(1 - \xi^2)$$

$$+ \frac{1}{2}\xi_i^2(\xi^2 - \xi\xi_i)(1 - \eta^2) \quad , \quad i = 2, 4, 6, 8$$

$$N_9 = (1 - \xi^2)(1 - \eta^2) \quad (I.5)$$

More information on the shape functions of 2D C^0 triangular and quadrilateral elements can be found in Oñate [On4].

References

[AA] Aitharaju, V.R. and Averill, R.C., An assessment of zig-zag kinematic displacement models for the analysis of laminated composites. *Mech. Composite Mat. and Struct.*, **6**, 1–26, 1999a.

[AA2] Aitharaju, V.R. and Averill, R.C., C° zig-zag finite element for analysis of laminated composite beams. *J. Engineering Mechanics, ASCE*, 323–330, 1999b.

[AB] Agarwal, B.B. and Broutman, L.J., *Analysis and Performance of Fiber Composites*. J. Wiley, 1980.

[AC] Adini, A. and Clough, R.W., Analysis of plate bending by the finite element method. *Nat. Sci. Found*, G7337, 1961.

[AD] Abel, J.F. and Desai, C.S., Comparison of finite elements for plate bending. *Journal of Structural Divison*, ASCE, **98**, ST9, 2143–48, 1972.

[AF] Arnold, D.N. and Falk, S.N., An uniformly accurate finite element method for the Mindlin-Reissner plate. Preprint Series No. 307, Inst. for Mathematics and its Applicat., Univ. of Minnesota, Abril, 1987.

[AFS] Argyris, J.H., Fried, I. and Scharpf, D.W., The TUBA family of plate elements for the matrix displacement method. *Aeronautical Journal of the Royal Aeron. Society*, **72**, 701–9, 1968.

[AG] Ashwell, D.G. and Gallagher, R.H., (eds.), *Finite element method for thin shells and curved members*. J. Wiley, 1976.

[Ah] Ahmad, S., *Curved finite elements in the analysis of solid shell and plate structures*. Ph.D. Thesis, Univ. of Wales, Swansea, 1969.

[AHM] Argyris, J.H., Haase, M. and Mlejnek, H.P., On an unconventional but natural formulation of a stiffness matrix. *Comput. Methods Appl. Mech. Engrg.*, **22**, 1–2, 1980.

[AHMS] Argyris, J.H., Haase, M., Mlejnek, H.P. and Schmolz, P.K., TRUNC for shells. An element possibly to the taste of Bruce Irons, *Int. J. Numer. Meth. Engng.*, **22**, 93–115, 1986.

[AIZ] Ahmad, S., Irons, B.M. and Zienkiewicz, O.C., Curved thick shell and membrane elements with particular reference to axisymmetric problems.

Proc. 2nd Conf. Matrix Methods in structural Mech, Air Force Inst. of Technol., Wright–Patterson A.F. Base, Ohio, 1968.

[AIZ2] Ahmad, S., Irons, B.M. and Zienkiewicz, O.C., Analysis of thick and thin shell structures by curved finite elements. *Int. J. Numer. Meth. Engng.*, **2**, 419–451, 1970.

[Ak] Akoussah, E., *Analyse non linéaire des structures à parois minces par éléments finis et son application aux bâtiments industriels*. Thèse de Doctorat, Université Laval, 1987.

[Al] Allman, D.L., A compatible triangular element including vertex rotations for plane elasticity analysis. *Computers and Structures*, **19**, 1–8, 1984.

[Al2] Allman, D.L., A quadrilateral finite element including vertex rotations for plane elasticity analysis. *Int. J. Numer. Meth. Engng.*, **26**, 717–730, 1988.

[Al3] Allman, D.L., Evaluation of the constant strain triangle with drilling rotations. *Int. J. Numer. Meth. Engng.*, **26**, 2645–2655, 1988.

[AL] Auricchio, F. and Lovadina, C., Analysis of kinematic linked interpolation methods for Reissner-Mindlin plate problems. *Comput. Methods Appl. Mech. Engrg.*, **190**, 2465–2482, 2001.

[AMR] Alfano, G., Marotti de Sciarra, F. and Rosati, L., Automatic analysis of multicell thin-walled sections. *Computers and Structures*, **59**(4), 641–655, 1996.

[APAK] Argyris, J.H., Papadrakakis, M., Apostolopoulou, C. and Koutsourelakis, S., The TRIC shell element: theoretical and numerical investigation. *Comput. Methods Appl. Mech. Engrg.*, **182**, 217–245, 2000.

[Ar] Argyris, J.H., Matrix displacement analysis of anisotropic shells by triangular elements. *J. Roy. Aero. Soc.*, **69**, 801–5, 1965.

[Ar2] Argyris, J.H., Continua and discontinua. In Proc. 1st Conf. Matrix Methods in Structural Mechanics. Volume AFFDL-TR-66-80, pp. 11–189, Wright Patterson Air Force Base, Ohio, October 1966.

[AR] Alwar, R.S. and Ramachandran, K.N., Theorical and photoelastic analysis of thick slabs subjected to highly localised loads. *Building Science*, **7**, 159–66, 1972.

[As] Ashwell, D.G., Strain elements with applications to arches, rings and cylindrical shells, en *Finite elements for thin shells and curved members*. D.G. Ashwell and R.H. Gallagher (eds.), John Wiley, 91–111, 1976.

[AS] Abramowitz, M. and Stegun, I.A. (eds.), *Handbook of Mathematical Functions*. Dover Publications, New York, 1965.

[ASR] Ashwell, D.G., Sabir, A.B. and Roberts, T.M., Further studies in the application of curved finite elements to circular arches. *Int. J. Mech. Science*, **13**, 6, 507–17, 1971.

[AT] Auricchio, F. and Taylor, R.L., A shear deformable plate element with an exact thin limit. *Comput. Methods Appl. Mech. Engrg.*, **118**, 393–412, 1994.

[AT2] Auricchio, F. and Taylor, R.L., A triangular thick plate finite element with an exact thin limit. *Finite Elements in Analysis and Design*, **19**, 57–68, 1995.

[ATO] Argyris, J.H., Tenek, L. and Olofsson, L., TRIC: a simple but sophisticated 3-node triangular element based on six rigid-body and 12 straining modes for fast computational simulations of arbitrary isotropic and laminated composite shells. *Comput. Methods Appl. Mech. Engrg.*, **145**, 11–85 ,1997.

[ATPA] Argyris, J.H., Tenek, L., Papadrakakis, M., Apostolopoulou, C., Post-buckling performance of the TRIC natural mode triangular element for isotropic and laminated composite shells. *Comput. Methods Appl. Mech. Engrg.*, **166**, 211–231, 1998.

[AU] Alam, N.M. and Upadhyay, N.Kr., Finite element analysis of laminated composite beams for zigzag theory using MATLAB. *Int. J. of Mechanics and Solids*, **5**(1), 1–14, 2010.

[Av] Averill, R.C., Static and dynamic response of moderately thick laminated beams with damage. *Composites Engineering*, **4**(4), 381–395, 1994.

[AY] Averill, R.C. and Yuen Cheong Yip, Development of simple, robust finite elements based on refined theories for thick laminated beams. *Computers and Structures*, **59**(3), 529–546, 1996.

[BA] Build Air Engineering and Architecture SA (www.buildair.com).

[Bab] Babuška, I., The stability of domains and the question of formulation of plate problems. *Appl. Math.*, 463–467, 1962.

[Ban] Bank, L., Shear coefficients for thin-walled composite beams. *Composite Structures*, **8**, 47–61, 1987.

[Bar] Barnes, M.R., Form finding and analysis of tension space structure by dynamic relaxation. *PhD Thesis*, Department of Civil Engineering, The City University, London, 1977.

[Bar2] Barbero, E.J., *Finite element analysis of composite materials*, CRC Press, 2008.

[Bat] Batoz, J.L., An explicit formulation for an efficient triangular plate bending element, *Int. J. Numer. Meth. Engng.*, **18**, 1077–89, 1982.

[Bat2] Bathe, K.J., *Finite element procedures*. Prentice Hall, Inc., 1996.

[BB] Bank, L.C. and Bednarczyk, P.J., A beam theory for thin-walled composite beams. *Composites Science and Technology*, **32**(4), 265–277, 1988.

[BBH] Batoz, J.L., Bathe, K.J. and Ho, L.W., A study of three node triangular plate bending elements. *Int. J. Numer. Meth. Engng.*, **15**, 1771–812, 1980.

[BBHH] Benson, D.J., Bazilevs, Y., Hsu, M.C. and Hughes, T.J.R., Isogeometric shell analysis: The Reissner-Mindlin shell. *Comput. Methods Appl. Mech. Engrg.*, **199**, 276–289, 2010.

[BBR] Braun, M., Bischoff, M. and Ramm, E., Nonlinear shell formulations for complete three-dimensional constitutive laws including composites and laminates. *Comput. Mech.*, **15**, 1–18, 1994.

[BBt] Batoz, J.L. and Ben Tahar, M., Evaluation of a new quadrilateral thin plate bending element, *Int. J. Numer. Meth. Engng.*, **18**, 1655–77, 1982.

[BC] Bauchau, O.A. and Craig, J.I., *Structural Analysis with Applications to Aerospace Structures*. Springer, 2009.

[BCC+] Bazilevs, Y., Calo, V.M., Cottrell, J.A., Evans, J.A., Hughes, T.J.R., Lipton, S., Scott, M.A. and Sederberg, T.W., Isogeometric analysis using T-splines. *Comput. Methods Appl. Mech. Engrg.*, **199**, 229–263, 2010.

[BCIZ] Bazeley, G.P., Cheung, Y.K., Irons, B.M. and Zienkiewicz, O.C., Triangular elements bending-conforming and non conforming solution. *Proc. Conf. Matrix Meth. in Struct. Mech.*, Air Force Inst. of Tech., Wright Patterson A.F. Base, Ohio, 1965.

[BD] Bathe, K.J. and Dvorkin, E.N., A four node plate bending element based on Mindlin-Reissner plate theory and mixed interpolation. *Int. J. Numer. Meth. Engng.*, **21**, 367–383, 1985.

[BD2] Bathe, K.J. and Dvorkin, E.N., A formulation of general shell elements. The use of mixed interpolation of tensorial components. *Int. J. Numer. Meth. Engng.* **22**, 697–722, 1986.

[BD3] Batoz, J.L. and Dhatt, G., A state of the art on the discrete Kirchhoff plate bending elements. In *Calcul des Structures et Intelligence Artificielle*, J.M. Fonet, P. Ladeveze and R. Ohayon (Eds.), Ed. Pluralis, 1988.

[BD4] Batoz, J.L. and Dhatt, G., *Modelisation des structures par élements finis*. Vol. **1**: *Solides elastiques*, Hermes, Paris, 1990.

[BD5] Batoz, J.L. and Dhatt, G., *Modelisation des structures par élements finis*. Vol. **2**: *Poutres et plaques*, Hermes, Paris, 1990.

[BD6] Batoz, J.L. and Dhatt, G., *Modelisation des structures par élements finis*. Vol. **3**: *Coques*, , Hermes, Paris, 1990.

[BD7] Bucciarelli, L.L. and Dworsky, N., *Sophie Germain: an Essay in the History of the Theory of Elasticity*. Reidel, New York, 1980.

[Be] Bell, K., A refined triangular plate bending finite element. *Int. J. Numer. Meth. Engng.*, **1**, 101-22. 1969.

[Be2] Bert, C.W., Simplified analysis of static shear factors for beams of non homogeneous cross sections. *J. Comp. Mat.*, **7**, 525–529, 1973.

[Be3] Beyer, W.H., *CRC Standard Mathematical Tables*. CRC Press, 28th ed., 1988.

[Bel] Belytschko, T., A review of recent developments in plate and shell elements. In *Computational Mechanics - Advances and Trends*, AMD, Vol. **75**, ASME, New York, 1986.

[BF] Bergan, P.G. and Felippa, C.A., A triangular membrane element with rotational degrees of freedom. *Comput. Methods Appl. Mech. Engrg.*, **50**, 25–69, 1985 .

[BF2] Bergan, P.G. and Felippa, C.A., Efficient implementation of a triangular membrane element with drilling freedoms. In *Finite Element Methods for Plate and Shell Structures*, T.J.R. Hughes and E. Hinton (eds.), **1**, 128–152, Pineridge Press, Swansea, 1986.

[BFS] Bogner, F.K., Fox, R.L. and Schmit, L.A., The generation of interelement compatible stiffness and mass matrices by the use of interpolation formulae. *Proc. Conf. Matrix Methods in Struct. Mech.*, Air Force Inst. of Tech., Wright Patterson A. F. Base, Ohio, 1965.

[BFS2] Bogner, F.K., Fox, R.L. and Schmit, L.A., A cylindrical shell discrete element, *AIAA Journal*, **5**, 4, 745–50, 1967.

[BFS3] Brezzi, F., Fortin, M. and Stenberg, R., Error analysis of mixed interpolated elements for Reissner-Mindlin plates, *Mathematical Models and Methods in Appl. Sciences*, **1**, 2 125–51, 1991.

[BG] Bert, C.W. and Gordaninejad, F., Transverse shear effects in bimodular composite laminates. *J. Comp. Mat.*, **17**, 282–298, 1983.

[BH] Benson, P.R. and Hinton, E., A thick finite strip solution for static free vibration and stability problems. *Int. J. Numer. Meth. Engng.*, **10**, 665–678, 1976.

[BH2] Bergan, P.G. and Hanssen, L., A new approach for deriving 'good' element stiffness matrices. In J.R. Whiteman (ed.), The Mathematics of Finite Elements and Applications, pages 483–497, Academic Press, London, 1977.

[BK] Batoz, J.L. and Katili, I., On a simple triangular Reissner-Mindlin plate element based on incompatible modes and discrete constraints. *Int. J. Numer. Meth. Engng.*, **35**, 1603–1632, 1992.

[Bl] Blanco, E., *Estudio numérico y experimental de la influencia de distintos parámetros en la respuesta de tableros oblícuos de puentes de sección transversal losa.* Ph.D. Thesis, E.T.S. de Ingenieros de Caminos, Univ. Politècnica de Catalunya, 1988.

[BL] Batoz, J.L. and Lardeur, P.A., A discrete shear triangular nine DOF element for the analysis of thick to very thin plates. *Int. J. Numer. Meth. Engng.*, **28**, 5, 1989.

[BLD] Barbero, E.J., Lopez-Anido, R. and Davalos, J.F., On the mechanics of thin-walled laminated composite beams. *Journal of Composite Materials*, **27**(8), 806–829, 1993.

[BLOL] Belytschko, T., Liu, W.K., Ong, J.S.J. and Lam, D., Implementation and application of a 9-node Lagrange shell element with spurious mode control. *Computers and Structures*, **20**(1), 121–128, 1985.

[BN] Bergan, P.G. and Nygard, M.K., Finite elements with increased freedom in choosing shape functions. *Int. J. Numer. Meth. Engng.*, **20**, 643–663, 1984.

[Bo] Bouabdallah, S., Détermination des facteurs de correction de cisaillement et des rigidités de torsion des poutres composites par éléments finis. Rapport interne, Division MNM, Université de Technologie de Compiègne (UTC), 1990.

[BOM] Botello, S. Oñate, E. and Miquel, J., A layer-wise triangle for analysis of laminated composite plates and shells. *Computers and Structures*, **70**, 635–646, 1999.

[BR] Buechter, N. and Ramm, E., *Comparison of shell theory and degeneration.* In Nonlinear Analysis of Shells using Finite Elements, CISM, Udine, Italy, June 1991.

[BR2] Buechter, N. and Ramm, E., 3D-extension of nonlinear shell equations based on the enhanced assumed strain concept. In *Computational Methods in Applied Sciences*, C. Hirsch (Ed.), Elsevier, pp. 39–59, 1992.

[BR3] Bischoff, M. and Ramm, E., Shear deformable shell elements for large strains and rotations. *Int. J. Numer. Meth. Engng.*, **40**, 4427–4449, 1997.

[BRR] Buechter, N., Ramm, E. and Roehl, D., Three-dimensional extension of nonlinear shell formulation based on the enhanced assumed strain concept. *Int. J. Numer. Meth. Engng.*, **37**, 2551–2568, 1994.

[BRI] Baldwin, J.T., Razzaque, A. and Irons, B.M., Shape functions subroutine for an isoparametric thin plate element. *Int. J. Numer. Meth. Engng.*, **7**, 431–440, 1973.

[BS] Babuška, I. and Scapolla, T., Benchmark computation and performance evaluation for a rhombic plate bending problem. *Int. J. Numer. Meth. Engng.*, **28**, 155–180, 1989.

[BS2] Budiansky, B. and Sanders, J.L. Jr., On the best first order linear shell theory. *Progress in Applied Mech.*, MacMillan, New York, 1963.

[BS3] Booker, J.R. and Small, J.C., Finite layer analysis of consolidation. Part I. *Int. J. for Num. and Anal. Meth. in Geomechanics*, **6**(2), 173–194, 1982.

[BS4] Booker, J.R. and Small, J.C., Finite layer analysis of viscoelastic layered materials. *Int. J. for Num. and Anal. Meth. in Geomechanics*, **10**, 415–430, 1986.

[BS5] Brunet, M. and Sabourin, F., A simplified triangular shell element with a necking criterion for 3-D sheet-forming analysis. *J. Mater. Process. Technol.*, **50**, 238–251, 1995.

[BS6] Brunet, M. and Sabourin, F., Analysis of a rotation-free shell element. *Int. J. Numer. Methods Engrg.*, **66**, 1483–1510, 2006.

[BSC] Belytschko, T., Stolarski, H. and Carpenter, N., A C_0 Triangular plate element with one point quadrature. *Int. J. Numer. Meth. Engng.*, **20**, 787–802, 1984.

[BSL+] Belytschko, T., Stolarski, H., Liu, W.K., Carpenter, N. and Ong, J.S.J., Stress projection for membrane and shear locking in shell finite elements. *Comput. Methods Appl. Mech. Engrg.*, **51**(1), 221–258, 1985.

[BT] Belytschko, T. and Tsay, C.S., A stabilization procedure for the quadrilateral plate element with one point quadrature. *Int. J. Numer. Meth. Engng.*, **19**, 405–19, 1983.

[BT2] Bauld, N.R. and Tzeng, L.-S., A Vlasov theory for fiber-reinforced beams with thin-walled open cross sections. *International Journal of Solids and Structures*, **20**(3), 277–297, 1984.

[BTL] Belytschko, T., Tsay, C.S. and Liu, W.K., A stabilization matrix for the bi-linear Mindlin plate element. *Comput. Methods Appl. Mech. Engrg.*, **29**, 313-327, 1981.

[BW] Bergan, P.G. and Wang, X., Quadrilateral plate bending elements with shear deformations. *Comput. Methods Appl. Mech. Engrg.* **19** (1–2), 25–34, 1984.

[BW2] Back, S.Y. and Will, K.M., Shear-flexible thin-walled element for composite i-beams. *Engineering Structures*, **30**(5), 1447–1458, 2008.

[BWBR] Bischoff, M., Wall, W.A., Bletzinger, K.-U. and Ramm, E., Models and finite elements for thin-walled structures. In E. Stein, R. de Borst, T.J.R. Hughes (Eds.), Encyclopedia of Computational Mechanics, Solids Structures and Coupled Problems, vol. 2, Wiley, (Chapter 3), 2004.

[BWS] Belytschko, T., Wong, B.L. and Stolarski, H., Assumed strain stabilization procedure for the 9-node Lagrange shell element. *Int. J. Numer. Meth. Engng.*, **28**(2), 385–414 1989.

[Ca] Carrera, E., Historical review of zigzag theories for multilayered plate and shell. *Applied Mechanics Review*, **56**(3), 287–308, 2003.

[Cal] Calladine, C.R., *The theory of shell structures*. Cambridge University Press, 1983.

[Cas] Casadei, F., *A bibliographic study of finite elements for the elasto-plastic analysis of 3D shell like structures subjected to static and dynamic loading.* Technical Report No. 1.06.C1.86.79, Commission of the European Communities, Joint Research Center, Ispra, Italy, 1986.

[CB] Connor, J. and Brebbia, C., A stiffness matrix for a shallow rectangular shell element. *Journal of Engng. Mech. Div.*, ASCE, **93**, 43–65, 1967.

[CBS] Carpenter, N., Belytschko, T. and Stolarski, H., Locking and shear scaling factors in C_0 bending elements. *Comput. and Struct.*, **22**, 39–52, 1986.

[CC] Cantin, G. and Clough, R.W., A curved cylindrical shell finite element. *AIAA Journal*, **6**, 1057–62, 1968.

[CD] Coull, A. and Das, Y.P.C., Analysis of curved bridge decks. *Proc. Institution Civil Engineers*, **37**, 75–85, 1987.

[CF] Clough, R.W. and Felippa, C.A., A refined quadrilateral element for analysis of plate bending. *Proc. 2d Conf. Mat. Meth. Struct. Mech.*, AFIT, Wright-Patterson, Air Force Base, Ohio, 399–440, 1968.

[CHB] Cottrell, J.A., Hughes, T.J.R. and Bazilevs, Y., *Isogeometric Analysis Towards Integration of CAD and FEA*. J. Wiley, 2009.

[CKLO] Cowper, G.R., Kosko, E., Lindberg, G.M. and Olson, D.M., A high precision triangular plate bending element. *Report LR–514, National Aeronautical Establishment*, National Research Council of Canada, Ottawa, 1968.

[CKZ] Cheung, Y.K., King, I.P. and Zienkiewicz, O.C., Slab bridges with arbitrary shape and support condition - A general method of analysis based or finite elements. *Proc. Inst. Civil Engrg.*, **40**, 9–36, 1968.

[CLO] Cowper, G.R., Lindberg, G.M. and Olson, M.D., A shallow finite element of triangular shape. *Int. J. Solids and Struct.*, **6**, 8, 1133–56, 1970.

[CMPW] Cook, R.D., Malkus, D.S., Plesha, M.E. and Witt, R.J., *Concepts and applications of finite element analysis*. 4th edition, Wiley, 2002.

[Co] Cox, M.G., The numerical evaluation of B-splines. Technical report. National Physics Laboratory, DNAC 4, 1971.

[Co1] Cohen, G.A., Transverse shear stiffness of laminated anisotropic shells. *Comput Meth. Appl. Mech. Engng.*, **13**, 205–220, 1978.

[Co2] Courtney, T.H., *Mechanical behaviour of materials*. McGraw-Hill, 1990.

[Co3] Cook, R.D., On the Allman triangle and a related quadrilateral element. *Computers and Structures*, **2**, 1065–1067, 1986.

[Co4] Cook, R.D., A plane hybrid element with rotational d.o.f. and adjustable stiffness. *Int. J. Numer. Meth. Engng.*, **24**, 1499–1508, 1987.

[Co5] Cook, R.D., Modified formulations for nine-DOF plane triangles that include vertex rotations. *Int. J. Numer. Meth. Engng.*, **31**, 825–835, 1991.

[Co6] Cowper, G.R., The shear coefficient in Timoshenko's beam theory. *J.A.M.*, 335–340, June 1966.

[CP] Crisfield, M.A., and Puthli, R.S., Approximations in the non linear analysis of thin plate structures. In *Finite elements in non linear mechanics*, P. Bergan et al. (Eds.), **1**, Tapir, Trondheim, Noruega, 373–92, 1978.

[Cr] Crisfield, M.A., *Finite element and solution procedures for structural analysis, I: Linear analysis*. Pineridge Press, 1986.

[Cr2] Crisfield, M.A., A four-noded thin-plate bending element using shear constrains. A modified version of Lyon's element. *Comput. Methods Appl. Mech. Engrg.*, **38**, 93–120, 1983.

[Cr3] Crisfield, M.A., A quadratic Mindlin element using shear constraints. *Computers and Structures*, **18**, 833–52, 1984.

[Cr4] Crisfield, M.A., Explicit integration and the isoparametric arch and shell elements. *Communications in Applied Numerical Methods*, **2**(2), 181–187, 1986.

[ChR] Chao, W.C. and Reddy, J.N., Analysis of laminated composite shells using a degenerated 3D element. *Int. J. Numer. Meth. Engng.*, **20**(11), 1991–2007, 1984.

[CRE] Cohen, E., Riesenfeld, R.F. and Elber, G., *Geometric Modeling with Splines: An Introduction*. A.K. Peters, Natick, MA, pp. 638, 2001.

[CSB] Carpenter, N., Stolarski, H. and Belytschsko, T., A flat triangular shell element with improved membrane interpolation. *Communications in Applied Numerical Methods*, **1**, 161–168, 1985.

[CSB2] Carpenter, N., Stolarski, H. and Belytschsko, T., Improvements in 3 node triangular shell elements. *Int. J. Numer. Meth. Engng.*, **23**, 1643–67, 1986.

[CT] Clough, R.W. and Tocher, J.L., Finite element stiffness matrices for analysis of plates in bending. *Proc. Conf. Matrix Meth. in Struct. Mech*, AFIT, Wright-Patterson, Air Force Base, Ohio, 515-45. 1965.

[CW] Clough, R.W. and Wilson, E.L., Dynamic finite element analysis of arbitrary thin shells. *Computers and Structures*, **1**, 33–56, 1971.

[Ch] Cheung, Y.K., Finite strip analysis of elastic slabs. *Proc. Am. Soc. Civil Eng.*, **94**, 1365–78, 1968.

[Ch2] Cheung, Y.K., The finite strip method in the analysis of elastic plates with two opposite simple supported ends. *Proc. Inst. Civil Engng*, **40**, 1–7, 1968.

[Ch3] Cheung, Y.K., Analysis of box girder bridges by the finite strip method. *Am. Conc. Inst. Public.*, SP 26, 357–78, 1969.

[Ch4] Cheung, Y.K., Folded plate structures by the finite strip method. *Am. Soc. Civil Eng.*, **96**, 2963–79, 1969.

[Ch5] Cheung, Y.K., The analysis of cylindrical orthotropic curved bridge decks. *Pub. Int. Ass. Struct. Engng.*, **29**, 41–52, 1969.

[Ch6] Cheung, Y.K., *The finite strip method in structural analysis*. Pergamon Press, 1976.

[ChC] Cheung, Y.K. and Chakrabati, S., Analysis of simply supported thick layered plates. *J. of Engineering Mechanics, ASCE*, **97**(3), 1039–1044, 1971.

[ChC2] Cheung, Y.K. and Chakrabati, S., Free vibration of thick layered rectangular plates by finite layer method. *J. of Sound and Vibration*, **21**(3), 277–284, 1972.

[ChC3] Cheung, M.S. and Cheung, Y.K., Analysis of curved bridges by the finite strip method. Research Report. Dept. of Civil Engineering, Univ. of Calgary, Canada, 1970.

[ChC4] Cheng, W.J. and Cheung, Y.K., Refined 9-dof triangular Mindlin plate elements. *Int. J. Numer. Meth. Engng.*, **51**, 1259–1281, 2001.

[ChC5] Cheng, W.J. and Cheung, Y.K., Refined quadrilateral element based on Mindlin-Reissner plate theory. *Int. J. Numer. Meth. Engng.*, **47**, 605–627, 2000.

[ChF] Chan, A.S.L. and Firmin, A., The analysis of cooling towers by the matrix finite element method. *Aeronaut. J.*, **74**, 826–35, 1970.

[ChF2] Cheung, Y.K. and Fan, S.C., Analysis of pavement and layered foundations by finite layer method. *Proc. of Third Int. Conf. on Num. Method in Geomechanics*, Aachen, Germany, 2–6 April 1979, 1129–1135, 1975.

[ChLC] Chinesta, F., Ladeveze, P. and Cueto, E., A short review on model order reduction based on Proper Generalized Decomposition. *Archives for Numerical Methods in Engineering*, **18**(4), 395–404, 2011.

[ChP] Cho, M. and Parmerter, R.R., Efficient higher order composite plate theory for general laminations configuration. *AIAA J.*, **31**, 1299–1306, 1993.

[ChT] Cheung, Y.K. and Tham, L.G., *The finite strip method*. CRC Press, 1998.

[ChTC] Chong, K.P., Tham, L.G. and Cheung, Y.G., Thermal behavior of formed sandwich plate by finite-prism-strip method. *Computers and Structures*, **15**(3), 321–324, 1982.

[Da] Dawe, D.J., Shell analysis using a simple facet element. *J. Strain Analysis*, **7**, 266–70, 1972.

[Da2] Dawe, D.J., Some higher order elements for arches and shells. In *Finite elements for thin shells and curved members*, D.G. Ashwell and R.H. Gallagher (Eds.), J. Wiley, 131–53, 1976.

[Da3] Dawe, D.J., Curved finite elements for the analysis of shallow and deep arches. *Computers and Structures*, **4**, 559–82, 1979.

[DB] Dvorkin, E.N. and Bathe, K.J., A continuum mechanics based four node shell element for general non-linear analysis. *Engineering Computations*, **1**, 77–88, 1984.

[DeB] De Boor, C., On calculation with *B*-splines. *Journal of Approximation Theory*, **6**, 50–62, 1972.

[Del] Delpak, R., *Role of the curved parametric element in linear analysis of thin rotational shells*, Ph.D Thesis, Dept. Civil Engineering and Building, The Polytechnic of Wales, 1975.

[Dem] Demmel, J., *Applied Numerical Linear Algebra*. Society for Industrial and Applied Mathematics (SIAM), Philadelphia, PA, 1997.

[DG] Dupuis, G. and Goël, J.J., A curved finite element for thin elastic shells. *Int. J. Solids and Struct.*, **6**, 11, 1413–28, 1970.

[DG2] Di Sciuva, M. and Gherlone, M. A global/local third-order Hermitian displacement field with damaged interfaces and transverse extensibility: FEM formulation. *Composite Structures*, **59**(4), 433–444, 2003.

[DG3] Di Sciuva, M. and Gherlone, M., Quasi-3D static and dynamic analysis of undamaged and damaged sandwich beams. *Journal of Sandwich Structures & Materials*, **7**(1), 31–52, 2005.

[Dh] Dhatt, G., Numerical analysis of thin shells by curved triangular elements based an discrete Kirchhoff hypothesis. *Proc. ASCE Symp. on Applications of FEM in Civil Engng.*, Vanderbilt Univ., Nashville, Tenn., 13–14, 1969.

[Dh2] Dhatt, G., An efficient triangular shell element. *AIAA J.*, **8**, 2100-2, 1970.

[Di] Dill, E.H., A triangular finite element for thick plates. *Computational Mechanics III*, S.N. Atluri *et al.* (Eds.), Springer Verlag, 1988.

[DiS] Di Sciuva, M., A refined transverse shear deformation theory for multilayered anisotropic plates. *Atti Accademia delle Scienze di Torino*, **118**, 279–295, 1984.

[DiS2] Di Sciuva, M., Development of an anisotropic, multilayered shear deformable plate element. *Computers & Structures*, **21**(4), 789–796, 1985.

[DiS3] Di Sciuva, M., Multilayered anisotropic plate model with continuous interlaminar stresses. *Composite Structures*, **22**(3), 149–168, 1992.

[DJS] Dörfel, M.R., Jüttler, B. and Simeon, B., Adaptive isogeometric analysis by local h-refinement with T-splines. *Comput. Methods Appl. Mech. Engrg.*, **199**(5–8), 264–275, 2010.

[DL] Donea, J. and Lamain, L.G., A modified representation of transverse shear in C_0 quadrilateral plate elements. *Comput. Methods Appl. Mech. Engrg.*, **63**, 183–207, 1987.

[DM] Dharmarajan, S. and Mac Cutchen, H., Shear coefficients for orthotropic beams. *J. Composite Materials*, **7**, 530–535, 1973.

[DMM] Dhatt, G., Marcotte, L. and Matte, Y., A new triangular discrete Kirchhoff plate shell element. *Int. J. Numer. Meth. Engng.*, **23**, 453–470, 1986.

[DMMT] Dhatt, G., Marcotte, L., Matte, Y. and Talbot, M., Two new discrete Kirchhoff plate shell elements. 4th Symp. on Num. Meth. in Engrg., Atlanta, Georgia, 599–604, 1986.

[DV] Dhatt, G. and Venkatasubby, S., Finite element analysis of containment vessels. *Proc. First Conf. an Struct. Mech. in Reactor Tech.*, **5**, Paper J3/6, Berlin, 1971.

[DZ] Drysdale, W.H. and Zak, A.R., Structural problems in thick shells. In *Thin Shell Structures*, Y.C. Fung and E.E. Schler (Eds.), pp. 453-464, Prentice Hall, Englewood Cliffs, New Yersey, 1974.

[EKE] Ertas, A., Krafcik, J.T., Ekwaro-Osire, S., Explicit formulation of an anisotropic Allman/DKT 3-node thin triangular flat shell elements. *Composite Material Technology, ASME*, PD-Vol.37, 249–255, 1991.

[EOO] Eijo, A., Oñate, E. and Oller, S., A four-noded composite laminated Reissner-Mindlin plate element based on the refined zigzag theory. Research Report No. 382, CIMNE, Barcelona, 2012.

[Fa] Fan, S.C., *Spline finite ship method in structural analysis*. Ph.D. Thesis, Dept. of Civil Engineering, University of Hong Kong, 1982.

[FB] Foye, R.L. and Baker, D.J., Design of orthotropic laminates. *Proc. 11th Annual AIAA Structures, Structural Dynamics, and Materials Conference*, Denver, Colorado, April 1970.

[FB2] Felippa, C.A. and Bergan, P.G., A triangular plate bending element based on energy orthogonal free formulation. *Comput. Methods Appl. Mech. Engrg.*, **61**, 129–60, 1981.

[FdV] Fraeijs de Veubeke, B., Displacement and equilibrium models in finite element method. In O.C. Zienkiewicz and G.S. Holister (Eds.), Stress Analysis, Chapter 9, pages 145–197, John Wiley & Sons, Chichester, 1965.

[FdV2] Fraeijs de Veubeke, B., A conforming finite element for plate bending. *Int. J. Solids and Struct.*, **4**, 95–108, 1968.

[FE] Flores, F.G. and Estrada, C.F., A rotation-free thin shell quadrilateral. *Comput. Methods Appl. Mech. Engrg.*, **196**, 2631–2646, 2007.

[Fel] Felippa, C., The amusing history of shear flexible beam elements. *IACM Expressions*, **17**, 15–19, 2005. Available from www.iacm.info.

[Fl] Flügge, N., *Stresses in shells*. Springer-Verlag, 1962.

[FM] Felippa, C.A. and Militello, C., Developments in variational methods for high performance plate and and shell elements. In Analytical and Computational Models of Shells, Noor *et al.* (Eds.), CED, **3**, ASME, 191–215, 1989.

[FM2] Feo, L. and Mancusi, G., Modeling shear deformability of thin-walled composite beams with open cross-section. *Mechanics Research Communications*, **37**(3), 320–325, 2010.

[FO] Flores, F. and Oñate, E., A comparison of different finite elements based on Simo's shell theory. Research Report No. 33, CIMNE, Barcelona, 1993.

[FO2] Flores, F. and Oñate, E., A basic thin shell triangle with only translational DOFs for large strain plasticity. *Int. J. Numer. Meth. Engng.*, **51**, 57–83, 2001.

[FO3] Flores, F. and Oñate, E., Improvements in the membrane behaviour of the three node rotation-free BST shell triangle using an assumed strain approach. *Comput. Methods Appl. Mech. Engrg.*, **194**, 907–932, 2005.

[FO4] Flores, F. and Oñate, E., A rotation-free shell triangle for the analysis of kinked and branching shells. *Int. J. Numer. Meth. Engng.*, **69**, 1521–1551, 2007.

[FO5] Flores, F. and Oñate, E., Wrinkling and folding analysis of elastic membranes using an enhanced rotation-free thin shell triangular element. *Finite Elements in Analysis and Design*, **47**, 982–990, 2011.

[FOO] Flores, R., Ortega, E. and Oñate, E., Simple and efficient numerical tools for the analysis of parachutes. Publication PI387, CIMNE, October 2012.

[Fr] Fried, I., Shear in C_0 and C_1 plate bending elements. *Int. J. Solids and Struct.*, **9**, 449–60, 1973.

[Fr2] Fried, I., Residual balancing technique in the generation of plate bending finite element. *Comput. and Struct.*, **4**, 771–78, 1974.

[Fra] Franson, B., A generalized finite strip method for plate and wall structures. Publication 77:1, Dept. of Structural Mechanics, Chalmers University of Technology, 1977.

[Fri] Friedrich, R., Finite strips: 30 years. A bibliography (1968–1998). *Engng. Computations*, **17**(1), 92–111, 2000.

[FT] Fam, A. and Turkstra, C., A finite element method for box bridge analysis. Computers and Structures, **5**, 179–186, 1975

[FV] Fezans, G. and Verchery, G., Some results on the behaviour of degenerated shell (DS) elements. *Nuclear Engineering and Design*, **70**, 27–35, 1982.

[FY] Fried, I. and Yang, S.K., Triangular nine degrees of freedom C_0 plate bending element of quadratic accuracy. *Quart. Appl. Math.*, **31**, 303–312, 1973.

[Ga] Gallagher, R.H., Shell elements. *World Conf. on Finite Element Methods in Structural Mech.*, E1–E35, Bournemouth, Dorset, Inglaterra, Oct., 1975.

[Ga2] Gallagher, R.H., *Finite Element Analysis Fundamentals*. Prentice–Hall, Englewood Cliffs, N.J., 1975.

[Ga3] Gay, D., *Matériaux composites*. Hermès, Paris, 1987.

[Ga4] Galerkin, B.G., Series solution of some problems in elastic equilibrium of rods and plates. *Vestn. Inzh. Tech.*, **19**, 897–908, 1915.

[GB] Guo, Y.Q., Batoz, J.L., Résultants du cas test sur la poutre console. Bulletin du Club ϕ^2AS, IPSI, **13**, 3A, 1989.

[GiD] GiD. The personal pre-postprocessor. www.gidhome.com, CIMNE, Barcelona, 2012.

[GL] Greimann, L.F. and Lynn, P.P., Finite element analysis of plate bending with transverse shear deformation. *Nuclear Engineering and Design*, **14**, 223–230, 1970.

[GM] Giannini, M. and Miles, G.A., A curved element approximation in the analysis of axisymmetric thin shells. *Int. J. Numer. Meth. Engng.*, **2**, 459–70, 1970.

[Go] Gould, P.L., *Finite Element Analysis of Shells of Revolutions*. Pittman Pub. Co., Marsfield, MA, 1985.

[Go2] Gould, P.L., *Analysis of shells and plates*. Springer Verlag, 1988.

[GO] García, J and Oñate, E., An unstructured finite element solver for ship hydrodynamics problems. *J. Appl. Mech.*, **70**, 18–26, 2003.

[GP] Gunnlaugsson, G.A. and Pedersen, P.T., A finite element formulation for beams with thin-walled cross-section. *Computers and Structures*, **15** (6), 691–699, 1982.

[GS] Grafton, P.E. and Strome, D.R., Analysis of axi-symmetric shells by the direct stiffness method. *A.I.A.A.J.* , **1**, 2342–7, 1963.

[GSW] Greene, B.E., Strome, D.R. and Weikel, R.C., Application of the stiffness method to the analysis of shell structures. In *Proc. Aviation Conf. of American Society of Mechanical Engineers*, *ASME*, Los Angeles, March 1961.

[GTD] Gherlone, M., Tessler, A. and Di Sciuva, M., C° beam element based on the refined zigzag theory for multilayered composite and sandwich laminates. *Composite Structures*, **93**(11), 2882–2894, 2011.

[Gu] Guo, Y.Q., *Analyse non linéaire statique et dynamique des poutres tridimensionnelles élasto-plastiques*. Thèse de Doctorat, UTC, 1987.

[HB] Horrigmoe, G. and Bergan, P.G., Non linear analysis of free form shells by flat finite elements. *Comput. Methods Appl. Mech. Engrg.*, **16**, 11-35, 1979.

[HB2] Hughes, T.J.R. and Brezzi, F., On drilling degrees-of-freedom. *Comput. Methods Appl. Mech. Engrg.*, **72**, 105–121, 1989.

[HC] Hughes, T.J.R. and Cohen, M., The Heterosis finite element for plate bending. *Computers and Structures*, **9**, 445–50, 1978.

[HCB] Hughes, T.J.R., Cottrell, J.A. and Bazilevs, Y., Isogeometric analysis: CAD finite elements NURBS exact geometry and mesh refinement. *Comput. Methods Appl. Mech. Engrg.*, **194**, 4135–4195, 2005.

[He] Hellen, T.K., An assessment of the semi-loof shell element. *Int. J. Numer. Meth. Engng.*, **22**, 133-151, 1986.

[HH2] Hinton, E. and Huang, H.C., A family of quadrilateral Mindlin plate element with substitute shear strain fields. *Comp. and Struct.*, **23**, 409–431, 1986.

[HH3] Huang, H.C. and Hinton, E., A nine node Lagrangian Mindlin plate element with enhanced shear interpolation. *Engng. Comput.*. **1**. 369–79, 1984.

[HH4] Huang, H.C. and Hinton, E., Lagrangean and Serendipity plate and shell elements through thick and thin. In *Finite element methods for plate and shell structures*, T.J.R. Hughes and E. Hinton (Eds.), Pineridge Press, 1986.

[HH5] Huang, H.C. and Hinton, E., A new nine node degenerated shell element with enhanced membrane and shear interpolation. *Int. J. Numer. Meth. Engng.*, **22**, 73–92, 1986.

[HK] Harvey, J.W. and Kelsey, S., Triangular plate elements with enforced continuity. *A.I.A.A.J.*, **9**, 1026–6, 1971.

[HL] Hughes, T.J.R. and Liu, W.K., Non linear finite element analysis of shells. Part I. Three dimensional shells. *Comput. Methods Appl. Mech. Engrg.*, **26**, 331–362, 1981.

[HL2] Haas, D.J. and Lee, S.W., A nine node assumed strain finite element for composite plates and shells. *Computers and Structures*, **26**, 445–452, 1987.

[HMH] Hughes, T.J.R., Masud, A. and Harari, I., Numerical assessment of some membrane elements with drilling degrees of freedom. *Computers and Structures*, **55**(2), 297–314, 1995.

[HO] Hinton, E. and Owen, D.R.J., *Finite element software for plates and shells*. Pineridge Press, 1988.

[HRS] Hughes, T.J.R., Reali, A. and Sangalli, G., Efficient quadrature for NURBS-based isogeometric analysis. *Comput. Methods Appl. Mech. Engrg.*, **199**, 301–313, 2010.

[HT] Hughes, T.J.R. and Taylor, R.L., The linear triangular plate bending element. *Proc. of 4th MAFELAP Conf.*, Brunel Univ.,Uxbridge, 1981, (Edited by J.R. Whiteman), 127–42, Academic Press, 1982.

[HTC] Hampshire, J.K., Topping, B.H.V. and Chan, H.C., Three node triangular elements with one degree of freedom per node. *Engineering Computations*, **9**, 49–62, 1992.

[HTF] Hetmaniuk, U., Tezaur, R. and Farhat, C., Review and assessment of interpolatory model order reduction methods for frequency response structural dynamics and acoustics problems. *Int. J. Numer. Meth. Engng.*, in press.

[HTK] Hughes, T.J.R., Taylor, R.L. and Kanok-nukulchai, W., A simple and efficient element for plate bending. *Int. J. Numer. Meth. Engng.*, **11**, 1529–43, 1977.

[HTT] Hughes, T.J.R., Taylor, R.L. and Tezduyar, T.E., Finite elements based upon Mindlin plate theory with particular reference to the four node bilinear isoparametric element. *J. Appl. Mech.*, **46**, 587–596, 1981.

[Hu] Hughes, T.R.J., *The finite element method. Linear Static and Dynamic analysis*. Prentice Hall, 1987.

[Hu2] Huang, H.C., *Static and dynamic analysis of plates and shells*. Springer-Verlag, Berlin, 1989.

[HWW] Henshell, R.D., Walters, D. and Warburton, G.B., A new family of curvilinear plate bending elements for vibration and stability. *J. Sound Vibr.*, **20**, 327–343, 1972.

[Hy] Hyca, M., *Predicting shear centre location in open cross sections of shear deformable thin walled beams*. XVIIth IUTAM Conference Grenoble, 1988.

[IA] Irons, B.M. and Ahmad, S., *Techniques of finite elements*. Ellis Harwood, Chichester, 1980.

[IA2] Iura, M. and Atluri, S.N., Formulation of a membrane finite element with drilling degrees of freedom. *Computational Mechanics*, **9**(6), 417–428, 1992.

[Ib] Ibrahimbegovic, A., Quadrilateral finite elements for analysis of thick and thin plates. *Comput. Methods Appl. Mech. Engrg.*, **110**, 195–209, 1993.

[IO] Idelsohn, S.R. and Oñate, E., Finite volumes and finite elements: two "good friends". *Int. J. Numer. Meth. Engng.*, **37**, 3323–3341, 1994.

[Ir] Irons, B.M., A conforming quartic triangular element for plate bending. *Int. J. Numer. Meth. Engng.*, **1**, 29–46, 1969.

[Ir2] Irons, B.M., The semiloof shell element, *Finite elements for thin shells and curved members*. Chap. **11**, 197–222, D.G. Ashwell and R.H. Gallagher (Eds.), J. Wiley, 1976.

[ITW] Ibrahimbegovic, A., Taylor, R.L. and Wilson, E.L., A robust quadrilateral membrane finite element with drilling degrees of freedom. *Int. J. Numer. Meth. Engng.*, **30**, 445–457, 1990.

[Je] Jetteur, Ph., Improvement of the quadrilateral JET shell element for a particular class of shell problems. Technical Report IREM 87/1, Ecole Polytechnique Federale de Lausanne, February 1987.

[Ji] Jirousek, J., A family of variable section curved beam and thick shell or membrane-stiffening isoparametric elements. *Int. J. Num. Meth. Eng.*, **17**, 171–86, 1981.

[JO] Jovicevic, J. and Oñate, E., Analysis of beams and shells using a rotation-free finite element-finite volume formulation. Monograph No. M43, CIMNE, Barcelona, 1999.

[Jor] Jordan, F.F., Stresses and deformations of the thin-walled pressurized torus. *J. Aerospace Science*, **29**, 213–25, 1962

[JP] Jang, J. and Pinsky, P., An assumed covariant strain based 9-node shell element. *Int. J. Numer. Meth. Engng.*, **24**, 2389–2411, 1987.

[JP2] Jang, J. and Pinsky, P. Convergence of curved shell elements based on assumed covariant strain interpolations. *Int. J. Numer. Meth. Engng.*, **26**, 329–347, 1988.

[JS] Jones, R.E. and Strome, D.R., Direct stiffness method of analysis of shells of revolution using curved elements. *A.I.A.A.J.*, **4**, 1519–25, 1965.

[JS2] Jones, R.E. and Strome, D.R., A survey of analysis of shells by the displacement method. *Proc. Conf on Matrix Methods in Struct. Mech.*, Air Force Inst. of Tech., Wright-Patterson, Air Force Base, Ohio, 1965.

[Ka] Kanok-nukulchai, W., A simple and efficient finite element for general shell analysis. *Int. J. Numer. Meth. Engng.*, **14** (2), 179–200, 1979.

[Ka2] Katili, I., A new discrete Kirchhoff-Mindlin element based on Mindlin-Reissner plate theory and assumed shear strain fields. Part I. An extended DKT element for thick-plate bending analysis. *Int. J. Numer. Methods Engrg.*, **36**, 1859–1883, 1993.

[Ka3] Katili, I., A new discrete Kirchhoff-Mindlin element based on Mindlin-Reissner plate theory and assumed shear strain fields. Part II. An extended DKQ element for thick-plate bending analysis. *Int. J. Numer. Methods Engrg.*, **36**, 1885–1908, 1993.

[KA] Kikuchi, F. and Ando, Y., A new variational functional for the finite element method and its application to plate and shell problems. *Nuclear Engineering and Design*, **21**(1), 95–113, 1972.

[KBLW] Kiendl, J., Bletzinger, K.-U., Linhard, J. and Wüchner, R., Isogeometric shell analysis with Kirchhoff–Love elements. *Comput. Methods Appl. Mech. Engrg.*, **198**, 3902–3914, 2009.

[KDJ] Kapuria, S., Dumir, P.C. and Jain, N.K., Assessment of zigzag theory for static loading, buckling, free and forced response of composite and sandwich beams. *Composite Structures*, **64**, 317–27, 2004.

[KK] Kantorovitch, L.V. and Krylov, V.I., *Approximate methods of higher analysis*. J. Wiley, 1958.

[Ki] Kirchhoff, G., Uber das Gleichqewicht und die Bewegung ciner elastlchen. Scheibe. *J. Reine und Angewandte Mathematik*, **40**, 51–88, 1850

[Kl] Klein, S., A study of the matrix displacement method as applied to shells of revolution. *Proc. Conf. on Matrix Methods in Struct. Mech.*, Air Force Inst. of Tech., Wright Patterson Air Force Base, Ohio, Oct. 1965.

[Ko] Koiter, W.T., A consistent first approximation in the general theory of thin elastic shells. *First IUTAM Sympos.*, W.T. Koiter (ed.), North Holland, **2**, 1960.

[Ko2] Kollár, L.P., Flexural-torsional buckling of open section composite columns with shear deformation. *International Journal of Solids and Structures*, **38**(42-43), 7525–7541, 2001.

[KP] Kollár, L.P. and Pluzsik, A., Analysis of thin-walled composite beams with arbitrary layup. *Journal of Reinforced Plastics and Composites*, **21**(16), 1423–1465, 2002.

[Kr] Kraus, H., *Thin elastic shells*. J. Wiley, 1967.

[KS] Kabir, A.F. and Scordelis, A.C., CURDI–Computer program for curved bridges and flexible bents. *Internal Report, Dept. Civil Enginering*, Univ. of California, Berkeley, Septiembre, 1979.

[KS2] Kim, R.Y. and Soni, S.R., Supression of free-edge delaminations by hybridization. In ICCM-V, San Diego, CA, July 1985.

[KSK] Kim, N.-I., Shin, D.K. and Kim, M.-Y., Exact solutions for thin-walled open-section composite beams with arbitrary lamination subjected to torsional moment. *Thin-Walled Structures*, **44**(6), 638–654, 2006.

[KV] Kandil, N. and Verchery, G., New method of design for stacking sequences of laminate. Proceeding of Computer Aided Design in Composite Material Technology, C.A. Brebbia, N.P. de Wilde and W.R. Blain (Eds.), Computational Mechanic Publications, Southampton, pp. 243–257, 1988.

[LB] Lardeur, P.A. and Batoz, J.L., Composite plate analysis using a new discrete shear triangular finite element. *Int. J. Numer. Meth. Engng.*, **27**(2), 343–359, 1989.

[LC] Loo, Y.C. and Cusens, Y.A.R., *The finite strip method in bridge engineering*. Viewpoint, 1978.

[LD] Lynn, P.P. and Dhillon, B.S., Triangular thick plate bending elements. In *Proceedings 1st Internacional Conference on Structural Mechanics in Reactor Technology*, M6/5, Berlin, 1971.

[Le] Lee, J.H., Flexural analysis of thin-walled composite beams using shear-deformable beam theory. *Composite Structures*, **70**(2), 212–222, 2005.

[Li] Livesley, R.K., *Matrix methods in structural analysis*. 2nd ed., Pergamon Press, 1975.

[LKM] Lim, P.T.K., Kilford, J.T. and Moffatt, K.R., Finite element analysis of curved box girder bridges. In *Developments in Bridge Design and Constructions*, K.C. Rockey *et al.* (Eds.), Crosby Lockwood, London, 264–86, 1971.

[LL] Liu, D. and Li, X., An overall view of laminate theories based on displacement hypothesis. *Journal of Composite Materials*, **30**(14), 1539–1561, 1996.

[LL2] Li, X. and Liu D., An interlaminar shear stress continuity theory for both thin and thick composite laminates. *J. Appl. Mech.*, **59**, 502–509, 1992.

[LL3] Lee, J.H. and Lee, S., Flexural-torsional behavior of thin-walled composite beams. *Thin-walled Structures*, **42**, 1293–1305, 2004.

[LS] Lee, C.Y. and Small, J.C., Finite layer analysis of laterally loaded piles in cross-anisotropic soils. *Int. J. for Num. Anal. Meth. in Geomechanics*, **15**(11), 785–808, 1991.

[LWB] Linhard, J., Wüchner, R. and Bletzinger, K.-U. Upgrading membranes to shells - the CEG rotation-free shell element and its application in structural analysis. *Finite Element Anal. Des.*, **44**, 63–74, 2007.

[Ly] Lyons, L.P.R., *A general finite element system with special reference to the analysis of cellular structures.* Ph D. Thesis, Imp. College of Science and Technol., Londres, 1977.

[Ma] MacNeal, R.H., A simple quadrilateral shell element. *Computers and Structures*, **8**, 175–83, 1978.

[Ma2] MacNeal, R.H., *Finite elements: their design and performance.* Marcel Dekker, New York, 1994.

[Ma3] Marguerre, K., Zur thorie der Gekrummten platte grosser formanderung. *Proc. 5th. Int. Congress Appl. Mech.*, J. Wiley, Londres, 98–101, 1938.

[Mar] Marshall, A., *Handbook of Composites.* G. Lubin (ed.), SPI-Van Nostrand Reinhold, New York, 1982.

[MB] Massa, J.C. and Barbero, E.J., A strength of materials formulation for thin-walled composite beams with torsion. *Journal of Composite Materials*, **32**(17), 1560–1594, 1998.

[Me] Melosh, R.J., A stiffness matrix for the analysis of thin plates in bending. *Journal of Aerosapce Science*, **28**, 1, 34–42, 1961.

[Me2] Melosh, R.J., Basis of derivation of matrices for the direct stiffness method. *A.I.A.A.J.*, **1, 7**, 1631–37, 1963.

[Me3] Melosh, R.J., Structural analysis of solids. *J. Structural Engineering*, *ASCE*, **4**, 205–223, August 1963.

[MG] Murthy, S.S. and Gallagher, R.H., A triangular thin shell finite element based on discrete Kirchhoff theory. *Comput. Methods Appl. Mech. Engrg.*, , **54**, 197–222, 1986.

[MII] Malkus, D.S. and Hughes, T.J.R., Mixed finite element methods-reduced and selective integration techniques: A unification of concepts. *Comput. Methods Appl. Mech. Engrg.*, , **15**, 63–81, 1978.

[MH2] MacNeal, R.H., and Harter, R.L., A proposed standard set of problems to test finite element accuracy. *Finite Elements in Analysis and Design*, **1**, 3–20, 1985.

[Mi] Mindlin, R.D., Influence of rotatory inertia and shear in flexural motions of isotropic elastic plates. *J. Appl. Mech.*, **18**, 31–38, 1951.

[Mo] Morley, L.S.D., The triangular equilibrium element in the solution of plate bending problems. *Aero Quart.*, **19**, 149–69, 1968.

[Mo2] Morley, L.S.D., On the constant moment plate bending element. *J, Strain Analysis*, **6**, 10–4, 1971.

[Mo3] Morley, L.S.D., Finite element criteria for some shells. *Int. J. Numer. Meth. Engng.*, **20**, 587–92, 1984.

[Mo4] Morley, L.S.D., *Skew Plates and Structures*. Macmillan, New York, 1963. International Series of Monographs in Aeronautics and Astronautics.

[Mo5] Morris, A.J., A deficiency in current finite elements for thin shell applications. *Int. J. Solids Struct.*, **9**, 331–45, 1973.

[Mo6] Mollmann, H., *Introduction to the theory of thin shells*. John Wiley, 1981.

[MS] Mebane, P.M. and Stricklin, J.A., Implicit rigid body motion in curved finite elements. *AIAA J*, **9**(2), 344–345, 1971.

[MS2] Moan, T. and Soreide, T., The analysis of stiffened plates considering non linear material and geometrical behaviour. *World Congress on Finite Elements in Structural Mechanics*, (Ed. J. Robinson), Robinson & Assoc., Verwood, 14.1–14.28, 1975.

[MSch] Milford, R.V. and Schnobrich, W.C., Degenerated isoparametric finite elements using explicit integration. *Int. J. Numer. Meth. Engng.*, **23**, 133–154, 1986.

[Mu] Murakami, H., Laminated composite plate theory with improved in-plane responses. *ASME, Journal of Applied Mechanics*, **53**, 661–666, 1986.

[Na] Naghdi, P.M., *The theory of shells and plates*. Handbuch der Physik, Vol. VI, A2 (Flügge ed.), Springer Verlag, Berlin, 1972.

[NB] Noor, A.K. and Burton, W.S., Assessment of shear deformation theories for multilayered composite plates. *ASME, Applied Mechanics Review*, **42**(1), 1–13, 1989.

[Ni] Niordson, F.I., *Shell theory*. North-Holland, Amsterdam, 1985.

[No] Novozhilov, V.V., *Theory of thin shells*. P. Noordhoff Ltd., Groningen, The Netherlands, 1959.

[No2] Noor, A.K., Bibliography of monographs and surveys on shells. *Applied Mechanics Review*, **43**(9), 223–234, 1990.

[NP] Noor, A.K. and Peters, J., Analysis of laminated anisotropic shells of revolution. In *Finite element methods for plate and shell structures, Vol. 2, Formulations and algorithms*, Pineridge Press, 179–212, 1986.

[Ny] Nygard,M.K., *The free formulation for non linear finite elements with applications to shells*. Ph. D. Thesis, Univ. Trondheim, Norway, 1986.

[NU] Nay, R.A. and Utku, S., An alternative to the finite element method. *Variational Methods Engineering*, **1**, 1972.

[OA] Obinata, G. and Anderson, D.O., *Model reduction for control system design*. Springer, 2001.

[OB] Oñate, E. and Bugeda, G., A study of mesh optimality criteria in adaptive finite element analysis. *Engineering Computations*, **10**(4), 307–321 1993.

[OC] Oñate, E. and Castro, J., Derivation of plate elements based on assumed shear strain fields. In *Recent Advances on Computational Structural Mechanics*, P. Ladeveze and O.C. Zienkiewicz (Eds.), Elsevier Pub, 1991.

[OC2] Oñate, E. and Cervera, M., Derivation of thin plate bending elements with one degree of freedom per node. *Engineering Computations*, **10**, 543–561, 1993.

[OCK] Oñate, E., Castro, J. and Kreiner, R., Error estimation and mesh adaptivity techniques for plate and shell problems. In *The 3rd. International Conference on Quality Assurance and Standars in Finite Element Methods*, Stratford-upon-Avon, England, 10–12 September, 1991.

[OCM] Oñate, E., Cendoya, P. and Miquel, J., Non linear explicit dynamic analysis of shells using the BST rotation-free triangle. *Engineering Computations*, **19**(6), 662–706, 2002.

[OCZ] Oñate, E., Cervera, M. and Zienkiewicz, O.C., A finite volume format for structural mechanics. *Int. J. Numer. Meth. Engng.*, **37**, 181–201, 1994.

[OEO] Oñate, E., Eijo, A. and Oller, S., Two-noded beam element for composite and sandwich beams using Timoshenko theory and refined zigzag kinematics. Publication PI346, CIMNE, Barcelona 2010.

[OEO2] Oñate, E., Eijo, A. and Oller, S., Simple and accurate two-noded beam element for composite laminated and sandwich beams using a refined zigzag theory. *Comput. Methods Appl. Mech. Engrg.*, **213–216**, 362–382, 2012.

[OEO3] Oñate, E., Eijo, A. and Oller, S., Modeling of delamination in composite laminated beams via 2-noded Timoshenko beam element and zigzag kinematics. Research Report PI383, CIMNE, Barcelona 2012.

[OEO4] Oñate, E., Eijo, A. and Oller, S., Two-noded troncoconical element for composite laminated axisymmetric shells based on a refined zigzag theory. Research Report PI376, CIMNE, Barcelona, 2012.

[OF] Oñate, E. and Flores, F.G., Advances in the formulation of the rotation-free basic shell triangle. *Comput. Methods Appl. Mech. Engrg.*, **194**, 2406–2443, 2005.

[OF2] Owen, D.R.J. and Figueiras, J.A., Anisotropic elasto-plastic finite element analysis of thick and thin plates and shells. *Int. J. Numer. Meth. Engng.*, **19**, 541–566, 1983.

[OFM] Oñate, E., Flores, F.G. and Marcipar, J., Membrane structures formed by low pressure inflatable tubes. new analysis methods and recent constructions. In *Textile Components and Inflatable Structures II*, E. Oñate and B. Kröplin (eds.), pp. 163–196, Springer, 2008.

[OFN] Oñate, E., Flores, F.G. and Neamtu, L., Enhanced rotation-free basic shell triangle. Applications to sheet metal forming. In *Computational Plasticity*, E. Oñate and R. Owen (Eds.), pp. 239–265, Springer 2007.

[OG] Oñate, E. and García, J., A stabilized finite element method for fluid-structure interaction with surface waves using a finite increment calculus formulation. *Comput. Methods Appl. Mech. Engrg.*, **182**(1-2), 355–370, 2000.

[OGI] Oñate, E., García, J. and Idelsohn, S.R., Ship hydrodynamics. Encyclopedia of Comput. Mechanics, E. Stein, R. de Borst and T.J.R. Hughes (Eds.), **3**, Chapter 18, 579–610, 2004.

[OHG] Oñate, E., Hinton, E. and Glover, N., Techniques for improving the performance of isoporametric shell elements. *Applied Numerical Modelling*, C. Brebbia and E. Alarcón (Eds.), Pentech Press, 1979.

[OK] Oñate, E. and Kröplin, B. (Eds.), *Textile Components and Inflatable Structures*. Springer, Netherlands, 2005.

[OK2] Oñate, E. and Kröplin, B. (Eds.), *Textile Components and Inflatable Structures II*. Springer, Netherlands, 2008.

[OL] Owen, D.R.J. and Li, Z.H., A refined analysis of laminated plates by finite element displacement methods-I. Fundamentals and static analysis; II Vibration and stability. *Computers and Structures*, **26**, 907–923, 1987.

[OL2] Owen, D.R.J. and Li, Z.H., Elasto-plastic numerical analysis of anisotropic laminated plates by a refined finite element model. *Comput. Methods Appl. Mech. Engrg.*, **70**, 349–365, 1988.

[On] Oñate, E., *Comparisons of finite strip methods for the analysis of box girder bridges*. M. Sc. Thesis, Civil, Eng. Dpt., Univ. College of Swansea, 1976.

[On2] Oñate, E., Edge effects in composite materials. *Analysis and design of composite materials structures*, Chap. 21, Y. Surrel, A. Vantrin and G. Verchey (Eds.), Pluralis, 1990.

[On3] Oñate, E., A review of some finite element families for thick and thin plate and shell analysis. In *Recent Developments in Finite Element Analysis*, T.J.R. Hughes, E. Oñate & O.C. Zienkiewicz (Eds.), CIMNE, Barcelona, 1994.

[On4] Oñate, E., *Structural Analysis with The Finite Element Method. Vol. 1: Basis and Solids*, Springer-CIMNE, 2009.

[OO] Oller, S. and Oñate, E., Advanced models for finite element analysis of composite materials. *Encyclopedia of Composites*, 2nd Edition, L. Nicolais and A. Borzacchiello (Eds.), J. Wiley & Sons, New Jersey, 2012.

[OR] Oden, J.T. and Ripperger, E.A., *Mechanics of elastic structures*. 2nd Edition, Hemisphere Publishing Corporation, 1981.

[OS] Oñate, E. and Suárez, B., A comparison of the linear, quadratic and cubic Mindlin strip element for the analysis of thick and thin plates. *Computers and Structures*, **17**, 427–39, 1983.

[OS2] Oñate, E. and Suaréz, B., The finite strip for the analysis of plate, bridge and axisymmetric shell problems. In *Finite element software for plate and shell analysis*, E. Hinton and D.R.J. Owen (Eds.), Pineridge Press, Swansea, 1984.

[OS3] Oñate, E. and Suárez, B., An unified approach for the analysis of bridges, plates and axisymmeric shells using the linear Mindlin strip element. *Computers and Structures*, **17**, 3, 407–26, 1986.

[OTZ] Oñate, E., Taylor, R.L. and Zienkiewicz, O.C., Consistent formulation of shear constrained Reissner-Mindlin plate elements. In C. Kuhn and H. Mang (Eds.), *Discretization Methods in Structural Mechanics*, pp. 169–180. Springer-Verlag, Berlin, 1990.

[OZ] Oñate, E. and Zárate, F., Rotation-free triangular plate and shell elements. *Int. J. Numer. Meth. Engng.*, **47**, 557–603, 2000.

[OZ2] Oñate, E. and Zárate, F., Rotation-free plate and beam elements with shear deformation effects. *Int. J. Numer. Meth. Engng.*, 2009.

[OZ3] Oñate, E. and Zárate, F., Rotation-free beam elements. A review. Research Report PI384, CIMNE, Barcelona 2012.

[OZF] Oñate, E., Zárate, F. and Flores, F., A simple triangular element for thick and thin plate and shell analysis. *Int. J. Numer. Meth. Engng.*, **37**, 25–2582, 1994.

[OZST] Oñate, E., Zienkiewicz, O.C., Suárez, B. and Taylor, R.L., A general methodology for deriving shear constrained Reissner-Mindlin plate elements. *Int. J. Numer. Meth. Engng.*, **33**(2), 345–367, 1992.

[Pa] Parekh, C.J., *Finite element solution system*. PhD Thesis, Department of Civil Engineering, University of Wales, Swansea, 1969.

[Pa2] Parisch, H., A critical survey of the 9-noded degenerated shell element with special emphasis on thin shell application and reduced integration. *Comput. Methods Appl. Mech. Engrg.*, **20**, 323–50, 1979.

[Pan] Panc, W., *Theories of elastic plates*. Sifthoff and Noordhoff, 1975.

[Paw] Pawsey, S.E., *The analysis of moderately thick to thin shells by the finite element method*. PhD dissertation, Department of Civil Engineering, University of California, Berkeley, 1970 (also SESM Report 70-12).

[PC] Pawsey, S.E. and Clough, R.W., Improved numerical integration of thick shell elements. *Int. J. Numer. Meth. Engng.*, **3**, 545–86, 1971.

[PC2] Phaal, R., Calladine, C.R., A simple class of finite elements for plate and shell problems. I: Elements for beams and thin plates. *Int. J. Numer. Meth. Engng.*, **35**, 955–977, 1992.

[PC3] Phaal, R., Calladine, C.R., A simple class of finite elements for plate and shell problems. II: An element for thin shells with only translational degrees of freedom. *Int. J. Numer. Meth. Engng.*, **35**, 979–996, 1992.

[PCh] Pilkey, W.D. and Chang, P.Y., *Modern formulas for statics and dynamics.* McGraw-Hill, New York, 1978.

[PF] Perruchoud, M. and Frey, F., How to modelize beams of slab–beam structures in a simple way. Rapport Interne IREM90/3, Dpt. Genie Civil, Ecole Polytechnique Federale de Laussanne, Abril, 1990.

[PFTV] Press, W.H., Flannery, B.P., Teukolsky, S.A. and Vetterling, W.T., *Numerical Recipes. The art of Scientific Computing,* Cambridge Univ. Press, 1986.

[PHZ] Pugh, E.D.L., Hinton, E. and Zienkiewicz, O.C., A study of quadrilateral plate bending elements with reduced integration. *J. Appl. Mech.,* **12**, 1059–1079, 1978.

[Pi] Pilkey, W.D., *Analysis and design of elastic beams.* Computational Methods, Wiley 2002.

[PK] Pluzsik, A. and Kollár, L.P., Effects of Shear deformation and restrained warping on the displacements of composite beams. *Journal of Reinforced Plastics and Composites,* **21**(17), 1517–1541, 2002.

[PN] Panda, S.C. and Natarajan, R., Finite element analysis of laminated shells of revolution. *Computers and Structures,* **6**, 61–64, 1976.

[POM] Pellegrini, L., Oñate, E. and Miquel, J., Development and validation of numerical procedure for analysis of sails in racing boats with the finite element method (in Spanish). Research Report IT356, CIMNE, Barcelona, 2000.

[PP1] Pipes, R.B. and Pagano, N.J., Interlaminar stresses in composite laminates under uniform axial extension. *Journal of Composite Materials,* **4**(4), 204–221, 1970.

[PP2] Pagano, N.J. and Pipes, R.B., Influence of stacking sequence on laminate strength. *Journal of Composite Materials,* **5**(1), 50–57, 1971.

[PP3] Pagano, N.J. and Pipes, R.B., Some observations on the interlaminar strength of composite laminates. *International Journal of Mechanical Sciences,* **15**, 679–688, 1973.

[PP4] Prinja, N.K. and Puri, A.K., An introduction to the use of material models in FE. NAFEMS, Publication, UK, Hardback, 88 pp., November 2005.

[PPL] Popov, E.R., Penzien, J. and Liu, Z.A., Finite element solution for axisymmetric shells. *Proc. Am. Soc. Civil Eng.,* EM, 119–45, 1964.

[Pr] Przemienieck, J.S., *Theory of matrix structural analysis.* McGraw-Hill, New York, 1968.

[PT] Papadopoulus, P. and Taylor, R.L., A triangular element based on Reissner-Mindlin plate theory. *Int. J. Numer. Meth. Engng.,* **5**, 1029–51, 1990.

[PT2] Piegl, L. and Tiller, W., *The NURBS Book.* Monographs in Visual Communication. 2nd edition, Springer-Verlag, New York, 1997.

[Ral] Ralston, A., *A first course in numerical analysis.* McGraw-Hill, New York, 1965.

[Raz] Razzaque, A., *Finite Element analysis of plates and shells.* Ph.D. Thesis, Civil Eng. Dpt., Univ. of Wales, Swansea, 1972.

[RB] Rowe, R.K. and Booker, J.R., A finite layer technique for calculating three-dimensional pollutant migration in soil. *Geotechnique*, **36**, 205–214, 1986

[RCh] Rossow, M.P. and Chen, K.C., Computational efficiency of plate elements. *ASCE*, **103**, St2, 447–51, 1977.

[Re] Reissner, E., The effect of transverse shear deformation on the bending of elastic plates. *J. Appl. Mech.*, **12**, 69–76, 1945.

[Re2] Reissner, E., A note on variational theorems in elasticity. *International Journal of Solids and Structures*, **1**, 93–95, 1965.

[Re3] Reissner, E., Reflection on the theory of elastic plate. *Appl. Mech. Rev.*, **38**(11), 1453–1464, 1985.

[Red] Reddy, J.N., On refined computational model of composite laminates. *Int. J. Numer. Meth. Engng.*, **27**, 361–382, 1989.

[Red2] Reddy, J.N., *Mechanics of laminated composite plates and shells. Theory and analysis.* 2nd Edition, CRC Press, Boca Raton, 2004.

[Ro] Rogers, D.F., *An Introduction to NURBS with historical perspective.* Academic Press, San Diego, CA, 2001.

[RTL] Rio, G., Tathi, B. and Laurent, H., A new efficient finite element model of shell with only three degrees of freedom per node. Applications to industrial deep drawing test. In *Recent Developments in Sheet Metal Forming Technology*, Barata Marques, M.J.M. (ed.), 18th IDDRG Biennial Congress, Lisbon, 1994.

[Sa] Savoia, M., On the accuracy of one-dimensional models for multilayered composite beams. *Int. J. Solids Struct.*, **33**, 521–44, 1996.

[San] Sander, G., Bornes superieurs et inféreures dans l'analyse matricielle des plaques en flexion-torsion. *Bull Soc. Royale des Sc. de Liege*, **33**, 456–94, 1964.

[San2] Sansour, C., A theory and finite element formulation of shells at finite deformations including thickness change: circumventing the use of a rotation vector. *Arch. Appl. Mech.*, **10**, 194–216, 1995.

[SB] Stolarski, H. and Belytschko, T., Membrane locking and reduced integration for curved elements. *J. Appl. Mech.*, **49**, 172–6, 1982.

[SB2] Sabourin, F. and Brunet, M., Detailed formulation of the rotation-free triangular element "S3" for general purpose shell analysis. *Engrg. Comput.*, **23**, 460–502, 2006.

[SBCK] Stolarski, H., Belytschko, T., Carpenter, N. and Kennedy, J.M., A simple triangular curved shell element. *Engineering Computations*, **1**, 3, 210–8, 1984.

[SBL] Stolarski, H., Belytschko, T. and Lee, S.-H., A review of shell finite element and co-rotational shell theories. *Computational Mechanics Advances*, **2**, 125–212, 1995.

[SCB] Stolarski, H., Carpenter, N. and Belytschko, T., Bending and shear mode decomposition in C0 structural elements. *Journal of Structural Mechanics, ASCE*, **11**(2), 153–176, 1983.

[SCB2] Sabourin, F., Carbonnière, J. and Brunet, M., A new quadrilateral shell element using 16 degrees of freedom. *Engineering Computations*, **26**, 500–540, 2009.

[SCLL] Soh, A.K., Cen, S., Long, Y.Q. and Long, Z.F., A new twelve DOF quadrilateral element for analysis of thick and thin plates. *Eur. J. Mech. A/Solids*, **20**, 299–326, 2001.

[SD] Smith, I.M. and Duncan, W., The effectiveness of nodal continuities in finite element analysis of thin rectangular and skew plates in bending. *Int. J. Numer. Meth. Engng.*, **2**, 253–258, 1970.

[SF] Simó, J.C. and Fox, D.D., On a stress resultant geometrically exact shell model. Part I. Formulations and optimal parametrizations. *Comput. Methods Appl. Mech. Engrg.*, **72**, 267–304, 1989.

[SFR] Simó, J.C., Fox, D.D. and Rifai, M.S., On a stress resultant geometrically exact shell model. Part II. The linear theory: Computational aspects. *Comput. Methods Appl. Mech. Engrg.*, **73**, 53–92, 1989.

[SG] Salerno, V.L. and Goldberg, M.A., Effect of shear deformation on the bending of rectangular plates. *Journal of Applied Mechanics, ASME*, 54–58, March 1960.

[Sh] Sharman, P.W., Analysis of structures with walled open sections. *Int. J. Mech. Science*, **27** (10), 665–677, 1985.

[SHTG] Stricklin, J.A., Haisler, W., Tisdale, P. and Gunderson, R., A rapidly converging triangular plate element. *A.I.A.A.J.*, **7**, 180–181, 1969.

[SJ] Singh, P.N. and Jha, P.K., *Elementary mechanics of solids*. Wiley Eastern, New Delhi, 1980.

[SK] Soni, S.R. and Kim, R.Y., Analysis of supression of free edge delamination by introducing adhesive layer. Internal Report, Univ. of Dayton Research Inst., Dayton Ohio, 45469 USA, 1986.

[SL] Sanders Jr., J.L. and Liepins, A., Toroidal membrane under internal pressure. *A.I.A.A.J.*, **1**, 2105–10, 1963.

[SLC] Soh, A.K., Long, Z.F. and Cen, S., A new nine d.o.f. triangular element for analysis of thick and thin plates. *Comput. Mech.*, **24**, 408–417, 1999.

[SM] Sawko, F. and Merriman, Y.P.A., An annular segment finite element for plate bending. *Int. J. Numer. Meth. Engng.*, **3**,, 119–129, 1971.

[SMO] Suárez, B, Miquel, J. and Oñate, E., A general thick finite strip method for plates and shells. Research Report, PI377, Barcelona, 2012.

[SNP] Stricklin, J.A., Navaratna, D.R. and Pian, T.H.H., Improvements in the analysis of shells of revolution by matrix displacement method: Curved elements. *A.I.A.A.J.*, **4**, 2069–72, 1966.

[Sp] Specht, B., Modified shape function for three noded plate bending element passing the patch test. *Int. J. Numer. Meth. Engng.*, **26**, 705–15, 1988.

[SP] Sze, K.Y. and Pan, Y.S., Hybrid stress tetrahedral elements with Allman's rotational DOFs. *Int. J. Numer. Meth. Engng.*, **48**(7), 1055–1070, 2000.

[SPH] Stanley, G.M., Park, K.C. and Hughes, T.J.R., Continuum-based resultant shell elements. In *Finite Element Method for Plate and Shell Structures. Vol. 1: Element Technology*, T.J.R. Hughes and E. Hinton (Eds.), Pineridge Press, Swansea, 1986, 1–45.

[Sr] Srinivas, S., A refined analysis of composite laminates. *J. Sound and Vibration*, **39**(4), 495–507, 1973.

[SR] Srinivas, S. and Rao, A.K., Flexure of thick rectangular plates. *J. Appl. Mech., ASME*, **40**(1), 298–299, 1973.

[Sta] Stanley, G.M., *Continuum-based shell elements*. Ph.D. Thesis, Dpt. Appl. Mechanics, Stanford Univ., 1985.

[Sta2] Stavsky, Y., Bending and stretching of laminated plates. *ASCE, J. Engng. Mech.*, **87**, 31–36, 1961.

[Ste] Stephen, N.G., Timoshenko's shear coefficient for a beam subjected to gravity loading, *J. Appl. Mech.*, **47** (1), 121–127, 1980.

[ST] Sheikh, A.H. and Thomsen, O.T., An efficient beam element for the analysis of laminated composite beams of thin-walled open and closed cross sections. *Composites Science and Technology*, **68**, 2273–2281, 2008.

[Su] Suárez, B., *A finite strip method based on Reissner-Mindlin theory for analysis of plates, bridges and axysummetric shells (in spanish)*. Ph.D. Thesis, School of Civil Engineering, Technical University of Catalonia (UPC), 1982.

[SV] Stenberg, R. and Vihinen, T., Calculations with some linear elements for Reissner-Mindlin plates. *European Conf. on New advances in Computational Structural Mechanics*, P. Lavedeze and O.C. Zienkiewicz (Eds.), 505–11, Giens, Abril, 1991.

[SWC] Sze, K.Y., Wanji, C. and Cheung, Y.K., An efficient quadrilateral plane element with drilling degrees of freedom using orthogonal stress modes. *Computers and Structures*, **42**(5), 695–705, 1992.

[Sz] Szilard, R., *Theory and analysis of plates: classical and numerical methods*. Prentice Hall, 1974.

[Ta] Taylor, R.L., Finite element analysis of linear shell problems. In J.R. White-man (ed.), *The Mathematics of Finite Elements and Applications VI*, 191–203, Academic Press, London, 1988.

[TA] Taylor, R.L. and Auricchio, F., Linked interpolation for Reissner-Mindlin plate elements. Part II. a simple triangle. *Int. J. Numer. Meth. Engng.*, **36**, 3057–3066, 1993.

[TChC] Tham, L.G., Chong, K.P. and Cheung, Y.K., Flexural bending and axial compression of architectural sandwich panels by combined finite-prism-ship method. *J. Reinforced Plastics and Composites*, **1**(1), 16–28, 1982.

[TD] Tessler, A. and Dong, S.B., On a hierarchy of conforming Timoshenko beam elements. *Computers and Structures*, **14**, 335–344, 1981.

[TDG] Tessler, A., Di Sciuva, M. and Gherlone, M., A refined zigzag beam theory for composite and sandwich beams. *J. of Composite Materials*, **43**, 1051–1081, 2009.

[TDG2] Tessler, A., Di Sciuva, M. and Gherlone, M., A consistent refinement of first-order shear-deformation theory for laminated composite and sandwich plates using improved zigzag kinematics. *J. of Mechanics of Materials and Structures*, **5**(2), 341–367, 2010.

[TDG3] Tessler, A., Di Sciuva, M. and Gherlone, M., A homogeneous limit methodology and refinements of computationally efficient zigzag theory for homogeneous, laminated composite and sandwich plates. *Num. Meth. Partial Diff. Eqns.*, **27**(1), 208–229, 2011.

[Te] Tessler, A., A C^0 anisoparametric three node shallow shell element. *Comput. Methods Appl. Mech. Engrg.*, **78**, 89–103.

[Tho] Thorpe, J., Ph. Thesis, University of Dundee, 1976.

[TH] Tessler, A. and Hughes, T.J.R., An improved treatment of transverse shear in the Mindlin type four quadrilateral element. *Comput. Methods Appl. Mech. Engrg.*, **39**, 311–35, 1983.

[TH2] Tessler, A. and Hughes, T.J.R., A three node Mindlin plate element with improved transverse shear. *Comput. Methods Appl. Mech. Engrg.*, **50**, 71–101, 1985.

[TH3] Teh, K.K. and Huang, C.C., Shear deformation coefficient for generally orthotropic beam. *Fib. Sci. and Tech.*, **12**, 73–80, 1979.

[TH4] Tsai, S.W. and Hahn, H.T., *Introduction to Composite Materials*. Technomic Publishing Co. Inc., Lancaster, Pennsylvania, 1980.

[Ti] Timoshenko, S.P., On the correction for shear of differential equations for transverse vibrations of prismatic bars. *Philosophical Magazine Series*, **41**, 744–746, 1921.

[Ti2] Timoshenko, S.P., *Strength of Materials. Part 1: Elementary Theory and Problems*. Van Nostrand Company Inc., New York, 1958.

[Ti3] Timoshenko, S.P., *Strength of Materials. Part 2: Advanced Theory and Problems.* 3rd. Edition, Van Nostrand-Reinhold, New York, 1956.

[TK] Tocher, J.L. and Kapur. K.K., Comment on basis for derivation of matrices for the direct stiffness method. *A.I.A.A.J.*, **6**, 1215–16, 1965.

[TM] Toledano, A. and Murakami, H., A higher-order laminate plate theory with improved in-plane response. *Int. J. Solids and Struct.*, **23**, 111–131, 1987b.

[To] Too, J.J.M., *Two dimensional, plate, shell and finite prism isoparametric elements and their application.* PhD Thesis, Department of Civil Engineering, University of Wales, Swansea, 1970.

[Ts] Tsai, S.W., *Composites Design.* 4th Edition, Think Composites, Paris, 1988.

[Tu] Tu, T., *Performance of Reissner-Mindlin elements.* PhD thesis, Rutgers University, Department of Mathematics, 1998.

[TW] Timoshenko, S.P. and Woinowsky-Krieger S., *Theory of Plates and Shells.* McGraw-Hill, New York, 3rd Edition, 1959.

[TZSC] Taylor, R.L., Zienkiewcz, O.C., Simó, J.C. and Chan, A.H.C., The patch test - A condition for assesing FEM convergence. *Int. J. Numer. Meth. Engng.*, **22**, 32–62, 1986.

[Ug] Ugural, A.C., *Stresses in plates and shells.* MacGraw-Hill, 1981.

[UO] Ubach, P.A. and Oñate, E., New rotation-free finite element shell triangle using accurate geometrical data. *Comput. Methods Appl. Mech. Engrg.*, **199**, 383–391, 2010.

[Va] Vargas, P., *A finite element formulation for composite laminated thin-walled beams with open section.* Ph.Thesis (in Spanish), Technical University of Catalonia (UPC), Barcelona, July 2011.

[VGM+] Versino D., Gherlone M., Mattone M., Di Sciuva M. and Tessler A., C^0 triangular elements based on the Refined Zigzag Theory for multilayered composite and sandwich plates. *Composites: Part B, Engineering*, pp. 13, 2012.

[Vi] Vinson, J.R., *The behaviour of thin walled structures, beams, plates and shells.* Kluwer Academic, Publishers, 1989.

[VK] Voyiadjis, G.Z. and Karamandlidis, D. (Eds.), *Advances in the theory of plates and shells.* Studies in Appl. Mech. 24, Elsevier, 1990.

[Vl] Vlasov, V.Z., *Thin-walled Elastic Beams.* Israel Program for Scientific Translations, 1961.

[Vl2] Vlasov, V.Z., General theory of shells and its application to engineering. *NASA TTF-99*, 1964.

[Vl3] Vlachoutsis, S., Explicit integration for three dimensional degenerated shell finite elements. *Int. J. Numer. Meth. Engng.*, **29**, 861–880, 1990.

[VOO] Vargas, P., Oñate, E. and Oller, S., A family of finite elements for composite laminated thin-walled beams with open section. Publication PI389, CIMNE, Barcelona, 2012.

[We] Wempner, G.A., Finite elements, finite rotations and small strains of finite shells. *Int. J. Solids Struct.*, **5**, 117–53, 1964.

[We2] Wempner, G.A., *Mechanics of solids with applications to thin bodies*. Sijthoff and Noordhoff, 1981.

[We3] Wempner, G.A., Mechanics and finite element of shells. *Applied Mechanics Review, ASME*, **42**(5), 129–142, 1989.

[Wh] Whitney, J.M., *Structural analysis of laminated anisotropic plates*. Technomic Publishing Company, 1987.

[Wh2] Whitney, J.M., The effect of transverse shear deformation in the bending of laminated plates. *J. Composite Materials*, **3**, 534–547, 1969.

[Wi] Wilson, E.L., Structural analysis of axisymmetric solids. *A.I.A.A.J.*, **3**, 2269–74, 1965.

[WJ] Weaver, W. (Jr.) and Johnston, P.R., *Finite elements for structural analysis*. Prentice Hall, 1984.

[Wo] Wood, R.D., A shape function routine for the constant moment triangular plate bending element. *Engineering Computations*, **1**, 189–198, 1984.

[WOK] Wempner, G.A., Oden, J.T. and Kross, D.A., Finite elements analysis of thin shells. *Eng. Mech. Div. Proc. ASCE*, **94**, EMS6, 1273–94, 1968.

[WR] Wilkinson, J.H. and Reinsch, C., *Linear Algebra. Handbook for Automatic Computation.* Volume II. Springer-Verlag, Berlin, 1971.

[WS] Willam, K.J. and Scordelis, A.C., Cellular structures of arbitrary plan geometry. *J. Struct. Div.*, ASCE, 1377-442, 1972.

[WTDG] Wilson, E.L., Taylor, R.L., Doherty, W.P. and Ghabussi, T., Incompatible displacement models. In *Numerical and Computer Methods in Structural Mechanics*, S.T, Fenves, *et al.* (Eds.), Academic Press, 1973.

[WV] Wong, C.C.K. and Vardy, A.E., Finite prism analysis of plates and shells. *Int. J. Numer. Meth. Engng.*, **21**, 529–41, 1985.

[WZ] Wood, R.D. and Zienkiewicz, O.C., Geometrically non linear finite element analysis of beams, frames, arches and axisymmetric shells. *Computers and Structures*, **7**, 725–35, 1977.

[Xu] Xu, Z., A thick-thin triangular plate element. *Int. J. Numer. Meth. Engng.*, **33**, 963–973, 1992.

[XZZ] Xu, Z., Zienkiewicz, O.C. and Zeng, L.F., Linked interpolation for Reissner-Mindlin plate elements. Part III. an alternative quadrilateral. *Int. J. Numer. Meth. Engng.*, **36**, 3043–3056, 1993.

[Ya] Yang, T.Y., *Finite Element Structural Analysis*. Prentice Hall, 1986.

[YJS+] Yang, D.Y., Jung, D.W., Song, L.S., Yoo, D.J. and Lee, J.H., Comparative investigation into implicit, explicit and iterative implicit/explicit schemes for simulation of sheet metal forming processes. In *NUMISHEET'93*, Makinouchi, A., Nakamachi, E., Oñate, E. and Wagoner, R.H. (Eds.), RIKEN, Tokyo, 35–42, 1993.

[Yo] Young, W.C., *Roak's formulas for stress and strain*. 6th Edition, McGraw-Hill, 1989.

[YSL] Yang, T.Y., Saigal, S. and Liaw, D.G., Advances of thin shell finite elements and some applications. Version 1, *Computers and Structures*, **35**(4), 481–504, 1990.

[YY] Yuqiu, L. and Yin, X., Generalized conforming triangular membrane element with vertex rigid rotational freedoms. *Finite Elements in Analysis and Design*, **17**(4), 259–271, 1994.

[ZBMO] Zienkiewicz, O.C., Bauer, J., Morgan, K. and Oñate, E., A simple and efficient shell element for axisymmetric shells. *Int. J. Numer. Meth. Engng.*, **11**, 1545–1559, 1977.

[ZCh] Zienkiewicz, O.C. and Cheung, Y.K., The finite element method for analysis of elastic isotropic and isotropic slabs. *Proc. Inst. Civ. Engng.*, **28**, 471–88. 1964.

[ZCh2] Zienkiewicz, O.C. and Cheung, Y.K., Finite element procedures in the solution of plate and shell problems. In *Stress Analysis*, O.C. Zienkiewicz and G.S. Holister (Eds.), Chapter 8. John Wiley & Sons, Chichester, 1965.

[ZCh3] Zienkiewicz, O.C. and Cheung, Y.K., Finite element methods of analysis for arch dam shells and comparison with finite difference procedure. In *Proc. Symp. on Theory of Arch Dams*, 123–140, Southampton University, 1964; Pergamon Press, Oxford, 1965.

[ZK] Zhang, Y.X. and Kim, K.S., A simple displacement-based 3-node triangular element for linear and geometrically nonlinear analysis of laminated composite plates. *Comput. Methods Appl. Mech. Engrg.*, **194**, 4607–4632, 2005.

[ZL] Zienkiewicz, O.C. and Lefebvre, D., Three field mixed approximation and the plate bending problem. *Comm. Appl. Numer. Meth.*, **3**, 301–9, 1987.

[ZL2] Zienkiewicz, O.C. and Lefebvre, D., A robust triangular plate bending element of the Reissner-Mindlin type. *Int. J. Numer. Meth. Engng.*, **26**, 1169-84, 1988.

[ZO] Zienkiewicz, O.C. and Oñate, E., Finite elements versus finite volumes. Is there really a choice? In *Nonlinear Computation Mechanics. State of the Art*, Wriggers, P. and Wagner, W. (Eds.), Springer, Berlin, 1991.

[ZO2] Zárate, F. and Oñate, E., Extended rotation-free shell triangles with transverse shear deformation effects. *Computational Mechanics*, **49**(4), 487–503, 2012.

[ZO3] Zárate, F. and Oñate, E., Enhanced rotation-free beam and plate elements with shear deformation effects. Research Report PI385, CIMNE, Barcelona, 2012.

[ZO4] Zárate, F. and Oñate, E., Finite elements versus finite volumes. Is there really a choice? In *Non Linear Computational Mechanics. State of the Art.* P. Wriggers and R. Wagner (Eds.), Springer 1992.

[ZPK] Zienkiewicz, O.C., Parekh, C.J. and King, I.P., Arch dams analysed by a linear finite element shell solution program. In *Proc. Symp. on Theory of Arch Dams*, Pergamon Press, Oxford, 1965.

[ZQTN] Zienkiewcz, O.C., Qu, S., Taylor, R.L. and Nakazawa, S., The Patch test for mixed formulations. *Int. J. Numer. Meth. Engng.*, **23**, 1873–1883, 1986.

[ZT] Zienkiewicz, O.C. and Too, J.J.M., The finite prism in analysis of thick simply supported bridges. *Proc. Inst. Civ. Eng.*, **53**, 147–72, 1972.

[ZT2] Zienkiewicz, O.C. and Taylor, R.L., *The Finite Element Method for Solid and Structural Mechanics*. Sixth Edition, Elsevier, 2005.

[ZTPO] Zienkiewicz, O.C., Taylor, R.L., Papadopoulus P. and Oñate, E., Plate bending elements with discrete constraints: New Triangular Elements. *Computers and Structures*, **35**, pp, 4, 505–2, 1990.

[ZTT] Zienkiewicz, O.C., Too, J.J.M. and Taylor,R.L., Reduced integration techniques in general analysis of plates and shells. *Int. J. Numer. Meth. Engng.*, **3**, 275–90, 1971.

[ZTZ] Zienkiewicz, O.C., Taylor, R.L. and Zhu, J.Z., *The Finite Element Method. Its Basis and Fundamentals*. Sixth Edition, Elsevier, 2005.

[ZXZ+] Zienkiewicz, O.C., Xu, Z., Zeng, L.F., Samuelsson, A. and Wiberg, N.-E., Linked interpolation for Reissner-Mindlin plate elements. Part I. A simple quadrilateral element. *Int. J. Numer. Meth. Engng.*, **36**, 3043–3056, 1993.

Author index

E. Oñate, *Structural Analysis with the Finite Element Method. Linear Statics: Volume 2: Beams, Plates and Shells*, Lecture Notes on Numerical Methods in Engineering and Sciences, DOI 10.1007/978-1-4020-8743-1, © International Center for Numerical Methods in Engineering (CIMNE), 2013

Subject index